D1037297

MANAGING WILDLIFE

MANAGING WILDLIFE

Presented by the
Alabama Wildlife Federation
Dan Dumont, Executive Director

By Greg K. Yarrow and Deborah T. Yarrow
Clemson University

1999 Sweetwater Press Edition

This edition published by Sweetwater Press by arrangement with Alabama Wildlife Federation.

Published in the United States by Sweetwater Press, Birmingham, Alabama.

Library of Congress Cataloging-in-Publication Data

Yarrow, Greg K. and Deborah T.
Managing Wildlife : on private lands in Alabama and the Southeast / edited by Dan Dumont ; by Greg Yarrow and Deborah Yarrow with the Alabama Wildlife Federation.
p. cm.
Includes bibliographical references and index.
ISBN 1-58173-157-4
1. Wildlife management—Alabama. 2. Wildlife habitat improvement-habitat improvement—Southern States. I. Dumont, Dan. II. Alabama Wildlife Federation. III. Title.
SK361.Y27 1998
639.9'0975—dc21 98-35971
CIP

Printed in Hong Kong

Cover photograph by Paul T. Brown

Editing, design, and production by Lenz Design & Communications, Inc.

Alabama Wildlife Federation
46 Commerce Street
Montgomery, AL 36104
Phone (334) 832-9453
Email: awf@mindspring.com
Web site address: alawild.org

Dedication

This book is dedicated to the owners and managers of private lands
whose stewardship has resulted in the rich and diverse forest and wildlife resources
of Alabama and the Southeastern United States.

The Alabama Wildlife Federation

Sponsors

We wish to gratefully acknowledge the generous contributions of the following who made this work possible:

The Curtis and Edith Munson Foundation

Alabama Forestry Commission

National Fish and Wildlife Foundation

Alabama Forest Resources Center

BOOKS·A·MILLION

Table of Contents

Chapter 1: Alabama's Landscape ... 1

Chapter 2: Developing a Wildlife Habitat Management Plan ... 17

Chapter 3: Wildlife Habitat Management ... 31

Chapter 4: Managing Forests for Wildlife and Timber Production ... 49

Chapter 5: Managing for White-tailed Deer ... 105

Chapter 6: Managing for Wild Turkey ... 137

Chapter 7: Managing for Bobwhite Quail ... 165

Chapter 8: Managing for Squirrels and Rabbits ... 185

Chapter 9: Managing for Mourning Dove ... 201

Chapter 10: Managing for Waterfowl ... 211

Chapter 11: Managing for Furbearer Species ... 229

Chapter 12: Managing for Non-game Wildlife ... 255

Chapter 13: Threatened and Endangered Species ... 285

Chapter 14: Attracting Backyard Wildlife ... 315

Chapter 15: Fish Pond Management ... 339

Chapter 16: Supplemental Plantings as Wildlife Food Sources ... 363

Chapter 17: Economic Considerations of Forest and Farm Wildlife Management ... 385

Chapter 18: Conservation Ethics and Opportunities ... 415

Chapter 19: Legal Considerations and Natural Resource Management ... 435

Appendixes ... 459

- Appendix A: References and Suggested Reading ... 460
- Appendix B: Book Steering Committee ... 477
- Appendix C: Nest and Denning Box Dimensions for Some Alabama Birds and Mammals ... 478
- Appendix D: Quick Reference and Wildlife Usage of Commonly Planted Wildlife Foods ... 479
- Appendix E: Cultivated Plants Important to Wildlife in Alabama ... 484
- Appendix F: Native and Naturalized Trees and Plants Important to Wildlife in Alabama ... 511
- Appendix G: Oaks Important to Wildlife In Alabama ... 519
- Appendix H: Pines Important to Wildlife In Alabama ... 522
- Appendix I: Upland Herbaceous Plants Important to Wildlife in Alabama ... 523
- Appendix J: Aquatic and Marsh Plants Important to Wildlife in Alabama ... 524
- Appendix K: Sources of Wildlife Planting Materials ... 526
- Appendix L: Soil pH Range for Select Southern Tree Species ... 532
- Appendix M: Sample Hunting Lease ... 534
- Appendix N: Sample Timber Sale Contract ... 538
- Appendix O: Sources of Technical and Educational Assistance ... 541
- Appendix P: Glossary of Terms ... 546
- Appendix Q: Threatened and Endangered Species List ... 557

Index ... 571

Foreword By Tom Kelly

It appears to be a universal characteristic of us outdoor types to believe we came here with an innate sense regarding sporting pursuits and wildlife management. While we are perfectly willing to consult dentists for a toothache or doctors for medical advice, and are generally reluctant to practice law without a license, this willingness to defer to experts does not extend to matters concerning game and fish.

As a general rule, any person who has shot one squirrel or caught one bream on a pole and line, feels fully capable of managing fish and game of all species on an ecosystem scale. He doesn't hesitate to disagree with professional wildlife biologists, to override the opinions of trained foresters or other natural resource managers, and, in general, to prove again, for the umpteenth time, the truth of the adage that "entrenched beliefs are never altered by exposure to facts."

This excellent, comprehensive manual makes no claim to omnipotence but it does reflect the opinions of some of the best minds in the Southeast, in a variety of disciplines, and sets forth in plain language practices which the private landowner or land manager can implement to improve the status of fish, game, and forests on his land.

You are, of course, entitled to disagree with anything written here and that may be part of the fun, but you should be aware of some historical facts before you do.

Go back 50 or 60 years, and take a look at the fish and game laws and the seasons and bag limits in Alabama and the Southeast in those "good old days."

With the single exception of the number of ducks allowed in the bag, every season is longer now than it was then, every bag and creel limit is as liberal and, in most cases, more liberal, and the truth is, the hunter, angler, and forest landowner of the 1990s is immeasurably better off than his grandfather was in the 1930s when a group of concerned sportsmen formed the Alabama Wildlife Federation.

Present conditions are very likely the result of private landowners doing the things the Alabama Wildlife Federation (and state game and fish and forestry agencies) have recommended for the past 60-plus years. It is highly possible that these guys know what they are talking about.

It might not be a bad idea to read on and see for yourself. Even *you* may learn a thing or two!

Tom Kelly
Spanish Fort, Alabama

Author of
Tenth Legion 1973
Dealer's Choice 1976
Better on a Rising Tide 1994
The Season 1996
The Boat 1998

The Lyons Press
31 W. 21st St.
New York, NY 10010

Preface and Acknowledgments

For over 63 years, the Alabama Wildlife Federation has been a recognized conservation leader at the national, regional, and state levels. Many of those accomplishments have been achieved quietly, however, and many Alabamians are unsure of who we are and what we do. The Federation is often confused with The Wildlife Society, the U.S. Fish and Wildlife Service, Auburn's Department of Zoology and Wildlife Sciences, and, most frequently, the Game and Fish Division of Alabama's Department of Conservation and Natural Resources. Although the comparisons are flattering, the Federation is none of those things and we have traditionally played a complementary role to all of these organizations. Still, the confusion persists. A significant proportion of the calls to the Montgomery office will attest to that. Over and over, the office staff is quizzed on a wide variety of technical and nontechnical wildlife topics, such as

What effects do clear-cuts have on wildlife?
How can I attract doves to my property?
What is the very best winter food plot planting for deer?
Are there panthers in Alabama?
Does stocking pen-raised birds increase quail populations?

Questions like these and thousands of others are asked each year. Our staff either refers the caller to one of the professional wildlife agencies listed above or rushes to find the answer, usually by referencing one of the many excellent publications provided by those agencies or through a hastily placed phone call to someone "in the know" at one of those agencies. This has been true throughout the history of the Alabama Wildlife Federation and it is still true today. In 1994, Dan Dumont, who serves as executive director and in-house counsel for AWF, realized that even though almost all of the information requested was available somewhere, there was no comprehensive guide, no "one-stop shopping place" for interested land owners, hunters, and other natural resource managers. The idea of producing such a reference continued to intrigue Dan and we discussed the possibilities one night at his hunting camp until the wee hours. Several weeks later, a more fully developed version of Dan's idea was presented to the AWF board of directors and Dan was given the go-ahead to produce a book with the support of then AWF President Dr. Jeff McCollum and Projects Committee Chair Dr. Tom Wells.

The first step was to form a Steering Committee to refine the basic outline, determine the form the book would take, secure funding to pay to have it written and published, and find someone to write it. That committee, which Dan chairs, was made up of leaders in the natural resources community, including professional foresters, wildlife biologists, sportsmen, landowners, agency representatives, university faculty, and others. (A complete list of committee members is included in Appendix B.) The committee went to work in earnest and quickly developed an outline to guide prospective authors. The first task was to decide just what we expected the book to be. The consensus was that the book should be aimed at nonprofessionals and take a cookbook approach to wildlife and habitat management rather than a textbook approach. It should be informative without lapsing into the technical jargon that often leaves landowners frustrated and confused. It should be arranged into logical divisions that, when taken as a whole, would provide a comprehensive guide to all aspects of forest and farm management for wildlife, but which, separately, could stand alone as succinct bits of useable information. In other words, the book could be read in its entirety, cover to cover, and provide an education in wildlife and natural resource management. Or it could be used to answer specific management questions without leading the reader through lengthy preparatory and supporting material. The committee felt that several elements were critical to the success of the book, including readability, "user friendliness," use of best available and most current information, good glossaries and indices for reader access, and profuse photographs and illustrations.

With those guidelines in place, a nationwide search was conducted to solicit proposals from prospective authors. At the same time Dan and the Steering Committee began soliciting funds to pay for the project. Proposals were made to a variety of funding sources and several private foundations, agencies, and one corporation committed sufficient funds to begin the project. Generous financial support was provided by the Curtis and Edith Munson Foundation, the National Fish and Wildlife Foundation, the Alabama Forestry Commission, International Paper, and the Alabama Forest Resources Center.

Meanwhile, about 10 serious proposals were received from prospective writers and considered by the Steering Committee. A first cut was made and the proposals reconsidered. At that point, all of the proposals were good and the choice was extremely difficult but the final decision was made to offer the contract to writers from Clemson University: Dr. Greg Yarrow, Associate Professor of Wildlife in the Department of Aquaculture, Fisheries and Wildlife; and Debbie Yarrow, a doctoral candidate in the Department of Forest Resources and Greg's wife. Dr. David Guynn, Jr., Professor in the Department of Forest Resources at Clemson University, agreed later to serve as a technical editor.

Greg Yarrow grew up in Hattiesburg, Mississippi, the son of a University of Southern Mississippi professor. After graduation from that school, Greg completed a master of science degree at Mississippi State University in wildlife Ecology studying white-tailed deer while working under the direction of Dr. Harry Jacobson. He went on to earn a doctorate degree at Stephen F. Austin University studying interactions and resource competition between deer and feral hogs. He worked in Extension Wildlife at Mississippi State University before accepting a split Extension and research position at Clemson University in 1988. He is a certified wildlife biologist and has been active in leadership roles in many professional and popular natural resources organizations. He works extensively with landowners, natural resource professionals, and young people in the course of his job. From those experiences he brings valuable insights to the writing of this book.

Debbie Yarrow spent her childhood in Alabama before leaving at the age of 16 to enter the University of Southern Mississippi. After graduation, she earned a master of science degree at Hollins College in Virginia and subsequently taught writing and communications at Virginia Polytechnic Institute, Stephen F. Austin University, and Clemson University. While at Clemson, she began pursuit of a Ph.D. in forest resources working with Dr. Guynn. Her research investigated public and private landowner attitudes and perceptions toward forest management on an ecosystem scale. She has published articles in both scientific and popular outlets on a variety of topics.

After the Yarrows were chosen to write the book, a subset of the Steering Committee was chosen to serve on the Technical Review Committee. This group, which I chair, included Dan Dumont; Dr. Jimmy Huntley, wildlife biologist for the National Forests in Alabama; Joe McGlincy, a wildlife biologist with Southern Forestry Consultants, Inc. in Bainbridge, Georgia and formerly with International Paper; Tim Gothard, Forest Management Chief with the Alabama Forestry Commission; Stan Stewart, a forest stewardship biologist with the Alabama Division of Game and Fish; Mickey Easley, a forester, biologist and manager of Wyncreek Plantation; Dr. Jeff McCollum, past president of the Alabama Wildlife Federation, TREASURE Forest landowner, and wildlife conservationist; and Ted DeVos, a forester and biologist with Regions Bank Trust Department. This group, with its wide array of backgrounds and interests, met at regular intervals to review submissions for technical accuracy, readability, and applicability. Regular communications were established between the Yarrows and this committee and a working relationship was formed.

Finally, edited material was submitted for comment to a panel of lay readers, some which were professional biologists and foresters, and others which were outdoor enthusiasts. A special thanks is given to the following readers: Tom Kelly, Dr. Tom Barnes, Dr. Bud Cardinal, Ted DeVos, Hugh Durham, Wayne Fears, Dr. Kathryn Flynn, Neil Johnston, Mason McGowin, Dr. Bill McKee, Daniel Powell, Corky Pugh, Jim Russell, Dr. David Thrasher, Glenn Waddell, Bob Waters, Dickie Williams, Chester "Mac" Baker, Chester "Bubba" Baker, Dr. Joseph Clark, and Chris Barnes. The reviewer's comments were considered and incorporated when appropriate. Interviews with land managers, natural resource professionals, landowners, and sportsmen were incorporated throughout the book. A special thanks is given to these individuals who include Dr. Doug Phillips, Stan Stewart, Dr. Dave Van Lear, Dr. Keith Causey, Dr. Harry Jacobson, Dr. George Hurst, Chuck Peoples, Dr. Lee Youngblood, Ted DeVos, Wayne Fears, Jeff Hadaway, the late William C. "Billy" Perdue, Keith McCutcheon, Frank Boyd, Chad Philipp, Mike Sievering, Tony Melchoirs, Bob McCollum, Ralph Costa, Barry Smith, Don Keller, Dr. Danny Everett, Dr. Jimmy Huntley, Mickey Easley, Ed McMillan, Dr. Bill McKee, and M. N. "Corky" Pugh. A special thanks is also given to Dr. Jim Dickson who contributed heavily to the text on the wild turkey in Chapter 6; to the Black Bear Conservation Committee which graciously provided much of the text on black bears in Chapter 12 from their publication *Black Bear Management Handbook*, to Auburn University, U.S.Fish and Wildlife Service, Champion International Corporation and Canal Wood Corporation for providing text in the species listing section of Chapter 13 from their publication *Threatened and Endangered Species of Alabama: A Guide to Assist with Forestry Activities;* to South Carolina Department of Natural Resources for information from their *Game on Your Land* publications; and Auburn University for providing information for Chapter 15 from their publication *Management of Recreational Fish Ponds in Alabama.*

Slides and photographs were taken or collected from dozens of sources and the best available slides selected to illustrate important points in the text. Special thanks is given to Paul Brown, a renowned wildlife photographer, for his contribution of photographs to the book and to the many other natural resource professionals, wildlife agencies, and organizations who also provided photographs. Thank you to Richard T. Bryant for photo editing the book. The result, I think you will agree, is a most appealing, readable, and useful addition to any home, hunting club, or natural resource professional's library.

Attractive enough to grace a coffee table and utilitarian enough to take to the field, this book is a resource that Dan Dumont, the writers and editors, and the Alabama Wildlife Federation are proud to make available to all lovers of the natural resources of Alabama and the Southeast.

Rhett Johnson

Director, Solon Dixon Forestry Education Center
School of Forestry
Auburn University
Past President, Alabama Wildlife Federation
Chair, Technical Review Committee

Introduction

By the time this book is published I will have ended my seven-year tenure as Executive Director of the Alabama Wildlife Federation. This has been a wonderful and rewarding period in my life, but it is time to return home to Mobile, Alabama.

When I took this position seven years ago, I immediately began getting calls from AWF members and others asking what they could do to improve the habitat on their land or the land they hunted on. I gave them my opinions and referred them to various other places for information, but it quickly became apparent that there was no single source of information available that answered all of their questions. I was familiar with an excellent book on piloting and navigation called "Chapman on Piloting" that is in its 61st edition and is considered "the bible" on the subject matter it deals with. I began to realize that if AWF could produce such a book for landowners and managers that helped them manage their forest and farmland for timber production *and* wildlife, we could assist a greater number of people and hopefully have a positive effect on a more significant amount of habitat.

With the full support of the Alabama Wildlife Federation's board of directors, I was given the green light to begin the production of this book about four years ago. I was fortunate in assembling a great team of writers and editors who have worked tirelessly on this project. The writers were Greg and Deborah Yarrow of Clemson. The technical editing committee was composed of Rhett Johnson, Tim Gothard, Stan Stewart, Ted DeVos, Jimmy Huntley, Joe McGlincy, Dr. Jeff McCollum, Mickey Easley, and myself. This group has spent countless hours editing and rewriting the draft chapters produced by the writing team, and each one helped fine tune the book based on his own unique background and experience. The result is a comprehensive book on forest and wildlife management for private landowners, which incorporates wildlife habitat management into commonly employed forest land management practices. The book is broadly based and designed to address differences in management practices across physiographic regions and ownership and management interests. While in most cases the book uses Alabama examples and refers the reader to Alabama agencies, it is applicable across the South, from eastern Texas to the Carolinas, in any given physiographic region. A reader outside Alabama can go to the comparable agency in his or her state for assistance.

We were also fortunate in having some very generous sponsors that enabled us to approach the project on the scale the subject matter deserved. I would like to personally thank Bruce Reid and the Hobbs family of Montgomery whose initial support through the Curtis and Edith Munson Foundation got the project started. They were joined by International Paper, the National Fish and Wildlife Foundation, the Alabama Forestry Commission, and the Alabama Forest Resources Center.

Books-A-Million, through their Sweetwater Press division, has been very supportive of this book throughout the editorial, design, production, and marketing phases of the project and we are most appreciative of their help.

Richard Lenz of Lenz Design & Communications and his team in Atlanta did a great job on the editing, layout, and design of the book.

Our goal was to produce a user-friendly book that could be read in its entirety or used as a reference by a manager preparing to perform a particular land management practice. Like Chapman, we hope this book will be improved upon and republished over the years as research, conditions, laws, and circumstances change. We encourage you to write to us if you find errors or would like to suggest improvements for or participate in future editions. Our email address is awf@mindspring.com

The measurement of the success of our efforts will be determined by whether or not you integrate our recommendations into your management plans to improve the wildlife habitat on land you own or manage. While the Federation makes no guarantees or warranties we stand ready to assist you, and we wish you every success in your stewardship of the places you know and love.

In closing, I would like to thank the membership, officers, and members of the board of directors of the Alabama Wildlife Federation for their support and encouragement of me, my staff, and our organization over the years. This is their book and I am very grateful and honored to have had the opportunity to be a part of it.

Dan Dumont

Executive Director
& In-House Counsel
Alabama Wildlife Federation
Chair, Steering Committee

Geology, soils, natural landforms, and dominant vegetation have been used by scientists to separate Alabama's unique landscape into five major natural regions that include the Plateau, Ridge and Valley, Piedmont, Upper Coastal Plain, and Lower Coastal Plain. Each region has a unique mixture of soils,topography, climate, forests, and vegetative communities that influence the type and quality of habitat for a variety of wildlife species. These differences determine the potential of a region to support certain species of wildlife.

DAN BROTHERS

CHAPTER 1

Alabama's Landscape

CHAPTER OVERVIEW

Alabama is blessed with a diversity of natural regions that support a variety of vegetative communities and a wide array of wildlife habitats. The inherent capability of these regions to produce quality wildlife habitat and support viable wildlife populations is determined by interacting factors such as geology (underlying rock formations), soil characteristics, topography, hydrology, vegetation, climate, and land use by man and animals. By understanding the physical and biotic characteristics of Alabama's regions, landowners can understand their property's capacity and limitations for supporting healthy and abundant wildlife populations.

CHAPTER HIGHLIGHTS PAGE

TOPOGRAPHY 2
SOILS 2
CLIMATE 3
LAND CLASSIFICATION 4
NATURAL REGIONS OF ALABAMA 5
Plateau Region 5
Ridge and Valley Region 7
Piedmont Region 8
The Fall Line 9
Coastal Plain Region 9
UPPER COASTAL PLAIN REGION 9
LOWER COASTAL PLAIN REGION 14
Wetlands 14
FOREST LAND OWNERSHIP 15
SUMMARY 15

Alabama's natural landscape and the diversity of **wildlife habitats**, the place where plants or animals live or grow, is a product of past geologic formations and climate. Over time, the various rock formations have weathered to produce a variety of soils in north Alabama, while marine deposits have formed the foundation for central and south Alabama soils. Geological forces have also produced a diverse topography across the state. The interaction of topography, soils, and climate has created a variety of natural regions in the state that differ in their productivity for wildlife.

TOPOGRAPHY

Topography defines which land will become floodplains and uplands. Floodplains are low enough in elevation to be regularly inundated with water by streams or rivers, and are generally more fertile and productive than uplands. Uplands, in contrast, are higher in elevation than floodplains, are rarely flooded, and usually have thinner and less productive soils.

Upland topography influences site quality and productivity in several ways. Topography can affect vegetative composition on certain sites by creating "microclimates" that alter soil moisture and nutrients. **Microclimates** are relatively small areas that have different moisture and temperature regimes than the surrounding environment. As a result, plant communities in these areas are usually different than the surrounding landscape. **Coves** are productive upland sites. These areas are found along stream heads and upland stream bottoms. Cove soils are normally moist, deep, and fertile, creating a favorable climate for plant growth. **Aspect**, which is the direction a slope faces, also influences the productivity of certain upland sites. Slopes facing the south and west tend to be warmer and drier than north- and east-facing slopes because of prolonged exposure to the sun. Length of exposure to the sun helps determine plant composition on a site by influencing soil moisture and the amount of **evapotranspiration**, or loss of water in plants. High moisture losses create stress on plants and reduce plant productivity. As a result, southern and western slopes contain fewer drought-tolerant plant species, while northern and eastern slopes generally support a broader diversity of plants. On gentle slopes, with gradients that have less than an 8 percent incline, aspect normally has little or no effect on vegetation growth and abundance.

Topographic position, which is the site's location on a slope, also affects upland productivity. In general, the lower portion of slopes is more fertile and productive than higher areas and ridges. An area at the bottom of a slope has the advantage of richer soils, because topsoil is gradually displaced downward from rainfall runoff. Over time enough soil has been displaced from upper regions of slopes to provide deeper, more fertile soils at the base of slopes.

SOILS

Wildlife managers are often able to predict the type and quality of wildlife habitat based on soils. Soil maps are helpful in locating fertile and productive soils that respond well to habitat management practices, or conversely, in ruling out poor, infertile sites. Sites that require special attention, such as fragile soils with thin topsoil layers, can also be identified with soil maps. To maintain soil fertility and productivity, these areas often require low impact management practices that maintain organic matter in the soil. In these soil-sensitive areas, the timing, intensity, or frequency of habitat improvement practices may have to be modified to reduce or eliminate soil damage caused by management practices. Understanding the capabilities and limitations of various soils, and where these soils are located on a particular tract, is a prerequisite to selecting the most appropriate wildlife habitat or forestry improvement practice.

Across Alabama there are over 300 different soil types that influence the variety and quality of wildlife habitats. Each of these soil types has an inherent capacity to support a certain variety and quality of vegetation. In some cases, the distribution and abundance of wildlife, as well as their physical condition and health, are also related to the soils of a region. Some of the most productive soils for wildlife habitat in the state are located in the Black Belt Region and in alluvial floodplains of river and stream bottoms. Because of their high fertility, these areas are primarily in agricultural production. Other regions of Alabama are not as productive as the Black Belt and alluvial floodplains. The deep, sandy soils of the coastal plains are low in fertility due to a high rate of nutrient leaching. The upper slopes of mountainous regions of the state may also be low in fertility and relatively unproductive because of thin or nonexistent topsoil layers caused by erosion. Other regions in the state are either medium to poor in productivity for wildlife habitat, depending on site characteristics and past land-use practices. The fertility, pH, and productivity of soils in a region limit the potential for management practices to improve wildlife habitat. Managers should not expect to see the same responses in vegetation and wildlife production from management practices conducted on poor soils as compared to more fertile sites.

The potential of an area to produce quality wildlife habitat is directly related to soil fertility. Soil fertility depends on soil structure (texture) and pH. Soils are classified by texture or soil particle size, which refers to the amount of sand, silt, and clay content. Sandy soils are coarse in texture, silty soils are intermediate, and clay soils are fine. Loamy soils contain more than one particle size. From a wildlife habitat perspective, texture matters because it determines the ability of soils to retain moisture and nutrients for plant growth. Soils with finer texture have a high water-holding capacity, while soils that are mostly sandy hold little water. Predominately sandy soils are dry sites with limited plant distribution, abundance, and quality. These areas are usually low in fertility and contain only the hardiest of drought-tolerant plants. Conversely, soils that are predominately fine in texture may not be desirable either because they are often poorly drained, which limits plant growth to species adapted to wet sites. Soils that contain intermediate particle sizes and retain some moisture but also allow some drainage are ideal for most wildlife plants. Some wildlife have also developed associations with certain soils based on texture. The gopher tortoise is only found in the Coastal Plain soils that have high sand content because it prefers to construct burrows in sandy or sandy loam soils.

Soil pH, an important component of soil chemistry, also affects the type and quality of wildlife habitat in a region. Soils are either basic (pH range of 7.1 to 14.0), acidic (pH range of 0 to 6.9), or neutral (pH of 7.0). Plants and trees are adapted to a certain range of soil pH. Plants having a wide pH range are generally more abundant and widely distributed than plants that have a narrow pH range requirement. The greatest variety of plants usually occurs in soils that have an intermediate pH value between 6.5 and 7.0. Soil pH preferences for southern tree species can be found in Appendix L. Soils which are highly acidic or strongly basic often suppress plant growth and support fewer plants that are poor in quality for wildlife. Strongly acidic soils tie-up vital nutrients necessary for plant growth and also increase the ability of plants to absorb aluminum and manganese. These two elements are toxic to many plants, even at low concentrations.

In general, acidic soils are associated with uplands dominated by pines. Basic soils are characteristic of limestone outcrops and calcareous soils such as in the Black Belt. Eastern red cedar is a primary indicator of basic soils. Neutral soils are most commonly associated with alluvial bottomland soils and hardwood stands. Soil pH's effects on wildlife have been documented. One study found that soils with a pH above 6.0 correlated positively with litter size of cottontail rabbits in five soil regions of Alabama. In this study, litter size increased from 3.6 to 4.7 when soil pH levels rose above 6.0.

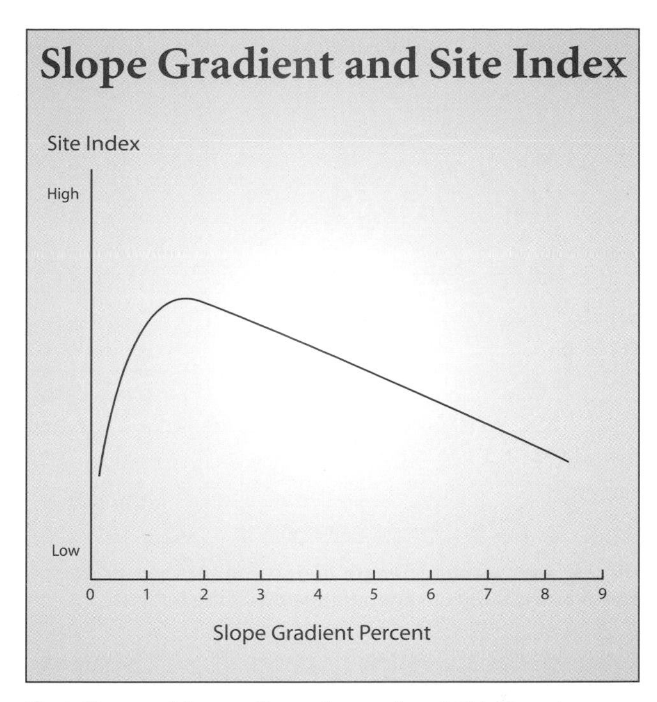

Fig. 1: Slopes with a gradient of more than 1.5% often decrease in site index (fertility) and productivity for wildlife. Areas at the bottom of slopes usually are higher in fertility due to soil accumulation from slope erosion.

Soil fertility also varies depending on the presence and amounts of phosphorus, nitrogen, potassium, and other nutrients. Sites that are deficient in these minerals usually have poor plant growth. Some wildlife that inhabit areas with poor soils have adapted to these sites by selectively eating vegetation or vegetative parts where nutrients are concentrated such as buds, fruits, acorns, or seeds. Soils high in nutrients are more likely to support healthy wildlife populations. Strong evidence exists that the quality of some wildlife is a function of soil fertility in their home ranges. When comparing harvested animals from Missouri counties with high, medium, and low soil fertility, one study noted that the largest raccoons, muskrats, opossums, and rabbits came from counties with high soil fertility, while the smallest animals came from counties with low fertility. For muskrats and opossums, fur quality was also related to soil fertility. The study noted that litters of fox squirrels from high-fertility counties were also larger. In Alabama, white-tailed deer have higher weights and better antler characteristics on sites with fertile soils, like the Black

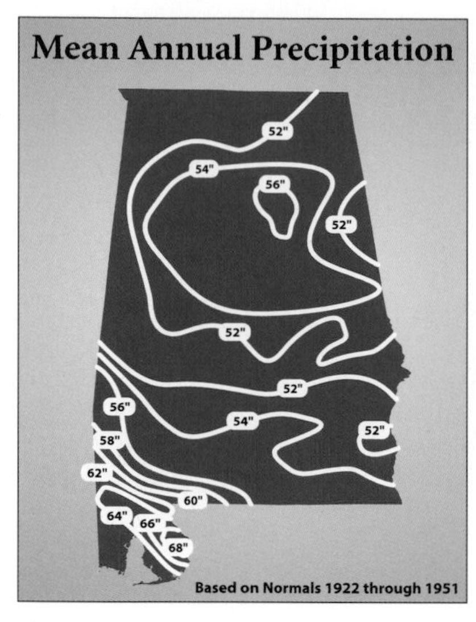

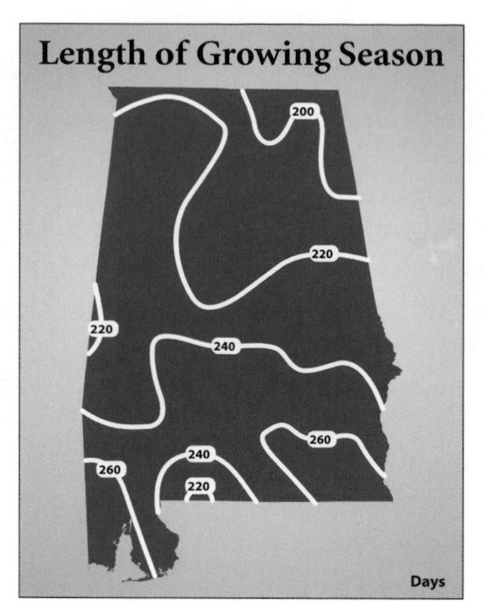

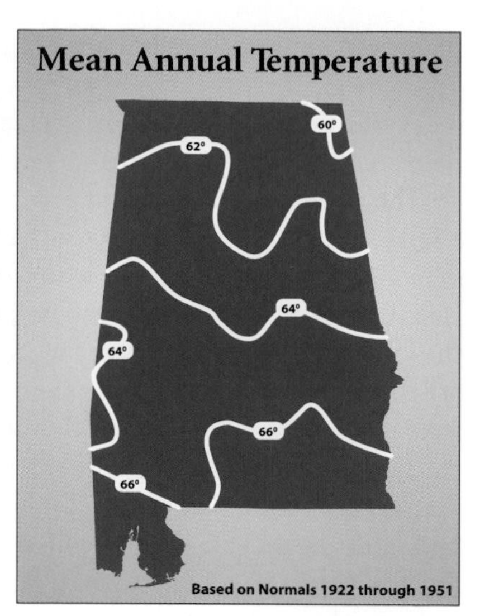

Fig. 2–4: Precipitation, length of growing season, and temperature combine with soil characteristics to determine the abundance and quality of vegetation and wildlife habitat.

Belt and alluvial floodplains associated with river bottoms, where calcium and phosphorus are abundant. The quality of bobwhite quail habitat is directly related to the soils of an area. Sands, sandy loams, and red clays are some of the best soils in Alabama for producing quail. These soils support weeds and grasses that grow in such a way that some bare ground is left between plants. Because quail are weak scratchers when searching for food, bare ground interspersed with plants provides ideal feeding habitat.

"The specific ways in which mineral-rich soils benefit animals are by no means well understood. Nevertheless, one thing stands out clearly: good soils yield the best crops, both in quantity and quality, of practically everything that lives upon them."

—*Durward L. Allen, 1974*

CLIMATE

Wildlife habitat is also influenced directly by climatic conditions such as precipitation, temperature, length of growing season, winds, and violent weather such as thunderstorms, tornadoes, and hurricanes. These factors, along with soil characteristics, interact to determine the composition, diversity, and quality of wildlife habitat. Of these factors, precipitation and temperature have the greatest influence on the growth of vegetation and quality of wildlife habitat. Alabama can be divided up into three general climatic regions: 1) the typical warm, humid climate found in the majority of central Alabama; 2) the wetter maritime climate of southern Alabama; and 3) the relatively cooler climate of the northern counties.

Alabama receives an average annual precipitation between 50 and 60 inches. In general, the most important precipitation period for wildlife in Alabama occurs during the warm season between March and September. For wildlife, it is critical to have adequate rainfall during this period. Droughts during summer and early fall cause vegetation to dry up, decreasing palatability, digestibility, and nutrient content of plants. During droughts, white-tailed deer often become nutritionally stressed because of the decline in food quality and availability. Although

RICHARD T. BRYANT

Because of Alabama's abundant rainfall, water is not a critical limiting factor for wildlife. Lakes like Guntersville (shown here), ponds, rivers, and streams provide a ready source of water for wildlife in the state.

not as well documented, other wildlife are also affected in a variety of ways during dry periods. Too much rainfall can have a negative effect on flora and fauna as well.

Temperatures across the state are fairly mild with an average annual temperature of 63°F. A record high temperature of 112°F was recorded in Centreville on September 5, 1925. A record low temperature of -27°F was recorded at New Market on January 30, 1966. Unusual extremes such as these take a toll on wildlife. Extremely hot temperatures magnify drought conditions, causing vegetation to decline rapidly in quality. Unusually low temperatures require wildlife to expend more energy to stay warm, and in extreme but rare cases cause wildlife fatalities in Alabama. Unseasonably cold or freezing temperatures during early spring can also inhibit flowering of fruit-producing shrubs and trees valuable to wildlife.

DAN DUMONT

The Appalachian Plateau, sometimes called the Cumberland Plateau, is located in north-central Alabama and is a diverse area composed of plateaus, mountain ridges, hills, and valleys. Most of the soils are derived from sandstones and shales, with sandy loam topsoil over sandy clay loam.

LAND CLASSIFICATION

Geology, soils, natural landforms, and dominant vegetation have been used by scientists to separate Alabama's unique landscape into five major natural regions that include the 1) Plateau, 2) Ridge and Valley, 3) Piedmont, 4) Upper Coastal Plain, and 5) Lower Coastal Plain. Each region has a unique mixture of soils, topography, climate, forests, and vegetative communities that influence the type and quality of habitat for a variety of wildlife species. These differences determine the potential of a region to support certain species of wildlife. The distribution of evergreen trees across the state illustrates this point well. The percentage of evergreens in Alabama increases from 16 percent (mostly Eastern red cedar) in the northern Plateau region to about 90 percent (mostly pine) in the Coastal Plain. The arrangement of conifers, trees that retain some or all of their leaves throughout the year (such as pines), across the state influences the distribution and abundance of some species of wildlife, especially non-game birds. Red-cockaded woodpeckers, brown-headed nuthatch, and pine warblers are either scarce or absent within 50 miles of the Tennessee state line because pines are uncommon in this area. Conversely, increasing numbers of conifers in the Coastal Plain limit the breeding range of the wood thrush, red-eyed vireo, American redstart, and other non-game birds that prefer deciduous trees. Many species of wildlife that depend on mast-producing hardwoods are also limited in pine-dominated landscapes.

The unique natural regions of Alabama support a variety of forest types that cover 22 million (67 percent) of the state's 33 million land acres. Dendrologists (who study trees) and botanists (who study plants) have identified approximately 300 native tree species and 3,400 plants in Alabama. The most common trees are loblolly pine, sweetgum, hickories, water oak, white oak, shortleaf pine, southern red oak, longleaf pine, yellow poplar, and slash pine. They make up the majority, 63 percent, of the woody biomass in the state. The forests of Alabama are classified by dominant tree species into five categories: 1) oak-hickory, 2) oak-pine, 3) oak–water tupelo–bald cypress, 4) loblolly–shortleaf pine, and 5) longleaf–slash pine. Collectively these categories consist of 7.4 million acres of pine, 4.5 million acres of mixed-pine hardwood, and 10 million acres of hardwood.

NATURAL REGIONS OF ALABAMA

Plateau Region

The Plateau Region covers about one-fourth of Alabama in the northern section of the state. This region contains three physiographically distinct areas: the Interior Low Plateau, Tennessee Valley, and Appalachian Plateau.

The **Interior Low Plateau**, also known as the Highland

Fig. 5: Selected Plants Important to Wildlife by Region in Alabama.

PLANTS	REGION	WILDLIFE THAT BENEFIT
Alabama Supplejack (*Berchemia scandens*)	PL,PE	Turkey, quail, deer, raccoon, squirrel, mallard, songbirds.
American Beautyberry (*Callicarpa americana*)	PE,CP	Quail, deer, squirrel, raccoon, opossum, fox, songbirds.
American Beech (*Fagus grandifolia*)	PL,PE	Squirrel, turkey, chipmunk, black bear, quail.
American Elder (*Sambucus canadensis*)	PE,CP	Robin, mockingbird, catbird, quail, turkey, dove, squirrel, rabbit, raccoon, opossum, chipmunk, deer.
American Holly (*Ilex opaca*)	PE,CP	Songbirds, dove, turkey, quail, small mammals, deer.
American Hornbeam (*Carpinus caroliniana*)	PE, CP	Quail, rabbit, beaver, turkey, deer, wild hog.
Blackberry (*Rubus* spp.)	PL,PE,CP	Deer, quail, raccoon, chipmunk, squirrel, many birds.
Blackgum (*Nyssa sylvatica*)	PL,PE,CP	Quail, wood duck, deer, black bear, turkey, fox, beaver, opossum, raccoon, squirrel, robin, mockingbird, brown thrasher, flicker, and other birds.
Black Cherry (*Prunus serotina*)	PL,PE	Deer, rabbit, quail, squirrel, raccoon, waterfowl, songbirds.
Blueberry (*Vaccinium* spp.)	PL,PE,CP	Turkey, quail, scarlet tanager, robin, cardinal, thrasher, deer, chipmunk, rabbit, fox, raccoon.
Dwarf Huckleberry (*Gaylussacia dumosa*)	CP	Quail, turkey, fox, squirrel, songbirds, deer.
Eastern Red Cedar (*Juniperus virginiana*)	PL,PE	Quail, turkey, rabbit, fox, raccoon, skunk, opossum, coyote, cedar waxwing.
Flatwoods Plum (*Prunus umbellata*)	PE,CP	Deer, raccoon, fox, wild hog.
Flowering Dogwood (*Cornus florida*)	PL,PE,CP	Turkey, deer, quail, rabbit, fox, black bear, chipmunk, wood duck, cardinal, grosbeak, robin, thrasher, cedar waxwing.
Gallberry (*Ilex glabra*)	CP	Dove, quail, turkey, deer, rabbit, black bear, hermit thrush, mockingbird, songbirds.
Greenbriar (*Smilax* spp.)	PL,PE,CP	Turkey, quail, deer, rabbit, black bear, opossum, raccoon, squirrel, songbirds.
Hackberry (*Celtis occidentalis*)	PL	Turkey, raccoon, quail, waterfowl, squirrel, songbirds.
Hawthorns (*Crataegus* spp.)	PL,PE,CP	Turkey, quail, wood duck, squirrel, fox, raccoon, rabbit, deer, sparrow, cedar waxwing, songbirds.
Hickories (*Carya* spp.)	PL,PE,CP	Squirrel, turkey, black bear, fox, rabbit, raccoon, deer, songbirds.
Eastern Hophornbeam (*Ostrya virginiana*)	PE,CP	Quail, squirrel, deer, rabbit, turkey, purple finch, rose-breasted grosbeak, downy woodpecker.
Japanese Honeysuckle (*Lonicera japonica*)	PL,PE,CP	Deer, rabbit, turkey, quail, songbirds.
Large Gallberry (*Ilex coriacea*)	CP	Deer, quail, songbirds.
Oaks (*Quercus* spp.)	PL,PE,CP	Deer, turkey, squirrel, quail, raccoon, wild hog, black bear, mallard, wood duck, woodpeckers, bluejay.

Fig. 5: Continued.

PLANTS	REGION	WILDLIFE THAT BENEFIT
Persimmon (*Diospyros virginiana*)	PL,PE,CP	Deer, opossum, raccoon, fox, skunk, turkey, quail, crow, rabbit, wild hogs, songbirds.
Possumhaw (*Ilex decidua*)	CP	Deer, quail, turkey, many birds.
Red Mulberry (*Morus rubra*)	PE,CP	Songbirds, squirrel, quail, turkey, opossum, raccoon, deer, small mammals.
Sassafras (*Sassafras albidum*)	PE,CP	Turkey, quail, raccoon, squirrel, black bear, songbirds.
Strawberry (*Fragaria* sp.)	PL,PE	Deer, rabbit.
Sugarberry (*Celtis laevigata*)	CP,PL,PE	Deer, quail, turkey, songbirds, small mammals.
Viburnum (*Viburnum* spp.)	PE,CP	Turkey, quail, squirrel, beaver, rabbit, raccoon, chipmunk, deer, songbirds.
Water Tupelo (*Nyssa aquatica*)	CP,PL,PE	Wood duck, quail, squirrel, raccoon, turkey, deer, fox, beaver, opossum, robin, mockingbird, brown thrasher, thrush, flicker.
Wild Grape (*Vitis* spp.)	PL,PE, CP	Deer, raccoon, black bear, fox, quail, wood duck, squirrel, songbirds.
Yaupon (*Ilex vomitoria*)	CP	Deer, turkey, quail, squirrel, raccoon, songbirds.

PL=Plateau Region, PE=Piedmont Region, CP=Coastal Plain

Rim and Chert Belt, is in the extreme northern portion of the state. Topography varies from hilly to nearly flat close to streams. Most of this region's geology is composed of compacted rock, often called "consolidated rock," that originates from limestone, and to a lesser extent sandstone and shale. The diverse geology of this area gives rise to a mixture of soils from silty to sandy loam topsoil that ranges from acidic to basic. The soils are fairly fertile, but their productivity depends upon the amount of rock fragments and topsoil depth. The dominant forest type is oak-hickory; however, some acidic soils support Virginia and shortleaf pine. Basic soils typically grow Eastern red cedar. Common trees include white oak, northern red and black oak with associated pignut, shagbark, mockernut, and bitternut hickory. The most productive soils in this area are often in cultivation or pasture.

The **Tennessee Valley** is a broad band of relatively level, fertile land that lies adjacent to the Tennessee River in north Alabama. Soils are mostly red clay derived from limestone and alluvial (river flooding) deposits with silty loam topsoils and clay subsoils. The more fertile sites are alluvial soils. Most of the land has been cleared for pasture and crop production, but some land remains in pines, hardwoods, and scattered patches of Eastern red cedar.

The **Appalachian Plateau**, sometimes called the Cumberland Plateau, is located in north-central Alabama and is a diverse area composed of plateaus, mountain ridges, hills, and valleys. Most of the soils are derived from sandstones and shales, with sandy loam topsoil over sandy clay loam. Dominant forests are pine-oak, and to a lesser extent oak-hickory. Common trees include white oak, northern red oak, black oak, chestnut oak, pignut, shagbark, mockernut, and bitternut hickory, loblolly pine, shortleaf pine, and Virginia pine.

The most productive habitats for wildlife in the Plateau Region are associated with alluvial soils adjacent to rivers, streams, and creeks. Because of the fertility of these soils, these areas produce some of the larger oaks, hickories, yellow poplar, and beech that are valuable mast producers for wildlife. These sites also produce an abundance of high quality **browse**, the woody plants that are consumed by many species of wildlife. Other areas within the Plateau Region are not as productive for wildlife. Habitat suitability of these areas varies from moderate to low depending on topsoil depth, amount of rock fragments in the soil, topography, and other site characteristics. Landowners and managers will see the best response in vegetation growth and diversity from wildlife habitat improvement practices that are conducted in alluvial bottomlands.

DENNIS HOLT

The diverse landscape of the Ridge and Valley region is transversed by major rivers and streams. Little River Canyon cuts into Lookout Mountain, a dominant feature of the Ridge and Valley region. Alluvial soils adjacent to these waterways are some of the most productive soils for wildlife in Alabama.

Ridge and Valley Region

The Ridge and Valley Region, also known as the "foothills of the Appalachians," is located southeast of the Plateau Region and extends from the northeast corner of the state southwest to the center of the state. This area contains a series of mountain ridges with rugged topography interspersed with hills and valleys, most notably the Coosa and Cahaba Valley. Soils are derived from consolidated rock containing a series of sandstone ridges and limestone valleys and range from loams containing gravel to clay. Over half of the region is open land with the remainder in oak-pine forests with small pockets of loblolly–shortleaf pine. Common trees include white oak, northern red oak, black oak, chestnut oak, loblolly pine, shortleaf pine, and Virginia pine.

After the Civil War, this region's timber resources were exploited to the point that entire mountain ranges were completely denuded of vegetation. As a result, productive soils were washed away, dramatically decreasing site productivity for forest and wildlife habitat. Today habitat suitability for many species of wildlife in the mountainous areas of the Ridge and Valley Region is moderate to poor. Steep sites have little or no topsoil, which limits plant growth and browse availability. Areas of low soil fertility and poor forage quality scattered throughout this region are responsible for a low carrying capacity for deer, reduced birth weights of fawns, lower body weights of adult deer, and poor antler development. Deer and other species of wildlife in this region depend heavily upon mast-producing trees for food. Because of the high variability in mast production from year to year and low browse availability, wildlife food sources are often a limiting factor in this region.

Piedmont Region

The Piedmont of Alabama is a somewhat triangular area with gently rolling hills that extends into central Alabama from Georgia. Sites consist of clay and rocky soils that are derived from granite, schist, and igneous rock underlying the topsoil. In the early 1800s nearly all the Piedmont was cleared for farming. Years of poor farming practices associated with growing cotton, tobacco, and corn resulted in extensive erosion, depletion of soil nutrients, and a loss of soil productivity in this region. As soils were depleted, croplands were abandoned and these lands eventually reverted back to forest. Unfortunately, many areas in the Piedmont are still composed of "worn out" soils that are low in productivity for wildlife habitat. Ridgetops in this region are primarily in pine, while hardwoods occur along lower slopes and in bottomlands. Forest communities in the Piedmont are mostly loblolly–shortleaf pine, large blocks of replanted loblolly pine, and to a lesser extent mixed oak–pine. Longleaf pine is found on drier sites in the western section of this region. Major tree species in this region include loblolly and shortleaf pine with a mixture of white oak, northern red oak, scarlet oak, black oak, chestnut oak, Virginia pine, and hickories.

The quality of wildlife habitat within the Piedmont varies depending upon the site. Ridgetops and slopes are not as productive as lower portions of slopes and bottomlands associated with streams and rivers. Pine ridgetops and upper slopes should be managed as pine sites because they are usually not suitable for the growth of high quality hardwoods. In pine-dominated uplands, thinning and prescribed burning are recommended wildlife habitat enhancement practices. In some areas topography may limit the feasibility of these management practices. Lower portions of slopes and bottomlands have fertile soils that should be managed for browse and mast (soft and hard) production.

Tuskegee National Forest in Macon County is located in the Fall Line Hills (left). A stream crossing the Fall Line (right).

The Fall Line

The Fall Line runs from the northwestern corner to the eastern portion of central Alabama. This physiographic boundary marks the transition zone separating the mountainous regions from the Coastal Plain. It clearly delineates the older geologic formations of the upland regions from the more recent marine deposits of the Coastal Plain Region. The Fall Line is an important physiographic feature and delineates the distribution and abundance of some wildlife, especially **herpetofauna** (reptiles and amphibians).

Coastal Plain Region

The Coastal Plain covers two-thirds of the state and is located directly south of the Fall Line, sloping gently toward the Gulf of Mexico. In ancient times oceans covered these lands, which is why shark's teeth and marine fossils are found in some areas of this region. The geologic materials of the Coastal Plain are more recent in origin than those of the upland regions and consist primarily of unconsolidated or loose material, unlike the hard rock formations in the mountainous regions and Piedmont. Most of the Coastal Plain is relatively flat, but it contains a few hilly areas with topographic relief of 200 to 300 feet in elevation. Forests of the Coastal Plain are dominated by loblolly, slash, longleaf, and shortleaf pine with scattered stands of hardwoods. Because of the adaptability and productivity of southern pines to the Coastal Plain, this region is often referred to as the "timber belt" of the state. The Tombigbee, Alabama, Chattahoochee, and Conecuh rivers run through this region and have adjacent floodplain forest communities of oaks, water tupelo, and bald cypress. The Coastal Plain has two primary divisions: the Upper Coastal Plain (Hilly Coastal Plain) and the Lower Coastal Plain (Flatlands Coastal Plain).

Much of the Coastal Plain soils and vegetation, except for hardwood drainages and other unique sites, are adapted to and dependent upon periodic fire. As a result many species of wildlife have also become dependent upon fire. Bobwhite quail depend on periodic burning in the Coastal Plain to maintain understory vegetation at early successional stages. Red-cockaded woodpeckers require periodic burning in pine stands to maintain an open canopy necessary for feeding and foraging. Pine stands should be burned every two to five years as a wildlife habitat improvement practice in pine–dominated stands in this region after regeneration has been accomplished and the stand is old enough to withstand fire. Frequency of burning depends upon soil fertility, soil moisture content, and other site characteristics (such as topography) that influence how quickly vegetation grows back after burning.

UPPER COASTAL PLAIN REGION

The Upper Coastal Plain originates from the oldest marine sediments from what is now the Gulf of Mexico. In general, the land types within this province have higher elevations and greater relief (degree of slope) than those in the Lower Coastal Plain. Dominant forest communities are loblolly–shortleaf pine and, to a lesser extent, oak-pine. Several unique areas are found in the Upper Coastal Plain such as the Fall Line Hills, Black Belt, Chunnenuggee Hills, and Red Hills. The **Fall Line Hills** area, also known as the "Central Pine Belt," is located in the top portion of the Upper Coastal Plain between the Fall Line and Black Belt Region. This region extends from the upper northwestern portion of the state where the terrain is moderately hilly to the gently rolling portions of central Alabama. Soils are mostly well drained and diverse, ranging from clay to sand. Because of the variety of soil types, the Fall Line Hills contains the greatest habitat diversity of any area in the state.

RICHARD T. BRYANT

RICHARD T. BRYANT

Because of the high soil fertility of the Black Belt, wildlife managers recognize the wildlife habitat value of this region. Calcareous chalk outcrops are a common feature. Most of the Black Belt is open land supporting row crops, cattle and commercial catfish ponds.

The sandy, well-drained sites often support forests dominated by longleaf pine and turkey oak. Upland sites are primarily in pine with hardwoods on lower drainages. Some areas of the Fall Line Hills also have pockets of calcareous or limey soils that are primarily in red cedar.

Because of the diversity of soils and habitat types in the Fall Line Hills, wildlife habitat improvement practices must be tailored to soil and site characteristics of each area. Habitat improvement practices should concentrate on sites that have the greatest productivity and potential to produce quality wildlife habitat.

The **Black Belt** is a crescent-shaped band located in the center of the Upper Coastal Plain that extends east-southeast across central Alabama. The name "Black Belt" is derived from the dark-colored soils, which are made up of heavy clays that vary greatly in color. Soils of this area were formed mainly from marine deposits of calcareous chalk, are high in fertility, and are alkaline at the surface. Because of this high alkalinity, many Black Belt soils are not well suited for pine production, although they are capable of producing some impressive stands of red cedar. Some areas with more acidic soils grow loblolly and shortleaf pine, while a few pockets of neutral soils support scattered hardwoods. Most of the Black Belt is open land in row crops and grasslands. Most of the commercial catfish production in Alabama is in this area of the state.

The high fertility of Black Belt soils produces nutrient-rich vegetation higher in quality than native vegetation found in most areas across Alabama. This makes the Black Belt one of the most important areas for providing high quality habitat for some wildlife species such as white-tailed deer. Because of the high fertility of the Black Belt, sportsmen and wildlife managers recognize it as an area that consistently produces greater weights and better antler characteristics in white-tailed deer than most areas of the state (see Chapter 5). The number of record book bucks harvested in the Black Belt far exceeds those taken from any other area of Alabama. This area also supports higher deer densities than other regions of the state.

The **Chunnenuggee Hills,** also known as the "Blue Marl Region," is a narrow band running from east to west across the entire state between the Black Belt and Red Hills. The terrain is gently to strongly hilly, with the southeastern portion containing most of the hills. Soils have developed from unconsolidated marine deposits that are high in clay with sandy loam topsoils over sandy clay to clay subsoil. Dominant forests are loblolly–shortleaf pine, except in the western portion of this area which is composed primarily of post oak. This area is often called the Post Oak Flatwoods section of the Chunnenuggee Hills.

The **Red Hills** is a band of relatively fertile soils approximately 30 to 40 miles wide stretching across the southern portion of the state from east to west. Rocky bluffs, deep ravines, and clear, rock-bottomed streams characterize this region. The dominant forest type is loblolly-shortleaf and, to a lesser extent, oak-pine. Ridgetops and upper ravine slopes support a mixture of pines and hardwoods, while coves and lower slopes are composed of hardwoods dominated by oaks, hickories, beeches, and magnolias. Some of these sites are the only home to the Red Hills salamander, a federally listed threatened species.

Alabama's Natural Diversity

Dr. Doug Phillips is a Conservation Educator at the Alabama Museum of Natural History on the campus of the University of Alabama. He has been a conservation leader and educator in Alabama for many years and is host and producer of the critically acclaimed Public Broadcasting series Discovering Alabama.

Alabama has all the major ingredients of America the Beautiful, from the mountains in the north to the Gulf of Mexico in the south. Within Alabama there are five major physiographic provinces that can be further broken down into 37 unique geologically diverse regions. This gives rise to a wide range of soil types that support an uncommonly rich assortment of vegetation, forest communities, and wildlife. The constant forces of nature have also played a role in shaping the natural diversity of Alabama's landscape. Together this makes Alabama a kaleidoscope of unique habitats and a very biologically rich state. This variety makes Alabama one of the most naturally diverse states in the country.

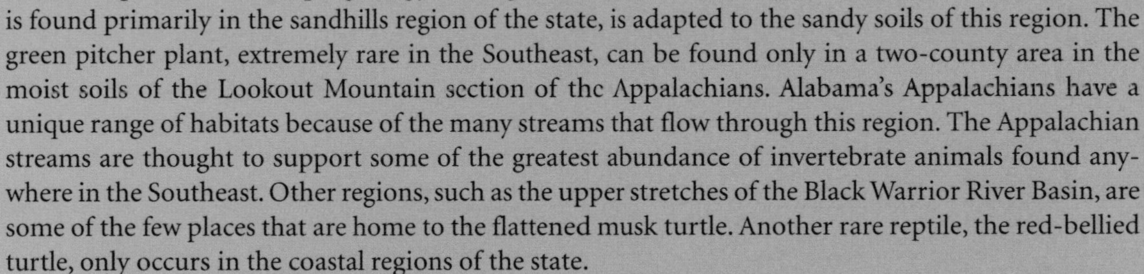

The large assortment of habitats fit together like pieces of a puzzle, providing transition areas or edge habitats that further increase Alabama's habitat complexity. Some species of wildlife depend on the unique geology and vegetation of an area. The gopher tortoise, which is found primarily in the sandhills region of the state, is adapted to the sandy soils of this region. The green pitcher plant, extremely rare in the Southeast, can be found only in a two-county area in the moist soils of the Lookout Mountain section of the Appalachians. Alabama's Appalachians have a unique range of habitats because of the many streams that flow through this region. The Appalachian streams are thought to support some of the greatest abundance of invertebrate animals found anywhere in the Southeast. Other regions, such as the upper stretches of the Black Warrior River Basin, are some of the few places that are home to the flattened musk turtle. Another rare reptile, the red-bellied turtle, only occurs in the coastal regions of the state.

As Alabama was settled, the natural landscape began to change as a result of various land-use practices such as agriculture, forestry, and industry. In the early 1800s a phenomena known as "Alabama fever" was at its peak and settlers were rushing into the state because the word was out that Alabama was rich in opportunity and natural resources. Woodlands, wildlife, and water were abundant. These resources later became the economic backbone of the state. However, as the land was settled, people forgot that it was the great abundance and diversity of natural resources that first attracted settlers to Alabama. By the early 1900s this forgetfulness had contributed to the overuse of natural resources and exhaustion of the land throughout much of Alabama. Although Alabama's natural landscape remains changed forever, much of the state's natural resource base has rebounded, largely because of conservation efforts initiated in the 1930s and 1940s by groups like the Alabama Wildlife Federation.

We should continue to remember that our diverse natural landscape, and the natural resources they support, are our economic and cultural foundation. They define our heritage and who we are as Alabamians. Alabama's landscape is becoming increasingly unique in that it is not a highly developed "industrial corridor" and instead can be recognized as a "corridor of nature." Our challenge for the future is to develop and initiate long-range land-use planning that incorporates economically sound decisions that sustain our diverse natural landscape. In essence we have to find ways to keep the best of the old while attracting the best of the new. Progressive leaders are striving to develop long-range conservation and economic development plans for the entire state. The plans call for evaluating the health and status of each physiographic region, determining areas that need to be protected, and assessing the suitability of each area for different land uses. With proper planning and foresight, Alabama can continue to balance economic development in a sustainable manner that conserves our natural landscape.

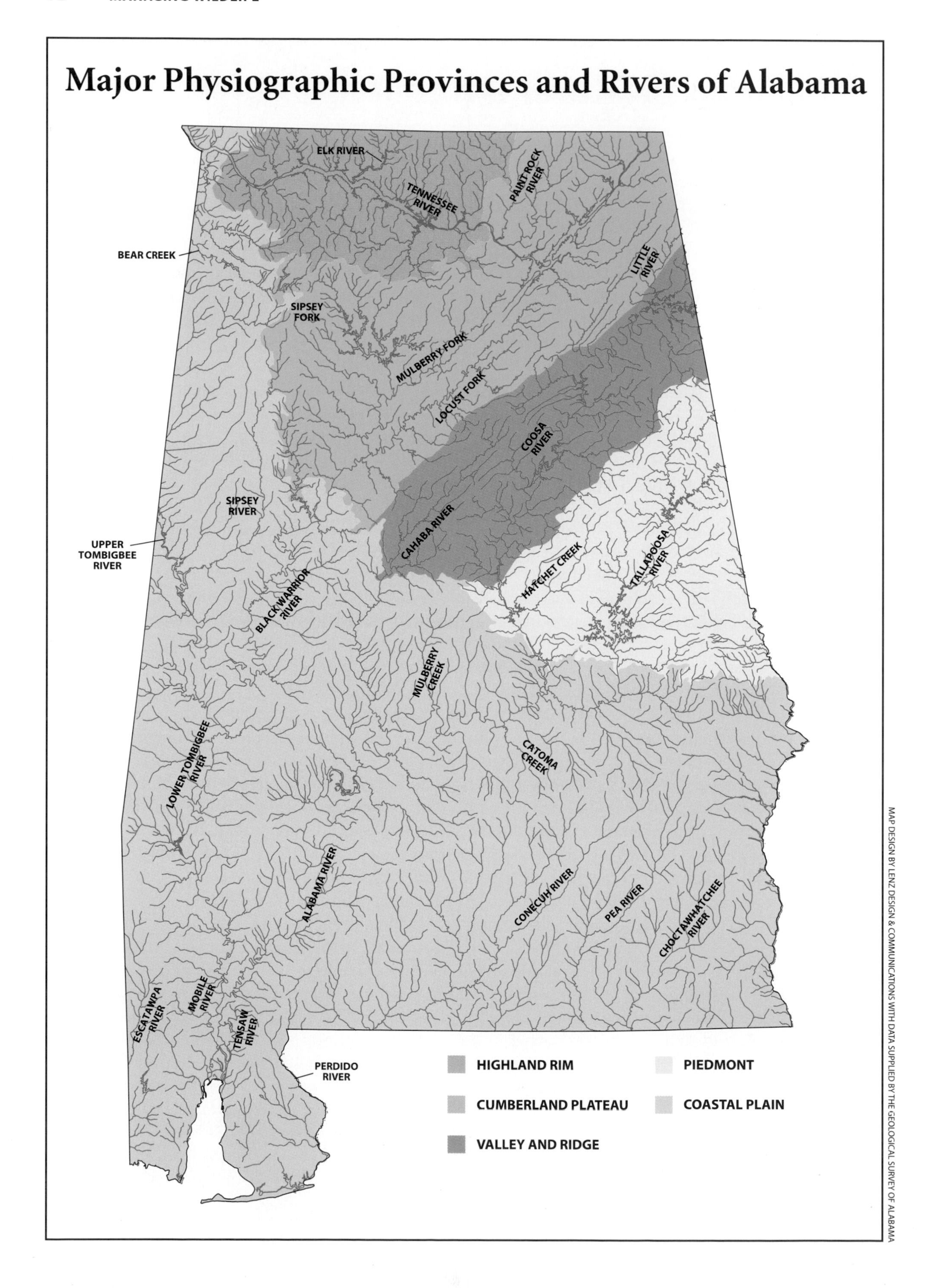
Major Physiographic Provinces and Rivers of Alabama
ELK RIVER
TENNESSEE RIVER
PAINT ROCK RIVER
BEAR CREEK
LITTLE RIVER
SIPSEY FORK
MULBERRY FORK
LOCUST FORK
COOSA RIVER
SIPSEY RIVER
CAHABA RIVER
UPPER TOMBIGBEE RIVER
HATCHET CREEK
TALLAPOOSA RIVER
BLACK WARRIOR RIVER
MULBERRY CREEK
CATOMA CREEK
LOWER TOMBIGBEE RIVER
ALABAMA RIVER
CONECUH RIVER
PEA RIVER
CHOCTAWHATCHEE RIVER
ESCATAWPA RIVER
MOBILE RIVER
TENSAW RIVER
PERDIDO RIVER
HIGHLAND RIM
CUMBERLAND PLATEAU
VALLEY AND RIDGE
PIEDMONT
COASTAL PLAIN
MAP DESIGN BY LENZ DESIGN & COMMUNICATIONS WITH DATA SUPPLIED BY THE GEOLOGICAL SURVEY OF ALABAMA

Physiographic Regions of Alabama

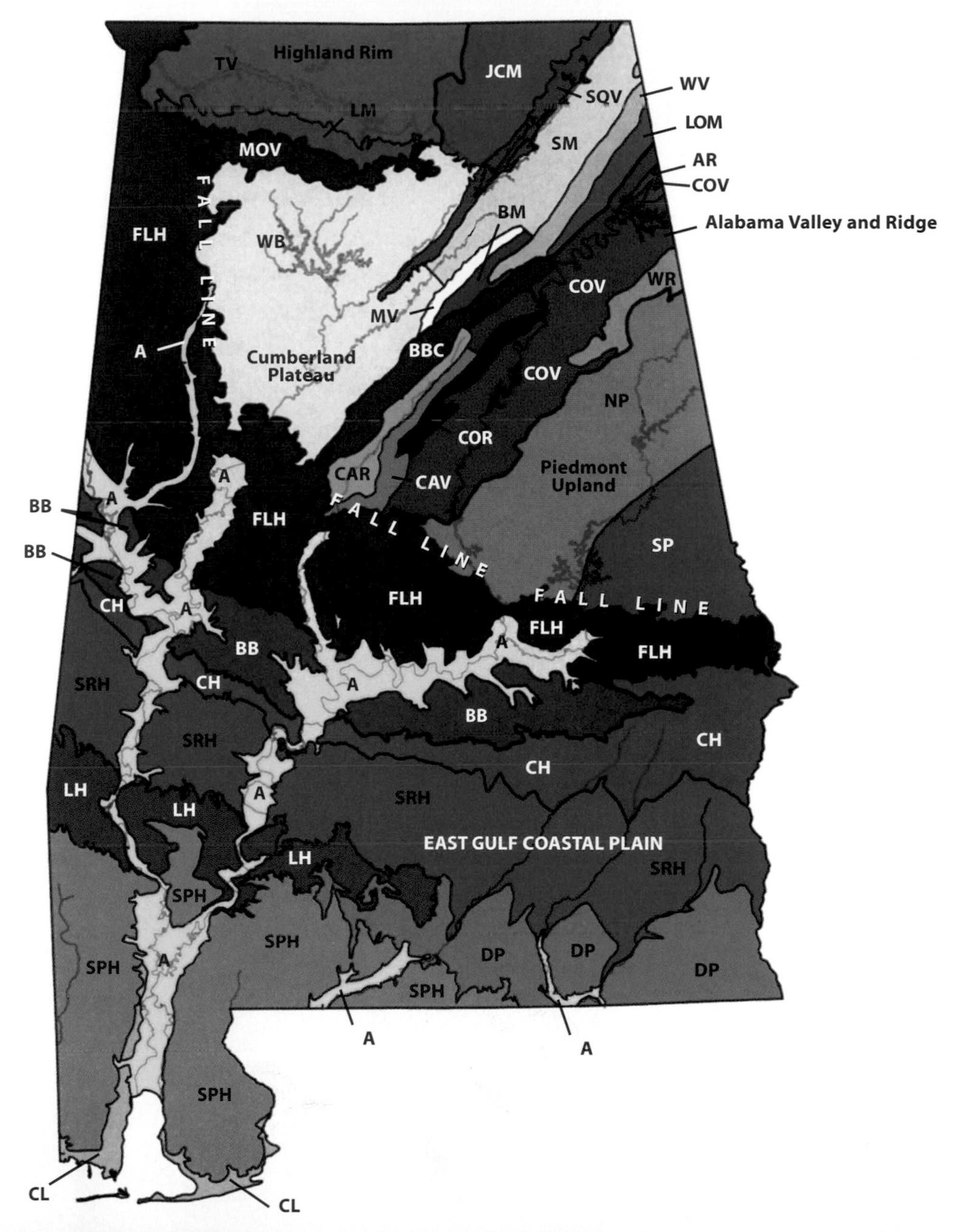

MAP DESIGN BY LENZ DESIGN & COMMUNICATIONS ADAPTED FROM *FISHES OF ALABAMA AND THE MOBILE BASIN*, METTEE M.F. ET AL., OXMOOR HOUSE, 1996.

Major Division	Province	Physiographic Section	District
Interior Plains	Interior Low Plateau	Highland Rim	Tennessee Valley (TV), Little Mountain (LM), Moulton Valley (MOV)
Appalachian Highlands	Appalachian Plateau	Cumberland Plateau	Warrior Basin (WB), Jackson County Mountains (JCM), Sand Mountain (SM), Sequatchie Valley (SQV), Blount Mountain (BM), Murphrees Valley (MV), Wills Valley (WV), Lookout Mountain (LAM)
	Valley and Ridge	Alabama Valley and Ridge	Coosa valley (COV), Coosa Ridges (COR), Weisner Ridges (WR), Cahaba Valley (CAV), Cahaba Ridges (CAR), Birmingham-Big Canoe Valley (BBC), Armuchee Ridges (AR)
	Piedmont	Piedmont Upland	Northern Piedmont Upland (NP), Southern Piedmont Upland (SP)
Atlantic Plain	Coastal Plain	East Gulf Coastal Plain	Fall Line Hills (FLH), Black Belt (BB), Chunnenuggee Hills (CH), Southern Red Hills (SRH), Lime Hills (LH), Dougherty Plain (SP), Southern Pine Hills (SPH), Coastal Lowlands (CL)
Alluvial-deltaic Plain (A) (not limited to one province)			

ROBERT P. FALLS, SR.

The Wire Grass Area of the Lower Coastal Plain in southeast Alabama is dominated by stands of longleaf pine. Periodic thinning and prescribed burning of this fire dependant land enhances timber production and supports many species of wildlife.

LOWER COASTAL PLAIN REGION

The Lower Coastal Plain, locally called the "pineywoods," is derived from marine sediments and is dominated by sandy loam, sand clay, or sandy soils. The majority of this region is low in elevation with little or no hills. Exceptions are the dry coastal sand dunes that are relatively infertile and support mostly sand pine. Much of the Lower Coastal Plain contains the natural range of loblolly, slash, and longleaf pine. Extensive clearing for agriculture, replanting of loblolly pine, and limited use of fire has reduced the native longleaf pine communities to isolated pockets surrounded by loblolly pine and hardwoods. Drier upland sites, or "sandhills," often contain longleaf pine and turkey oak communities.

Wetlands

Wetlands occur in various forms in each region of the state and are essential habitats for many wildlife species. Wetlands increase in size and number from north to south and are most common in the Coastal Plain. Unfortunately, over 50 percent of the original wetlands in Alabama have been lost through drainage and development since the time of the earliest settlers.

Wetlands include river bottom lands, swamps, marshes, bogs, and similar areas that are inundated or saturated with water during most of the year and have soils and vegetation that are characteristic of wet areas. Alabama's wetlands can be grouped into three broad categories: 1) salt and brackish water marshes, 2) freshwater marshes, and 3) forested wetlands.

Salt and brackish marshes are found adjacent to the Gulf of Mexico where the vegetative communities are composed of salt-tolerant plants such as cordgrass. Most of the salt and brackish marshes in Alabama are along the coast and associated with Mobile Bay and the Mobile/Tensaw Delta. Tidal influences have a major effect on the composition of most salt and brackish water marshes.

Freshwater marshes are relatively open areas, like bogs and wetland meadows. These areas are scattered across the state and are located where there is a consistent source of fresh water. Dominant vegetation includes grasslike plant species such as sedges, grasses, rushes, and cattails.

Most forested wetlands occur in the floodplain of rivers and streams. The most notable forested wetlands are adjacent to major rivers such as the Black Warrior, Sipsey, Tombigbee, and Alabama. The biggest block of forest wetlands in Alabama is about 288,000 acres and extends 50 miles north of Mobile to the junction of the Alabama and Tombigbee rivers. Another 100,000 acres extends up both rivers another 50 miles in a swamp that is ½–1 mile wide. The dominant forest commu-

BETH MAYNOR

The Grand Bay Savannah in the Coastal Lowlands of the Lower Coastal Plain in southwest Alabama contains stands of loblolly, slash, and longleaf pine.

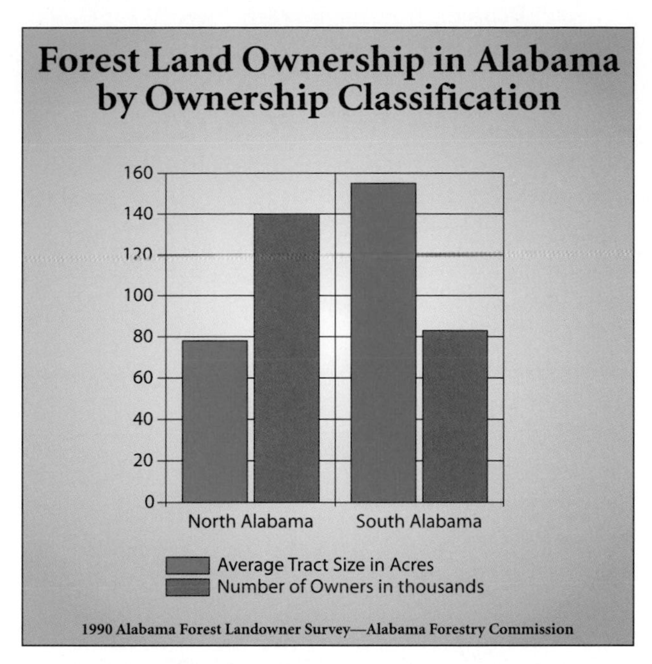

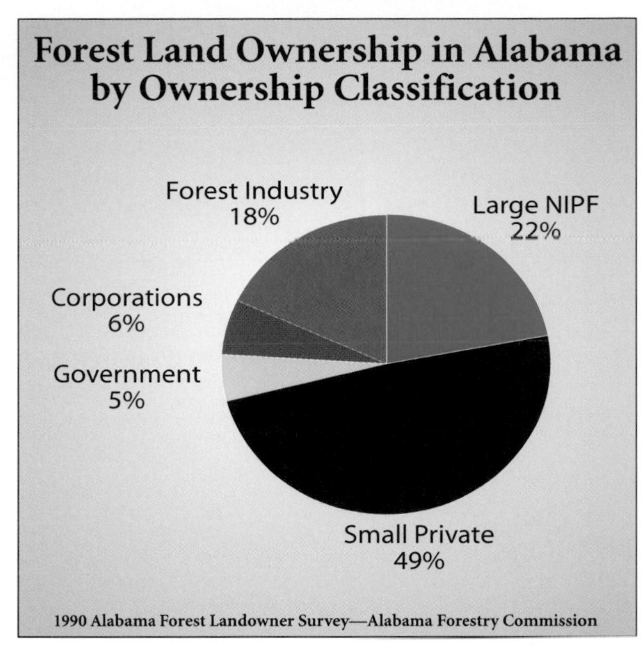

Fig. 6–7: Most of the forestland, and consequently a large portion of the wildlife habitat in Alabama, is owned by non-industrial private landowners. The number of forest landowners is greater in the northern portion of the state but tract size is smaller than those of south Alabama.

nity in forested wetlands is oak–tupelo gum–bald cypress and includes water tupelo, bald cypress, Nuttall oak, Shumard oak, and overcup oak in areas with prolonged flooding. Seasonally flooded forested wetlands are more commonly composed of elm, ash, hackberry, oaks, bitternut hickory, boxelder, and red maple. Other hardwoods, such as willows, maples, sycamore, cottonwoods, and beech, are also often present.

FOREST LAND OWNERSHIP

Approximately 95 percent of the forest land in Alabama is privately owned by non-industrial and industrial landowners. As a result, access to wildlife and wildlife management on these lands is almost exclusively controlled and conducted by private landowners. These individuals and corporations have been the key to past wildlife management successes that have sustained many of the wildlife populations that Alabamians enjoy today. The future of Alabama's wildlife resource depends on the continued management of private lands for wildlife and other natural resources.

The typical non-industrial forest owner in Alabama retains a relatively small forested tract averaging 80 acres. This ownership group controls nearly 50 percent (11 million acres) of all the privately owned forest lands in the state. The remaining forest land in Alabama is owned by large (greater than 500 acres) non-industrial private landowners who own 22 percent (4.8 million acres), forest industry with 18 percent (3.9 million acres), other corporations with 6 percent (1.3 million acres), and federal and state government with 5 percent (1.1 million) of the state's woodlands.

Within the state there is a significant difference in land ownership patterns. North Alabama contains a large number of owners with small tract sizes, while south Alabama has fewer owners with larger tracts. Cullman, Jefferson, and Marshall counties have the largest number of owners with typically the smallest average tract size in Alabama. Montgomery, Lowndes, and Chambers counties have the least number of owners with the largest average tract size.

SUMMARY

Alabama's diverse landscape is primarily a result of the five natural regions that support a variety of wildlife habitats across the state. Although some general assumptions can be made for each natural region, each land tract is unique. Interacting factors such as geology, soil characteristics, topography, climate, and use by man and animals influence the current wildlife habitat on a tract as well as the potential to improve the area for wildlife. Landowners should evaluate these factors in order to understand the capabilities and limitations of each site for producing quality wildlife habitats.

White-tailed deer are the most popular big game animal in Alabama. They are a valuable aesthetic, recreational, and economic resource to the state. Deer are now abundant in all of Alabama's 67 counties, making them one of Alabama's most successful wildlife management stories.

PAUL T. BROWN

CHAPTER 2

Developing a Wildlife Habitat Management Plan

CHAPTER OVERVIEW

A wildlife habitat management plan is the foundation of a successful wildlife management program. Written plans are a compass, as well as a step-by-step guide, for implementing habitat improvement practices that accomplish management objectives. This chapter discusses a straightforward approach for developing a successful wildlife habitat management plan.

CHAPTER HIGHLIGHTS PAGE

STEPS TO WRITING A PLAN 18
Step 1. Identifying Objectives 18
Step 2. Resource Inventory 19
Step 3. Designating Management Compartments 22
Step 4. Selecting Habitat Improvement Practices 26
FORMAT FOR A SIMPLE WILDLIFE HABITAT MANAGEMENT PLAN 26
IMPORTANT TOOLS FOR DEVELOPING A WILDLIFE HABITAT MANAGEMENT PLAN 27
ADDITIONAL WILDLIFE HABITAT MANAGEMENT PLAN CONSIDERATIONS 28
RECORD-KEEPING AND EVALUATION 29

Who would consider building a house without a blueprint? Or taking a trip without a road map? Successful wildlife managers and landowners carefully plan and target management activities to accomplish their objectives, minimize expenses, and ensure the long-term productivity of their property for wildlife and other resources. Management plans are a written guide for how, when, and where to implement wildlife habitat improvement practices. Developing or contracting a wildlife biologist to develop a sound wildlife management plan for forest or farmland is time and money well invested.

Components of an effective wildlife management plan include land management objectives and priorities, a resource inventory, site-specific habitat improvement recommendations, a schedule for conducting management practices, record keeping, and evaluation of management efforts and their impacts on wildlife habitat. A carefully developed plan provides a logical approach for using an assortment of habitat improvement practices. Some government cost-sharing programs also require that a management plan be written before cost-sharing funds are allocated.

STEPS TO WRITING A PLAN

No two wildlife management plans are exactly the same. Plans vary depending on management objectives and priorities, habitat and site characteristics, financial resources, existing land uses (such as forestry or farming), and the individual(s) writing the plan. Assistance for developing and writing a wildlife management plan is available from a variety of sources such as private consulting firms, the Game and Fish Division of the Alabama Department of Conservation and Natural Resources (ADCNR), the Alabama Forestry Commission (AFC), and the USDA Natural Resources Conservation Service (NRCS). Natural resource professionals from government agencies can provide advice and guidance in developing wildlife habitat management plans at no cost to the landowner. Some forest industry landowner assistance programs also provide guidance in developing management plans. In addition to agency and forest industry assistance programs, natural resource consultants also provide management plan expertise and services for a fee. Consultants should be professionally trained and registered foresters and/or certified wildlife biologists. At a minimum, a wildlife management plan should be reviewed by a certified wildlife biologist. A list of consultants can be obtained by contacting the AFC (334) 240-9300.

Wildlife habitat management plans can be prepared in a variety of ways depending on available resources. Large timber companies often have sophisticated computer programs for recording, analyzing, and displaying land management information. The advantage of using a computer-based recording and reporting system is that a large amount of diverse information can be quickly updated and is easily accessed in a variety of formats (maps, charts, text) for making management decisions. The disadvantage is that these systems are usually cost-prohibitive for the average landowner, although some private consultants offer these services. More typically, management plans include a written and visual description (sketch) of the land and other resources with recommendations for habitat improvements.

Whatever the approach, it is important that management plans are usable and flexible documents which guide forest and farm owners toward improving their land for wildlife. The following are important steps in developing a sound wildlife habitat management plan:

Step 1. Identifying Objectives

Landowners who neglect to identify and prioritize their management objectives and set priorities before conducting habitat improvement practices are often disappointed with their efforts and results because they never clearly defined what was important or what they wanted to accomplish. As one wildlife manager stated, "If you have no idea where you are going, how do you know when you get there?"

The first and most important step in developing a management plan is to clearly define, in writing, wildlife habitat management objectives and expectations. Objectives should be as specific as possible and include wildlife species to be managed as well as the expected outcome. For example, one objective may be to manage farm or forest land for quality deer with an expected outcome of healthy deer with large antlers and heavy body weights. Habitat improvement practices that improve the abundance and nutritional quality of native and planted deer foods can then be developed and implemented to help meet management objectives.

Objectives should also be measurable. This helps in evaluating the success or failure of habitat improvement efforts. Plans that lack measurable objectives are often ineffective, because there is no way to know if management objectives were ever reached. The ability to determine whether or not management objectives were accomplished helps identify successful habitat improvement practices. Plans can then be modified to include only those habitat improvement practices that successfully meet management objectives.

Landowners also need to consider how their wildlife management objectives fit with other land-use objectives such as farming or timber operations. Rarely do forest and farm owners have only one land-use or management objective. Landowners should prioritize their land-use/management objectives to have a clear understanding of where wildlife habitat enhancement efforts fit with other land management operations. In most cases, wildlife habitat improvement practices are compatible with other land management activities. If wildlife and habitat improvements are a top priority then some concessions and modifications may have to be made in timber, agricultural, or other land uses. Opportunity costs, or potential revenue foregone from other land management operations in favor of wildlife, should be a consideration when prioritizing land management objectives. However, if improving lands for wildlife is a secondary objective, then some concessions in wildlife habitat improvements may have to be made to accommodate other land uses. Defining and prioritizing land management objectives, as well as expected outcomes, helps landowners determine the best approach to managing their lands for wildlife and other resources.

A Wildlife Management Plan—Your Road Map to Improved Wildlife Habitat

Joe McGlincy is a consulting natural resource professional with Southern Forestry Consultants in Bainbridge, Georgia. He has many years of experience writing wildlife and forest management plans for private landowners across the Southeast.

Just as you would not think of going on a long trip to an unfamiliar place without a road map, a landowner should not begin a wildlife management project without a plan. The plan will be your road map for wildlife habitat enhancements on your property. Well-designed plans include strategies developed to assure that landowner objectives are met. Wildlife management is not an exact science and usually requires a little "trial and error." Landowners who implement a wildlife plan can evaluate the practices, continuing those that work and modifying or discontinuing those that do not. Don't leave home without your map and don't launch into a wildlife management project without a plan!

Step 2. Resource Inventory

A resource inventory is the process of identifying, locating, and recording land and other physical characteristics that have a potential to support wildlife or meet other land management objectives. An inventory helps to determine what you already have or will need to meet your objectives. It should include, for example, an assessment of the property and existing habitat, wildlife present on the property, equipment (tractors, disks, planters, etc.), facilities (lodging, barns, skinning and equipment sheds, etc.), labor requirements (by you and others), estimated management expenses and income, cost-sharing options, and sources of technical assistance. Information derived from a resource inventory and/or timber appraisal, in combination with management objectives, is the foundation for selecting and implementing habitat improvement recommendations.

Management plan objectives should be revisited and examined after a resource inventory and may need to be modified, depending on inventory results. A land survey may have revealed management limitations that would make accomplishing certain objectives difficult or unrealistic. The resource inventory may have also identified management opportunities that were not apparent when the objectives were first developed. Now is also an opportune time to reexamine personal resources. In light of the resource inventory, are objectives realistic in terms of time and money needed to achieve them? A review of management objectives, inventory information, and financial resources is prudent before selecting the type and level of intensity of habitat improvement practices.

THE PROPERTY

A survey of the property will determine availability

The Process of Developing and Implementing a Wildlife Habitat Management Plan

- Identify Management Objectives for Property
- Revisit and Modify Objectives (if necessary)
- Conduct Resource Inventory
- Designate Management Compartments
- Record Objectives and Descriptive Information by Compartment
- Select Habitat Improvement Practices and Schedule of Activities by Compartment
- Implement Management Practices by Compartment
- Refine Management Practices Based on Results
- Record Keeping and Evaluation

A typical profile of the Bama, the official state soil of Alabama, consists of a layer of dark brown fine sandy loam topsoil about 5 inches thick; a subsurface layer of pale brown fine sandy loam about 6 inches thick; and a subsoil red clay loam and sandy clay loam to a depth of 60 inches or more. Bama soils occur in 26 counties in Alabama on more than 360,000 acres in the state, mainly in the western and central part of the state, paralleling major river systems. Bama soils are well drained, have desirable physical properties, and are located on high positions of the landscape. These characteristics make them well suited to most agricultural and urban uses. They are well suited to cultivated crops, pasture, hay, and woodland. Cotton and corn are the principal cultivated crops grown on these soils.

and quality of existing habitat and the potential for improvement. Information in Chapter 3 is helpful in understanding general habitat components that should be present for wildlife. Habitat components important for individual wildlife species can be found in other chapters. This information should be reviewed prior to conducting a property inventory.

A property inventory is a two-step process that includes 1) identifying physical features (such as land-use and vegetative types, water sources, terrain, soils, and other natural and man-made features) from various maps and aerial photographs; and 2) a more detailed in-the-field survey of land features that are not easily identified from maps or aerial photographs. Information from maps, aerial photographs, and field observations should be included as a sketch or computer-generated base map and written description in the management plan.

Most land features can be identified using topographical quadrangle maps from the Geological Survey of Alabama, recent aerial photographs from the county USDA Farm Services Agency office, soil surveys and soil maps from the county USDA Natural Resources Conservation Service office, and property blueprints (plats) from the county tax assessor's office. These items are invaluable tools for developing a wildlife habitat management plan. The sketch map and written description should include information from maps, surveys, and aerial photographs such as property location, soil types and capabilities, topography, current land use, vegetative cover types, streams and other water sources, boundary lines, rights-of-ways, road systems, and other important features. If there is too much information to include on one sketch map, separate maps should be drawn. One map could include major features such as soil and vegetation cover types, while a second map could include other pertinent information. Transparent acetate sheets can also be used as overlays on sketch maps to provide additional information on sketches.

The next step is to add additional information to the sketch and written description that could not be identified from resource maps or aerial photographs. This is accomplished by walking over the property with the sketch map and noting unique features that might enhance or restrict wildlife habitat management efforts. Special attention should be given to the presence, arrangement, and condition of natural vegetation that provides food and cover for wildlife of interest. Landowners and managers should also note existing timber and mast-producing trees and other vegetation on the property, as well as other areas that could support additional trees, shrubs, grasses, and legumes that benefit wildlife.

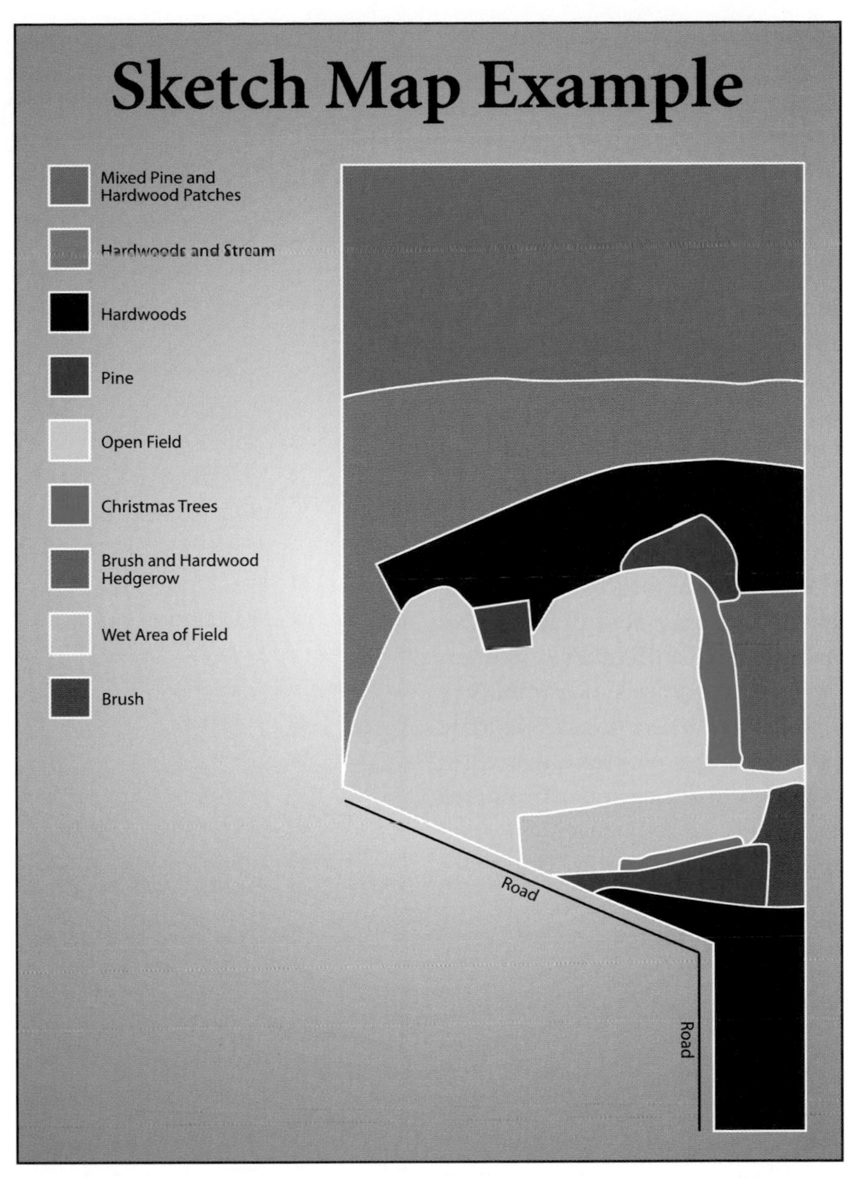

Fig. 1: A sketch map provides a visual illustration of features of the land such as vegetation or habitat types as well as other physical features on the property. Sketch maps should begin with a delineation of similar habitat types.

On-the-ground inventories should be made at least twice, at dawn and dusk, because these are peak activity periods for many species of wildlife. Walking over the property during these times helps determine what wildlife species are present on the land. Other signs of wildlife, such as scats or droppings, tracks and travel lanes, feeding areas, beds, nests, dens, burrows, and sounds can also help identify wildlife species that use your land. Special attention should be devoted to determining if threatened or endangered species are present on the property. Chapter 13 provides a helpful overview of threatened and endangered species that are most likely to occur on private lands in Alabama. Sites that support threatened or endangered species should be noted on the sketch map. These areas will require special attention and specific management considerations.

EQUIPMENT AND FACILITIES

Most farm and forest owners have some equipment and facilities that can be used for wildlife habitat improvement practices. A farm tractor can be used for establishing food plots, creating and maintaining fire lanes, and disking natural openings. If no equipment is available, some habitat improvement practices can be contracted out to local farmers and others who own tractors, disks, and planting equipment. Consultants usually have a list of vendors that own management equipment and provide habitat improvement practices. Every effort should also be made to integrate wildlife habitat improvement practices with existing farm and forestry operations to lower costs. The key is advanced planning and coordination with other land management activities.

Existing facilities, such as an old house or barn, are useful for housing management personnel and storing equipment. Labor personnel can stay on the property, while management activities are being conducted, for extended periods to reduce travel and expenses. Old barns and sheds can also be used to store seeds, fertilizers, lime, equipment, and other management tools and materials. Barns and old houses can also be refurbished and used as lodging for hunters or other guests.

Features to Include on a Sketch Map

BOUNDARIES:

- Property
- Land-Use/Cover Types
- Hardwood Forests
- Pine Forests
- Mixed Forests
- Wetlands
- Farmland (cultivated fields, old fields, pastures, food plots, bicolor/partridge pea patches, shrub plantings, and other plantings)
- Residential and other developed areas
- Water Sources (rivers, ponds, streams)

MAN-MADE FEATURES:

- Roads
- Trails
- Fences
- Firelanes/Firelines
- Houses
- Buildings
- Utility Rights-of-Way
- Easement Areas

NATURAL FEATURES AND SPECIAL SITES:

- Soil Type(s)
- Steep Slopes
- Ridges
- Seeps, Springs, Natural Meadows, and Glades
- Rock Outcrops
- Caves
- Waterfalls
- Rare, Endangered, or Threatened Species Habitat
- Special Recreation Areas
- Vistas, Special Scenic Areas
- Archaeological and Historical Sites

FINANCIAL CONSIDERATIONS

Management expenses depend on objectives, availability of labor and equipment, current land conditions, and whether or not wildlife habitat enhancement practices can be integrated with other land management operations such as forestry or farming. Where possible, wildlife habitat improvement practices should be planned and coordinated with other land management practices to reduce costs and disturbance to wildlife.

Management practice costs should be criteria for selecting the level and intensity of wildlife habitat improvement practices. In general, intensive management practices are more costly. Prescribed burning and disking may have similar effects on enhancing vegetation growth, but in general, an area can be burned at a lower cost than it can be disked. Management costs per acre are lowered as they are applied over a larger area. In other words, management costs per acre are lower on large land tracts than on small tracts of land. Management costs can also be reduced if they qualify for cost-sharing assistance. For more information on cost-sharing assistance programs see Chapter 17.

Step 3. Designating Management Compartments

Farms and woodlands are seldom uniform in the distribution of plant species, soils, productivity, and management potential. Because of these differences, a variety of management strategies are necessary for enhancing wildlife habitats across an individual farm or forest ownership. Land tracts should be divided up into management units called "compartments" to make the process of recommending and conducting habitat improvement practices over a large and diverse area easier and more efficient. Compartments are areas that have

Compartment Record Sheets

Compartment record sheets are vital components of a wildlife management plan. They are standardized information forms (8½ x 11 inch, three-hole punch) to record compartment management objectives, compartment descriptions, management recommendations, schedules of management activities, and records of management activities and impacts. Below is an example compartment record sheet that can be modified to meet landowner needs.

Compartment no. ______________________

Management objectives (includes priorities for wildlife, timber, and other land uses)

Wildlife ______________________

Timber ______________________

Other ______________________

Location of compartment ______________________

Description of compartment (Narrative description of compartment) ______________________

Size of compartment ________ acres

COMPARTMENT CHARACTERISTICS

Soil type and capabilities ______________________

Site index ______________________

Drainage ______________________

Aspect ______________________

Tree species composition ______________________

Volume/basal area of timber ______________________

Trees per acre ______________________

Mean DBH (diameter of tree at breast height) ______________________

Mast-producing trees ______________________

Fruit-bearing shrubs and herbaceous plants ______________________

Den trees and snags ______________________

Specific wildlife habitat information ______________________

ACTIVITIES TO BE CONDUCTED (IN A CALENDAR YEAR FROM START TO FINISH)

1. ______________________
2. ______________________
3. ______________________
4. ______________________
5. ______________________
6. ______________________
7. ______________________
8. ______________________
9. ______________________
10. ______________________

RECORD OF WILDLIFE, TIMBER, AND OTHER MANAGEMENT ACTIVITIES

Activity ______________________

Year ______________________

Impact of management activities ______________________

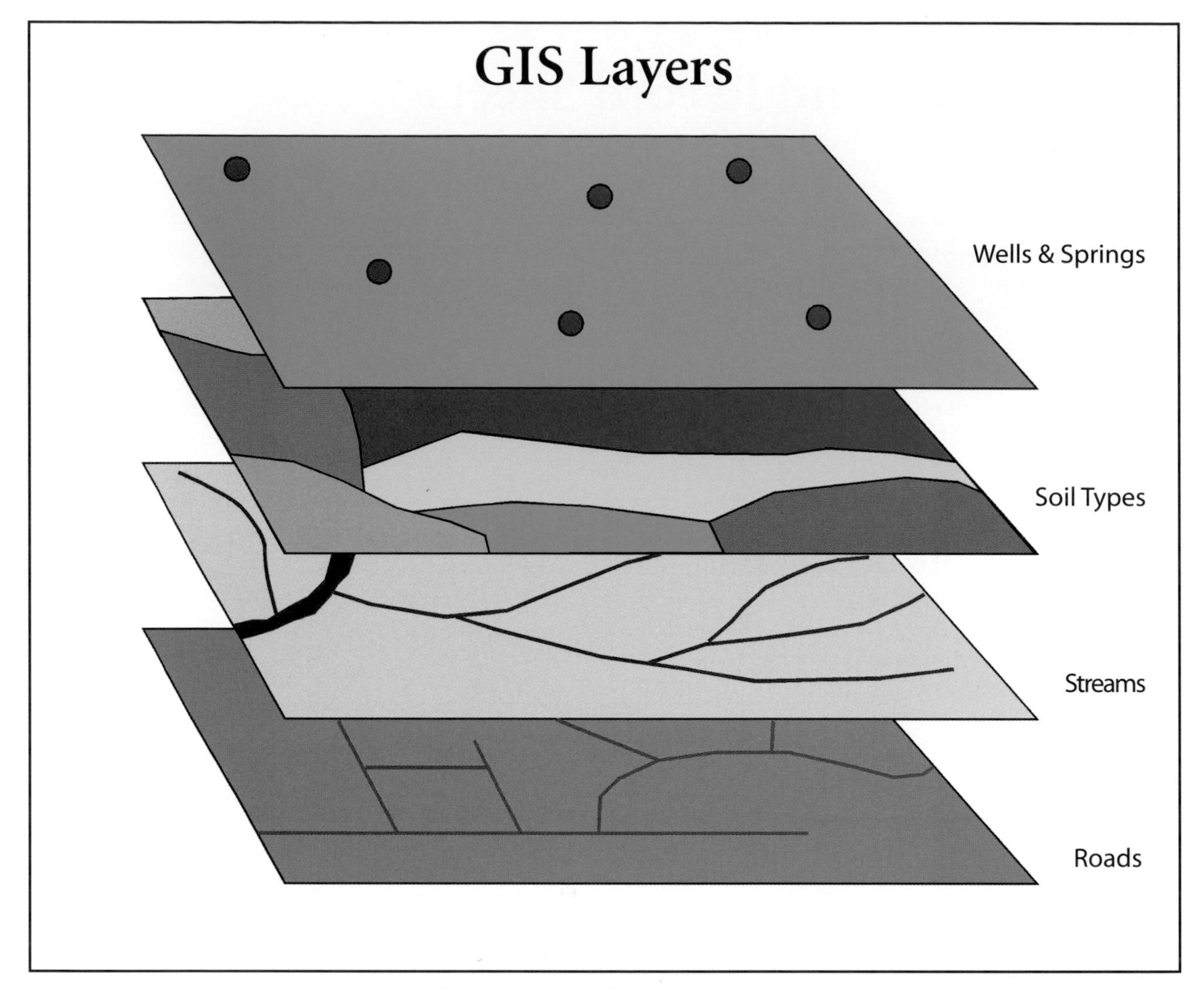

Fig. 2: Every landscape consists of a variety of characteristics or features that can be mapped to produce a more clearly understood model for planning land management activities.

Geographic Information Systems (GIS)

A geographic information system (GIS) is a method of entering, storing, manipulating, analyzing, and displaying information about land characteristics, use, and management on a computer. Maps and other information can be updated more quickly and accurately with a GIS than by conventional methods. GIS can also be used to predict or "model" impacts of certain land management activities before they are conducted, making it a powerful tool for wildlife habitat management decisions. Recent technological developments have made GIS programs available on desktop computers, helping to revolutionize land planning and management. Several GIS software packages are available for Windows-based desktop computers such as 1) Desktop Mapping System™ (DMS), R-Wel Inc., P.O. Box 6202, Athens, GA 30604, (706) 353-4166; 2) ArcView™ for Windows, Environmental Research Institute Inc. (ESRI), 380 New York Street, Redlands, CA 92373, (714) 793-2853 ext. 1375; 3) ArcCAD™, 3800 SW Cedar Hills Boulevard, Suite 225, Beaverton, OR 97005, (503) 646-5393; 4) PAMAP GIS/AVS, PAMAP Technologies, 3440 Douglas St., Victoria, B.C. Canada, V8Z 3L5, (604) 381-3838; and 5) EPPL7/GIS, InterTechnologies Group, Suite 330, 658 Cedar St., St. Paul, MN 55155, (612) 296-1206.

Fig. 3: This image is a Landsat Thematic Mapper false-color infrared satellite image of the South Carlton area in southwest Alabama, just north of the confluence of the Tombigbee and Alabama Rivers. The digital Landsat imagery, purchased from Space Imaging Eosat, Inc., is being used by the Geological Survey of Alabama to produce land use/land cover maps and GIS data sets.

Fig. 4: Aerial photography is another tool used in managing property for wildlife. Above, an aerial photograph of a section of Monroe County, Alabama taken by Marengo Aerial Service.

Compartment Information

Information to Include in a Written Inventory of Compartments (description of features identified on the sketch)

- Compartment Number (identifies compartment on land tract)
- Management Objectives (wildlife, timber, and other land uses)
- Location of Compartment
- Description of Compartment:
  - Size of Compartment (number of acres)
  - Soil type(s) and Capabilities
  - Site Index
  - Drainage
  - Aspect
  - Dominant Vegetation
- Timber Inventory:
  - Timber Species Distribution
  - Age Classes
  - Stand Density (number of trees per acre or basal area)
  - Average Sizes
  - Timber Volume/Basal Area and Grades
  - Timber Management History
  - Special Trees (number of mast-producing trees, den trees, snags)
- Game and Non-Game Habitats
- Wildlife Feeding Areas and Plant Composition
- Brush Piles and Windrows
- Nesting Sites and Water
- Unique Areas
- Special Places and Historical Sites
- Threatened and Endangered Species Habitat
- Areas Presenting Special Problems and Opportunities:
  - Stream Banks
  - Streamside Management Zones (SMZs)
  - Steep Slopes
  - Glades
  - Coves
  - Wetlands

similar characteristics such as vegetation, soils, topography, productivity, or other features. Compartments may be a pine plantation, hardwood stand, swamp, riparian forest, old homesite, or any particular field or field system. Because of their uniqueness, compartments can usually be identified from aerial photographs and maps.

After dividing a land tract into compartments, each compartment's potential for producing quality wildlife habitat should be evaluated using information from the resource inventory. This information should be used to develop site-specific management objectives and recommendations for each compartment. Wildlife habitat improvements should focus first on compartments that have the greatest potential for providing wildlife habitat.

Step 4. Selecting Habitat Improvement Practices

After the current conditions and management potential of each compartment are assessed, habitat improvement practices discussed in the book should be reviewed and selected for each compartment. As you read through the book, write down practices that would be most appropriate for your land, and then discuss them with a natural resource professional (a certified wildlife biologist or registered forester). When considering management alternatives, be sure to consider the impacts of practices, timing, costs, and the potential for each practice to complement or conflict with on-going land management operations on your tract and adjacent tracts managed by others.

FORMAT FOR A SIMPLE WILDLIFE HABITAT MANAGEMENT PLAN

The following is a suggested format for organizing a management plan into a three-ring, loose-leaf notebook. This notebook makes it easier to add materials to the plan. At a minimum, the plan should contain the six sections below. These sections can be marked in the notebook with colored index tabs for easy access:

- **General Description of the Entire Property:** Includes a brief description of the entire property such as location in the county, number of acres, past and current land uses, general forest and vegetation conditions, and number of compartments.
- **Land-Use and Management Objectives:** Includes a priority listing of wildlife and other land-use and management objectives. This section should also include a brief index of each

Sources of Information and Assistance for Developing Management Plans

Alabama Department of Conservation and Natural Resources (ADCNR)
Division of Game and Fish
64 North Union St.
Montgomery, AL 36109
(334) 242-3456

Alabama Cooperative Extension System
109 Duncan Hall
Auburn University, AL 36849-5612
(334) 844-4444

Alabama Forestry Commission (AFC)
513 Madison Ave.
Montgomery, AL 36130-0601
(334) 240-9378

Geological Survey of Alabama
420 Hackberry Lane
P.O. Box O
Tuscaloosa, AL 35486-9780
(205) 349-2852

USDA Farm Services Agency
P.O. Box 235013
Montgomery, AL 36123-5013

USDA Natural Resources Conservation Service
P.O. Box 311
Auburn, AL 36830
(334) 887-4560

U. S. Fish and Wildlife Service
P.O. Drawer 1190
Daphine, AL 36526
(334) 441-5181

compartment's management objectives.

- **Sketch Map:** Provides a visual description (sketch) of the property. May include several maps such as 1) a base map that shows boundaries, roads, and other man-made features; 2) a type map that differentiates cover types (timber stands, agricultural fields, and open fields); 3) a soils map that shows the location of different soil types; and 4) a compartment map that indicates where habitat improvement practices have or will take place, etc.
- **Compartment Record Sheets:** Contains descriptive information and wildlife habitat improvement recommendations for each compartment. Also includes a schedule of recommended management activities for the compartment for a 10-year period.
- **Field Notes Section:** Provides a commentary on impacts of management activities and wildlife observations taken directly from logbooks and archived in the three-ringed binder. The most appropriate place for storing field notes is by compartment.
- **Resource Materials Section:** Contains copies of aerial photographs, as well as topographic and soil maps used to draw the base map. This section should also include reference materials such as bulletins, leaflets, and articles on wildlife habitat management. The names, addresses, and telephone numbers of resource professionals that helped prepare the management plan and who will be conducting management practices should be included here.

IMPORTANT TOOLS FOR DEVELOPING A WILDLIFE HABITAT MANAGEMENT PLAN

- **Aerial photographs** are used to locate and identify natural and man-made features such as vegetation and forest types, land use, water sources, roads, rights-of-ways, buildings, and other features. They are also useful in delineating management compartments. Aerial photographs are available in black and white, color, or color infrared and in various scales. An ideal scale for management plans is 1"= 660'. Aerial photographs can be obtained from the county USDA Farm Services Agency or the county USDA Natural Resources Conservation Service office, or they can be contracted to be flown by private forestry firms.

- **Topographical maps** help to locate property in relation to physical features such as elevation, roads, water sources, and other land characteristics. Topographic maps can be obtained from the Geological Survey of Alabama at (205) 349-2852 or local map vendors.
- **County soil surveys** provide a description and map of soil types in a county. Soil surveys also provide soil suitability and productivity ratings for growing timber, producing wildlife habitat, and other land uses. They often have a description of the vegetation on various soil types. Soil surveys can be obtained from the county USDA Natural Resources Conservation Service office.
- **A field notebook and tape recorder** are useful for recording observations during the field inventory. Information recorded in the field can be transferred later from field notes and a tape recorder to the management plan.
- **Field guides** are useful for identifying wildlife, trees, shrubs, vines, and herbaceous vegetation during the field inventory. Color photograph guides with detailed descriptions are ideal.
- **Landscape architecture templates** are useful for drawing trees, shrubs, and other natural and man-made features on sketch maps. Templates can be found at most draftsman supply stores.
- **A camera** can be used to document wildlife habitat conditions before and after management practices.
- **Information from earlier land management plans** is invaluable in describing, recommending, and scheduling wildlife habitat improvement practices.

ADDITIONAL WILDLIFE HABITAT MANAGEMENT PLAN CONSIDERATIONS

- Where possible, integrate wildlife habitat improvement practices with other land management such as forestry or agriculture. If conducted properly, most silvicultural practices are also good wildlife habitat improvement practices and vice versa. Examine existing forest and farm management plans and modify them to include practices that also benefit wildlife. Wildlife habitat improvement practices should be an integral part of a total forest or farm master plan.
- Landowners should know the types and condition of wildlife habitat and current management practices on neighboring lands. In any case, neighboring land-use activities affect the landowners activities, and in some situations, land-use and management objectives may complement those of adjacent properties. Management practices, such as prescribed burning, can be conducted jointly with neighbors. Adjacent property may also provide habitat components not found on your land. Whenever possible, planning, development, and implementation of wildlife habitat improvement practices should be coordinated with neighboring landowners.
- Management plans (forestry, farming, and wildlife) should be shared with user groups such as hunters, horseback riders, and other outdoor enthusiasts, especially if these groups pay an access fee. Informing user groups of land management objectives and future management activities reduces potential conflicts and misunderstandings. User groups that are aware of management activities may also be willing to donate labor, such as hunters who may be interested in establishing and maintaining food plots or wildlife openings.
- Game population objectives and harvest strategies should also be included in the management plan.
- Nuisance wildlife problems (such as beavers or depredating deer) and control methods should be included in the management plan. Controlling nuisance wildlife requires a detailed plan of action.
- Technical assistance should be sought from consultants or agency professionals to design and implement a wildlife management plan. Natural resource professionals should ideally be Alabama registered foresters or wildlife biologists certified by The Wildlife Society.

RECORD-KEEPING AND EVALUATION

Management plans are dynamic documents that should be evaluated and updated periodically. Evaluations should be made annually for each compartment so that effective practices can continue to be implemented, while those that produce few or no results can be modified or discarded. Recording impacts of management efforts on compartment sheets is important in helping to evaluate the effectiveness of certain management practices. Keeping a logbook of observations and changes that occur in compartments can also provide valuable information for evaluating management efforts. Recorded observations should include estimates of vegetative responses to management practices as well as wildlife responses, such as deer and turkey use of food plots. Food plots that are not heavily utilized by wildlife in one area should be discontinued and moved to more suitable sites after an appropriate amount of time to allow wildlife to accept them. There is no substitute for good record-keeping as a basis for evaluating the effectiveness of wildlife management practices.

In a speech given to the National Wildlife Federation in April of 1997 in Tucson, Arizona, Nina Leopold Bradley recounts that her father, Aldo Leopold, was a great believer in the value of record-keeping. She stated, "My father's life-time habit of keeping records fortified his expanding factual and scientific knowledge of the natural system, and his continuing sense of wonder. From childhood on he wrote in a journal, detailing his emotions and his encounters with foxes, deer, and all manner of birds, plants, and weather. Year after year, the seasons come and go with their innumerable associated events. Spring brings the return of migrating birds, summer the blooming of prairie, fall the fruiting of plants, winter the freezing and thawing of the land. Phenology is the term for the science that records and studies these events. As my father kept his phenological records he summarized as follows:

"'Keeping records enhances the pleasure of the search and the chance of finding order and meaning in these events.'"

Aldo Leopold, the father of wildlife management.

SUMMARY

A wildlife habitat management plan is the foundation of a successful wildlife management program. Plans serve as a guide for how, when, and where to conduct wildlife habitat improvement practices. At a minimum, written plans should include management objectives, descriptive information about current land conditions and available resources, habitat improvement recommendations, timetables for conducting management practices, and a method for keeping records to evaluate the success of management efforts.

Wildlife habitat management plans should be viewed as flexible and dynamic documents that can be updated and revised periodically as habitat conditions and landowner objectives change. If forest or farm plans already exist, wildlife plans should be integrated into these plans to provide a coordinated and comprehensive total land management plan.

Information found in this chapter and throughout this book can provide guidance and information that can help landowners develop and write their own wildlife habitat management plan. Management plans written by landowners should be reviewed by a natural resource professional before any management practices are conducted. Wildlife habitat management plans can also be developed and written with the assistance of ADCNR biologists and AFC foresters. Natural resource consultants are also available to write plans and provide additional services for a fee.

Timber management is an integral part of wildlife management. This view illustrates an open longleaf pine stand, which provides excellent habitat for a variety of wildlife.

LEWIS C. ROGERS/SOUTH CAROLINA DNR

CHAPTER 3

Wildlife Habitat Management

CHAPTER OVERVIEW

Forests and farmlands in Alabama can be managed to provide most of the survival needs for a wide variety of wildlife. One should understand what the primary needs of wildlife are and how these requirements can be met through management. We will define wildlife and examine important habitat and landscape features that are necessary for supporting wildlife. Finally, we will relate several approaches to managing wildlife across the Alabama landscape and look at traditional management practices as well as the newly emerging ecosystem management concept.

CHAPTER HIGHLIGHTS PAGE

WILDLIFE HABITAT 32

- **Food** 32
- **Cover** 35
- **Water** 37
- **Space** 38
- **Arrangement of Habitat Components** 38

HABITAT SUCCESSION 40

- **The Importance of Dead Wood** 42

TRADITIONAL APPROACHES TO WILDLIFE MANAGEMENT 44

AN ECOSYSTEM APPROACH TO WILDLIFE AND NATURAL RESOURCE MANAGEMENT 45

- **Ecosystem Management Considerations** 47

Before attempting to describe wildlife habitat management, it's necessary to clarify the term wildlife, as wildlife means different things to different people. To a backyard bird enthusiast, wildlife may denote chickadees, nuthatches, and cardinals. To a hunter, it may suggest white-tailed deer, wild turkey, bobwhite quail, or gray squirrels. To a cattle farmer, it may mean coyotes.

Although a universally accepted definition of wildlife does not exist, wildlife biologists usually agree that **wildlife** are undomesticated, free-ranging animals. Because of society's appreciation of and emphasis on certain familiar animals, biologists have focused research and management efforts on those animals that have the greatest appeal to society.

Fish have not traditionally been categorized as wildlife by natural resource agencies. In 1940, President Franklin D. Roosevelt combined the federal Bureau of Biological Survey with the Bureau of Fisheries, creating the U.S. Fish and Wildlife Service. The new name implied that wildlife and fish were two separate categories, a distinction that persists today.

The majority of wildlife in Alabama and across the Southeast are native to the region. However, a growing number of wildlife species have been introduced intentionally or by accident. These wildlife are considered "non-native" or "exotic" and include species like the European starling, the house sparrow, and wild boar. Some non-native plants, like kudzu or bermuda grass, have become so invasive that they dominate thousands of acres that once produced valuable native wildlife foods. On the other hand, many trees and plants that are routinely planted to improve wildlife habitat for game species, like sawtooth oak, bicolor lespedeza, and chufa (a sedge-like plant), are also non-natives. Some exotics have become pests to humans and may serve as reservoirs for disease, often displacing native wildlife. For example, starlings and house sparrows compete with the native Eastern bluebird for nesting sites. This sometimes creates a dilemma for biologists and land managers.

Biologists, natural resource managers, and landowners will continue to debate the advantages and disadvantages of exotic introductions. Regardless, exotics have become and will remain a part of the plant and animal communities in the Southeast. Future efforts should be directed at reducing or eliminating exotics that have negative consequences for man and native wildlife. Beneficial exotics that are non-invasive can be useful during wildlife habitat management but should be used with care.

WILDLIFE HABITAT

Each species of wildlife requires a specific environment or **habitat** in which to live. To properly manage land for the benefit of wildlife, forest and farm owners must be aware of those parts of the environment that provide the habitat that wildlife need to survive and reproduce.

Wildlife have specific habitat requirements that must include **food**, **cover**, **water**, and **space**, all arranged in an accessible fashion. To sustain a wildlife species, these four components must be present in sufficient quantity and quality within that species' home range. Each species of wildlife requires different types of habitat components. For example, in most habitats white-tailed deer need at least 640 acres of land with a mixture of various trees, shrubs, and herbaceous plants that provide food and cover. A source of standing or flowing water within the 640 acres is also important. In contrast, during its entire life, a white-footed mouse requires only 2 to 3 acres with enough herbaceous vegetation to provide cover and seeds for food. Free-standing water is not required by white-footed mice since they obtain the majority of their water from succulent vegetation, dew, and the water produced as a by-product of their bodies' metabolism. Specific habitat requirements for many wildlife species of interest are discussed throughout the book.

Food

Obviously, food is an important habitat component for wildlife. Animals that have adequate food and proper nutrition throughout their lives grow larger and remain healthier than animals with poor nutrition during part or all of their lives. Generally, wildlife in good physical condition have higher reproductive rates, produce healthier offspring, are more resistant to disease, and can escape predators better than animals in poor condition. Consequently, nutrition directly affects birth and death rates and is key in the overall survival of any wildlife population.

The availability of food varies by season of the year and geographic location. Food can be abundant in one area during one season of the year and in critically short supply in the same area during other seasons. In Alabama, many plant-eating wildlife species or **herbivores** experience some nutritional stress during late summer and late winter. During these periods many natural food sources have been depleted or the quality of remaining foods has deteriorated. Lignification (hardening of plant cells) makes plants less palatable, digestible, and nutritious. Cold weather can also cause stress and force animals to consume more food to maintain body heat.

Increasing and improving food supplies are usually two of the main objectives of a wildlife management plan, and landowners should work to provide high quality food during the late summer and winter nutritional stress periods.

Diet selection in wild animals is driven by the quantity and quality of available food and physiological demands (pregnancy, raising young, etc.) of the animal. For instance, coyotes are adapted to eating a diet of small animals—mice, voles, and other rodents—during much of the year. However, when insects, fruits, and berries are abundant in summer, they can account for more than 80 percent of a coyote's diet. Farmers and home gardeners with holes eaten in their watermelons by coyotes can attest to this. Likewise, wild turkeys and bobwhite quail are fruit, seed, or grain eaters during much of the year, and a large portion of their diet will be dewberries, blackberries, and soft mast from other fruit-producing plants when they are available. However, wild turkeys and quail also consume large amounts of insects during the reproductive season to meet their high protein requirements during this time.

For predators, food availability means plenty of prey animals. Predators generally do not experience problems with diet quality because most animal matter is nutritionally complete and easy to digest. Even though predators expend a large amount of energy searching for, chasing, capturing, and killing their food, this extra expenditure of energy is offset by the higher nutritional content of animal matter.

Herbivores may become nutritionally stressed by a lack of food or by a lack of highly nutritious food. For example, in years when acorns are abundant, white-tailed deer are healthier because much of their diet consists of high-energy acorns. Dur-

Native Versus Exotic Species

Native species are those that naturally occur on an area, while **exotics** are species that persist outside their natural range of distribution. Over time many exotic plant species have been able to survive in the wild without cultivation; these species are called **naturalized plants**. The concern over exotic species is related to documented cases in which introduced species have become invasive, competing with native plants and animals. Some exotic animals also pose disease risks for native wildlife and domesticated animals. In addition, some invasive exotics have become economic pests for man. A few examples of exotic pests include the gypsy moth in oak forests, water hyacinth and hydrilla in wetlands and lakes, tall fescue in old fields, the zebra mussel in waterways, and kudzu across the southern landscape.

On the other hand, there are some introduced plants and animals that have contributed positively to man. Examples include most agricultural crops, the ring-neck pheasant, ornamental plants, and many plants that are used as supplemental plantings for wildlife (see Appendix D).

The debate on native versus exotic species will continue. Landowners and managers should evaluate each species on its own merits. Invasive plants should not be established for wildlife, especially if there are other choices. Native plant species should be a first consideration since they and have evolved with adaptations to local environmental conditions and with natural controls. One drawback to native plantings is that sources of seeds and seedlings are difficult to find and often expensive. However, as more wildlife managers begin to incorporate native plantings into their management plans, native plant materials should become easier to locate and less costly. (For a listing of sources of native plant materials see Appendix F.)

RICHARD T. BRYANT/ARISTOCK

ROBERT P. FALLS

RICHARD T. BRYANT/ARISTOCK

DR. H.S. BANTON

From top: Honeysuckle, kudzu, Johnson grass, and Chinese tallow tree are examples of invasive exotics that have become prevalent in Alabama.

From left to right, top to bottom: Strawberry bush, acorns, and greenbriar are examples of preferred and staple deer foods. Poison ivy, yaupon, and Eastern red cedar are examples of emergency or staple foods.

ing years when acorn crops are scarce or when many of the acorns are rotten, deer still have plenty of food to eat (tree and shrub leaves, forbs and grasses, etc.), but these plants contain less available energy than acorns.

Herbivores, like deer, do not feed randomly in the environment but show definite feeding patterns. These patterns are called **food preferences.** Food preferences can be determined by ranking the amount of a particular food item found in the diet relative to its availability in the field. Wildlife foods can be classified into four categories:

- **Preferred**—foods that are more abundant in an animal's diet compared to their abundance in the field;
- **Staple**—foods eaten on a regular basis and which meet nutritional needs, usually an animal's second choice;
- **Emergency**—foods eaten to meet short-term nutritional needs, usually eaten when staple foods are absent; and
- **Stuffers**—foods with low nutritional value that are consumed because there is nothing else to eat.

Even when food is available, wildlife can die of starvation or become malnourished. Although the woods and fields may look green and are covered with lush plants, this doesn't necessarily mean deer and other herbivores in the area have adequate food. Managing food supplies for herbivores requires matching the animal's food habits with what the land can provide. Some of the preferred foods of specific wildlife species can be found in later chapters.

It is important to understand the difference between starvation and malnutrition. Starvation is dying directly from a lack of food. If something like tularemia, a bacterial disease, reduces or eliminates rabbit or small mammal populations in an area, a red fox living in the same area may starve. However, wildlife in the Southeast rarely die from starvation.

On the other hand, wildlife in the Southeast may die indirectly from malnutrition because there is insufficient food of the required quality and quantity. Herbivores, like white-tailed deer, rarely die in the Southeast because they cannot find enough to eat but will suffer if foods are low in quality. Habitat that produces poor quality foods may lead to malnutrition and affect wildlife health, and in extreme cases, it can cause death. Over time, poor habitat may lead to nutritional deficiencies that cause the health of individual animals to decline, making them more susceptible to disease and environmental stress. Poor food quality makes body maintenance, reproduction, and caring for young difficult and in many cases impossible. In white-tailed deer, for example, the nutritional level of milk remains the same regardless of the doe's condition. However, the quantity of milk produced declines with malnutrition. In severe cases, a pregnant doe that is malnourished may abort a fetus or absorb it back into her body. Either way, malnourished does can reduce deer populations dramatically.

Landowners and managers can reduce or eliminate the potential problem of malnutrition by improving the quantity and quality of desirable native wildlife foods through habitat improvement practices. Land management practices that enhance native wildlife foods can be found in chapters throughout the book. Landowners can also supplement native foods by establishing food plots as described in Chapter 16, "Supplemental Plantings as Wildlife Food Sources."

Determining What Wildlife Eat in Your Area

There are several ways to determine what wildlife prefer and eat in your area. Determining wildlife food preferences can be done in three ways: 1) directly observing feeding animals, 2) examining vegetation or prey for signs of use (browsing, feeding on carcasses), 3) examining stomach contents of harvested game or trapped animals, or 4) examining scat or droppings.

PAUL T. BROWN

Observing feeding deer is one way to determine food habits. Deer tend to select the most nutritious foods by "mouthing" or tasting food items.

Cover

Cover is a vital habitat component that provides the protection necessary for an animal's survival and reproduction. Some people think of cover as something animals hide under. Actually, wildlife cover has many components. It provides cover that shelters animals from adverse weather conditions, and it provides escape or screening cover for protection while they feed, rest, and care for their young.

Cover is multi-dimensional in nature and relates to the functional needs of the animal. Consider when a deer is flushed in an open field. It will flee into the closest woods then often stop and resume feeding once inside the trees. The deer gets enough horizontal cover, space, and foliage between you and it to feel secure.

Deer use different types of cover depending upon their needs and activities. They may feel comfortable enough to forage in the protective cover of an open pine or hardwood forest, in an open field that has an overgrown fence line, or in a wide, weed-filled ditch. However, when resting, they prefer to bed in relatively thick areas such as a 3- to 10-year-old pine plantation or a thicket. A diverse mixture of vegetative types and ages will provide the majority of cover that different wildlife species require.

Wildlife need cover for nesting, escaping predators, breeding, rearing young, and resting. For bobwhite quail, the general cover requirements are a mixture of 30 to 40 percent open pine woodland or grassland, 20 to 50 percent cropland, and 5 to 20 percent brushy or thicket cover. Many biologists believe that declines in quail numbers across the Southeast are related to changes in the availability of cover. One potential cause is large-scale prescribed burning, called **compartment or block burning**. Compartment burning reduces critical nesting cover and makes quail and other ground-dwelling wildlife more vulnerable to predation. Large expanses of open forest created by compartment burning may be ideal for timber management and is suitable habitat for deer and turkey, but quail suffer because the loss of escape cover makes them easy targets for predators and nesting cover is eliminated over large areas. An alternative to compartment burning is **patch burning**, where small areas are burned in a mosaic pattern across a forest stand. Patch burning retains valuable cover for ground-dwelling wildlife like quail, but at the same time it provides the benefits associated with prescribed burning. A drawback to patch burning is that it is more labor intensive and costly than burning large compartments.

Some animals are not very selective about what they use for cover. The opossum can live in almost any type of habitat including cities and garbage dumps. Other wildlife are more selective about the type of cover they require. Given the interrelated cover needs of a variety of wildlife, it is almost impossible to provide cover for one species without influencing the cover needs of other species.

Often the cover requirements of wildlife may be similar. Cottontail rabbits, cardinals, and ground-nesting

Fig. 1: Different Types of Brush Piles for Cover

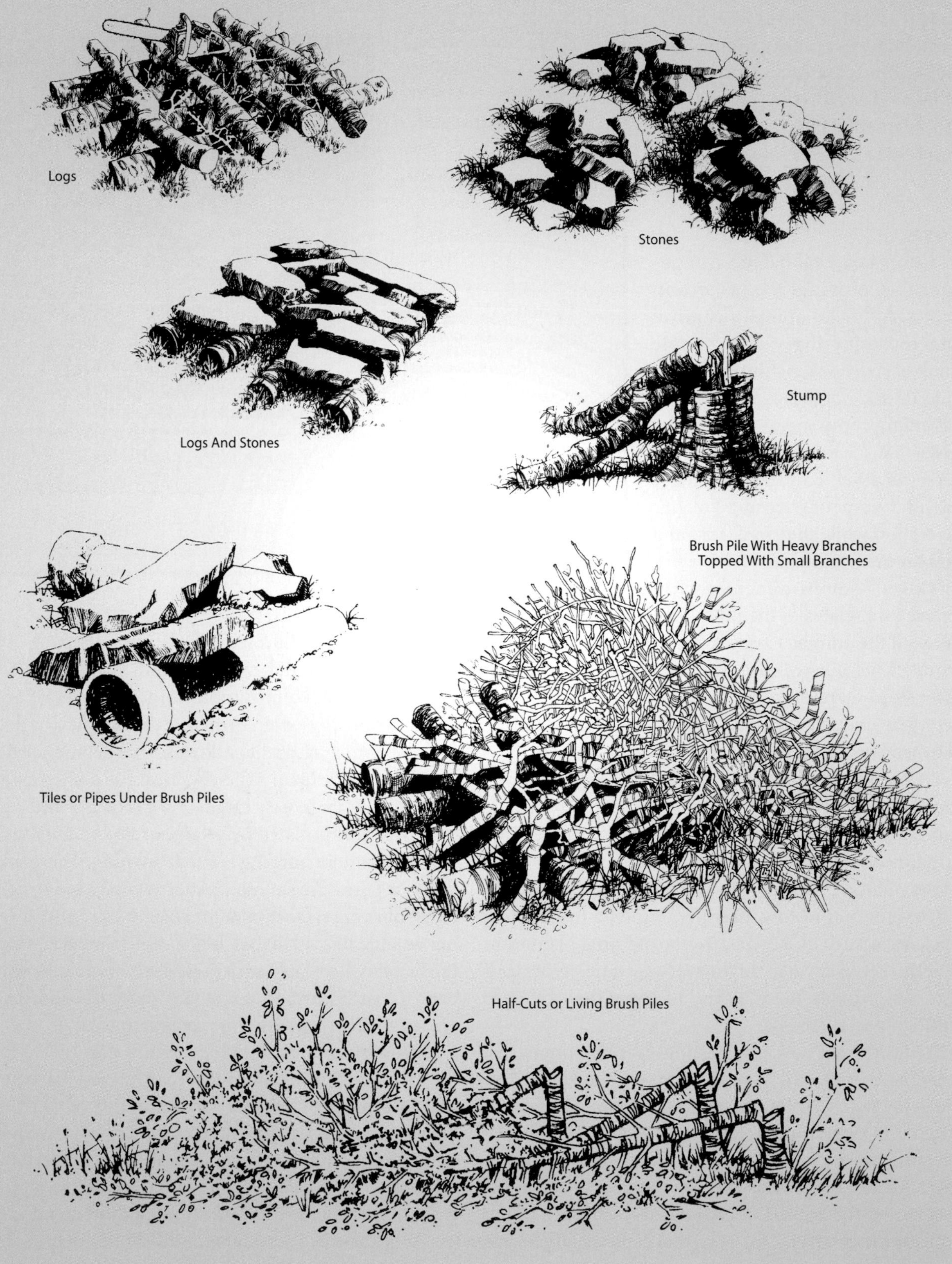

Logs

Stones

Logs And Stones

Stump

Brush Pile With Heavy Branches Topped With Small Branches

Tiles or Pipes Under Brush Piles

Half-Cuts or Living Brush Piles

wildlife will all benefit from cover established for quail. Gray squirrels may require different types of cover such as the vertical cover found within the canopy layers of a forest. The key is understanding the cover requirements of the desired wildlife species and providing these requirements in sufficient quantity. This may involve distributing a variety of cover types across the landscape. A diversity of cover types contributes to the patchiness of vegetation and often enhances the different types of wildlife found on a given tract of land.

Cover can also be important in connecting habitats that have been broken or fragmented by land-use practices like agriculture, silvicultural operations, or human development. Travel corridors can help to connect habitats that have been separated. **Travel corridors** are areas of continuous or unbroken habitat that permit animals to travel securely from one area to another. These pathways can vary in size and may consist of simply an overgrown ditch or fencerow running through a field, or a river that traverses the landscape and is bordered on each side by forest (see Streamside Management Zones [SMZs] Chapter 4). Providing corridors can be an integral component of management plans where land-use practices have separated large sections of wildlife habitat. For more on travel corridors refer to Chapter 4.

Water

Wildlife require water for digestion, metabolism, reducing body temperature, and removing metabolic wastes. More than 80 percent of an animal's body is composed of water. Wildlife can survive only a few days without water. Clearly, water is crucial for survival. Fortunately, the supply of freestanding water in Alabama is not usually a concern. Springs, creeks, farm ponds, and other sources, including about 60 inches of annual rainfall, provide adequate standing water for most species of wildlife.

Some wildlife species can obtain all the water they require through a diet of green plants, from dew on leaves, or as a metabolic by-product of their bodies breaking down fat and starches. The water requirements of wildlife vary, and people sometimes overestimate the importance of freestanding water, although the availability and proper distribution of standing water usually enhances wildlife populations. As a general guideline, water sources should be no more than ½ mile apart.

Basic Biology and Habitat Requirements

Stan Stewart is a wildlife biologist with the Game and Fish Division, Alabama Department of Conservation and Natural Resources in Montgomery, Alabama. He provides wildlife technical assistance to landowners across the state.

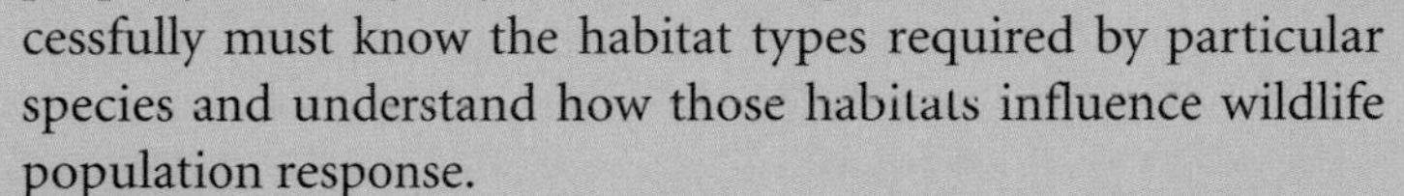

A common experience in my work is seeing effort and money ineffectively invested on practices to enhance wildlife. This happens because the practitioner does not understand basic habitat requirements and behavior of wild animals. A mechanic can repair an automobile because he is familiar with the parts of an engine and understands how the different parts work together. He observes how the engine is running, identifies the faulty parts, and replaces or corrects those that are not running properly. Similarly, anyone who manages wildlife successfully must know the habitat types required by particular species and understand how those habitats influence wildlife population response.

Before a practice is employed, a manager should consider how the practice relates to habitat structure and if it meets a need of wildlife. Does the practice supply a missing or faulty part of an animal's environment? As an example, the first and often only practice I see performed by landowners interested in quail is the cultivation of food patches. Food is assumed to be the missing part of the quail habitat. Other habitat needs are not considered. But, in my experience, quail populations most often do poorly because of deficient nesting cover. This critical habitat type is seldom provided or allowed to develop extensively in today's landscape. A food patch cannot increase the quail population when the faulty part of the habitat is nesting cover.

I would advise anyone to invest time learning the basic biology and habitat requirements of a species prior to investing time and money on management practices.

Shrubby borders increase the benefits of the "edge effect" by providing both food and cover for wildlife.

Space

Each wildlife species requires a certain amount of space in which to live, move about, avoid or escape predators, locate a mate, obtain sufficient food and water for survival, and rest. This space is often referred to as the **home range** of an animal. Space requirements are behavioral and social responses that help to ensure an animal's well being.

Wildlife space requirements vary by species, but generally, the amount of space required is determined by the quantity and quality of food, cover, and water found in an area. Other factors affecting space needs of wildlife include the following:

- **Body size**—larger animals require more space than smaller animals.
- **Dietary preferences**—carnivores require more space than herbivores.
- **Carrying capacity**—the number of animals a given area can support. The lower the carrying capacity of the land the larger the home range required.
- **Mobility**—some wildlife travel longer distances than others.
- **Territorial and social behavior**—some animals will share home ranges while others will not.

Arrangement of Habitat Components

Food, cover, water, and space requirements vary from one species of wildlife to another. Within an individual species these requirements can also vary based on age and sex, physiological condition (pregnancy, nursing young, antler growth, etc.), time of year, and geographic location. The location of the habitat necessary to meet these requirements is also important, since many wildlife species (except those that migrate) rarely travel more than 1 mile from where they were born. To provide all the habitat requirements of a particular species, each component must be properly arranged within the area that is used by that species.

The intermixing of habitat components such as forest, pasture, cropland, and other plant communities is called **interspersion**. Interspersion creates **horizontal habitat diversity**—the presence of a variety of habitats across the landscape. The greater the mixing of habitat types, the greater the degree of interspersion and horizontal habitat diversity. Many animals, game species in particular, benefit from high interspersion and

Different species and age classes of trees and other vegetation within a forest stand provide a layering effect and increase vertical habitat diversity.

Measuring Interspersion and Edge on Your Property

There is a simple way to measure the amount of edge and interspersion on a tract of land. The first step is to obtain an aerial photograph of the property. Draw two imaginary lines connecting each corner of the property boundary and count the number of times the habitat changes along each line. Next, add the two numbers together to get an **interspersion index** value. The higher the value, the better the tract of land for quail, rabbits, and other wildlife that like areas with high interspersion. You can also circle each point where three vegetation types occur to identify how many "headquarter" areas are on the property. **Headquarters** are areas that are highly preferred by many wildlife species.

horizontal habitat diversity. The intermixing can be thought of as a puzzle where each piece of the puzzle represents a different type of habitat. All of the pieces must be present and in the right place for the puzzle to be complete. Properly arranging the pieces of the puzzle helps ensure that the needed habitat components are available within an animal's home range on a year-round basis.

Within a forest or plant community, the different layers in which plants grow is also an essential part of habitat arrangement. This is called **vertical layering**. Vertical layering increases **vertical habitat diversity** and enhances the ability of a forest or plant community to maintain a variety of wildlife species.

Some wildlife species may use the ground level (herbaceous) layer or understory for food but also need the tallest level (tree canopy) or overstory for cover, making both low and high layers vital. The middle layer between the tree canopy and the herbaceous layer is comprised of shrubs and small trees and is called the **midstory**. Mature forest communities have different types of vertical layering. Some may have one distinct layer of tall trees. Others may have a variety of layers comprised of grasses, broadleaf weeds (forbs), shrubs, small trees, and large trees. Forests with multiple layers provide more habitats for wildlife than forests with single layers.

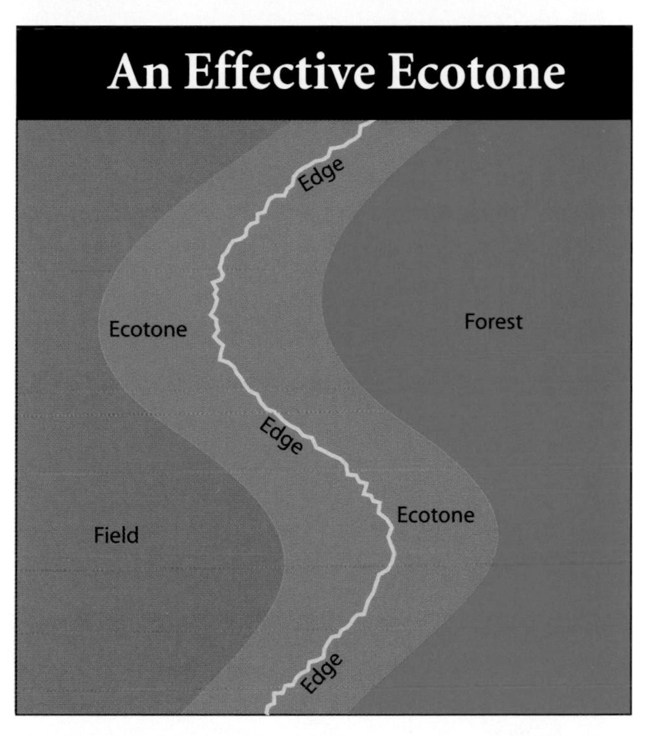

Fig. 2: The transition zone where two or more habitat types come together is called an ecotone. Ecotones provide a gradual change and a greater diversity of food and cover as compared to an abrupt change from one habitat type to another.

The boundary where two or more different plant communities or habitats join, such as where a forest meets cropland or where two different forest types meet, is called an **edge**. Sometimes the landscape abruptly changes between plant communities. At other times no sharp or distinct differences exist, but rather there is a gradual change from one plant community to another. This mixture or transition between two adjacent habitats is called an **ecotone**. Gradual edges share characteristics of both plant communities and offer a greater selection of food and cover for many species of wildlife. The greater the width of the ecotone, the greater the value to wildlife. Wildlife that prefer these transition areas are called **edge-adapted species** and include most game species, such as rabbits and quail, and many species of non-game wildlife, such as the American robin and the common flicker.

Large amounts of edge and highly interspersed landscapes are not beneficial for all wildlife species. Some wildlife, such as the Swainson's warbler or red-cockaded woodpecker, need one type of unbroken habitat to meet their survival needs. Landowners who are inter-

Headquarters, also called "coverts" or "prime habitat corners," are points where more than three vegetation or habitat types meet.

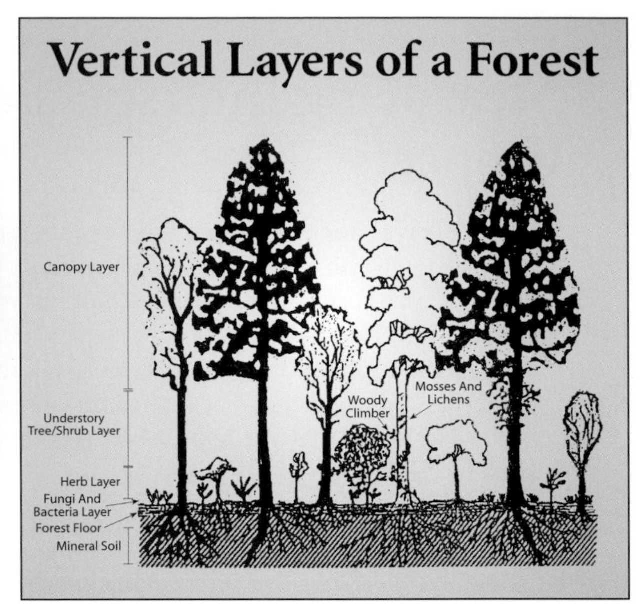

Fig. 3: Multiple canopies or layers within a forest stand provide many habitat types that are used by a variety of wildlife, especially birds.

ested in providing habitat for a wide variety of wildlife should consider trying to maintain a balance between edge habitat and blocks of unbroken forest.

Headquarters, also called "coverts" or "prime habitat corners," are areas where more than three vegetation or habitat types meet. Like edge, headquarters are very attractive to wildlife because they provide multiple habitat components in a small area and should be encouraged in management plans.

HABITAT SUCCESSION

Matching land management with the needs of wildlife requires an understanding of how habitats change over time. This natural change or process is called **plant succession**, which is an orderly and predictable change in plant communities over time. Most forest owners are aware of these variations as they see their newly regenerated forest grow from seedlings to mature trees. Farmers also know about succession such as when a fallow field gradually becomes overgrown with weeds, shrubs, and then trees. Often this is referred to as "going back to woods."

Plant succession and changes in plant communities directly affect wildlife. Consider an abandoned soybean field in Alabama. During the first year, annual plants that are propagated from seeds stored or banked in the soil, such as ragweed, foxtail grass, and doveweed, become established. At this time, ground-nesting birds like doves, quail, and some songbirds find the area desirable because they feed on seeds produced by these **annual** plants which grow, produce seed, and die each year.

As time goes by, the field will become filled with heavy-seeded and berry-producing plants, weeds,

Openings within a forest stand are first dominated by early successional herbaceous plants favoring wildlife such as insects, insectivorous birds, and herbivores such as deer.

grasses, and legumes. Many of these plants are **perennial**, meaning that they survive for more than one year. This type of plant community favors quail, cottontail rabbit, and numerous other small mammals like voles or white-footed mice. In turn, these small mammals provide abundant food for red foxes, coyotes, hawks, and owls. Wild turkeys find this stage of plant growth (usually about three years of growth and often called a **rough**) ideal for nesting. Likewise, deer will readily feed on the encroaching succulent herbaceous and woody growth like American beautyberry, blackberry, and the tender tree seedlings that sprout during spring and early fall.

During the third to eighth year, broomsedge and other grasses, blackberries, pines, and hardwood tree sprouts form briar thickets and other dense, brushy areas, creating habitats that continue to favor cottontail rabbits, quail, deer, and many species of songbirds. Deer browse (woody plants) and cover are generally greater during this period than at any other time. Nesting sites for brush-nesting songbirds are also abundant during the early part of this stage. After about 8 to 10 years, shrubs and trees grow larger and begin to shade the ground. As a result the forbs and grasses that once grew in the field are suppressed and decrease in abundance.

As the forest matures it forms a plant community which is called a **climax community or climax forest.** This last community may survive for hundreds of years or until something happens to disturb it. In much of Alabama, depending upon disturbances, this relatively stable and long-term plant community is composed of a mixture of oaks and hickories. An undisturbed pine forest will eventually be naturally replaced by a mixture of oaks and hickories. During the final stages of succession, cavity-nesting wildlife such as raccoons, gray squirrels, and owls all flourish and other species like turkey benefit too.

RICHARD T. BRYANT/ARISTOCK

Dead and dying trees, such as those attacked by the pine bark beetle, create openings within a forest stand and increase habitat diversity.

Each species of wild animal thrives only in those plant successional stages that best meet its requirements for survival. Some animals thrive only in a specific stage of plant succession. For example, cottontail rabbits need shrubby fields. Other species like white-tailed deer need a wider variety of plant successional stages.

ED ORTH/ALABAMA ENVIRONMENTAL COUNCIL

Dead wood is an important component of wildlife habitat and should be retained within a forest stand. Fallen trees quickly become infested with fungi and insects that provide food for wildlife.

Wildlife habitat management involves manipulating succession to improve habitats for desired wildlife species. An area of land may be "set back" to an earlier stage of succession by mechanical or chemical methods, by fire, or by natural catastrophes like tornadoes, insects (like pine beetles), and disease. In general, the more intensive the management activity, the further succession is set back. Many silvicul-

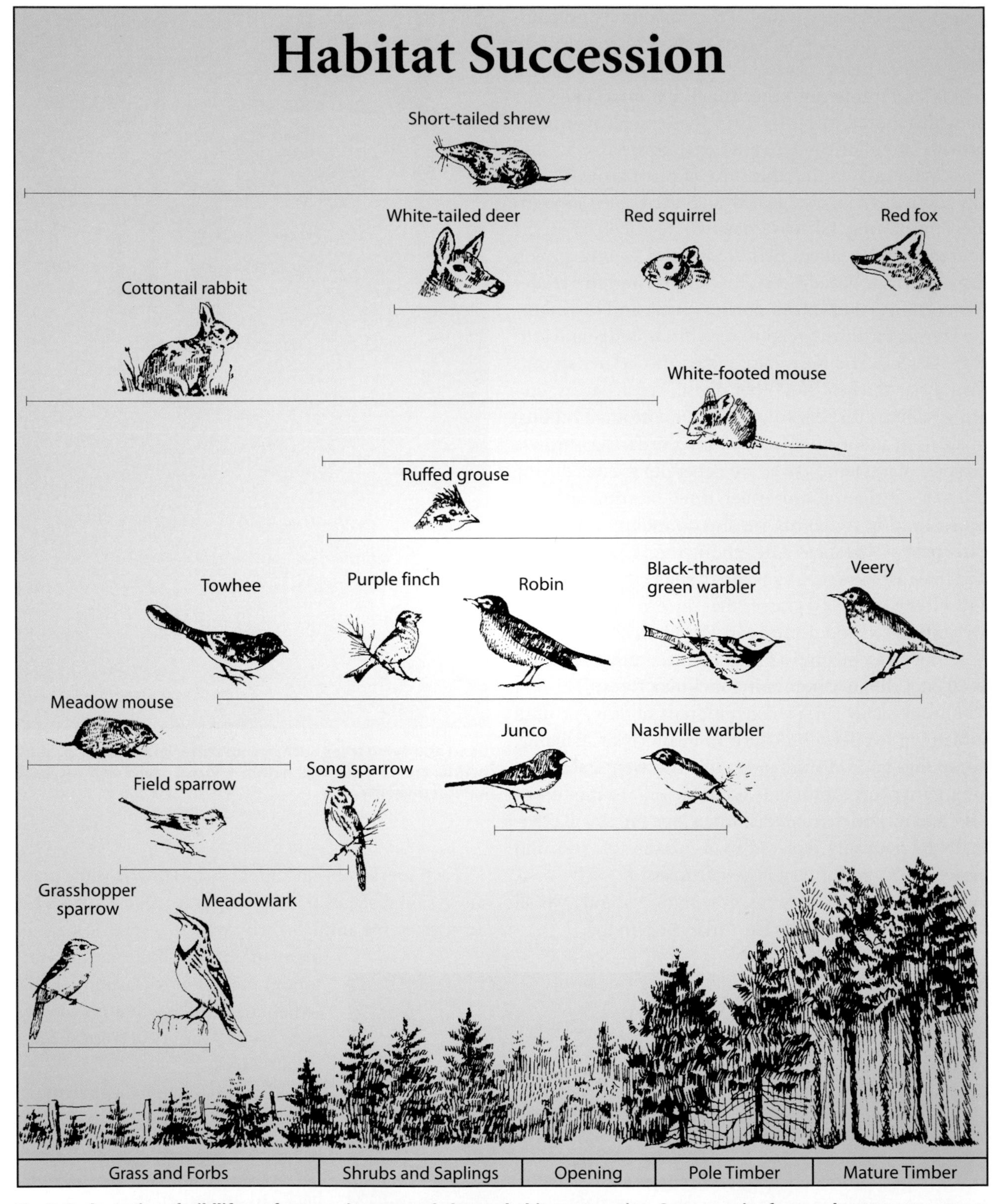

Fig. 4: Each species of wildlife prefers certain stages of plant or habitat succession. Some species favor only one or two stages of plant succession, while others can be found over a wide range of successional stages. Wildlife management involves understanding the habitat preferences of wildlife and tailoring management practices to meet those needs.

tural practices that impact succession and wildlife habitat are described in detail in Chapter 4. The key to habitat management for wildlife is understanding what successional stages are important for the wildlife you desire and choosing the appropriate methods to provide those successional stages.

DR. DANNY EVERETT

Transitions of vegetation types from forest to shrubs to grasses along field borders increase habitat diversity for wildlife.

The Importance of Dead Wood

Air, water, and soil are the media that support life, according to ecologist Dr. Eugene Odum. Soil is the foundation upon which all forest life depends. Soil is the product of the combined effects of geology, vegetation, rainfall, and temperature. Soil is composed of minerals and decaying organic materials from trees and other plants and living organisms such as beetles, fungi, and bacteria. The fertility of soil is determined by the relationship among these components. The more fertile the soil, the more productive it is for producing vegetation for animal life. A fallen tree is literally soil for the future. Therefore, dead wood is an integral part of the ecology of forests and wildlife communities. Dead parts of living trees and dead trees, whether standing or fallen, are required by many wildlife species. Standing dead trees are called **snags**; they provide cover for cavity-nesting birds like woodpeckers and for mammals like flying squirrels. Snags also provide a ready source of insects for woodpeckers and other insect-feeding wildlife. For more information on snags see Chapter 4.

A fallen tree, or **log**, also fulfills essential wildlife needs. A fallen tree quickly becomes infested with fungi and insects. As the tree decomposes, nutrients are recycled into the soil, and a microhabitat favorable for growing new plants and tree seedlings is often created. Insects, salamanders, toads, snakes, mice, shrews, and other smaller animals also find logs an excellent part of their habitat. Insect-foraging wildlife, such as feral hogs, skunks, opossums, and turkeys also find logs an abundant source of protein-rich insects. The accumulation of organic material, including damp, rotting wood and leaves, favors the production of mushrooms and other fungi. Mushrooms and fungi are foods for insects, turtles, birds, mice, squirrels, and deer. Mushrooms also provide a ready source of phosphorus to wildlife. Among other things, phosphorus is a key mineral required for antler growth in deer. During winter periods when food is scarce, highly nutritious mushrooms can compensate for nutrient deficiencies in deer browse. Turkeys and rufous-sided towhees, among other wildlife, will often nest adjacent to partially elevated logs. Depending on their size, hollow logs can shelter a variety of forest mammals such as shrews, chipmunks, bear, foxes, and coyotes.

The removal of all or most dead wood has been shown to have a long-term detrimental impact on flora and fauna. Research has shown that a 20-acre clear-cut site where all the wood has been removed was relatively unattractive to over 100 mammals, amphibians, and reptiles for over

CLEMSON UNIVERSITY

In some cases wildlife management involves decreasing animal numbers to maintain habitat quality and ensure viability of other land uses. As pictured here, deer can destroy valuable crops. An electric fence has been erected in this field to protect soybeans from deer depredation.

Standing dead trees with cavities provide den sites for raccoons and other wildlife.

40 years. During timber harvests, standing and fallen dead trees should be retained. Limbs and small stems that remain after timber harvest, called **slash**, should also be left on the harvested site to provide cover for wildlife. For the best distribution of logs on harvested areas, non-commercial sections of butt logs (logs with deformed or swollen ends) should not be piled at the log-loading site. Instead, they should be cut from the saleable portion of the log and left where the tree was cut. Natural disturbances that knock down trees, such as windstorms or fires, leave dead wood on the forest floor. Although this process rarely appears neat and clean, leaving dead wood replicates the recycling of nutrients in natural processes.

TRADITIONAL APPROACHES TO WILDLIFE MANAGEMENT

Landowners have several options when considering what to do with the wildlife and the wildlife habitats on their lands. One option is **preservation,** where wildlife and the habitats that support them are not managed and are left alone without human disturbances. The goal of preservation is to leave ecosystems in a natural state. However, undisturbed ecosystems are not necessarily stable. Natural changes in plant communities, such as fire and windstorms, constantly create a multitude of habitats for different species of wildlife. As the undisturbed landscape changes over time, conditions may not remain suitable for the continued existence of some wildlife. For example, as a newly established forest matures, the presence and abundance of bobwhite quail will decline because the habitat no longer meets all their needs.

The other option for landowners is to actively manage their lands for specific wildlife. **Wildlife management** is the deliberate act of manipulating wildlife populations and their habitats for the benefit of people and wildlife. This can either be done to increase, decrease, or maintain wildlife varieties and/or numbers. For example, wildlife managers can increase wildlife populations in many cases by improving the quality and quantity of habitats. Many examples of habitat improvements can be found throughout this book. Other situations require a reduction in wildlife numbers, such as in areas where too many deer are damaging (depredating) agricultural crops or orchard trees, or causing a significant number of deer-vehicle accidents. Deer can also damage and degrade their own habitat when their numbers exceed the carrying capacity of the land they inhabit.

For game species, wildlife managers strive to stabilize wildlife populations at an optimum level, making sure that enough "breeding stock" remains to replace those animals that are harvested during regulated hunting or trapping seasons. This approach to management is called **sustained yield.**

Landowners who are interested in managing their lands for wildlife have three basic approaches, depending on their objectives. The first approach is to provide a variety of habitat types in an attempt to support as many different wildlife species as possible. This approach is called **multiple species management**. Under this system, a landowner's objective is to manage his or her property to provide a mixture of areas in different stages of plant succession. Management should provide areas with large amounts of edge interspersed with some unbroken tracts of forest, and forested areas with vertical layering of trees, shrubs, and broadleaf weeds.

The second approach to managing wildlife, which is the method the majority of wildlife managers use, is called **featured species management.** The goal of this approach is to concentrate on providing habitat for one or two selected species. Although the emphasis may be on one or two species, other wildlife with similar habitat requirements also benefit. For example, a landowner might choose to manage for wild turkey or bobwhite quail exclusively; however, other wildlife with similar habitat requirements, like white-tailed deer, will also benefit. The key to featured species management is to identify the habitat requirements of the featured species and to select habitat management practices that provide the requirements that may be lacking.

The third approach to wildlife management is **limiting factor management**. This approach is similar to featured species management, except that it centers specifically on reducing or eliminating factors that have been identified as restricting wildlife population growth and survival. Examples of limiting factors include predation, disease, critical food (quantity and quality) shortages, loss of cover, or negative human impacts like nest disturbance and poaching. Limiting factors are like holes in a keg filled with water, where water represents wildlife and the keg represents the ability of an area to support wildlife. Limiting factors can quickly reduce or eliminate the ability of a wildlife species to survive in a particular area. This type of management concentrates on plugging-up the holes in the keg, allowing it to remain full of water. Examples of limiting factor management include predator and disease control, habitat improvement practices to increase food and cover, and restrictive measures to reduce public access to wildlife that are vulnerable to disturbances and poaching. Half a century ago in the Southeast, just one insect, the screwworm, created a limiting factor for deer and cattle and decimated populations. Although research and progressive management eradicated that factor, new problems continue to emerge.

AN ECOSYSTEM APPROACH TO WILDLIFE AND NATURAL RESOURCE MANAGEMENT

Dramatic changes in natural resource conservation are evolving as we learn more about man's impact on natural systems. The focal point of these changes centers around examining the interdependent relationships of plants, animals, people, and the ecological processes (such as fire and windstorms) that link all living creatures with the physical environment. The ecosystem approach to wildlife and natural resource management looks at the natural variability in ecosystems and directs man's management activities to mimic or replicate that natural variability. Unfortunately, land managers cannot always do this. Past activities and land uses, the complications of trying to manage larger areas, and balancing human impacts and needs all affect managers'

The Importance of Biodiversity

An ecological approach to management has its foundation in managing, conserving, and protecting biodiversity. **Biodiversity** is defined as the variety of life and its processes. It includes the variety of living organisms, the genetic differences among them, the communities and ecosystems in which they occur, and the ecological and evolutionary processes that keep them functioning. There are many different kinds of organisms on earth, many of which are easily recognized like white-tailed deer, bobwhite quail, longleaf pine, white oak, red maple, and a host of others. While maintaining these species is perhaps the best known aspect of maintaining biodiversity, most people never realize the sheer enormity of the number of living organisms. Scientists have identified about 1.8 million species and theorize that 10 to 100 million species are suspected to exist on earth.

So why include biodiversity as a component of land management? The answer is simple: to maintain the health of our ecosystems, the systems that all life depends upon. Scientists have long realized that biodiversity maintains a healthy environment, providing an insurance policy for our own survival. When thinking about the concept of maintaining diversity as part of land-use and management practices, it is wise to remember Aldo Leopold's statement, *"If the land mechanism as a whole is good, then every part of it is good, whether we understand it or not. If the biota, in the course of eons, has built something we do not understand, then who but a fool would discard seemingly useless parts? To keep every cog and wheel is the first precaution of intelligent tinkering."* The best way to "keep all the parts" is by using management practices that have minimal adverse impacts on the land and that maintain biodiversity and the integrity of the ecosystem.

NANCY WEBB

Management at the landscape or multiple land ownership level is essential in order to provide for the habitat needs of the black bear which ranges over large areas.

decisions. In essence, this new approach to management is guided toward understanding the natural forces of change in ecosystems and how those forces are affected by human activities.

Ecosystems vary across landscapes based upon geography, watersheds, and region. Natural processes and human actions shape the diversity and productivity of ecological systems over time. No ecosystem is static; they are dynamic and malleable. The forest stands in the landscape today, if left undisturbed, will change over the next 50 to 100 years. An ever-shifting mosaic of natural systems like weather and growth cycles, combined with the actions of people will determine what those changes will be.

An ecosystem approach to management has been termed new perspectives forestry, managing for biodiversity, an ecosystem approach to management, holistic management, an ecological approach to management, or simply ecosystem management. To reflect current thinking and remain consistent, **ecosystem management** is the term we will use. The focus of ecosystem management centers around a philosophical blend of Gifford Pinchot's ideas of sustained forest yield combined with Aldo Leopold's and John Muir's ideas of land or environmental ethics as outlined in Chapter 18, Conservation Ethics and Opportunities. The ultimate goal of managing land at this level is to provide for **sustainable use** of natural resources, which means that the desired ecological conditions as well as the desired economic benefits from the land can be maintained over time. Scientists and land managers recognize a fundamental need to sustain productive soils, provide high-quality air and water, and maintain vigorous native plant and animal populations. Likewise, **sustainable development** meets the needs of people today without compromising the ability of future generations to meet their own needs from the land.

Limiting factors are like holes in a keg filled with water, where water represents wildlife and the keg represents the ability of an area to support wildlife. Limiting factors can quickly reduce or eliminate the ability of a wildlife species to survive in a particular area.

A primary theme of ecosystem management is the reduced emphasis on commodity production and increased emphasis on an approach that maintains the ecological integrity of natural systems, or maintaining overall health and sustainability. Ecosystem management integrates long-range land use and management on the local as well as the landscape level. This approach considers the needs of people, including their economic and social values, in harmony with environmental values. At the core of this new thinking is managing natural resources at a larger landscape level.

The black bear provides one example of the implications of

managing large southern forested landscapes for wildlife. Black bears require more than one habitat or forest type to meet their needs. Female home ranges usually cover large areas of 6 to 19 square miles. Male home ranges are considerably larger. When young males disperse in search of a new home, they can travel hundreds of miles. Therefore, only by managing at the landscape level can we truly address the needs of black bears. Effectively managing large, free-ranging mammals like these will require cooperation between federal, state, and local governments, non-governmental organizations, and private landowners. The Alabama Black Bear Alliance, composed of the Alabama Wildlife Federation, The Alabama Nature Conservancy, and other conservation groups, agencies, forest products companies, and private landowners is an example of a cooperative effort to protect black bear and bear habitat in Alabama. (See black bears, Chapter 12)

Ecosystem Management Considerations

Ecosystem management is a dynamic process that requires a strategy that develops, enhances, and protects the ecological and socioeconomic values of natural resources. The challenge is how to implement an ecosystem management approach while still providing for the needs of people. Since over 95 percent of the land in Alabama and 70 percent across the South is in private ownership, the responsibility falls to many private individuals to manage their property in a sustainable manner. When landowners combine their private values with responsible stewardship, keeping in mind the public's values for those resources, a landscape-level private stewardship plan can work. For instance, some individuals may appreciate the use of the forest for hunting, hiking, or bird-watching. These are private values. The citizens of Alabama want wildlands to ensure the survival of wintering waterfowl and songbirds. These are public values. A landscape-level stewardship plan can simultaneously satisfy both of these values.

Landowners should develop and implement a natural resources management plan that is consistent with and supports the general goals of a landscape-level stewardship plan. At the present time, most planning efforts on private lands are directed toward single resource management, either for forest products or wildlife. Sometimes an individual will receive conflicting information from resource professionals with a different focus of interest, like a forester and a wildlife biologist. Land management plans should integrate all the principles of managing forests, wildlife, soil, and water resources.

Planning implementation across multiple ownerships will be essential for ecosystem management to work. Because ecosystem management entails bringing people together to sustain multiple uses of larger areas of land, a wider organizational structure will also be needed to inform landowners about ecosystem management efforts throughout the state. The Mobile River Basin initiative is one example of a landscape-level management approach where public and private conservation agencies and organizations have joined forces to protect the largest watershed east of the Mississippi River.

SUMMARY

Forest and farmlands in Alabama and across the Southeast can be managed for income from timber and agricultural products while also providing suitable habitat for a wide variety of wildlife. Wildlife require food, cover, water, and space to live. Understanding how these habitat components can be arranged and enhanced through land management practices is essential to benefitting wildlife populations and sustaining healthy ecosystems. Private landowners who own the vast majority of forest and agricultural lands in the South may be asked to work cooperatively with others in an ecosystem approach to land management in the future. To take no action on their land has repercussions just as noticeable as if direct changes were made. This chapter reviewed concepts like preservation, sustained yield, multiple and featured species approaches, limiting factor management, as well as the newly emerging philosophy of ecosystem management. All these approaches to wildlife management are options that landowners and managers should consider and understand. By making informed choices, landowners can better shape the land and the wildlife that depend on it for a sustainable existence.

Silvicultural practices are applied to establish, maintain, or enhance a site or stand based on an identified management objective. For wildlife, silvicultural practices are a catalyst for positive or negative habitat change. The results depend on the specific practice used, the time of year applied, and how often performed.

PAUL T. BROWN

CHAPTER 4

Managing Forests for Wildlife and Timber Production

CHAPTER OVERVIEW

This chapter examines forest management practices commonly conducted in Alabama and the dramatic effect they have on wildlife habitat. Silvicultural practices, like natural forces, may improve habitat for some wildlife while decreasing the quality of habitat for other wildlife. By understanding the effects of various silvicultural practices on vegetative communities that provide food and cover for wildlife, land managers and owners can tailor management plans to meet specific timber and wildlife management objectives.

CHAPTER HIGHLIGHTS PAGE

FOREST STAND VALUE FOR WILDLIFE **50**
SILVICULTURAL PRACTICES AND EFFECTS ON WILDLIFE HABITAT **51**
OTHER FOREST HABITAT IMPROVEMENTS FOR WILDLIFE **83**
WOODLAND GRAZING **91**
MULTIPLE FOREST STAND MANAGEMENT CONSIDERATIONS **100**
LANDSCAPE LEVEL MANAGEMENT **100**
MANAGEMENT BY LAND CLASSIFICATION **101**
MONITORING AND REVISION OF FOREST STAND MANAGEMENT ACTIVITIES **102**

Forest management for timber production is typically conducted on the stand level. A **stand** is a group of trees that are similar in age and species composition. For timber production, management recommendations or "prescriptions" are developed for each stand based on landowner objectives, site potential, and current vegetation. Forest management practices, also called **silvicultural practices,** impact the quality and availability of forest wildlife habitat. In many cases, entire vegetative communities, and consequently wildlife habitat, are dramatically altered by forest management practices. Silvicultural practices that alter forest habitat in favor of certain groups of wildlife may negatively affect other wildlife species. The challenge is understanding the impact of each silvicultural practice on forest wildlife habitat and using that knowledge to benefit desired wildlife species. Although we have learned a great deal about the relationship of forest management practices to wildlife habitat, we still do not understand how many plant and animal relationships are affected by forestry practices. However, what we do know has many practical applications.

This chapter will examine some of the factors that determine a stand's value for wildlife and the impacts that forest management practices may have on wildlife habitat. Forest habitat improvement recommendations for specific wildlife species can be found throughout the book. Proper planning and an understanding of the effects of various silvicultural practices on wildlife habitat allow stand management to be tailored to maximize both timber production and wildlife habitat enhancement.

FOREST STAND VALUE FOR WILDLIFE

A forest stand's value for wildlife depends upon the productivity of the land, the vegetative structure of the stand, and wildlife habitat requirements. The underlying factor that determines the value of a site for growing timber and providing quality wildlife habitat is the productive capacity of the land. For forest stands, site index is one way to measure the capacity and suitability of a piece of land to grow timber, and it can also give a general indication of the ability of an area to support quality wildlife habitat. **Site index** is defined as the height a tree will attain on a given site at a certain age, usually 50

RICHARD T. BRYANT/ARISTOCK

With proper planning and an understanding of the effects of various silvicultural practices on wildlife habitat, forest management can be tailored to enhance both timber production and wildlife habitat.

years. For example, an area with a site index of 60 to 65 for red oak should produce red oaks 60 to 65 feet tall when the trees are 50 years old. On a better site, site index might range from 80 to 90 for red oak. As site index increases, so does the potential for growing larger quantities of marketable timber and valuable wildlife plants.

An exception to this rule would be in the Black Belt Region of Alabama, where many soils are extremely fertile yet support few trees due to high alkalinity. Consequently, site index for tree growth in the Black Belt is not a good indicator of soil fertility or wildlife habitat productivity. Estimates of site index for a particular soil type can be found in county soil surveys available from local county Farm Service Agency offices or in *Considerations for Forest Management on Alabama Soils*, available from the Alabama Forestry Planning Committee (334) 240-9305.

Plant Succession

Another important factor in determining the value of a stand for wildlife is the existing stage of plant growth. As discussed in the last chapter, vegetation goes through a series of changes known as **succession**. Through each stage of succession, certain characteristics of a forest change, favoring some wildlife species over others. Silvicultural practices may alter forest succession in ways that favor certain wildlife and negatively impact others. For example, clear-cutting and site preparation for tree planting can set back a stand to an early stage of succession that favors the habitat of bobwhite quail and cottontail rabbits.

Wildlife such as gray squirrels and forest interior birds require mature forests and are negatively impacted by practices that set back forests to early succession. It is impossible to manage for all wildlife species within a single stand. This is only possible on a multiple stand or landscape level. If a landowner's objectives are to provide habitat for a combination of early, intermediate, and late successional wildlife species, a mosaic of different stands that are diverse in age, tree composition, and structure is necessary. The distribution and relationship between stands is also important in providing all the habitat components within the home range (Chapter 3) of a wildlife species. Silvicultural practices should be used to create a mixture of habitats beneficial to a variety of wildlife species.

Site Index

Site index is closely related to a combination of factors such as the parent material of the soil (or type of rock from which the soil derived), topsoil depth, soil texture, restrictive layers (see soil hardpans, Chapter 16), and drainage. Topography also influences site productivity, especially in hilly or mountainous terrain. The lower portion of a slope is more productive than the upper portion due to greater soil depth, a result of natural weathering that moves soil from higher elevations to lower elevations. Slopes facing north or east are more productive than slopes facing south or west due to the natural track of the sun in our hemisphere and its effect on soil moisture. North-facing slopes receive little direct sunlight because the sun tracks to the south. East-facing slopes receive direct sunlight only in the morning when humidity is naturally higher. These factors lead to higher levels of soil moisture on north- and east-facing slopes than on slopes facing south or west. Slopes facing northeast are usually the most productive areas, whereas slopes facing southwest are the poorest. Forest stands on northwest and southeast slopes are usually moderately productive. Coves and drainages are also productive areas in forest stands. Topography remains an indicator of site productivity as it determines the potential of a piece of land to retain water and to supply water for trees and other vegetation.

SILVICULTURAL PRACTICES AND EFFECTS ON WILDLIFE HABITAT

Silvicultural practices are applied to establish, maintain, or enhance a site or stand based on an identified management objective. Most often silvicultural practices are used to improve the condition and value of a site or stand for timber production. There is one certainty with the use of any silvicultural treatment: change. For wildlife, silvicultural practices are a catalyst for positive or negative habitat change. This change may be minute or large, and short-lived or long-term. The results depend

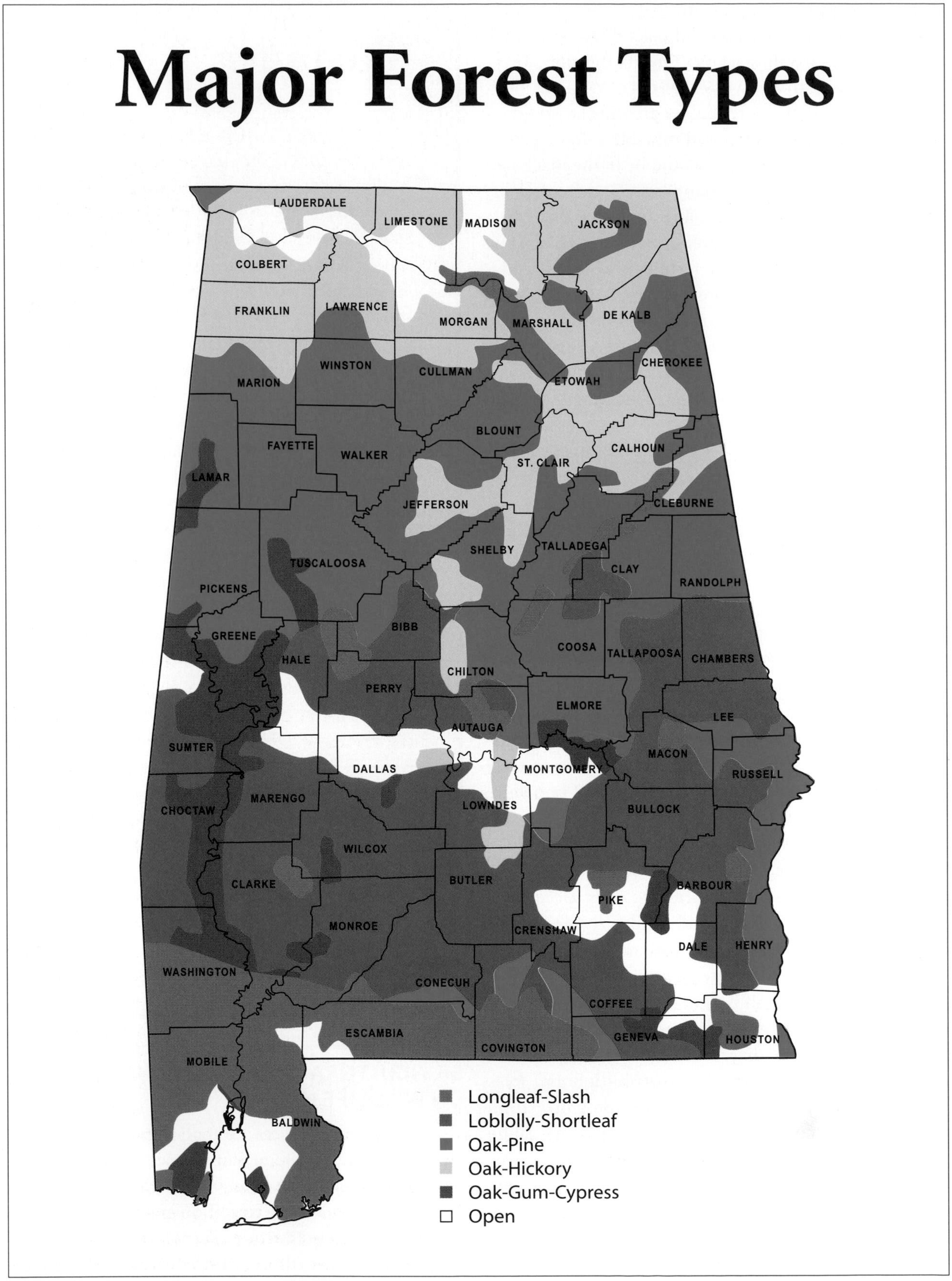

Major Forest Types

- Longleaf-Slash
- Loblolly-Shortleaf
- Oak-Pine
- Oak-Hickory
- Oak-Gum-Cypress
- Open

Fig. 1: Each forest type in Alabama has specific characteristics that require special consideration when integrating forest and wildlife management.

on the specific practice used, the time of year applied, and how often performed. Fortunately, silvicultural practices used for timber production can also provide significant benefits to a variety of wildlife species and can be modified to minimize negative effects. Once wildlife management objectives are identified, silvicultural practices can be selected and applied for maximum benefit. Common treatments include timber harvest, site preparation, tree establishment, prescribed burning, fertilization, and herbicide use. The following section provides an overview of typical silvicultural practices and their effects on wildlife habitat.

MORRIS GILLESPIE/NATURAL RESOURCES CONSERVATION SERVICE

Even-aged forest stands typically have single-storied canopies such as this loblolly pine stand. Thinning and burning in this stand has opened up the canopy layer and allowed sunlight to reach the forest floor, stimulating the growth of herbaceous and woody plants valuable to wildlife.

Forest Harvest and Regeneration

When managing forest stands for timber and wildlife, the first decision a forest owner should make is whether to manage the existing stand or harvest and regenerate (re-establish) a new stand. In some cases, past timber harvests have removed only high quality trees, a practice called **high grading** or "taking the best and leaving the rest." This practice has resulted in some stands containing a majority of poor quality trees with little or no value for wildlife or forest products. Improving these stands' value for both timber and wildlife may require total harvest and regeneration.

Several harvest and regeneration methods are commonly used in Alabama. The type of method chosen depends on the predominant tree species, site characteristics, the condition of the existing forest, landowner objectives, and economics. Some of the more common methods used to harvest and regenerate stands include clear-cutting, seed-tree, shelterwood, group selection, and single-tree selection. Clear-cutting, seed-tree, and shelterwood regeneration methods remove all or a significant portion of a stand in one or two harvests. This favors the establishment of a stand of trees that are the same age. These **even-aged** stands typically have single-storied canopies until they are thinned and allowed to develop mid-stories. Consequently, wildlife that require multi-layered forest canopies, like some species of songbirds, are less abundant. Early successional stages in even-aged forests provide food and cover requirements for many species of wildlife. If even-aged forests are harvested before they reach biological maturity, wildlife species that prefer older forests generally do not frequent these areas. When managing a forest for timber production, trees are harvested based primarily on size or economic maturity rather than on age or biological maturity.

Group and individual tree selection systems produce **uneven-aged** forests, often called **all** or **mixed age stands**. Group selection, also known as **small patch clear-cutting** removes small groups of trees periodically and produces uneven-aged forests composed of small even-aged groups. This system is often used when commercially valuable shade intolerant trees such as pine, oak, poplar, and ash are preferred, but large harvest areas are undesirable. Individual tree selection, as the name implies, removes single trees in a random pattern and produces uneven-aged conditions throughout the forest. This method does not produce large openings in the forest and is sometimes used when a continuous tall forest and regular income are desired.

Many landowners feel that uneven-aged management results in a more aesthetically pleasing forest than even-aged management. One reason for this feeling is that only a small portion of the forest is harvested at any one time using uneven-aged management. In addition, harvest sites are small and scattered over the forest. In comparison, even-aged management eventually removes all or larger portions of the forest in one cutting. However, to produce a comparable volume of wood as that of most even-aged systems, uneven-aged management requires frequent harvests over a short period of time. When smaller harvests occur regularly, the impacts of uneven-aged management can affect the site similar to clear-cutting. Landowners interested in uneven-aged

Fig. 2: Even-aged Versus Uneven-aged Management

EVEN-AGED MANAGEMENT

ADVANTAGES

- Provides most stages of succession including early successional stages
- Allows for horizontal habitat diversity across the landscape
- Reduces relative cost of timber management and harvest compared to uneven-aged management
- Favors shade-intolerant trees that are generally higher in timber value

DISADVANTAGES

- Lacks vertical diversity within forest stands
- Eliminates mature trees and snags are not prevalent
- Reduces species, age, and size diversity of trees in stand
- Increases forest and habitat fragmentation which may be detrimental to some forest interior wildlife species
- Aesthetically less pleasing than uneven-aged management

UNEVEN-AGED MANAGEMENT

ADVANTAGES

- Provides vertical diversity from multi-storied canopies in forest stands
- Provides for a diversity of trees species, ages, and sizes
- The scale of stand disturbance is less than even-aged management, although entries into stands may be more numerous
- Retains more mature trees and snags than even-aged management
- Results in a more continuous forest canopy which is more aesthetically pleasing than even-aged management
- Used for shade-tolerant tree regeneration and management

DISADVANTAGES

- Little horizontal diversity and early successional stage habitats
- Timber management and harvest costs are higher than even-aged management

management may have to compromise in the short-term by receiving less income from timber harvests in exchange for the benefits uneven-aged forests provide to a variety of wildlife.

The following section discusses harvest and regeneration methods and their effects on wildlife habitat. Often these practices, such as clear-cutting, appear to destroy wildlife habitat when in fact they enhance habitat for certain wildlife species for several years. With an understanding of the effects a particular harvest and regeneration method has on vegetative communities, the proper method can be used to meet specific wildlife habitat improvement objectives.

CLEAR-CUTTING

Clear-cutting is a common method used to harvest and regenerate forests in Alabama. This practice removes an entire stand of trees in one harvest. Without a doubt, clear-cutting is the one silvicultural practice that alters forest wildlife habitat the most. It completely removes the existing forest and sets back plant succession to its earliest stage—bare ground, grasses, and shrubs. Because of the drastic change in the appearance of a forest after clear-cutting, this method has become one of the most controversial and misunderstood silvicultural practices in use today.

Clear-cuts can be conducted in strips, patches, or blocks, depending on the objectives of the landowner. Contrary to the belief of some individuals, this practice is a valuable wildlife management technique for many wildlife species if shape, size, and distribution are taken into account. For southern pines, clear-cuts are most often replanted with seedlings and occasionally by distributing seed directly on a recently cleared forest site. Planting seedlings and direct seeding are often referred to as **artificial regeneration.**

Direct seeding involves mechanically distributing seed over a site and is often used to quickly regenerate pine on rough terrain or large areas at low cost. The most common problem associated with direct seeding is regulating the density and spacing between seedlings. Most attempts at direct seeding result in stands with too few or too many seedlings. Too few seedlings, often due to heavy seed depredation by birds and rodents, create stands that have limited value for timber production. Too many seedlings result in dense stands that retard timber growth and reduce or eliminate the occurrence of valuable wildlife plants due to shading and lack of sunlight on the forest floor.

Planting pine seedlings by hand or machine is more

Fig. 3: Spacing of Pine Seedlings and the Relative Effects on Wildlife Habitat and Timber Production

SPACING (FEET)	TREES/ACRE	RELATIVE EFFECTS
20 x 20 15 x 15	109 194	Maximum herbaceous plant production with partial or no crown closure. Excellent for wildlife, poor for timber.
12 x 12 10 x 12	303 363	Excellent herbaceous plant production with delayed and partial crown closure. Excellent for wildlife, poor to fair for timber production.
8 x 12 8 x 10	454 545	Good herbaceous plant production and delayed crown closure. Good compromise between wildlife habitat improvement and timber production.
6 x 8 6 x 6	908 1210	Short-lived herbaceous growth due to rapid crown closure. Poor for wildlife and excellent for certain types of timber production.

common than direct seeding. This approach offers the ability to control seedling density and increase the production of understory vegetation valuable to wildlife. Planting pine seedlings on wide spacings can delay crown closure and prolong the availability of understory plants valuable to wildlife. This practice, however, causes more limbs and therefore later thinnings, lower quality, and lost revenue.

Pines can also be re-established by **natural regeneration** that relies on seed or seedlings already present in the stand. Seed-in-place and seedling-in-place methods involve timing the clear-cut so that sufficient seedlings and seeds are present in the stand. The **seed-in-place method** calls for preparing the land for optimal seed germination and seedling survival by controlling understory vegetation through a series of prescribed burns or herbicide treatments. The existing stand is then harvested in the fall after the seeds have dispersed. If a new stand of seedlings does not become established, the area can then be artificially regenerated by direct seeding or planting seedlings.

The **seedling-in-place method** can be used to regenerate a new stand if each acre of existing mature pines has several thousand established seedlings in the understory. With this method, stands of mature pine trees should be harvested in late summer or fall after the seedlings have survived their first growing season. Care should be taken to minimize damage to seedlings when the mature timber is harvested.

Strip or patch clear-cutting removes narrow strips of mature pines, relying on seed from adjacent stands

RICHARD T. BRYANT/ARISTOCK

Light thinning and single tree selection can produce forest stands with trees that are different in age and species composition such as this hardwood stand. Vertical wildlife habitat diversity is enhanced with uneven-aged management.

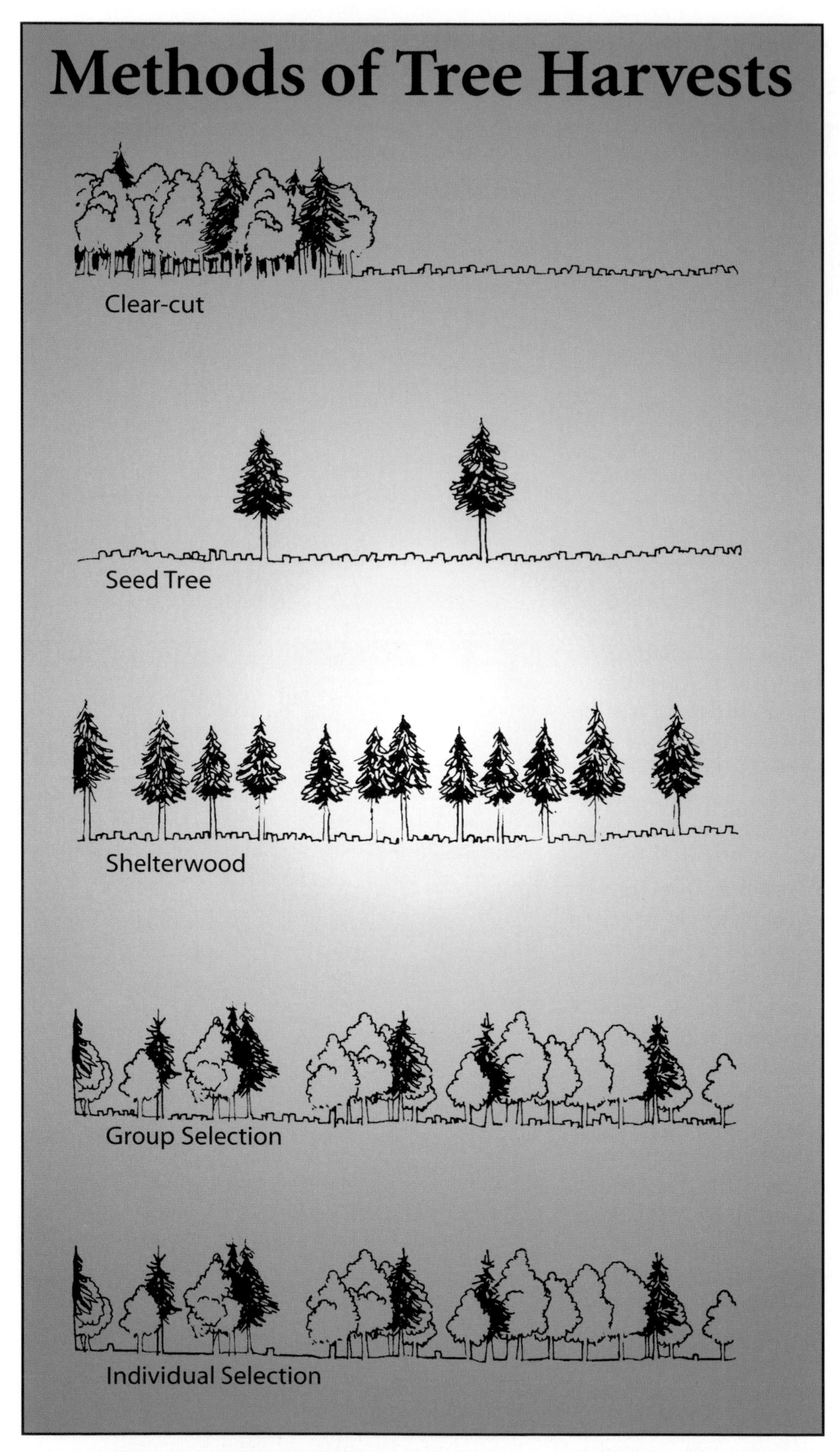

Fig. 4: A variety of tree harvest methods are available for woodland owners. Clear-cuts, seed-tree, and shelterwood harvests result in even-aged forest stands. Group and individual tree harvests favor uneven-aged forests.

to reforest the opening. This is also an excellent method to increase horizontal habitat diversity for wildlife within a stand. Strips should be cut perpendicular to the prevailing winds and located on the down-wind side of the stand that will provide the seed. This arrangement will insure that seed from trees in the uncut strips will blow into the clear-cuts. Before seeds fall, the cleared strips should be burned to expose mineral soil, enhance germination of fallen seeds, and reduce competition from woody vegetation.

Southern hardwoods are most effectively regenerated by natural regeneration of the stand from root and stump sprouting. Less frequently, natural regeneration can occur from seedlings that already exist in the stand and from germination of seeds that have fallen from trees. In some cases, hardwood regrowth from roots and stump sprouts can be enhanced by using various site preparation techniques. Clear-cutting provides ideal conditions for fast-growing, desirable hardwood species that are adapted to full sunlight conditions. It also can create an even-aged stand that can eventually produce a variety of high-quality mast-producing hardwoods for wildlife. Site quality seems to be an important factor in determining the species composition of regenerated upland hardwood stands. On productive sites seedlings of fast-growing species such as yellow poplar, ash, black locust, black cherry, red maple, and a mixture of other species are common. On average sites, oaks are often the most abundant with an occasional yellow poplar, pine, hickory, red maple, or other tree species. In the hilly and mountainous areas of Alabama, light-seeded species are usually not as prevalent. Desirable species like oaks, that come from advanced regeneration (understory trees) and stump sprouts, are more prevalent in these areas.

Because of the high success rate and relatively low cost of re-establishing hardwood stands by natural regeneration, planting seeds or seedlings has had limited use. Consequently, planting hardwood seeds and seedlings usually should not be considered as an economic option in regenerating clear-cuts. In some cases, however, like old agricultural fields and pastures where there are no natural sources of seed or sprouts, artificial regeneration is the only alternative for establishing a hardwood stand. However, artificial regeneration of hardwoods has often been unsuccessful due to low survival rates of seedlings and poor germination of seeds. In addition, planting and maintenance of hardwood seedlings has proven to be time-consuming and expensive. Recently the Southern Hardwood Laboratory in Stoneville, Mississippi developed a successful method to plant acorns and other hardwood seeds.

Clear-cutting is the most frequent method used to harvest and regenerate mixed pine and hardwood stands. Generally, the same guidelines used to regenerate pine and hardwood stands are adopted to regenerate mixed stands. Re-establishment of a mixed stand can be

Planting Acorns for Wildlife

Acorns can be collected for planting from a wide range of oaks or purchased from a seed distribution vendor (see sources of planting material, Appendix K). Seed maturity varies by species and year, but generally acorns mature and drop from about the first of October to the first of January.

Before storing or planting, acorns should be checked for viability, and those that are rotten, have insect damage, or are underdeveloped should be discarded. One method of checking acorns for soundness and viability is using the "float test" which involves placing collected acorns in a container of water. Acorns that float to the top are not viable and should be discarded. Acorns can be kept in cold storage at 35° F or in polyethylene bags about 1 foot deep in the ground. Properly stored red oak acorns will remain viable for up to three years, while white oak acorns can only be stored for a few months.

DR. H. S. BANTON

With proper planning and care, species like this overcup oak can be planted from acorns.

Acorns can be planted at any time by hand or by machine at a depth of 2 to 6 inches. Hand seeding can be accomplished with a dibble bar or a modified bar that gives a consistent depth and eliminates the need to bend over to put the acorns in the holes. Machine planting is more suitable for open fields. A modified soybean planter can handle acorns ranging in size from water oaks (small) to Nuttall oaks (large); it automatically drops acorns every 30 inches. Variable speed auger planters have also been successful in machine planting acorns. On old fields, expect a germination rate of about 35 percent. With this in mind, planting 1,500 acorns per acre will provide 525 seedlings, with between 150 to 375 successfully reaching 10 years of age. For more information about planting acorns or hardwood seedlings for wildlife, contact the Alabama Forestry Commission's Hauss Nursery in Atmore, Alabama (334-368-4854) or the Southern Hardwood Laboratory in Stoneville, Mississippi (601-686-3190).

Natural Versus Artificial Regeneration

NATURAL REGENERATION

ADVANTAGES
- Lower establishment costs
- Less labor and equipment required
- Seeds adapted to the site
- Less soil disturbance
- More visually appealing

DISADVANTAGES
- Less control over spacing and stocking rates
- Longer rotations needed
- Greater risk of seed loss
- Lose option of using genetically improved seedlings
- Pre-commercial thinning sometimes required
- Irregular stands that result are often difficult to harvest
- High value trees at risk—possible lower stumpage value

ARTIFICIAL REGENERATION

ADVANTAGES
- Control of seedling spacing
- Use of genetically improved seedlings

DISADVANTAGES
- Higher establishment costs than natural regeneration
- Greater chance for site disturbance and increased potential for soil erosion
- Less visually appealing

accomplished either by natural regeneration or by a combination of natural and artificial regeneration methods. For example, forest stands that have been recently clear-cut can be allowed to re-establish on their own, or pines can be supplemented by hand planting, usually at rates lower than used for even-aged pine forests. The lower planting rates allow space for hardwoods to grow naturally along with planted pines. In some cases where seed trees are present, pines will also naturally regenerate in a newly cut stand. The result will be a mixture of pines and hardwoods. Recently the price of pine and hardwood pulpwood is more or less equal in most areas of the state, and many landowners are happy to have mixed stands. This eliminates the expense of practices designed to eliminate hardwoods in the stand like herbicide treatments and extensive site preparation.

CLEAR-CUTS AND WILDLIFE HABITAT

The visual and biological changes caused by clear-cuts drastically alter forest stand structure, but they also add another dimension to the forest landscape by increasing edge and creating open areas. In general, early successional and edge-adapted wildlife species are favored by this change. Large clear-cuts are attractive to wildlife species such as mourning doves, rabbits, small rodents, reptiles, and raptors. On the other hand, small patch clear-cuts can improve both vertical and horizontal habitat diversity favoring wildlife that live in both forested and open areas. After clear-cutting and during the early stages of forest regeneration, an abundance of highly nutritious weeds, grasses, woody vegetation (leaves, twigs, buds), and fruits provide food and cover for a variety of early successional wildlife species. During the first three years after a clear-cut, the abundance and nutrient content of many wildlife food plants increases. Herbaceous plant production often peaks at two to three years after a total harvest and remains relatively high for five to six years. As the forest stand grows and tree crowns block out sunlight, herbaceous vegetation declines drastically in quantity and nutritional quality. In naturally regenerated pine-hardwood stands, both herbaceous and low-growing woody plants are most abundant two to five years after cutting, but then decline. In rich cove hardwood stands, forbs, grasses, and woody plants become most abundant one to three years following a clear-cut. In general, forests grow more rapidly on fertile sites and therefore have a shorter grass or forb stage following clear-cutting.

In most areas of Alabama deer utilize clear-cut areas intensively year-round, especially during the spring and summer, when succulent new woody growth, herbaceous forage, and fleshy fruits are most abundant. Use of clear-cuts by deer drops off in winter, if acorns are readily available in surrounding areas. When acorns and other mast foods are not available in adjacent areas, clear-cuts will continue to be frequented by deer during the winter months.

Clear-cutting temporarily provides game birds with an increase in food supply; however, the effects of this harvest practice on game birds vary with the species. In pine forests, for example, quail populations reach their greatest number two seasons after cutting and site preparation. This period coincides with the availability of nesting cover and the maximum abundance of legumes and other quail foods brought about by clear-cutting. After the fourth or fifth year, clear-cuts become densely vegetated and support few quail for hunting. Turkeys will also utilize clear-cuts in search of foods and for nesting sites. The open areas also provide important bugging areas (places to feed on insects), especially for young turkey and quail.

Other wildlife species also benefit from this practice. For instance, in bottomland hardwoods the American woodcock prefers clear-cut stands as wintering habitat. In pine plantations and natural pine stands, small mammal populations peak after cutting when annual plants produce their second crop of seed. Generally speaking, clear-cutting is followed by increases in small mammal abundance and diversity that persist until tree canopies close. Raptors, like hawks and owls, and terrestrial predators, such as foxes, that prey on small mammals also concentrate on clear-cut sites. Rabbits prefer young pine plantations over undisturbed natural areas since there is an abundance of food and cover.

Not all wildlife benefit from clear-cuts. Interior forest species and cavity-nesting wildlife that depend on dead or dying tree snags, or large tracts of mature forests may be negatively affected by clear-cutting. Other site-sensitive species, like the Red Hills salamander, have also been reported to be negatively impacted by clear-cutting. Fortunately, with proper planning, clear-cuts or other silvicultural practices can be conducted in a fashion that reduces or eliminates the negative impacts on these species.

Clear-cutting can improve habitat for edge-adapted wildlife, but it is not recommended for wildlife species that prefer large expanses of mature forests. Factors that influence the value of clear-cuts to many wildlife species include the size, shape, distribution, and retention of valuable habitat components within the area being cut.

As a general rule, the smaller the property or forest stand, the smaller the clear-cut. The influence of clear-cut size on deer has been studied in the South, and results have shown that deer overuse and eliminate the better browse and herbaceous plants in clear-cuts of 1 acre or less but fail to effectively utilize available forage in clear-cuts over 50 acres. In addition, long and narrow clear-cuts tend to attract more deer than square or circular cuts. Consequently, clear-cuts that are less than 50 acres and irregularly shaped are preferred more by deer and other wildlife over large, square clear-cuts. Some animals, like deer, are reluctant to move more than 100 yards from cover; because of this, a high percentage of large clear-cuts remain unused by deer until vegetation reaches a height that provides visual cover. If clear-cuts are over 50 acres in size, the shape of cuts becomes even more important. To maximize the benefits to deer, clear-cuts over 50 acres should be long, narrow (200 yards or less in width), and irregular in shape. Unevenly shaped cuts create more edge compared to other cuts of the same size. For example, a square clear-cut of 50 acres has 37.8 feet of edge per acre, while an irregular clear-cut of 50 acres may have 118 feet of edge per acre. Relatively long and narrow cuts that follow natural contours of the land are preferable. Large square cuts that provide less edge should be avoided in managing for edge-adapted wildlife. However, some neotropical migrant birds require little or no edge and prefer forest interior habitats. For this reason, square clear-cuts that minimize edge may be more desirable than irregular clear-cuts when neotropical migrant birds are considered. In general, small patch clear-cuts can be a good compromise in retaining or improving both vertical and horizontal habitat diversity.

Another reason clear-cuts should be small and well dispersed is to increase habitat diversity and avoid forming large areas of a single cover type of the same age. Maximum populations of game species favored by early successional habitats generally occur when about 50 percent of the land is composed of small clear-cuts scattered across a management unit in a patchwork fashion. Wildlife managers can choose to create many openings of various sizes or few openings in long, narrow, contour strips. Either design benefits most large game species. Small game and non-game edge-adapted species prefer a greater number of small openings that provide more edge than larger openings.

A general guideline for laying out clear-cuts for wildlife on large areas is that the portion of an area to be regenerated in any decade should equal 1/R, where R equals the desired stand rotation age (length of time until harvest) in decades. For example, if an area is to be regenerated in loblolly pine with a rotation age of 40 years (four decades), no more than ¼ of the area should be regenerated during any 10-year period. To distribute clear-cuts uniformly in time and space, multiple cuts should be made in each 10-year period and scattered

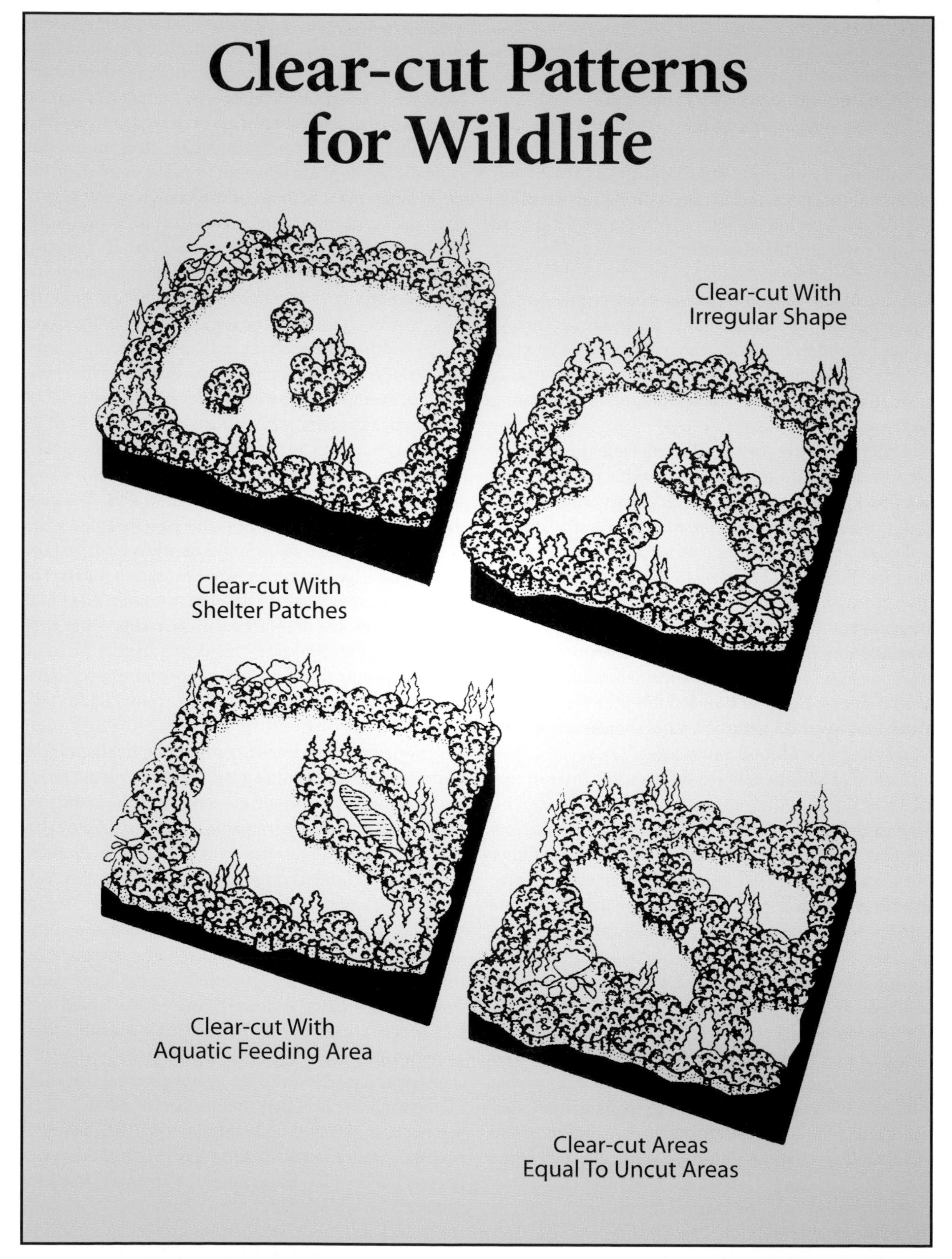

Fig. 5: Clear-cuts can be conducted in a variety of ways to enhance areas for many species of wildlife while minimizing the negative impact to others. Factors affecting the value of clear-cuts to wildlife that benefit from openings include size, shape, and distribution of clear-cuts across the landscape.

JOE MCGLINCY

JOE MCGLINCY

Size of clear-cuts can affect wildlife use. Dove and bluebirds prefer large clear-cuts for feeding such as those pictured on the left. Deer and turkey prefer smaller clear-cuts, such as those on the right, that provide easy access to nearby escape cover in adjacent forest strips or blocks. Varying the size and shape of clear-cuts and including SMZs and buffer strips, as depicted on the left, greatly enhance their value for wildlife.

throughout the management area. Small clear-cuts in adjacent stands should be scheduled six to eight years apart to increase stand age-class diversity.

Despite the advice that smaller patch clear-cuts are recommended, there will be situations and landowner preferences that call for creating larger clear-cuts. Strips of uncut upland timber at least 100 yards wide should be left between clear-cuts over 50 acres in size. These areas, called **buffer** or "leave" strips, increase habitat diversity within clear-cuts and provide travel lanes for wildlife between and through the newly opened sections of forest. Buffer zones that extend into clear-cuts in a finger-like fashion tend to increase wildlife use of clear-cuts. These buffer zones are similar to SMZs but should not be confused with them. They both provide separate values. If possible, buffer zone areas should be chosen that contain mast-producing hardwoods to provide an additional food source for wildlife in newly cutover management units. In addition, buffer zones also provide protective cover for many species of wildlife that routinely use clear-cuts. Wild turkey nests for example are commonly found in, or adjacent to, leave strips.

SEED-TREE

Seed-tree harvest and regeneration is closely related to clear-cutting, except that a few seed-producing trees are left after the harvest. The seed-tree method naturally regenerates a stand by the cutting of all but a few seed-bearing trees. Seeds of these remaining trees are distributed by wind. After windblown seeds germinate and the seedlings are well established, the remaining seed trees can then be removed. The number of seed trees left after an initial harvest varies, but usually ranges from 10 to 15 mature seed-producing trees per acre, depending on the tree species. The seed-tree method is normally used for trees that have light seeds that are easily spread by wind, like loblolly pine and yellow poplar. This method of regeneration can be successful if initial harvests are timed with seed fall, and if the site is prescribed burned or disked ahead of seed fall to expose seeds to the soil and reduce competition from woody vegetation. The success of the seed-tree method depends upon adequate seed production. Mature seed-bearing trees do not always produce a consistent seed crop each year and a poor crop may result in a partial or total regeneration failure. If so, regeneration will be delayed until the next good seed crop.

The value of seed trees that remain in a newly regenerated stand will determine whether or not the seed trees will be removed later. If market conditions make it uneconomical to remove seed trees, then they can either be left in the stand until market conditions improve, they can be killed and left to serve as denning and roosting trees within the younger forest, or they can be left alive for possible harvest when the small trees are thinned. As a general rule, seed-tree cuts should not be conducted in wet or sandy soils. These sites do not promote strong root systems and make the remaining seed trees vulnerable to wind damage. The few remaining seed-trees that are left in a stand are also more prone to mortality caused by lightning strikes.

Seed-tree cuts have almost the same effect as clear-cuts on wildlife habitat. The same guidelines that are used for clear-cut size, shape, and arrangement of harvests to enhance wildlife habitat are also appropriate for seed-tree cuts.

SHELTERWOOD

A **shelterwood** harvest naturally regenerates pines and hardwoods through a series of two or three cuts

Oak Regeneration for Wildlife Using Shelterwood and Prescribed Fire

Dr. David Van Lear is professor and forest ecologist at Clemson University with a national reputation involving the use of prescribed fire for silvicultural and wildlife improvement practices. Together with his graduate research assistant Pat Brose, he has developed a unique approach to regenerating oak stands using shelterwood harvest and prescribed fire.

Regenerating (re-establishing) oaks on fertile upland sites has always been a particular problem in the South. This is especially important since acorns provide the bulk of fall and winter diets of many wildlife species. Often these sites are dominated by yellow poplar, red maple, and other less desirable wildlife trees. Part of the problem may be the absence of fire as a management tool to favor oaks. Results of our ongoing research suggest that oak regeneration can be enhanced through a shelterwood harvest followed in a few years by a prescribed burn to limit advanced regeneration of other species that compete with oaks.

We have found that shelterwood cuts that leave 15 to 20 oaks per acre, followed three to five years later by a winter burn, improve oak regeneration. Spring burns are even more effective in "leveling the playing field" to make regenerating oaks more competitive. High intensity summer burns help to control red maple and other resilient species. Burning should only be conducted three to five years after a shelterwood harvest so sprouting oaks can develop an adequate root system to withstand fire and make them more competitive. Because of their "carrot-type" roots, oaks are better able to resprout after burning than other competing trees.

Yellow poplar is a prolific seeder and should be removed in the initial shelterwood cut. Overstory oaks in the shelterwood provide acorns for wildlife and periodic burning increases the likelihood that the next stand will be dominated by oaks. The shelterwood and burn technique offers a viable alternative to regenerate fertile upland sites in oaks, resulting in a stand that is valuable for both timber and wildlife.

over time, rather than by one clear-cut. This harvest practice establishes a new stand by gradually removing the existing trees so that seedlings become established under the protection of older trees.

Like clear-cuts, a shelterwood cut can be made in strips, patches, or blocks. The first cut opens the forest canopy by about 50 percent, leaving between 20 and 40 seed-producing trees per acre for re-establishing the forest stand. This cut actually resembles a heavy thinning which reduces the density of trees in a stand to a **basal area**, a method of measuring the stocking of timber in a stand, of about 30 or more square feet per acre. The second cut often removes half of the remaining stand. Sunlight then reaches the forest floor and encourages successful germination of seeds and the establishment of seedlings, while still leaving a partially stocked overstory stand. Openings created by shelterwood cuts also allow sunlight to stimulate the production of legumes, grasses, and forbs that are important wildlife foods. The third cut removes the rest of the mature trees. The volume of timber removed in each shelterwood cut depends upon site productivity and tree species. For example, on productive sites longleaf pine can be cut to a basal area of 60 square feet per acre at the first cutting and 25 to 30 square feet per acre at the second cut.

Shelterwood cuts continue to provide both cover and food for wildlife until the sheltering trees are removed and the trees beneath them begin to mature and shade-out understory vegetation. In addition to perpetuating timber stands that produce beneficial wildlife foods, shelterwoods are often preferred by landowners for aesthetic reasons since a portion of the mature forest remains standing for 6 to 12 years as compared to a clear-cut which removes all commercial timber at once.

This harvest practice may be particularly appealing to landowners who want to naturally re-establish longleaf pine stands since the large, heavy seeds of these trees are not dispersed far from the parent trees. It is also recommended for forest owners who want to increase the oak component in a hardwood stand. To increase the number of oaks in a stand before the final cut, oak seedlings must already be present in the understory. For harvesting and regenerating an oak stand, a three-step shelterwood system is planned. The first cut should reduce the existing stand by 70 to 80 percent or to a basal area of 20 square feet. Two years later, a second cut should be made to reduce the stand another 40 to 50 percent or a basal area of 10 square feet. The final cut should occur five years after the initial cut. An alternative option used by some landowners who own bottomland hardwoods is to make three cuts at 10-year intervals.

Fig. 6: Examples of Shade-Tolerant and Shade-Intolerant Trees

SHADE TOLERANT (THRIVES IN SHADE)	SHADE INTOLERANT (THRIVES IN SUN)
Red mulberry	Most oaks
Magnolia	Southern pines
Dogwood	Butternut
Redbud	Yellow poplar
Honeylocust	Sassafras
American holly	Sumac

Studies conducted for over 15 years at the Bent Creek Research and Demonstration Forest in Asheville, North Carolina, have successfully demonstrated methods to naturally regenerate oak stands in the Appalachian region of the South. Some of these findings should have applications in regenerating oak stands in northern Alabama. The Bent Creek techniques involved opening up dominant and midstory tree canopies through a combination of shelterwood cuts and selective herbicide treatments. This allowed enough sunlight to reach emerging oak seedlings. As the oak seedlings grew, the remaining dominant trees in the stand were harvested. For more information about the Bent Creek Research and Demonstration Forest projects contact the Bent Creek Research and Demonstration Project Leader at (704) 667-5261.

Using this harvesting method can potentially provide the greatest flexibility of any forest harvest and regeneration system because it allows for both vertical and horizontal forest habitat diversity, benefitting a wide array of wildlife. The degree of vertical or horizontal habitat diversity achieved by shelterwood cuts depends upon the intensity of the harvest. For example, horizontal diversity is favored if shelterwood harvests remove a greater number of trees per acre, leaving few seed trees. Horizontal diversity is also favored if the remaining mature trees in the stand are removed in successive cuts in a relatively short period of time, usually within two to five years.

Conducted properly, shelterwood cuts cause less site disturbance than clear-cutting; consequently, wildlife species that are negatively affected by site disturbances from clear-cutting are less impacted by shelterwood cuts. However, intensive shelterwood management results in greater disturbance to wildlife due to more entries into the forest and the greater use of forest roads. Early forest successional stages are often not as prevalent in shelterwood cuts as in clear-cuts, which somewhat reduces the value of shelterwood cuts for edge-adapted wildlife species.

Although an even-aged system, shelterwood harvests provide some multi-storied canopies and a variety of tree age classes which benefit various non-game birds and tree-dwelling wildlife. Since intensive site preparation for planting is not necessary with shelterwood or seed-tree cuts, rotting stumps from harvested trees provide shelter for a wide variety of wildlife such as mice, chipmunks, rabbits, foxes, and other species.

Landowners who choose shelterwood harvest methods should use the same guidelines as for clear-cuts for size, shape, and arrangement of harvest to enhance wildlife habitat. For wildlife preferring a vertically diverse forest, the three-stage shelterwood cuts have less of a detrimental impact than two-stage cuts. This is because the three-stage approach creates smaller horizontal openings and retains a greater portion of vertical diversity within the forest stand. Vertical habitat diversity is lost, though, when the final shelterwood cut eliminates the entire overstory tree canopy.

For pines, successful shelterwood cuts can result in regeneration rates as high as 10,000 seedlings per acre. When this occurs, stands should be pre-commercially thinned. Seedlings should be thinned as early as possible to allow sunlight to reach the forest floor and enhance the production of understory vegetation valuable to wildlife. For timber production, ideal stocking rates should be between 900 to 1,200 seedlings per acre; however, this rate promotes rapid crown closure and results in poor wildlife habitat. A good stocking rate compromise for both timber and wildlife would be around 500 to 550 seedlings per acre or an average spacing of approximately 8 to 10 feet.

GROUP SELECTION

Group selection can also be used in creating and managing uneven-aged forests. This silvicultural system is most often applied in mixed hardwood stands where natural regeneration of shade intolerant species is desired from seedlings already in the stand (advanced regeneration), stump sprouts, or seeds. Group selection cuts involve removing clusters of trees in an area less than 2 acres. This form of harvesting resembles a small-scale clear-cut in appearance. As a general rule, openings should be at least two times as wide as the height of the surrounding trees to allow sufficient sunlight to reach the forest floor and stimulate regeneration and growth of shade-intolerant tree species and herbaceous vegetation for wildlife.

This is also an ideal harvest method in a forest stand that has a variety of trees that require different amounts of sunlight. Shade tolerant trees will thrive along the edges of the harvest area that receives less sunlight. Shade intolerant species will do well in the interior where sunlight is abundant. The mixture of trees resulting from group selection is beneficial to many wildlife species since it can increase habitat diversity and edge in the forest.

Group selection also allows a wide variety of understory vegetation to grow, which increases habitat diversity within a forest stand. Small openings in the forest create favorable conditions for the growth of grasses, forbs, shrubs, and other early successional plants. Consequently, those wildlife species that are adapted to living among early succession plant communities benefit. Group selection cuts that are well distributed over a stand increase edge habitat and forest stand habitat diversity. In uneven-aged stands, group selection increases the horizontal habitat diversity that is often lacking. To encourage the growth of early successional plants and shade intolerant trees, group selection openings should be 1 to 2 acres in size. Openings smaller in size favor shade-tolerant trees. Over time, as trees become established in openings created by group selection cuts, early successional ground layer vegetation is shaded out, and horizontal diversity decreases. Some studies have indicated that this process may take longer in partially shaded openings created by group selection cuts as compared to larger openings that receive full sunlight. Consequently, partially shaded openings remain more favorable to wildlife like deer for a longer period of time. However, as horizontal habitat diversity decreases, vertical diversity increases. Emerging trees provide a multi-canopied layer within the forest stand.

There are advantages to group selection harvest. Some landowners find this harvest practice more aesthetically appealing than clear-cuts since much of the stand remains after harvest. Many tree insects and diseases attack only one species of plant, and a multi-species stand will slow the spread of infestations. Group selection harvests can help provide a steady flow of income from a series of light harvests.

While group selection is reasonably effective and often seems aesthetically pleasing, it too has drawbacks. Openings still have to be large enough to allow adequate sunlight for effective natural regeneration to occur. This may require cutting additional immature trees. In pine stands the constant presence of young trees significantly impedes use of prescribed fire to control brush and improve wildlife habitat. The major disadvantage of the group selection method is that many small forest stands are created, making timber management more difficult and costly to administer. On large forest ownerships this may be a problem, making group selection a more practical alternative for small forest landowners. Other disadvantages include 1) lower short-term income from timber harvest as compared to larger cuts (clear-cuts), 2) greater management costs per acre as compared to larger harvests, and 3) shade tolerant species of lower timber value are often favored.

SINGLE-TREE SELECTION

The most labor intensive harvest and regeneration method is **single-tree selection** or **individual-tree selection**. This method of timber harvest favors uneven-aged forest stands and involves removing only individual trees, as opposed to removing all trees or groups of trees at the same time. Individual trees that are ready for market as well as low-value trees are removed from the forest stand. Under this system, forest stands can produce timber at 10- to 25- year intervals, with new trees constantly emerging from natural regeneration to take the place of harvested trees. Single-tree selection appeals to many landowners, especially those concerned with the appearance of their forests, since almost the entire stand is still remaining after harvest. Single-tree selection cuts are also desirable in site sensitive areas, like wetlands and riparian zones, since site disturbance is held to a minimum and forest communities remain relatively unchanged.

This harvest method maximizes vertical habitat diversity by creating additional forest canopy layers. Consequently, those wildlife species that require multi-layered forest canopies benefit the most from single-tree selection cuts. Trees that are low in quality for wildlife and timber should be removed by single-tree selection to make room for higher quality trees that benefit wildlife and are valuable for timber.

Different tree harvest and regeneration methods can lead to a variety of wildlife habitats. Clear-cutting (loblolly pines top left) produces an even-aged stand that can be maintained to produce an abundance of understory vegetation valuable to wildlife. Uneven-aged forests resulting from natural regeneration (longleaf pine stand top right, and thinned hardwood stand bottom left) produce multiple forest canopies that increase vertical habitat diversity. Intensively managed pine plantations (bottom right) have little value for wildlife prior to thinning.

Some quail plantations conduct single-tree selection on a seasonal basis, in conjunction with prescribed burning, to create favorable quail habitat. Poor quality pines and other low value trees are removed across a forest stand every one to two years to maintain a basal area that allows for a continuous open canopy and production of quail foods. Single-tree selections conducted on long rotations maintain ideal habitat conditions for many wildlife species for an extended period of time.

Single-tree selection has generally not proven to be an effective way to regenerate most forest stands in the South. The major drawback to single-tree selection is that the method fails to provide for a quick and orderly removal of overhead competition. Most desirable hardwoods and pines are relatively intolerant of shaded conditions that are prevalent with single-tree cuts. They can begin to grow under partial shade, but they must be released from overhead competition early in life. An exception is the American beech, which develops well under single-tree cuts. Single-tree selection also promotes the development of shade-tolerant species, many of which are not desirable for timber or wildlife.

Unfortunately, proper application of single-tree selection in the past has been rare. Most attempts have resulted in **high grading** through the use of diameter limit cutting, an approach that removes trees larger than a specified diameter. Since larger trees most often represent the best trees in a stand, when improperly applied this approach removes high quality trees and leaves poor quality and less desirable tree species to make up the next stand. To use diameter limit cutting with any success, stands must be evaluated to insure that desirable, quality trees for timber and/or wildlife management objectives are present and ready to fill the void before harvest takes place. If suitable replacement trees are not present, diameter limit cutting should be avoided and other silvicultural treatments should be used to create a suitable crop of replacement trees. This can be accomplished through silvicultural treatments such prescribed burning, herbicide use, tree planting, and selective harvests that remove both merchantable and non-merchantable trees to favor the development of more desirable trees consistent with the specific timber and wildlife objectives for the stand.

Site Preparation

Practices used to prepare a site for establishing a new forest are collectively called **site preparation**. Mechanical methods, herbicides, fire, or a combination of these are commonly used. They may be applied to open fields or forest land, before or after timber harvest, and to facilitate artificial or natural regeneration. For timber production these practices prepare a site for seed germination or seedling growth and temporarily control vegetation that will compete with the newly established stand. Overall, site preparation efforts seek to improve the chance of regeneration success and to promote future growth and development of desirable trees.

Before choosing a particular site preparation method, timber and wildlife management objectives should be identified. In addition, each site should be evaluated to identify any specific needs or constraints that must be considered when choosing the best method or methods to use. This can be accomplished by weighing the cost and effectiveness of each site preparation practice against its effects on competing vegetation, soils, and wildlife habitat.

Some site preparation techniques involve significant soil disturbance, while others disturb the soil very little or none at all. When topography is rolling or hilly or when soils are sandy and easily moved, site preparation methods that disturb the soil may create erosion problems. Others may create problems when site preparation methods are improperly applied. The degree of soil disturbance and potential for erosion are important considerations because loss of topsoil can lead to a decline in site productivity for both timber and wildlife habitat. On some sites, the loss of 2 inches of topsoil can reduce tree growth by as much as 50 percent. This loss in site productivity will also result in a decrease in the quantity and quality of vegetation valuable to wildlife for food or other habitat needs.

Site preparation techniques also differ in their influence on vegetation. Some practices are intensive and remove most of the vegetation on a site, disturb soil, and leave bare ground that is quickly replaced by forbs, grasses, and legumes. These conditions favor ground-feeding and ground-nesting wildlife. During the first two years after intensive site preparation, quail foods can increase more than six-fold, resulting in increased quail numbers. Turkey, dove, and other species that use early successional habitats can also find beneficial food and cover during the first few years following intensive site preparation. Other practices are less disruptive, do not disturb the soil significantly, leave less bare ground, and promote higher levels of woody vegetation such as trees, shrubs, and vines. These areas provide food and cover favored by white-tailed deer. The length of time site-prepared areas remain valuable to a particular species of wildlife and how long they remain unsuitable to wildlife species that do not benefit from early successional stage habitat depends on how rapidly plant succession restores these sites to tree and shrub vegetation.

Little research has been conducted on the effects of site preparation on small mammals, amphibians, and reptiles. As with other wildlife species, the effects would probably depend on the intensity and type of site preparation, the size of the area involved, and site conditions. The few studies that have been conducted on pine plantations in the Lower Coastal Plain found that site preparation had little effect on long-term small mammal populations. One study did find that rabbits used less intensively prepared sites more frequently. Other studies have suggested that minimum site disturbance and the presence of logging debris favors reptiles and amphibians.

Timing of site preparation also has an impact on the composition of plant communities and wildlife habitat. Site preparation conducted after October favors plants that produce large seeds like those of legumes. Earlier site preparation favors plants that produce small, wind-carried seeds such as ragweed. Summer or fall site preparation results in areas being sparsely vegetated the following winter with a lower abundance and variety of plants.

The following section describes site preparation methods and some of their effects on wildlife habitat. An understanding of the various site preparation alternatives and how they affect wildlife habitat will help land managers choose a method that best complements their particular objectives.

MECHANICAL SITE PREPARATION

Mechanical site preparation involves the physical disturbance of a site by machinery to reduce undesirable plant competition during the re-establishment of a forest stand. Most mechanical site preparation methods utilized in the South are intended to reduce hardwood shrub or tree competition for establishing pine stands. Several mechanical methods may be used to remove or "knock-back" the vegetation that remains after a timber harvest to enhance the establishment of the new stand. These methods include shearing vegetation and piling debris (windrowing), root raking, burning, drum chopping, disking and bedding, or various combinations of these techniques. On soil-sensitive areas where there is a potential for erosion or site degradation, like stands on steep terrain, a new technique called fell and burn has proven effective for low-cost site preparation with little soil disturbance.

Site Preparation Tools and Methods

CLEMSON UNIVERSITY

RICHARD T. BRYANT/ARISTOCK

ANDREW NIX

ALABAMA FORESTRY COMMISSION

CLEMSON UNIVERSITY

CLEMSON UNIVERSITY

Many tools and methods are used to prepare a newly harvested forest for regrowth. The type of method and type of tool can have an influence on forest stand regeneration and wildlife habitat. Intensive site preparation initially favors grasses and other herbaceous plants, while less intensive methods favor woody plants. Top left, heavy disks break up the ground and roots for tree planting; top right, spray paint marks trees to be removed or retained; middle left, tree planters plant seedlings in open land; middle right, rollers create beds for tree seedlings to increase survival; bottom left, shearing blades remove stumps and roots; and bottom right, a rolling drum chopper knocks down and cuts apart unwanted woody shrubs and seedlings that remain after tree harvest.

Windrow Management for Wildlife

Rhett Johnson, a wildlife biologist and forester, is Director of Auburn University's Solon Dixon Forestry Education Center in Andalusia, Alabama and a past president of the Alabama Wildlife Federation. He has been a driving force in Alabama for illustrating how timber and wildlife management can be compatible on private forestlands.

Windrows can provide a valuable habitat component for a wide range of wildlife species on areas that are intensively site-prepared for timber regeneration. Windrows are piles of stumps and timber slash arranged in linear rows, usually on the contour, across a newly cleared site. The number of windrows needed depends on the size of the clearing and the amount of timber slash left after a harvest. It is not unusual for windrows to occupy up to 10 percent of the site. As part of site preparation, windrows are often burned to enhance tree planting and return nutrients back into the soil. Burning of windrows, in combination with intensive site preparation, often leaves little cover for small mammals, reptiles, and amphibians.

Some windrows should be retained to maintain wildlife cover on areas that have been intensively site-prepared. They can provide protective cover in the same manner as brush piles, except that they are linear in shape and stretch across the entire length of harvested sites. Unlike brush piles, windrows provide protective travel corridors for small mammals across opened timber stands.

Because of the accumulation of topsoil and seeds, windrows quickly become surrounded by a variety of herbaceous and woody plants that provide both cover and food for wildlife. In some cases, production of these plants can be enhanced by fertilizing windrows and several feet adjacent to them. Many species of birds find these areas attractive feeding and nesting sites. Small mammals such as rabbits thrive in these areas because of the close proximity to food and cover. Deer heavily utilize these areas and the adjacent cover for bedding.

Over time the structure of windrows breaks down as regenerated stands become well established. As they deteriorate they leave linear openings through the forest stand. Because of the fertile nature of these sites, they can be easily managed as linear wildlife openings in either planted or native wildlife plants. These strips left from decomposed windrows increase habitat diversity within a stand and provide excellent feeding and brood-rearing habitat for turkeys and other wildlife.

JOE MCGLINCY

Unburned windrows provide cover and travel corridors for many wildlife species. They also quickly become established in herbaceous and woody plants that provide a source of food, additional cover, and variety in even-aged stands.

The most intensive site preparation technique used to convert mixed hardwood stands to pine plantations is a combination of shearing, raking and windrowing, and disking and bedding called "three-pass" site preparation. With this method all standing trees and brush are pushed down by a bulldozer using a pointed, sharpened blade called a K-G or V blade. The remaining material or slash is pushed into windrows with a root-rake, which is a bulldozer blade with teeth along the bottom. In some cases, after the site has been windrowed, the ground is disked to break up any roots left in the soil to prevent hardwood resprouting and competition with pine seedlings. On poorly drained soils, such as some of those in the Lower Coastal Plain, soils are then pushed into linear mounds called **beds** to improve drainage, survival, and establishment of new stands.

Although pine seedling growth can be enhanced by the three-pass method, the potential for soil erosion is increased. This can be a serious problem in hilly terrain. On some sites, soil erosion losses of 50 tons or more per acre can occur during the first year after three-pass site preparation. This severely impacts future timber growth as well as the productivity of the site to provide quality wildlife habitat.

Intensive site preparation techniques, like the three-pass system, temporarily increase herbaceous plant diversity and the total number of desirable wildlife plants during the first two to three years as compared to a mature uncut forest. During the first three to four years after site preparation, these areas are used by a wide array of early successional wildlife as well as other species like deer. If windrows are not burned, they can provide cover for rabbits, snakes, deer, and other wildlife as well as food plants like wild grapes, blackberry, and pokeweed that often grow up around windrows. Windrows that become established with hardwoods can also provide a break in the pine canopy, making the habitat more diverse. If properly arranged, they can also serve as wildlife travel corridors through and between clear-cuts and other habitat types. Although intensive mechanical site preparation can temporarily increase herbaceous vegetation for wildlife, it also uproots large mast-producing hardwoods and destroys other woody plants, hardwood sprouts, and fruit-producing shrubs and vines that provide food and cover for wildlife. Over time some of these plant species will resprout, but their lifespan is usually short since they are shaded out by faster growing pine seedlings.

A more desirable mechanical site preparation method for some wildlife, like deer, is roller chopping. With this technique, a large water-filled drum with sharpened fins is pulled behind a bulldozer. The chopper crushes any remaining trees and brush, while the blades on the chopper cut the slash into small pieces. Soil erosion rates are lower using chopping when compared to more intensive site preparation methods. In most cases the remaining slash is burned afterwards. The use of prescribed fire following chopping will temporally reduce woody vegetation and fruit-bearing plants but will increase the nutrient content of the soil and growth potential of re-sprouting plants including legumes and other herbaceous plants that are preferred by wildlife. Studies have shown that shrubs providing browse and fruits for wildlife recover faster after site preparation if the sites are also prescribed burned. Chopping and burning, as compared to more intensive site preparation methods, produces herbaceous plants for a longer period because the slower growth of pines results in delayed crown closure. A final benefit of chopping and burning is that it is generally one-third the cost of more intensive site preparation methods.

Fell and burn is a mechanical site preparation technique useful in promoting mixed pine–hardwood stands. This method combines mechanical felling of trees that remain after harvest with prescribed burning and low density pine tree planting. It can be applied to many sites but is particularly applicable in steep terrain dominated by hardwoods of low timber value where timber production and wildlife habitat are important objectives.

Following a commercial clear-cut, all standing trees greater than 1 to 2 inches at ground level are cut during the spring using a chainsaw or shear. This promotes full sunlight conditions that favor shade-intolerant species such as oak and pine and encourages quality stump sprouts. Spring felling is used to provide partial hard-

CLEMSON UNIVERSITY

Fell and burn is a low-impact site preparation alternative for steep areas prone to erosion. It combines cutting and felling of trees that remain after timber harvest followed by burning. Many hardwoods often grow back on fell and burn sites planted to pine, resulting in mixed pine–hardwood stands.

wood control, reduce the vigor of new sprouts, and provide an early competitive advantage to the new pine seedlings. The felling step should be used whenever the density of trees that remain following harvest is greater than 20 square feet of basal area per acre. When residual trees amount to less than 20 square feet per acre, adequate sunlight will result and residual felling may be an unnecessary expense. Stump height during the harvest and felling process is important and should be no more than 8 inches to produce quality stump sprouts less susceptible to decay. Prescribed burning is conducted in the summer to provide additional hardwood control and to prepare the site for planting.

The fell and burn technique can produce mixed stands of pine and hardwood that are excellent habitat for a variety of wildlife. In addition, stand establishment costs can be as much as one-half the expense of intensive site preparation methods. Tradeoffs include the difficulties of conducting summer burns, excessive debris that may make tree planting more difficult and costly, and the potential for lower income from timber production with mixed pine–hardwood stands.

Trends in modern forest management, especially where wildlife habitat is to be managed, are toward lower costs and lower intensity in site preparation techniques. Landowners are finding that the slower tree growth resulting from less intensive site preparation is acceptable if investment costs are reduced, especially if wildlife habitat is also improved. In general, the less intensive mechanical methods of site preparation are better for a wider array of wildlife, while the more intensive systems promote quicker seedling growth at the expense of preferred wildlife food plants. Intensive systems allow pine seedlings to achieve crown closure quickly, therefore shading and retarding the growth of understory wildlife food plants. In most cases, less intensive site preparation methods are adequate for pine establishment, more beneficial to wildlife, and less costly overall.

CHEMICAL SITE PREPARATION

Herbicides are also used as a site preparation method to reduce unwanted vegetation that competes with newly planted trees. In many cases, after herbicides are used, sites are also prescribed burned to further reduce competition and improve conditions for seedling planting and establishment. This practice is called **brown and burn**. Herbicides have an advantage over other site preparation methods in that soil disturbance and erosion are kept to a minimum. This can conserve soil on site-sensitive areas including forest stands with steep terrain. Herbicides can also have a more prolonged impact on controlling vegetation than other site preparation techniques. Chemicals that are activated through the soil or translocated to the roots can kill tree roots, which delay the regrowth of vegetation. Although herbicides are most often used for site preparation, they also can be applied later in the life of a stand to release desirable trees from unwanted competition.

Widespread broadcast application of herbicides by mist-blowers or from helicopters can drastically reduce the abundance and diversity of existing vegetation, primarily woody plants. The initial effect of most herbicides incorporated in site preparation is a reduction in both herbaceous and woody plant abundance after the first year's application. During the first growing season after herbicides are applied, bare ground areas are increased by 50 to 90 percent. These areas are highly favored by ground-feeding wildlife, such as quail and doves. Turkeys also use newly created bare-ground areas as bugging sites for insects during the first two years after herbicide treatment.

Studies have shown that approximately one year following herbicide application, herbaceous vegetation returns, although the composition and diversity of plants may be different depending on the herbicide or combination of herbicides used. Newly established pine stands site-prepared with herbicides to reduce hardwood competition change from sites that are dominated by woody plants to stands that are primarily composed of grasses and forbs. Some early successional wildlife species, like quail, mice, and cotton rats, prefer herbaceous plants and benefit from a reduction in woody vegetation.

Studies have shown that a combination of herbicides and prescribed burning can result in more deer forage (grasses, weeds, vines, and woody browse) than is produced by intensive mechanical site preparation such as crushing and burning, or shearing-raking-bedding. In addition, herbicides can be used to kill residual, non-merchantable trees that will create snags, denning, and perching trees that are used heavily by woodpeckers, bluebirds, chickadees, squirrels, raccoons, and other wildlife. These species need trees with cavities to reproduce and raise young.

In certain areas of the South prescribed burning is not feasible. Some natural resource managers are experimenting with forest herbicides to promote valuable wildlife plants over other less preferred plants. For example, Imazapyr (trade name Arsenal®) has been shown to be effective in controlling most hardwood species that compete with pines but does not kill many favored wildlife plants like legumes, blackberry, or dewberry. Several studies have also shown that combinations of certain herbicides may be effective in promoting valuable wildlife plants and also controlling hardwood competition.

Favoring Wildlife Plants With Herbicides

The following is a list of common herbicides used in forest and open land settings. The list specifically includes those plants that are **not** killed by the herbicide treatment. These herbicides can be used to kill vegetation that competes with the plants listed below.

HERBICIDE COMMON NAME	HERBICIDE TRADE NAME(S)	PLANTS NOT CONTROLLED[1]
Forest Herbicides		
Glyphosate	Accord[2], Roundup[2]	red maple, black cherry, ash, hickory, dogwood, greenbriar, Virginia creeper, trumpet creeper
Hexazinone	Pronone 25G[3], Pronone 10G[3], Power Pellets[3]	yellow poplar, Eastern red cedar, sassafras, blackgum, hollies
	Velpar L[4], Velpar ULW[4]	American beauty-berry, bermuda grass, white snakeroot, broomsedge, Johnson grass, sicklepod, trumpet creeper, morning glory
Imazapyr	Arsenal[5]	elm, locust, redbud, buckeye, wax myrtle, sicklepod, coffeeweed, tropical croton, blackberry/dewberry, honeysuckle
Metsulfuron	Escort[4]	bermuda grass, croton, Johnson grass, trumpet creeper, broomsedge, all trees
Picloram	Tordon K[6]	ash, Eastern red cedar
Picloram + 2,4-D	Tordon 101[6], Tordon 101R[6], Tordon RTU[6]	ash, Eastern red cedar
Sulfometuron methyl	Oust[4]	bermuda grass, croton, Johnson grass, trumpet creeper, broomsedge, all trees
Triclopyr (amine)	Garlon 3A[6]	black cherry, Eastern red cedar
Triclopyr (ester)	Garlon 4[6]	black cherry, Eastern red cedar
Open land Herbicides		
Atrazine	AAtrex 4L[7], AAtrex Nine-O[7]	bermuda grass, broomsedge, Johnson grass
Dicamba + 2,4-D	Banvel 720[8]	red maple, ash, hickory, elm, dogwood, persimmon
Dichlorprop	Weedone 2,4-DP[9]	red maple, ash, elm, dogwood, persimmon, willow
Dichlorprop +2,4-D + Dicamba	Weedone 170[9]	red maple, ash, elm dogwood, persimmon
Fluazifop-butyl	Fusilade 2000[10]	all broadleaf weeds and trees
Sethoxydim	Poast[11]	all broadleaf weeds and trees
Trifuralin	Treflan[6]	woody plants, briars

[1] Not all plants listed are good for all wildlife species. [2] Trademark of Monsanto Company, [3] Trademark of Pro-Serv, Inc., [4] Trademark of Du Pont Company, [5] Trademark of American Cyanamid, [6] Trademark of DowElanco Company, [7] Trademark of Ciba-Geigy Corporation, [8] Trademark of Sandoz Corporation, [9] Trademark of Rhone Poulenc, Inc., [10] Trademark of ICI Americas, [11] Trademark of BASF Corporation

Herbicides are also useful in hardwood management to control species composition. Hardwoods of little or no value for timber or wildlife can be selectively killed using herbicides, allowing more growing room for higher quality hardwood trees.

The method of application can also influence the impact that forest herbicides have on wildlife habitat. For example, there are several alternatives to spraying an entire stand with herbicides. One alternative for planted seedlings is to confine herbicide applications to strips where seedlings will be or are planted. This technique is called **band spraying**. Areas between the strips are free to grow in natural vegetation benefitting wildlife and reducing overall site preparation costs. Other herbicide application methods such as directed sprays, stem applications, and individual tree injection can be used to selectively remove certain plants and trees of little value for wildlife or timber. If this method is used, field crews should be trained to identify and spray only vegetation less preferred for timber or wildlife value like sweetgum, leaving beneficial tree and brush species likes oaks, blueberry, huckleberry, dogwood, and other valuable shrub and tree species like those listed in Appendix D–J.

Some scientists are concerned about the negative effects on wildlife and man from indiscriminate use of herbicides. Dr. George Fokerts of Auburn University writes in *Aquatic Fauna in Peril*, "Herbicides used to kill hardwoods and control the growth of herbaceous plants will cause further harm to associated watersheds. Not only do these chemicals destroy plants which help to retain nutrients on site, but some have been shown to be toxic to algae and thus are bound to interfere with processes in aquatic habitats. There have been no thorough tests of how herbicides applied to forests will affect aquatic systems." Obviously more research will be necessary to answer these questions.

Intermediate Forest Stand Practices

Intermediate stand management practices are treatments applied to a forest stand between establishment and final harvest. A variety of treatments are commonly used to:

- Improve tree growth,
- Improve stand quality,
- Reduce tree mortality,
- Obtain periodic income,
- Improve wildlife habitat and aesthetics, and
- Protect the stand against damage from wildfire and insects.

Specific practices used include pre-commercial thinning, commercial thinning, prescribed burning, release, and forest fertilization. These treatments can also increase the quantity and quality of wildlife habitat.

PRE-COMMERCIAL THINNING

Pre-commercial thinning is used to reduce the number of seedlings or saplings in a stand that is too "thick" or dense to meet stand objectives. This type of thinning is conducted before trees have commercial or market value, hence the term pre-commercial. Most often this technique is used to improve the growth of desirable trees in dense natural stands. Naturally regenerated pine stands are the most likely candidates for this practice, but it can also be beneficial in naturally regenerated hardwood stands that have an abundance of stump sprouts. Occasionally it is used to benefit pine plantations that were planted at high densities.

If tree density is not reduced on overstocked sites, the length of time necessary for trees to reach merchantable size will increase significantly, compared to an adequate or understocked stand on the same site. Another impact of overstocking is poor tree vigor, which increases the susceptibility of trees to insect and disease attack and mortality. Heavy stocking rates, however, promote heavier, tighter-ringed trees with fewer limbs and knots and increase height growth over diameter growth. In addition, overstocked stands have closed canopies and limited understory vegetation, which provides little or no food for wildlife which depend on understory plants. The only value that overstocked stands provide to some wildlife is protective cover for resting and escape from predators.

Pine stands that contain more than 1,500 seedlings or saplings per acre should be pre-commercially thinned as soon as possible, preferably at age two to three years but up to age five. Thinning early provides the opportunity to avoid growth loss and to conduct the practice at the lowest possible cost. Seedlings or saplings remaining after thinning should be 8 to 10 feet apart. The greater the distance between saplings, the longer the period before crown closure and loss of understory plants valuable to wildlife.

Several methods are used to pre-commercially thin. On small tracts, hand thinning using a brush axe, weed-eater style equipment, or chainsaw may be the only option. On larger areas, seedlings less than 5 feet tall can be thinned with a mower or bushhog. Larger trees may require a heavy duty chopper. Heavy duty forestry or wildland disks can also be used. When conducted with mowers, choppers, or disks, strips are made in the stand leaving rows 8 to 10 feet apart to provide a source of crop trees. Leave strips should be as narrow as possible. Openings created during the process can be maintained for wildlife by periodic mowing, disking, or planting in wildlife foods. Once the crop trees are large enough to tolerate fire, prescribed burning can also be incorporated for timber and wildlife benefit. Pre-commercial thinning is rarely pleasing to the eye. When conducted properly it often appears that the entire stand has been destroyed.

Although pre-commercial thinning is a labor-intensive expense, it often pays for itself. Compared to an overstocked stand with no treatment, a pre-commercially thinned stand will often provide increased income at an earlier age due to improved tree growth. It will also improve the vigor of remaining crop trees.

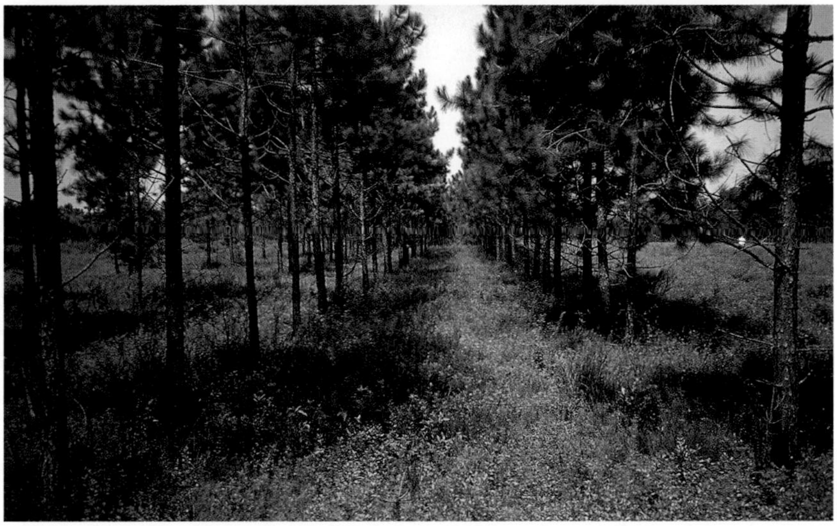

CLAY SISSON

Thinned planted pines between agricultural fields provide an attractive buffer and corridors where wildlife can travel across previously open areas.

COMMERCIAL THINNINGS

Commercial thinning is used for many of the same reasons as pre-commercial thinning. The difference is that the trees removed are older, larger, have market value, and therefore generate income. Commercial thinnings are used extensively in pine management, especially when growing high value trees suitable for solid wood products such as lumber, plywood, and utility poles. Though most common in pine management, commercial thinning can also be used with care in hardwood stands to improve their value for both timber and wildlife. In hardwood stands, this practice is often called timber or wildlife stand improvement cutting.

Pine stands managed for timber production are thinned using a variety of methods. Low thinning generally removes smaller trees and leaves the best trees. Intermediate thinning removes some small and some large trees. Thinning from above removes larger trees. Row thinning removes entire rows in pine plantations. The most common method used in pine plantations is a combination of row thinning and low thinning. Every third to fifth row is removed, then individual trees in the rows that remain are selectively thinned with an emphasis placed on leaving the better quality or larger, dominant individuals for the future crop. Individual trees within the remaining rows that are harvested are usually diseased, poorly formed, or relatively good individual trees too close to a tree of similar or better quality.

After crown closure, tree growth begins to slow due to competition with adjacent trees for sunlight, nutrients, and water. Thinnings open up forest stands and provide the remaining trees with more growing room. The increase in growth rate of the remaining higher quality trees is a result of reduced competition with lower quality trees that have been taken out of the stand. The desired outcome of thinnings for timber production is an improvement in tree vigor, growth, and value.

In stands where crown closure has completely shaded out the forest floor and where the forest is composed of a single canopy layer of trees, there is generally a lack of vertical and horizontal wildlife habitat diversity. Thinnings open up canopies, allowing sunlight to reach the forest floor and stimulating the production of understory vegetation valuable for food and cover to many species of wildlife. The immediate impact of thinning is an increase in the productivity of ground-layer herbaceous vegetation. The production and nutritional quality of herbaceous vegetation can be further enhanced if stands are prescribed burned after thinning.

In general, heavy thinnings produce a tremendous response in the growth of understory vegetation. The more open the forest stand, the greater the production and availability of herbaceous plants. After thinning, forage and herbaceous plant growth usually peaks in two to three years. Soft mast and browse production from woody plants peaks in about five years.

To balance the benefits of thinnings for tree growth and wildlife habitat, most pine stands should be thinned to a basal area between 60 and 70 square feet per acre. Pines should initially be thinned between the ages of 12 and 20 years, and every 5 to 10 years thereafter until final harvest. The number of times a stand needs to be thinned depends upon several factors such as the type of pine, growth rate, and site conditions. For example, on most sites the faster growing loblolly pine should be thinned every 5 to 6 years, and the slower growing longleaf and shortleaf pine every 8 to 10 years. To provide maximum benefit to wildlife, most managers agree that pines should be thinned as soon and as often as possible. Thinning in pine stands with a low or very low site index should be heavier than in sites that are highly productive. Poor quality sites need the additional benefit of added sunlight to assist both tree growth and understory growth. Early thinnings on poor quality pine sites should leave a residual basal area of less than 60 square feet per acre.

Thinning activities should be dispersed throughout a forest. The result is a "patchy" pattern of open understory intermixed with dense shrubby growth across the forest landscape. Removing select trees along a forest edge is a useful technique to create a transition zone that minimizes the contrast between cover types. Such gradual edge or ecotone zones are ideal for many species of wildlife. For example, to create a cover condition ranging from open land to weed/shrub to forest, remove 75 percent of the tree cover from the first 50 feet of forest edge, 50 percent of the trees from the next 50 feet, and 25 percent of the trees from the next 50 feet.

In traditional pine thinnings for timber production, both merchantable and nonmerchantable, poor quality trees are marked for removal during the operation. In most cases however, non-merchantable trees of poor quality are simply left in the stand, since their value for timber is negligible. Many non-merchantable trees in a forest stand provide food and cover for wildlife. Tree and plant species such as oaks, hickory, beech, persimmon, serviceberry, blackgum, American holly, hawthorn, dogwood, wild grapes, and others provide nuts, soft fruit, and browse for wildlife and should be retained in the stand. In addition, at least two to four den trees per acre of adequate size should be retained to provide nesting and denning sites for birds and small mammals.

Studies have shown that deer foods such as herbaceous plants, fruits, and woody browse can be increased dramatically by thinning a closed canopy forest. Blackberry, woody vines, and other preferred deer browse increase dramatically after thinning. Prescribed burning a recently thinned pine forest can further stimulate herbaceous plant production and can increase the nutritional value of plants for deer. Pine stands that are periodically thinned and burned also provide increased forbs, grasses, and legumes that are utilized by quail, turkey, and other wildlife. If thinning is not used as a management strategy, mast-producing trees should be retained throughout a stand whenever possible.

Although each stage of a forest's growth and the corresponding vertical and horizontal structure favors certain groups of birds. Thinned forest stands have a greater number and diversity of birds than unthinned stands. Thinnings encourage the development of a shrub and midstory layer that provide ideal habitat conditions for many species of birds. More heavily thinned stands favor species like the chestnut-sided warbler, gray catbird, rufous-sided towhee, Kentucky warbler, and indigo bunting. Other bird species favored by thinnings include the American redstart, black-and-white warbler, hooded warbler, and Eastern wood-pewee. Some species like red-eyed vireo and the scarlet tanager can do well in both thinned and unthinned timber. Other species like the ovenbird and the black-throated green warbler are negatively affected by forest stand thinning. In order to maximize bird species diversity in a forest, land managers should incorporate a variety of thinning practices in combination with retaining unthinned stands to provide habitat conditions favorable to a wider variety of species. This requires developing a thorough wildlife habitat improvement plan as discussed in Chapter 2.

Thinnings conducted in conjunction with prescribed burning are also a valuable tool used in maintaining the endangered red-cockaded woodpecker's habitat in mature pine stands. Red-cockaded woodpeckers excavate cavities for nests in old living pines and prefer open foraging areas within mature pine forests with no midstory shrub or tree layer. Thinnings maintain an open area for foraging, while frequent prescribed burning prevents the establishment of a forest midstory.

PRESCRIBED BURNING

Natural fire has influenced southern forests for thousands of years. Fire continues to impact our forests today, but, rather than by chance, it is deliberately used to bring about desired results. This planned use of fire to achieve specific objectives is called **prescribed burning** or **controlled burning**. Low intensity fires are used as a forest management tool to prevent forest fires by reducing the accumulation of combustible material on the forest floor. Low intensity fires are also used to eliminate hardwood competition in pine forests and prepare sites for re-establishment by artificial or natural regeneration. Prescribed fire is also an excellent wildlife habitat improvement practice. This intentional burning enhances horizontal diversity within a forest stand by stimulating the growth of many forbs, grasses, and other herbaceous vegetation that provide valuable food and cover for early successional wildlife. The seeds of most legumes require some sort of scarification, the process of breaking down hard seed coats, before the seeds will germinate. Prescribed burning effectively acts as a scar-

Fig. 7: Fire-Dependent Animals and Plants of Concern in Alabama[1]

ANIMALS	SCIENTIFIC NAME	FEDERAL STATUS
Red-cockaded woodpecker	*Picodies borealis*	Endangered
Eastern indigo snake	*Drymarchon corais couperi*	Threatened
Gopher tortoise	*Gopherus polyphemus*	Threatened
Dusky gopher frog	*Rana capito sevosa*	Species of Concern
Flatwoods salamander	*Ambystoma cingulatum*	Species of Concern
Florida pine snake	*Pituophis melanoleucas mugitus*	Species of Concern
Black pine snake	*Pituophis melanoleucas lodingi*	Species of Concern

PLANTS	SCIENTIFIC NAME	FEDERAL STATUS
Green pitcher plant	*Sarracenia orephila*	Endangered
Alabama canebrake pitcher	*Sarracenia alabamensis*	Endangered
Chaffseed	*Schwalbea americana*	Candidate Species
Wiregrass	*Aristida striata*	Species of Concern
Three-awn grass	*Aristida simpliciflora*	Species of Concern
Chapman's aster	*Aster chapmanii*	Species of Concern
Coyote-thistle aster	*Aster eryngifolius*	Species of Concern
Panhandle lily	*Lilium iridollae*	Species of Concern
Birds-in-a-nest	*Macbridea carolina*	Species of Concern
Panic grass	*Panicum nudicaule*	Species of Concern
Butterwort	*Pinguicula planifolia*	Species of Concern
Meadow beauty	*Rhexia parviflora*	Species of Concern
Panhandle meadow beauty	*Rhexia salicifolia*	Species of Concern
White-topped pitcher plant	*Sarracenia leucophylla*	Species of Concern
Wherry's sweet pitcher plant	*Sarracenia rubra* spp. *wherryi*	Species of Concern
Royal catch-fly	*Silene regia*	Species of Concern
Mohr's goat's-rue	*Tephrosia mohrii*	Species of Concern
Kral's yellow-eyed grass	*Xyris longisepala*	Species of Concern
Rough-leaved yellow-eyed grass	*Xyris scabrifolia*	Species of Concern

[1]Listing as of 1997 by the Alabama Natural Heritage Program. Candidate Species are those proposed to be listed under the Endangered Species Act as threatened or endangered. Species of Concern include plants or animals whose numbers are so low the threat of extinction is of concern and may be listed as Candidate Species in the future.

ifying agent, increasing the chance of seed germination. Studies have shown that prescribed burning also serves as a fertilizing catalyst that releases nutrients otherwise bound-up in dead organic material on the forest floor. The burning process frees these nutrients for use by plants, increasing their palatability, digestibility, and nutritional value for wildlife. The only nutrient lost is nitrogen, since it is vaporized as a gas during burning. This loss, however, is only temporary since nitrogen-fixing (producing) legumes grow almost immediately after a burn, generally producing more nitrogen than was lost. Studies have indicated both the high value of fire-adapted legumes (seeds and browse) for wildlife food and their additional value in providing a natural source of nitrogen essential for plant growth.

FIRE ADAPTED LANDSCAPES

When the first Europeans arrived in Alabama they reported landscapes that are quite different from what we see today. Open park-like stands of longleaf pine were common and covered millions of acres in the Coastal Plain as well as high elevations in the central part of the state. While most tree species were killed by periodic hot fires caused by lightning or intentional burning by

Native Americans, longleaf pines were well adapted to periodic fires. Plant and wildlife communities evolved and adapted to sites that were frequently burned and became unique to these areas.

Since the early settlements of Alabama's Coastal Plain and other areas where fire-adapted communities existed, man over time has limited the use of fire. This has changed the ecology of many areas in favor of plant and animal communities that do not require periodic fire. Still, a number of today's species depend on periodic burning in managed forest stands. For example, the gopher tortoise, protected by the Endangered Species Act over a portion of its range, digs deep burrows to avoid summer heat, to protect against cold winters, and to escape fires. These burrows in turn are used by a number of declining reptiles and amphibians including indigo snakes, gopher frogs, and pine snakes. Red-cockaded woodpeckers, another endangered species, also depend upon frequent summer and spring fires that kill hardwood sprouts and maintain preferred longleaf pine stands with open understories. Of Alabama's plant and animal species listed by the U.S. Fish and Wildlife Service as threatened, endangered, or candidates for listing status, eight animals and at least 18 plants are considered dependent on fire for the maintenance of their habitats. The Alabama Forestry Commission, in cooperation with the U.S. Fish and Wildlife Service and the Alabama Natural Heritage Program, uses fire to manage several bogs containing remnant populations of endangered pitcher plants. Clearly the use or omission of fire has played a major role in the forest ecology of many stands across Alabama.

CONTROLLED BURNING EFFECTS

Although there is an initial reduction of low-growing hardwoods and shrubs after prescribed burning, hardwoods resprout quickly after burning. Periodic burning maintains hardwoods and shrub browse in a palatable state at a height that is easily accessible by deer. However, because periodic prescribed burning retards midstory hardwood and shrub growth, vertical habitat diversity is reduced.

Periodic burning also increases insect abundance. Insects are extremely important sources of protein for young quail and turkeys and are heavily used by these birds during the first one to four weeks of life. Insects continue to comprise a significant portion of quail and turkey diets through adulthood. Other benefits of prescribed burning include controlling certain tree diseases, enhancing the appearance and value of forest stands, improving access, and controlling unwanted insects like ticks and other parasites that plague people and wildlife alike.

The fact that hardwoods are easily damaged or killed by fire, because of their thin bark layer, makes prescribed burning an effective tool in reducing hardwood competition in pine stands. This same fact, however, prevents prescribed burning from being a viable wildlife habitat improvement tool in hardwood stands, although in some cases wildlife managers have been successful in using low-intensity "cool burns" in hardwood stands. The possibility of using low-intensity fire in older hardwood stands to stimulate herbaceous plant growth yet not damage mature hardwoods needs investigating. As a rule, however, prescribed burning in hardwood stands should be avoided if timber production is a high priority.

To increase benefits from prescribed burning, timing is critical. Spring and summer burns are the most effective for controlling hardwoods in pine stands, but they are also the most risky since pine trees can also be killed by hot fires. Burning during the first of May is probably the most beneficial. During this time of year, young hardwoods have utilized most of their stored nutrients in putting on new leaves; therefore, resprouting and regrowth is less likely. In some cases, several burns may be needed to effectively control unwanted hardwoods.

Until recently, many wildlife managers discouraged spring and summer fire for wildlife. These periods coincide with nesting season for many ground-nesting wildlife such as quail, turkey, chuck-wills-widow, and Bachman's sparrows and may negatively affect nesting success. Prescribed burning during spring can make newly born wildlife more vulnerable to predation since protective cover is often drastically reduced. In some cases, direct mortality can also occur if young wildlife and nesting birds are unable to escape prescribed fires. However, if spring and summer burns are properly conducted, protective cover can be retained. This can be accomplished by plowing firelines, often called "ring-arounds," around high-valued cover sites before burning. Ring-arounds should be at least 5 acres in size to provide adequate nesting cover protection for species like bobwhite quail. A less intensive method of protection is to provide unburned corridors that encompass high-value cover sites. This can have the same effects as ring-arounds at a lower cost. Plantations that have been experimenting for several years with spring and summer burns have found that overall plant diversity and abundance, especially of native warm-season grasses, may be enhanced over time by burning during the growing season. Other studies by Auburn University have indicated that spring burns for wild turkey brood habitat have no benefits over traditional winter burning and should be avoided if possible to reduce any risk

Prescribed burning in pine stands is one of the most beneficial and cost-effective management tools for wildlife habitat improvement and timber production. Low-intensity burning enhances horizontal diversity within a forest stand by stimulating the growth of many forbs, grasses, and herbaceous plants that provide valuable food and cover for wildlife. Prescribed burning breaks apart the outer wax-like barrier of seeds and enhances germination. Burning also acts a fertilizing agent by releasing many nutrients that are bound in dead organic matter. Pines are fire-adapted because of their thick bark layer and can be burned periodically when the trees are more than 15 feet tall, usually at 8 to 10 years of age. Top left: Longleaf pines can be burned earlier than other pines during the seedling stage. Top right and bottom: Low-intensity backing fires can be effectively conducted in older, naturally regenerated pines or pine plantations. Bottom left: Fallow fields can be burned to remove dead and dying vegetation and promote new plant growth valuable to wildlife.

ANDREW NIX

RICHARD T. BRYANT/ARISTOCK

ALABAMA FORESTRY COMMISSION

Firebreaks serve to contain prescribed burns to a specific tract or compartment. Top: Secondary woods roads can be utilized as firebreaks. Middle: Firelines plowed through the woods can also effectively keep fires contained within or excluded from a specific area. Bottom: Wider firebreaks, called firelanes, can also be planted as wildlife food plots and used to access timber stands for wildfire control, timber harvest, and travel by sportsmen.

posed to turkey nests.

Traditional prescribed burning for most wildlife species has been conducted during late winter, mid-January to mid-March. Burning during this time does not conflict with ground-nesting birds like quail and turkey. Late winter burning enhances the spring production of plants, especially legumes and hardwood sprouts, and also insects valuable to wildlife. Late winter, low-intensity prescribed burns leave "patches" or small pockets of unburned vegetation necessary for nesting and escape cover. Escape cover is important since many raptors are migrating north during this time. Terrestrial predators are also stressed due to the low abundance of prey at this time and are actively searching for food.

Ideal burns for wildlife create a mosaic of burned and unburned areas that stimulate the growth of new vegetation while retaining adequate cover. The "patchy" result of these burns intersperses a variety of plant species and cover types across forest stands that help meet the habitat requirements of many wildlife species. Various burning techniques can be utilized depending on the situation, but backfires are recommended for most wildlife habitats because they tend to leave unburned patches.

To obtain the best growth of vegetation for wildlife, prescribed burning should be conducted in close conjunction with timber thinnings. The combination of burning and newly opened canopies resulting from thinning creates several benefits: 1) more sunlight reaches the forest floor, 2) seeds are scarified so they can sprout, and 3) the release of nutrients by burning can have an almost immediate effect in stimulating the production of legumes, forbs, grasses, and other herbaceous wildlife plants. Ideally, prescribed burning should be conducted before and immediately after pine thinnings for maximum herbaceous plant production. Prescribed burning after crown closure has minimal benefits for wildlife.

Prescribed burning to improve wildlife habitat should be conducted at the earliest age that pine trees could tolerate fire. Young pines are more susceptible to damage, however, than mature trees and a greater degree of care should be taken to protect young stands from fire damage. For loblolly and shortleaf pine, stands can be burned as early as 8 to 10 years of age. Longleaf is the only southern pine that can be prescribed burned at the seedling stage or what is often called the "grass stage." Longleaf pines between the seedling stage and at a height less than 8 to 10 feet tall should only be burned very cool in midwinter, since they are susceptible to damage during this time. All four major southern pine species (loblolly, longleaf, shortleaf, and slash pine) can be prescribed burned when trees are more than 15 feet tall, usually at 8 to 10 years of age.

The frequency of prescribed burning is important in maintaining the availability and quality of herbaceous wildlife plants in pine stands. In the Middle and Lower Coastal Plain of Alabama, prescribed burning every one to two years benefits wildlife that prefer ground-layer vegetation like quail and turkey. Burning on a one- to two-year cycle increases seed production of legumes and grasses but reduces soft mast production and fruiting of shrubs and small trees. In the Upper Coastal Plain, Piedmont, and upper regions of the state where vegetation generally is not as thick, prescribed burning for early successional wildlife species should be conducted every two to three years. Delaying the frequency of prescribed burning to every three to five years increases woody sprouts and fruit production and benefits wildlife species like deer and other wildlife associated with shrub layer vegetation. Woody plants that are severely burned can resprout and provide highly nutritional browse for deer. Many shrubs reach their peak fruit production in the second to fourth growing season after burning. An ideal approach to prescribed burning for improving wildlife habitat is to divide a tract of land into burn units. Adjoining units should not be burned in the same year, therefore creating differing stages of plant growth and a mosaic of habitat types. Unique areas within burn units that provide valuable wildlife habitat (snags, plum thickets, abandoned orchards, old homesites) should be protected from fire. Other selected areas within burning units should also be set aside to provide nesting areas for wildlife species like quail. Studies have shown that quail prefer to nest in vegetation that has had one or

Planning and Conducting a Prescribed Burn for Wildlife

Landowners who follow the recommendations below should meet the requirements of the Alabama Prescribed Burning Act. This Act recognizes prescribed burning as a legitimate management tool. The Act also reduces the liability associated with prescribed burning if the person conducting the burn is a certified burn manager.

1. **Evaluate Area(s) to be Burned.** Timber stands and other areas should be evaluated before the burning season to determine current conditions, needs, and actions necessary prior to burning.

2. **Prioritize Areas that Need Burning.** If several areas need to be burned, give higher priority to those sites that have the greatest need for burning. Identify areas that require special considerations and specific weather conditions for burning such as sites that have heavy fuel loads (lots of dead organic matter), small trees, steep terrain, a potential for smoke problems, and wildlife food and cover plants that need to be protected from burning.

3. **Develop a Written Burning Plan.** A written prescribed burning plan should be prepared with the assistance of an Alabama Forestry Commission (AFC) forester or a consultant who is a certified prescribed burn manager. Forms for developing burning plans are available from local AFC offices. Plans should be completed and approved by a certified burn manager before the burning season.

4. **Conduct a Burn.** Be prepared to burn when weather conditions are favorable during the burning season. Permits must be obtained from local AFC offices prior to burning. To obtain a burning permit call the AFC toll-free number located on the inside cover of your local phone book.

AFC personnel will help you decide if weather conditions are conducive for burning. Prescribed burning should be conducted by or under the supervision of an AFC forester or other certified burn manager. Individuals interested in becoming certified burn managers must attend an approved prescribed burning course. Dates and times of course offerings can be obtained from local county Extension System offices or by contacting the Auburn University School of Forestry, 108 M. White Smith Hall, Auburn, AL 36849-5418, (334) 844-1054. To complete the certification process you must send proof of training (registration receipts and meeting agenda) and a $50 processing fee to the Alabama Forestry Commission, Attn: Prescribed Burn Certification Program, P.O. Box 302550, Montgomery, AL 36130-2550.

Methods to Apply Herbicides in TSI and WSI

Girdle: Cutting and removing a band of bark and cambium layer from around the stem.

Frill Girdle: Making a continuous cut around the stem going through the cambium and into the sapwood and applying herbicide to the cut.

Cup Girdle: Spacing cuts at intervals around the base of a stem, usually with an axe, for application of herbicide.

Basal Spray: Applying herbicides to small stems, less than 3 inches in diameter, by spraying the bark or lower stem.

Spot Treatment: Application of herbicide to individual trees using directed spray or spot gun.

Stump Treatment: Application of herbicide by spray or brush to a freshly cut stump.

Injection: Application of a specific amount of herbicide beneath the bark of individual trees using a tree injector or hypo-hatchet.

more growing seasons without fire, or what is commonly called a two-year (or greater) rough. These areas should be scattered throughout burn units.

Even though prescribed burning is one of the most effective and cost-efficient wildlife habitat improvement tools, its future as a management practice may not be as bright. Because of increasing concerns about air pollution and global warming, EPA air pollution (particulate size) regulations, local ordinances regulating burning, smoke management problems near populated areas and highways, and fears of litigation, it has become increasingly difficult for the average forest owner to burn his lands. In addition, some timber companies are beginning to restrict or eliminate prescribed burning as a management policy on their lands because recent research by Weyerhaeuser Company has documented reduced pine growth (volume) the first and second year following a burn. Even though the impacts of future regulations and public policy affecting burning are unknown, landowners should continue to use prescribed burning as a wildlife habitat and timber management tool.

TIMBER AND WILDLIFE STAND IMPROVEMENT PRACTICES

Many forests are composed of **cull** or poor quality trees that have little or no value for timber. These trees compete with higher quality trees for growing space, water, and nutrients, ultimately reducing the stand's potential to produce timber, and to provide food and cover for wildlife.

Stand improvement practices for timber (TSI) or wildlife (WSI) can be used to reduce or eliminate poorer quality trees from forest stands in favor of higher valued timber and wildlife trees. The ultimate goal of TSI or WSI is to make growing space available for more desirable or beneficial trees. The techniques of TSI and WSI are the same, but the objectives are different. With TSI practices, the emphasis is on "releasing" high valued trees from competition with poorer trees to increase the growth, volume, and market value of trees in a forest stand. WSI practices, on the other hand, remove trees that are of little or no value to wildlife in favor of trees that provide food and cover. In some cases, trees that are favored in WSI practices are those that are removed in TSI. At other times both practices favor the same trees.

Timber and wildlife improvement practices that remove inferior trees are effective with all four of the major southern pine species, with mixed pine–hardwood stands, and with hardwood stands. In hardwood stands, reducing competition increases crown size of the remaining trees. The larger the crown, the more mast produced for wildlife. In addition, reducing competition prevents delays in mast production resulting from overcrowding.

During the fall when mast is present, prime mast-bearing oaks and other species can be identified and retained in the stand. Non-producing and otherwise inferior trees surrounding highly valued wildlife trees can then be marked for removal. The removal of inferior trees around the mast-bearing trees will allow the more valuable trees to increase their crown and produce a larger mast crop. TSI practices that remove inferior trees will also allow understory soft mast trees and shrubs to bear more fruit. Trees to leave are those that provide food and cover for wildlife. (For a listing of trees that produce wildlife foods see Chapter 16, "Supplemental Plantings as Wildlife Food Sources.") Mature trees that provide denning, nesting, and roosting sites for wildlife should also be left intact in the forest stand.

Stand improvement practices can be conducted mechanically by cutting and removing poor quality trees (also called improvement cuts), killing unwanted trees with herbicides, or a combination of both techniques. Mechanical removal is usually the cheapest method and involves cutting unwanted trees with a chainsaw or fellerbuncher like that used in thinning operations. Trees can be left where they fall on the forest floor or stacked in brush piles to provide shelter for small mammals and herpetofauna (reptiles and amphibians). Where practical, trees that are cut can also be sold or used for firewood.

Herbicides can also help reduce the number of poor quality trees in a forest stand, but the treated trees are usually left standing for some period of time. Although dead, these herbicide-treated trees can provide excellent denning, nesting, and roosting sites for wildlife until they decay and fall. The effectiveness of a herbicide treatment depends on the time of year that herbicides are applied, tree species, vigor and size of tree, and weather conditions. Generally, the best time for herbicide use is during the spring and summer growing season.

Using Herbicides to Manage for Red-cockaded Woodpeckers

Herbicides have been used effectively to reduce midstory hardwoods from longleaf pine stands to improve habitat for the federally endangered red-cockaded woodpecker. The red-cockaded woodpecker prefers open stands of mature pine for foraging areas. A combination of herbicides and periodic prescribed burning can be used to effectively control hardwoods and maintain pine stands in suitable red-cockaded woodpecker habitat.

JOE MCGLINCY

Herbicides and prescribed burning can be used to maintain mature pine stands in a park-like fashion for red-cockaded woodpeckers that require open pine stands for foraging.

FOREST FERTILIZATION

Although this is a relatively new practice, fertilizers are applied to some pine forests in the South. Most applications are on industrial forests where the goal is maximum fiber production. The purpose of fertilizing stands is to accelerate tree growth and increase the volume of merchantable timber. In addition, studies have shown that fertilization improves tree vigor, making them more resistant to some diseases. Unfortunately, for the average non-industrial forest owner in Alabama, forest fertilization may be impractical because of costs and uncertain results. The response of stands to fertilization varies depending upon tree species, age of the stand, site conditions (soil fertility, competing vegetation), and time of application. In general, the best potential for a significant response from fertilization is on intensively managed forest stands that are on nutrient deficient sites. Because forest fertilization is expensive, a thorough understanding of site conditions is required in order to gain the maximum benefit in timber growth. Industrial forest managers usually will have the soil and leaves of trees tested for nutrient content before recommending fertilization.

The most common nutrients applied to industrial forest stands are phosphorus (P) and nitrogen (N). Phosphorus-deficient sites are commonly treated before stand establishment by incorporating concentrated super phosphate or diammonium phosphate into the soil. The typical rate of application is 25 pounds of phosphorus per acre. Fertilization with phosphorus, in combination with weed control, gives southern pine seedlings, like slash pine, a jump-start in growth.

Studies on the effects of fertilization on forest wildlife food plants have been limited to pine habitats. However, some wildlife managers have also reported using nitrogen fertilizers to increase acorn production and palatability in hardwoods. The effects of fertilization on wildlife plants and trees vary with the type of fertilizer, application rate, and soil type. Generally, forest fertilization in pine stands increases understory herbaceous forage production by 20 to 30 percent for several years depending on the site. Total soft mast yield, in mast like wild grapes and blackberries, has been shown to be greater in some fertilized pine plantations as compared to unfertilized plantations. Supplementing phosphorus and nitrogen has also been shown to increase protein levels of some wildlife food plants. For example, studies at Auburn University have shown dramatic increases in protein levels in honeysuckle patches that were fertilized for deer. In general, deer prefer most plants that have been fertilized over those receiving no fertilizers, even if the plants are not generally preferred by deer. But once again the bottom line for private forest owners contemplating large-scale forest fertilization may be money, as it may seem cost-prohibitive given the indefinite payoffs to wildlife.

White-tailed deer, especially those whose range encompasses nutrient-deficient soils, may benefit more than most wildlife species from forest fertilization. Studies have indicated strong correlations between soil nutrients, especially phosphorus, and deer body weights. A low phosphorus content and a wide calcium to phosphorus ratio in soils are believed to suppress the digestibility of vital nutrients in deer forage that already occur in low levels. Fertilization with phosphorus on poor or depleted forest soils should result in improved digestibility of many browse plants and improved growth and reproduction for deer, rabbits, and other herbivores.

The effect of forest fertilization on wildlife plants is not all positive. After forest fertilization, the variety of plants has been reported to decrease, since invasive plants out-compete other plants for nutrients. Decreasing the availability of a variety of plants reduces the overall value of the site for a wide range of wildlife species. In addition, studies have indicated that fertilizers that contain nitrogen depress legume growth in favor of grasses and forbs. Legumes are important food plants for a variety of wildlife species.

Fertilizing Honeysuckle for Deer

Japanese honeysuckle is a woody evergreen vine that is an ideal year-round browse for white-tailed deer, especially during the fall and winter in areas where acorns are lacking. The leaves and seeds are also eaten by rabbits, quail, and wild turkey. A study conducted by Auburn University found that fertilization and liming could be cost-effective methods to increase year-round production and nutritional quality of honeysuckle browse. Protein contents of honeysuckle in the study were increased from 11 to 17 percent by fertilizing.

Honeysuckle patches were limed and fertilized in the spring (April) and fall (September/October) at a rate of 3.5 tons of lime per acre, and 300 pounds of triple 13 (13-13-13) per acre, and top-dressed with 300 pounds of ammonium nitrate per acre. For the purpose of the study, honeysuckle patches were mowed to a uniform height in order to measure changes in growth and yield from fertilization. Mowing of natural honeysuckle patches that grow on forestlands, however, is ***not*** recommended to maintain or enhance honeysuckle production. Honeysuckle forage production averaged over 2,480 pounds per acre, and protein content increased significantly at a cost of approximately $134 per acre. In comparison, a cultivated deer food plot containing biennial red clover would only produce about 1,115 pounds of forage per acre at a cost of $150 per acre. In addition, red clover would not be available to deer from September through December, which is a critical period due to the reduction in nutrition and palatability of most vegetation.

Landowners or hunting club members can fertilize honeysuckle patches for deer using the following steps:

1. Locate honeysuckle patches and estimate approximate size in acres.
2. Determine lime and fertilizer rates based upon size of patch and recommendations above. For example, patches ¼ of an acre should receive 500 pounds of lime, and 75 pounds of 13-13-13, and be top-dressed with 75 pounds of ammonium nitrate.
3. Lime and fertilizer should be broadcast evenly over the entire patch by hand or using a cyclone seeder (for fertilizer) in April and again in September.

Currently there is work underway to examine the potential usefulness of applying industrial residues, like ash and municipal waste, to increase timber production. Studies have indicated that such waste can increase tree growth and understory plant production. Although using waste as fertilizer has some promise in forest management, plant species diversity may be decreased by waste application. In addition, there is concern about the effects that heavy metals and other toxins found in waste might have on wildlife. However, preliminary results suggest little heavy metal build-up in the tissues of small mammals captured in waste-treated forest stands. This is an area to watch as future research results become available.

PRUNING

Pruning is the removal of live and dead limbs from the main trunk or stem of a tree for the purpose of improving tree form and quality. Pruning concentrates growth of the tree upward rather then outward and decreases knots and defects on the lower bole of the tree. Although not a common practice on non-industrial private forests, some industrial timber companies have recently begun pruning intensively managed pine plantations. Because pruning is labor intensive, it is questionable whether the benefits in tree growth and quality outweigh the associated costs.

Removing limbs reduces vertical habitat structure. Consequently, those wildlife species that depend upon vertical diversity within a stand could suffer from the effects of pruning. Eliminating live limbs close to the ground also reduces forage and cover for wildlife associated with the lower levels of a stand. Removing dead limbs also reduces the availability of perching sites for raptors and diminishes the insect prey of insectivorous birds. Studies have shown that restricting pruning to the most valuable timber producing trees may help lessen the impact that pruning has on wildlife, especially songbirds.

Pruning may help increase horizontal diversity within a forest stand if pruning is conducted in conjunction with timber thinnings. Prunings, like thinnings, allow sunlight to reach the forest floor, helping to stimulate the production of early successional stage vegetation. In pine forests, the changes are even greater if the area is prescribed burned after thinning and pruning. However, pruning has little likelihood of increasing understory vegetation production if a closed forest canopy is not opened by thinning.

OTHER FOREST HABITAT IMPROVEMENTS FOR WILDLIFE

In addition to improving habitat during normal timber operations, land managers can use other practices to enhance forest wildlife habitat. The following section highlights some examples of these practices.

Openings

Openings in a forest can serve to increase horizontal habitat diversity for wildlife. Openings can either be planted and cultivated as wildlife food plots or established and maintained as "natural" openings in a variety of native grasses, forbs, and other herbaceous vegetation. A mixture of planted and natural openings may be the best approach in providing the greatest variety of food and cover sources for wildlife within a stand.

Ideally, forests should have at least 1 acre of openings for every 25 acres of forest. Openings created during normal silvicultural operations like logging decks, woods roads, selection cuts, or salvage cuttings to remove disease- or insect-infested trees all provide excellent opportunities for establishing permanent herbaceous openings for wildlife. Wildlife openings can be created at any time, but openings are less disruptive to timber management if they are created in conjunction with forest management activities. The key to integrating openings into forest stands is prior planning and incorporation into the overall forest management plan before silvicultural practices are implemented.

Natural openings can be valuable for wildlife. Once sunlight reaches the ground, openings quickly become covered in a mixture of native grasses, forbs, and other non-woody plants. These areas provide excellent sites for nesting, brood rearing, and feeding for a variety of wildlife such as quail, turkeys, and deer. Herbaceous seed and green forage produced in these areas are key wildlife food sources. They also become "bugging" sites: areas where wildlife feed on insects, since they attract a variety of insects that provide protein for quail, turkey, and other insect-eating (insectivorous) wildlife. For smaller species of wildlife, like quail and small mammals with limited home ranges, openings should be at least ½ acre in size. For larger species of wildlife, like deer and turkeys, openings should be from 2 to 5 acres in size.

To maintain openings for herbaceous plants, these areas should be mowed, disked, or prescribed burned every one to two years. These practices suppress woody plants and keep herbaceous vegetation at a palatable, nutritional, and productive stage for wildlife. Research also verifies that periodic mowing and disking increases insect production.

Riparian Forests and Streamside Management Zones

Riparian areas or streamside forests are land areas adjacent to a body of water, usually a stream or river. The boundary of a riparian area is difficult to determine because the land may be a transition between aquatic and terrestrial habitats, and because of seasonal and annual variation in flooding, soil moisture, and vegetation. These wetter forests protect water quality and enhance wildlife habitat. Historically, portions of the forest protected the rivers and streams, but deforestation for agriculture and urban expansion has dramatically affected their ability to filter pollutants and hold stream banks in place. The result of deforestation has included some water becoming unfit for human or livestock consumption, recreation, or fisheries production. Many of these problems are ultimately linked to contamination from nutrients, sediments, animal wastes, or other pollutants associated with urban or agricultural runoff.

These forest buffers are essential for water quality, helping to filter sediments and suspended solids from surface runoff, transform or alter the chemical composition of certain compounds like nitrates, change toxic chemicals like pesticides to non-toxic forms, store excessive nutrients like nitrogen in plant tissue, and provide a source of energy to downstream ecosystems in the form of dissolved organic compounds. Streamside forests also shade waterways to allow optimal light and temperature conditions to foster the growth of aquatic organisms. Water temperatures that are too hot can kill fish and other aquatic life.

Riparian forests are extremely valuable wildlife habitat because of the 1) predominance of woody plants, especially hardwoods, that provide mast and cavities; 2) presence of surface water and soil moisture; 3) diversity of habitat types forming an ecotone or edge habitat; and 4) pathways or travel corridors. These areas provide for movement of plants and animals. Studies have shown that many wildlife species require these unique habitats to meet a part of their life needs. In general, forested floodplains have higher wildlife populations than surrounding upland forests because of their unique attributes and the high level of productivity associated with these areas. These lands are often the most productive sites within a forest stand. One study found bottomland hardwood forests supported 2 to 5 times the diversity of game animals as surrounding pine forests and 10 times as many birds. At least 49 species of mammals, 38 amphibian species, and 54 reptile species use riparian forest habitats in the Southeast. These forests also serve as wintering, migrating, feeding, and breeding areas for many migratory species.

To provide riparian forests to benefit wildlife, many managers have established streamside management zones adjacent to watersheds. **Streamside management zones (SMZs)** are strips of vegetation along watersheds (such as creeks and streams) that are left intact when adjacent forests are harvested. In the past, these areas were left mostly untouched because of their inaccessibility for timber harvesting operations. Today, they are retained because of their value in preventing soil erosion, maintaining water quality, and enhancing wildlife habitat.

Maintaining adequate forest cover on each side of a stream or drainage can increase forest habitat diversity, especially within a pine-dominated landscape. SMZs provide travel corridors for wildlife between habitat types. They also provide food and cover for deer, quail, turkeys, squirrels, and other wildlife. Hardwoods adjacent to smaller creeks or drainages that flow into SMZs should be left intact to serve as a network of corridors to link forest stands and other habitats. Development, cultivation, and past land management practices have fragmented many forests and divided the landscape into isolated areas or islands. SMZs can join isolated forest fragments, which enhances the movement of wildlife to and from these otherwise unconnected areas. A free flow of animals and genetic material is critical for maintaining biological diversity.

Streamside management zones also aid in preventing soil erosion and protecting water quality during silvicultural operations. Because of their buffering effect in halting soil erosion, "Alabama's Best Management Practices (BMPS) for Forestry 1993" manual recommends that SMZs be a minimum of 35 feet in width on each side of a stream and larger if terrain and soil dic-

JOE MCGLINCY

Riparian forests, which are located adjacent to waterways, should be retained as streamside management zones (SMZs). They provide travel corridors between habitat types for many wildlife species, serve to increase habitat diversity within a pine-dominated landscape, and protect water quality in streams and rivers.

tate. This is the **minimum** width thought necessary to comply with the prohibitions on stream siltation set out in the Clean Water Act. If wildlife are an important management consideration, then the minimum width recommendation should be increased. SMZs must be wide enough to retain their distinct habitat conditions for wildlife. When these zones are too narrow, sunlight pours in from both sides and changes the vegetation on the forest floor. Some wildlife biologists and managers say that a good rule of thumb is that one should not be able "to see through a SMZ from one side to another on a winter day."

Some guidance on SMZ width may be found from studies conducted on large streams in the Southeast. For example, wild turkey populations were found to thrive in areas with substantial pine plantations, if the plantations were managed so that mature mixed hardwood–pine forests extended at least 270 feet on either side of a stream. Squirrels and squirrel nests were also found to be in greater abundance in SMZs larger than 165 feet in width. Amphibians and reptiles are reported to be more prevalent in SMZs with widths ranging from 99 to 313 feet. Small mammals (primarily rodents), however, seem to prefer narrow SMZs less than 80 feet in width. The narrow width may be more conducive to small mammals since these areas are often more brushy due to the greater exposure to sunlight. Other research has found that the minimum width to maintain habitat for neotropical migrant songbirds is 150 feet on each side of a stream, and minimum widths of 300 feet tend to maximize songbird diversity.

Forest landowners should use the results of the preceding studies as a general guideline. Based on these studies, a minimum SMZ width of 150 feet on each side of a stream would benefit most wildlife species. However, the key point to remember is that the necessary corridor width will vary depending on the species of wildlife. The preceding guidelines may provide some general information, but more research is needed to determine the optimal buffer for different species.

One misconception about SMZs is that they are permanent set-aside areas where timber management must be foregone to provide undisturbed buffer zones for wildlife. Although this is one option, SMZs can also be "management" zones, as the name implies, where timber is carefully removed on a limited basis using less intensive harvesting methods like group or single-tree selection. With careful planning, SMZs can benefit wildlife and at the same time continue to provide timber revenue for the landowner. Landowners interested in establishing SMZs should have them marked before negotiating or soliciting bids for timber sales. They should also mandate in timber harvest contracts what type of harvest, if any, will be allowed within SMZs.

Mast Trees and Shrubs

Mast is the fruit of forest trees and shrubs. Mast is classified as either hard or soft, depending on the tree or shrub. Hard mast is simply fruit with a hard exterior such as acorns or hickory nuts. Soft mast like persimmons, plums, and black cherries, has a flesh-like exterior that covers a pulp-like interior. Mast is an important food source for many species of wildlife. Acorns can comprise over 85 percent of the rumen (stomach) contents of white-tailed deer during fall and winter months. Deer, like other species of wildlife, depend upon acorns and other fruits as a source of carbohydrates that provide a source of high energy. Carbohydrates are stored as fat reserves that are tapped as sources of energy during winter months. The reproduction of some wildlife species like wild hogs and squirrels has been closely correlated with mast production. In years of low acorn availability, the number of young born to these species declines. For deer, weight and antler diameter have also been positively correlated with mast production. Mast production in one year strongly influences deer condition the following year.

The variety of fruits is just as essential to wildlife as abundance. Fruit-producing trees and shrubs produce during different times of the year, which increases the availability of food to wildlife year-round. Rarely will one species of tree or shrub produce fruit consistently from year to year, so it is wise to maintain many different types to ensure a constant wildlife food source.

Knowledge of fruiting patterns of mast-producing trees helps in making decisions concerning their management. Fruiting habits vary by tree species, locality, and between individual trees of a particular species. A desirable wildlife management practice is to favor a wide variety of mast-producing species such as beech, hickory, black gum, dogwood, wild grape, and oaks. Where stand conditions permit, forest owners should manage for both white and red oaks because the fruiting pattern of white oaks differ from those of red oaks. White oaks flower and bear fruit in one growing season and their acorns are found on the current year's growth. Red oaks flower and set fruit in one growing season, but the acorns are not mature until the following season. Consequently, a total mast failure of all oaks in the same year is rare if there is a good mixture of oak species in the forest stand. Poor years of acorn production do occur, however, primarily due to poor pollination in the flowering stage, flower abortion, or a lack of rainfall during the acorn de-

Mast producing shrubs should be identified within forest stands before silvicultural practices are implemented. These valuable shrubs and trees should be clearly marked with flagging tape or marking paint and pointed out to logging crews so that they are retained for wildlife. Examples of valuable mast trees include (top left) flowering dogwood, (top right) black cherry, (middle left) serviceberry, (middle right) persimmon, and (bottom right and left) plum and plum thickets.

velopment stages in late summer.

In most forest stands, a few high quality oaks produce a disproportionate share of the mast. The age of a tree has a lot to do with mast production. Trees in the middle of their lifespan produce more fruit than either young or old trees. However, even during bumper acorn production years, not all acorn-bearing trees produce. For example, in good years about 90 percent of red oaks are productive, and only 60 percent of the trees in the white oak group are productive. The most dependable acorn producers in Alabama are the water, willow, and laurel oaks. For more information on oaks, see Appendix G.

Stands containing low-producing mast trees should be selectively thinned to remove the poorer quality trees. Thinnings should be conducted during late summer or early fall, and those trees with visibly larger crops should be retained in the stand. The amount of mast produced by most species is related to the size of a tree's canopy and the amount of sun reaching the canopy. The best acorns come from healthy, vigorous trees with large trunk diameters (at least 14 inches DBH or greater) and well-developed crowns. Poor quality trees that overtop or dominate potentially good mast producers should be removed. To meet the food needs of most mast-using wildlife, managers should try to maintain a mixture of at least five red and white oaks per acre.

The first step toward ensuring that stands have a variety of mast-producing trees and shrubs for wildlife is to understand which plant species are the most beneficial for wildlife. (See Appendixes D–J that contain listings and descriptions of woody plants that should be maintained in stands for wildlife.) A variety of color identification guides are also available that provide pictures and descriptions of trees and plants that wildlife depend on. Professional foresters and wildlife biologists are also excellent resources in helping to locate significant mast-producing trees and shrubs for wildlife.

After identifying key mast-producing plants, efforts should be made to protect these species during normal silvicultural practices. Valuable trees and shrubs should be marked in a special manner with marking paint so that they will be protected and maintained. All forest management and logging crews should be aware of the markings and be informed that these trees and shrubs are to be protected. Timber sales agreements (see Appendix N) can include provisions that specify that marked trees should not be harmed during timber harvest and that the logging contractor is liable for damages to these trees and shrubs. Forest owners may want to specify a dollar value before the harvest begins. After any silvicultural operation, it's prudent to have a professional forester or wildlife biologist make a follow-up inspection of mast-producing trees and shrubs to make sure that damage has not occurred.

Snags

A **snag** is a standing dead or dying tree. Because snags are usually of poor form, these trees rarely have any value in the wood products market, but they can be invaluable when retained for wildlife. They provide both food and cover, making them important to the distribution and abundance of many wildlife species. Snags are produced naturally by fire, disease, insects, lightning, flooding, and drought. Physically girdling the base of trees or applying forest herbicides can also create snags.

Wildlife need snags for a variety of purposes. Woodpeckers and other small birds feed on insects that are found on and within snags. Birds of prey frequently use snags as hunting perches. Because of this fact, snags should not be located adjacent to small game food plots. Many songbirds that occupy edge or open habitats use snags as singing perches. Examples of birds that frequently use snags include the Eastern bluebird, Carolina chickadee, red-bellied woodpecker, red-headed woodpecker, loggerhead shrike, northern flicker, and red-tailed hawk. Woodpeckers often use undecayed portions of snags as drumming sites for territorial marking. Cavity-dwelling mammals, like raccoons and squirrels (gray, fox, and flying), use snags for denning sites. Studies have shown that gray squirrel reproduction and survival of young is much higher in nests that are constructed in tree cavities as opposed to the more exposed leaf nests built in the open canopy of trees.

If cavity nesters and denning wildlife are to be given consideration in timber management, at least three to five snags per acre should be left during timber harvesting to provide a continuous supply of potential cavity trees. Even snags left in clear-cuts increase the diversity of birds found in newly cut areas. Tall and large snags should be left since they are most often used by cavity-nesting and denning wildlife. Where possible, some living trees that are over 60 years of age should remain standing. These trees often provide the best wildlife shelter.

Dead or dying trees are classified as either hard or soft snags, depending on their rate of decay. The decay rate determines the durability of a snag and how long it will be available for wildlife. Trees vary on their resistance to decay. Not all snags provide the same type of habitat for wildlife. Soft snags, which are "crumbly" and structurally weak on the interior, don't last long in a forest stand, usually only two to three years. However, they provide excellent foraging areas for insect-eating wildlife. Hard snags, on the other hand, are more structurally

Snag Durability

Some trees are more resistant to decay and will provide standing dead nesting and denning sites for a longer period of time compared to other trees. To have snags in various degrees of decay over a period of time, it may be advantageous to select potential snags from more than one category below. Trees killed in the non-resistant category will provide punky (soft and sponge-like) wood relatively soon, and about the time they are falling down, trees in the resistant category will just be getting punky. Another factor affecting snag durability is the degree of lean in the tree. Trees that lean significantly are less likely to stand for a long period of time.

RESISTANCE TO DECAY	
EXTREMELY RESISTANT	Black locust, Red mullberry
RESISTANT	Red cedar, Osage-orange, Catalpa, Sassafras, Black cherry, Black walnut, White oak, Chestnut oak
MODERATELY RESISTANT	Honeylocust, White pine
NON-RESISTANT	Willows, Cottonwood, Buckeye, Butternut, Yellow poplar, Sweetgum, Elm, Hackberry, Beech, Hickory, N. red oak, Black oak, Pines

RICHARD T. BRYANT/ARISTOCK

RICHARD T. BRYANT/ARISTOCK

Standing snags (living left; dead right) have little value for timber, but are important for many species of wildlife. Snags can provide both cover and food. Cavity-dwelling mammals like raccoons and squirrels use snags for denning and raising young. Many species of birds also use snags for nesting, perching, insect foraging, and as territorial drumming sites (woodpeckers).

Brush Piles Provide These Advantages for Wildlife:

- Concealment cover from predators. An overhead canopy and surrounding brush hide nests from outside view.
- Protection from predators. The tight network of strong twigs and small openings eliminates entry of many predators.
- Protection from weather. Nests are sheltered from chilling rains, wind, and excessive sunlight.
- An area for various seeds to sprout. The accumulation of twigs and grass are ideal sites for seed germination and young plant growth.
- Food and cover for wildlife. Living or half-cut brush piles covered with vines or woody vegetation allow for easy access to food sources.

Brush piles are an effective method of providing cover for many wildlife species where natural cover is lacking.

sound and can persist for longer periods of time. Living trees that provide cavities for wildlife are also considered hard snags.

Snags can be created through wildlife and timber stand improvement techniques by using either physical or chemical methods mentioned earlier in this chapter. Herbicide application is easier and usually kills trees faster than girdling; however, girdled trees persist for long periods in the stand as compared to herbicide-treated trees that decay faster. Toxicity potential should be checked on the labels of all herbicides that are used to create snags.

Brush Piles

When cover is a limiting factor for wildlife, brush piles may help protect certain animals like rabbits, small mammals, reptiles, and amphibians from predation and harsh weather. Materials necessary to construct brush piles often can be a by-product of silvicultural operations. Timber harvest, timber stand improvement practices, and release cuttings all create slash (woody limbs and debris) suitable for brush piles. Correctly constructed brush piles that are evenly distributed across a forest stand can provide nesting and protective cover with the same value as a dense thicket of natural vegetation. Studies have indicated that brush piles are an excellent method for improving rabbit habitat or "rabbitat." Where cover is significantly lacking, brush piles can actually help increase rabbit populations. Every effort should be made not to dislodge or pile up soil in the process.

Half-cuts or **living brush piles** can also provide cover for many species of ground-dwelling wildlife. These can be constructed by cutting partially through the trunks of a group of four to six small trees that are close together, then bending each treetop downward so that they all come together to form a pile. The idea is to provide cover without killing the trees. To ensure that leaves will remain on trees through at least one season, cuts should be made during the spring after the sap has risen and the leaves have matured. Half-cut trees that have fruit or forage-producing vines growing on them like honeysuckle, wild grape, greenbriar, and Alabama supplejack can add extra food value to living brush piles. Often these vines will continue to flourish long after the trees have died. Trees that make optimal half-cut candidates include willow, live oak, hackberry, elm, blackjack and post oak, and ash.

Size and distribution of brush piles do make a difference. Brush piles should be large enough to provide adequate cover for wildlife and should also be placed within an animal's home range. For example, for rabbits and quail, two to three brush piles per acre are recommended where cover is lacking. Brush piles should be located close to food sources and at least 200 feet from other escape cover. To provide sufficient cover for small mammals, they should be 25 to 50 feet in length, 12 to 15 feet in width, and 4 to 5 feet high.

RICHARD T. BRYANT/ARISTOCK

CLEMSON UNIVERSITY

Wildlife habitat can be improved on farm and forest lands. Left: Woods roads can be widened significantly and planted to grasses or left in native vegetation. Middle: Old home sites should be retained since many of these areas contain fruit-producing shrubs and trees that were planted in days gone by. Right: One side of a woods road can be widened and maintained in native herbaceous plants valuable to wildlife.

Natural and man-induced changes do occur, so brush piles should be inspected annually to replace material that has decayed or has been disturbed by land management practices. Those that have completely lost their value for wildlife can be burned and replaced by new piles. For quail, only living brush piles and thickets such as blackberry should be used. Dead wood brush piles attract too many predators such as rats.

Firebreaks

Firebreaks are constructed primarily for two purposes: for protection against wildfires and to facilitate prescribed burning. Firebreaks vary in size from a few feet in width to greater than 20 feet. Smaller width firebreaks are called **firelines** and are made with a single chisel fire plow pulled by a tractor or bulldozer. Larger width firebreaks are called **firelanes**, are established with a bulldozer blade, and offer forest managers greater flexibility in use and management. In addition to acting as fire barriers, wide firelanes can provide access for wildfire-fighting equipment, as well as silvicultural and wildlife management activities. An added benefit of firelanes is that they provide recreational access into forest stands.

These pathways enhance forest habitat for many wildlife species. For instance, they can "break up" and diversify wildlife habitat in large monoculture pine plantations. They also provide transitions between different stand types and protect vulnerable forest stands (hardwood stands, SMZs, and young pine stands) from fire damage.

The location of firelanes should be noted in forest wildlife management plans and determined well in advance of conducting silvicultural practices. To reduce the potential for soil erosion, firelanes should follow natural contours of the land and be stabilized with water bars in steep terrain. To further reduce the potential for erosion and enhance forest wildlife habitat, firelanes should be maintained in non-woody vegetation. In steep terrain, every effort should be made to maintain firelanes in native grasses and forbs to reduce soil disturbance and the chance of erosion. Native plants can be maintained by mowing or light disking every two years. If firelanes in steep terrain are planted, they should be established in perennial rather than annual wildlife food plants to minimize soil erosion. Perennial plants require less maintenance and consequently lower site disturbance. Before planting firelanes as wildlife food plots, be sure to read Chapter 16, "Supplemental Plantings as Wildlife Food Sources," and Appendix E–K for ideas on appropriate choices of plants that benefit wildlife.

If firelanes are maintained in herbaceous vegetation, caution should be used to prevent the growth in these areas from becoming too thick and consequently building up a fuel load of dead and dying plant material that would carry a fire across a firebreak. To reduce the chance of this happening, firelanes may be mowed and raked, or disked annually.

Forest Roads

Forest roads provide access to stands for management and timber harvest. They also are valuable in providing recreational access for hunters and other outdoor enthusiasts. In addition to providing access, forest roads can also be improved for wildlife. Borders of secondary forest roads can be widened to a width of 20 to 25 feet or more on each side. This practice, often called **day-lighting**, creates linear wildlife openings on each side of the roadway. Maintenance of roadside openings in native herbaceous plants valuable to wildlife usually requires annual disking, mowing, or prescribed burning. Opened forest road borders can also be used as sites for planting wildlife food plots. Daylighting roads provides an additional benefit by decreasing drying time of roads after heavy rains or flooding.

Tertiary roads, which are lightly used vehicle paths through the woods, can also be used as openings for native and planted wildlife foods. These roads should be at least 25 feet wide to allow enough sunlight to reach the forest floor. Unlike secondary roads, all of the tertiary woods roads can be maintained in native vegetation or planted in a wildlife food. However, once these roads are established in native or planted wildlife foods, vehicle travel should be kept to a minimum.

Roads created during silvicultural operations in some cases may actually decrease the value of a forest stand for some wildlife. For example, black bears and wild turkeys may suffer from uncontrolled road access that increases negative human interactions like poaching. In some cases this may be a factor limiting population growth. Some scientists also believe that forest interior wildlife, like some neotropical migrant birds, are negatively affected by roads. Roads tend to break up contiguous forests (habitat fragmentation) into smaller areas that may not be suitable for interior forest wildlife.

Another disadvantage of forest roads is the potential change they can create in rainfall runoff patterns, increasing the speed of runoff water and soil erosion. Roads along streams are particularly undesirable, due to increased runoff and stream sedimentation. To reduce forest stand degradation and soil erosion into streams, roads should be constructed according to "Alabama's Best Management Practices for Forestry 1993" manual, available from the Alabama Forestry Commission.

Allowing more access by the general public sometimes encourages uninvited visitors, littering, and hunting from vehicles. Entry points of forest roads should be controlled with a lockable gate or highly visible cable. Wire cables running through white PVC pipes are an excellent way to increase visibility of cabled barriers, but wire and cable gates should be avoided due to potential injury to all-terrain vehicle riders.

Protection of High-Value Habitats

Sites that may be marginal for forest management but which are valuable for wildlife should be protected during silvicultural operations. For example, bays, swamps, beaver ponds, and other wetland areas are often inaccessible and costly or illegal to drain. These areas provide important habitats for a wide variety of wildlife such as black bear, deer, turkey, squirrel, waterfowl, and many species of non-game wildlife.

During and shortly after the small farm era in Alabama, good wildlife habitat was created around small field borders, wide fence rows, ditch banks, and old abandoned buildings. Remnants of old homesites that still remain on forested lands provide a unique and valuable habitat for wildlife. Shrubs, vines, and fruit trees that were planted in the past still surround many of these sites, providing excellent food and cover sources for many species of wildlife. Plum thickets also provide excellent cover and feeding areas for bobwhite quail. Old homesites and other old farm structures in or adjacent to forest stands should be protected from fire, logging, and other silvicultural operations.

MULTIPLE FOREST STAND MANAGEMENT CONSIDERATIONS

For wildlife that require a diversity of habitats, one of the best opportunities for improving habitat diversity across multiple forest stands is designing management units that maximize differences between stands. For example, a mosaic of even- and uneven-aged stands may be interspersed across the management unit. Habitat diversity can also be enhanced in even-aged management units if adjacent timber stands are maintained in different age classes. This is easily accomplished by staggering timber harvests so that there is a five to seven year age difference between adjacent forest stands. Staggering harvest periods may also be a wise plan for tax and investment reasons.

Proper arrangement of forest stands within a management unit is also an important consideration. Wildlife that require a variety of habitat components in more than one stand should have a variety of stand types located within their home range. Arranging diverse forest stands in a manner that meets the habitat needs of wildlife enhances the overall value of a management unit for wildlife.

The size of stands within a management unit varies, depending on soils, topography, and management objectives. For timber management, large stands are generally more economical to manage and harvest than smaller stands. Interior forest wildlife also prefer large expanses of mature forests. For other wildlife, like early successional and edge-adapted species, large expanses of pine monoculture are not very beneficial. Pine stands of various age classes should be kept relatively small and well interspersed throughout the management unit.

In some cases, large timber harvests or concurrent harvests in adjacent stands create large open spaces that separate or isolate valuable wildlife habitat. If some species of wildlife are to continue using these areas, some type of travel corridor may be required through the newly harvested areas. **Forested corridors** can serve the same

Champion's Forest Patterns Land Classification System

Champion, using GIS data and other land surveys, manages its land holdings by classifying property tracts into one of four categories:

Protected Areas

Land with unique natural or cultural value that has been removed from timber production. These areas are either part of Champion's Special Places in the Forest™ program or other efforts designed to maintain and protect the area's unique values. Examples include rare geologic features, significant historical sites, natural heritage sites, threatened and endangered species habitat, unique high-value recreation areas, and exemplary natural communities.

Restricted Management Areas

Land where non-timber values, such as wildlife habitat, water quality, or recreation potential, are the bases for management decisions, with timber production a secondary objective. Examples include streamside management zones, cypress domes and ponds, gum ponds, river and creek floodplain swamps, and critical wildlife and plant life areas.

General Management Areas

Land where timber production is the primary objective, but where management intensity is constrained by factors such as regulations, economic and wildlife considerations, slope stability concerns, inaccessibility, political sensitivity, aesthetics, or lower site productivity. Selective harvesting or clear-cutting with natural regeneration is the typical management prescription for these areas. Examples include bottomland hardwood stands, upland hardwood/softwood areas with natural regeneration, wildlife corridors, and stands with extended rotation ages.

High Timber Yield Areas

Land with the highest timber productivity suitable for intensive timber management. Consideration is given to all other resource values, but timber production is the primary focus. The objective for these areas is to optimize wood and fiber production through the application of the latest scientific forest management techniques, with harvest schedules based on the optimum age for target products. Examples include plantations and other intensive solid wood and fiber production areas. Management typically includes vegetation control, fertilization, genetically improved planting stock, and other intensive practices.

purpose as SMZs, except that they are not necessarily associated with watersheds. These passages through clear-cuts are heavily used by wildlife as travel routes linking unconnected stands.

Corridors also have some drawbacks. Predators learn to target these corridors, resulting in increased predation on some wildlife species. Current research findings on turkey predation along riparian habitats support this notion. In some cases, narrow corridors have given nest predators like raccoons easier access to turkey hens or their nests. In these cases the immediate cause of turkey death is predation, but the ultimate problem is insufficient corridor width. The message is simple: the wider the corridor the better. Increased width decreases the likelihood that predators will find their prey. A corridor that is at least 300 feet wide provides sufficient cover to reduce nest predation.

LANDSCAPE LEVEL MANAGEMENT

A **landscape** is essentially a mosaic of different habitats across a large area where wildlife live and die. Landscapes contain a series of habitat patches, small areas of similar habitat, in an overall matrix that may or may not be connected by corridors. Landscapes have patterns that consist of repeated habitats occurring in various sizes, shapes, and arrangements. The best way to envision a landscape is by looking down from an airplane window at the land passing below or by examining aerial photographs to see how individual areas fit into the larger picture.

Historical evidence indicates that the abundance of wildlife in pre-colonial forests resulted from a mixture of interconnecting forest types with an abundance of mature forests interspersed with openings and early successional forests with few people. This mixture provided a wealth of vertical and horizontal habitats across the landscape. Over time landscapes changed as a result of natural disturbances like fires, ice storms, floods, tornadoes, and hurricanes that helped to insure a diversity of interconnecting habitats across the landscape. These natural disturbances ultimately gave the landscape a mosaic appearance of interconnected habitats. Like a giant patchwork quilt, the beauty and function of the landscape can only be fully appreciated when the entire area is viewed and understood in its entirety.

As Alabama was settled, the natural landscape changed even more to accommodate man's need for raw materials. Eventually, much of the forest landscape was harvested, and large areas were converted to agricultural production or planted in monocultures to produce wood and fiber. Other man-made disturbances like highways, power and gasline corridors, and urban sprawl also disrupted the natural appearance and function of the landscape. Eventually habitat patches that were once connected became isolated islands created by man's activity, disrupting the flow of life across the landscape.

Today the isolation of habitat patches, called **fragmentation**, is thought to be a major cause in the decline of some wildlife species like neotropical migrant birds that require relatively large tracts of connected habitat. Other species, like the black bear and cougar, are also thought to be in decline because of man-made habitat fragmentation. Because of these concerns, managers of public lands, such as the U.S. Forest Service, have begun to consider natural resource management beyond the forest stand or multiple stand level to the management of entire ecosystems at the landscape level. The primary purpose of this large-scale planning is to maintain habitat connectivity and possibly to reconnect habitat patches that have been isolated due to fragmentation by man. Some large industrial timber companies have also begun to incorporate landscape level planning into their management philosophies and strategies. Because of the large scale that landscape level management entails, it is unlikely that small private forest owners will be able to participate in landscape level management, at least solely on their own lands. A more likely scenario is that private forest owners might be given the opportunity in the future to become partners with public land agencies or large industrial forest landowners in helping to establish corridors that connect critical habitat patches, or to form their own private landowner cooperative management agreements.

A Farmland Matrix

The matrix in this figure is farmland. Historically, the matrix was forested, not in agriculture. The farmland matrix was formed when people settled and cleared areas for small farms because they used horses and could not farm larger areas of land. With mechanization and modern agriculture, an increase in the size of the agricultural fields has resulted. At first farmland occurred as gaps in the forest. Eventually as these gaps grew, farms replaced forest as the primary matrix, and the forest became the secondary patches. In this picture you can see how corridors serve the purpose of connecting patches of isolated forestland and how other forest patches are isolated and unconnected. Even though scientists are still studying and debating the necessary size of corridors for use by different species, they agree that more corridors increase the connection between habitat patches and ensure the viability of the land and all that lives there.

DENNIS HOLT

Agriculture and industrial and urban development often create isolated patches of forested wildlife habitat. Forested corridors used to connect isolated forest blocks are one way to ease the pressure of changing land-use patterns on wildlife.

Fig. 8: Woodland Forage Plants Utilized by Cattle and Wildlife

PLANT	CATEGORY	VALUE TO CATTLE	VALUE TO WILDLIFE
Ash (*Fraxinus* spp.)	w	good	good
Bahia, Paspalum (*Paspalum* spp.)	ws,p	good	good
Barnyard Grass (*Echinochloa* spp.)	a,ws	good	good
Beggarweed (*Desmodium* spp.)	p	good	excellent
Blackberry (*Rubus* spp.)	w	fair to poor	excellent
Bluestems (*Andropogon* spp.)	ws,p	excellent	fair
Foxtail (*Seteria* spp.)	ws,a	good to fair	excellent
Hackberry (*Celtis* spp.)	w	good	excellent
Honeylocust (*Gleditsia triacanthos*)	w	good	good
Indian grass (*Sorghastrum nutans*)	ws,p	excellent	good
Lespedeza (*Lespedeza* spp.)	a/p	good	excellent
Oaks (*Quercus* spp.)	w	fair	excellent
Orchard grass (*Dactylis glomerata*)	cs,p	excellent	good to fair
Panic Grasses (*Panicum* spp.)	ws,a/p	good to fair	excellent
Purpletop (*Tridens flavus*)	cs,p	fair	fair
Redroot (*Ceanothus* spp.)	w	good	good
Sedge (*Carex* spp.)	p	good to fair	excellent
Sensitive Brier (*Schrankia uncinata*)	p	good	fair
Sweet Clover (*Melilotus* spp.)	a	excellent	good
Sumac (*Rhus* spp.)	w	good to poor	good
Timothy grass (*Phleum pratense*)	cs,p	excellent	good
Wild Bean (*Strophostyle* spp.)	a/p	good	excellent
Wild Lettuce (*Lactuca* spp.)	a	fair	excellent

a=annual, p=perennial, cs=cool-season, ws=warm-season, w=woody

Woodland grazing, if conducted properly, can complement wildlife and timber management objectives. Unfortunately, cattle are often left to roam freely in forests for extended periods with little regard to stocking rate or impact on wildlife plants. This has created the perception among many biologists that cattle and wildlife do not mix.

MANAGEMENT BY LAND CLASSIFICATION

In an attempt to manage land in the best possible manner for a variety of timber and non-timber values such as wildlife, several forest industries have developed land classification systems to help determine management objectives and decisions. One example is Champion International Corporation's Forest Patterns classification system. Using this system, lands are evaluated, classified, and managed using a four-tier landscape approach. The intensity of management practices changes depending on the classification. If non-timber values are the focal point for management decisions on a particular area, the area is placed in either a **protected** or **restricted management** classification. The difference between these categories depends on whether timber can be removed without risking important non-timber values. If timber cannot be removed, the area is placed in the **protected** classification.

On areas where timber values drive management decisions, the area is placed in either the **general management** or **high timber yield** classification. The difference between these categories depends on the intensity of the forest management practices. If practices are not constrained by factors such as regulations, or biological, economic, or high impact visual concerns, the area is placed in the high yield classification.

With the exception of the protected category, classification types should be viewed as dynamic, changing over time where appropriate as management practices improve. Champion's classification system is one example of how private forest owners may vary the intensity of management based upon existing land capabilities for a variety of timber and non-timber resources.

MONITORING AND REVISION OF FOREST STAND MANAGEMENT ACTIVITIES

To evaluate the success or failure of silvicultural practices for wildlife, forest owners should monitor the impact that management activities have on wildlife habitat over time. Each year forest owners should record this information, and every five years re-evaluate and if necessary revise their forest management activities. Forest owners should ask themselves if silvicultural practices have accomplished their objectives of improving forest wildlife habitat. If not, what changes need to be made in forest stand management? Also, does new knowledge exist that can improve management decisions or land management practices?

WOODLAND GRAZING

To increase the revenue potential from timberlands, some forest owners have taken a multiple-use approach that integrates not only timber and wildlife management, but also includes management for cattle production. The combination of both timber and cattle management on the same land is referred to as **woodland grazing**. The native grasses that grow underneath pine stands in Alabama have the potential of providing forage for livestock, and an opportunity for landowners to receive additional income from the annual sale of beef. Annual income from forest stands is helpful, because most landowners don't receive revenue from timber stands until a commercial thinning or harvest is conducted later in the life of the stand.

Cattle grazing in the piney woods of the Southeast sis nothing new. Spanish explorers introduced cattle to the region over 450 years ago and subsistence livestock grazing of pine woodlands was a major land use in the Deep South well into this century. Woodland grazing fell out of favor in the 1930s because uncontrolled, free-ranging livestock was causing considerable damage to young timber stands as well as degrading creeks and streams. One of the earlier projects of the Alabama Wildlife Federation was the "stock law war" of the 1930s, which pitted timber interests and grazing interests against one another. The need to exclude both cattle and fire from newly regenerated pine stands also forced many

Plants Toxic to Cattle

Woodland grazing programs should strive to reduce or eliminate plants that are toxic to cattle. Below is a list of plants known to be toxic to cattle that should discouraged in a woodland grazing program.

TOXIC PLANTS	COMMENTS
Johnson grass (*Sorghum* spp.)	Livestock have occasionally been killed by prussic acid (hydrocyanic acid) poisoning from these plants during dry summers or immediately after a frost.
Indian hemp (*Apocynum* spp.)	
Milkweed (*Asclepias* spp.)	
Milk vetch (*Astragalus* spp.)	
Wild indigo (*Baptisia* spp.)	Suspected of being poisonous.
Water hemlock (*Cicuta maculata*)	Poisonous to both livestock and humans.
Jimson weed (*Datura stramonium*)	
Larkspur (*Delphunium* spp.)	
Dutchman's breeches (*Dicentra* spp.)	
Horsetail (*Equisetum* spp.)	
White snakeroot (*Eupatorium* rugosum)	
Spurge (*Euphorbia* spp.)	
Bitterweed (*Helenium* spp.)	
Ground cherry (*Physalis* spp.)	Poisonous but seldom fatal.
Bracken fern (*Pteridium aquilinum*)	
Black nightshade (*Solanum* spp.)	
Buckeye (*Aesculus* spp.)	
Bittersweet (*Celastrus scandens*)	Thought to be toxic but few poisonings reported.
Black cherry (*Prunus* spp.)	Leaves and fruit sometimes poisonous.
Elderberry (*Sambucus canadensis*)	Cattle have been poisoned by eating new shoots, leaves, and buds.

Woodland Grazing and Wildlife

Dr. Jimmy Huntley is a retired Wildlife Biologist with the U.S. Forest Service in Montgomery. His training and years of field experience in applied wildlife and forest management have provided insight and direction in integrated resource management. As a part of that work, he has been involved with examining the feasibility of combining woodland cattle grazing with timber and wildlife management on forest lands in Alabama.

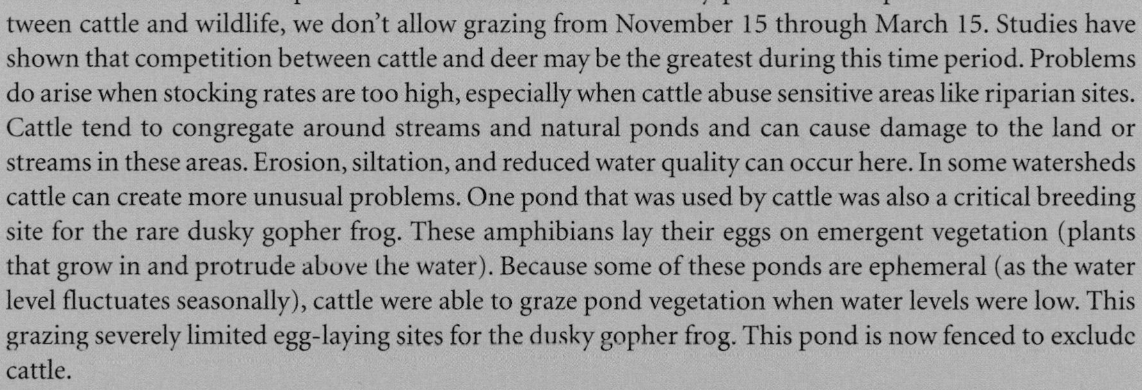

We have a woodland grazing program on about 2,000 acres, called an "allotment," on the Conecuh National Forest. Our program is part of our multiple-use management efforts that examine how woodland grazing may be a management option for some private forest owners in the Coastal Plain of Alabama.

Before allowing cattle on the National Forest we estimate the forest's capability to support both wildlife and cattle by determining the amount and types of available forage. This information is used as the basis for selecting a minimum cattle stocking rate that will not overgraze the allotment area. We look primarily at the percent of grass coverage, and to a lesser extent the percent of forb coverage. Grasses, primarily bluestems and panicums, make up most of the woodland forage consumed by cattle. If cattle stocking is kept low, we don't have a concern with competition from wildlife. To reduce any potential competition between cattle and wildlife, we don't allow grazing from November 15 through March 15. Studies have shown that competition between cattle and deer may be the greatest during this time period. Problems do arise when stocking rates are too high, especially when cattle abuse sensitive areas like riparian sites. Cattle tend to congregate around streams and natural ponds and can cause damage to the land or streams in these areas. Erosion, siltation, and reduced water quality can occur here. In some watersheds cattle can create more unusual problems. One pond that was used by cattle was also a critical breeding site for the rare dusky gopher frog. These amphibians lay their eggs on emergent vegetation (plants that grow in and protrude above the water). Because some of these ponds are ephemeral (as the water level fluctuates seasonally), cattle were able to graze pond vegetation when water levels were low. This grazing severely limited egg-laying sites for the dusky gopher frog. This pond is now fenced to exclude cattle.

Based upon our work in the Conecuh National Forest and the Forest Service's experience in other areas, we see two approaches to woodland grazing for private forest owners in the Coastal Plain. The first option is grazing in open pine stands without improved pastures. These areas must have an open forest canopy that allows sunlight to reach the ground for sufficient grass and forb growth. Grazing rates in open pine stands should be low; on most areas this will range between 20 and 30 acres per head of cattle. Sensitive sites and unique areas should be protected by fencing. Depending on the price of cattle and other factors, open woodland grazing may not be profitable.

The second option is to develop improved pastures adjacent to and within moderately stocked pine stands. Ideally, improved pastures should be established after timber harvest and site preparation, but before pine seedlings are planted. Old fields or pastures that are being planted in pines make excellent areas for improved pastures. Open areas should be disked and seeded with a grazing grass such as bahia, then fertilized, and then planted in pine seedlings with wide spacings, at least 10 x 10 feet. Several years after these areas are established in pines, they should be pre-commercially thinned to maintain a wide tree spacing that continues to allow sunlight to reach pasture grasses. If grazing is controlled and kept at a moderate level, these areas provide excellent brood habitat for turkeys and "bugging areas" for quail. Woodlands with improved pastures can often sustain a stocking rate as low as 2 acres per head of cattle. Woodlands that incorporate improved pastures can be valuable for wildlife, cattle, and timber production.

An Example of a Grazing System for Timber, Wildlife, and Cattle Production

The following is a description of a rotational grazing system for a hypothetical landowner who owns 2,890 acres in a longleaf pine flatwoods area. The values presented in this example are based upon estimated annual forage production and should not be used to establish stocking rates in other situations.

Description of Property

A forest owner is interested in establishing a rotational grazing system for 92 head of cattle on 2,890 acres of longleaf pine. In order to accomplish this, the entire tract is divided into four units based upon specific stand characteristics that are described below.

Stand 1. This 940-acre stand contains 8- to 12-year-old longleaf pines originally planted at about 700 trees/acre. Estimates of current cattle stocking rates are 18 acres per animal, which should not decline greatly for the next three to four years. Based on forage production estimates, this unit could support 52 cows for one year.

Stand 2. This 848-acre stand contains 12- to 16 year-old longleaf pines originally planted at 700 trees per acre. Estimates of current cattle stocking rates are 20 acres per animal for one year, but this will begin to decline sharply in the near future. Based upon forage production estimates, this unit will support 42 cows for one year.

Stand 3. This 474-acre stand contains 3- to 6-year-old longleaf pines originally planted at 700 trees per acre. Estimates of current cattle stocking rates are 12 acres per animal for one year. This rate will not decline significantly for the next four to five years.

Stand 4. This 628-acre stand contains 22- to 26-year-old longleaf pines that were originally planted at about 1,000 trees per acre. The trees have never been thinned. Canopy cover is complete, and forage production is so low that cattle currently cannot be grazed.

Recommendations

Stands 1, 2, and 3 should be placed in a three-pasture rotational system and grazed on a four-month cycle during the height of the growing season (April through July), and on an eight-month cycle the remainder of the year. This means that grazing during the four-month growing season will only be allowed once every three years for each stand. This allows stands to maintain forage production without being overgrazed. The remaining eight-month grazing period will also be rotated to occur once every three years for each stand. Estimated forage production within these three stands allows 15 percent of the native forage to be retained for wildlife, with the remainder being more than enough to support 92 head of cattle. The conservative stocking rate and rotational grazing will ensure that grazing pressure will remain low, reducing the possibility of overgrazing, tree damage, or wildlife habitat degradation. All three stands should be thinned as soon as timber is merchantable, to maintain or improve native forage production. Thinnings that remove 30 to 40 percent of the merchantable timber are preferred for optimal native plant production. Prescribed burning should be conducted as soon as the stands can be burned, every one to two years, and in conjunction with timber thinnings.

Stand 4 should be thinned immediately to a basal area of 60 to 70 square feet to improve timber growth and to open the tree canopy for sunlight penetration. Prescribed burning should be conducted following thinning and thereafter every one to two years. Cattle should be excluded from this stand until grass and herbaceous vegetation has responded to thinning and burning. After native forage production is well established, estimates of yield should be determined to develop initial cattle stocking rates. Stand 4 should then be incorporated into a four-pasture rotational grazing system with stands 1–3.

cattlemen out of business or into grazing their livestock on pastures.

Research since World War II has shown that when pine stands are properly managed and livestock numbers carefully regulated, woodland grazing can be a viable component of a multiple-use forest management plan that also includes wildlife. However, because of poor woodland grazing practices and overgrazing of wildlife habitat in the past, wildlife managers have been reluctant to allow cattle grazing on pinelands where wildlife are a management objective.

Cattle congregate around streams and ponds causing erosion, siltation, and reduced water quality. Providing supplemental water for cattle can help prevent these problems.

A forest owner considering woodland grazing has basically two options. The owner can either allow cattle to graze on native grasses or establish pasture grasses underneath pine stands.

Native Grazing Ranges

Alabama has three native grazing ranges: longleaf/slash pine–wiregrass, loblolly/shortleaf pine–hardwood, and upland hardwood–bluestem. Since most woodland grazing efforts have been focused on pine stands, the following discussion will center on these stand types as potential livestock grazing ranges.

Longleaf/slash pine–wiregrass stands are commonly found in the Alabama Coastal Plain. Pineland threeawn (wiregrass) is the dominant grass in the understory. Other grasses that occur on pine-wiregrass range include dropseeds, bluestems, Indian-grass, paspalums, and panic grasses. If wiregrass dominates the range, cattle will make most of their weight gains from early spring until midsummer. Prescribed fire should be applied on a two-year rotation during the late-winter to early spring to keep wiregrass nutritious and palatable. On ranges that contain mostly dropseed and bluestems, burning should be on three-year intervals in order to maintain a good variety of grasses.

Loblolly/shortleaf pine–hardwood stands are the most common ranges in Alabama. Found throughout the state, this stand-type can consist of pure pine stands of loblolly and shortleaf pine or mixed pine–hardwood stands. Grazing values on these areas can vary from nonexistent in dense pine-hardwood stands to excellent in pine stands that are thinned and burned. Common grasses on pine-hardwood ranges include bluestems, panic grasses, and paspalums and spike grasses. Livestock can graze pine-hardwood ranges from early spring until the fall or first frost. Prescribed fire on a three-year rotation works best for maintaining a variety of grasses in the understory.

Pine Pastures

Shade tolerant pasture grasses can also be established underneath pine stands in a woodland grazing program. Bahia grass is a good choice because it is readily utilized by some wildlife, especially wild turkeys. Pasture grasses that are invasive and have little value to wildlife, such as bermuda, should be avoided because they quickly compete with native wildlife plants, and once established are difficult to eradicate. Fescue is also not a good choice because it provides little food for wildlife. Fescue grows in thick mats and may become so dense that it inhibits movement by ground-dwelling birds like quail. Stands of fescue also seem to produce limited amounts of insects that are vital to young quail and turkeys. Fescue also is known to produce a fungus that is toxic to cattle and horses.

Native warm-season grasses can also provide excellent forage for cattle and food and cover for ground-nesting birds. Warm-season grasses, commonly called "bunch grasses" because they grow in upright bunches with patches of bare ground in between, include many of the native grasses that were once common to the Coastal Plain of Alabama. Some of these species include big and little bluestem, Indiangrass, switchgrass, side-oats grama, Eastern gammagrass, and others. Warm-season grasses have an advantage over many grasses because they are more drought resistant, require little or no fertilizer, and produce about twice the forage of traditional cool-season grasses. Warm-season grasses have a few drawbacks that need to be considered. Finding and purchasing seed is sometimes difficult and may cost more than traditional pasture grass seed. Native warm-season grasses may also take two full years to become established. Once established, warm-season grasses cannot be grazed as short as fescue, requiring more rotation among pastures and closer management of cattle stocking rates. However, for managing woodlands for both cattle and wildlife, the advantages of native warm-season grasses often outweigh the disadvantages in the long-term.

To establish pine pastures, pine seedlings are planted in wide rows (8 x 12 feet or wider) or double-row (6 x 12 x 40 feet) configurations in the winter, just before sowing grass seeds. After the grasses are well established and the pines are 3 to 6 feet tall, cattle can be stocked in pine pastures at rates similar to open pastures. Cattle should graze stands for seven months from spring until early fall. Annual fertilization on pine pastures to enhance grass growth is required for good beef production. In addition, pines will also benefit from grass fertilization and will respond with increased growth.

Timber should be managed by thinning stands to a basal area of no more than 70 to 80 square feet per acre in order to keep stands open for desirable forage production. Prescribed burning should be used during the winter to reduce fuels and brush encroachment. Firelanes, woods roads, and other open areas in stands should be seeded in grass as well. In addition, both bahia and fescue can be seeded with clover to provide better cattle nutrition, lengthen the grazing season, and reduce nitrogen applications. Select varieties of clover are also heavily used by many species of wildlife.

Pine pasture management offers a good opportunity for landowners who are reforesting poor or marginal cropland. Land conditions are usually ideal for establishing pines and grasses on these sites, and research has shown landowner cash-flow can be significantly improved by initiating a woodland grazing operation on reforested cropland. Ideally, pine pastures should be limited to marginal cropland or abandoned open land that is converted to pines to prevent destroying or eliminating native plant communities that already exist under pine stands. Converting marginal cropland and open sites to pine pastures is also more cost-effective than establishing pine pastures in established pine plantations.

Grazing of native range is seasonal. Cattle will lose weight or, at best, maintain their weight if grazed on native range year-round. During the dormant season cattle diets must be supplemented by feeding hay or moving cattle to winter pastures. Year-round supplemental grazing can be provided by planting or seeding grass and clover mixtures on firebreaks, woods roads, and small fields within the woods. These grazing mixtures should be planted and fertilized as if they were pastures.

Woodland Grazing Considerations

A grazing management plan developed by both a wildlife and range specialist can serve as a guide for landowners. The plan should identify the amount and location of forages on the property, as well as physical features like existing fences, watering sites, and working pens. In addition, the plan should specify the maximum number of animals that can optimally graze on the property without damage to other natural resources.

Determining forage yields is the most important step in establishing a sound woodland grazing program that protects young trees, wildlife habitat, and watersheds from damage caused by overgrazing. Cattle stocking rates should be based upon the amount of available forage and the estimated use by both cattle and wildlife. One rule of thumb is to reserve at least 15 percent of the available forage on a pine stand for wildlife. In order to arrive at a stocking rate, you must be able to estimate the amount of forage that is available within pine stands. Forage availability can be estimated using procedures outlined in the publication *Management of Southern Pine Forests for Cattle Production,* U.S. Forest Service, General Technical Report, R8-GR4. It can be obtained by writing the Southern Forest and Range Experiment Station, U.S. Forest Service, 701 Loyola Ave., New Orleans, LA 70113 or calling (504) 589-6787.

Grazing recommendations should take into account the class of grazing animals, the seasons of use, and the key grazing plant species. Various classes (e.g. bulls, heifers, commercial brood cows) of livestock may require different pastures, supplemental feeding schedules, and amounts of forage each day. The grazing management plan should create a schedule for setting aside different areas of the property during the year so that grasses can recuperate and produce seed. It's important to identify plants that are preferred by cattle so that their use can be monitored closely to prevent overgrazing.

Grazing intensity should be adjusted so that half of the current year's growth on all grazing plants (particularly key species) will be left at the end of each grazing season. Moderate use of forage will provide grasses with sufficient leaf area to produce the energy and nutrients necessary for maintenance of adequate growth. As forest stands mature and canopy cover increases, forage production will decline. Consequently, a planned reduction over time in the number of grazing animals is required to reduce the risk of overgrazing. Thinning, harvesting, and regeneration practices will open up pine stands and replenish forage so that stocking levels can be maintained.

Understory shrubs often compete with herbaceous plants for sunlight, water, and nutrients. Shrubs can be controlled with herbicides, bushhogging and prescribed burning, or roller chopping and prescribed burning. All of these practices help to control woody shrubs and enhance forage production. However, care should be taken to leave adequate cover for wildlife. If roller chopping is used, pine trees need to be maintained at wider than normal spacings to accommodate mechanical equipment and prevent root damage to the trees. Roller chopping should be performed every four to seven years depending upon the density and species composition of shrubs. If timber production is a primary objective, bushhogging and prescribed burning every year or two is more desirable to control woody shrubs. This will eliminate the potential of tree damage and soil compaction associated with roller chopping. Bushhogging can also be conducted with equipment usually already available in forest or woodland grazing operations and is cheaper than roller chopping.

Prescribed fire benefits both wildlife and livestock by keeping grasses nutritious and palatable. Thinning of pine timber as early as economically practical is essential in keeping forage at grazable levels. There is an inverse relationship between forage and pine basal areas. As pine basal area increases, forage levels decrease. A good rule of thumb for thinning pines to stimulate forage production is to thin to a residual basal area of no more than 70 to 80 square feet per acre or less, at intervals of 6 to 10 years.

Pine regeneration sites should be protected from grazing for at least one growing season to allow newly planted pines and forages to become established. Pines are most susceptible to grazing damage during the first year following planting. Typically, pines are planted in the winter months; therefore, cattle should be excluded from the site until at least the following winter or possibly longer if the native forages do not grow well. Care must be taken not to overgraze stands during this time, which would retard forage production and tree growth.

Supplemental feeding troughs, mineral feeders, and watering sites should be located away from newly regenerated areas. Cattle are naturally drawn to these areas, increasing the potential for damage. As a rule, all items that tend to congregate cattle should be located away from regeneration sites. Site damage can be avoided by keeping supplemental feeding locations and water troughs away from new plantations.

Riparian areas, streams and creeks, should be protected from use by livestock. Overgrazing of vegetation adjacent to watersheds can quickly lead to erosion and stream sedimentation, reducing water quality and the ability of a stream to support aquatic life. In addition, direct degradation of stream banks by trampling livestock can quickly ruin an otherwise productive creek or stream. Protecting riparian areas from livestock abuse can be accomplished by lowering stocking rates, or limiting the duration and frequency of riparian use through seasonal or rotational grazing. In some cases, fencing of riparian areas may also be an option to exclude cattle. Cost-share options should be investigated by landowners interested in this practice. Fencing, however, can be expensive and may not be economical in some situations. Livestock may also be attracted away from riparian areas by creating favorable conditions in adjacent upland areas such as shade, alternate water sources, rubbing posts, and mineral and feeding stations. No matter what is done, some livestock prefer riparian areas to upland sites.

Wildlife Considerations

The most important consideration in making cattle grazing compatible with timber and wildlife management on pine forests is the ability to adjust cattle stocking to available forage levels. Adjustments must be made in cattle stocking to allow sufficient forage for wildlife. As mentioned earlier, reserving 15 percent of the total available forage is one method to retain an adequate amount of grasses for wildlife. A recommended livestock grazing system with wildlife as a consideration is the multi-pasture, rest-rotation or deferred rotation system that helps to sustain native grass production for wildlife. This system only allows cattle on native ranges during a portion of the year, as opposed to year-long grazing which can be detrimental to native forage production. Studies have shown that moderate levels of forage used by livestock will not be detrimental to timber production or native wildlife habitat. Research in the Southeast has demonstrated that dietary overlap with deer is not significant at low to moderate cattle stocking rates. At these stocking levels, cattle diets consist of mostly grasses, while deer prefer forbs and woody browse. However, when stocking rates are high, cattle not only eat grasses but also consume forbs and woody browse valuable to deer and other wildlife.

Cattle grazing, in some situations, can be used to improve wildlife habitat. The use of livestock as a manipulative tool in wildlife habitat management to create specific habitat conditions is called **strategic grazing.**

This system involves short-duration grazing alternated with periods of no grazing, which allows time for plants to grow and reproduce. Strategic grazing can have the same effect as mowing, by keeping plants at a palatable and nutritious stage of growth. In addition, some studies have indicated that moderate levels of strategic grazing increase plant species diversity. It can also keep areas open enough for use by turkeys. Strategic grazing has been used in Texas to improve bobwhite quail and Canada goose habitat and has also been used in Florida to improve grass and legume production for wildlife. A review of numerous studies that evaluated the effects of cattle grazing on wildlife habitat indicates that most strategic grazing systems tend to benefit wildlife when compared to year-long or season-long grazing.

Determining Native Forage Yield and Stocking Rates for Woodland Grazing and Wildlife

Determining what the native forage yields are in a timber stand is an important first step in deciding on cattle stocking rates. This can easily be accomplished by using grass and herbaceous sampling plots to determine native forage production on potential woodland ranges.

The following step-by-step method can be used to determine native forage yields and cattle stocking rates in pine stands that are managed for timber, wildlife, and livestock:

1. **Select Sampling Areas.** Use a map or aerial photograph to locate woodland grazing units. Determine the number of vegetation sampling plots needed for each unit. As a general rule of thumb, 10 sampling plots are needed on forest units that are more than 200 acres. In woodland grazing units less than 100 acres, 5 plots are sufficient.

2. **Establishing Plots.** Sampling plots should be evenly distributed across the grazing unit and can either be square (3.1 x 3.1 feet) or circular (circumference of 10.96 feet and a diameter of 42 inches) in shape. Instead of measuring each sampling plot for size, portable wire circles or rectangular metal or wooden frames the size of a sampling plot can be used for plot borders.

3. **Determine Forage Weights.** Forage weights can either be estimated or determined by clipping and weighing the forage. For the latter, clip all herbaceous vegetation at ground level and place in a paper bag. Plots should be raked by hand before clipping to remove pine needles and dead vegetation. Weigh the bag and forage in grams and subtract the weight of the bag to obtain actual weight of forage.

4. **Calculating Forage Yields.** Record the weight of each plot and then get an average for all plots within the sampled grazing unit. Convert green weights to dry weights by multiplying green weight by 1) 0.40 if clipped in the late spring, 2) 0.50 if clipped in the summer, 3) 0.60 if clipped in the fall, and 4) 0.80 if clipped in the winter. Then multiply the results in grams by 10 to obtain the average yields in pounds per acre.

5. **Determine Total Available Forage.** Multiply dry-weight yield per acre times the number of acres in the grazing unit to determine total available forage. At least 15 percent of the available forage should be deducted for wildlife. This amount should be deducted before determining the carrying capacity for livestock.

6. **Determine Initial Cattle Stocking Rate.** To determine a conservative initial stocking rate of livestock for grazing during the forage growing season, take the total available forage for the entire grazing unit and divide by 2,250. For year-long grazing use 3,000 instead of 2,250 to determine initial stocking. These figures are estimates of the total pounds of forage eaten per animal during seasonal or year-long grazing. This will give the total months of animal use available.

During the first few years of grazing, annual monitoring of range use should be conducted to determine if stocking rates are too high or low. Adjustments will usually be required as conditions change, trees grow larger, and forage yields decrease as the canopy closes and blocks sunlight from reaching the forest floor. The best method to monitor stocking rates is by measuring forage yields and use. Native forage utilization can be estimated by installing a few (5 to 10) ungrazed enclosed plots (exclosures) which are placed in grazing units before cattle are allowed on an area. Forage use is estimated by comparing the amount of forage inside and outside the exclosures. Exclosures can be monitored monthly during cattle grazing and evaluated at the end of a grazing period. These exclosures can easily be constructed by using three to four steel fence posts and wrapping wire fencing around them. For the best comparisons in forage use, these movable exclosures should be placed in recently burned areas. If annual evaluations of native forage production and use indicate that the stocking rate is too high, then the landowner has two options. First, cattle numbers can be reduced, or secondly, native forage can be stimulated to produce more by heavier tree thinnings or more frequent thinnings.

SUMMARY

After reading this chapter you should be able to clearly envision a forest stand and its value for timber production and wildlife habitat. More importantly, you should have a better understanding of the different silvicultural practices that are conducted on stands in Alabama and how these practices affect wildlife habitat. Silvicultural practices affect the horizontal and vertical structure of a stand for the benefit or detriment of wildlife. An understanding of how forest harvest and regeneration, site preparation, intermediate stand practices, and other forest management activities impact wildlife habitat is important for forest owners who are interested in improving wildlife habitat on their lands.

Woodland grazing of cattle in southern pine forests has been around since the time of America's first explorers. Until the early part of this century, cattle grazing in pine forests was a common practice. Unfortunately, unrestricted cattle grazing quickly became a problem that caused overgrazing of wildlife habitat, forest regeneration problems, and degradation of waterways. However, past research and the experience of some managers has indicated that regulated short-duration cattle grazing in pinelands can be compatible with timber and wildlife management. In some cases, carefully controlled grazing can also enhance wildlife habitat. The key to a successful woodland grazing program is understanding the forage production capabilities of pinelands, the management practices that enhance forage production, and the appropriate stocking rates of livestock.

JOE MCGLINCY

With proper planning and an understanding of the effects that silvicultural practices have on wildlife habitat, land managers and owners can tailor timber management to enhance both timber and wildlife objectives.

Land managers who fail to take these considerations into account should be discouraged from initiating a woodland grazing program. Nevertheless, those who wish to incorporate cattle grazing into a multiple-use management plan have the opportunity to increase revenue from their timberlands.

The information in this chapter and other chapters specific to certain wildlife species will be helpful for forest owners who are preparing wildlife and timber management plans.

The white-tailed deer is one of the most abundant and sought after game species in North America

PAUL T. BROWN

CHAPTER 5

Managing for White-tailed Deer

CHAPTER OVERVIEW

This chapter describes the white-tailed deer in Alabama as well as techniques to enhance forest and farm habitats for this species. An understanding of the biology, life history, and habitat requirements of the white-tailed deer is necessary in guiding the selection and implementation of appropriate habitat management practices. Game population management strategies that use hunting as a management tool are also discussed where appropriate.

CHAPTER HIGHLIGHTS PAGE

BIOLOGY AND LIFE HISTORY ... 107
DISEASE AND PARASITES ... 110
HABITAT REQUIREMENTS ... 111
HOW MUCH AREA DO DEER REQUIRE? ... 116
HABITAT IMPROVEMENTS ... 116
ESTABLISHING FOOD PLOTS ... 120
SUPPLEMENTAL FEEDING ... 123
THE USE OF SALT OR MINERAL BLOCKS ... 124
HERD MANAGEMENT ... 126
DETERMINING THE HEALTH OF A HERD ... 128
ESTIMATING POPULATION TRENDS ... 128
QUALITY DEER MANAGEMENT ... 129
RECORD-KEEPING ... 129
MEASURING AND RECORDING ANTLER CHARACTERISTICS ... 130
HOW TO AGE JAW BONES ... 131
DEER MANAGEMENT ASSISTANCE PROGRAM ... 134
CROP AND ORNAMENTAL DAMAGE ... 134
QUALITY DEER MANAGEMENT ASSOCIATION ... 135

Hunting white-tailed deer (*Odocoileus virginianus*), the most popular big game animal in Alabama, is a form of recreation filled with tradition. Each year over 211,000 Alabamians head for the woods in pursuit of the whitetail, harvesting more than 331,000 deer. In addition to the recreational opportunities they provide, there is considerable interest in the aesthetic, economic, and educational values of deer. Alabama's deer herd is a valuable resource that with proper management will continue to thrive. Today, deer are abundant in all of Alabama's 67 counties, and biologists estimate the total number of deer in the state to be approximately 1.4 million. This represents an average of about one deer per 16 to 17 acres across Alabama. Deer were not always this plentiful in the state.

In pre-colonial times, the extensive mature forests of the Southeast did not provide the optimum habitat diversity necessary to maintain high deer populations. Deer were locally abundant, though, in areas where lightning fires and other factors had opened up the dense forest canopy. The natural enemies of deer such as cougars and wolves also played a significant part in regulating deer numbers and in keeping them in balance with their habitat. Later, during post-colonial settlement, extensive cotton production and subsistence and market hunting became severe limiting factors on deer, and herds began to decline drastically.

By the early 1900s, only a few thousand deer remained in Alabama. Because of public concern about conservation, game laws were enacted and law enforcement efforts were strengthened. But, as deer numbers slowly started to climb in some areas, the screwworm epidemic in the 1940s had a devastating effect on some southeastern deer populations. Scientific advances eliminated the screwworm problem by 1958, and once again deer herds began to expand their range.

After World War II, a deer-restocking program was initiated across Alabama and much of the Southeast. Deer were trapped from areas that had adequate populations and were relocated in 56 of the 67 counties that had relatively few deer. Not all deer that were trapped and relocated were from Alabama or even the Southeast. The Bankhead National Forest and part of Lauderdale County in northwest Alabama were restocked with a northern subspecies of white-tailed deer (*Odocoileus virginianus borealis*) brought in from Michigan and Wisconsin. The restoration and management efforts of the Alabama Department of Conservation and Natural Resources (ADCNR), the solid support of private landowners and the general public, increasing forest management efforts, and the elimination of the screwworm enabled the spectacular comeback of the white-tailed deer that we see in Alabama today.

Since the vast unbroken stands of pre-settlement forest and the large predators are gone, fewer natural factors now act to limit deer numbers. However, deer can become their own worst enemy if steps are not taken to limit population growth. The most valuable, choice food plants are overbrowsed or eliminated. Deer are then forced to utilize lower preference foods with little nutritive quality and the problem becomes more complicated with each successive fawn crop. Natural mortality increases significantly until the population often experiences a dramatic reduction in numbers. Unfortunately, by the time this occurs, the habitat has usually been severely damaged and may require many years to recover.

Man, ultimately, has assumed the role of primary predator. Through regulated hunting, adequate numbers of animals are removed from growing populations each year. Die-offs and the resultant waste of the resource are avoided. By balancing the harvest with the annual fawn crop, relatively stable populations can be maintained.

ANDREW NIX

The successful restocking effort of the Alabama Department of Conservation and Natural Resources after World War II has led to the spectacular comeback of the white-tailed deer in Alabama.

BIOLOGY AND LIFE HISTORY OF THE WHITE-TAILED DEER

White-tailed deer are polygamous breeders: that is, one male mates with several females during a breeding season. A dominant buck may breed with seven or more does in a season. The breeding period, often called the rut, occurs in most of Alabama from January through the middle of February with the peak period occurring during the last two weeks of January. In other southeastern states and parts of Alabama the rut occurs at different times depending upon a variety of factors, such as herd density, buck-to-doe ratios, and genetic background. Some Alabama deer have an earlier breeding season than other herds. On the Bankhead National Forest, the Choccolocco Division of Talladega National Forest, the Lauderdale Wildlife Management Area, and in east Alabama along the Chattahoochee River, the breeding season begins approximately two months earlier than other parts of the state. Earlier breeding dates on these areas are probably a reflection of the genetic origin of the deer. These areas were restocked in the past with deer from Michigan, North Carolina, and Wisconsin. Regardless of origin, all deer in the Southeast breed during the fall and winter but occasionally the periods may extend from November to March.

The gestation or pregnancy period for white-tailed deer is slightly less than seven months or between 190 and 210 days. Most fawns in Alabama are born between late August and early September, except on the areas mentioned above, where fawning occurs between the end of May and the end of August. Some fawns in Alabama are born as late as November. Fawns born later in the year have a harder time "catching up" to fawns born during the normal fawning period, especially as they approach the winter when food supplies are sometimes scarce and of low quality. In severe winters, with a shortage of food, late-born fawns often do not survive. Fawns that do survive rarely reach their full potential in weight or antler characteristics. Only in penned conditions where there is optimal nutritional conditions (available protein and carbohydrates) do late-born deer eventually rebound in quality. Even then, this takes several years. Late fawning has serious implications for managers interested in optimizing antler growth and conformation. Late fawning contributes to a disproportionate number of bucks having spiked antlers during their first few years, although this is not the only cause. In poor to adequate habitat, late-born bucks may never reach their full antler growth potential.

TES RANDLE JOLLY

"Rubs" and "Scrapes" serve as signposts indicating bucks' breeding territory. Scent on the bucks' forehead, from glands on the forehead or from rubbing against the tarsal glands on the back legs, is deposited on trees.

PAUL T. BROWN

Mother and fawn in summer coat. Most fawns in Alabama are born between late August and early September. Adequate nutrition for the doe is important at this time since nursing fawns place a heavy energy demand on the mother.

Keys to Healthy Deer: Factors Influencing Antler Size and Conformation

Once a spike always a spike? Not necessarily true. Extensive research with penned deer has shown that there is no way to look at a yearling buck (one and a half years old) and predict from his first set of antlers what he will produce as a mature buck. In fact, several yearling spiked bucks used in breeding studies went on to develop antlers that were larger in size (mass, number of points) than bucks that had forked antlers as yearlings. These findings put to rest the myth that all spike bucks are inferior and should be removed or "culled" from a herd by selective harvest. Another interesting point brought out in penned deer breeding studies is that does contribute as much to antler quality as bucks, genetically passing along antler characteristics from their ancestors to their offspring. Unfortunately, there is no practical method to determine which does carry the genetic traits to potentially produce a set of quality antlers.

Managers and landowners interested in improving antler characteristics of bucks on their lands should consider the following four factors:

1. **Age.** Antler size is directly related to the age of a buck. Bucks in Alabama reach their greatest antler potential between four and six years of age. After this period, antler quality starts to decline. Most bucks in Alabama are harvested at one and a half years of age and they never become old enough to reach the age of greatest antler development. The message is simple: to increase the potential for quality antlers, do not harvest younger bucks. Allow them to pass by so they can reach older age classes.

DR. KEITH CAUSEY

Age, nutrition, and genetics determine antler characteristics of bucks. These two bucks are one-and-a-half-year-old brothers and still exhibit significantly different antler characteristics.

2. **Nutrition.** The limiting nutritional factor for deer in Alabama is usually a lack of protein. Bucks need an adequate amount of dietary protein (13–16 percent) during antler development. If adequate levels of dietary protein are not available, bucks may never reach their true genetic or age-related antler growth potential. In addition to protein, adequate levels of calcium, phosphorus, and vitamins A and D should be present in deer diets. Unfortunately, native vegetation in most areas in the state averages only 8 percent crude protein. Exceptions include the Black Belt Region and fertile bottomland habitats along watersheds, which provide sufficient levels of nutrients needed by deer for adequate antler growth. Most areas in the state are less fertile and simply do not supply enough nutrients for deer to reach their full antler growth potential. On these areas, habitat improvement practices that improve the nutrient content of native deer foods, along with supplemental plantings, can go a long way toward enhancing antler growth of deer. On areas that have too many deer, overbrowsing of native deer foods reduces the availability of nutritional foods causing a drop in the quality of deer. Harvest strategies should focus on maintaining deer populations at levels the land can support with buck-to-doe ratios approaching one to one.

3. **Genetics.** Genetics definitely affects the potential for bucks to grow quality antlers. However, without proper nutrition and adequate age, even bucks with superior genes will never be able to reach their full genetic potential for antler growth. Most biologists agree that Alabama deer have the genetic potential to grow quality antlers but are limited by age and nutrition, particularly protein. To improve antler quality, managers should concentrate on increasing the average age and nutrition of deer, and the genetics will take care of itself.

4. **Herd Management.** Deer numbers should be kept within the carrying capacity of the land so adequate nutrition is available. Herd management should also focus on allowing deer to grow older to reach their biological potential in growth, body size, and conformation.

Both sexes are capable of breeding at one and a half years of age, but doe fawns receiving adequate nutrition are capable of breeding during their first year. Does bearing young for the first time often have a single fawn. Thereafter, they may have from one to four fawns, but usually two. In good habitat, the average is about one and one-half fawns per breeding doe each year. Young are produced once a year.

Fawns usually weigh 5 to 7 pounds, with males being larger than females. Typically, males slightly outnumber females at birth. Newly born fawns are able to stand on their spindly legs during their first day of life but do not ordinarily begin following their mother until they are about one month old. During this early period they remain secluded in heavy grass or brushy cover and are frequently visited by the doe for feeding. The fawns are eventually weaned at three to four months of age, but remain with their mother until the following year. They leave her a few weeks before she gives birth to next year's fawns.

Fawns are not usually abandoned, as many people believe when they find them unattended between feeding periods. The doe is almost always browsing or resting nearby. Fawns that are found in the woods should neither be molested nor taken home to be raised on a bottle. Bucks do not make safe pets, especially during the breeding season when they become aggressive. Many people have been seriously injured by "domesticated" bucks that have lost their fear of humans.

Male fawns have a type of antler growth usually referred to as buttons when they are five to seven months old. Yearling bucks get their first noticeable antlers at the age of one to one and a half years and have been known to have up to 10 or more antler points when raised in penned conditions and provided good nutrition. This is rarely the case, however, in the wild where yearling bucks are most often spikes to "six pointers." Generally, older animals have heavier, better-developed antlers than animals of a younger age, if nutrition has been comparable. With good nutrition, bucks usually attain their best antler size between four and six years of age.

Bucks shed their antlers each year. The time of antler-drop may vary somewhat, but most antlers in Alabama are shed between the middle of February and late March. People are often bewildered by their inability to locate shed antlers in the woods. This is because antlers have a high protein content as well as an abundance of calcium phosphate and are quickly consumed by rodents. For those with a keen eye, collecting "sheds" (antlers) dropped in the woods is a recreational pastime.

Once a deer sheds his antlers, new growth starts immediately, though signs of antler growth are sometimes not apparent for several weeks. Rapid growth allows the antlers to mature in three to four months. Throughout the summer, the antlers are fed by a very rich blood supply and are covered with a hair-like membrane commonly known as velvet. While "in velvet," a deer's antlers are particularly vulnerable to injury. Cuts or bruises suffered at this time often result in freakish or deformed antlers. By late summer or early fall, annual growth is completed and the antlers become solid and hard. The velvet dries and sloughs off or is rubbed off on trees or other objects. Healthy bucks then maintain polished antlers throughout the breeding season.

Deer coloration changes with the seasons. The short, reddish summer coats are gradually shed over a period of several weeks in the late summer or early fall and are replaced by the heavier, gray or gray-brown winter coat. The winter coat is replaced through a similar process the following spring. Fawns normally lose their spotted coloration after three or four months or at the time their

Fig. 1: This map illustrates the remnant distribution of the deer population in Alabama before restocking efforts began.

DAN BROTHERS
GARY W. GRIFFEN

Left, during the spring and summer, antlers are rapidly growing and are covered with a hair-like membrane known as "velvet." Damage to antlers during this time usually causes deformities that are visible when the antlers harden during late summer and early fall.

winter coat replaces their first summer coat. Minor color variations also occur. Biologists with ADCNR have reported that deer in the river floodplains of southwestern Alabama appear darker in color than those from the upland areas.

In Alabama, the whitetail is the only native species of deer. There is, however, a small herd of imported fallow deer, mostly in Wilcox and Dallas counties, though this herd is too small and isolated to furnish many hunting opportunities.

DISEASE AND PARASITES

It should not be too surprising that an animal as abundant as the white-tailed deer has many diseases and parasites that affect populations in the Southeast. Some of these diseases and parasites can cause deer mortality, lower the health and productivity of a deer herd, or cause minor irritations in the least serious situations. The common causes of many of the diseases and parasites that affect deer are viruses, bacteria, protozoan, and internal and external parasites. Landowners should be aware of the presence and significance of some of the most common diseases and parasites as part of sound deer management programs.

One of the most fatal diseases that infiltrate deer herds is epizootic hemorrhagic disease (EHD) also known as "blue tongue." This disease is a virus that can cause extensive internal hemorrhaging and ultimately death. In the past there have been sporadic outbreaks of EHD, which in severe cases have killed thousands of deer. In the early 1970s scattered outbreaks were reported across the Southeast. EHD is an ever-present virus that is usually spread to deer by blood-sucking arthropods such as gnats. It appears most often in late summer or early autumn. Deer can be carriers of the disease without exhibiting any symptoms, however, and when deer are stressed, EHD may become acute. External symptoms include hoof loss, rapid heart rate, and excessive salivation. Dehydration can result and sick or dead deer are often found near water in hot weather. Deer that are affected with EHD initially lose their fear of man, do not eat, and become weaker over time. When field dressing EHD-infected deer, it is apparent that many of the organs have lesions where extensive hemorrhaging has occurred. Although EHD can be transmitted to other animals, it does not pose a threat to man.

A fairly common deer disease that hunters may encounter is skin tumors, commonly referred to as warts. This disease is caused by a virus that is spread by insects, contaminated vegetation, or other deer. Although unsightly, the skin tumors rarely affect the health of a deer, unless they impair eyesight, hearing, or breathing by covering the nose. There are no human health threats from deer warts.

Anthrax has also infected deer across the U.S., and in the past has been responsible for significant deer die-offs. The disease is a bacterium that is spread to deer through soil and water that has been contaminated by other infected animals. Deer that succumb to anthrax lose their coordination and often have bloody discharges from their nose and anus. These animals usually die within 48 hours after such symptoms appear. If anthrax is observed in living or dead deer, the animals should not be touched and health officials should be contacted

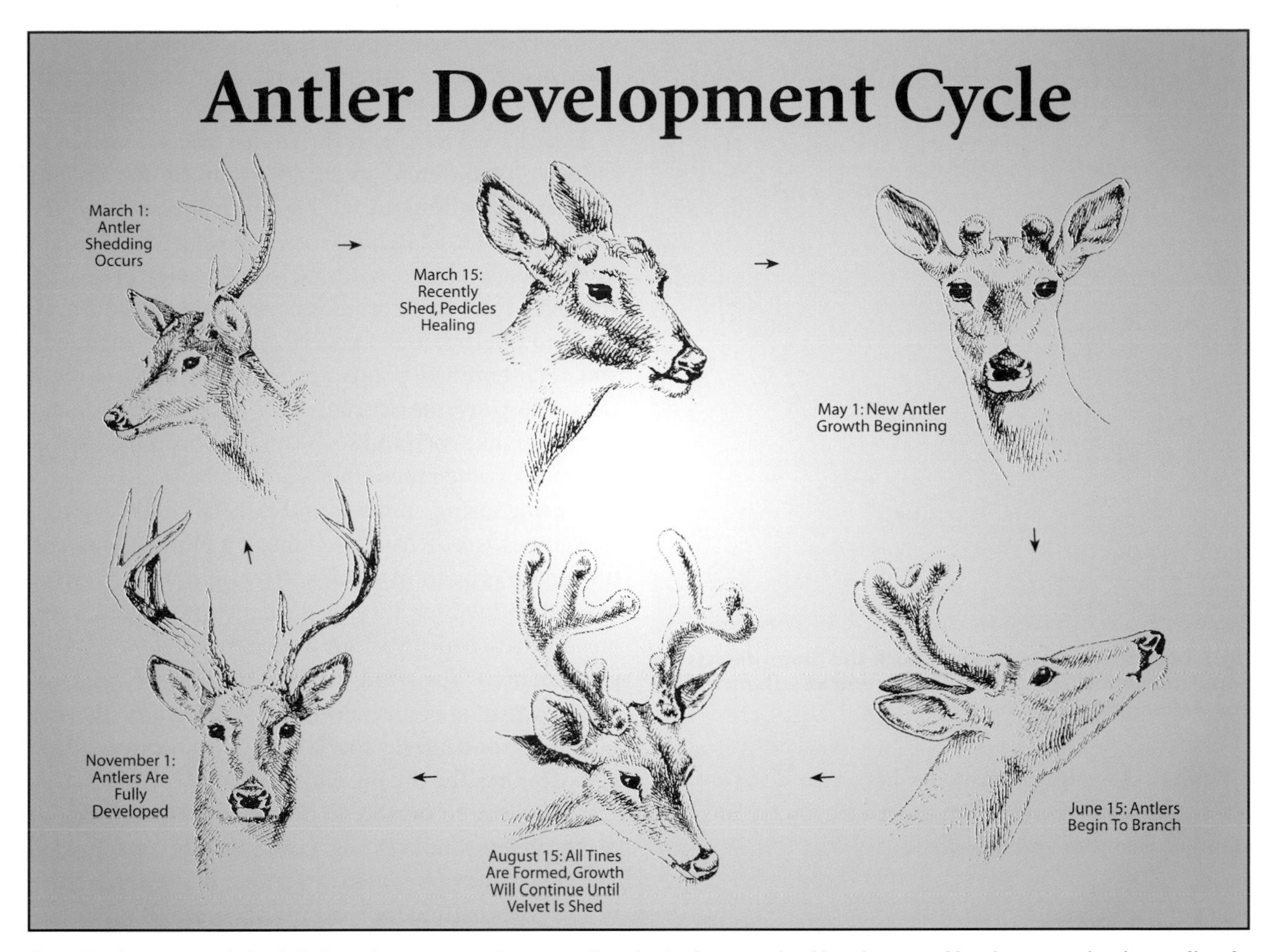

Fig. 2: Bucks grow and shed their antlers in a regular, annual cycle. Antlers are shed in winter and begin regrowing immediately through the spring and summer until the velvet is shed in late summer or early fall.

immediately since humans are highly susceptible to the disease. Anthrax can be controlled by vaccination and using sanitary practices.

There are a host of parasitic protozoans, worms, and arthropods that affect deer. The parasitic worms produce the most significant problems. Usually high numbers of parasitic worms indicate poor health and malnutrition associated with overpopulation. The presence and abundance of parasitic worms in deer is often used as a management tool. The type and number of stomach worms can be directly related to deer density. Abomasal (one of the four stomachs) worm counts are often used by managers as an index of the relative deer density in an area and the general health of a deer herd.

Nasal bots are often encountered by hunters who examine the oral cavity (area containing the nasal cavity) of deer. Nasal bots are larvae of flies (*Cephenemyia* spp.) which have a "grub-like" appearance ranging in color from white to tan. Adult flies deposit eggs on the skin around the nose or mouth of deer, and when deer lick these areas larvae within the eggs are released and migrate to the nasal passages. Inside the naval passages they grow until they exit the deer, drop to the ground, and eventually become adult flies. Nasal bots are not harmful to deer. Although unsightly, hunters should not be concerned with consuming deer that have nasal bots.

HABITAT REQUIREMENTS OF WHITE-TAILED DEER

Though white-tailed deer are extremely adaptable animals, their essential requirements include food, cover, and water. Water is readily available in Alabama and is rarely a limiting factor for deer populations. Of these three, food and cover are the two most important con cerns in deer management. Abundant forest land within the state provides suitable cover except where large tracts of land are in agricultural production. Because the transition zones between cover types are used heavily by deer, an interspersion of brushland, woodland, and non-forested land provides a desirable diversity in the types and amounts of food and cover. Many timber harvest operations today create an "edge effect" and add diversity to

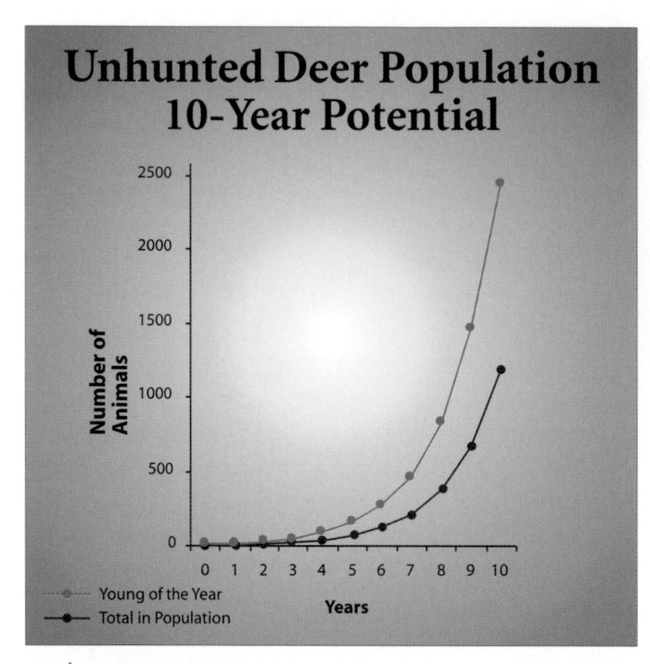

Fig. 3: Beginning with a doe and a buck, this graph depicts the population growth potential in 10 years of an unhunted deer population, including natural mortality.

a habitat. An edge is created where two different age classes or forest types meet, as in a newly regenerated stand adjacent to an older stand.

Cover and Deer

Cover is rarely a limiting factor for deer in the Southeast like it can be for northern deer exposed to extremely cold temperatures. However, deer in the Southeast do prefer areas that provide cover during most of the year. Deer use cover for concealment and protection during the summer fawning period, thermal cover for protection against extremely hot weather, concealment while resting, and concealment in areas subject to heavy hunting pressure. Ideal cover types include dense, unthinned pine stands; dense three- to four-year-old openings created by forestry practices; abandoned fields that have grown up in tall grasses, shrubs, and trees; evergreen thickets; and virtually any area that contains thick vegetation. Dense areas that provide both cover and food, like honeysuckle and greenbriar thickets, are used heavily by deer, especially during the summer months when deer movements are at a minimum. In the Southeast, most land management practices provide enough cover to meet the needs of deer, except in agricultural or forestry operations that create extremely large openings with little or no cover.

Deer have a large and varied diet and are known to eat over 700 different species of plants at one time or another, depending on the season of the year and the availability of food items. Whitetails were once thought to be exclusively browsers, selecting twigs primarily. However, food habit studies have revealed that leaves, bark, and herbaceous material such as grasses, weeds, and soft-stemmed plants are common in their diets. Deer also consume acorns, other nuts, fruits, agricultural crops like corn and soybeans, mushrooms, lichens, and mosses when these foods are available.

During spring, summer, and early fall, deer eat grasses, legumes, weeds, fruits, certain agricultural crops, and the tender growth of shrubs, trees, and vines. During late fall and winter, they feed on acorns, grasses, evergreen leaves, fruits, various agricultural crops, and the green stems of dogwood, greenbriar, blueberry, and sassafras. Acorns are especially important because they are high in carbohydrates that help build fat reserves and keep deer healthy during winter months.

Deer seem to be able to determine which foods or plants are most nourishing. Foods readily consumed in one area may not be eaten in another due to differences in soil types, succulence, deer numbers, and other factors. Studies have shown that deer prefer plants on burned rather than on unburned areas, and on fertilized rather than on non-fertilized areas. Burning releases nutrients bound-up in forest floor litter like leaves and dead wood and returns them to the soil. The improved

Cover is not a limiting factor for deer in the Southeast, although deer will seek out areas of dense vegetation and use them for relief during temperature extremes, for refuge in areas that are heavily hunted, and for concealment during fawning periods.

SEASONALLY IMPORTANT NATIVE DEER FOODS[1] IN ALABAMA

Spring	Summer	Fall	Winter
Alabama supplejack	Alabama supplejack	acorns	acorns
American beautyberry	American beautyberry[2]	Alabama supplejack fruit	Alabama supplejack fruit
assorted grasses	blackberry[2]	American beautyberry fruit	beechnuts
blackberry	blueberry	assorted forbs[*] (vetch, beggar's tick, milk pea, annual lespedezas, wild sweet clover, and others)	blueberry
blueberry	dogwood	beechnuts	greenbriar
dogwood	fringetree[2]	blueberry	holly
fringetree	greenbriar	dogwood fruit	fruit
greenbriar[*]	hawthorn fruit	fringetree[2]	honeysuckle[*]
hawthorn[2]	honeysuckle[*]	gallberry fruit	mushrooms
honeysuckle[*]	plum fruit	greenbriar	possumhaw fruit
huckleberry	red mulberry	holly fruit	yaupon[2]
oak	sassafras	honeylocust fruit	
persimmon	wild grape[2]	honeysuckle[*]	
plum	yaupon	mushrooms	
possumhaw	yellow poplar	pecan nut	
red mulberry		persimmon	
sassafras		plum fruit	
viburnum		strawberry bush	
water and black tupelo		water and black tupelo fruit	
wild grape		yaupon[2]	
yaupon[*]			
yellow poplar			

[1]Leaves and stems [2]Leaves and fruit

[*]Plants that are high in crude protein content

Seasonally important deer foods include (from left to right, top to bottom) blackberry, vetch, French Mulberry (American beautyberry), acorns, greenbriar, Japanese honeysuckle, Mexican clover, huckleberry, and poison ivy.

The Magic of the Black Belt

Deer hunters have long known about the quality of bucks routinely produced in the Black Belt counties of central Alabama. The deer populations in these counties are among the highest in the state and the number of record bucks taken far exceeds those taken from any other portion of Alabama.

The Black Belt is a crescent-shaped region extending some 300 miles across central Alabama, northeastern Mississippi, and into Tennessee. A relatively flat, fertile plain, it ranges in width from 20 to 25 miles throughout most of its length. In Alabama, the Black Belt covers large portions of Sumter, Greene, Marengo, Hale, Dallas, Perry, Lowndes, Montgomery, Bullock, and Macon counties, and smaller portions of Russell, Butler, Wilcox, and Pickens counties.

The name "Black Belt" is derived from the dark color of the soil, which is made up of heavy clays, ranging in color from gray to red to black. The soil is rich in lime and other minerals, making deer forage and other foods highly nutritious. Because of this high nutritional value, the natural browse, acorns, and agricultural crops grown in the Black Belt provide deer with above average amounts of protein and minerals that increase body weights and antler size.

Alabama's Black Belt

Fig. 4: Deer weights and antler characteristics have been reported to be greater in the Black Belt because of the region's high soil fertility.

nutrient content of native foods following prescribed burning can also be beneficial for deer. Well-fertilized agricultural crops or nursery stock are often browsed severely in areas where deer densities are high. Preferences for individual food items are basically a reflection of the palatability and the nutritional content of the food that is available at a particular time and place.

Research has shown that the nutritional health of deer is directly related to soil fertility. On sites that are relatively poor in soil nutrients, deer tend not to reach their full body or antler potential. The protein content of deer forage is the most important nutrient enabling deer to reach their maximum potential in body and antler size. For deer to reach their genetic potential in size and antler characteristics, their diet should contain at least 17 percent protein. Unfortunately, very few native foods contain 17 percent crude protein. Most native

DAN BROTHERS/ADCNR

Deer diets are based upon the availability and quality of foods in a particular area. Deer have an uncanny ability to select the most nutritious plants and parts of plants that contain the highest protein and mineral content, such as new growth like buds and leaves.

foods contain between 6 and 10 percent. However, because deer are ruminants (having four stomachs), they are able to increase the protein content of ingested foods through a process called microbial digestion. Microbial digestion involves naturally occurring microorganisms (microbes) found in the rumen (one of the four stomachs) that aid in the breakdown of plant parts during digestion. During this process microbes also contribute protein to deer diets. Microbial protein, in combination with plant protein, increases the overall protein levels absorbed by deer. Consequently, in quality habitat and with the aid of microbial digestion, deer are able to come close to a dietary protein level of 17 percent.

In Alabama, one of the most fertile sections of the state is the Black Belt Region of central Alabama. Deer harvested within the nutrient-rich soils of the Black Belt have historically been heavier animals with greater antler mass than deer from surrounding areas. Although the Black Belt comprises only 3.5 percent of the land area in Alabama, this small area has produced 85 percent of all the whitetails listed in the *Alabama Whitetail Records.* Bottomlands associated with river drainages also contain highly fertile soils.

Deer are fond of a variety of agricultural crops including grains and vegetables. Damage to commercial agricultural crops is often extensive and severe. Nurseries and orchards also sustain losses to deer browsing. Orchard losses are usually greatest to smaller trees that often have to be replaced several times because of severe damage. Another type of deer damage often overlooked is the destruction to natural or planted trees on forest land. Deer depredation is usually most severe where deer population densities are high and in areas where small agricultural fields, orchards, or forest regeneration areas are interspersed with forested deer habitat. It is no coincidence that these areas are highly attractive to deer, because most are established and maintained by fertilization. Research and field observations have documented that deer prefer fertilized areas more than unfertilized sites, because fertilization increases the palatability, digestibility, and nutrient content of vegetation. Even one- to two-year-old abandoned agricultural fields are attractive to deer because of higher soil fertility and increased nutrient content of vegetation.

There is a relationship between food supply and population density, but its effect on reproduction and antler development is often misunderstood. Many people fail to realize that overpopulation can occur when deer numbers are not extremely high. If there are more deer than available food, the herd is overpopulated—not only the habitat but also the individuals within the population suffer. Deer with inadequate nutrition are prone to parasites, have lower body weights, show obvious signs of poor antler development, and experience lowered reproductive rates.

As a deer population continues to increase beyond its food supply, animal quality declines even further and natural mortality becomes more significant. There are usually rising complaints of deer damage to agricultural crops, forest regeneration, and severely overbrowsed areas. Deer-vehicle accidents also increase in frequency. Additionally, the ever-present threat of a mass die-off from malnutrition, parasitism, or disease becomes more likely.

HOW MUCH AREA DO DEER REQUIRE?

The home range size of white-tailed deer will vary, depending upon the quality of the habitat. In general, the better the habitat, the smaller the home range. Research with radio-collared deer has shown that in relatively good habitat, bucks will spend most of their life on 640 acres, whereas does require substantially less, around 300 acres. Home range size also varies by season, depending on food availability, the rut, and fawning.

HABITAT IMPROVEMENTS FOR WHITE-TAILED DEER

Land management practices exert a direct influence upon the value of an area as habitat for deer. Habitat manipulation through timber harvests, controlled burning, and agricultural or wildlife plantings has been shown to be of major value in providing a proper combination of food and cover necessary to maintain healthy deer populations. Deer plantings, for example, can provide proper combinations of food and cover necessary to maintain healthy deer populations, especially in areas where quality native foods are sparse. In addition, optimal deer habitat is nearly always a by-product of forest management or some other land-use practice. Fortunately, forest management techniques directed toward providing deer habitat can be compatible with timber produced for income. Silvicultural practices can enhance both timber production and deer habitat with understanding and proper planning.

Forest Management

Pure stands of unmanaged pine generally provide poor deer habitat because of the low abundance and quality of understory forage and the scarcity of mast-producing hardwoods. Dense stands and closed canopies reduce

DAVID K. NELSON

WENDY NEEFUS/ANIMALS ANIMALS

Forest stands should have a balanced mixture of oaks from the red oak group (represented by laurel oak acorns on the left) and from the white oak group (represented by white oak acorns on the right). Generally, deer prefer acorns from the white oak group because of their lower tannic acid content as compared to red oaks.

browse and fruit yields. These stands are mainly valuable to deer as cover. Management efforts in these forests should be directed toward increasing browse production. These stands should be thinned to a basal area of 50 to 60 square feet per acre to open the overstory and to encourage the production of desirable understory vegetation.

Mixed pine–hardwood stands generally provide good deer habitat and yield important mast, fruit, and browse production. These stands should also be thinned frequently to renew understory forage and hasten early mast yields. Where possible, retain a mixture of valuable mast-producing hardwoods. Stands should have a basal area of at least 20 square feet of mature mast species per acre. A balance between the white and red oak groups is desired to provide consistent mast production for deer and other wildlife (see Appendix G for a listing of white and red oaks). This is necessary since acorn production is not always dependable from year to year with white or red oaks alone. Red oak acorns take two years to mature, whereas white oak acorns mature in one year. Providing a mix of red and white oaks increases the chance that acorns will be available for deer each year.

Bottomland hardwood forests, generally composed of oak, gum, and ash, can also provide quality deer habitat. These areas normally have fertile soils that can produce high quality browse if plant succession is set back by some type of disturbance. Unfortunately, they are often subject to flooding which may reduce available food supplies. Mast production in this habitat is generally ample, but as in the mixed pine–hardwood type, both white and red oaks should be retained. If possible, it is wise to keep half of the trees on this forest type in a variety of mature mast producers. Large expanses of mature, closed canopy hardwood forests can present seasonal problems for deer with regard to food availability. Although these areas may produce mast that is available during fall and winter, they are often void of forage and browse during the rest of the year. In years of poor mast production, closed canopy hardwood stands offer little food value for deer. To alleviate this problem, stands can be opened up by thinning or timber harvests that remove only portions of the stand, stimulating the production of understory forage plants for deer.

A general recommendation for stand size and mixture for timber production and deer habitat on large upland tracts is to have stands that are ideally 50 to 150 acres in size, with 70 to 80 percent of the stand in pines and 20 to 30 percent in mixed pine–hardwood and bottomland hardwoods. Landowners who have smaller tracts can create a similar effect by intermingling forest types and ages where possible to increase habitat diversity and edge for deer. Regeneration areas should be well distributed and spaced at ¼- to 1½-mile intervals. The method of timber harvest, whether it be seed-tree, clear-cut, or shelterwood is not a critical factor in deer management, but when practical, clear-cuts should be as small as economically feasible. Distributing small 5- to 10-acre clear-cuts throughout an area improves deer habitat. In addition, harvesting and regenerating in linear, irregular shapes benefits deer by maximizing edge effect.

To stimulate the production of valuable deer forage in pine stands, a regeneration harvest, intermediate thinning, or other silvicultural practice is usually needed at least every 6 to 10 years. On better sites, forage production peaks about 2 to 3 years after regeneration and then declines steadily for the next 5 or 6 years. At about 8 to 10 years of age, the tree canopy closes, shading out sunlight and causing deer forage production to decline even further. From then until thinning at 15 to 25 years, forage production is at its lowest. Any subsequent silvicultural operation such as an intermediate cutting temporarily boosts forage production.

The conversion of extensive mixed-hardwood or

pine-hardwood stands to pure pine reduces the amount of available hard mast (acorns). Conversion of bottomland oak forests to forests that do not produce mast should especially be avoided and may be a Clean Water Act violation. In fact, the intermingling of mast-rich tree species of bottomland hardwood forests should be encouraged, especially when they occur as wide linear bands along streams, creeks, and rivers. Prior to a timber harvest, areas that provide critical habitat, such as grape or honeysuckle thickets, soft or hard mast trees and shrubs, live oak clumps, or greenbriar thickets in bottomland hardwoods should be identified and protected. Also, natural openings, savannahs, open or wet prairies, fields, and old house sites should be retained since these areas provide natural foods and cover for deer. Old house sites are often surrounded by fruit-producing trees, shrubs, and vines that are valuable deer foods.

Preparing sites for regeneration by drum chopping or shearing can increase the benefits of newly regenerated stands for deer. These site preparation techniques provide the greatest amount of forage and soft mast production. When using herbicides for site preparation, proper herbicide selection, rate, and timing of applications can also minimize damage to desirable food plants for deer. Care should be taken to prevent overuse and poisoning of aquatic habitats.

Before altering a site, valuable wildlife plants such as dogwood, blueberries, persimmon, crabapple, and wild grape should be marked with paint or flagging to identify trees to be saved. When replanting, wide seedling spacing helps delay canopy closure. Planted pine seedlings should be spaced somewhere between a rate of 300 (12 x 12–foot spacing) to 545 (8 x 10–foot spacing) per acre. Wider seedling spacing allows a greater variety and quantity of understory food supplies to exist over a longer period of time; however, landowners should be aware that timber quality will be reduced with extremely wide spacings that encourage limb retention which reduces the market value of pines.

Where practical, the planting of desirable mast-producing hardwood seedlings on suitable sites can also enhance deer food availability. However, with the exception of sawtooth oaks, most oaks don't begin to produce a significant amount of mast until around 25 years of age. Therefore, it is important to mark patches or groups of mature oaks to be retained when planning a timber sale.

PRESCRIBED BURNING

Prescribed burning is a practical and economical deer management tool. Prescribed burning in pine stands benefits deer by increasing browse yields and improving the palatability and nutrition of understory plants. Burning acts as a fertilizing agent by releasing many nutrients that are bound up in the forest litter. Herbaceous vegetation and woody browse plants respond immediately after a fire with increased production (in pounds of forage per acre) and protein content. For maximum effect, prescribed burning should be conducted in conjunction with timber thinnings that open up pine canopies for sunlight penetration. Prescribed burning provides little benefit to deer and other wildlife if stands have closed canopies and no initial understory vegetation.

Generally, fire should not be used in hardwood stands. Hardwoods are intolerant of fire due to their thin bark. If burned they will eventually die or be damaged so severely that mast production and timber quality is drastically reduced. In some mixed pine–hardwood stands, where the quality of hardwoods for wildlife and timber is low, managers do use prescribed burning to enhance deer habitat. If poor quality hardwoods are damaged or killed in these situations, there is little economic or wildlife value lost in the short run. The increase in browse yield by prescribed burning more than offsets the loss of mast production in poor hardwood stands. Control of these poor quality trees can actually improve some stands for timber and wildlife by "releasing" or reducing competition with the more valuable trees that remain in the stand.

Initial burns in pine stands should be made as early in the life of the stand as possible. Chapter 4 describes in detail when prescribed burning should first be conducted depending on the species of pine within a stand. The most practical burning regime for deer in pine stands is burning on a three-year rotation. This length of rotation provides optimal herbaceous growth and maintains browse within easy reach of deer. Woody browse plants grow several feet in the two to three years without fire, then are killed and resprout again after burning. On some sites the frequency of burning may have to be altered by a year or two depending on the productivity of the area. Fertile sites in general will have to be burned more often than poorer sites since understory plant growth is more rapid. Observation of the response of deer forage plants to burning over time can give a clearer indication of how often to burn. Burning should be on rotations longer than three years to enhance the production of soft mast such as blackberry. For deer, burning should be conducted during late winter, February or early March, especially if ground-nesting wildlife like wild turkey are a consideration. If greater control or "knock down" of hardwood browse plants is desired, burning should be conducted in early spring when hardwoods are most susceptible to fire. Burning in late summer can also be effective.

Prescribed burns should only be conducted by

RICHARD T. BRYANT/ARISTOCK

Prescribed burning is one of the most beneficial deer habitat and forest management practices. Burning every two to three years helps keep native vegetation palatable and within reach of deer. This woods road serves as a firebreak between the stand that has been recently burned on the right and the unburned stand on the left.

trained professionals who are knowledgeable about the safe use of fire to improve deer habitat. A permit must be obtained from the Alabama Forestry Commission before burning. For more information on prescribed burning see Chapter 4 or contact your local office of the Alabama Forestry Commission.

FERTILIZATION OF NATIVE DEER FORAGE

A recent study conducted at Auburn University indicated that the fertilization of selected plants to increase their nutrient content might be a viable option in improving deer habitat. The Auburn study examined the effects of fertilization on Japanese honeysuckle, a preferred deer forage. Honeysuckle was fertilized with a complete fertilizer (13-13-13) which doubled plant growth and increased plant protein content from 11 percent to 17 percent. The Auburn study recommends that 300 pounds of 13-13-13 per acre be applied to honeysuckle in early spring and again in the fall. For the fall application, ammonium nitrate may be substituted at a rate of 300 pounds per acre. Add lime based on soil test results.

Research at Auburn has also suggested that fertilizing white oaks increases acorn production, the size of acorns, and their palatability. The effect of fertilization on red oaks and other mast-producing trees and shrubs is yet unclear. One study examining the effects of forest fertilization in pine plantations did find an increase in soft mast production, primarily blackberry fruit, in six- to seven-year-old plantations that were fertilized with 300 pounds of urea per acre. The same study also suggested a potential to increase soft mast production by combining fertilization with timber thinning.

In general it is not economically feasible to attempt to increase acorn production over large areas with fertilization. However, fertilization of select mast-producing trees may be feasible if conducted on a small scale. To maximize results, candidates for fertilization should be dominant trees that have a history of being

Habitat and Hunting Management Intensities for White-tailed Deer

Management Practices	Low Intensity	Medium Intensity	High Intensity
Timber Harvest/ Regeneration Types	Any method	Clear-cut/Plant, Natural	Clear-cut/Plant, Natural
Timber Harvest Rotation Length	15–25 years	20–45 years	35–70 years
Thinnings	Once or none	Every 4–5 years	Every 5 years
Fire	None or one	Dormant season every 3 years	Dormant season every 3 years and growing season every 9–10 years
Site Preparation	Shear/pile & disk, broadcast herbicide, banded or broad spectrum herbicide	Chop/burn, selective spot treatment herbicide	Fire, banded herbicide, release after minimal site prep
Windrows	Burn	Leave alternate windrows	Leave unburned
Seedling Spacing	6 x 6, 6 x 8	8 x 8, 8 x 10	10 x 10, 12 x 12
Hardwoods	Retain in drains	Retain in drains and mast-producing clumps	Retain all mast producers
Food Plots	1–3% of land in winter plots	3–5% of land area in winter and summer foods	Greater than 5% of land area in a variety of seasonal foods
Fertilization	None/food plots	Food plots, honeysuckle patches	Food plots, honeysuckle patches, mast producers, complete stand after fire
Feeding Supplements	None	Mineral supplements	Minerals and pellets, soybeans during off-season
Deer Harvest	Harvest does, all legal bucks	Harvest does, limit buck harvest to forked antlers	Harvest does, bucks must have at least 8 points and a 16-inch main beam, 16-inch inside spread

good mast producers. These trees can be identified in the stand by monitoring acorn production of individual trees over several years. Even without fertilization, a variety of mast-producing trees should be retained and protected within a stand for deer and other wildlife. It is especially important that these trees or shrubs be identified and marked before any silvicultural practices are performed to prevent damage or removal.

ESTABLISHING FOOD PLOTS FOR DEER

Supplemental Plantings

Deer readily feed on supplemental plantings. These plantings can help compensate for yearly and seasonal fluctuations in food supplies, especially mast. They are less important, however, if the habitat is enhanced through coordinated and sustained silvicultural practices that improve deer habitat. Food plots for deer should never be viewed as a substitute for properly managed native habitat. As a general rule, at least 4 to 5 percent of forest land should be planted in deer food plots. On most areas this is about one food plot for every 25 acres of land, but some research in the Southeast has indicated that maintaining as little as 0.5 percent of an area in year-round food plots can significantly increase deer quality.

Food plots also provide an excellent method to attract deer for adequate harvests or viewing. Research on radio-collared deer indicates that deer will shift their center of activity (where they spend most of their time), to areas where food plots are planted. In other words, deer alter normal foraging patterns to frequent food plots, making these locations ideal for increasing the chance of observing and harvesting deer.

Natural openings of 1 to 3 acres in size should also

be retained and managed within timber stands. These openings should be irregularly shaped, preferably linear, and strategically located throughout an area to provide maximum diversity and edge. Openings should not be developed adjacent to major roads or other access routes. Unused logging roads, skid roads, and trails can be "opened up" and managed as natural plantings in native vegetation or seeded to provide supplemental food. Widening roads and trails by removing at least 5 to 20 feet of adjacent trees on each side allows sunlight to reach the ground and stimulates deer forage growth. This also allows the road to dry quicker and reduces road maintenance. Natural openings can be maintained by periodic mowing, disking, or prescribed burning every two years to stimulate new plant growth.

Plants favored by deer that are often planted in food plots include American jointvetch, barley, clovers (osceola, regal ladino, red, burseem, Redland II, arrowleaf, ball, bur, crimson, subterranean, and white), corn, soybeans, oats, rye, ryegrass, vetch, and wheat. Some of the most preferred plantings for deer (and turkey) on uplands in north and central Alabama include mixtures of wheat, Redland II clover, and crimson clover. On uplands in south Alabama, a mixture of wheat and crimson clover is preferred. On bottomlands across the state, regal ladino and osceola clover have proven to be dependable producers. In general, clovers are a good choice for deer food plots since they are available during late winter to early summer, retain their nutrient value, and can be planted in various combinations with small grains. In addition, perennial clovers require less time and labor since they should remain productive for about three to five years before needing to be replanted. For more information on establishing plantings for deer see Chapter 16 "Supplemental Plantings as Wildlife Food Sources" and Appendix D–J.

Fertilizing Oaks to Increase Acorn Production

Research suggests that fertilizing individual oak trees may increase acorn production and availability for wildlife. The best time to apply a fertilizer is immediately after a heavy frost or between January and April before spring. Pelleted or liquid nitrogen fertilizers (ammonium sulfate, ammonium nitrate, or urea) should be applied directly on the soil surface under the tree canopy at an annual rate of about 6 pounds of nitrogen per 1,000 square feet of surface area. To prevent fertilizer from burning exposed roots, fertilizers should not be applied within 3 feet of the tree trunk. Every three to five years, in the spring or fall, a slow-release complete fertilizer (such as 12-6-6 or 12-12-12) should be applied at a rate of 6 pounds per 1,000 square feet. If soils are compacted under tree canopies then fertilizers may be applied as subsoil treatments in angular holes (gives more exposure of fertilizers to tree roots than holes straight down) 12 to 18 inches deep, 2 feet apart.

WHEAT, REDLAND II CLOVER, AND CRIMSON CLOVER

This combination is especially recommended on upland soils of medium to high fertility in central and northern Alabama. Wheat is a cool-season annual grass; Redland II clover is a cool-season, short-lived perennial legume (usually surviving for two years in Alabama); and crimson clover is a cool-season annual legume. Some landowners and hunting clubs like to add cold-tolerant oats. Since oats are a cool-season annual cereal grain, they add variety to the combination. In some parts of the state, deer seem to prefer oats to wheat. However, according to research, wheat is more nutritious for deer.

The best time to plant wheat, Redland II, and crimson clover is between September 15 and September 30 after a rainfall so that there will be enough moisture in the soil for the seeds to germinate. Plantings should be avoided before September 1 and after October 15. Before planting, the site should be prepared using a disk harrow with a

LEWIS O. ROGERS/SOUTH CAROLINA DNR

Proper food plot preparation helps to ensure good seed to soil contact, resulting in successful germination of seeds. Here a grain drill is being used to plant seeds in a properly prepared seedbed.

Suggestions for Managing Openings for Deer

Because a diversity of wildlife openings is desirable, rotation of the following alternatives should be practiced.

1. Allow some openings to grow up in native vegetation and maintain these in an early stage of plant succession by annual mowing.
2. Establish annual crops such as soybeans, cowpeas, black-eyed peas, corn, or one of the grain sorghums in other openings.
3. Plant some openings in a combination of wheat, oats, and rye for winter grazing.
4. Perennials such as clover can be planted and maintained with annual late summer mowing and periodic fertilization.
5. Establish sawtooth oak, Chinese chestnut, Sweet Hart™ chestnut (available through the National Wild Turkey Federation at 1-800-THE-NWTF) and other mast-producing trees at the edge of openings to benefit deer and other wildlife species.

slight angle. Wheat seed should then be broadcast at a rate of 1½ bushels (90 pounds) per acre over the prepared seedbed. Cold-tolerant oats can also be added, if desired, by planting 45 pounds of oats and 45 pounds of wheat per acre. Seeds should be broadcast evenly across the entire plot with a cyclone-type seeder or planted with a seed drill. After broadcasting, seeds should be covered with about 1 inch of soil with a disk harrow. A cultipacker or other type of roller should then be used to firm up the soil covering the seeds. A cultipacker is a tractor attachment used to compact freshly plowed land to pack the dirt tighter in the seedbed. A tightly packed seedbed enhances seed to soil contact which is a critical factor in facilitating proper germination.

The next step is to immediately broadcast 15 pounds of inoculated Redland II and 30 pounds of crimson clover seed per acre over the covered wheat (and oat) seed. Attempt to broadcast the seed evenly over the entire plot, though this may be difficult because the seeds are so small. The seeds may be broadcast using the same methods as wheat, but sand may have to be mixed in with the seed to make sure seeds are evenly distributed and don't come out too quickly. Redland II and crimson clover seeds should then be covered with about ¼ inch of soil with a drag attachment, and then the soil firmed up using a cultipacker or other roller. Be careful not to cover seeds too deeply.

WHEAT AND CRIMSON CLOVER

Wheat and crimson clover combinations are recommended for providing fall, winter, and early spring food for deer (and turkeys) on upland soils of medium to high fertility, especially in southern Alabama. Both plants should be established in the same manner as mentioned previously, first the wheat and then the clover. With this mixture, cold-tolerant oats may also be added by reducing the seeding rate to 45 pounds wheat seed and 45 pounds of oat seeds per acre.

TES RANDLE JOLLY

Food plots provide additional nutrients for deer and also help hunters meet harvest management objectives by concentrating deer.

REGAL LADINO CLOVER

Regal ladino is a cool-season perennial legume developed at Auburn University that is well suited to most bottomland soils in Alabama. It provides high quality, year-round forage for deer, especially between May and September when antlers are developing.

Once established, a regal ladino

clover plot should last about five years before competition from other native plants makes it necessary to replant. Weed competition can be reduced during land preparation by disking deeply several times prior to planting. This practice exposes the roots of weeds and kills these plants. Seeds of weeds will also be stimulated to sprout and can be killed by the next disking.

Regal ladino clover can be planted between August 15 and November 1. The optimal time to plant is between September 15 and October 15, when there is enough moisture in the soil for the seeds to germinate. This will provide enough time for the clover to develop a good root system before the first freeze.

Planting sites should be prepared in advance by running a disk harrow with a slight angle over the ground. The seedbed should then be firmed up with a cultipacker or other type of roller. Once the seedbed is firm, inoculated seeds should be broadcast evenly over the seedbed at a rate of 5 pounds per acre. Seeds should then be covered with ¼ inch of soil by making a second trip over the plot with a cultipacker.

Once regal ladino plots are established, they should be maintained by reducing the spread of competing plants. Taller weeds and other plants should not be allowed to produce seed and should be clipped two or three times a month in July, August, and September with a rotary mower or bushhog, especially if rainfall is heavier than normal. Soils should be tested every three years, and lime and fertilizer applied according to the soil test recommendations. After five years, clover plots will usually have to be replanted.

Lablab For Deer

Ed McMillan is a landowner, wildlife manager, and AWF board member who has been experimenting with a new plant variety called Lablab on approximately 200 acres of his family's Cedar Creek Lodge in Dallas County, Alabama. As he explains below in his own words, this warm-season legume from Africa that has been extensively used for cattle grazing in Australia, can be grown as a summer planting for Alabama deer.

Lablab is a drought-resistant, annual, warm-season legume that is similar to soybeans, but looks like a vine and withstands heavy browsing from deer. Deer eat the leaves of the plant, which have about a 22 percent protein content. Because of the nutritional value of Lablab, it may be one planting that helps keep deer in good condition through most of the summer. We plant Lablab in late June by broadcasting seeds at 35 to 45 pounds per acre on a well-prepared seedbed. Depending on the results of our soil tests, we use 13-13-13 fertilizer at an average rate of 300 pounds per acre and 2 tons of lime per acre. Usually we only lime our fields about every three years. Once deer become accustomed to eating Lablab they will consume it consistently. Unlike, soybeans or cowpeas, Lablab will continue to grow after it has been browsed close to the ground. Depending on the deer pressure, we usually get from six to eight weeks of deer utilization from our Lablab plantings. In some of our fields, we also grow a mixture of Lablab, cowpeas, and soybeans. In fields where grass and weeds become a problem, we incorporate a pre-emergent herbicide such as Trelan® at a rate of 1½ pints per acre.

SUPPLEMENTAL FEEDING

There have been several commercial feed rations developed specifically for deer. Usually these feed mixtures can be purchased at the local feed and seed store or from specialty sportsmen's outlets. The rations are usually in pelleted form and have been specially formulated to supply high levels of protein, minerals, and vitamins to deer. Soybeans and a mixture of soybeans and corn can also be used. The main drawback is that ration feeds and other forms of supplemental feeds for deer are expensive, especially if placed in feeding troughs distributed over large areas. In most cases, the enhancement of native deer forage through habitat improvement practices and the planting of supplemental food plots are more cost-effective methods of providing high quality deer food. If ration feeds are used, however, they should be made available to deer from the end of deer season until the acorns begin to fall, and remember it is illegal to hunt over feed in Alabama. All supplemental feeds must be removed prior to the hunting season, and all traces

Ideal Cool-Season Planting Mixtures For Deer

Small Grain and Clover Combinations

Mix 1	Mix 2
2 bushels wheat	1 bushel wheat
1 bushel oats	1 bushel grain rye
5 pounds crimson clover	1 bushel oats
7 pounds red clover	5 pounds crimson clover

Mix 3	Mix 4
2 bushels grain rye	15 pounds red clover
5 pounds ladino clover	10 pounds crimson clover

The amount in these mixtures is for 1 acre of food plot. (Source: Alabama Cooperative Extension Service)

of the feed must be gone 10 days prior to hunting if you are to hunt on or around areas that have had feed stations in place. Before initiating a supplemental feeding program, it is wise to check with the local Conservation Enforcement Officer for a clear interpretation of the use and management of supplemental feeding sites.

THE USE OF SALT OR MINERAL BLOCKS

Deer use salt or fortified mineral blocks (that contain additional nutrients and minerals) when they are available, especially in late spring and early summer. However, on most soil types across the state, additional salt and other minerals are usually not needed. In fact, research at Auburn University has shown that well-fed deer receive no apparent benefits in increased body weights or antler size from mineral and vitamin supplementation, but on nutrient deficient or depleted soils, the addition of mineral blocks could increase mineral availability to deer, and in some situations improve antler growth. The most important minerals for antler growth and development are calcium and phosphorus, as opposed to salt (sodium chloride). Salt should comprise no more than 35 percent of a mineral block, the remainder should contain calcium, phosphorus, and other trace elements and vitamins.

Salt blocks alone provide little benefit for deer even though the blocks are an attractant. If salt blocks are going to be put out for deer, fortified mineral blocks should be used rather than the unfortified blocks used by many cattle growers. Another option is using commercially prepared granular mineral supplements that are specifically formulated for deer. Deer should not have to travel more than ½ mile to mineral stations, which should be placed on the ground and well away from food plots, roads, agricultural fields, and other places where they can easily be seen by people. Granular mineral mixes can be placed in a dirt hole or in a covered feeding trough in an area that is frequented by deer. Mineral blocks and granular mixtures that are kept out of the rain by placing them in covered troughs last longer than exposed mineral supplements. As a reminder, it is illegal in Alabama to shoot deer over mineral blocks; however, Alabama law does allow hunters to hunt deer over pure salt blocks that contain no other minerals. All fortified salt blocks should be taken up at least 10 days prior to hunting.

TES RANDLE JOLLY

One method to provide additional salt for deer is to incorporate it with dirt using a shovel.

Seasonal Deer Forage Plantings in Alabama

Plant[1] Species	Planting Date	Seeding Rate[2] Pounds/Acre	Best Sites	Planting[3] $ Cost/Acre	Forage $ Cost/Ton	Production Total Pounds/Acre
Oct.–Mar.						
Wheat	Sept. 1–Nov. 15	90	Loams to clays.	89	45	4,027
Rye	Sept. 1–Nov. 15	85	Well-drained sandy to clay loam soils.	93	52	3,544
Oats	Sept. 1–Oct. 31	80	Sandy loam to clay loam soils. Subject to winter kill.	90	65	2,805
Jan.–Apr.						
Ryegrass	Aug. 1–Oct 15	25	Best on clay loam soils.	88	64	2,782
Crimson clover	Sept. 1–Oct. 31	20	Best on well-drained soils. Tolerates lower pH (5–6.5).	53	62	1,715
May–June						
Ladino clover	Sept. 1–Nov. 15	3	Moist, but not wet, bottoms and fertile uplands.	47	34	2,755
June–July						
Red clover	Sept. 1–	10	Well-drained, fertile soils.	56	28	4,008
July–Aug.						
American jointvetch	May 1–July 31	30	Poorly drained or wet soils. Drought sensitive.	103	244	851
Aug.–Sept.						
Grain sorghum	Apr. 15–July 15	20	Well drained productive soil.	48	—	—
July–Sept.						
Davis soybean	May 1–June 30	100	Best on bottomland, deep loam. Subject to overgrazing.	61	37	3,298
Quail Haven soybean	May 1–June 30	100	Best on bottomland, deep loam.	68	59	2,341

[1]Plants listed are high forage producers and are heavily utilized by deer during the period listed. All listed plants are adapted statewide.

[2]Rates listed are for broadcast application as recommended by the Alabama Cooperative Extension System.

[3]Costs are 1994 figures that include seed, fertilizer, lime, and use of equipment. Forage costs are based on total annual production.

For more information on these and other plants for deer, see Appendix D–J.

Supplemental Feeding Using Soybeans

Dr. Keith Causey, Ireland Professor of Wildlife Sciences at Auburn University, has studied Alabama white-tailed deer for most of his career. In this interview Dr. Causey discusses deer herd management implications from a supplemental soybean feeding program during the 1996–1997 hunting season on a 3,300-acre hunting club in the Lower Coastal Plain in southwest Alabama.

A total of 110 deer were harvested during the 1996–1997 hunting season on an area where we incorporated supplemental soybean feeding. Twenty-one males and 89 antlerless deer were taken. Hunters did a very good job avoiding yearling and two-year-old males with only six yearlings and two two-year-olds taken. Likely the six yearlings were mistaken for antlerless deer. Thirteen of the males that were harvested were more than three years old.

After comparing the 1996–1997 harvest data to the previous year's (1995–1996) harvest information, some interesting facts surface. Most striking are the increases in average body size among several of the age and sex classes of harvested deer.

Mature bucks (three years and older) increased in body size 15 pounds on average. Average weights for 1996–1997 were 171 pounds with a range from 147 to 214 pounds. In 1995–1996 this age-class averaged 157 pounds with a range in weight from 126 to 177 pounds.

Weights of yearling females increased 10 pounds on average as did the two-and-one-half-year age-class. Some three-year-old and older female weights increased an average of 14 pounds, with a range from 74 to 132 pounds. Male and female fawn body weights remained almost constant with only a slight increase over the previous year. Yearling male weights also showed only a 4-pound increase over the 1995–1996 weights. It seems logical to assume the increase in average weights in the older males and females over the past year was in response to the soybean feeding program.

Lack of response the second year among the fawns might logically be explained by the removal of the soybean food source prior to fawns reaching the size and age to access this high energy food source. Given the dispersal patterns of yearling male deer, one might assume that many of these young males might have been recent arrivals from adjacent habitats where soybeans were not available to them for most of their first 18 months. Of course this is only speculation.

Based upon what we have observed so far, I have recommended maintaining the current harvest strategy and continuing soybean supplementation through the 1997–1998 hunting season with careful monitoring of soybean consumption throughout the year. When we have collected three to four years of data, we will be able to fine-tune our management recommendations.

DEER HERD MANAGEMENT

A regulated deer harvest is an essential part of sound management for white-tailed deer. Maintaining harvest numbers is necessary to keep deer populations in balance with their food supply. Where food is abundant and deer are healthy, a sustained harvest maintains healthy conditions and prevents overpopulation. Sustained annual harvests can range from 30 percent to 40 percent of the total deer population in a healthy deer herd. In areas where deer are approaching overpopulation and food supplies are becoming critical, herd reduction is overdue and heavy harvests become necessary to prevent further damage to the habitat, to reverse declining deer quality, and to prevent increasing natural mortality or major die-offs. Where deer are too abundant they may become their own worst enemy, as well as the enemy of other wildlife, by over-browsing areas to the point of damaging habitat productivity and native plant communities. Studies comparing forest plant communities where deer were abundant, compared to adjacent areas where they had been excluded for over 20 years, indicated a significant reduction in the number and diversity of ground and midstory vegetation as well as the vertical structure (number of vegetative layers from the ground to the dominant tree canopy) in stands where deer were abundant. Deer must be maintained below the carrying capacity of the land to sustain healthy and productive habitats for deer and other wildlife.

Regulated either-sex harvests are necessary for proper herd management where deer have become well established. Buck hunting alone cannot control a growing population. Harvesting deer of both sexes will not extirpate them any more than it will quail, squirrels, or other game species, provided the harvest is regulated and does not exceed the annual production. When few deer are lost to causes other than legal hunting, a reasonable harvest of both bucks and does assure a healthy population for the future. At a minimum, harvest rates for antlerless deer should be one deer per 60 acres. Higher rates of antlerless deer should be harvested in overpopulated areas.

Many people believe that deer populations contain a high proportion of old barren does (animals that are too old to have young) and use that as a justification for harvesting antlerless deer. Antlerless deer harvests can be justified for a number of reasons, but this is not one of them. Does that have never produced fawns or have stopped producing entirely are almost nonexistent. Furthermore, when barren does are found, it is almost always a result of some physiological malfunction rather than age. Obviously, there is a limit to the age at which does remain productive, but very few individuals reach such an advanced age. The effect of these few individuals is essentially insignificant to the productivity of the population.

Inbreeding in wild deer herds often concerns deer enthusiasts, but is rarely a problem because of the genetic diversity and the number of animals found in most deer herds. In addition, individual deer movements in and out of their home ranges insures sufficient interaction with a variety of other deer. Theoretically, inbreeding could occur where deer herds become isolated over a long period of time, such as on an island or in an enclosure. Certain physical traits that are often rare under natural conditions, such as true albino deer (white coats and pink eyes), usually can increase in frequency by inbreeding within a population.

Proper management of deer herds is accomplished by regulating harvests to keep deer populations in balance with their food supplies and manipulating habitats through various land management practices to make an area more favorable to deer. Another factor that must be considered, however, is the compatibility of deer populations with commercial agriculture, forestry, and other land-use interests. Problems often arise when adjoining landowners have competing or conflicting interests and objectives. It should be a realistic objective in deer management to annually produce and harvest an optimum number of healthy animals that keeps the population balanced at or below a level compatible with other uses of the land. This is often called the **social carrying capacity** for deer.

Effects of an Over-Populated Deer Herd

- Choice deer foods are eliminated.
- Decreased fawn crops.
- Mortality is higher, especially among fawns and older deer.
- Average weights, especially of the one-and-one-half-year age class, decrease.
- Bucks have smaller antlers.
- Harvestable bucks make up a smaller percentage of the deer herd.
- In agricultural areas, crop damage is more serious.
- Parasites and diseases of deer are more prevalent.
- Newly regenerated forest stands are heavily browsed.
- Nurseries and other fertilized plants are heavily browsed.

How Many Deer to Harvest?

Most wildlife biologists agree that once a deer herd is established, at least one-third of the fall population must be harvested each year to prevent overpopulation and a decline in deer quality. Research has shown that slightly more than one-third of the fall population can be harvested each year with no apparent effect on the following year's population. Landowners should rely on the advice of trained biologists to estimate local deer populations and to recommend a harvest rate that would approximate a one-third herd reduction. In most cases, equal numbers of both sexes should be removed during the hunting season. In situations where the sex ratio is skewed heavily in favor of does, harvests should include a greater portion of does than bucks.

DETERMINING THE HEALTH OF A DEER HERD

Many landowners and sportsmen want to know how big a deer should be at a certain age. In addition, they want to know how many deer they should harvest. Since it is difficult to calculate the total number of deer in the same manner as you would calculate the number of trees, biologists look at the physical condition of the deer at a certain age. The best age class to determine herd health is the one-and-one-half-year-old class, because this age category represents deer that are still growing, making them more sensitive to changes in food supplies and other environmental factors. On most sites in Alabama, one-and-one-half-year-old bucks should average a live weight (before field dressing) of 100 to 110 pounds, while one-and-one-half-year-old does should weigh between 80 and 100 pounds. If average weights of one-and-one-half-year-old deer are lower than this, there may be too many deer and not enough quality food. In this situation there are three options: 1) reduce the number of deer, 2) increase the food supply, or 3) a combination of both. Biologists with the ADCNR can examine harvest records and formulate recommendations for landowners.

NANCY WEBB

One method of estimating deer densities on a tract of land is to conduct nightly spotlight counts. Local Alabama Department of Conservation and Natural Resource conservation enforcement officers should be contacted in advance since spotlighting from public roads is illegal in Alabama.

ESTIMATING DEER POPULATION TRENDS

Biologists use several techniques to estimate the number of deer on a given tract of land. The following are four different techniques that have been used in the past to give an index to deer density.

1. **Night Spotlight Counts.** A route is established that can be driven safely at night and that covers all the major habitats on a farm or forest tract. The route should begin shortly after dark and be driven no faster than 8 miles an hour. When a deer is spotted, the age (fawn, yearling, adult) and sex of the animal should be recorded if possible. The route should be run at least three to four times on successive nights to get an average number of deer sighted. An index is then calculated by dividing the average number of animals seen on all nights by the acreage or mileage covered. For example, if an average of 60 deer were counted on a 10-mile route, then the index of the deer population would be six deer per mile. This method gives an index in a relatively short period of time. The disadvantages are that it is labor intensive; it requires a fairly extensive road system; and at least three people are needed to conduct the counts, one to drive and two to observe either side. In Alabama, spotlighting from public roads is prohibited by law.

2. **Fecal Pellet Counts.** Establish several permanent transects (imaginary lines) that bisect a farm or forested tract. Along each of these lines, establish 10 to 40 permanent sampling plots (12 x 50 feet). Mark these plots so they can be visited year after year. Within each plot remove all deer fecal pellets. Recheck the plots after two weeks and count the number of deer pellet groups. The index of the deer population will be the number of pellet groups per transect. For example, if three transects were established on the property with 10 stations per transect and 5 pellet groups were on transect 1, 3 on transect 2, and 10 on transect 3, the index to the population would be an average value of six pellet groups per transect (18 divided by 3). This number can be compared to other fecal pellet counts to assess relative deer abundance. The disadvantages of this technique are that it is labor intensive and fecal pellets of deer in the Southeast deteriorate rapidly, making the index less accurate.

3. **Track Counts.** Disk a long, narrow strip of ground 4 to 9 feet wide during August. The next morning count the number of deer that walked through the strip. It is important to follow the tracks of each animal as it entered and left the strip to ensure that each deer is counted only once. Repeat the procedure at least three times to ensure that counts were not made on a day when the deer were not moving. Try to separate the number of fawn tracks from adult tracks, using a size difference, to get a percentage of fawns produced. The index is calculated by counting the number of deer crossing each linear mile. This roughly estimates the number of deer per square mile. The disadvantages of this technique are that it is labor intensive; it depends upon deer movement; and it is invalidated if rainy weather washes tracks away.

4. **Biological Monitoring of the Harvest.** This technique does not give a direct index of population levels, but it is very useful in monitoring the health of deer herds, which is related to population density. The three most common measurements taken are yearling (one-and-one-half-year-

old age class) buck and doe field-dressed (without internal organs) weights; main antler beam diameter of yearling bucks; and lactation (presence of milk) rates of yearling does. These three measurements are good indicators of the condition of a deer herd. Slight shifts from year to year can indicate subtle changes, but this should not be a concern. If these measurements steadily decline, this would indicate that there are too many deer and habitat quality is in decline. An increasing deer herd will reach a point at which the nutrition (food supply and quality) is inadequate to support healthy animals. During this time biologists will also notice that most preferred deer foods are absent, and a heavy browse line appears on other vegetation not usually preferred by deer. Average or normal yearling field-dressed body weights, buck beam diameters, and lactation rates for specific areas of Alabama can be obtained from the regional ADCNR biologist.

QUALITY DEER MANAGEMENT

Quality deer management (QDM) is an approach that has become popular with many hunting clubs, landowners, and other deer enthusiasts in the Southeast who are interested in increasing antler quality and improving the overall health of a deer herd. Fundamental to QDM is the voluntary use of restraint in harvesting young bucks combined with an adequate harvest of antlerless deer to maintain a healthy population in balance with existing habitat conditions. QDM allows bucks to reach an older age class, increasing the potential for quality antlers. QDM strives to maintain a balanced buck-to-doe ratio that approaches one to one. This level of deer management involves the production of quality deer (bucks, does, and fawns), quality habitat, and quality hunting.

Involvement in QDM extends the role of hunters from mere consumers to that of active managers. Guidelines are formulated according to desires, goals, and limitations. Participating hunters enjoy tangible and intangible benefits from following the guidelines that they establish together. Pleasure can be derived from each hunting experience whether or not a shot is fired. What is important is the ultimate chance to harvest a quality buck. When a quality buck is taken, the pride is shared by all members of the hunting club because they helped produce the deer by allowing the animal to progress into an older age class represented by individual deer with larger bodies and antlers.

The success of quality deer management involves the following:

- **Goals and Objectives.** Establish clear and obtainable goals and objectives.
- **Professional Assistance.** Obtain the assistance of an ADCNR biologist or other professional wildlife biologist for recommendations on deer harvest strategies.
- **Harvest Records.** Maintain harvest records. They profile a deer herd over time. Harvest records are the basis for making recommendations and evaluating the success of a management approach.
- **Age.** Don't harvest young bucks; let them pass and mature into older age classes, increasing the potential for quality antler development. Remember, every time a trigger is pulled a management decision is made.
- **Nutrition.** Maintain and enhance the nutritional content of deer foods through habitat management and supplemental plantings.
- **Balanced Sex Ratio.** Keep the sex ratio at one male to three or fewer females with adequate antlerless harvests.

"Quality deer management is first and foremost an attitude—a means of expression. The hunter views a deer not as a resource for recreation and food, but as a part of nature to which he or she willingly belongs. The self-imposed restriction of taking antlerless deer while allowing young antlered bucks to pass provides the hunter opportunities to study deer, learn their behaviors, and sharpen hunting skills. The success of the hunt is no longer measured merely as a bagged animal."

— Dr. David C. Guynn, Jr., Founding Board Member, Quality Deer Management Association

RECORD-KEEPING

Detailed and accurate record-keeping is essential to a successful deer management program. Records should document the date and sex of each deer harvested along with biological data on age, weight, antler development, and reproductive condition. Reproductive information simply involves determining if harvested does are lactating (producing milk). The percentage of does that are lactating indicates that a doe has produced at least one fawn and gives an overall snapshot of the reproductive success of a deer herd for a particular year.

Data from the yearling (one-and-one-half-year) age class provides the most reliable indicator of a deer herd's condition and health. Deer are aged by the replacement pattern and wear of their teeth. Accurate determination of age, therefore, is essential to obtaining useful records. A mandible (lower jawbone) should be removed and properly labeled from each animal harvested. Labeled jawbones and the corresponding biological information

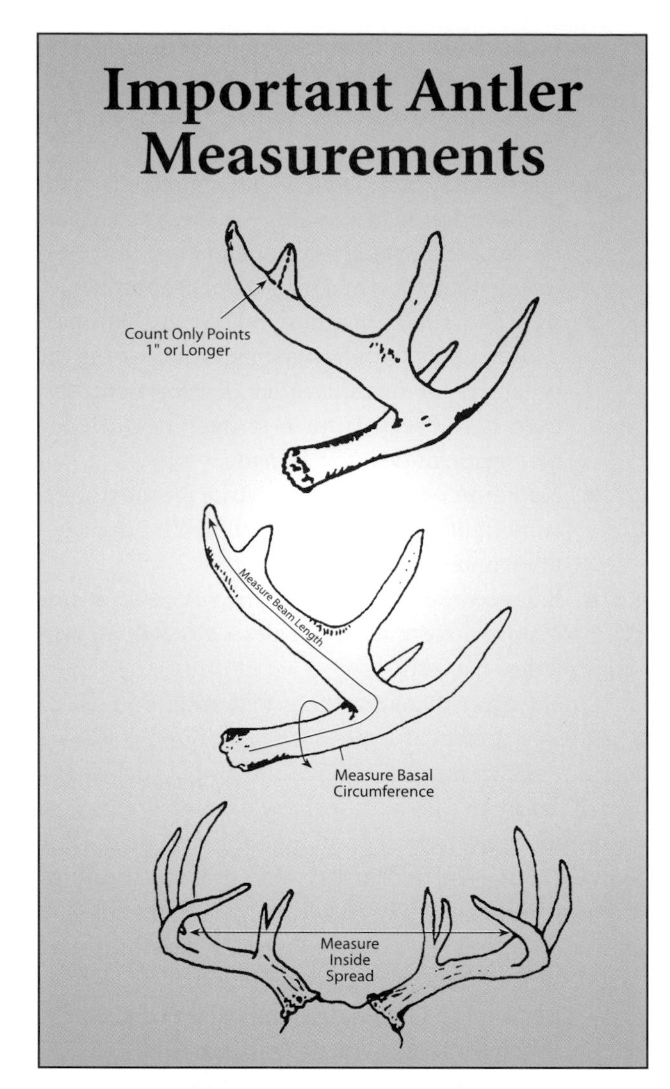

Fig. 5: Measuring and recording antler characteristics from harvested bucks helps to determine trends and changes in antler development and quality from year to year. Important measurements are basal circumference, inside spread, main beam length, and number of points.

Inside antler spread is an important indicator of antler quality. Many hunting clubs interested in quality deer management use an antler spread of "outside the ears" as a minimum guideline for harvesting bucks.

from harvested deer should be given to a wildlife biologist for aging and interpretation.

Knowledge of trends in annual harvests, the biological condition of harvested deer, and occurrences of parasites is necessary. In addition, population estimates and browse-use surveys provide biologists and land managers with a sound basis for making future deer harvest recommendations. This approach helps to assure the maintenance of quality habitat and a healthy deer herd.

After records have been kept for at least three years, trends in recorded measurements should be examined. During the first three years, average live weights of both sexes in the various age classes (one-and-one-half-years, two-and-one-half-years, three-and-one-half-years, etc.) will either remain the same, increase, or decrease. If they have increased, managers should continue harvesting the same number of deer each year for the next three to five years, insuring that adequate numbers of both males and females are harvested. If the average live weights of both sexes in various age classes remains the same or decreases, managers should increase the yearly harvest by 10 percent for the next three to five years. Sportsmen and landowners should continue adjusting harvest rates until the average live weight of one-and-one-half-year-old bucks is between 100 and 120 pounds and that of one-and-one-half-year-old does between 80 and 110 pounds.

MEASURING AND RECORDING ANTLER CHARACTERISTICS

Measuring and recording antler characteristics helps to determine changes in antler development and quality from year to year. Knowing which antler characteristics to measure as well as how is important. A flexible cloth measuring tape (the type used for sewing) should be used for measuring antlers. All measurements should be recorded to the nearest ⅛ of an inch. The following are important antler measurements that should be recorded.

- **Basal circumference** is the distance around the base of one of the antlers. The base is the area immediately above the hair line.
- **Inside spread** is the longest inside distance between the two main antler beams.
- **Main beam length** is the distance from the base to the tip of the longest main antler beam.
- **Number of points** is the number of tines (points) 1 inch and longer branching off both main beams.

HOW TO AGE DEER JAW BONES

Deer are aged by examining teeth on one side of the lower jaw (mandible) and looking at the pattern of tooth wear and replacement. The best approach is to remove the lower jaw bone using a jaw spreader/remover and shears. Jaw bones can be removed without ruining the deer for mounting. After one side of the lower jaw has been removed, the teeth can be examined and aged according to the guidelines below. Deer are grouped in one-year increments beginning at one half year of age. Reliability of this method decreases with increasing age of the deer.

Fawn Or One-Half Year

Less than six teeth in the lower jaw bone. Body size should also give an indication that the animal is a fawn.

Yearling Or One-And-One-Half Years

Six lower jaw teeth. The first three milk teeth, which are generally whiter in coloration, still intact. Third tooth has three cusps on the top of the tooth.

Two-And-One-Half Years

First three milk teeth have been replaced with permanent two-cuspid teeth. The cusps of the fourth tooth are sharp and the enamel (white) is wider than the dentine (brown). Little or no wear.

Three-And-One-Half Years

The fourth tooth is showing wear and the dentine is as wide or wider than the enamel. In the fifth tooth, the enamel is wider than the dentine.

Four-And-One-Half Years

The dentine is wider than the enamel in both the fourth and fifth teeth. On the sixth tooth, the enamel is wider than the dentine.

Five-And-One-Half Years

The dentine is wider than the enamel on all six teeth. The fourth tooth is showing more wear than all the other teeth. Only a small amount of enamel remains in the center of the fourth tooth.

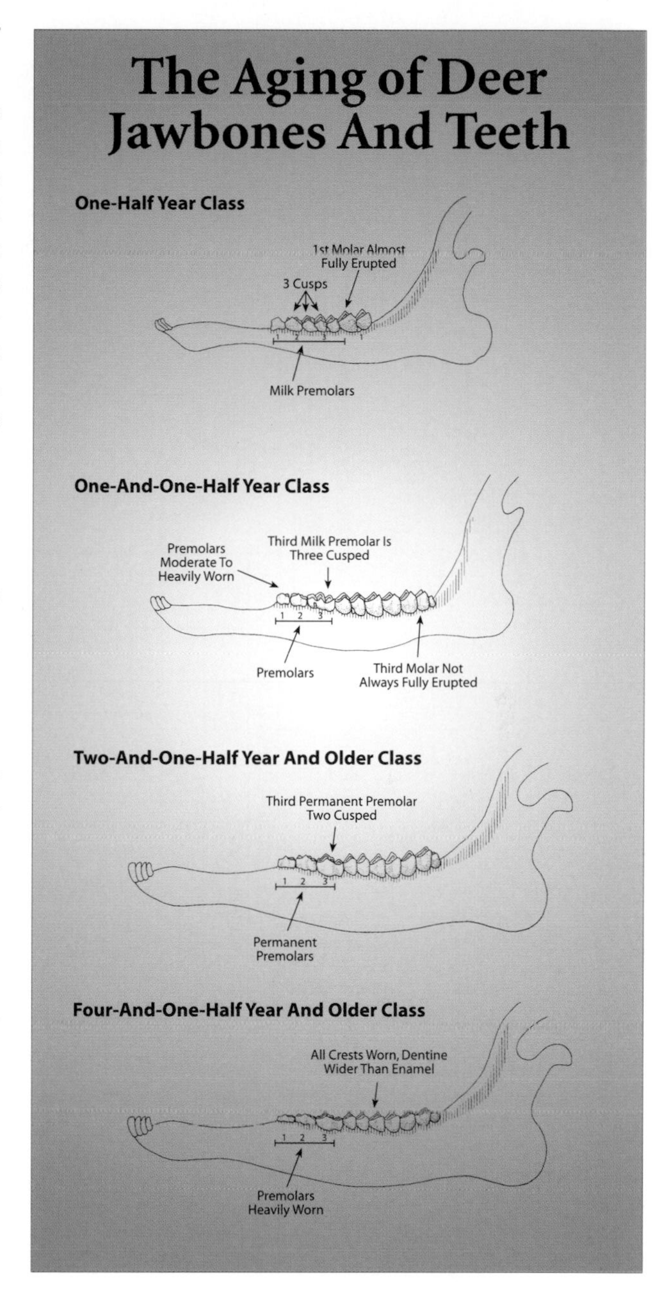

Fig. 6: The pattern of tooth wear and replacement from lower jawbones is a method used by biologists to estimate ages of deer.

Six-And-One-Half Years And Older

In the fourth tooth, all the enamel is completely gone. The fifth and sixth teeth have only a small amount of enamel left. Deer over six and a half years old will have the fourth, fifth, and sixth teeth worn completely smooth, almost down to the jawbone.

ALABAMA COOPERATIVE DEER MANAGEMENT ASSISTANCE PROGRAM - DEER HARVEST REPORT

CLUB OR LANDOWNER: ______________________________

COUNTY(S): ______________________________

HUNTING SEASON: ______________________________

FOR GAME & FISH DIVISION USE ONLY

Data File Name ____________________

DEADLINE: DATA SHEETS MUST BE RETURNED BY MARCH 31 TO AVOID LATE FEE PENALTY.

Type Hunting	Buck Harvest	Doe Harvest
Archery	______	______
Firearms	______	______
Subtotal	______	______
Total	______	

Complete this table before returning forms to the DMP Biologist. Include all deer taken on the property.

DEER NUMBER	JAWBONE MARKED	FOR OFFICE USE ONLY	HARVEST DATE		SEX	WEIGHT		MILK PRESENT	ANTLER MEASUREMENTS						TAG #	GUN OR BOW	COMMENTS/ HUNTER NAME
									# POINTS	CIRCUMFERENCE (inches)		LENGTH (inches)		INSIDE SPREAD			
			Month	Day	MALE/ FEMALE	LIVE	DRESSED	Y/N		LEFT	RIGHT	LEFT	RIGHT	(inches)			
1	√	EXAMPLE	11	25	M	176			8	3 3/4	4 1/8	18 1/2	19	15 5/8		B	F. BROWN
2	√	EXAMPLE	12	1	F		91	Y							41097	G	J. SMITH
1																	
2																	
3																	
4																	
5																	
6																	
7																	
8																	
9																	
10																	

Fig. 7: The Alabama Department of Conservation and Natural Resources can provide landowners and sportsmen with forms to record deer harvest information as part of the department's Cooperative Deer Management Assistance Program.

Deer Management Considerations

Dr. Harry Jacobson, formerly a Professor of Wildlife at Mississippi State University, has conducted research on free-ranging and captive white-tailed deer for over 25 years. He is well-known for his research that has led to a better understanding of the biology and ecology of deer. His work instigated improvements in deer management across the South.

Before attempting to manage deer populations, landowners need to identify their specific objectives. For example, are objectives focused on managing for a healthy deer herd, a certain number of days of recreation, or for maximum harvest? Once objectives are identified, a management plan can be developed for improving habitat quality and outlining deer harvest strategies. When developing a deer management plan, landowners and managers should remember that deer are a product of their environment: given the proper distribution of food and cover, they will thrive. Deer herds must be in balance with their habitat to reach their potential in body weight, antler development, and reproduction.

To achieve most deer management objectives there must be a large enough area to work with, at least a block of 1,000 acres, but preferably 2,000 acres or more. If individual land ownerships are small in size, larger management units can be created by forming associations or cooperatives with adjoining neighbors.

Nutrition, age, and genetics are three factors that determine the health and quality of deer. Habitats that provide an abundance of high quality foods year-round are important. Harvest strategies to keep deer numbers in balance with their habitat and allow bucks to reach older age classes increase the chance of deer reaching their genetic potential in weight and antler development. In most areas of the South, managers should not be concerned with genetics as a factor in the deer management equation. Either you have the genetics or you don't. There is little you can do if don't have the genetic potential of other areas; be satisfied with what you have and try to manage deer to meet the genetic potential in a given area. Managers should not concern themselves with genetic introductions. However, with all of this said we need to recognize the importance that genetics plays in the expression of certain antler characteristics. Genetics does play a final role in the type of antler traits a buck will ultimately exhibit and will have bearing on trophy scores with some scoring systems.

Immunocontraceptives is one population control method currently being researched. It is hard to tell what the future will bring to this approach to controlling nuisance deer populations. In my opinion, trying to control populations by altering a wild animal's physiology is unacceptable because it changes the conditions under which those species have evolved. Some species like the white-tailed deer evolved under predation and hunting that makes them what they are today. By changing the animal you take out the evolutionary selection factor—creating something other than a white-tailed deer. When we start manipulating the physiological processes of a wild animal we are in effect domesticating that animal.

Our studies here at Mississippi State University indicate that sterilization creates a tremendous bioenergetics drain on individual bucks because it prolongs the rut. Therefore, whatever gains are made in population control from sterilization may be offset by additional energy demands that may increase food consumption and ultimately depredation to agricultural crops and other plantings. Sterilized deer undergo more energetic stress compared to deer that breed naturally in the wild.

As a final comment it is important to understand that records and record-keeping are vital in deer management. Unless you can look back and see the effects of a management decision, in terms of change in condition of an animal or population, then you are really not managing but simply shooting in the dark.

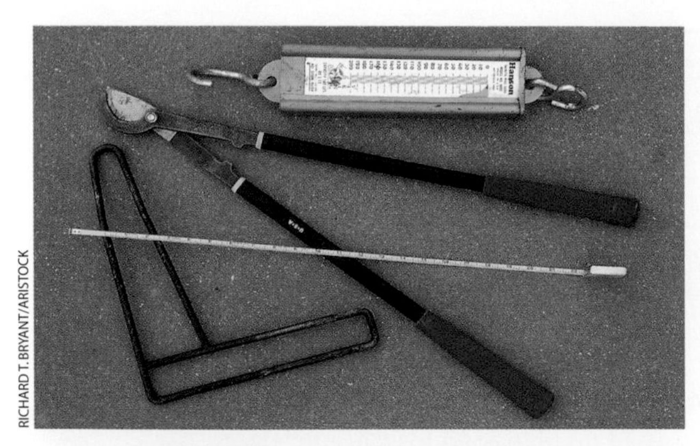

RICHARD T. BRYANT/ARISTOCK

Important instruments and accessories for collecting information from harvested deer include a hanging scale to record weight, a triangular bar to remove lower jaw bones for aging, shears for cutting jawbones, and a measuring tape. Some of these items can be obtained from a local hardware store or by ordering directly from Forestry Suppliers, Inc., 800-354-3565.

ALABAMA'S DEER MANAGEMENT ASSISTANCE PROGRAM

The Game and Fish Division of the Alabama Department of Conservation and Natural Resources provides technical assistance to landowners and hunting clubs who request help in managing their deer herds. For a small fee the ADCNR provides assistance through the Alabama Cooperative Deer Management Program, or simply the "Deer Management Program" (DMP) as it is commonly called. Alabama's DMP has over 1,900 participating cooperators (hunting clubs and landowners) managing more than 3.7 million acres. Most DMP participants are concentrated in the Black Belt Region of the state.

Landowners and hunting clubs that participate in the program are required to keep accurate harvest records and collect a lower jawbone from every deer harvested during each hunting season. The harvest records contain information such as date harvested, antler development, live weight (with nothing removed), and the sex of each deer collected.

Both the harvest records and the lower jawbones are numbered consecutively, and the number placed on a jawbone must be the same as the number on the harvest record of the deer from which the jawbone was taken. Landowners and members of hunting clubs must remember that a jawbone without a corresponding harvest record is of little value. Likewise, a harvest record without a lower jawbone is of even less value.

After the hunting season, ADCNR biologists determine the age of each deer by examining the jawbones. Ages are then recorded on the harvest record form. Harvest records of deer in the various age classes are then analyzed, with bucks considered separately from does. After evaluating harvest records for the various age classes, the ADCNR biologists make specific recommendations to the landowners or to the hunting club regarding management of deer herds on their particular lands.

The ADCNR recommendations are based on the stated objectives of the landowner or hunting club. For instance, a landowner who wants to produce quality bucks gets an entirely different set of recommendations than a landowner who merely wants to protect his or her land from becoming overpopulated with deer. These landowners would receive differing sets of recommendations even though the deer densities and the habitat conditions on the two tracts of land may be almost identical. The following four categories are deer management options that are available through the ADCNR Deer Management Assistance Program:

1. **Present Deer Density and Physical Condition.** This plan prevents deterioration of the current condition of deer.
2. **Higher Deer Density.** This plan increases deer per unit of range. Under this option physical condition of deer may become below average.
3. **Better Antler Development and Physical Condition.** Average to good antler development is a focus.
4. **Trophy Deer.** Low deer density with a good portion of heavier antlered bucks in above average physical condition is the goal.

The state's deer management program has operated successfully since 1984 and is getting more popular each year. It provides a valuable service to people who are interested in improving the quality of deer herds through management. The program is evolving as new information is gained and some changes can be expected in the future as DCNR Game and Fish Division Biologists and managers integrate this information with the desires of the hunting public. Landowners and hunting clubs who are interested in managing their deer herds should contact their local ADCNR biologist or ADCNR, Division of Game and Fish, 64 North Union Street, Montgomery, AL 36130, (334) 242-3469.

CROP AND ORNAMENTAL DAMAGE BY DEER

In many areas where deer have become too numerous, agricultural crop damage and damage to ornamental plantings is commonplace. In these situations, sportsmen should be sensitive to the damage that deer are causing landowners and make concerted efforts to lower deer numbers through hunting. This is the most feasible long-term solution to reducing deer damage. For short-term relief of deer damage, forest, farm, and homeowners should

Alabama Whitetail Records

Each year the Alabama Whitetail Records, in conjunction with the Alabama Wildlife Federation, sponsor a big buck recognition program. Deer antlers are scored by an ADCNR biologist or a certified Boone and Crockett scorer. High-scoring racks are then entered into the Alabama Whitetail Records Book. For more information about the annual big buck program, contact the Alabama Wildlife Federation or call Dennis Campbell at (205) 822-3383. For a copy of the Alabama Whitetail Records Book write Campbell's Publishing, P.O. Box 31072, Birmingham, AL 35231.

consider temporary electric fences. Two designs that have been tested by Auburn University researchers and proven to be effective are the 5-wire Penn State design and the poly-tape electric. The effectiveness of fencing designs can be increased with slight modifications. For example, sometimes deer jump through poly-tape fences without getting shocked. Peanut butter placed directly on poly-tape can cause deer to smell and lick electrified poly-tapes and receive a shock. Other options that have been used with varying degrees of success include repellents, frightening devices, and a host of homemade remedies. For more information about methods to reduce deer damage, contact the Extension Wildlife Damage Specialist, Department of Zoology & Wildlife Science, 331 Funchess Hall, Auburn University, AL 36849-5643, (334) 844-9233.

QUALITY DEER MANAGEMENT ASSOCIATION

The Quality Deer Management Association (QDMA) is a non-profit organization devoted to promoting ethical hunting and the sound management of deer. Below are the general objectives of the QDMA:

- Improve the quality of deer herds and the hunting experience through sound management of buck/doe ratios, buck age structure, and deer densities that are compatible with habitat conditions and the land-use objectives of landowners;
- Promote hunter education through seminars, workshops, demonstrations, and the publication of a journal, *Quality Whitetails;*
- Enhance the public image of deer hunters and deer hunting by providing a code of conduct for QDMA members to follow; and
- Promote and financially support deer research projects.

For more information about QDMA, write the Quality Deer Management Association, P.O. Box 707, Walterboro, SC 29488, or call 1-800-209-DEER (3337).

DR. H. S. BANTON

Soybeans, considered an "ice-cream food" by deer, are one of the most severely damaged crops in Alabama and the Southeast.

SUMMARY

Game species are an important recreational and economic resource in Alabama. Each year thousands of Alabamians head to the field to hunt game. This activity generates approximately $1.2 billion in the state. Hunting also contributes significantly to the economy of local communities and to landowners who lease their lands for hunting. To make private lands more attractive for white-tailed deer, forest and farm owners should first have a clear understanding of the biology, life history, and habitat requirements of deer. This information is helpful when selecting and matching management practices that will provide the habitat needs of whitetails. Descriptions of deer, their habitat needs, and species-specific habitat improvement recommendations found in this chapter should provide a solid foundation for farm and forest owners who want to improve their lands for white-tailed deer.

The majestic wild turkey is king of the game birds in Alabama. From an estimated statewide population of 13,487 in 1941, wild turkey now number over 500,000 in Alabama, making them part of a remarkable wildlife management success story.

PAUL T. BROWN

CHAPTER 6

Managing for Wild Turkey

CHAPTER OVERVIEW

This chapter describes the biology, life history, and habitat requirements of the wild turkey in Alabama. Habitat enhancement practices that meet the needs of wild turkeys on Alabama forests and farms are throughly discussed. Special emphasis has been placed on integrating forest and farm management practices to enhance wild turkey habitat on private lands in Alabama. Landowners, managers, and sportsmen who are interested in the wild turkey and its management will find this chapter contains a wealth of information.

CHAPTER HIGHLIGHTS PAGE

SUCCESS OF TRAPPING AND RELOCATION 138
DESCRIPTION AND LIFE HISTORY 138
THE MATING SEASON 139
BROOD-REARING 142
THE FALL AND WINTER PERIOD 143
ANNUAL POPULATION LEVELS 143
FACTORS THAT LIMIT POPULATIONS 143
HABITAT REQUIREMENTS 148
FOODS AND FEEDING 149
HABITAT IMPROVEMENT PRACTICES 152
HUNTING 163
THE NATIONAL WILD TURKEY FEDERATION 163

By early explorers' accounts, wild turkeys (*Meleagris gallopavo*) were abundant in pre-colonial America. Mature forests dominated the landscape, with occasional openings created by natural events like wind storms and fires. Native Americans periodically burned the forests, fostering new growth. The wild turkey thrived during this period. Early settlers cleared small patches in the forest and grew crops for sustenance for their families and their livestock, with some impact on wild turkeys. These birds remained abundant in the state through the early 1800s. For example, Phillip Henry Gosse in 1859 in his *Letters from Alabama on Natural History* noted the enormous number of turkeys that were common along the Alabama River from Selma to Mobile. But in the latter half of the 1800s and early 1900s, as the human population grew, the impact on wild turkey habitat and populations increased dramatically. By 1900, the statewide turkey population had reached a critical state of decline. Wholesale cutting of the forests and indiscriminate and unregulated subsistence and market hunting took its toll. Gobblers and hens were hunted year-round, often over bait. Better access to Alabama forests and turkey flocks was provided by new roads and railroads. The wildlife inventory of 1941 showed an estimated statewide population of only 13,487 wild turkeys scattered over a range of less than 10,000 square miles. These remnant flocks were generally found on areas of limited human activity. At that time over three-fourths of Alabama's wild turkeys were located in the Lower Coastal Plain, mostly in the western portion on large land holdings and in major river bottoms with limited human intrusion. In 1941 only five counties (Baldwin, Choctaw, Clarke, Sumter, and Washington) were thought to have over 1,000 turkeys each.

Fortunately, better times were ahead for Alabama wild turkeys. Wildlife management got its real start when the Alabama State Legislature created a State Game and Fish Commission in 1907 for the purpose of administering and enforcing the first game laws. One of these initial regulations was a "gobbler law," the first of its kind in the U.S., which established the first open season on gobblers from December 1 to March 1, with a bag limit of two gobblers per day.

According to *The Wild Turkey in Alabama* (1948) "the enactment of the original game laws (in 1907) stimulated a number of owners of large tracts of land in Southwest Alabama to close their holdings to hunting and provide turkeys a limited amount of refuge. The bountiful supply of forest game that now exists in that portion of the state is largely due to the foresight and courage of Colonel E. F. Allison, Messers, Tony Slade, Howard Douglas, Captain Ed Powell, Fred T. Stimpson, and others who actively championed the cause of wildlife conservation during its darkest hour." The abundant turkey populations seen today in Alabama are a result of trapping turkeys from many of these areas, such as the Fred T. Stimpson Game Sanctuary, and releasing them to other parts of the state where turkeys numbers were low or non-existent. Funds from the Pittman-Robertson Act of 1937, which earmarked a federal tax on arms and ammunition, were used with matching state monies to aid in trapping and relocating turkeys.

Other factors also contributed to the return of the wild turkey in Alabama. After World War II, the rural human population decreased as many people left small farms for employment in the cities. The forests, which had been devastated in the early part of the century, were now maturing again. The establishment of the National Wild Turkey Federation also increased the awareness of the need for protection and management of the wild turkey, which has helped turkey flocks thrive.

SUCCESS OF TRAPPING AND RELOCATION

The combination of these factors began to dramatically increase wild turkey numbers in the state. By the mid 1970s an estimated 250,000 turkeys abounded statewide on about 31,000 square miles of habitat in all counties of Alabama, almost four times the number that occupied the same area in 1940. Today it's estimated that wild turkeys in Alabama number over 500,000 birds on over 35,000 square miles of habitat. Each year some 54,000 hunters harvest more than 43,600 turkeys, mostly during the spring season.

DESCRIPTION AND LIFE HISTORY

There are five subspecies of the wild turkey in the U.S., portions of southern Canada, and northern Mexico. The Eastern wild turkey (*Meleagris gallopavo silvestris*) is the native subspecies in the eastern U.S. and Alabama. In southern Alabama, it is generally thought that the Eastern wild turkey may have interbred with the Florida wild turkey (*Meleagris gallopavo osceola*).

The Eastern wild turkey is the largest game bird native to Alabama. Gobblers average about 17 pounds with some weighing 21 pounds or more, depending upon age and winter food supplies. Hens average between 8 and 11 pounds.

Alabama's abundant wild turkey population is a result of cooperative trapping and relocation efforts between the Alabama Department of Conservation and Natural Resources, wildlife conservation organizations, and private landowners. Right: Wild turkeys are placed in carrying containers for relocation by ADCNR personnel. Left: Cannon nets are used to trap wild turkeys for relocation in other parts of Alabama.

The gobblers have colorful caruncles, or bumps of skin, as well as a snood (an upward protrusion from the base of the beak) about the head and neck, which easily change colors. The greatest display of these color variations occurs during the spring mating season. The exposed feathers of the gobbler have iridescent qualities that cause several different colors to be seen as the sunlight reflects from different angles. Feathers of the hen have less iridescence. The breast feathers of the gobbler have a velvety black tip; while those of the hen are buff colored.

Gobblers have a "beard" which grows from the middle of the breast and varies in length, depending on age, but seldom exceeding 11 inches. Rarely, hens will also grow beards, but these are not as thick or as long as that of an adult gobbler. The "hairs" of turkey beards are actually modified feathers called filaments. Beard filaments on one- and two-year-old gobblers have amber tips.

Gobblers also have spurs on their legs, but hens do not. The length of the spur can be used as a rough aging method. Spurs less than ½ inch are from young gobblers called jakes or first-year gobblers. Spurs ½ to 1 inch long are from two year olds, spurs 1 to 1¼ inch long are from three year olds, and spurs longer than 1¼ inch are from four year olds and older birds.

THE MATING SEASON

The most complex period of an animal's life is the time surrounding its reproductive cycle. For the wild turkey, this period occurs from about March through June, although some related activities may begin as early as mid-February and late nesting attempts may take place well into the summer. During this critical period, there are many environmental and behavioral mechanisms affecting turkeys. Since wild turkeys flock together through the winter, the first events of the reproductive cycle begin while they are still in this loose association. When day length increases and the weather begins to get milder, internal mechanisms trigger changes in the behavior of both gobblers and hens. These behavioral changes intensify as the reproductive instinct increases.

The first noticeable behavioral changes occur in the gobblers. Mature males will begin to gobble several times

Physical characteristics distinguish the hen (left) from the gobbler (right). Gobblers have beards, colorful carnucles, spurs on the legs, and greater iridescence on the feathers when exposed to direct sunlight.

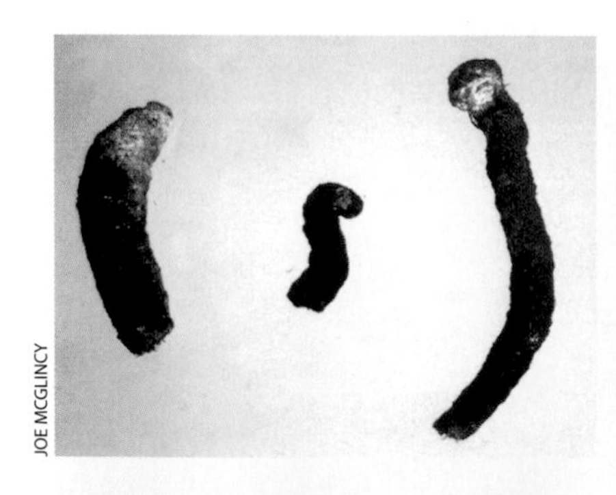

JOE MCGLINCY

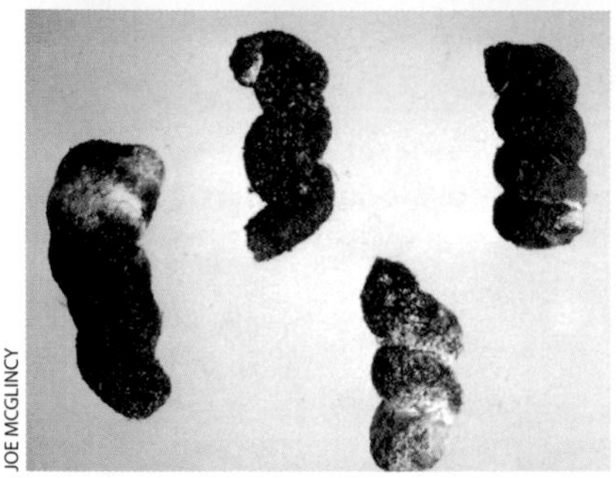

JOE MCGLINCY

A close look at turkey droppings reveals differences in the J-shaped gobbler droppings (top) and the spiraling mound-shaped droppings of hens (bottom).

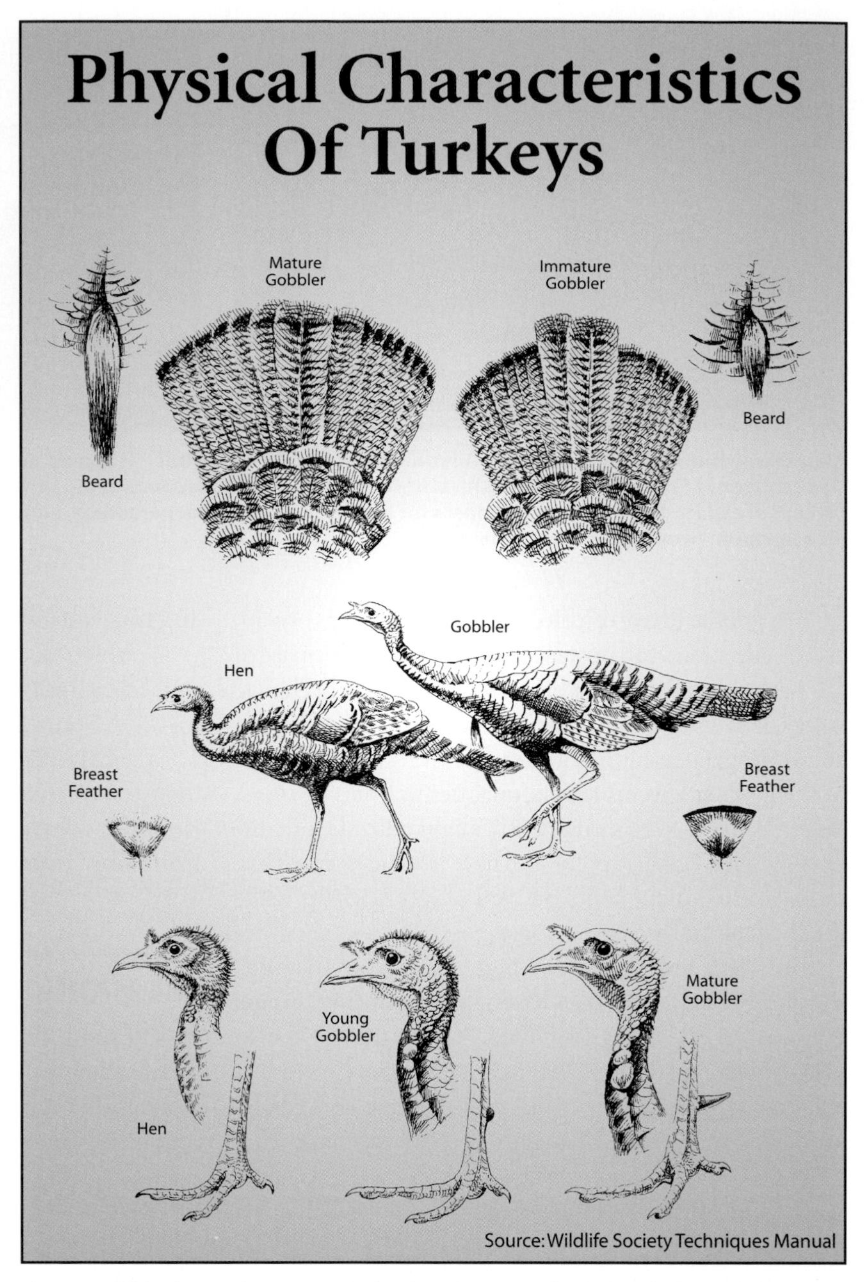

Fig. 1: Feather formation, type, and coloration can also help distinguish between mature and immature gobblers, as well as between hens and gobblers. Top: Beards of mature gobblers are well developed and tail feathers have completed growth. Middle: Breast feathers of gobblers have a velvety black tip, those of the hen are buff-colored. Bottom: Hen necks are covered more in modified feathers, gobblers' necks are bare. Hens lack spurs on their legs; mature gobbler spurs are longer and more developed than immature gobbler spurs.

before leaving their roosts at daylight. They will spend brief periods "strutting" about with tails raised and fanned, body feathers puffed out, and wings dragging the ground. While indulging in this type of display, they will make low "drumming" sounds and will occasionally gobble. Dominance among males is established by challenges to other males which may result in brief fights. The "boss" gobblers are established through this process, and they will begin to assert their presence toward the hens in the area before the winter flocks have broken up. Although turkeys become sexually mature at about one year of age, the young gobblers or "jakes" are more often spectators than participants in the task of servicing the hens, or actual mating.

Gobbling, strutting, drumming, and fighting among males increases as spring progresses. The antisocial behavior of the gobblers and the hens' increasing instinct to begin nesting causes the winter flock to partially disband. Breeding males are often accompanied by a harem of hens numbering as many as six or eight, but more often two or three. Several young gobblers often range together as the various contingents of the winter flock go their separate ways.

About three to four weeks after gobbling begins, the area used by the combined winter flock increases in size due to dispersal of individuals and smaller groups. Flocks that spent the winter in major hardwood bottoms now scatter into the mixed pine–hardwood uplands. Often a winter flock that may number 20 or

more birds uses no more than a 50-acre home range in which to roost, feed, and loaf during early March. After the flock breaks up into its breeding range, individuals and small groups may disperse over several miles. Flock break-up in Alabama usually occurs around the third week in March.

Egg laying usually begins in late March to early April. While hens are laying their clutches, averaging about 10 eggs per clutch, they occasionally visit an available gobbler, even though one mating would probably fertilize the entire clutch. About two weeks is required for the hen to lay her clutch of eggs. Egg laying varies, but most hens lay one egg each day, usually in the middle of the day. Incubation may begin irregularly when the clutch is nearly completed. The hen begins continuous incubation within a day or two after laying the final egg, leaving the nest only for brief periods to feed and drink water.

During the nesting phase, particularly when the hen roosts on the ground during incubation, the hen and her nest are particularly vulnerable to predation and disturbance. If the nest is predated or abandoned by the hen soon after the onset of incubation, renesting is likely. Hens that lose their nest later in the incubation period may go into a non-reproductive phase until the following spring. Landowners and managers should keep disturbances, such as silvicultural operations and other land-use activities, to a minimum during nesting to prevent possible abandonment of the nests by hens.

It requires about 28 days of incubation to hatch the clutch of eggs, and all fertile eggs normally hatch within a one-day period. Most hens begin incubation sometime during the first three weeks in April. Considering early nest failures that lead to additional nesting, the peak

JOE MCGLINCY

Wild turkeys remain in flocks throughout the winter. The beginning of the reproductive cycle begins in late winter while they are still in this loose association.

GARY W. GRIFFEN

Hens prefer to build nests adjacent to logs and stumps, or under low over-hanging vegetation. Egg-laying begins in March to early April and the eggs hatch in about 28 days. Young turkey poults are extremely vulnerable to predation. About ½ to ¾ of the newly hatched poults die before they reach two weeks of age. This hen is yelping or "imprinting" so that her poults will recognize her voice.

of the hatch appears to occur during the last half of May. Occasionally young poults have been observed in early April and as late as early September. Usually one half or more of the nests are unsuccessful due to predation or abandonment.

Soon after first nest incubation has peaked, the greatest gobbling intensity occurs, as gobblers can no longer easily locate a receptive hen. Peak gobbling intensity and duration occurs in April with the exact dates varying depending on weather, although some sporadic gobbling may continue well into June in some years. Late gobbling is often associated with unsuccessful hens returning to gobblers for mating before attempting to start a new nest. Gobblers may continue gobbling for several weeks after the last hen has gone to her nest.

BROOD-REARING

The first poults to appear may hatch several hours before the eggs that were laid last. If poults attempt to leave the nest before the hen is ready, she uses her head and neck to retrieve them and push them back into the nest. Once she leads her brood from the nest she does not return. This is probably an instinctive response to avoid predators after the young have hatched. Poults develop rapidly and begin flying to roost on low branches when they are about 10 days old. By 18 days of age they are strong fliers. At about two to three weeks of age, poults start roosting in trees, which greatly increases survival.

With the small poults following close behind, the hen moves slowly and deliberately, no object escaping her incredibly perceptive eyes. Poults are especially vulnerable while they are flightless. During the first days of their lives, the hen and brood inhabit only a few acres. Nights are spent on the ground with the young crouched under her belly feathers and half-spread wings. She "broods" them during rains and on chilly days, as they are susceptible to chilling and exposure to the raw elements. If confronted by an intruder, she will try to lure it away from her brood, but if that fails she will pretend to attack. Communication between the hen and her poults is excellent. Poults obey her vocal commands completely and stay hidden until her call announces safety.

Each day the family range expands. During spring and summer months, the borders of agricultural fields, pastures, old fields, and new forest regeneration sites provide brood-rearing habitat that provides large quantities of protein-rich insects, tender green vegetation, and ripening seed heads needed for rapid growth. By mid-October, hen poults weigh 5 to 7 pounds and males 6 to 10 pounds. Growth slows and may halt during winter, but resumes in the spring.

Studies in Alabama have shown that one-half to nearly three-fourths of newly hatched poults die before they are two weeks old, primarily from predation. Poult survival is directly related to the proximity of nesting sites to brood-rearing habitat. Poults follow the hen, sometimes as much as 2 miles, to brood-rearing habitat to feed on insects. During this time poults are extremely vulnerable to predation. The farther the distance between nesting sites and feeding areas, the greater the risk of predation on poults. In managing forest and farm land for turkeys, it is important to provide nesting habitat as close to brood-rearing habitat as possible to reduce predation on poults.

Poult survival is also directly related to the type of brood-rearing habitat selected by hens. Research conducted by Auburn University on radio-tagged hens and poults shows that brood habitat is characterized by lush grass and forb vegetation often with or near some shrub or tree cover. Hens with broods used several types of openings, including pastures, hayfields, grainfields, and old fields that have been abandoned for three to five years. In northwestern Alabama, improved pastures, cutover hardwoods, wildlife openings, and rights-of-way were preferred brood habitat by hens that were successful in raising their broods. Upland hardwood sites were avoided by successful hens. Studies at Auburn University have also shown that hens successful in raising poults to two weeks of age selected brood habitat that had fewer large trees, less shrub-level vegetation, and denser herbaceous vegetation than hens that were unsuccessful in raising their brood.

MARCIA W. GRIFFEN

Young turkey poults quickly learn feeding and behavioral traits from the hen. Brood range expands with the passing of each spring and summer day to include borders of fields, pastures, old fields, and new forest-regeneration sites. These areas provide large quantities of protein-rich insects, tender green vegetation, and ripening seed heads for rapid growth of poults.

CLAY SSSON

RICHARD T. BRYANT/AEISTOCK

DR. H. S. [illegible]ANTON

Turkey poult survival is directly related to the type of brood-rearing habitat selected by hens. Ideal brood habitat consists of lush grasses associated with fallow wheat fields (left), fallow fields planted the summer before (middle), or grassy glades (right).

Throughout the poult-rearing period, the mature gobblers range either alone or in small groups. The hens unsuccessful at nesting usually join other hens, some with and others without poults. By late summer, it's common to see several hens with their broods using an area together.

THE FALL AND WINTER PERIOD

The flocking behavior of wild turkeys continues through the winter months. Fall flocks usually consist of adult hens and their poults, with old gobblers being segregated into distinct groups or ranging by themselves. By winter, many of the young-of-the-year gobblers have separated from family flocks to form young gobbler flocks.

Through the fall and winter, turkeys spend most of their time in woodlands searching for acorns, beechnuts, dogwood berries, and other types of mast. When turkeys range through farmland, they will also utilize waste grain in corn and other grain fields, if they are not harassed or alarmed by human activity. Through this period people commonly see several groups of turkeys feeding together on grain fields. Turkey flocks of 30 or more are often observed in good winter range. Old gobblers, young gobblers, and hens frequently feed together, but they roost separately and generally exhibit some degree of segregation on common feeding grounds. On occasions when a group of gobblers enter the feeding area, the hen group or young gobblers may move off or leave the feeding site.

The total amount of range used by turkey flocks during the fall and winter varies with the available habitat and with the abundance of food sources. In poor mast years, more range is required than in ample mast years, provided that the turkey population density is relatively unchanged. On a day-to-day basis, a turkey flock may use 50 to 100 acres, but the flocks may suddenly move a mile or more to a new area. Such sudden movements are often associated with the birds' traveling to a particular stand of preferred mast-producing trees, like wild cherry.

ANNUAL TURKEY POPULATION LEVELS

Brood size, or the number of young poults comprising a brood, varies greatly, both within and between years. Generally turkeys have between 4 and 10 poults per brood. The number of broods observed each year varies even more than brood size. As a result of these two factors, the fall population is dependent on the number of broods produced and average brood size. Turkey populations in rather small sections have been known to fluctuate as much as 50 percent from one year to the next, as a result of poor nesting success and low brood survival. The compounding factors of low reproduction and a loss of habitat can make it difficult to maintain a sizeable population of turkeys. However, one good year of turkey reproduction and survival can offset the declines caused by several successive years of poor reproduction.

Like most species of wildlife, mortality is always high among very young and very old turkeys. Among the younger groups, most mortality has occurred by late summer. When the poults have reached 12 weeks of age they have an increasing likelihood of surviving through, the first year. In a successful production year, the late summer poult-to-hen ratio is three to one or better. This ratio includes all hens and not just those with poults. The average lifespan of gobblers and hens is approximately one- to one-and-a-half years. Recently established populations tend to have a higher rate of productivity than the older, more established turkey populations.

FACTORS THAT LIMIT TURKEY POPULATIONS

Habitat Loss

Scarcity of quality habitats constitutes the main limiting factor for the wild turkey. Unless the turkeys' requirements for food, water, and cover are met on a year-round basis, turkeys have great difficulty surviving. In favorable habitat, it is possible to maintain one flock per 640 acres, but as habitat quality diminishes, the flock's range increases.

Changing land uses are the primary factors that reduce the quality and quantity of habitat and wild turkey abundance. Economics often dictate that farms and forest lands that provide nesting, brood-rearing, escape, and feeding areas for turkeys be converted to uses that make a profit but are not always compatible with turkey habitat needs. For example, unimproved pastures that provide excellent brood-rearing habitat are often planted in trees or merely abandoned when the price of cattle drops. Mast-producing hardwoods that provide food for adult turkeys and brood-rearing habitat for poults are lost when harvested and the land is replanted in pine plantations. Marginal agricultural lands that once supported soybeans and other crops eaten by turkeys have been converted to pine plantations during the USDA-sponsored Conservation Reserve Program. These are just a few examples of how changing land-use patterns can dramatically affect the wild turkey. Even though land-use change is often inevitable, landowners interested in wild turkeys should be sensitive to the habitat needs of turkeys when altering the ways they use and manage their land.

The variety of habitats used by wild turkeys for roosting, brood-rearing, nesting, feeding, and escape vividly demonstrates the need for a rich mosaic of habitats to provide for wild turkeys alone, much less the many other species of valuable wildlife.

Weather

Turkeys, like other birds, survive in certain areas because they are adapted to the normal regional climate and available habitat. Severe deviations from the normal weather patterns cause decreases in turkey population size. In the Southeast, weather is not normally a limiting factor, except in situations with prolonged rains and flooding during the nesting and brood-rearing seasons. Spring flooding along major drainages can be especially devastating, wiping out turkey nests, and drowning poults. Drought, the other extreme, may cause problems by drying up needed water sources and withering vegetation that feed both the turkey and the insects on which they rely for food. Cold temperatures during the spring have little effect on nesting or brood-rearing success. In general weather, particularly rainfall, must deviate significantly from normal patterns for an extended period of time before it affects turkey populations in Alabama.

Poaching

Turkeys are quite susceptible to poaching (the illegal killing of game) and need plenty of protection. One study indicated that poaching is a significant mortality factor in some areas of Alabama. Landowners and managers can reduce the potential for poaching turkeys by strictly limiting access to turkey habitat. Woods roads and other access points should be closed, particularly during nesting and brood-rearing season. Firelane entry points should also be blocked by creating a dirt mound barrier that is impassible by vehicles. In addition, natural openings and food plots that attract turkeys should not be visible from public roads.

Diseases and Parasites

Under wild conditions, diseases and parasites usually have the most impact when populations become too dense for the habitat to support. In these cases, disease often causes the population to decline substantially below the capacity of the habitat.

Occasionally, periodic declines in normal population numbers are blamed on diseases and parasites. The diseases and parasites that are most detrimental to wild turkey populations are those introduced with the release of "pen-reared wild turkeys." Like domestic poultry, pen-raised turkeys can be immune to infectious diseases but still carry the disease and infect wild turkeys. Blackhead (histomoniasis), avian pox, and coccidiosis are examples of diseases spread by both domestic poultry and pen-reared turkeys.

Contact between pen-raised or domestic stock and wild turkeys should be avoided. In addition, locations of turkey food plots should be rotated every two or three years to prevent fecal buildup and the chance of infectious diseases being transmitted to concentrations of turkeys that use these supplemental planting sites. Past wild turkey declines in some sections of Alabama appear to have been partly due to contacts between wild and domesticated flocks.

Landowners and managers who use corn in supplemental feeding programs should be careful that the grain is not contaminated with aflatoxins. Aflatoxins are poisons produced by the fungi *Aspergillus*, which if ingested at high levels, can be fatal to animals that consume contaminated corn. The effect of aflatoxin on domestic animals and humans is well documented. Depending on the dosage, health problems in animals can range from acute to chronic liver damage, increased susceptibility to infectious diseases, cancer, and death. Information on the risks to wildlife health caused by aflatoxin-contaminated corn is inconclusive, but there have been several documented cases of migratory waterfowl and sandhill crane deaths caused by aflatoxins. In general, birds are more susceptible than mammals. Landowners, managers, and sportsmen should use caution when buying "wildlife corn" since these sources often contain aflatoxin levels too high for other uses. Suspect corn should be tested to determine if aflatoxins are present and at what level. To discourage further toxin production that would cause aflatoxin content to increase over time, corn should not be piled up at feed stations, but should be spread evenly over the ground. Feed stations should be moved periodically to prevent build up of soil-borne pathogens.

LEWIS O. ROGERS/SOUTH CAROLINA DNR

Avian pox is a major viral disease of wild turkeys in the Southeast that produces lesions on the head. Avian pox has been reported to cause mortality in local populations of wild turkey in Alabama.

DR. GEORGE HURST/MISSISSIPPI STATE UNIVERSITY

Nest predation is a serious limiting factor of wild turkeys in Alabama. Over 50 percent of wild turkey nests and eggs are destroyed by predation each year.

Predation

Wild turkeys are part of a complex ecosystem that has evolved over thousands of years. Studies have determined generally which animals prey on wild turkeys, but the overall effects of predation on wild turkey populations by different predators and turkey densities in different habitats is unclear.

Overall wild turkey mortality is substantial. About one-half of all hens and approximately one-third of all gobblers die each year. Mortality for hens is highest during the nesting and brood-rearing season when hens do not roost in trees but stay on the ground overnight with their nest or brood. Studies in Mississippi have revealed that 92 percent of hen deaths were caused by predation, and most (69 percent) of the mortalities occurred during the nesting and brood-rearing season. Usually gobblers are hardy once they mature and hunters become their main predator. However, a number of animals, including free-ranging dogs, coyotes, bobcats, foxes, and owls, prey on grown turkeys.

Since turkeys nest on the ground, eggs and young poults are predated heavily by a number of animals. Studies in Alabama have suggested that raccoons are the number one predator on turkey nests and eggs. Free-ranging dogs are the primary predator on turkey poults and hens in some areas. Skunks, like raccoons, are also important predators and consistently prey on nests and young poults. Eggs and poults are consumed also by opossums, crows, foxes, hawks, owls, and snakes. In a six-year study conducted by Auburn University, over 50 percent of turkey nests and 70 percent of poults were lost as a direct result of predation. The study concluded that predation, especially on nests and broods, is a serious limiting factor on turkeys in Alabama and that control of mammalian predators should be an integral part of turkey management.

Feral hogs also can destroy turkey nests. Wild hogs

DR. GEORGE HURST/MISSISSIPPI STATE UNIVERSITY

DR. KEITH CAUSEY

PAUL T. BROWN

The primary nest, egg, and chick predators of wild turkeys are opossums, raccoons, and free-ranging dogs. Raccoons are the number one nest predator in Alabama.

in many river bottoms of the state potentially affect turkeys and turkey management in a number of ways. Hogs root the ground and can disrupt almost anything in their path. Certainly they create problems with management activities such as maintaining food plots for turkeys. They also compete with turkeys for some foods. For example, acorn production in mixed hardwood forests fluctuates from year to year. In years of very low production, hogs compete with turkeys and other native wildlife for a limited acorn supply. In addition, hogs compete with turkeys for food in planted plots, such as chufas, and grain crops. On the other hand, hogs can uncover foods which turkeys later find. Also, hogs may promote small areas of grass and forb growth with their rootings, which may be beneficial to young poults. The best solution to managing feral hogs and turkeys on the same land is to keep hog numbers at a modest or low level through regulated hunting or trapping. Feral hogs may be legally trapped in Alabama and once reduced to possession are no longer considered a game animal.

Studies of nesting turkeys have not yet implicated the fire ant as a substantial nest predator in the Southeast. But recent studies in Texas show reduced reproduction of several wildlife species in areas with very high fire ant densities as compared to areas where fire ants were controlled. Also, fire ants consume an enormous quantity of invertebrates, so they may be a serious competitor with young turkey poults for invertebrate prey. Wide-scale control of fire ants, however, is probably not practical at this time.

Studies have shown that substantial harvest of predators may enhance turkey survival. However, control of predator populations is difficult, usually cost prohibitive, and controversial. Killing certain predators, such as raptors (hawks and owls), is also illegal. Trapping of select furbearers, especially important turkey predators like raccoons, is a legitimate and legal control method that should be considered in areas where raccoon depredation is heavy (see Chapter 11). With the expansion and proliferation of coyotes in Alabama, managers are also concerned about their impact on turkeys. Although coyotes are opportunistic and will eat poults and adult turkeys, studies in Alabama have shown that coyotes may be only a minor turkey predator. Landowners should be more concerned about free-ranging dogs. Free-ranging dogs and cats should not be tolerated by landowners interested in managing wild turkeys (or quail and rabbits), especially since they have been implicated as major nest and poult predators. Landowners and adjoining neighbors should restrain dogs and cats, especially during nesting and brood-rearing seasons. Finally, creating high-quality habitat (especially nesting and brood-rearing) that helps turkeys withstand predation and reproduce more consistently is probably the best use of a landowner's time.

Factors Limiting Wild Turkeys

Dr. George Hurst is Professor of Wildlife in the Wildlife and Fisheries Department at Mississippi State University and is nationally known for his research on wild turkey for over 20 years. He has written extensively on the wild turkey and is widely recognized as a leading authority on wild turkeys in the South.

During my years as a wild turkey researcher I have seen a series of factors affect turkey populations. From my experience I consider the following factors, in descending order of importance, to be the most important in limiting the growth of wild turkey populations: 1) habitat alteration, 2) predation, and 3) disease. These factors interacting together can be devastating to wild turkeys.

Today we are in the midst of growing our third, and in some cases, fourth forest in the South. In the process, we have lost a lot of mature pine, pine-hardwood, and hardwood forests to pine plantations. Large blocks have been cleared and in some counties in the South over half of the mature forests have been lost. During the first few years after timber harvest, these areas provide some turkey habitat; however, between 4 and 15 years of age these areas are virtually useless for turkeys. Much of our forest landscape is at this stage now. The fact that these changes have taken place on such a large scale has been detrimental to wild turkeys. Large-scale habitat changes have been driven by timber market factors and demand for forest products.

Mature bottomland forests have been converted to pine plantations or agricultural production. Some row crops, such as soybeans and corn, do provide food for turkeys and have not been as detrimental as other agricultural plantings such as cotton. In my opinion, government farm programs in general have also had a negative impact on turkey habitat. The Conservation Reserve Program has converted small soybean fields, pastures, and hayfields to dense loblolly pine stands and overgrown fields that are not ideal for wild turkeys. These thickets attract and concentrate turkey predators because they contain an abundance of prey such as rats, mice, and rabbits. The dairy buyout program, in combination with falling cattle prices, has reduced the number of dairy and cattle producers, causing a decline in the number of pastures, ryegrass fields, and hayfields that are excellent turkey habitat. Many of these areas have been converted to pine plantations or left fallow to grow up in thick vegetation not suitable for turkeys. These large-scale habitat changes have been bad news for the wild turkey.

Predation is the second most important factor affecting turkeys. We are now seeing an all-time high in the populations of turkey predators as compared to the past. This in part is probably due to reduction in predator trapping and hunting combined with an increase in predator habitat. Predators like raccoons, opossums, skunks, bobcats, foxes, and great horned owls have historically been natural turkey predators; however, as their populations have increased so has their impact on wild turkeys. Because of this, predators may not only be a limiting factor in preventing the expansion of turkey populations, but they may also be a regulating factor that keeps turkey numbers below what the habitat can support (carrying capacity). Predator control and management must be a component of turkey management.

Diseases, in my view, are the third most important factor that affect turkey populations. When turkey densities are high, disease is most likely to take its toll on turkey populations. Turkeys flocks that have been infected with disease are usually slow to rebound to their original numbers. Diseases such as avian pox, histomoniasis (blackhead), and others can infect turkey flocks and cause mortality. A new disease affecting wild turkeys, called lymphoproliferative, may have been responsible for wild turkey die-offs in Alabama and Mississippi. This disease is a virus which has previously not been documented in the U.S. To reduce the potential for fecal build-up and the chance of disease, turkey managers should rotate food plots every one to two years to prevent an accumulation of fecal matter from infecting turkeys. In addition, the practice of spreading litter from commercial poultry houses on pastures should be eliminated since this also exposes wild turkeys to a variety of domestic fowl diseases.

Fig. 2: Predators of Wild Turkeys

Mammals	Birds	Reptiles	Insects
free-ranging dog	Cooper's hawk	rat snake	fire ant
raccoon	broad-winged hawk	Eastern coachwhip	
skunk	crow	pine snake	
opossum	goshawk	king snake	
fox (red and gray)	great horned owl	other snakes	
bobcat	golden eagle	alligators	
coyote	red-tailed hawk		
feral house cat	red-shouldered hawk		
feral hog	barred owl		
man			

HABITAT REQUIREMENTS

To support wild turkeys the land must provide suitable habitat on a year-round basis. Generally speaking, suitable turkey habitat includes a scattering of mature mast-producing hardwoods, mainly oaks, with smaller hardwoods that will eventually replace those becoming over-mature. A mixture of midstory trees, like dogwood and wild cherry, which provide food and cover are also needed. Turkeys make efficient use of green plants and seed heads found in pastures, fields, roadsides, and some regeneration sites. These areas can also provide the insects needed by poults to obtain the high quantity of protein necessary for their first few weeks of growth. Quality turkey habitat will support one bird per 20 to 30 acres or one flock to about 640 acres.

When management efforts are successful and turkeys set up residence and begin to reproduce, they can increase their numbers at an amazing rate. Often it appears that an area will support one bird per 5 or 10 acres as turkey populations are expanding and quickly becoming established. This is usually not a lasting situation, as all species tend to expand rapidly when they first become established. In Alabama, wild turkeys have been established in all suitable habitat. Although considered a big game species, wild turkey populations often show the boom or bust cycles typical of small game species. When weather conditions and food supplies are abundant, two or three years of successful reproduction can easily double population size. Likewise, poor weather and food supplies and resulting poor reproductive success can lead to drastic reductions in turkey population levels.

FOODS AND FEEDING

Turkeys are opportunistic feeders, meaning they consume a wide variety of foods throughout the year. They feed by picking, scratching, clipping, stripping, and ingesting whole food items such as salamanders. During the spring, green grasses and leaves are consumed in large quantities. During the summer and early fall, picking and stripping methods are used to eat ripened seed heads of grasses and other plants. Scratching in the leaves for acorns and berries is the primary feeding method used from late fall through early spring.

Young poults, particularly during the first two weeks of life, require a high protein diet that is furnished mostly by insects. During the first week of life, the diet of poults is usually composed of 90 percent animal matter. Young poults will eat anything slow enough to catch and small enough to swallow. Grasshoppers, crickets, stink bugs, beetles, flies, wasps, ants, moths, tadpoles, small frogs, millipedes, snails and spiders make up the bulk of the animal matter consumed. In addition, grit is an important material needed in the gizzard to aid in grinding the food to a digestible stage. This is why turkeys may peck around sandy or rocky areas seemingly devoid of plant life. After about a month of age, the diet of young turkeys is predominately plant material. Wild turkeys benefit from supplemental plantings such as chufa, a nutsedge that makes an excellent supplementary winter food source. Several varieties of clovers, as well as wheat, rye, oats, corn, soybeans, cowpeas, vetch, and bahia grass also make excellent plantings for turkeys.

Key Habitat Types for Wild Turkey

NESTING

Provides visual protection of hen and eggs from predators in forest stands having open overstories and well-developed understories. Characterized by abundant grass, herbaceous and shrub vegetation up to 3 feet in height. Often turkeys nest near logging slash or brush with vines over nests or nests at the base of a tree. Frequently near brood-rearing habitat and openings. Can be created by retaining three- to five-year-old abandoned fields, three- to seven-year-old timber regeneration sites, utility rights-of-way, edges between forest and fields, or other areas that provide well-developed herbaceous and shrub vegetation.

RICHARD T. BRYANT/ARISTOCK

BROOD-REARING

Provides poults with insects that are easily accessible with enough cover for hiding but low enough (about 2 feet) to allow the hen unobstructed vision for detecting predators. Characterized by openings having grass and forb vegetation interspersed with forest. Several types of openings are used by broods for insect feeding including improved pastures, burned pine stands, hayfields, grainfields, cutover hardwoods, wildlife openings, utility rights-of-way, and old grass-dominated fields. Landowners should have several openings for insect foraging within the 75 acre weekly home range of broods.

CLAY SISSON

ROOSTING

Provides protection against predators and is necessary throughout the year. Roosting characteristics are not specific; however, turkeys prefer to roost between 30 and 100 feet above the ground. There also is a preference for roosts that are near or above water. Heavily hunted turkeys use denser, more remote roosts. During winter when temperatures are below freezing, roosts tend to occur on south-facing slopes away from prevailing north winds. Landowners should retain mature hardwoods and pines across the yearly range as potential roosting sites.

RICHARD T. BRYANT/ARISTOCK

FALL AND WINTER

Provides food and roosting cover for young and adult turkeys. Turkeys increase their use of forested cover during this time and decrease their use of open areas. Hardwood stands containing a diversity of mast-producing trees, interspersed with pines and field edges should be retained by landowners.

CLEMSON UNIVERSITY

Examples of key habitat types for wild turkey in Alabama: nesting and emergency escape cover (top), brood-rearing (2nd from top), roosting (3rd from top), fall and winter mast-producing sites (bottom).

ESCAPE COVER

Sometimes provides emergency cover for turkeys pursued by avian predators or from heavy hunting pressure. Found in timber-cutting slash and dense vegetation.

FIG. 3: SEASONALLY IMPORTANT FOODS[1] OF WILD TURKEYS IN ALABAMA

Spring	Summer	Fall	Winter
green vegetation (such as new shoots)	insects	acorns	acorns
insects	side seed grass	insects	insects
acorns	green vegetation	green vegetation	pecans
soft mast (blueberry and blackberry)	soft mast (huckleberry blackberry, blueberry, gallberry fruit, poison oak, wax myrtle, wild grape, black cherry, noseburn)	crabgrass	hophornbeam seeds
snails	crabgrass	panic grass	green vegetation
panic grass	panic grass	side seed grass	soft mast (gallberry, wax myrtle, dogwood)
wood sorrel	bahia grass	beggarweed	pine seeds
chufa	paspalum	soft mast (American beautyberry, plum, persimmon, dogwood, blueberry, wax myrtle, black tupelo)	beechnut
clovers	chufa	pine seeds	chufa
	soybeans	saw palmetto seeds	corn
	browntop millet	pecans	wheat
		beechnut	oats
		chufa	ryegrass
		corn	clovers
		soybeans	
		clovers	
		browntop millet	

[1]Hard and soft mast and seeds of grasses provide carbohydrates that are a ready source of energy for turkeys. Insects, seeds, and cultivated plants provide a source of protein. Invertebrates such as snails provide protein and a source of calcium, which is important for the hen before and during egg laying. Essential vitamins and minerals are obtained from a wide assortment of green vegetation.

Top (left to right): Acorns, vetch, clovers. Bottom (left to right): blackberry, wild strawberry, insects, dogwood berries.

Water

Water was long thought to be an important part of wild turkey habitat. Most habitat recommendations in the past have included having water sources readily available from permanent streams, springs, and ponds distributed at about four sources per square mile. Recent information from radio-instrumented turkeys has shown that turkeys in a relatively humid range such as Alabama can get moisture from their diet of insects, fruits, and succulent vegetation. Observers noted that radio-tagged hens with their broods in south Alabama made no attempt to travel to permanent water sources.

Cover

Cover requirements of wild turkeys change with the season of the year in relation to the events of the turkey's life cycle. Nesting cover and brood-rearing cover are perhaps the most critical types of cover for turkeys. Nesting cover provides visual protection for the hen and her eggs and in forest stands is characterized by open overstories and well-developed understories. Understory vegetation should be composed of abundant herbaceous and shrub vegetation up to 3 feet in height. Nests are often located under or adjacent to logging slash, brush, vines, or the base of a tree. Other habitats, such as three- to five-year-old abandoned fields, three- to seven-year-old timber regeneration sites, utility rights-of-way, edges between forest and fields, or other areas that provide well-developed herbaceous and shrub vegetation can also provide ideal nesting cover.

Brood-rearing cover should provide poults with enough visual protection for hiding while feeding on insects, but be low enough (about 2 feet) to allow the hen unobstructed vision for detecting predators. Several types of openings yield cover for broods to feed on insects including improved pastures, hayfields, grainfields, cutover hardwoods, wildlife openings, utility rights-of-way, and old grass-dominated fields. Food plots left in wheat, oats, and clover through the summer provide excellent brood-rearing habitat because browned-out wheat is tall enough for cover, short enough for hens to see over, and open enough at the ground level to allow easy movement and feeding by the poults.

Mature timber for roosting is also an important habitat component. Turkeys need several suitable roosting sites scattered over the flock's range, as turkeys seldom roost in the same place on successive nights. Aside from the many varied vegetative cover types found in turkey range, natural land contours that allow turkeys to avoid intruders can also be considered a form of cover. Rolling pasture land is a prime example of contour cover, as turkeys feel safer when they can retreat over a hill to get out of sight of whatever may have startled them.

A variety of sites are selected for roosting in Alabama. In upland areas with some topographical relief, turkeys often roost on the upper portions of slopes, just off of ridgetops. On upland sites turkeys often roost in pines. On bottomland sites turkeys roost in mature cypress or hardwoods, often over water. Broods will use pine plantations for roosting during spring and summer.

The Importance of Nesting and Brood Habitat for Turkeys

Chuck Peoples conducted his graduate research at Auburn University where he investigated nesting and brood habitat selection by wild turkeys. He is currently a Wildlife Biologist at the National Wild Turkey Federation in Edgefield, South Carolina.

To be successful, wild turkeys must have adequate nesting and brood-rearing habitat. My graduate research at Auburn University examined areas where wild turkeys nest and raise their broods. We found that typical nesting sites contained vegetation that was in a transition from herbaceous plants to woody shrubs or saplings. Typical areas included fallow fields, utility rights-of-ways, small forest openings created by fallen trees, and areas of pine forest under a periodic burning regime. Our investigations found that the size of nesting habitat and distance of nests to edge were important factors determining the success of hens in producing poults. On average, successful nesting and hatching occurred on 8-acre or larger blocks. We also found that the farther the nests were from edge habitat, the greater the chances of successful nesting. Large blocks of nesting cover located away from edge habitat reduced predation by hampering the ability of predators to locate nests.

Brood-rearing habitat that provides an abundance of insects and protective cover from predators was also found to be key to the survival of turkey poults. These areas are the "grocery stores" for poults, providing high protein insects that are essential dietary items during the first few weeks after hatching. These sites are typically characterized by knee-high herbaceous cover found in fields and other open areas. Brood habitat can be created and maintained in openings with fall disturbances that include mowing and burning, disking, or planting crops such as winter wheat and clover. Left unmowed through the following summer, these areas also provide an abundance of insects.

Ideal nesting and brood-rearing habitat should be located adjacent to hardwood drains. These areas provide travel routes for hens and poults and are considered the "interstates" of the turkey world. Hens may move 2 to 3 miles along hardwood drains with a brood of poults to suitable brood-rearing habitat right after the young are hatched. Hardwood drains should be left undisturbed in as wide a corridor as possible to provide suitable cover and travel access for turkeys. As a general rule, hardwood drains should be wide enough that you cannot see through from one side to another during winter. Hardwood drains that are narrow often become too thick and dense with vegetation, decreasing visibility, mobility, and the hens' ability to detect and avoid predators.

HABITAT IMPROVEMENT PRACTICES

Whether the area considered for turkey management is 320 or 3,200 acres, the starting point and basic considerations are the same. The first step is to make an inventory of what habitat components are already available to turkeys and evaluate the quality and quantity of available habitat. Recent aerial photographs of the property, as well as of adjacent lands that may be used by turkey flocks, can help in assessing existing turkey habitat. Topographical maps are also helpful in evaluating an area for turkey habitat. Making use of good habitat components on adjacent property is especially helpful to the small landowner, who may not be able to support a flock of turkeys year-round on his or her land alone. Turkeys know no property boundaries and will roam wherever necessary to obtain food, water, cover, and protection.

Turkey managers should carefully study the habitat distribution on the aerial photographs. It may first appear to be a patchwork quilt, but close examination will reveal what each habitat type offers turkeys under its present condition. Then consider the combination of adjacent habitat components. Remember that during the critical reproductive period (from early spring through summer), hens, poults, and gobblers like the areas where spring's first green plants reappear. They also prefer open

Fig. 4: Hard Mast Producers for Wild Turkeys

	Average Number of Seeds Per Pound/Average Number Pounds Per Tree	Minimum Age of Production	Interval Between Large Crops (Years)
White Oaks			
White oak	120/8	20	4–10
Post oak	380/6	25	2–3
Chestnut oak	100/—	20	2–3
Swamp chestnut oak	85/—	20	3–5
Red Oaks			
Blackjack oak	210/—	25	2–3
Bluejack oak	100/—	25	2–3
Dwarf live oak	—/—	4	—
Laurel oak	560/11	15	1
Live oak	352/—	25	1
Northern red oak	380/—	25	2–3
Runner oak	360/—	4	—
Southern red oak	540/—	25	1–2
Water oak	395/45	20	1–2
Willow oak	462/40	20	1
Other Hard Mast			
American beech	1,600/—	40	2–3
Eastern Hophornbeam[1]	30,000/—	25	—
Soft Mast			
American beautyberry	32,500/5	2	1–2
Black cherry	4,800/—	5	1–5
Black tupelo	3,380/—	—	—
Dogwood	4,500/2	6	1-2
Huckleberry	4,300/—	15	—
Gallberry	29,000/—	5	1–2
Persimmon	1,200/—	10	2
Plum	4,240/—	5	1–5
Wild grape	14,500/—	4	—

[1]Seeds high (15%) in crude protein.

(Source: Southern Fruit-Producing Woody Plants Used by Wildlife, Halls 1977; and USDA Silvics of North America, Volume 2, 1990)

areas for feeding on insects. During fall and winter, they utilize the woodlands in search of acorns and other mast crops. With these and other factors in mind, like possible sources of human disturbance, future timber operations and other land management practices, a habitat development plan should begin to take shape.

Hardwood Management

To the wild turkey, oaks are vitally important. In most areas of Alabama, as much acreage as possible should be left in hardwood stands that offer a variety of different oak species. Most oak species begin to bear acorns at 20 to 25 years of age, but, the best mast-producing years are from age 50 to 100 or, in size, from 14 to 24 inches diameter at breast height (DBH).

DR. GEORGE HURST/MISSISSIPPI STATE UNIVERSITY

An acorn basket placed under oaks and other hardwoods is one method of estimating hard mast production.

A mix of oak species fosters better overall acorn production. Acorn production fluctuates from year to year based on species and individual trees. White oak acorns mature in one growing season, while red oak acorns take two growing seasons to mature. A diversity of white and red oaks increases the chance that some form of acorn crop will be available each year. The presence of a variety of mast-producing species is especially important during years when the acorn crop is poor. If feasible, hickory and beech stands should be managed on a slightly longer timber rotation than oak, from 80 to 150 years, which is the period of their best mast production. Other mast-producing trees like pecan in river bottoms, dogwood, and black gum provide essential winter foods, particularly during years of oak mast failure. During silvicultural practices, individual high-value hardwoods that produce abundant mast should be left for turkeys. These trees should be identified and clearly marked prior to the timber sale and before timber harvest. Their benefits to wildlife over the long-term outweigh the immediate income from timber harvest, in most cases.

Because many wildlife species utilize acorns as a primary food source, land managers should plan to provide a continuous crop of mast-producing trees. Before a given percentage of the available hardwoods is harvested, an equal percentage should be entering the best period of mast production (50 years or older). An alternative to this would be to have a higher percentage entering the 30- to 40-year-old class, so that the collective production of the acorns would about equal the harvested acreage of mature oaks. Harvest rotations for oaks should not be shorter than 60 years or exceed about 100 years.

Swamps, river bottoms, creek bottoms, and drains should be targeted for mast production. Often these areas contain a high volume of poplar, elm, gum, cypress, and other less valuable mast species. These species should be removed, if possible, in favor of oaks, beech, and other mast-producing species; however, on many wetter sites it will be impractical to convert the stand to mast producing species. In the Piedmont Region oaks are often more abundant on upland ridges, slopes adjacent to bottomland sites, and around old abandoned house sites. These areas should be maintained as key mast-producing areas.

Good hardwood management can accelerate acorn production for turkeys and other wildlife. Hardwood forests naturally develop dominant and codominant oaks that have the potential over time to be excellent mast producers. Management should focus on identifying these trees early in the life of a stand and providing them with ample opportunities to reach their full acorn-producing potential. This can be accomplished by removing poorer quality trees that compete with dominant and codominant oaks. Reducing competition helps maximize crown development of the dominant and codominant oaks to facilitate acorn production and quality.

In small hardwood stands (less than 50 acres), the best timber harvest method is group selection of overmature or poor mast producers to allow the younger, more vigorous trees to reach the canopy. A stand should be harvested or thinned no more often than every five years to minimize habitat disturbance. Where large stands occur or where clear-cutting is necessary, the size of the clear-cut should not exceed 50 acres and preferably not be more than 25 acres. A wise first step toward getting a hardwood regeneration area back into mast production is thinning the young stand at about 20 to 35 years to facilitate full crown development of the most desirable trees.

Wild turkeys also need the soft mast produced by dogwood, grape, wild cherry, huckleberry, blackberry, dewberry and wild strawberry. These species help assure the needed variety of food sources in year-round turkey range. Of these, dogwood is probably the most important, as it provides a fall food source that can help offset the effects of a poor acorn crop. During thinning and/or selective cutting, care should be taken not to remove or damage dogwood and wild cherry trees. Other understory tree species quickly rebound after silvicultural operations, but dogwood and wild cherry are important enough to warrant special attention.

In managing overstory hardwood stands, tree crowns should not be so dense that they completely shade out the understory. Overstory trees should develop a full crown, but some spacing between trees helps assure that sufficient sunlight reaches the forest floor, allowing some understory to develop. At the turkey's eye level, the forest floor should be fairly but not completely open. Their keen vision is a key part of their defense, and they feel safe as long as they have a view of at least 50 feet or more.

Pine Management

Turkey populations, for the most part, are determined by landscape (or habitat) patterns. Generally, mature pine or pine-hardwood stands are good turkey habitat, but they represent a decreasing portion of Alabama's timberland. When these stands are harvested, turkeys use the cutover area for a couple of years, especially if some residual mature trees or mature adjacent stands are left unharvested. Recently cut stands have abundant grass and forb growth and ample soft mast, such as blackberries and dewberries, as a result of overstory removal and increased sunlight. But within two or three years, vegetation normally becomes too dense for wild turkeys. Hens often nest in thickets, including edges of clear-cuts, but other turkey use is restricted. Pine plantations represent a substantial and increasing portion of Alabama's commercial timberland. Dense pine plantations are effective habitat for deer and several species of small mammals and birds, but pine plantations remain too dense for turkey use until the plantation reaches about 10 years of age.

At about age 10, when pine canopies close, understories are sparse enough for some wild turkey use, but shaded understories are usually devoid of turkey food items, such as grass seed, forbs, and fruits of shrubs. In middle-age pine plantations from 10 to 20 years old, thinning of up to one-half of the pine volume, especially in conjunction with prescribed burning, opens up dense stands to sunlight that enhances understory growth valuable for turkeys.

Older pine and pine-hardwood stands grown for sawtimber products are better than young pine plantations for wild turkey habitat. Tall pines are used for roosting and also produce an abundance of seed that is readily eaten by turkeys. In fact, turkeys do well in mature pine forests and pine seeds are a primary source of food when available. Mature pines or pine-hardwoods combined with openings can provide excellent turkey habitat. Mature pine stands that have an open understory with some herbaceous vegetation and some fruit-producing shrubs, broken up with small agricultural fields (particularly those with waste grain) or herbaceous fallow fields are also outstanding turkey habitat.

Hardwoods should be distributed throughout pine-hardwood stands, preferably pure hardwood groupings, or in streamside zones or hardwood corridors traversing pine stands. Streamside management zones comprised of strips of mature hardwoods or mixed hardwood–pines along waterways or moist sites can provide fine winter habitat for turkeys and improve year-round suitability of pine plantation landscapes for wild turkeys and a host of other wildlife species (See Chapter 4 on SMZs and travel corridors for wildlife).

Managing strictly for short rotation pine pulpwood on large acreages in contiguous blocks does not produce turkey habitat. Dense stands of even-age pines do not allow hardwoods to develop, even in the under and midstory, where dogwood and wild cherry combined with other soft mast species could provide turkey foods. Managing for pine sawtimber, on the other hand, provides a good potential for turkey management if wildlife values are kept in mind.

If wildlife is a management objective, the conversion of large stands (200 acres and up) from hardwood and mixed forest types to pine plantations should be avoided. Clear-cuts in blocks over 1,000 acres often destroy range that would support a large flock of turkeys year-round.

For multiple-use purposes, pine management should always be confined to those sites best suited for pine. The slopes or zones between bottomland and upland stands generally support good mast-producing trees that should be maintained where turkeys are a management objective. Trying to grow mast trees on every site is poor forest management and will result in many areas with very low-grade hardwood. It is also poor forest and wildlife management to convert a good hardwood site to pine. Clear-cutting and regeneration of pine stands should be limited to areas of no more than 200 acres and preferably less than 100 acres. Consideration should be given to the proportion of land area relative to the

ALABAMA FORESTRY COMMISSION

Hardwood corridors and strips of mature hardwoods or hardwood-pine mixtures along waterways that are maintained as streamside management zones (SMZs) provide excellent winter habitat for turkeys. These areas also improve year-round suitability of pine-dominated landscapes for wild turkeys and other wildlife.

flock's home range. Scattered oaks, hickories, dogwoods, and other mast-bearing species in natural pine stands should be maintained.

Clear-cuts should be irregular in shape. This is easily accomplished if drainages are left intact during silvicultural operations. Irregularly shaped clearings increase edge, a habitat type beneficial for turkeys and many other wildlife species. In some areas, a recently regenerated area may provide short-term nesting and brood range before the ground cover becomes too dense. Gobblers will frequent the edges of these areas, usually following the hens.

Areas managed on a sawtimber rotation of 30 years and longer usually increase in value to turkeys over time. To speed the habitat development of these areas, thinnings should begin at the sapling stage and continue in 5- to 10-year intervals until the stand is about 30 to 40 years old. This practice allows a diverse under- and midstory to develop fairly early in the rotation. Regularly thinned stands will provide open-wooded habitat that turkeys prefer as well as soft and hard mast-producing trees.

Landscapes dominated by pine plantations are less than ideal for turkeys when adjacent stands are converted to pine plantations every few years. The overall effect is the same as one massive clear-cut. Within a short period, several thousand acres of adjacent sites can be converted to young plantations that are approximately the same age. Landowners interested in maintaining turkey habitat should maintain enough stands of saw timber-size trees to prevent elimination of turkey range over a several thousand-acre block. A minimum of 50 percent of the timber should be pole size or larger, and these stands should not be isolated as islands in vast, young pine stands. Sawtimber stands should connect to other fairly large timber to avoid creating a natural barrier that turkeys will not or cannot pass through. Such a management approach may cause a turkey flock to use more acreage to satisfy its annual needs, but at least the flock will not be extirpated from its home range. Enough suitable year-round turkey habitat should be maintained in a pine management unit to support a minimum of 20 birds per 3,000 acres.

Prescribed Burning

Longleaf, shortleaf, loblolly, and slash pine, or mixtures of these species have survived wild fires throughout the Coastal Plain for many thousands of years. They are fairly fire resistant, except in the early seedling and sapling stages. Prescribed burning by skilled managers to control understory growth in pine habitat can be an excellent turkey management practice.

Unlike pine, hardwood bark does not have good insulating qualities and therefore prescribed fire should be excluded from pure hardwood stands. It is imperative to exclude fire from oak and mixed hardwood transition zones between pine upland and gum swamps. Repeated burning of transition zones will eventually convert these areas to pine. Where scrub oaks occur within longleaf pine stands, extra care should be taken to protect them from fire at desired locations. The mast produced by these oaks is very important to turkeys and other wildlife.

The turkey's need for a relatively open forest understory is easily accommodated by the use of prescribed burning. In addition to maintaining an open understory, prescribed burns increase the availability of many desirable food sources. Wild turkeys eagerly consume the new tender growth of forbs, grasses, and legumes stimulated by burning. Insects are often prevalent on recently burned areas, as they are attracted to the newly abundant flowering legumes. Hens and poults make excellent use of these areas in search of insect and plant materials.

Prescribed burning should be conducted during January and February and no later than March 15. Some variation in opinion exists on how often to burn as well as how much area to burn. Two primary theories on burning will be described so that the one best suited to a given situation can be used.

One type of burning technique employs low intensity burns over one-third to one-half of the entire pine woodland each year. This burning should be accomplished with a cool, low-burning fire (a fire that can be stepped over) that only partially burns understory thickets and debris. Because of the mosaic of burned and unburned areas within a stand, these burns are often called "dirty burns" or "wildlife burns." These burns help increase the quantity and quality of the food supply each year over the turkey range. The same site generally should be burned every two to three years, except for those sites where vegetative regrowth warrants burning on shorter intervals.

The second type of burning technique is a **compartment burn,** which is achieved with a slightly hotter fire than with spot fires. For forest management purposes, compartments may encompass several forest stands ranging up to several thousand acres in size and comprising about one-fourth to one-third of the total management unit. Compartment burns are most commonly used by larger landowners because large areas can be burned for a lower cost per acre. Large scale compartment burns, without protecting nesting and brood habitat, are detrimental to turkey and quail. For turkey, burns should be conducted only within individual stands and not across multiple stands. Cool fires should be used, whenever possible, to provide a mosaic of burned and unburned areas that provide both nesting and brood-rearing habitat for turkeys. If hot fires are used, nesting habitat should be protected by establishing firelines around blocks of 10 acres or more. Burning should be conducted on a two- to four-year rotation, altering the location of ring-arounds in the year of the burn.

Managing Openings

Wild turkeys require open areas in which to feed on green vegetation, seeds, and insects. Insect populations are highest in open grassy areas and therefore turkeys use openings regularly. Openings may include pastures, fields, cropland, orchards, logging decks, roadside, powerlines, gaslines, newly regenerated areas, and other areas that provide a break from continuous woodland.

Studies have shown that turkeys use open areas of almost any size, but prefer openings from 5 to 20 acres. Apparently, these openings produce more forage, seed, and insects than smaller openings, and the turkey flock is not able to exhaust the food supply as readily.

From a development standpoint, it may be too costly or difficult to bulldoze openings over 5 acres. Unless developed for other purposes, establishing large openings may not be practical. Several closely spaced, smaller openings, created by silvicultural operations or other land management practices, may be a better approach.

About 10 percent of the overall acreage should be maintained as some type of open land. Turkeys will fare better if a higher percentage is left in openings that are well distributed among the wooded tracts. A tract of land may contain as much as 40 to 50 percent pasture, cropland, orchards, or other openings, yet still be an excellent place for turkeys. Prime turkey habitat can contain large openings if year-round food sources and mast-producing oaks are available in adjacent woodlands. Nesting hens are attracted to open lands because these areas can provide excellent brood habitat.

Before incurring the expense of developing new openings, it would be wise to inventory what is already available, but unused. Electric and gas rights-of-way are relatively open and may cost very little to develop. Bush hogging will remove brush from areas that can be planted and maintained. Leaving the unworkable sites in brush will provide nesting cover. It may be relatively easy to develop several on ¼- to 2-acre patches along the rights-of-way with a tractor, bushhog, and disc harrow. Power companies should be asked not to spray herbicides or mow during the nesting season, as this may disturb nesting. Because hens may nest in fields subject to early mowing, these areas should not be mowed until mid-July. On any opening to be mowed from late March through June, it would be wise to drive over the mowing route on a tractor without a mower and look closely for nesting hens. Land managers should at least take time to drive the area within 50 to 60 yards of woodland borders. By locating the nests before mowing, enough cover can be left around the nest sites to afford some protection if mowing cannot be delayed until nesting is over. On lands subject to early cultivation, it would be wise to mow the area in late fall, after the growing season, or keep the soil turned, so that cover would be too sparse for a hen to attempt nesting.

Logging decks, logging roads, and private roads are excellent areas to develop as openings at a low cost. Old logging decks and logging roads may contain stumps that should be removed before planting, but stump removal on these sites will cost less than developing an opening from scratch. Where timber sale contracts are involved, the timber operator could be required to remove stumps and leave the decks and roads workable. Removal of stumps is not absolutely necessary if they

JOE MCGLINCY

DR. GEORGE HURST/MISSISSIPPI STATE UNIVERSITY

RICHARD T. BRYANT/ARISTOCK

RICHARD T. BRYANT/ARISTOCK

Wild turkeys require openings for feeding on green vegetation, seeds, and insects. Any opening can be managed and maintained for turkeys. These are examples of managed openings for turkey: woods roads widened and planted in turkey foods such as bahia grass and clovers (top left), gas and powerline rights-of-ways planted for turkeys (top right), open bahia grass fields with scattered trees (bottom left), and agricultural openings created especially for food plots (bottom right).

are not too numerous and are left at nearly ground level. A light harrow can be used to break up or disturb the ground between the stumps without too much trouble or risk to equipment, and the stumps will be easier to remove after they have rotted for a few years.

Privately controlled roads may be excellent for roadside plantings. Brushy areas along the edge of roads also serve as possible nesting cover. To make these roads attractive for plantings, loggers can be instructed to clear-cut both sides of the roads approximately 30 to 50 feet from the roadbed. This process, called **daylighting**, allows the roadsides to be planted as food plots. Possible plantings include greenfield combinations such as wheat, clover, and vetch, summer grasses such as bahia, sorghum, corn, millet, or chufa. In areas with dense pine plantations, these roads become important travel corridors. Planting along exposed roads however can draw turkeys to areas where uncontrolled access makes them vulnerable to poachers. Care should be taken to restrict uncontrolled access to roadsides improved for turkeys.

Closely grazed pasture land is well suited to turkey use. Grazing sometimes keeps pasture grasses, such as fescue, in a tender growing condition that is more palatable to turkeys. Turkeys will also readily eat seeds and grain found in livestock droppings.

Turkeys also use corn fields, bean fields, and other cropland for fall and winter food sources. For wild turkeys and other animals, clean borders around cropland are wasted space that should be converted into winter grazing or cover. Planting the 10- to 20-foot-wide fringe around cultivated openings and orchards in food or cover crops is a practice beneficial to turkeys. Leaving these areas undisturbed for two or three years will be beneficial to many species of wildlife including quail.

Supplemental Plantings

Supplemental plantings are an effective method of increasing the quantity and quality of available food for wild turkeys; however, they should not be viewed as a substitute for managing native turkey foods by silvicultural and other land management practices. These plantings can be compatible with the primary land use, like planting ryegrass and winter wheat for cattle graz-

Fig. 5: Habitat and Harvest Management Intensities for Wild Turkey

Management Practices	Low Intensity	Medium Intensity	High Intensity
Timber Harvest/ Regeneration Types	Clear-cut/Plant	Clear-cut/Plant, Shelterwood or Seedtree	Small clear-cut/Plant, Shelterwood or Seedtree
Basal Area (sq. ft./acre)	> 120	50–90	55–85
Timber Harvest Rotation Length	15–25 years	20–45 years	35–70 years
Thinnings	Once or none	Every five years	Every four or five years
Fire in uplands	None or one every three years in winter	Every two to three years in winter	Every 2–3 years in winter, summer burn every 9–10 years, ring-arounds to protect nesting habitat
Site Preparation	Shear/pile & disk, broadcast broad spectrum herbicide	Chop/burn, selective herbicide, banded or spot treatment herbicide	Fire, banded selective herbicide, release after minimal site preparation
Windrows	Burn	Leave alternate windrows	Leave unburned
Seedling Spacing	6 x 6, 6 x 8	8 x 8, 8 x 10	10 x 10, 12 x 12
Hardwoods	Retain in drains	Retain in drains and mast producing clumps	Retain all mast producers
Food Plots	1–3% of land in chufas, clovers, and small grains	3–5% of land area in chufas, clovers, and small grains	Greater than 5% of land area in chufas, clovers, and small grains
Fertilization	None/Food plots	Food plots	Food plots, mast producers
Openings	1–3% of land maintained in pastures, fields, cropland, roadsides, right-of-ways	3–10% of land maintained in pastures, fields, cropland, roadsides, right-of-ways	Greater than 10% of land area maintained in pastures, fields, cropland, roadsides, right-of-ways
Turkey Harvest	Harvest all gobblers	Harvest some jakes	Harvest only adult gobblers

ing or grain harvest and soybeans and peanuts for their commercial value.

Of all supplemental foods specifically planted for turkeys, chufa, a nut sedge, is perhaps the most favored. Chufa does best on sandy loam soil on "new" ground or sites that have not been cultivated in the recent past. Chufa should be planted in plots 1 acre or larger. Five acres of chufa per section of land (640 acres) can produce a food source that may effectively offset an acorn failure. In wildlife plots of 2 acres or larger, chufa should be planted on part of the opening with the remainder planted in other turkey foods or maintained as natural openings. Turkeys that have never scratched for chufa may need help in finding it. This can be accomplished by running a disk across the plot to expose some of the tubers. To make it easier for turkeys to locate chufa, pull up and scatter several clumps from the patch from about November until the turkeys discover them.

Clovers are also an excellent source of supplemental food for turkeys. They have a high vitamin A content, valuable to pre-nesting hens, and also support high densities of insects and other invertebrate animals that are valuable for young turkey poults. Because of the high insect production in clover patches, fallow fields, and unimproved pastures, these areas are often referred to as bugging areas. Ladino clover holds up well and lasts longer into spring and summer than does crimson or white clover. Ladino clover is best suited to fertile clay

RICK CLAYBROOK

Chufa is a warm-season sedge that produces underground nutlets or tubers that are relished by turkeys.

and silty loams, while crimson and white clover are more widely adapted. The Tillman variety is a hardy clover that does well in places where others might fail, especially near the Gulf Coast. Combination plantings of crimson clover, vetch, and wheat or rye also do well together. Clovers may be planted on roadsides, powerlines and gaslines, and around edges and between rows in some croplands and orchards.

Clovers can also be utilized for both livestock and turkeys. When incorporating clover in a pasture program, clovers should be seeded around pasture edges which turkeys frequent most. Bahia grass and clover can be planted together in early spring. This combination will provide a year-round food source but requires periodic maintenance. While bahia grass is a good food source for turkeys, it is considered a pest plant in some areas because it may spread to unwanted sites. Otherwise, bahia is an effective pasture grass that works well in combination plantings if it is not allowed to get too dense.

Field corn provides an excellent fall and winter food source for turkeys. Only part of the crop should be made available at any one time. By periodically knocking down several rows of stalks, it will last much longer. Deer, squirrels, quail, raccoons, and other wildlife use corn during the winter, and therefore, plantings should be no smaller than 1 acre per plot. Where corn is planted in large acreages for silage or other purposes, leave several scattered patches standing near woodland margins for winter use by turkeys and other wildlife. Corn left on the ground in harvested fields will serve to feed many animals as long as it lasts.

Similarly, winter wheat, another traditional farm crop, is a good choice to plant for turkey food plots. If the patch is 1 acre or less, the entire opening should be planted in wheat, but on larger patches combine wheat with other crops. The same recommendations for winter wheat also apply for oats and rye; however, oats are a poor choice on low sites that have periods of standing water.

Grain sorghum, millet, milo, and Egyptian wheat should be planted in strips because a dense stand may be difficult for turkeys to traverse. The planted strips should be 3 to 4 feet wide, leaving about 2 feet between rows to allow easy traveling. Quail, doves, and other birds also use sorghum. If sorghum is planted for silage or other uses, leave unharvested strips around field borders for fall and winter food. Patches planted strictly for turkeys should range from about 1 acre up to 10 acres, depending on the expected use by other wildlife, particularly deer, which will

Volunteer Chufa

Dr. Lee Youngblood is a veterinarian and is a vice president of the Alabama Wildlife Federation from Selma, Alabama. He has been experimenting with various techniques to re-establish chufa without re-planting. For the past 15 years, he has worked to perfect a simple and effective method of establishing volunteer chufa patches on his 1,500 acre hunting club. Each year he successfully volunteers about 20 acres of chufa in 1 to 2 acre patches across his property. In his own words, he describes how his volunteer chufa program works.

For years my hobby has been experimenting with different techniques to improve wildlife habitat on my land. Through trial and error I've discovered some techniques that really produce results. One technique that I've had a lot of success with over the years has been re-establishing chufa without reseeding in plots that were planted the previous year. I've been able to grow beautiful chufa plots two to three years consecutively without having to replant. In some cases I have extended the original planting up for five years. The advantage of having volunteer chufa is that it is cheaper (there is no seed cost), and with my method weed control is more effective.

The theory behind getting volunteer chufa to grow is that it responds favorably to repeated disking. The more frequently you disk, the more prolifically the plants respond. To begin with you must have had a successful chufa plot established from seeds of the prior year. A person can't get chufa to volunteer if it was not planted in the plot the previous year. Don't worry about there not being enough residual chufa in the plot from the year before because turkeys and other wildlife will leave enough leftover chufa.

I like to disk my one year-old chufa plots the following spring and summer at least three times. In early April I start off with a light disk harrow and disk enough for bare ground to show. Afterwards the spring rains will cause the disked chufas to sprout and begin growing. Around mid-May I'll disk the plot again a little heavier than previously. About the second week in June I'll disk the plot a third time. Where crabgrass and other unwanted weeds are a problem I'll apply 1 to 2 quarts of Treflan® per acre; however, the majority of the time repeated disking will discourage most weed competition. After the last disking I'll also apply a fertilizer, according to a soil test, or 300 pounds of 13-13-13 per acre. On sandy soils, where chufa grows best, fertilizer application should only be released on the surface of the ground and not disked into the plot. Rains will eventually cause the fertilizer to be released and made available for chufas. Several weeks after chufa plots are established, I sometimes come back and apply nitrate at a rate of 200 pounds per acre. I only do this if my chufas are not green or growing like they should.

One of the biggest mistakes in re-establishing chufa is to not continue disking the second or third time. Sometimes it may be tempting not to re-disk a chufa stand that has come up nicely. Resist the temptation! Go ahead and re-disk. It will come back better than before. In some situations stands which come back may appear too thick. This is not as big a problem if initial chufa plots are seeded in rows and not broadcast.

As a last bit of advice, suppose after disking and re-disking you fail to establish a stand of volunteer chufas. (From my experience this is rare and will only happen if there were no chufas the year before, or if dry weather conditions prevented chufas from resprouting.) If this is the case, all is not lost. Chufa seed can always be broadcast in a plot up until the middle or latter part of June and still provide an excellent food plot for turkeys.

consume a large percentage of the crop.

For more information on plant varieties and cultural requirements for wild turkey plantings see Chapter 16 and Appendix O–J.

PLANTING CHUFAS FOR TURKEYS

Chufa is a warm-season sedge that produces underground nutlets or tubers that are relished by turkeys. Since chufa is also preferred by other wildlife, plots sizes should be no less than 1 acre. Chufa does best if planted on sandy and loamy soil and on areas that have never been planted before, but it can also be established on other areas. Forest clearings such as log landings and wide, abandoned or closed roads that receive full sunlight are good locations for planting chufa. Chufa plots should be well distributed across the landscape at a rate of about one plot per 100 acres of turkey habitat.

Chufa should be planted between May 15 and June 30. The best dates for planting are between June 15 and June 30, when there is enough soil moisture for germination. If chufa is planted before June 15, the nutlets mature about the time acorns drop in the fall. If both chufa and acorns are available at the same time, turkeys will feed first on acorns, leaving the chufa nutlets for raccoons and other wildlife. However, if chufa is planted after June 15, chances are good that most of the nutlets will be available to turkeys after the acorns are gone.

Openings for chufa plantings should be prepared by disking plots several times at different time periods beginning a month to several weeks before planting. This will help reduce weed competition with chufa. Soils should be tested in advance for determining proper fertilizing and liming rates. As a general rule, fertilization rates are about 300 to 400 pounds of 10-10-10 fertilizer per acre applied at planting. Once the chufa plants reach 8 to 12 inches high they can be top-dressed with about 100 pounds of ammonium nitrate per acre. In acidic soils, lime should be added to maintain soil pH between 5.0 and 7.5. Chufa can be planted by broadcasting tubers at a rate of 30 to 40 pounds per acre, or by row planting at a rate of 20 pounds of tubers per acre in rows 36 to 42 inches apart. Row planting takes fewer tubers to establish a plot and allows for cultivation between rows to reduce weed and grass competition. Row planting, however, is more susceptible to damage by nuisance animals like raccoons.

When the tubers mature in the fall, some portions of the chufa plots should be pulled up by hand or run over with a disk harrow to expose the tubers, making them easier for turkeys to find. This is especially important during the first year or two until turkeys become accustomed to finding the nutlets.

Chufa production can be extended each year by disking and fertilizing the previous year's plot during the spring. However, to obtain the best yields, plots should be replanted every year. After two or three successive years, chufa plots should be relocated to a different site to help reduce nematode infestation and the chance of disease transmission. The nutlets are less likely to be damaged by weevils, during the first two or three years. Wild hogs and raccoons will heavily utilize chufa plots and can decimate plantings where present in high concentrations. If wild hogs and raccoons are a potential problem, chufa plots should be located in upland areas away from streams and wetlands. As a final note, if livestock have access to the property, chufa plots should be protected from grazing by fencing.

Direct Feeding

The direct feeding of grains or other foods has limited value in sound turkey management. Feeding may concentrate human or natural predators and increase the potential for disease transmission, such as histomoniasis, at long-term feeding sites. If you choose to feed, feed sites should be rotated to prevent pathogen build up. Corn is often used as a direct feeding grain since it is high in carbohydrates and energy; however, it offers little of the other nutrients needed by turkeys. In addition, aflatoxin, a fungus often found in corn, can be a problem with turkeys, often causing death. Some corn that has higher levels of aflatoxins than can be legally marketed for livestock is often sold as deer or "wildlife" corn. Because of the potential problems associated with aflatoxins, aflatoxin corn should not be used for wild turkeys. As a final note, landowners and hunters should be aware that hunting turkeys over or in the proximity of direct feeds like corn is illegal in Alabama. Any trace of feed must be gone at least 10 days prior to hunting. This does not mean that you should stop feeding 10 days before the season opens. It means that all traces of feed must be gone at least ten days prior to hunting in the area. If any grain is present within 10 days prior to hunting, even a hunter without knowledge of the existence of the feed will be in violation of the law. For further clarification of what constitutes turkey baiting contact your local ADCNR law enforcement officer.

HUNTING

The sport of hunting wild turkey in the spring is a revered tradition in Alabama. The excitement of fooling an old long beard by imitating a hen is truly unparalleled. The thrill of calling an excited gobbler is

TES RANDLE JOLLY

Chufa can be planted by broadcasting seed or by planting in rows as pictured here. Row plantings take fewer seeds to establish a plot and allow for cultivation between rows to reduce weed and grass competition.

beyond description. The Alabama season, bag limit, and harvest are some of the most liberal of any state. The spring season runs from shortly after mid-March until about the end of April, and there is a fall season in some counties. The bag limit is one gobbler per day with a season limit of five. In recent years the state harvest has approached 50,000 turkeys each year. It is amazing to consider that this number is several times the estimated state turkey population of just a few decades ago, before the time of better management by landowners.

THE NATIONAL WILD TURKEY FEDERATION

The National Wild Turkey Federation (NWTF) is a non-profit organization dedicated to the wise conservation and management of the American Wild Turkey. Comprised of over 600 state and local affiliates, the NWTF supports educational and research programs concerned with enhancing the wild turkey in the U.S. We would like to thank Lynn Boykin, Dr. James Earl Kennamer, and Dr. Jim Dickson for their contributions to this book. For more information about the NWTF, write the National Wild Turkey Federation Inc., Wild Turkey Bldg., P.O. Box 530, Edgefield, SC 29824-0530 or call 803-637-3106.

SUMMARY

Historically, wild turkey populations in Alabama have fluctuated from being abundant in pre-colonial times, to extremely low populations in the early 1940s, to an estimated 500,000 birds in the state today. The comeback success story of the wild turkey in Alabama is the result of sound wildlife management practices that have included habitat improvement practices and protection. These same successful management techniques are described in this chapter and can be applied by any forest or farm owner. The key to having turkeys on your land is understanding the biology, life history, and habitat requirements of the wild turkey; tailoring management efforts to meet these needs; and understanding management options and their impacts.

The bobwhite quail has been considered the premier game bird in Alabama. A combination of factors has caused quail numbers to decline in Alabama and across the Southeast.

GLENN D. CHAMBERS

CHAPTER 7

Managing for Bobwhite Quail

CHAPTER OVERVIEW

This chapter describes the bobwhite quail in Alabama, its life history, biology, and habitat requirements. Common myths about this premier game bird are also discussed to examine possible causes for its decline. Because of their early successional-stage habitat requirements, quail management can be the most intensive of any game species in Alabama. This chapter provides an in depth review of forest and farm management strategies for bobwhite quail.

CHAPTER HIGHLIGHTS PAGE

BIOLOGY AND LIFE HISTORY 166
FACTORS LIMITING POPULATIONS 167
MYTHS AND MISCONCEPTIONS 168
HABITAT REQUIREMENTS 170
OTHER HABITAT COMPONENTS 172
HABITAT IMPROVEMENTS 173
FARM MANAGEMENT FOR QUAIL 174
FOREST MANAGEMENT FOR QUAIL 178
HUNTING QUAIL 181
ORGANIZATIONS DEDICATED TO QUAIL 183

The bobwhite quail (*Colinus virginianus*) has long been a favorite of hunters across Alabama. Although not nearly as abundant now as in the past, each year approximately 18,000 hunters take to the field and harvest 350,000 quail. A historical review reveals that quail populations have increased and declined as a result of natural factors and by land-use and management activities of man. Pre-colonial quail populations were probably low, due to vast, essentially unbroken tracts of mature forests that provided poor quail habitat. Early land-use practices associated with pioneer settlements created a patchy farming pattern that provided ideal quail habitat. Consequently quail numbers began increasing around 1800 and peaked in the early 1900s.

From the early 1900s to the mid-1940s, quail population densities remained high and quite stable. However, since the mid-1940s, quail numbers have declined throughout the South. This downward trend is largely associated with deteriorating habitat conditions resulting from many factors including a change to "cleaner" and more mechanized farming methods. Other factors include the joining of small patchwork fields to make large, unbroken fields suitable for intensive cultivation and the development of pastures for cattle. Much quail habitat has been lost as a result of abandoned farmland reverting to forests. In addition, intensive timber production has, to a lesser degree, reduced the amount of available quail habitat. Many forests do not support quail because the tree canopies are closed and herbaceous plants that provide valuable food and cover are shaded out. In addition, fire is not as common in today's forest as in years past and that has resulted in the growth of dense midstory shrubs and trees which eliminates herbaceous plants needed by quail. Ideal quail range generally includes approximately 50 percent open woodland and 50 percent fields.

A loss of ideal habitat in combination with predation are probably two of the most important reasons why quail numbers are so low in many parts of the Southeast today. The most valuable information gained through 75 years of research on bobwhite quail is the realization that every situation is different and dynamic, and that every quail population is dependent upon delicate balances unique to its own habitat. Ongoing research is seeking to find the weak link in the life-cycle of bobwhites so that managers can apply this knowledge to increase quail numbers. The challenge for managers today is to incorporate what knowledge is currently available to help quail through the many obstacles that threaten their survival.

BIOLOGY AND LIFE HISTORY

When days get longer and warmer, and the greening of foliage and flowering of plants begins, the bobwhite's whistle is one of the early signals of the coming of spring and summer. Shortly after the first bird is heard, groups of quail or **coveys** begin to break up around the third to fourth week in April, and courting pairs are often observed during this time. After initial mating has taken place pairs may stay together until the nesting and raising of the chicks is complete.

Recent research reveals that bobwhite quail have a complex mating system. During the course of a nesting season that includes initial nesting and re-nesting attempts, and continuous mortality, most birds will be associated with more than one mate. Re-mating must occur for re-nesting attempts whether both members of the pair are still alive or not. Some adults, however, develop a strong pair bond.

After choosing a nesting site, the pair gathers available dead plant material (dry grasses, stems, and pine needles) and constructs the nest in a slight depression in the soil. The female generally lays the first egg within a few days after the nest is finished, and usually will continue to lay one egg daily until the clutch is complete. Clutch size averages about 13 eggs in Alabama, with original nesting attempts containing a slightly greater number of eggs and re-nesting attempts slightly less than the average.

Females normally incubate around 75 percent of the clutches; a process that lasts for 23 days. Hens sometimes lay a clutch that is incubated by the male. Studies have shown that a male incubates about one out of four nests. Quail almost never share incubation duties, and in most instances a pair of quail hatches only one brood of young per year. Occasionally, during periods of favorable nesting conditions, quail have been known to raise two broods. This may happen in several ways: a hen may hatch a brood that is cared for by the male, and the hen subsequently re-nests; the hen may lay a clutch that is incubated by the male, and the hen lays and incubates another clutch; or the hen hatches and raises a brood to three to four weeks of age early in the summer, then incubates and hatches a second brood. If the nest is destroyed or abandoned prior to hatching, quail will attempt to nest again until achieving a successful nest or until the nesting season is over. Few hens are able to hatch a brood on their first attempt and generally bobwhite nesting success is low. Typically less than half of all nesting attempts are successful. However, 70 to 80 percent of hens that survive the summer eventually hatch a brood due to re-nesting attempts. Generally

PAUL T. BROWN

TED DEVOS

Female bobwhite quail (left) begin searching for nest sites in mid to late April and nest as early as May. Nesting habitat (right) composed of dry grasses and briars is critical for successful hatching, which can occur from May to September.

less than 25 percent of these broods will have been raised by males.

Quail nest in Alabama from early May through September with most of the hatching occurring from June to August. There are commonly two or three hatching peaks since quail usually re-nest several times over a period of three to four months. In most of the deep South, the greatest peak in hatching is in late June and early July. Most nest failures can be attributed to predation but others may fail due to harsh weather conditions, land-use activity, or other environmental factors. Nest failures, however, are not necessarily bad in that they spread out hatching throughout the summer and therefore reduce the total effect of any mass mortality of the young due to natural causes. Late-hatched birds have a greater chance of surviving until the hunting season, and a high percentage of late-hatched birds is generally associated with good fall hunting.

Newly hatched chicks are covered in natal down and are totally flightless for about two weeks. This time of life is especially critical for the newly hatched young; losses to predation and adverse weather (with the young becoming wet and hypothermic) may take 80 percent or more of the brood. Research indicates 60 to 70 percent mortality of chicks in the first two weeks of life, and 70 to 85 percent mortality prior to one month of age. After this period, it appears that mortality is similar to grown quail.

Chicks may be raised by one or both parents. Often broods that have lost their parent(s) may become adopted by unrelated adults. The adult and brood may cover up to 100 acres, but 15 to 20 acres is common depending on the habitat. At about six weeks of age the juveniles are fully feathered and their diet begins shifting from insects to seeds and fruit. At 16 weeks of age the young are considered full grown. By early fall, the young of the year will normally make up approximately 80 percent of the population.

FACTORS LIMITING QUAIL POPULATIONS

On most areas of the Southeast, quail populations over the years have shown a cyclical pattern of gradual increases for several years followed by severe declines. This pattern is thought to be influenced by changes in land-use practices, long-term weather trends, predation, disease, and quite possibly the inherent cyclical nature of quail populations. All these factors working together for an extended period of time may be the cause of the declines in quail seen in many areas across the Southeast, as compared to higher populations in the 1970s.

Natural predation, hunting, disease, exposure, and other mortality factors take about 80 of every 100 birds from one fall to the next. Most adult quail predation occurs in late winter through early summer when cover conditions are weakest and many migratory raptors are present. Quail are also vulnerable during the breeding season when mortality of nesting hens and especially chicks is quite high. Avian predators such as the Cooper's and sharp-shinned hawks take their share of quail from late winter through summer. Predation rates vary depending on the number of hawks migrating through an area and the quality and quantity of escape cover.

Other predators that prey on eggs and chicks include raccoons, opossums, skunks, free-ranging house cats and dogs, bobcats, foxes, coyotes, snakes, and cotton rats. The extent of quail predation depends upon the quantity, quality, and distribution of escape cover and predator densities. In areas of high quail densities, fall mortality can also be quite high and natural diseases such as pox can be prevalent.

Predators also prey on animals that compete with quail for the necessities of life. For example, during intensive efforts to increase quail populations on one Georgia plantation by killing or trapping all hawks, owls, foxes, cats, and skunks, a decline in the quail population occurred. It was later learned that due to this intensive predator removal, the cotton rat population had increased and the rats were destroying a majority of quail nests, resulting in fewer quail being produced. Nest predation may serve to spread the hatch through the summer and adult predation may help remove slow or weak birds.

Weather has a significant impact on quail production. During nesting season, when dry weather prevails, cover is thinner and food production and quality are lower. During these times predation on quail nests is heavier. When rainfall is adequate during the summer, cover is more prevalent, and natural and cultivated food production is higher. During these times nest predation on quail is lower because predators have an abundance of alternative foods or **buffer foods** that helps to reduce predation on quail. Nest losses are possible with heavy rains and flooding. Poor quail production has also been related to droughts, which may cause chicks to have a hard time breaking out of dry egg membranes and may dramatically reduce the amount of insects available to chicks. Weather extremes, either too wet or dry, likely decrease quail production.

To reduce quail mortality, the most practical solution is to improve habitat conditions so that an area can produce and carry a larger number of quail. Predator control, through legal trapping, should also be a component of quail management. Most field observations have noted that those lands that are managed to provide quality habitat for quail experience less predation than other areas. As conservationist Aldo Leopold noted, "If a habitat can't support game in spite of predators, it simply isn't good game habitat." One thing is for certain though: the debate about the impact of predators on bobwhite quail populations will continue until research clearly documents the role that predation has on quail.

MYTHS AND MISCONCEPTIONS ABOUT BOBWHITE QUAIL

Over the decades, the popularity of the bobwhite quail has spawned a number of popular beliefs and tales. Many are local while others are heard throughout the range of the bobwhite. Nearly all have been handed down through successive generations of hunters, landowners, and other quail enthusiasts. While these beliefs add to the appeal of the bobwhite and reflect a popular interest, many have actually taken these stories to heart and ultimately wasted time and money on misguided efforts to increase quail numbers. The following are some common myths concerning bobwhite quail.

Myth 1—Wild quail populations can be increased by stocking or releasing pen-raised quail.

Studies at the Tall Timbers Research Station at Tallahassee, Florida and elsewhere have found contrary evidence. In relationship to the number of quail released, only a small percentage survive a year after release, and only a few nest and produce broods. In some recent studies, however, up to 20 percent survival has been observed from fall to spring, and 5 percent to 7 percent have been documented to survive annually with full-fledged nesting and brood-rearing.

Myth 2—Many quail found in the Southeast today are descendants of a subspecies of bobwhite commonly called Mexican quail.

Between 1925 and 1928 approximately 400,000 Mexican quail were released across the Southeast. However, research has documented that relatively few of the Mexican quail actually survived, and that the characteristics and traits of those which did interbreed with the native bobwhites were quickly diluted and have become obscure with time.

Myth 3—Quail declines in the Southeast are due to increasing predators.

It's true that raccoons, opossums, snakes, foxes, hawks, cats, and other predators take their share of quail and quail eggs, but other factors also are thought to be influencing quail decline. Some of these factors include loss or deterioration of habitat, changing land-use patterns, and other mortality factors.

Pen-Raised Quail

Ted DeVos conducted research on the ecology of northern bobwhites at Tall Timbers Research Station near Tallahassee, Florida. As a graduate student at Auburn University, he conducted research on wild and pen-raised bobwhites in Alabama and Georgia. Currently Ted is a biologist and forester in charge of timber and wildlife management on lands managed by Regions Bank in Montgomery, Alabama.

As native bobwhite quail numbers have declined across the South, many wildlife managers and landowners have turned to releasing pen-raised quail on designed shooting courses. Pen-raised birds provide an opportunity for hunters to find a lot of birds in a short amount of time. These birds are purchased from a reputable producer and released either one or two months before the hunting season in what is called a preseason program or just prior to shooting (put and take). The advantages of releasing pen-raised quail early is that they usually become accustomed to the wild and develop flight and behavior characteristics similar to wild quail. The disadvantage of releasing pen-raised birds early is that a greater percentage will be lost to predators before hunting season opens.

Releasing pen-raised birds should **not** be viewed as a method to restock areas lacking sufficient wild quail numbers, since annual survival of released birds is usually quite low. Fall released birds which have survived to spring have been documented to produce offspring in the wild with pen-raised parents or pen-raised wild bird crosses. Either way the offspring have similar behavior and survival characteristics of wild birds.

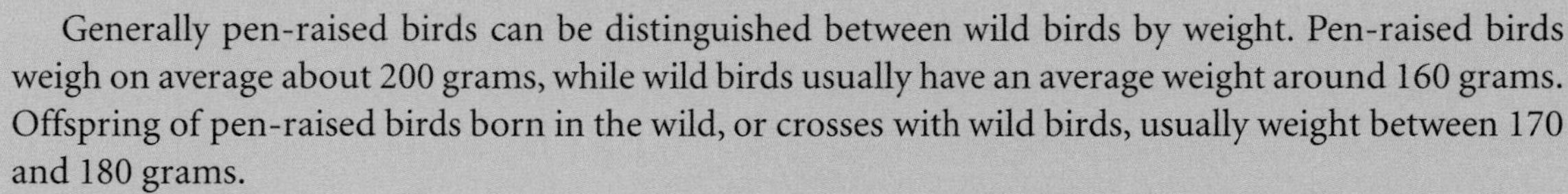

Generally pen-raised birds can be distinguished between wild birds by weight. Pen-raised birds weigh on average about 200 grams, while wild birds usually have an average weight around 160 grams. Offspring of pen-raised birds born in the wild, or crosses with wild birds, usually weight between 170 and 180 grams.

Pen-raised quail are one alternative to provide consistent viewing and hunting opportunities for sportsmen. They provide an opportunity to train bird dogs, and when properly released, can generate three to five covey rises per hour throughout the season. In a compromise between pen-raised birds and native quail, some Alabama managers have dedicated a particular block or area to release-bird shooting, while relying on native birds for a majority of the property. In this manner, sportsmen who prefer consistent quail shooting can enjoy the opportunities that released birds provide, while others can hunt wild quail in a more traditional fashion.

Myth 4—Releasing wild hens in areas where there is an abundance of unmated cocks will increase the quail population.

This idea is erroneous or impractical for several reasons. First, a hen may be trapped that is already mated and in some phase of nest-building or incubation. If this happens one potential hatch of young birds may have been destroyed. Also if the hen mates with an excess cock and begins nesting in unfavorable habitat, the chances of a successful hatch are extremely low.

Myth 5—Shoot as many quail as possible during the hunting season to prevent inbreeding.

This idea has been unconditionally disproved. Birds hatched during summer may move to several different coveys during fall covey formation. This movement between coveys continues during the early winter months, and during covey breakup in the spring, individual birds may move a distance of several miles. Because of this intermixing, inbreeding is not a concern with wild bobwhite quail. On the contrary, shooting too many birds out of coveys is a common management mistake. No more than 20 percent of each covey should be taken.

HABITAT REQUIREMENTS

Two factors should be understood by any landowner wishing to manage for bobwhites. Those factors are that quality quail management is expensive and intense. Assistance from an experienced professional is recommended. More money has been wasted by landowners attempting unproductive quail management techniques than on any other type of wildlife management. Understanding the habitat components that bobwhite quail require is the first step toward successful management. In certain seasons of the year and during different stages of their life cycle, quail require specific types of food and various types of cover. Having a wide variety of available foods throughout the year, in combination with specific cover types all arranged for easy access, can provide ideal quail habitat.

One of the most important factors known to determine the potential quality of quail habitat is soil. Fertile sands, sandy loams, and light red clays are some of the best soils for producing quality quail habitat. These are the same soils that are in high demand for agricultural production. The beauty of these light soils is that the woodland grasses and weeds grow in a way that leaves some bare dirt in between the clumps. Quail are weak scratchers and need bare ground to find food and walk through the vegetation. This is the reason that pasture grasses, like fescue and bermuda, are so poor for quail and that fire is so important for quail management. The heavier clay soils in central Alabama can cause additional problems, drying hard and becoming sticky when wet.

Food

Food is an essential requirement for quail survival that managers can influence and is an important component of quail management. Landowners benefit by knowing the important quail food plants and becoming familiar with favored food items.

Quail primarily feed in fields and open forests. Their diet is mainly vegetative and composed primarily of seeds, small fruits, and green forage. Animal matter may be consumed year-round, but it makes up the majority of the diet only during the warmer months. Insects provide an especially critical source of protein in the diet of young quail and are eaten almost exclusively during the first two weeks of life. Seeds of various legumes, like beggarweed and lespedezas, are probably the most preferred native quail foods, with grasses and sedges of secondary importance. Soft and hard mast and cultivated grains are also eaten readily. Foods must be close to cover to be of the greatest value to quail.

Cover

Ideal cover types required by quail are often deficient in Alabama. Managers need to know what constitutes optimal quail cover. In any instance, the amount of additional cover needed depends on the quality of what is already available. Preferred cover types are relatively dense above the ground yet open at ground level.

In general, quail prefer or select a diversity of cover types found in forests, brush, grass openings, and cultivated lands. Research has shown that quail require a variety of cover types for different functions and activities throughout the year. These cover types include the following:

NESTING COVER

Nesting cover is one of the most crucial habitat components determining the success of nesting and hatching. Since most quail live less than one year, population increases depend on annual nesting success. Studies suggest that lack of nesting cover may be a factor that is limiting quail in some areas of the South. Quail nesting cover is characterized by moderately dense mixtures of grasses and broadleaf weeds with nearly bare ground around grass clumps. Good nesting cover contains scattered broomsedge. Most nests in Alabama are constructed at the base of a broomsedge clump. Ideal nesting cover consists of herbaceous plants that grew during the summer before the current nesting season. Cover that is at least one year old at the onset of the nesting season provides reasonably secure sites for nesting. The cover may attract nesting birds for a few years, but soon becomes too dense at ground level for ease of movement. Periodic burning during late winter or early spring every two to four years maintains herbaceous plants in the proper stage for nesting. When prescribed burning for quail, at least one-fourth of the cover should remain unburned so that adequate nesting sites are available when nesting begins in the spring. The unburned cover should be on well-drained upland sites and scattered in blocks, especially in and around field edges. Vegetation should be disked during winter or early spring about every three to five years to maintain it in a stage suitable for nesting. Only a portion of the nesting areas should be disked in any one year so that adequate cover is always available.

BROOD-REARING COVER

Brood-rearing cover, like nesting cover, is an important habitat component that provides protection for chicks during feeding and contains insect-rich mixtures of annual legumes and forbs along with bare ground for ease of movement. The insects meet the chicks' high demand for protein, and the plants provide an overhead protective

FIG. 1: SEASONALLY IMPORTANT FOODS OF BOBWHITE QUAIL IN ALABAMA

Spring	Summer	Fall	Winter
insects	insects	acorns	acorns
green vegetation	clovers	insects	insects
soft mast (blackberry)	vetches	annual lespedezas	annual lespedezas
panic grass	green vegetation	bush lespedeza	bush lespedeza
clovers	soft mast (huckleberry, blackberry, blueberry, poison oak, wild grape, black cherry, plums)	panic grass	cowpeas
wheat	panic grass	Florida beggarweed	soft mast (gallberry, dogwood)
partridge pea	bahia grass	soft mast (American beautyberry, plum, wild grape, persimmon, dogwood, blueberry, black tupelo)	pine seeds
wood sorrel	wheat	pine seeds	partridge pea
early producing forb seeds	browntop millet	corn	corn
		soybeans	oats
		clovers	clovers
		millets	soybeans
		ragweed	sorghum
		milkpea	green vegetation
		butterfly pea	
		sunflower	
		vetches	
		sorghum	

canopy. Good brood habitat is found in fallow fields that are growing annual weeds, particularly ragweed and partridge pea. Brood habitat can be maintained in these fields by disking or mowing annually during fall to early spring and allowing them to become fallow during summer. Millet patches, old fields mowed in winter, and burned pineywoods also provide ideal brood-rearing habitat.

ROOSTING COVER

Roosting cover is composed of grasslands with vegetation approximately 2 feet high with scattered hardwood sprouts.

Quail eat a variety of food items throughout the year. Examples of foods that are seasonally important to quail include (top row from left to right) butterfly pea, milk pea, blue curls, (second row from left to right) black cherry, grasshopper, (third row from left to right) sumac, partridge pea, panic grasses; (fourth row from left to right) bicolor lespedeza, and acorns.

ESCAPE COVER

Escape cover consists of shrubby or woody areas like brushy fencerows or field dividers. Some of the better escape cover plants include wild plum, wild cherry, sumac, sweetgum, greenbriar, palmetto, viburnums, sassafras, blackberry, honeysuckle, and grapevines. In most areas sufficient escape cover may be developed simply by protecting areas from disturbances such as fire, disking, mowing, and overgrazing. Escape cover is often deficient in open, fire-maintained habitats in the Southeast.

THERMAL COVER

Thermal cover is composed of dense ground cover under a woody canopy that provides protection from extreme temperatures.

OTHER HABITAT COMPONENTS

Although quail are commonly observed in the vicinity of open water and occasionally drink surface water, drink-

TED DEVOS

RICHARD T. BRYANT/ARISTOCK

TED DEVOS

RICHARD T. BRYANT/ARISTOCK

Cover is a critical habitat component for quail. Land management practices should strive to maintain and provide nesting, brood-rearing, roosting, escape, and thermal cover for quail. Habitat types important for quail include shrubby nesting cover (top left), multi-use burned woodlands (top right), agricultural and fallow field cover for brood use and fall feeding areas (bottom left), and protective thicket cover (bottom right).

ing water is not essential since they obtain most of their water requirements from dew, insects, and succulent green vegetation. Water availability for quail is not a limiting factor in the Southeast, as it can be in the Southwest, since rainfall here is over 60 inches each year and there is an abundance of ponds, lakes, streams, and rivers.

Quail also prefer areas of loose bare ground where they can periodically dust themselves to get rid of external parasites. These "dusting" areas are usually located where land management activities have disturbed the soil, such as along dirt roads, forest regeneration and harvest sites, firelanes, or along the edges of cultivated fields. Other odd areas are also utilized by quail such as bare soil or rock surfaces where quail dry off after heavy rains, or elevated sites for singing posts used by cocks trying to attract hens.

HABITAT IMPROVEMENTS

The only way to permanently increase quail populations on any area is through the maintenance and development of quality quail habitat. Quail must be recognized as a product of good habitat management.

Landowners, hunters, and land managers should reflect back to a time when quail were abundant and think about where they consistently found quail. Most of these areas had similar habitat components that included adequate cover and important food plants, either cultivated or native. If either of these components is lacking, develop plans to provide these components across the farm or forest land.

TED DEVOS

Quail prefer areas that have loose bare ground where they can periodically dust themselves in the dirt to get rid of external parasites. Bare dirt is an important component of nearly all quail habitat types, not only for dusting but also for ease of movement and aid in locating foods.

FARM MANAGEMENT FOR QUAIL

Modern and intensive farming methods, coupled with the high cost of farm machinery, dictate that wise landowners seek the most return for their investments. This in turn has led to the abandonment of tenant-type farming, where the common practice was to leave nonproductive field edges and temporarily fallow fields in native vegetation. While this could not be considered progressive agriculture, it did provide ideal habitat for quail. The present-day practice of plowing and planting fields to the edge has done little to increase total agricultural production, since returns are poor on crops planted in field borders that are often shaded from the sun. These practices, coupled with the removal of fencerow habitat, have reduced the ability of most agricultural land to support an abundance of quail.

Just as important as developing new food and cover for quail is the preservation of existing food and cover. Land-use practices such as intensive farming, extensive pasture management, and commercial and residential development are destroying quail habitat at an alarming rate. Much of the farmland developed in the past was of marginal value to the farmer, and a hefty portion of this land was cleared at a cost that often exceeded returns from farming. For example, there may be only 5 acres of land set aside in 2 miles of an overgrown fence row 20 feet wide, but those 5 acres could provide food and cover for several coveys of quail.

Factors That Affect Quail Populations on Farms

TYPE OF CROPS IN FIELDS

Crops that have both a food and cover value for quail, like soybeans and corn, are more valuable to quail than other crops like cotton, which provide only overhead cover during the summer.

CHARACTERISTICS OF FIELDS

Fields that provide adequate food and cover (nesting, brood-rearing, and escape) are important habitats for quail. Field size should be relatively small to provide optimum edge. Borders and idle areas of fields should be managed for feeding, nesting, and brood-rearing cover, and for winter-roosting cover. Large fields should be broken up with pine strips and fallow ground.

CHEMICAL USE

Insects, including those that are agricultural pests, are also important food sources for quail. Not only do certain insecticides reduce the availability of insects for quail, in some cases they make quail sick and vulnerable to predators or kill birds that ingest chemically treated bugs. Farmers should read insecticide labels carefully and avoid using those that are toxic to quail. Field borders that provide valuable habitat for quail should not be treated with insecticides.

HIGH VALUE AREAS

Most farms have areas that already provide excellent habitat for quail in odd areas. These sites should be retained for quail. Some of these areas include plum and sumac thickets, blackberry and honeysuckle patches, and other areas that provide food and cover for quail.

Developing Food

Managing for quail food plants is generally no more complicated than being familiar with the locally preferred plants and maintaining these in conjunction with farming practices. This can be accomplished by manipulating plant succession to favor native grasses and legumes through prescribed burning, mowing, or disking. Food can also be provided from agricultural food crops such as soybeans or supplemental plantings intended specifically for quail.

Since the cost of various management practices can differ, it would probably be better to consider the least expensive and also the least labor-intensive method of increasing quail food supplies. Prescribed burning is usually the cheapest method of manipulating the land to produce desirable quail food plants, followed by disking. Old fields that contain broomsedge require periodic disking to prevent thick mats of broomsedge from forming, and stimulate the growth of native quail food plants such as beggarweed and partridge pea. Disking strips 50 to 75 feet wide in winter and early spring can provide hens and young chicks with unrestricted access to these areas for feeding. The new succulent vegetation that grows in these strips will also attract insects. These areas can be managed by re-disking every other year and applying fertilizer to the strips every third year. Studies at Tall Timbers Research Station have shown that seasonal disking favors certain plants over others. For example, disking in winter produces heavy-seeded quail foods such as ragweed and partridge pea, while disking in April increases the production of important seed-producing grasses such as panic grass. Disking in June was found to favor green vegetation that attracts insects and a number of major seed plants that quail readily feed upon in the fall. Disking in the summer, however, stimulates plants such as Florida pussley, poor-joe, and

LEWIS O. ROGERS/SOUTH CAROLINA DNR

LEWIS O. ROGERS/SOUTH CAROLINA DNR

Disking (left) and prescribed burning (right) are two important habitat improvement practices that stimulate the growth of grasses and forbs that maintain the correct cover conditions for quail. These practices also maintain insect habitat, which is essential for growing quail chicks.

blackberries. In general, seasonal disking can provide a diversity of seed-producing plants for quail.

Strips should always be established close to adequate cover. If an increase in seed production is desired, fertilizers recommended for legumes (which require no nitrogen) can be used at rates dictated by a soil test. Fertilizers may be applied to strips shortly after the disking is completed to enhance native plant production. The results of fertilization should be closely observed to see which plants benefit, since in some instances undesirable grasses, like bermuda grass, fescue, and crabgrass, may be encouraged by the increased soil fertility, especially if used in conjunction with summer disking or mowing. These weeds will often out-compete native quail plants and reduce seed production, or they may choke out the plants completely. On poor soils, some quail managers have had success increasing seed production of native quail plants by raising the pH levels of acidic soils by applying lime. This enables plants to better absorb soil nutrients. In another study in the Piedmont of Alabama, researchers more than doubled the coverage of quail food plants, primarily legumes, by broadcasting basic slag at the rate of 1 ton per acre.

Another method of providing food for quail is through the proper manipulation or harvest of select row crops. Corn provides carbohydrates while native plants in and around field borders provide additional seed and insects. Soybeans, smaller grains, and sorghums are also beneficial crops.

Present-day methods of planting corn in thick stands, especially for silage, have reduced the overall value of these fields for quail since dense stands and herbicide applications seriously reduce the volunteer growth of annual weeds and other plants preferred by quail. All agricultural harvest methods, however, leave behind some grain that will provide quail food for a short time after harvest. If possible, a couple of rows of corn or other grain crops should be left standing around the edges of fields after harvesting is complete. Portions of these set-aside strips may be mowed at intervals during the winter months to supply food for coveys.

Soybeans and grain sorghums are also valuable to birds when a few rows are left standing on field edges. The vegetative part of these plants offer some additional cover, and in most cases, seeds will slowly shatter and provide a source of food over a longer period of time. All of these food supplies should be left as close to available escape cover as possible. Plowing agricultural fields in late summer or fall should be avoided whenever possible, except where wet winters preclude disking. In this case, fall disking may be the only way to stimulate winter legumes. This practice reduces available feeding cover for quail. If necessary, however, unplowed strips containing crop residue should be left around field edges.

In areas where disking, burning, or available row crops do not provide sufficient food for quail, supplemental plantings of various high-quality quail food plants may be established. These plantings may be divided into two general types. One is for fall and winter use to concentrate birds for hunting, and the second is for spring and summer use for brood-rearing. Both annual and perennial plants are available. A combination of the two is probably best. Fall plantings help to hold birds during the hunting season and may draw some birds from adjoining unmanaged land. Late winter food helps to make an area attractive to the birds year-round.

Numerous plantings have been used to provide food for quail. Some of the better ones for fall use include the annual lespedezas (common, Korean, and Kobe) and Florida beggarweed. Perennial late winter foods normally planted in Alabama include bicolor lespedeza and large partridge pea. All of these plants require proper seedbed preparation and fertilization. In some areas of Alabama, bicolor patches are difficult to establish due to high numbers of deer that can easily overbrowse a bicolor planting. Past research has indicated that two other shrub lespedezas, *Lespedeza thunbergii* and *Lespedeza japonica*, may be more resistant to deer depredation. Amquail, an improved variety of *Lespedeza thunbergii*, has also been proven to be somewhat resistant to deer browsing. Bicolor lespedeza can become invasive, especially on heavy clay and loamy soils that are frequently burned. Because burning tends to enhance bicolor growth, it should be excluded from bicolor plantings on these soils. Otherwise, bicolor should be planted sparingly or not at all. *Lespedeza thunbergii*, a relatively new plant for quail, has not exhibited the tendency to spread and become invasive like bicolor, although fire does have the same effect of enhancing production. Specific planting recommendations for these quail food plants are given in Chapter 16 and Appendix D–J.

Size of perennial quail plantings should be at least 1⁄10 acre (15 feet x 300 feet) and normally not larger than ¼ acre. Plantings should be in narrow strips about 15 to 20 feet wide paralleling field borders, forest edges, roadways, grown-up ditches, or other areas adjacent to suitable escape cover. One planting for every 15 to 20 acres should be adequate.

Current research suggests that in some situations, quail do not readily use small annual food plots scattered across the landscape but prefer larger plots (up to 1 acre or more) that are strategically placed close to key quail cover (nesting, brood-rearing, and escape). More research is needed to determine the ideal size, distribution, and combination of quail plantings on a variety of lands. For hunting purposes, only low-growing plants such as partridge pea, annual lespedeza, and browntop millet should be planted in plots more than a ¼ acre in size. Shooting and bird dog work are difficult in larger plots that have tall plants such as sorghum or corn. These taller plants should be established in 1⁄10 acre plots.

AMQUAIL FOR QUAIL

Amquail is a perennial, warm-season legume that is an improved variety of *Lespedeza thunbergii* that produces an abundance of high quality seeds readily eaten by quail from late December to mid-March. Unlike lespedeza bicolor and other shrub lespedezas that can be heavily damaged by deer browsing, Amquail has proven to be fairly resistant to deer browsing.

Amquail can be established by planting seed or by transplanting one-year-old seedlings. Although usually more expensive, the best results have been obtained by planting one-year-old seedlings. Seeds or seedlings should be planted in strips 15 to 20 feet wide, for about 100 to 300 feet in length adjacent to quail escape cover. For optimal hunting, strip lengths should be closer to 100 to 200 feet with a 20-foot break. This usually works out to be about 1⁄20 acre in size. Ideal locations for planting include the edges of agricultural fields, or along woodland borders, ditch banks, idle crop fields, utility rights-of-way, forest openings, or as strips in thinned forests. Plantings should be distributed across the landscape at about one plot per 10 to 20 acres.

The best dates for planting scarified (where the seed coat has been removed to improve germination) seeds are from March 1 to April 15. Seedlings should be planted from December 1 to March 1. Seeds and seedlings should be planted on well-prepared seedbeds that are fertilized and limed according to a soil test. As a general rule, avoid using fertilizers that contain nitrogen, since legumes like Amquail do not respond well to nitrogen.

Amquail seeds can be planted in rows or by broadcasting. Plots planted in rows should contain six rows 300 feet long spaced 3 feet apart. Approximately 20 to 30 seeds per foot should be placed in a shallow furrow and covered with about ½ inch of soil. Seeds can be planted by hand or by using a single-row garden planter. Broadcast rates are 15 pounds per acre.

There are basically three methods for planting Amquail seedlings: planting in a furrow, planting with a dibble bar, or planting with a mechanical tree planter. Furrow planting should only be done on level ground to reduce the risk of erosion. Deep furrows should be plowed with either a turning plow or moldboard plow to a depth equal to the root length of the seedlings.

After placing seedlings about 2 feet apart in the furrow, cover the roots by plowing a second furrow adjacent to the first. This entire process should be repeated until all six rows are planted. About 12 to 14 man-hours are needed to plant a plot by this method. To facilitate planting, roots may be trimmed somewhat for easy handling.

A dibble bar can also be used to plant Amquail seedlings in the same manner that pine seedlings are planted. It takes about 8 to 10 man hours to plant an entire plot by this method.

On level ground, like abandoned crop fields, a mechanical tree planter pulled by a tractor is recommended, especially if a large number of plots are going to be planted. At least two people are needed, one to drive the tractor and the other to place seedlings in the ground. About one to three man-hours are needed to plant a plot by this method.

Once Amquail plots become established they should be maintained by mowing to a height of about 4 to 8 inches above the ground in late February after their first growing season. Mowing causes the plants to become bush-like, concentrates growth on seed production, and discourages unwanted plants from becoming established in the plots. After the initial and successive mowings, plots should be mowed in late February every three to five years and should be fertilized according to soil test recommendations.

Developing Cover

The four essential quail cover types that must be present are nesting cover, brood-rearing cover, escape cover, and winter-roosting cover. All four must be located within close proximity to each other and adjacent to food sources. Ideal nesting cover consists of moderately dense grass–broadleaf weed mixtures interspersed with nearly bare ground. Mixes of annual and perennial herbaceous grasses, forbs, and legumes that are between one and two years of age provide ideal nesting sites. In Alabama, good nesting cover usually is "brushy" and contains scattered broomsedge, legumes, and other herbaceous plants in addition to young hardwood sprouts. Nests are often located adjacent to openings around field edges, disked strips, roadways, and fence rows. Nesting cover should not be developed in areas that have wet soils or are prone to flooding.

Brood-rearing cover provides another important component of good farmland quail habitat. It contains mixtures of annual legumes, forbs, and grasses that provide enough bare ground for movement but also support high numbers of insects that are eaten by chicks to supply their high demand for protein. Patches of annual weeds like partridge pea and ragweed are ideal brood habitat. Burned woodlands, winter mowed or disked fallow fields, and annual planted patches are also good brood-rearing areas.

One of the best ways to provide both nesting and brood-rearing cover on farms is to create and manage transition zones. **Transition zones** are areas where two different habitat types meet and mix together to form a third habitat type. They create an edge that can be managed to provide components of nesting and brood-rearing habitat. In most instances transition zones can be established along the edges of cultivated fields or pastures (protected from grazing) that border woodlands, fence rows, ditch banks, and roads. Usually, agricultural field borders are unproductive for crops, making these areas excellent candidates for transition zones. These strips may cover all the unproductive field edge but should never be less than 30 feet wide. The types of plants that will invade these areas depend on soil type, fertility, and pH.

Developing transition zones is perhaps the easiest and least expensive quail management practice on agricultural land because nature accomplishes the work. These zones may be established by simply removing strips of land from their previous use and allowing them to grow up in a mixture of annual and perennial legumes, grasses, and weeds. After several years transition zones will become too thick for quail and will need to be maintained by periodic disturbances such as burning, plowing, or disking in late winter. A general rule is that when more than 50 percent of the soil is covered in dead vegetation, the land needs to be burned or disked. In Alabama this will occur sometime between two and six years after transition zones become established. Fields having transition zones around three or four sides may be maintained on one side annually, starting approximately two years after the transition zones are established. Disturbance should be performed on a rotational basis that leaves some transition zones intact so that nesting and escape cover is always available adjacent to disturbed areas. When managed in rotation, some strips will be composed predominately of annual weeds that provide brood habitat, while others areas provide a mixture of grasses and legumes for nesting.

The need for transition zones in quail management largely depends upon the type of habitat adjoining cultivated areas. Transition zones are less valuable in

JERRY JOHNSON/NATURAL RESOURCES CONSERVATION SERVICE

TED DEVOS

Weeds, in nearly all forms, are beneficial for quail. Producing weedy agricultural fields (left) and leaving strips of weeds between summer and winter plantings and wood lines (right) can provide much of a quail's annual requirements.

situations where early successional habitat types or ground cover immediately adjoin cultivated areas. They have the greatest value where unusually dense or sparse ground cover exists.

The third important quail habitat type, escape cover, should also be present on farmlands to provide emergency hiding places for quail pursued by predators. It consists of shrubby or woody areas like brushy fencerows or field dividers. Some of the better escape cover plants include wild plum, wax myrtle, pine thickets, wild cherry, sumac, greenbriar, palmetto, viburnums, sassafras, blackberry, honeysuckle, and grapevines. In most areas sufficient escape cover may be developed simply by protecting areas from disturbances such as fire, disking, mowing, and overgrazing. In many large agricultural openings, quail will not use the majority of the field because of a lack of escape cover. For example, large fields and pastures contain areas within the center that are not utilized by quail. Generally, quail will not venture more than 100 feet from the nearest escape cover. To provide escape cover and access routes into these areas, large fields may be broken into smaller tracts by providing travel lanes across or into these fields. This may be accomplished by leaving undisturbed strips in native vegetation and brush, or planting a tree corridor across fields. Brush strips should be maintained by mowing, disking, or burning one side of each strip every three to four years in the early spring. To receive some eventual economic return from land removed from agricultural production, these escape cover patches may be established in pine trees. Plant 8 to 12 rows on a spacing between 8 x 10 feet or 12 x 12 feet. These plantings will provide long-term escape cover as well as future income from timber sales. These strips should also have 30- to 50-foot transition zones established on their edges growing into fallow field habitat. Barring soil constraints, longleaf pine is the best pine species to use for corridors.

In general it is difficult to maintain quail populations on land established in temporary or permanent pastures because of a lack of cover and the "matty" nature of pasture grasses. In general fescue, bermuda, and other mat-forming pasture grasses make extremely poor quail habitat. Temporary pastures usually have little value for quail since they are planted in the fall and are usually heavily grazed by livestock the following spring and summer. Permanent improved pastures may be of minimal value to quail if they are properly managed and will provide a small amount of nesting and brood-rearing cover for quail if they are only moderately grazed. Pasture borders can be allowed to develop into transition zones if these areas are winter disked and protected by fencing. Other areas in the pasture can be disked in winter or early spring to enhance plant growth; however, care should be taken to leave the remaining areas fallow or undisturbed to provide enough cover for nesting and brood-rearing. Summer disking should be avoided since this promotes bermuda and bahia grass growth and the machinery may destroy quail nests or cause them to be abandoned. If the pasture is cut for hay, a strip 50 feet wide should be left undisturbed along field edges. This will help reduce the number of quail nests destroyed.

FOREST MANAGEMENT FOR BOBWHITE QUAIL

High quail populations and good quail hunting are also associated with properly managed pine woods, and some of the highest quail populations in the Southeast are associated with "broken" woodlands maintained in an open condition. Silvicultural practices that create open forests that contain an abundance of annual and perennial herbaceous vegetation, small agricultural openings, as well as thickets of woody shrubs and vines provide most of the essential habitat components.

In much of the South, quail have been closely associated with timber stands that have been routinely burned almost every year. As a result, many of these stands are dominated by fire-tolerant pines with few or no hardwoods. In the Coastal Plain of Alabama, open stands of loblolly and longleaf pine can provide excellent quail habitat, provided all habitat components are present. Longleaf is generally better suited than other pines to quail management, since it can be burned at an earlier age, and the seeds are preferred by quail over the seeds of other pines. Some longleaf pine stands are being burned as early as four years of age for quail. Other pines can also be managed for quail if open canopies are maintained and frequent disturbances like fire, disking, or mowing are practiced. An open stand of pines with few hardwoods, a very sparse midstory of fruit-producing trees like dogwood, and a patchy and brushy herbaceous plant layer is the best forest stand for quail, in contrast to a forest that is dominated by dense stands of only one species of tree.

For the purpose of quail management, newly regenerated pine stands should be planted at a spacing of 6 x 12 feet (600 trees per acre) or even lower at an 8 x 12 foot (450 trees per acre) spacing. This delays crown closure and maintains herbaceous plants for a longer period of time. The drawback of this open spacing is lower and later income production, and shorter trees with more limbs.

Mowing and disking between rows also helps to

maintain quail habitat for a longer period. Studies have shown that populations of quail will expand to huntable levels on newly regenerated areas during the first two to four years after a timber harvest, if the surrounding forest or farms already support quail. As the forest grows, tree canopies close and reduce or eliminate most of the valuable quail plants. To offset this loss, 3- to 5-acre openings or 100-foot-wide corridors can be retained within the regenerating stand and can be managed for native and/or cultivated foods, nesting cover, winter roosting cover, and brood habitat. Ideally, these openings should be scattered across the stand and comprise about 25 percent or more of the area, or an average of one opening per 15 to 20 acres of forest. If permanent openings take too much land from timber production, food patches may be established in timber stands that have been opened by thinning. Other openings such as logging decks, woods roads, and firelanes can also be managed as permanent openings for quail.

Pines should be thinned as soon as possible after they become merchantable. For quail management, stands should be thinned to a basal area that is equal to the site index minus 25 and no more than a basal area of 80 square feet. For example, an area with a site index of 90 should be thinned to a basal area of 65 square feet (90 site index - 25 = basal area of 65 square feet). To create stands with a patchy understory that is dense (for escape cover) and open (for feeding, nesting, and brood habitat), the intensity of thinning can be varied across the stand. Thinnings should be conducted in conjunction with prescribed burning for a maximum response in herbaceous growth. When thinning a pine or mixed stand, a good rule of thumb is that at least 50 percent of the ground has sunlight exposure during midday.

Of all the techniques used in forest management, prescribed burning is probably the cheapest and most effective method to improve quail habitat. Properly conducted burns act as a fertilizing catalyst by releasing nutrients and minerals that are bound up in forest litter and placing them back into the soil. The removal of forest litter during burning creates bare ground, making it easier for quail to find seeds and other food. Germination of primary quail plant seeds is enhanced by burning. The new succulent vegetation that grows after a burn also attracts a variety of insects for quail during spring and summer. Burned pinewoods are probably the first or second most preferred brood habitat. The best time to burn for quail is from February through March. Care must be taken to leave an adequate quantity of upland blocks unburned to provide some protection for quail when the majority of cover is burned off. This timeframe for burning coincides with the northerly migration of hawks, making good cover that much more important. Late spring burning may disrupt nesting, is detrimental to legume production, and kills shrubby patches needed for escape cover. Preliminary work by Tall Timbers Research Station, however, indicates that burning in the month of May for hardwood control in pine stands may not be as detrimental to quail as was once thought. In the Tall Timbers study, areas that were burned in May produced better hunting success than comparable areas burned in February, possibly due to greater insect production from May burns. Before attempting spring or summer burns for quail, managers should wait for additional research results that clarify the impacts that spring and summer burning have on quail production.

Prescribed burning must be carried out in a manner that stimulates seed and insect production for good brood-rearing habitat, but at the same time leaves enough thickets and shrubs for nesting and escape cover. Since quail prefer to nest in one-year-old weeds and grasses, one alternative is to burn every other year. Burns should be conducted in 10 to 50 acre units. Adjacent units should be burned the following year to leave adequate nesting cover. The overall effect should create a checkerboard appearance of blocks of alternating one- and two-year-old "rough" vegetation. Thickets and other escape cover should be protected within burn units by plowing a fire line or "ring-around" around these areas. Ring-arounds should be between 5 and 20 acres in size. If large areas are burned, efforts should be made to retain at least 25 to 40 percent of an area in unburned nesting cover scattered in patches that are at least 5 acres in size. This provides nesting areas large enough in size to deter unusually high nest predation. Current research indicates that nesting success is higher in larger (20 to 30 acre), unburned blocks. Another approach to creating a mosaic pattern of burned and unburned patches is to burn at night with a cool backfire. This technique, called a "dirty burn," leaves many patches unburned, creating cover adjacent to feeding areas and brood habitat next to nesting areas. Hardwood control, however, is minimal with cool night fires.

Extensive annual burning that leaves little cover should be avoided. This type of burning destroys most of the available nesting cover with little cover remaining for spring nesting or broods and leaves birds without protective cover for much of the spring. Only late in the summer does enough vegetation grow back for some late nesting use.

Disking can also enhance native quail foods and create nesting and brood-rearing habitat in open forest stands. This practice is also called **strip disking**, a pro-

Fig. 2: Habitat and Harvest Management Intensities for Bobwhite Quail

Management Practices	Low Intensity	Medium Intensity	High Intensity
Timber Harvest/ Regeneration Types	Clear-cut/Plant	Clear-cut/Plant, Single tree selection, Patch clear-cut thinnings	Clear-cut/Plant, Single tree selection, Patch clear-cut thinnings
Basal Area (sq. ft./acre)	Over 100	70–100	< 70
Timber Harvest Rotation Length	15–25 years	20–45 years	50–150 years
Thinnings	Once or none	Every five years or more	As often as necessary
Fire	None or once every three years	Every one to two years in winter	Rotate every year in winter, growing season as needed to control hardwoods
Site Preparation	Roller chopping and burning	Roller chopping, burning, selective herbicide	Shearing, piling, burning, root raking, and windrowing
Windrows	Leave unburned	Leave alternate windrows	Burn or no windrowing, scatter slash throughout the stand
Seedling Spacing	6 x 6, 6 x 8	8 x 8, 8 x 10	10 x 10, 8 x 12, 12 x 12
Hardwoods	Retain producers of small mast such as water oak	Retain in drains and mast-producing clumps	Retain in drains and scattered throughout upland pine stands
Food Plots	1–3% of land in shrub lespedeza and agricultural food crops	3–5% of land area in shrub lespedeza and agricultural food crops	Greater than 5% of land area in shrub lespedeza and agricultural food crops
Fertilization	None	Food plots every two to three years	Food plots annually
Openlands	1–3% of land maintained in nesting and brood habitat	3–10% of land maintained in nesting and brood habitat	Greater than 10% of land area maintained in nesting and brood habitat
Cover	Exclude fire from drains	Cool patch burns	Mosaic of burns with ring-arounds
Quail harvest	No strategy	Harvest only 5–6 birds per covey	Harvest only 3–4 (20–30%) birds per covey. No late afternoon or late season hunting.

cess in which a disk is pulled through open areas in stands, removing mat-forming grasses in favor of seed-producing legumes and grasses. Disked strips should be as wide as tree spacing will allow. Disking improves brood-rearing habitat for quail (especially if conducted in fall and spring) giving quail better access for their broods during the summer. Seasonal disking favors certain plants over others. Breaking up the land in the winter produces heavy-seeded quail foods such as ragweed, doveweed, and partridge pea. Disking in April encourages important grasses such as panic grass, while disking in June favors a variety of green vegetation that attracts insects as well as other seed-producing plants that quail use during the fall. A schedule that rotates disking during winter, spring, and summer across forest stands is probably the best approach to providing adequate sources of native foods throughout the year. Disking every year favors food production and brood-rearing, while disking every few years is preferred for nesting habitat.

As mentioned earlier, some studies have suggested that fertilization of strips shortly after disking can increase seed production of certain quail food plants. As on farmlands, this is also an option in open forest stands that have been disked. Fertilizers containing nitrogen should not be used since they suppress legume production and favor woody plants and matty grasses over herbaceous vegetation. Managers should monitor the results of fertilization closely to see which plants benefit, since in some instances undesirable grasses and woody plants may be encouraged by the increased fertilization. These plants will often out-compete native quail plants and reduce seed production or choke out the plants completely. The best approach is to experiment with various combinations of fertilizers, disking, and prescribed burning techniques to see what works on different sites.

Mowing with a bushhog can also be done in open forest stands to maintain vegetation at a level desirable for quail. In early spring this can keep the ground open enough for brood habitat and the vegetation at a growing stage that is attractive to insects. Mowing can also be a beneficial tool in maintaining grassy woods roads, road edges, and firelanes in brood habitat. Summer mowing should be conducted in hardwood thickets to control excess brush growth.

Estimating Quail Numbers

Harvesting game during the hunting season is based on the assumption that a surplus of animals is produced each year. To obtain a fairly accurate estimate of quail numbers before the hunting season, some plantations conduct a quail census. One technique involves several people walking in a straight line over quail habitat, with a distance of about 20 yards between individuals. This is repeated several times until the entire sampling area is covered. Sampling areas should be large enough to represent the quail habitat of an area. At a minimum, sampling areas should be a least 100 acres in size. All quail that flush are counted and recorded. Care should be made to not count the same quail more than once. For the best estimate, sampling should be done on three successive days and an average taken of the number of quail counted. Research has shown that about 50 percent of the birds on an area are counted by this method; therefore, the average number of quail counted is doubled to give an estimate of the total number of quail on the sampled area. If an average of 10 quail were counted on 100 acres during three sampling periods, the estimated number of quail on the sampling area would be 20 birds (10 x 2), or one quail per 5 acres.

HUNTING

Studies conducted at the Tall Timbers Research Station near Tallahassee, Florida indicate that harvest levels and hunting strategies can have a lot to do with how many birds remain for the following year's breeding stock. The common harvest strategy of shooting only two birds per covey rise and leaving the remaining birds is a good start. Once a covey is shot down to around eight birds these remaining birds may regroup with other small coveys to form a **bevy**, which hunters often mistake for a new covey. Late season hunting on these "new coveys" can seriously reduce the number of quail remaining for next year's brood stock. If at all possible, late season hunting in February, especially in areas of moderate to low numbers, should be avoided. One study also revealed that hunting in the afternoon may increase non-hunting mortality of quail. Birds that are separated from coveys by late afternoon hunting may not have time to regroup as a covey before nightfall, which puts individual quail at greater risks for predation. To reduce this risk, sportsmen should only hunt during the morning or early afternoon hours. Hunters should also avoid attempting to fill game bags in years when quail reproduction and numbers are low. The best advice is to leave areas with low quail populations alone and concentrate on other areas that have more quail.

Most intensively managed hunting plantations keep detailed records from year to year on harvest rates and estimated populations of quail. Quail managers estimate populations just before hunting season. This estimate gives a general indication of how many quail are present and what percentage should be harvested so that a viable breeding stock will remain for the following year. The number of quail to harvest depends upon specific site conditions and a variety of other factors. Successful quail managers eventually develop a knack for determining how many quail to harvest. Harvest rates are usually between 30 and 40 percent of the estimated population; however, rates closer to 20 percent are more prudent.

One Manager's Pen-Released Quail Program

Mickey Easley has long been respected in Alabama as a wildlife biologist and game manager who has built a reputation of excellence by combining practical, on-the-ground experience with his graduate training in Wildlife Ecology at Mississippi State University. He currently is the wildlife biologist and manager of Wyncreek Plantation in Hardaway, Alabama.

Native bobwhite quail numbers on Wyncreek Plantation, like many other places in the South, have declined. In the face of dwindling wild quail numbers, we have gone to a pen-released program in order to enhance our quail shooting opportunities. The new generation of hunters is a different breed. These hunters are fast-paced and want to see a lot of birds in a short period of time. Because of this change in hunter attitudes, many traditional hunting preserves are quickly becoming shooting preserves. Approximately 85 percent of the plantations in Alabama have gone to some form of releasing pen-raised quail, and each year more acres are converted to this type of program. Although we still manage our habitat for wild birds, we see pen-released quail as an acceptable alternative to wild bird hunting. Habitat management is still a high priority in a sound pen-release program.

Our pen-released program consists of several hunting courses scattered across the property. Each course has food and cover patches, primarily in *Lespedeza thunbergii* and Egyptian wheat. These sites provide critical cover from avian predators. Most of the quail are released approximately one month before the hunting season in several release stages. The first release is usually around 1,500 birds that eventually make up about 30 to 50 coveys. Three weeks later we usually release an additional 500 to 600 birds in select areas, and this generally meets our hunting needs until December. From December through January we supplement birds as needed about every two to three weeks.

Although we provide an abundance of native and planted food for released quail, we also supplementally feed the birds throughout the year. We feel this helps to acclimatize released birds to feeding in the wild and allows them to have a ready source of food, especially in January and February when native food sources are lean. We supply grain sorghum to our birds either in feeders or by broadcasting (scattering) the seed. We prefer to broadcast sorghum seeds in protective cover across our shooting courses to prevent predators from zeroing in on feeders. By scattering food out, we also eliminate some of the problems associated with bird dogs going directly from one feeding area to another. In addition, scattering bird feed across a larger area reduces the build-up of quail scent and keeps bird dogs from becoming too "birdy" when they come to feeding areas that routinely concentrate pen-released birds. We use grain sorghum, rather than corn or wheat, since it more closely resembles native seeds.

Releasing Pen-Raised Birds for Shooting

Releasing pen-raised quail was widely practiced by some state wildlife agencies and private conservation groups in the Southeast during the late 1940s and early 1950s, and is still done today on private land when a higher harvest is desired. After thorough evaluation, however, biologists and managers found stocking to be expensive and unsuccessful in increasing quail numbers over the long-term.

Most wildlife biologists believe that native quail can quickly occupy all of the suitable habitat available if there is a native population nearby. Some pen-raised birds are released into the wild into areas not even suitable for occupation by native birds and this is fruitless. Since a surplus of wild quail is produced each year naturally, released pen-raised birds will probably be the first to be eliminated as part of the surplus. In other words, it is not logical to expect a pen-raised bird to survive in habitat that will not support native quail, because those birds that are naturally adapted to withstand the pressures of nature will do so better than those raised artificially. The danger of introducing diseases into native bobwhite quail populations from pen-raised birds also poses a risk.

Releasing birds shortly before the opening of hunting season can be effective for short-term shooting, but not for population restoration. From an economic standpoint, releasing birds may seem costly at first, because only a small percentage are normally recovered. At an initial cost of $2.50 per bird and a 20 percent harvest, a pen-raised bird in the bag cost around $12.50. This may be relatively cheap compared to the cost of managing wild birds, which is estimated to be $500 to $600 per bird on intensively managed quail plantations.

Studies of radio-tagged pen-raised quail released into the wild, in intensively managed quail habitats, reveal that about 50 percent survive for the first two months with 20 percent surviving into the spring. These surviving birds readily breed and raise young throughout the summer. A study by Tall Timbers Research Station followed several thousand pen-raised birds that were released two and one-half months before hunting season. By the end of the hunting season 18 percent of the released birds were harvested. Less than 1 percent of the birds released in the first year of the study were harvested the following fall, suggesting that year-to-year survival is extremely low. However, this compares fairly well to wild quail that only have about a 4 percent harvest return from fall to fall. **Habitat improvements are still vital for maintaining wild quail populations and increasing survival of pen-raised birds.** Whether areas are managed for wild quail or pen-released birds, habitat management should be an integral part of any quail operation. Although quail habitat management is expensive, these practices also enhance the land for other wildlife as well.

ORGANIZATIONS DEDICATED TO BOBWHITE QUAIL

Quail Unlimited

Quail Unlimited, Inc. is a national conservation organization dedicated to improving quail and upland game bird populations through habitat management and research. Quail Unlimited, Inc. was organized to re-establish and manage suitable upland game habitat on both public and private lands across the country, and to educate the public to the needs for quail and other wildlife habitat management. The organization has state and local affiliate chapters across many parts of the Southeast. Quail Unlimited, Inc. also publishes *Quail Unlimited* magazine. For more information write Quail Unlimited, Inc., P.O. Box 610, Edgefield, SC 29824-0610, or call 803-637-5731.

Tall Timbers Research

Tall Timbers Research, Inc. is a non-profit research, resource management, conservation, and educational organization created to study fire's natural role in ecosystems and to promote prescribed burning in land management. Research and long-term monitoring provide insight into ecology, conservation, and management of native species, including but not limited to the bobwhite quail. The organization facilitates the exchange of information on fire ecology and wildlife management through conferences, special educational events, and publications. For more information write Tall Timbers Research, Inc., Route 1, Box 678, Tallahassee, FL 32312-9712, or call 904-893-4153.

Albany Area Quail Management Project

The Albany Area Quail Management Project (AAQMP) is a non-profit research organization dedicated to studying the ecology and management of bobwhite quail. The AAQMP is a cooperative effort between Auburn University, quail plantations, and quail enthusiasts to research areas of concern and interest to improve quail management on private lands. The Alabama Quail Hunters Association is a statewide organization dedicated to the propagation and management of quail. It sponsors seminars for quail hunters and those interested in quail management in Alabama. For more information contact Jim Bradford, 4960 Meadow Brook Road, Birmingham, AL. 35242, or call (205) 991-8635. To receive the AAQMP newsletter or obtain more information, write the AAQMP Leader, 331 Funchess Hall, Auburn University, AL 36849, or call 334-844-9247.

SUMMARY

Bobwhite quail numbers in Alabama, like in other areas in the Southeast, have declined. Wildlife biologists speculate that no one factor is to blame for the decline, but that a combination of limiting factors, such as loss of habitat or habitat quality, predation, and disease, are the most likely causes. Landowners, managers, and sportsmen interested in bobwhite quail can strive to reduce these and other limiting factors by following the management recommendations outlined in this chapter.

Gray squirrels are a common sight in most areas of Alabama that have nut-producing trees.

DENNIS HOLT

CHAPTER 8

Managing for Squirrels and Rabbits

CHAPTER OVERVIEW

Squirrels and rabbits are the most abundant small game species found in Alabama. Because of their adaptability and simple life requirements, management for squirrels and rabbits primarily involves providing adequate food and cover. This chapter describes gray squirrels, fox squirrels, and rabbits in Alabama and the various habitat improvement practices that ensure an abundance of these species on forest and farm lands in the state.

CHAPTER HIGHLIGHTS PAGE

BIOLOGY AND LIFE HISTORY OF SQUIRRELS .. **186**

Limiting Factors **187**

Habitat Requirements **188**

Habitat Improvements **189**

Harvest Requirements **191**

BIOLOGY AND LIFE HISTORY OF RABBITS **192**

Limiting Factors **193**

Habitat Requirements **194**

Habitat Improvements **194**

Harvest Requirements **199**

The gray squirrel (*Sciurus carolinensis*) and fox squirrel (*Sciurus niger*) have been a part of Alabama forests for thousands of years. Native Americans and early settlers used squirrels for food and, to some degree, for their fur. Historical accounts in the Southeast reveal high gray and fox squirrel numbers. For example, in 1834, two 50-man teams participated in a hunt to reduce the impact of a squirrel population explosion. During this outing the two top hunters shot 900 and 783 squirrels, respectively. The days of such hunts are gone forever, but so are the vast forests that supported such high populations. Squirrel populations still periodically soar however, and when they do, many migrate from overpopulated areas.

Both the gray and fox squirrel are present in Alabama. The gray squirrel prefers large tracts of dense, mature hardwoods, especially oaks and hickories with an understory of smaller trees and shrubs. The fox squirrel, on the other hand, prefers open park-like stands in pine, mixed pine, and oak, or oak-gum-cypress stands. Of the two, the gray squirrel is by far the most abundant and it is the only species numerous enough to manage for hunting purposes. For this reason, and the fact that the gray and fox squirrels' habitats and life history are similar, most of the emphasis in this section will be devoted to gray squirrels. Gray squirrel habitat improvement practices should also benefit fox squirrels.

RICHARD T. BRYANT/ARISTOCK

Gray squirrels utilize a variety of mast-producing trees and shrubs, including this dogwood tree and its berries.

BIOLOGY AND LIFE HISTORY

Reproductive Cycle

Female gray squirrels, called "sows," may produce two litters of young a year, if food is plentiful and other factors are not too adverse. Litter size can vary from one to six young, but is usually between two and four. The two major breeding periods in Alabama are from December through January and from May through June but some breeding may occur throughout the year. Though some juveniles produced from the winter breeding season may produce a litter their first summer, usually only yearlings and adults breed. Adult females often have larger litters than yearling females.

Male squirrels, known as "boars," carry their testicles inside their bodies during the non-breeding periods. This accounts for the notion that some hunters have about young boars being "clipped" by other dominant squirrels to prevent overpopulation. When the breeding season approaches, the boar's testicles enlarge and become exposed. Several males usually locate a sow after she has emitted a scent that attracts them, just prior to her entering breeding condition. Males may chase a sow for several days before she is ready to be bred. The dominant male of the area will usually assert himself over the other males through aggressive confrontations, and gradually the less dominant boars will leave in search of other prospects. Both the sow and her chosen mate will go through a brief courtship until she is ready to breed.

Gray squirrel pregnancy lasts about 44 days. The first litters are born in February after the winter breeding period, or in June after the spring and summer breeding period. As mentioned earlier, the number of young per litter depends on numerous factors, but the average litter is about three.

At birth the young are hairless, about 4½ inches long, and weigh about ½ an ounce, roughly the weight of a 50-cent piece. At birth, the squirrel's eyes and ears are closed and no teeth have erupted through the gums. After three weeks, the ears of the young squirrel open, followed by the eyes at five weeks. By the fifth week, the rapidly growing squirrels are about 10 inches in length, with the tail measuring about an additional 4½ inches. From the fifth through the ninth or tenth week, the young squirrel grows about 1 inch a week. Weaning of the young usually begins in the seventh week. Tooth development, which began in the third week, progresses, and by the seventh week, some buds and leaves near the nest can be eaten. However, the bulk of the diet is still milk from the mother.

PAUL T. BROWN

RICHARD T. BRYANT/ARISTOCK

Both fox (left) and gray squirrels (right) are present in Alabama. Fox squirrels prefer open park-like stands in pine, mixed-pine, and oak-gum-cypress forests. Gray squirrels prefer large tracts of dense, mature hardwoods, especially oaks and hickories, with an understory of smaller trees, shrubs, and vines.

During the eighth week, the young become more active outside the nest and begin to eat more solid foods. At about the time gray squirrels are weaned in the fall, trees are producing acorns and dogwood fruit is plentiful. During the spring young squirrels readily eat swelling buds, an important food source for young squirrels.

Squirrels use both tree cavities and leaf nests. Young squirrels of both sexes practice building lodging facilities, with varying results in quality of workmanship. Nests often become infested with a high number of fleas, mites, ticks, and other insects. However, other insects, including several kinds of beetles, help rid the nest of these biting pests.

Females that breed in the summer may return to the den tree cavity where they reared their winter litters, but they often seek new quarters in order to get away from the previous season's offspring. While most litters are reared in tree cavities, some summer litters are reared in leaf nests. Leaf nests are safer during the summer period than during winter because the surrounding green foliage protects the nest from aerial view by raptors (birds of prey) like hawks, owls, and other predators.

Squirrels born in the summer become active outside their nests by September or October. Because the autumn season provides a longer break between the breeding periods, the family group will likely stay together somewhat longer before the adult female leaves or drives her young away, usually in December. Females that are bred in May often leave their litters at 8 to 14 weeks after birth.

LIMITING FACTORS

Like most wildlife species, squirrels have a high mortality rate during their first year, with about 60 percent dying in the same year in which they were born. Those that survive their first year have the potential to live about six years in the wild. In captivity, squirrels have been known to live 15 years.

Because of the continually changing habitat conditions and other factors that help regulate reproductive rates, squirrel populations are very cyclical. When a squirrel population expands beyond the carrying capacity of a forest stand, a mass movement and relocation of squirrels to other areas takes place, leaving few squirrels behind. When this occurs, squirrels have been observed swimming across lakes and rivers. These large-scale movements involve thousands of animals, many of which die during the journey.

A population explosion is a unique limiting factor because the end result probably helps improve the overall gene pool of the gray squirrel over a broad area. In addition, because the remaining squirrels are greatly reduced in numbers, there are more available resources (food and shelter) to go around until a new population builds up. Squirrel populations will usually take about five years to build back up to the levels they achieved before a mass movement occurred.

Predators of the gray squirrel include rat snakes; red-tailed, red-shouldered, marsh, and Cooper's hawks; great horned and barred owls; red and gray foxes; bobcats;

Plants Important to Squirrels

Fruits of Trees, Plants, and Other Items that are Important Food for Squirrels

Hickory	Red mulberry
Oak	Black walnut
Beech	Maple
Fungi (mushrooms)	Yellow poplar
Animal matter	Corn
Flowering dogwood	

Seasonal Fruits of Trees and Plants that Squirrels Prefer

American hornbeam	Hackberry
Wild plum	Magnolia
Hophornbeam	Ash
Bald cypress	Osage orange
Basswood	Palmetto
Blackberry	Persimmon
Black Cherry	Pine seeds
Blueberry	Sweetgum
Carolina silverbell	Sycamore
Chinquapin	Tupelo gum
Wild grape	Witchhazel
Greenbriar	

Common Cavity Trees Used by Squirrels for Dens

Ash	Maple
Bald cypress	Oak
Beech	Sweetgum
Blackgum	Sycamore
Hickory	

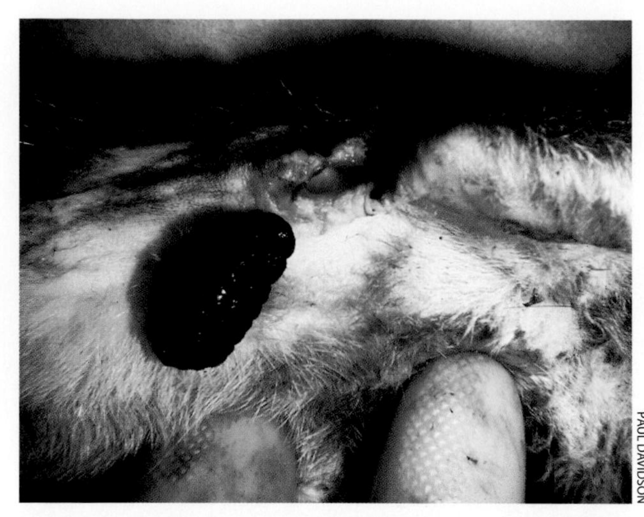
PAUL DAVIDSON

The bot or warble fly ("wolves") ranks as the squirrel's most serious pest. The bot fly lays eggs on the squirrel's fur, the eggs hatch, and the larvae burrow under the skin, creating a small hole for breathing. The larvae develop into a grub, falling out of the squirrel and on to the ground. Although unsightly, the meat of harvested squirrels is not affected by the bot fly larva.

raccoons; house cats, and dogs. However, considering that population build-up to the point of migration is fairly common, these natural predators do not limit overall population growth.

Scabies or mange, caused by the scabies mite, can be fatal to squirrels. Often squirrels will scratch themselves until their bodies are bloody and hairless. They then become weakened and susceptible to predation, secondary infections, and the effects of harsh weather.

The warble or bot fly (commonly called wolves) ranks as the squirrel's most serious pest, however. This fly lays its eggs on tree bark. When the eggs hatch, the larva attach to the first squirrel that passes by. The larvae then burrow under the squirrel's skin around the shoulders and legs and develops into a large grub, which keeps a hole open in the skin in order to breathe. Although this parasite does not affect the quality of the meat, hunters frequently don't clean or eat squirrels with wolves. Early fall is the period when squirrels are most bothered with them. Other parasites like ticks, fleas, and lice are only a minor nuisance to squirrels.

Weather may be the most limiting factor affecting populations, aside from man-caused habitat modification. Too little rainfall may severely reduce the production of valuable squirrel foods. Heavy rains, when litters are very young, may result in drowning of baby and juvenile squirrels. Cold weather during the early spring may reduce the food supply for squirrels by killing fragile buds that would normally have produced the fall mast crop.

HABITAT REQUIREMENTS

Food Habits

Squirrel reproduction and survival fluctuate with the changing availability of heavy-seeded mast, particularly acorns. When heavy mast is not available, squirrels feed on other fruits and berries, parts of flowers, buds, bark, roots, mushrooms, and some animal matter. About 1½ pounds of mast per week is required for each squirrel from September through March.

RICHARD T. BRYANT/ARISTOCK

RICHARD T. BRYANT/ARISTOCK

Hollow den trees are essential for protection of squirrels against extremely cold weather and for rearing young. Survival of litters is usually more that two times higher in den trees as compared to leaf nests. There should be at least two to six quality den trees for each acre of squirrel habitat.

The order of squirrel preference for hard-mast varieties is hickory nuts, beechnuts, and white and red oak acorns. This preference order varies from place to place because of differing food availability. When heavy-seeded mast crops fail, competition for food may become intense. Younger members of the population are forced out of the home range by well-established adults. Mortality of sub-adults increases and reproduction may decline or stop altogether. Mast failure during a population peak is the prime cause of the large squirrel movements and relocation described earlier.

Cover Requirements

Hollow den trees are essential to squirrels for winter shelter and rearing of young. Survival of litters is usually about two and one-half times higher in den trees as compared to leaf nests. Adult females with young will not tolerate other squirrels in the same den tree. Considering that only about 50 percent of the hollow trees identified from the ground are suitable as den trees, plenty of hollow trees are usually needed to provide shelter for squirrels. As a general rule of thumb, there should be at least two to six quality den trees for each acre of squirrel habitat. Suitable den trees should be identified, marked, and protected from silvicultural and other management operations. Adult squirrels usually "set up house" in at least two den trees located in their home range. Home range size varies from about 1½ to 8 acres for individual squirrels, with many ranges overlapping.

In some forested stands, a sufficient number of suitable den trees may not be available for squirrels. If this is the case, artificial nest boxes for squirrels can be constructed with a little effort and expense. Nesting boxes can be an effective substitute for natural cavities and are commonly utilized by squirrels. Components of good nesting boxes can be found in Chapter 14.

Water Requirements

Gray and fox squirrels use a variety of available water sources, including streams and pools, dew on plants, tree crotches, and succulent plant materials. Enough moisture can usually be taken in from succulent plants to satisfy their water requirements. However, squirrels still prefer to have water available from sources no farther than ¼ mile from their nest or den.

HABITAT IMPROVEMENTS

A large part of improving squirrel habitat involves providing several year-round food sources. Managing timber for pine pulp production (even-age pine

Fig. 1: Management Intensities for Gray and Fox Squirrels

Management Practices	Low Intensity	Medium Intensity	High Intensity
Timber Harvest/ Regeneration Types	Clear-cut less than 50 acres, shelterwood	Group or individual selection	No cutting of hardwoods except to improve squirrel habitat
Timber Harvest Rotation Length	20–45 years	45–60 years	60–100 years
Thinnings	Fox Squirrel: None Gray Squirrel: None	Fox Squirrel: Once Gray Squirrel: None	Fox Squirrel: Every four to five years Gray Squirrel: Release of good mast producers by removing inferior trees
Fire	Fox Squirrel: None Gray Squirrel: None	Fox Squirrel: Every five years Gray Squirrel: None	Fox Squirrel: Every two to three years Gray Squirrel: None
Fertilization of Mast-Producing Trees	None	Every two to three years	Annually
Site Preparation	Shearing, piling, burning, root raking, and windrowing	Roller chopping, burning, selective herbicide	Roller chopping and burning
Windrows	Leave unburned	Leave alternate windrows	Burn
Seedling Spacing	6 x 6, 6 x 8	8 x 8, 8 x 10	10 x 10, 12 x 12
Hardwoods	Retain in drains	Retain in drains and mast-producing clumps with corridors	Retain all mast producers and connect with corridors
Squirrel Harvest	No strategy	No strategy	Adjust harvest to abundance of squirrels and acorn availability

management) is seldom compatible with optimal squirrel habitat. To ensure that ample food sources are maintained, timber stands should be composed of a mixture of hardwoods or mixed pine–hardwood species. For better squirrel habitat, timber stands should be managed on at least a 60- to 100-year rotation. This will help ensure that stands produce a sufficient amount of mast and den trees for many years into the future. Timber stand improvement practices, like thinning and selective harvests, should be conducted to remove inferior trees that are poor mast-producers and often of little or no value for timber markets. When a stand is thinned, 40 to 60 percent of the management unit should be left in trees of mast-producing age, which for most oaks begins at 25 to 30 years of age. Small selection cuts that create openings of ¼ to 1 acre are less disruptive to squirrels than clear-cuts. Timber stand improvement practices can help to improve and accelerate crown development and increase mast yields of trees that remain in the forest stand.

Clear-cutting large areas and other timber harvest practices that remove mast producers and den trees are detrimental to squirrels. For all practical purposes, these areas are lost for squirrels for at least 25 years. All harvest cuts should be held to a maximum size of 20 acres and should be 500 feet or less in width, if possible. These harvest and regeneration areas should be scattered throughout the management unit so that the young stands are evenly distributed among older stands. This helps to minimize the impact to squirrels of losing mature trees over a large area. If clear-cuts must be greater than 20 acres, travel lanes that are at least 50 to 100 yards wide and are composed of mature trees should be retained until young trees in the clear-cut are old enough to produce mast.

RICHARD T. BRYANT/ARISTOCK

Mixed hardwood stands with an understory of mast-producing shrubs, such as along this streamside management zone (SMZ), provide excellent habitat for squirrels.

Hardwood and mixed hardwood–pine stands composed of mature timber of mixed age and species composition that are adjacent to waterways should be spared from timber harvests and left as streamside management zones (SMZs). SMZs are listed as an integral part of Alabama's Best Management Practices (BMPs) for forestry. A good rule of thumb is that SMZs should be wide enough that you can't see through them in the winter. Particular care should be taken not to leave small islands of timber surrounded by cleared land, since this lowers the value of these areas for squirrels. Isolated clusters of hardwoods several acres in size should be connected by hardwood corridors. Mature hardwood or mixed hardwood–pine stands should be selectively cut, not more than every five years, removing overly mature or inferior mast-producing trees.

As a final consideration, cattle should be excluded from woods intended for squirrel management since their presence reduces plant reproduction in the understory and adds yet another competitor to the list of species that use many of the foods squirrels need for survival. Prescribed burning should also be excluded from squirrel habitat because it reduces understory growth and is generally not considered a management practice that improves squirrel habitat. In addition, valuable hardwoods will be damaged or killed by fire.

HARVEST REQUIREMENTS

Hunting popularity of the gray squirrel in Alabama is second only to deer hunting. As a result, some local areas experience unusually high hunting pressure. In these cases, care should be taken to reduce hunting pressure in relation to the availability of squirrels and squirrel foods. On small forested tracts it is possible to assess the relative abundance of both squirrels and mast. Mast abundance is usually determined in September. Binoculars are helpful in scanning tree canopies for the presence or absence of mast such as acorns. Also, squirrels or their signs, such as nests or feeding debris, should be apparent during these habitat inventories.

If there is evidence of a high squirrel population in a year with low mast production, this indicates that the squirrels should be heavily hunted. This will help to reduce their numbers in a short period of time, before the low food supply is exhausted. If the squirrels go unhunted, competition for low food sources will become intense and large movements and relocations may result. Hunting during the early part of the season is most effective in protecting food supplies for the remaining squirrel population during years of poor mast availability.

If food is plentiful but squirrels are scarce, the hunting pressure on an area may be too heavy and should be reduced. When both food and squirrels are plentiful, the

KAREN LAWRENCE/ARISTOCK

The cottontail rabbit has an enormous ability to reproduce and relatively simple habitat requirements, which make it one of the most sought after game species.

hunting pressure should be heavy. Squirrel reproduction during heavy mast years is usually high, but mast production during the following year may be poor.

The key to the continued abundance of a huntable gray squirrel population in Alabama is the availability of a mix of mature forest types. These stands should contain heavy mast producers like hickory, oak, and beech, and plenty of hollow den trees. Fox squirrel habitat should contain an open park-like woodland in a pine, oak-pine, or oak-gum-cypress stand containing a sparse understory. With proper planning, gray and fox squirrel habitat can be maintained and enhanced on farm and forest land across Alabama for years to come.

MANAGING FOR RABBITS

Description and Historical Overview

The Eastern cottontail rabbit (*Sylvilagus floridanus*) is one game species familiar to virtually everyone in Alabama. Its requirements for life are relatively simple and this, coupled with an enormous reproductive capacity, makes it one of the most popular game species in Alabama and across North America.

The cottontail was not abundant in the United States when settlers arrived, but with the introduction of agriculture, cottontail numbers increased. Changing land-use practices in recent years, particularly conversion of patchy farming to forest production and intensive farming, have resulted in a population decline. This decline will continue unless landowners and managers are conscientious in providing the basic elements the cottontail requires to survive.

Four species of rabbits are found in Alabama. The Eastern cottontail is the most abundant and is distributed statewide. The large swamp rabbit (*Sylvilagus aquaticus*) or "cane cutter" primarily inhabits lowlands throughout the state. The marsh rabbit (*Sylvilagus palustris*) inhabits the marshes and swamps of extreme south Alabama and is the smallest of the state's rabbits. The New England cottontail (*Sylvilagus transitionalis*) has been reported in isolated portions of northeast Alabama but its current status is unknown.

Although these rabbits represent four distinct species, they are all commonly referred to as cottontails by the average hunter. Likewise, the terms cottontail and rabbit as used in this text apply not only to the Eastern cottontail but to the other species as well. Because cottontail rabbits are the most numerous species in the state, this section will be devoted primarily to cottontails and, to a lesser extent, swamp rabbits.

Biology and Life History

When the cold days of winter begin to merge with the warmer spring-like days of late February and March, males or "buck" cottontails begin to seek females or "does" as mates. This signals the start of the breeding season which extends through September.

Near the end of a 28-day pregnancy period, the heavy-bodied doe will begin selecting a well-drained location, usually in an old field, open woodland, garden, or even an open lawn, on which to construct her nest. A nest cavity is constructed that is approximately 5 inches wide and 7 inches long. After the nest has been excavated, the doe will line the nest cavity with layers of grass, leaves, small roots, or green grass that will later be topped with fur pulled from her underside.

At birth, baby cottontails weigh approximately 1 ounce and are blind, deaf, and completely helpless. A litter of baby cottontails may vary in number from one to seven, but the average litter size is around three or four. The young remain hidden by dead litter or leaves in the nest. Within 14 days they are strong enough to leave their mother and fend for themselves. The doe cottontail may breed the same day the litter is born and raise young three or four times during the breeding season. A new nest is generally constructed for each successive litter.

Under ideal conditions, one pair of rabbits can multiply into 25 in less than a year. This efficient reproduction is necessary to offset the high annual mortality in cottontail populations. Their potential lifespan is 8 to 10 years, but few rabbits in the wild reach even one year of age. In fact, the average life expectancy for cottontails in the wild is only four to six months. Only about 50 percent of those born can be expected to leave the nest; of those that do, fewer than one-half survive until fall.

Limiting Factors

Rabbits, like quail, are one of the most heavily preyed upon game species. They are also susceptible to a variety of diseases and parasites, some of which can be deadly. During certain seasons of the year, some Alabama predators, both terrestrial and avian, subsist mainly on a diet of cottontails. Foxes are probably the most important predators on rabbits, and at times rabbits compose as much as 50 percent of the fox's diet. The bobcat has a distinct preference for rabbits, and nearly 75 percent of its diet may consist of cottontails. The larger hawks and owls also consume rabbits readily, and up to 40 percent of a great horned owl's diet may be cottontails. Cottontails, their nest, and young, are hunted effectively by practically every wild predator, including coyotes, crows, dogs, feral house cats, foxes, hawks, minks, owls, snakes, and skunks.

Predation may be unpleasant to observe, but the process preserves the balance of nature. Predation also has benefits to rabbit species; with survival of the fittest, rabbit populations remain stronger and healthier.

PAUL T. BROWN

Rabbits are one of the most heavily preyed upon game species and provide food for a large variety of predators.

Diseases and parasites are common in most wild species and are generally insignificant. There are, however, some that should be of interest to hunters and other cottontail enthusiasts. While most cause the rabbit little harm, some are responsible for the needless waste of rabbit meat each year.

Probably the best known or most obvious cottontail afflictions are wolves, warbles, or bots. These terms are used interchangeably to describe the larvae of a large fly. The larvae have a grub-like appearance and can be found under the skin of infected rabbits, generally on the neck and chest.

Wolves, as they are most commonly called, are the larval stage of the bot fly. Eggs are laid on the hair of the rabbit by the adult fly and later hatch into immature larvae which bore into the rabbit's skin. These larvae grow until they are approximately 1½ inches long, at which time they emerge through the rabbit's skin, fall to the ground, bore into the soil, and later emerge as adult flies. These parasites are more abundant during warmer weather and generally leave infected rabbits by late fall or early winter. Infected rabbits rarely die, and the meat is edible if larvae are removed before the meat is cooked. Having the larvae in rabbit meat is similar to having an inchworm in an apple. By removing the areas surrounding the blemished spot, the rest can be eaten.

Fibroma or "rabbit horn" disease is a virus-induced black growth occasionally found on the skin of cottontails. This warty-appearing growth is normally confined to the skin and is usually removed along with the hide. Infected rabbits can be handled, cleaned, and eaten in perfect safety. The disease-producing virus is spread by ticks, mosquitos, biting flies, and other biting insects.

Tularemia or "rabbit fever" is the single most important and deadly disease affecting cottontails. The

The Facts About Tularemia

Although the seriousness of tularemia to humans has been reduced, there are several steps that can be taken to reduce the risk of contracting the disease. Even without these added precautions, annually there are only about 10 cases of tularemia for every 1 million people.

The following are precautions that should be considered in reducing risks of exposure to tularemia:

1. Rabbits should be hunted after the coming of cool weather when there is a better chance that most sick rabbits have died.
2. While in the field, if a rabbit shows no natural wariness or alertness, and no strong inclination to run, don't attempt to harvest the rabbit. Sick rabbits are often sluggish and "tame."
3. Always wear rubber gloves when dressing rabbits, even in winter. Avoid getting rabbit blood on your hands in the field.
4. After removing your rubber gloves, wash hands well with strong soap and hot water if available, and treat all cuts, scratches, and abrasions on your hands with iodine. This can be done in the field as well as at home.
5. Cook rabbits thoroughly. Never eat rare or undercooked rabbit meat.

organism that causes tularemia, a bacterium, *Pasteurella tularensis*, is capable of infecting a number of wild birds and mammals, including man, but it more commonly occurs in rodents and rabbits. Rabbits are especially prone to tularemia infections when populations are high and the disease can quickly decimate local rabbit populations. Infected rabbits die within a week or 10 days following the onset of the illness. Signs of the disease in rabbits include sluggishness, slow reactions, and refusal to run. In advanced stages, internal organs such as the lungs, liver, and spleen may be covered with small white spots. Positive diagnosis of tularemia requires laboratory work by disease experts.

Humans are susceptible to tularemia. At present there are no vaccines to prevent the disease, but tularemia responds quickly to antibiotics and should no longer be considered a serious threat if treated properly.

Habitat Requirements

The cottontail rabbit occurs over a wide variety of habitat types and conditions. Ideal rabbit habitat includes an abundance of well-distributed patches of brushy cover mixed with grass and weedy fields. Cottontails thrive where cropland, idle fields, hay fields, and newly (two- to eight-year-old) cutover forest land are present in one area. Cottontails move very little in good habitat making their daily and seasonal home ranges small. Most cottontails rarely range over more than 10 acres in search of food and cover during their entire lifetime.

Alabama cottontails eat an extremely wide variety of plant foods. Some of the more general items include grasses, sedges, sprouts, leaves, fruits, buds, and bark. During the summer, cottontails primarily consume grasses, legumes, succulent annuals, weeds, and an occasional garden vegetable. The winter diet includes small grains, as well as twigs, bark, and buds of shrubs and trees. In agricultural areas, wasted grains left in fields, such as corn and soybeans, provide a source of high-energy food when suitable adjacent escape cover is available.

Cover requirements for cottontails closely resemble those of quail in that they both require a diversity of habitats with an interspersion of woodlands, brush, grass, and cultivated lands. Since rabbits are fair game for nearly all predators and the primary prey for some, suitable escape cover must be present. Ideal cover includes brushy fencerows, drains, honeysuckle thickets, blackberry patches, and brush piles. Fields containing dense stands of low grasses provide preferred nesting cover. Hay fields and lightly grazed pastures are also preferred nest sites for cottontails.

Cottontails occasionally drink surface water from sources like streams and ponds, but they obtain most of their water needs by feeding on succulent vegetation and drinking dew. In Alabama, the availability of surface water does not influence rabbit densities or distributions.

Habitat Improvements

Cottontails are extremely prolific. They will eat a wide variety of plants, and in general, they require little management for average populations. Most areas of Alabama can support a stable population of about one rabbit per acre by providing green food plants and plenty of cover, especially during the winter. These habitat components should be provided within the 10-acre home range of rabbits.

In spite of the relative ease and effectiveness of rabbit management, some hunters and landowners believe the quickest, easiest, and best way to ensure a high rabbit population is through stocking. Studies have shown however, that only about 4 percent of released rabbits are ever harvested and that this figure only increases to 25 percent when rabbits are stocked just before hunting.

Since cottontails may spend most of their lifetime on 1 acre or less, it is not necessary to tie up large areas of land in rabbit management. Some management on odd field corners, on non-productive crop land, and on idle land will do much to increase cottontail populations.

Adequate escape cover must be present if huntable rabbit populations are to become established. Rabbits will use almost any type of cover as long as it is sufficiently large and protective. Cover strips, briar patches, or brush piles should be at least 20 feet wide. Cover development falls into two major categories: vegetative and artificial. Vegetative cover includes natural thickets such as blackberry, honeysuckle, fallow areas, bicolor lespedeza, or any naturally growing thicket which is thick enough to provide protection from coyotes, foxes, hawks, owls, dogs, and other predators. Artificial cover includes brush piles, fencerows, or piles of rock with drain pipes for access.

Studies have indicated that brush piles are an excellent method for improving rabbit habitat. Brush piles can be easily built by piling limbs, logs, or solid debris over rolls of old wire or crisscrossed logs. These piles should be 12 to 15 feet in diameter and stand 5 to 6 feet tall. They should be well distributed in close proximity to food sources and placed approximately 50 to 100 yards apart. Most are useful for only three to five years before replacement becomes necessary. Native vegetation can be encouraged adjacent to brush piles by fertilizing with 3 to 5 pounds of 13-13-13 around each pile.

Adequate cover may also be provided by allowing unused areas such as stream banks, drainage ditches, fence rows, pond edges, or edges of fields to revert to natural vegetation. These areas usually provide long narrow strips of cover that can be further improved by planting food strips adjacent to them. Most management practices found in Chapter 7 "Managing for Bobwhite Quail," are also suitable practices for rabbit management.

DEVELOPING TRANSITION ZONES

Transition zones are simply a third habitat type developed between two existing and different habitat types. These areas provide a mixture or blend of food and cover from several different habitat types. In most instances transition zones can be developed along an adjoining edge between fencerows, roads, ditch banks, timbered areas, and cultivated fields.

Improved farming methods over the years have helped landowners receive the most return from their investments. This in turn has led to the abandonment of tenant-type farming, where the common practice was to leave non-productive field edges in native vegetation. While this was not considered progressive agriculture, it did provide ideal habitat for the cottontail. The present-day practice of plowing and planting fields to the edge has done little to increase total agricultural production, since growth and yield are relatively poor on crops planted in the "shaded-out" field borders. These practices, however, have reduced the potential for cropland borders to meet the habitat needs of rabbits.

Transition zones between forest and field are extremely important because rabbits are an "edge" species. The amount and quality of edge present usually determines the abundance of cottontails on a particular area. Properly managed and maintained, these edge areas will provide much of the rabbits' needs year-round.

Transition zones may be established in field corners, edges, or borders. These zones may be located where woodlands meet crop fields, open pastures, or along fence lines and roadways. These transition strips may cover all the unproductive field edge but should never be less than 15 feet wide. The species and composition of the vegetation which invade these areas will depend upon the soil type, fertility, pH, plant succession, and timing of soil disturbances.

The easiest, least expensive rabbit habitat improvement practice that can be conducted on agricultural land is the establishment of transition zones because nature does the work. These zones may be established by simply removing strips of land from their previous use and protecting against grazing by livestock.

To maintain transition zones in a mixture of legumes, grasses, and weeds, they must be burned, plowed, or disked in late winter, preferably in early February. If this is done in the spring, it will destroy many rabbit nests. It is not necessary to disturb these areas every year. A general rule of thumb is that when more than 50 percent of the soil is covered in dead vegetation, transition zones need maintenance. In Alabama this will usually occur sometime between two and six years after establishment. Fields having transition zones around three or four sides may be maintained on one side annually, starting approximately two years after zones are established.

The value of transition zones in cottontail management depends largely upon the type of habitat adjoining cultivated areas. Transition zones have less value in situations where early successional habitat types or ground cover immediately adjoin cultivated areas. They have more value where unusually dense or sparse ground cover exists.

TED DEVOS

MARK BAILEY

RICHARD T. BRYANT/ARISTOCK

RICHARD T. BRYANT/ARISTOCK

DAN JAMES

RICHARD T. BRYANT/ARISTOCK

DAN DUMONT

RICHARD T. BRYANT/ARISTOCK

Important habitat types and foods for rabbits include field edges with annual weeds (top left), burned and unburned woodlands (top right), dense shrubs, plums, blackberry thickets, and field borders (second row), legumes such as vetch and clover (third and fourth row left), and grassy roadsides next to cover (above). Good cover near high-quality food sources is probably the most important component of excellent rabbit habitat.

Other key types of vegetative cover also deserve consideration. For example, large fields and pastures contain areas within the center which are not utilized by rabbits. Generally, rabbits will not venture far into the open. To provide access routes into these areas, large fields may be broken into smaller tracts by providing travel lanes across or into fields. This may be accomplished by leaving undisturbed strips that grow up over time in natural vegetation.

Strips should be at least 60 feet wide, or wider if practical. They may be established by connecting adjacent forest stands, cultivated fields, or other areas that provide adequate cover. Mowing, disking, or burning one side of the strip every two or three years in the early spring will maintain desirable plant growth. In order to receive some economic return from cropland removed from agricultural production, these areas may be established in pine seedlings on a spacing between 8 x 10 feet and 12 x 12 feet. These plantings will provide a permanent cover type as well as some financial return in the future. If pines are not used, brush piles or other suitable escape cover can be located randomly through transition zones and greatly increase rabbit use of the area.

DEVELOPING FOOD

Food sources for rabbits can be enhanced by improving the quality and abundance of native food plants or establishing supplemental plantings. Native grasses, legumes, and other forbs can be allowed to become established in ½ acre or more of open land. These areas should be well distributed across the landscape and located near cover. Natural openings can be easily maintained in early stages of plant growth by periodic mowing, disking, or prescribed burning.

Quarter-acre food patches or strips of food planted beside adequate escape cover can be especially beneficial to rabbits, especially in winter. The patches also concentrate rabbits, making them easier to find while hunting. If possible, there should be at least one food patch for every 2 to 5 acres of rabbit habitat. White and crimson clover and bahia grass provide a preferred food during the spring. Any green succulent vegetation such as alfalfa, vetches, millets, wheat, barley, ryegrass, winter peas, various annual grazing mixtures, and wasted grain from agricultural harvests will provide a supplemental winter food source. Fescue should not be planted for rabbits because of low utilization and problems with fungus-infected varieties, such as tall fescue KY 31, which is toxic to rabbits and causes lower reproductive rates. General recommendations concerning the planting and care of these crops, including when to plant, the kind and amount of fertilizers, and other pertinent information can be found in Appendixes D–K.

Forest Management for Rabbits

Although high rabbit populations and good rabbit hunting are generally associated with mixed cultivated areas and farmlands, properly managed forest types can also provide rabbit hunting opportunities. With an ever-increasing amount of land being converted to pine production in Alabama, use of timber management practices that enhance rabbit habitat is essential to maintaining adequate cottontail and swamp rabbit populations.

Newly regenerated timber sites can provide excellent habitat for rabbits for the first five to seven years. Several studies have indicated that rabbits prefer less intensively prepared (drum chopping) sites over those that were more intensively prepared for regeneration. Intensive site preparation (shearing, raking, and piling) dramatically reduces the amount and distribution of woody rabbit cover over an area. To offset this problem, windrows that contain timber harvest slash and debris should be retained and not burned. Over time these areas will become overgrown with woody vines, such as honeysuckle and blackberries, providing excellent cover and food sources for rabbits. Rabbits can also use windrows as travel corridors to move through open regenerated areas. Some managers have also established supplemental plantings adjacent to windrows or disked these areas annually to promote green vegetation valuable to rabbits.

Of all the techniques used in forest management, prescribed burning is probably the least expensive and most effective method known to improve rabbit habitat in certain timber types. Although the proper use of fire is an excellent wildlife management tool, there are some general guidelines which may be useful in burning, particularly to enhance habitat for rabbits and other small game. Burning should be conducted only in forests managed for pine. Pine seedlings, young pines and even mature hardwoods may be damaged or killed by fire. Prescribed burning in pine stands for intensive cottontail management should be conducted in alternate strips between mid-February to late March. Following this procedure, the total acreage should be burned every two years. Burning in April or later may do more harm than good. Burning should be limited to cool burns in the early morning or late afternoon. With rare exceptions, only a backfire (into the wind) should be used. Normally backfires should be burned into a steady breeze of five to eight miles per hour.

Relative humidity is vital while burning because it affects the combustibility of the ground litter. As the humidity falls, the fire is likely to become hotter. Ambient temperature impacts fire because greater combustibility of material and better burning conditions

Fig. 2: Habitat and Harvest Management Intensities for Rabbits

Management Practices	Low Intensity	Medium Intensity	High Intensity
Timber Harvest/ Regeneration Types	Clear-cut/Plant	Clear-cut/Plant, Natural	Clear-cut/Plant, Natural
Timber Harvest Rotation Length	35–70 years	20–45 years	15–25 years
Thinnings	Once or none	Every five years	Every four or five years
Fire	None	Every three–five years	Every two–three years
Site Preparation	Roller chopping and burning	Roller chopping, burning, selective herbicide	Shearing, piling, burning, root raking and windrowing
Windrows	Burned	Leave alternate windrows	Leave unburned
Seedling Spacing	6 x 6, 6 x 8	8 x 8, 8 x 10	10 x 10, 12 x 12
Hardwoods	Retain in drains	Retain in drains and clumps	Retain all hardwoods
Food Plots	1–3% of land in clovers and agricultural food crops	3–5% of land area in clovers and agricultural food crops	Greater than 5% of land area in clovers agricultural food crops
Fertilization	None	Food plots every two–three years	Food plots annually
Openlands	1-3% of land maintained grass and brush	3–10% of land maintained in grass and brush, establish brush piles	Greater than 10% of land area maintained in grass and brush, establish brush piles
Rabbit Harvest	Heavy harvest	Heavy harvest	Heavy harvest

are generally associated with higher temperatures. Higher ambient temperatures equal higher rates of woody stem mortality. Fires burn faster uphill than down, and the use of fire on steep slopes may remove ground litter necessary to prevent soil erosion. Fire breaks should be used around areas that are designated to be left unburned. Prescribed burning should only be conducted by those trained in prescribed burning techniques. A burning permit must be obtained from the Alabama Forestry Commission prior to burning.

Cottontail management is compatible with short or long rotation timber management. Openings created from frequent pulpwood harvests provide excellent rabbit habitat during the first five to seven years after cutting. To provide a continuum of rabbit habitat, timber harvests should be conducted on five- to seven-year intervals and be irregular in shape, less than 100 acres in size, and scattered across the management unit. On smaller land tracts, timber harvests can be 3 to 5 acres and still provide enough food and cover.

Properly managed long rotations in which merchantable trees are removed and natural reproduction periodically thinned can provide ideal rabbit habitat. In most instances, a properly managed forest will provide cottontail habitat if the forest is kept open and thinned. Management for open timber stands, especially when used in conjunction with annual or semiannual prescribed burns, will greatly reduce natural escape cover for rabbits. For this reason efforts should be made to protect escape cover from fire and to prohibit fire in areas where escape cover is being developed, either naturally or artificially. In addition, brush piles should be readily available and protected in areas that are burned. Research at Auburn University has shown that rabbits heavily use brush piles as an alternative cover source after prescribed burning when herbaceous cover is temporarily unavailable.

Since swamp rabbits usually inhabit bottomland hardwood stands associated with river bottoms and drainages, special management considerations need to be given to these areas. These sites should be retained as streamside management zones at a width of at least 300 feet on each side of a stream's bank to provide adequate habitat for swamp rabbits. Food and cover can be enhanced in these areas by creating scattered openings with group selection timber harvest or small clear-cuts within SMZs. These openings should be 1 or more acres in size and at least 50 feet wide. One opening for every 10 acres of forest should support a relatively high population of swamp rabbits. Two or three brush piles should be created in each opening from slash remaining from timber harvests. Half-cuts or living brush piles can also provide cover for rabbits in openings. If these openings are accessible and can be managed on a long-term basis, they can either be planted in supplemental foods (see Appendixes C–J) or maintained in native rabbit forage. Native food plants can be maintained by disking, mowing, or prescribed burning every one to three years to stimulate the production of green succulent vegetation for food. Care should be taken to protect brush piles and thickets that provide cover for rabbits during these disturbances.

Rabbit Harvest

In Alabama, hunters annually harvest over 400,000 rabbits, making it the fourth most popular game species in the state. Rabbit populations are not harmed by heavy hunting pressure. Rabbits, like quail and most other small game, have a high annual mortality (up to 80 percent) which occurs regardless of whether or not they are hunted. In other words, rabbits cannot be stockpiled regardless of the carrying capacity of the land. If hunters do not remove the annual surplus, then weather, disease, parasites, predators, or starvation will. Normally up to 65 percent of the fall rabbit population may be taken without affecting the spring breeding population.

SUMMARY

The gray, fox, and southern flying squirrel are the three tree squirrel species found in Alabama. Gray squirrels prefer large tracts of dense, mature hardwoods, especially oaks and hickories with an understory of smaller trees and shrubs. In contrast, fox squirrels prefer open, park-like stands in pine, mixed pine and oak, or oak-gum-cypress stands. The secretive flying squirrel is active at night in the same habitats preferred by both gray and fox squirrels. For all three species, management efforts should provide a mixture of mast-producing trees for food production during fall and winter months. Understory shrubs also are important in providing fruits during other parts of the year. Retaining den trees for shelter and raising young is a critical component of sound squirrel management. Rabbits, like squirrels, must have adequate food and shelter to thrive. Like bobwhite quail, rabbits require early successional-stage vegetation that provides dense thickets for cover and protection from predators, as well as succulent herbaceous vegetation for food. With proper planning, forest and farm management practices can be tailored to meet the needs of both squirrels and rabbits by following the suggestions in this chapter.

One Rabbit Hunter's Observations

Jeff Hadaway is an avid rabbit hunter from Lafayette, Alabama. His years of rabbit hunting have provided some unique observations concerning rabbit habitat and hunting.

I have hunted rabbits all across Alabama, and the places that are the most productive are so thick that you usually have a hard time walking through them. These areas provide excellent cover to protect rabbits from a variety of predators. Prime rabbit habitat is thick in weeds (grasses and forbs) and briars with a midstory of shrubs and young tree seedlings that provide both cover and food for rabbits. Newly regenerated pine stands, up until about five to seven years of age, are ideal habitat for "sagers" or cottontail rabbits. To enhance these areas for rabbits, small openings (1 to 2 acres in size) such as logging decks, utility right-of-ways, old woods roads, and other areas can be planted in clovers. Natural vegetation can also be fertilized to enhance growth and palatability for rabbits.

Some of my favorite places to hunt cottontail rabbits are abandoned farms, because they provide an ideal mix of food and cover. For the bigger canecutters, I prefer to hunt thick swamps and wetland areas. Old beaver ponds surrounded by tall grasses provide other prime places to hunt canecutters. For a good day of rabbit hunting, using a few beagles, I like to hunt an area that is 200 acres or more in size. This allows enough space to cover new areas, moving from one site to another after a rabbit has been killed. This also prevents overharvesting and ensures that enough rabbits remain across the entire tract to sustain healthy rabbit populations for future hunting.

The familiar cooing of the mourning dove is a sign that spring is near in Alabama. Each fall thousands of Alabama hunters take to the field to shoot doves for table fare in an annual ritual that is a southern social tradition.

MAC CONE

CHAPTER 9

Managing for Mourning Dove

CHAPTER OVERVIEW

Alabama is fortunate to have an abundance of resident and migratory mourning doves. This chapter will help familiarize you with the mourning dove, and its life history, biology and habitat requirements. Methods to enhance farms and open forest lands for doves are throughly described in this chapter, as well as recommendations for harvest strategies. Whether you are a landowner, manager, sportsman, or dove enthusiast, this chapter will be a handy reference on the mourning dove and its management.

CHAPTER HIGHLIGHTS PAGE

BIOLOGY AND LIFE HISTORY **202**
COMMON DISEASES THAT AFFECT DOVES **203**
FOOD NEEDS OF THE DOVE **203**
HABITAT IMPROVEMENTS **204**
September and October Dove Fields **204**
November–January Dove Fields **205**
OTHER HABITAT IMPROVEMENTS FOR DOVES **206**
ADDITIONAL CONSIDERATIONS TO IMPROVE DOVE HABITAT **206**
WHAT IS A LEGAL DOVE FIELD? **206**

Pioneers settling in Alabama encountered a small pigeon-like bird in and around forest openings. Although the passenger pigeon was larger and more numerous during this time, mourning doves (*Zenaida macroura*) were abundant enough to provide a source of pleasure with their low, mournful cooing and some variety in table fare for the early settlers. Unlike the passenger pigeon, which has long since become extinct from the pressures of unregulated market hunting and habitat loss, mourning dove populations have increased along with the progress of civilization and it has become one of our most popular game species.

An interesting fact about the scientific name of mourning doves, *Zenaida macroura*, is that it was first penned by Prince Charles Lucien Bonaparte, a nephew of Napoleon. Lucien Bonaparte (1803–1857) was an avid ornithologist who at the age of 22 published *American Ornithology*, in which he named and described doves as well as other birds. Bonaparte named the genus *Zenaida* after his wife Zenaide, and the species name *macroura* to describe the dove's long tail.

The mourning dove is widely distributed within the United States, occurring within all the contiguous 48 states, southern Canada, Central Mexico, and Cuba. Northern segments of the population are highly migratory and usually start their migrations south by mid-October and return in early spring. Southern doves in general do not exhibit as much seasonal movement.

Nearly 3.5 million doves are harvested each year by Alabama hunters, with most of these being resident birds, or those that live in the same general area where they were hatched. Resident doves usually make up the largest percentage of birds harvested during the first half of the hunting season, while doves migrating through Alabama account for a larger percentage of the harvest later in the season. This high harvest makes doves the state's most popular game bird, and dove hunting generates millions of dollars annually in Alabama. Many landowners in Alabama, particularly farmers, use dove hunting as an additional source of income through commercial pay-to-shoot dove fields.

Smaller and more streamlined than its relative the rock dove or domestic pigeon, the mourning dove normally attains a length of 11 to 13 inches. Its small body size lends itself to strong and swift flight. The wings produce a whistling sound audible at close range and often a clicking sound when doves are in large flocks during winter.

The dove's fondness for roadsides, open woodlands, suburbs, and farmland has also made it a familiar sight to most people. The "ocah, coo, coo, coo" call, for which mourning doves were so justly named, is most prevalent in spring, although it can be heard during other times of the year as well. The call is usually repeated every four to six minutes and, from a distance, only the last three coos may be audible.

Far from flashy in plumage, drab soft grays and browns give the bird an overall somber appearance to match its melancholy wistful call. Adults are slaty-brownish above with dark spots on the wings and back and a long, pointed, white-edged tail. The neck is reddish brown with an iridescent sheen, while the body's underside is pale tan with grayish wing linings. Upon close examination, the adult male can be distinguished from the female by prominent blue-gray feathers on the top of the head, a pinkish or rose-colored breast, and obvious iridescence along the sides of the throat.

The mourning dove has two characteristics that are rather unique in the bird world. While most birds must tilt their heads back to swallow water, the dove is capable of thrusting its bill into the water and drinking in a fashion similar to horses and cattle. Members of the dove family also produce a curd-like secretion called "pigeon milk." This substance is produced by the lining of the crop during the time of incubation and aids in the rearing of young as a nourishing food for newly born hatchlings.

BIOLOGY AND LIFE HISTORY

Courtship begins with the coming of spring. The male takes to the air with slow laborious wing beats that may carry him upward for 100 feet or more. Upon reaching the apex of his flight, he spreads his wings and glides earthward in a wide, sweeping circle. On the ground he struts with nodding head and cocked tail, uttering a series of calls to the mate he wishes to attract.

In Alabama, this courtship behavior is common from February to October, although it may be observed during any month, and reaches a peak in late spring and early summer. Once a male has attracted a female, the pair mate and remain together during the entire breeding season. The chores of nest-building are shared. When a suitable nest site has been selected, generally in pines, cedars, dogwoods, oaks, or low-growing shrubs, the male begins to collect small twigs and sticks and presents them to the female to construct the nest. Nest sites are usually about 15 feet off the ground and are generally along field and pasture edges or adjacent to other open areas. In treeless areas, doves may construct ground nests similar to those built in trees. Ground nesting, however, is not common in Alabama.

Usually only two eggs are laid in each nest. The first egg is laid a short time after completion of the frail, platform-like nest and is followed by a second egg in about 24 hours. Incubation begins with the laying of the first egg and is the reason why one egg hatches a day earlier. This early incubation of the first egg causes a noticeable difference in the size of nestlings. If all goes well, the eggs hatch after 14 days and the young birds are fed pigeon milk and partially digested seeds until they are ready to leave the nest.

DR. RALPH E. MIRARCHI

Dove nests are fragile structures constructed by both the male and female and are usually located about 15 feet off the ground in trees adjacent to fields, pastures, and other open areas.

The young doves, or squabs, develop rapidly and are capable of limited flight at around 12 days of age. Soon after the young leave the nest, the adults begin preparation for a second brood, frequently using the same nest. However, they may start their nest-building procedures anew or move to another nest constructed earlier. The entire process of raising the young may be carried out several times during the breeding season, with each cycle lasting about four weeks. In Alabama, from three to seven broods may be produced each year, and nesting doves have been observed during the winter months. Young doves that are born during the spring become reproductively mature at 90 to 100 days old and may reproduce during the same year.

This high reproduction rate is needed to balance the high mortality rate imposed by nature. The lifespan of the dove is generally between one and three years; however, doves seldom live longer than one year. About 70 percent of the doves in a fall population are usually young that are hatched in the same calendar year. Predation, disease, starvation and other factors take their toll on dove populations.

KAREN LAWRENCE/ARISTOCK

Usually only two eggs are laid and the female assumes full incubation duties. After 14 days the young birds hatch and are fed pigeon milk and partially digested seeds.

COMMON DISEASES THAT AFFECT DOVES

A fairly common disease in doves is fowl pox, characterized by wart-like growths on the skin of the head and feet. This disease may indirectly cause mortality in mourning doves by blinding, or through the formation of obstructive lesions on the bill or in the throat that restrict feeding and cause death by starvation.

Another important disease affecting doves is trichomoniasis, which is caused by a small, single-celled organism known as a protozoan that infects the mouth, throat, and crop areas. Doves are highly susceptible to this disease, and an outbreak in 1950 in the Southeast caused the death of an estimated 50,000 birds.

Severe winter weather may at times induce heavy mortality in dove populations. Ice storms and unusually cold weather are especially deadly when they occur in the southern portion of the range, as southern populations are not adapted to such weather extremes. If enough hard winter has ensued to produce snow, ice storms severe enough to snap trees in two, or late frosts that nip tree buds or blooms, it is fairly certain doves have also been impacted to some degree.

FOOD NEEDS OF THE DOVE

Mourning doves are primarily vegetarians, dining on waste grains and other seeds. However, they will occasionally eat small traces of animal matter, primarily insects. A majority of the dove's diet is composed of

Native and Cultivated Grasses Preferred by Doves

- barley
- barnyard grass
- bristlegrass
- browntop millet
- bull paspalum
- bullgrass
- crabgrass
- dove proso millet
- foxtail grass
- grain sorghum
- Japanese millet
- Johnson grass
- oats
- other millets
- wheat

native and cultivated grasses.

Legumes such as cowpeas, soybeans, and peanuts may also be taken, but are not as preferred as grass seeds. Other native foods are Carolina cranesbill, dove weed (tropic or woolly croton also known as goatweed), morning glory, pokeweed, ragweed, and sweetgum seed. Doves do not have strong feet and rarely scratch the ground for their food; therefore, foods must be on relatively open ground and plainly visible.

Like most seed-eating birds, doves require grit to help grind their food. Grit is normally composed of small bits of sand or gravel, but small snail shells and hard insect parts may also be ingested. In addition to food and grit, doves require a daily supply of fresh water to prevent dehydration and to soften and aid in the digestion of food.

HABITAT IMPROVEMENTS

Landowners wishing to attract this migratory, highly mobile, and gregarious species for hunting or bird-watching will find it relatively easy to attract to their land. It is possible to concentrate the birds for hunting by planting fields or other idle areas in a preferred dove food.

In most instances, the best methods of attracting doves can easily be combined with normal farming practices or grain-harvesting procedures. Ample dove concentrations in Alabama are frequently found near corn or peanut fields, and in fields from which wheat or corn has been combined, harvested, or sown.

Since doves prefer to walk and feed on ground free of dense vegetation, removal of excess vegetation by burning or light disking may be necessary. This practice also exposes waste grain and makes it readily available to the birds. Any field in which doves are being concentrated for hunting should be at least 5 acres in size.

Management practices that are useful in September may not be practical or legal in January. Therefore, the following management recommendations and suggestions are presented as a guideline for landowners wishing to have doves throughout the year.

September and October Dove Fields

Browntop millet is a highly preferred food frequently used to attract doves during the early fall season before corn is ready for harvesting. Browntop should be planted in a well-prepared, fertilized seedbed, in rows 36 to 42 inches apart, at the rate of 10 to 12 pounds of seed per acre. Fertilizer should be applied as determined by a soil test and turned under prior to planting. Six hundred pounds of 6-12-12 per acre or the equivalent can be used. Browntop can be planted from mid-May to mid-July, and young stands should be enhanced with an additional 180 pounds of ammonium nitrate per acre once they are established.

Cultivation treatments should be carried out as often as necessary to keep the areas between rows free of weeds and provide open feeding areas. Broadcasting about 20 pounds per acre of browntop millet also works well, especially if properly fertilized, and when the plants mature, the seeds are combined or the millet is cut, raked, and bailed as hay. This should be done two weeks prior to hunting season opening or the proposed date of the dove shoot. Browntop matures in approximately 60 days.

Dove proso millet is another highly preferred grain that doves readily consume. Proso may be planted using the same method of fertilizer application as browntop. Eight to 10 pounds of seed per acre should be used. Fields of proso may be left unharvested or harvested much the same as browntop two weeks prior to hunting. Dove proso matures in approximately 90 days.

Sunflower is also an excellent choice to plant for doves. Any of the varieties that produce small to medium-sized seeds high in oil content are good choices. In Alabama, Peredovik is the most popular variety. Sunflowers should be planted no later than July 15. Since the seeds mature in 90 to 100 days, they should be planted early enough so the seeds are mature at least two

STEVEN HAYSLETTE
STEVEN HAYSLETTE
STEVEN HAYSLETTE
RICHARD T. BRYANT/ARISTOCK
STEVEN HAYSLETTE
TES RANDLE JOLLY
BOB MCCOLLUM

Mourning doves primarily eat waste grains and other seeds. A majority of the dove's diet consists of native and cultivated grasses. The following dove foods are wooly croton (top left), signal grass (top middle), spiny pigwood (top right), ragweed (middle center), bristle grass (middle right), sunflower (bottom left), and pokeberry (bottom right).

weeks before the opening of dove season. Sunflower can be planted in continuous rows spaced 36 to 42 inches apart or in an alternating two-row pattern. The planting rate for continuous rows is 5 to 7 pounds per acre, spacing plants about 1 foot apart. If planted in an alternating two-row pattern, plant two rows 36 to 42 inches apart, then skip the width of two rows and plant two more rows 36 to 42 inches apart. For this method planting rates should be about 3 pounds of seed per acre. Planting in alternating rows provides more bare ground for dove feeding. If grasses and weeds are a problem between rows, they should be controlled by using herbicides or by repeated disking between rows.

Woolly and tropic croton, commonly called dove or goat weed, are native annual weeds that frequently come up on their own in fields and pastures. Woolly croton is most often found in pastures and abandoned fields. Tropic croton is frequently found growing in cotton fields, produces mature seeds during the late summer and early fall and is usually responsible for large dove concentrations sometimes found in cotton fields. Where doves are desired, agricultural measures that attempt to control croton should be restricted.

November–January Dove Fields

Alabama corn fields that have been harvested in late summer probably have more late fall and winter dove shooting than all other types of dove fields. However, for those who wish to provide an additional food source for doves, milo, grain sorghum, and peanuts are readily eaten by doves in the fall and winter. Maturity for these plants is between 90 and 120 days.

For those wishing to hold doves for the entire season, browntop and milo may be planted in the corn fields in alternating strips of about 10 rows each. To have doves

Fig. 1: Native and Planted Foods Preferred by Mourning Doves

Native	Planted
barnyard grass	browntop millet
bristlegrass	dove proso millet
bull paspalum	soybeans
common ragweed	corn
Carolina cranebill	sunflower (Peredovik variety)
croton	grain sorghum
pine seeds	Japanese millet
pokeberry	wheat
sweetgum	barley
lespedezas	benne (sesame)
	buckwheat
	cowpeas

Guidelines for planting and maintaining most of these dove foods can be found in Chapter 16 and Appendixes D–J.

Grain crops such as browntop millet, grain sorghum, sunflower, and corn are examples of preferred dove foods and provide excellent shooting opportunities for hunters. Prior planning and crop variety are important in establishing a successful dove field. Seedbeds should be well prepared and seeds planted far enough in advance so they will mature at least two weeks before dove season. Fields should have enough bare ground so doves can readily find fallen seeds.

during the summer, or to concentrate them for the early fall hunting season, 1 to 5 acres in wheat planted in the fall and left unharvested will provide food during the summer months.

OTHER HABITAT IMPROVEMENTS FOR DOVES

In addition to providing resting, roosting, and nesting cover, Alabama forests, particularly pine stands, may be managed during the early regeneration stage to provide concentrations of doves for hunting. Regeneration areas may be established in browntop millet by planting open strips between seedlings. These areas normally provide good dove hunting until pine stands become too thick and begin to shade out ground vegetation. Dove plantings may also help reduce competition to young pines from other vegetation.

ADDITIONAL CONSIDERATIONS TO IMPROVE DOVE HABITAT

- Favor shrubs, clumps of pine trees, and bushy fence rows for nesting.
- Plow or disk fallow fields to encourage volunteer growth of native grasses and weeds as dove food.
- Leave cultivated field corners in early successional vegetation to favor preferred dove food-producing plants.

WHAT IS A LEGAL DOVE FIELD?

Legal dove fields are those that are planted in a normal or bona fide agricultural method. In Alabama, the Department of Conservation and Natural Resources uses the Auburn Cooperative Extension System as an authority on what is considered a bona fide or normal agricultural practice. The local Extension office has several publications that explain these practices, which include bush hogging

A Lifetime of Hunting Doves

Billy Purdue was a life-long dove hunter from Mobile who hunted almost every day of the season for over 60 years. He was an internationally known competitive shooter and won hundreds of competitions in the United States and Europe. Competitive shooting was his passion and he pursued it worldwide. He won the North American Clay Target Championship in 1949 and in 1952 took the European pigeon shooting championship in Madrid. He was inducted into the Skeet Shooting Hall of Fame in 1972 and the Mobile Sports Hall of Fame in 1992. His insights on dove fields and hunting provide a unique perspective honed from many years in the field. He passed away shortly after giving this interview. He was 75.

Most of the fields that I like to hunt doves in are large open areas with a fresh grain supply left behind from recent harvest. Fields should have a lot of visibility because doves prefer fields that are open with few trees. Ideally fields should also have water nearby. If the fields have these ingredients, they will be heavily utilized by resident doves. Migratory birds that see native birds feeding in a field are more likely to stop over and also feed. Migratory birds will only augment native bird populations if an adequate food supply exists.

In warm weather, during the first part of the hunting season, small grains such as milo or browntop millet work best for doves. In winter, dove fields should be in corn because corn meets the necessary energy needs of doves during cold weather.

At least a week before dove season, I like to scout my fields to see if they are being used by doves. This is essential in determining which field you will hunt on opening day. There is no substitute for scouting—you just cannot wish the birds into the field. To scout a field you have to look to see if doves are feeding, or look for signs in the fields such as feathers and droppings. Remember to scout during normal feeding times.

I like as few shooters as possible in the fields that I hunt. Doves should be able to land in a portion of the field and feed undisturbed. When that happens you have the right number of hunters. The problem with saturating a field with hunters placed only a gun range apart is that doves will eventually become shy and avoid the field. With fewer shooters, you can hunt a field every week, but when there are too many shooters, you have to wait 10 days or more before the next hunt.

Some biologists believe that the number of migratory and resident doves in Alabama has declined over the years. I tend to agree and have also noticed a decline in the last couple of years, especially in south-central Alabama. I attribute the decline to changing land-use patterns with fewer acres being planted in small grains and more acres in pastures or pine trees. I think we have also lost a lot of our traditional food sources that tended to attract a large number of birds in the past due to changes in farming practices. Despite the apparent declines in doves, however, properly prepared fields can still attract adequate numbers of doves for good shooting. Fields with fresh feed in areas that traditionally have birds will have doves almost every day in the fall and winter in shootable numbers until the season closes, if they are not over-hunted.

standing corn, wheat, millet, and milo, or planting wheat or other small grain in a bona fide agricultural operation or procedure. These procedures may be used in conjunction with the management practices previously recommended and should help in attracting doves.

Federal regulations require that all small grains used for dove fields be planted in a normal agricultural manner. This means that all broadcast seeds, including wheat, must be covered with soil and not exposed.

Practices considered as baiting doves which are illegal include top sowing of all small grains (including wheat) without covering seeds, the use of scratch feed and salt, or returning to the field grain that has been harvested and stored. Manipulation of standing crops for dove hunting by bush hogging, mowing, burning, or partially harvesting a field is allowed as long as the grain has been grown

Doves are weak scratchers and prefer fields with areas of open ground where food is plainly visible.

in place and no grain is removed and redistributed on the field. As regulations are subject to change during any given year, detailed regulations are not included in this section. For the latest information, contact the Alabama Department of Conservation and Natural Resources conservation officers concerning the legality of current management practices for doves. For further clarification of what is and is not legal, the Alabama Wildlife Federation's publication *Dove Hunting in Alabama—Information for the Hunter and Landowner* is an excellent source of information regarding the legality of prepared fields for dove hunting. It can be obtained by writing the Alabama Wildlife Federation, 46 Commerce Street, Montgomery, Alabama 36104, or by calling (334) 832-WILD.

Since doves are migratory, they are placed under federal jurisdiction and protected by federal regulations and an international treaty with Canada and Mexico. The migratory nature of doves between states has caused considerable conflict in northern states where doves are considered songbirds and southern states where they are favored game birds. Nearly half of the states now provide year-round protection of the dove as a songbird. Many of the northern states that do not allow dove hunting would like to see the dove protected nationwide, while southern states argue for a liberal harvest. Studies have shown that hunting mortality accounts for only 10 to 15 percent of the total annual mortality (70 percent) of doves. The remaining 55 to 60 percent is caused by natural mortality. Therefore, many doves not harvested during dove season will ultimately die from natural causes before the next spring regardless of hunting.

States that allow dove hunting must operate within a framework set out by the U.S. Fish and Wildlife Service. An opening and closing date are established within which the participating states may choose a long, continuous season or a split season with a reduced number of hunting days. The season is normally split in Alabama to provide a greater harvest and to create equal hunting opportunities in both the northern and southern parts of the state.

The dove hunter who would like to attract and hold large numbers of birds should restrict shooting to no more than one hunt each week. In addition, sportsmen should restrict hunting to either morning hours, where legal, or early afternoon, but not both. Shooting during the last couple of hours before sunset should be avoided so that doves can feed undisturbed. Landowners in a community may get together and plan shoots on a rotational basis, to provide plenty of hunting without overshooting any one field. Some sportsmen prefer to shoot adjacent fields at the same time to keep the birds flying between two fields. It should be remembered, that doves are highly mobile as well as migratory, and once doves are concentrated in an area, plans should be made to shoot soon since they may not remain long in one area. As a final note, many dove hunters believe that a sufficient number of hunters are needed in fields to keep doves flying and to prevent them

Small Grain Planting Recommendations[1]

Although currently under review, top sowing of all small grains without covering the seed is not considered a normal agricultural practice. Most small grain is normally planted into prepared seedbeds by broadcasting or drilling. A bona fide attempt must be made to cover the seed by disking, raking, or cultipacking. Some incidental seed may remain on the surface following a bona fide covering attempt. The only recommended methods of planting small grain without a prepared seedbed are 1) no-till drilling, or 2) aerial seeding small grains into standing row crops, such as cotton or soybeans, just prior to defoliation. Recommended seeding rate for small grain is no more than 200 pounds per acre. Seeds should be uniformly distributed at approximately 50 seeds per square foot.

[1]Recommendations from the Alabama Cooperative Extension System as normal agricultural practices for small grains.

Fig. 2: Habitat and Harvest Management Intensities for Mourning Doves

Management Practices	Low Intensity	Medium Intensity	High Intensity
Timber Harvest/ Regeneration Types	No cutting	Group Selection	Clear-cut
Timber Harvest Rotation Length	35–75 years	20–45 years	15–25 years
Fire	None	Every two–three years	Annually
Site Preparation	Roller chopping and burning	Roller chopping, burning, selective herbicide	Shearing, piling, burning, root raking, and windrowing
Windrows	Burn	Leave alternate windrows	Leave unburned
Seedling Spacing	6 x 6, 6 x 8	8 x 8, 8 x 10	10 x 10, 12 x 12
Fertilization	None	Food plots every two–three years	Food plots annually
Openlands	None	Disturb abandoned fields by burning, mowing, or disking	Disturb abandoned fields by burning, mowing, or disking annually and managing agricultural crop fields for hunting
Dove harvest	Follow legal guidelines	Follow legal guidelines and restrict shooting to one hunt per week	Follow legal guidelines, restrict shooting to one hunt per week, limit the number of hunters to one per 3 acres of field or less

from landing in fields during shooting. While this does occur in some cases, to ensure hunter safety, never exceed one hunter for each acre of field or one hunter for each 100 yards of linear field edge. One hunter per 3 acres will actually provide the most pleasant and enjoyable shooting.

SUMMARY

Dove hunting is a rich tradition that signals the opening of the fall hunting season in Alabama. Each year approximately 3.5 million mourning doves are harvested by Alabama hunters who relish the social atmosphere of the hunt as much as the table fare provided by doves. In terms of dollars generated in the states economy from hunting activities, dove hunting is the most popular game bird, generating millions of dollars annually in Alabama. Like most seed-eating birds doves eat a variety of foods such as waste agricultural grains and seeds from native grasses and forbs. Habitat management efforts should be tailored to provide field openings that contain a mixture of agricultural grains or native seed-producing plants. Specific dove field planting and management recommendations for managing for doves were described in this chapter.

Permits Required to Hunt Doves in Alabama

Beginning in 1996, permits are required to hunt doves and other migratory game birds in Alabama. The free permits can be obtained from sporting license vendors in the state. The permits are part of the federal Migratory Bird Harvest Information Program that is attempting to obtain dove harvest information from hunters. This information will be used to set future bag limits and hunting seasons for doves. To obtain a permit, hunters must complete a harvest information card with their name, address, and answers to a few short questions about past hunting success. After completing the card, hunters will be given a permit that affixes to their hunting license. Permits must be available for inspection by law enforcement officers when hunting doves or other migratory game birds in Alabama.

The wood duck is one of the most beautiful birds in the world. Wood ducks and mottled ducks are the only species of wild ducks that raise young in Alabama.

RICHARD T. BRYANT/ARISTOCK

CHAPTER 10

Managing for Waterfowl

CHAPTER OVERVIEW

Alabama inland and coastal waters provide breeding (wood duck and mottled duck), migratory, and/or wintering habitat for approximately 26 different North American waterfowl species. This chapter describes some of the most common waterfowl species that reside or migrate through the state, as well as their life history, biology and habitat requirements. Methods to create and improve wintering waterfowl habitat on farm and forest lands are throughly described in this chapter.

CHAPTER HIGHLIGHTS PAGE

BIOLOGY AND LIFE HISTORY OF WATERFOWL 212

TWO COMMON WATERFOWL SPECIES IN ALABAMA 213

MANAGING MIGRATORY WATERFOWL ON A LARGE SCALE 218

WATERFOWL HABITAT PROTECTION 219

HOW WATERFOWL HARVEST REGULATIONS ARE DETERMINED 220

PREDATOR AND DISEASE CONTROL 220

HABITAT IMPROVEMENTS FOR WATERFOWL ON FARM AND FOREST LANDS 220

HARVESTING AND REGULATIONS 225

ORGANIZATIONS DEDICATED TO WATERFOWL 226

Alabama and its Gulf Coast waters provide breeding, migratory, and/or wintering habitat for approximately 26 different North American waterfowl species during some periods of the year, primarily in the fall and winter. Some species do not commonly appear and can only be observed during certain years, depending on the availability of wetlands and weather conditions.

Most waterfowl species are migratory and spend only a portion of the year in Alabama. Depending on the weather, duck concentrations are generally highest during late December and January when the fall migration from the northern U.S. and Canada peaks. The greatest number of species is found throughout Alabama from late February through early March.

Waterfowl migration routes, called **flyways**, are categorized by region. There are four major flyways on the North American continent. These include the Atlantic, Mississippi, Central, and the Pacific flyways. Alabama is considered part of the Mississippi Flyway.

Few waterfowl species breed within the boundary of Alabama. Wood ducks and mottled ducks are the only species that raise young each year in the state, except for domestic ducks, resident geese, and some hooded mergansers that nest in hollow trees. The prairie pothole region of the north-central U.S. and Canada are the primary breeding grounds for most North American waterfowl. In an average year, about 50 percent of the U.S. ducks are produced on less than 10 percent of the available breeding area. Other areas that provide substantial amounts of breeding habitat include eastern Canada, the northern latitudes of the midwestern U.S., central Alaska, and northern California.

BIOLOGY AND LIFE HISTORY OF WATERFOWL

Ducks, geese, and swans are all members of the same general family (*Anatidae*) which is characterized by large bills, webbed feet, rounded bodies, and long necks. Other duck traits include brightly colored male and drab female coloration, two body **molts** (the loss and replacement of feathers) a year, elaborate courtship displays, and brief annual pair bonds, although only the female incubates eggs and cares for the young.

North American ducks are separated into three major groups: diving, dabbling, and perching. Members of the diving duck group, appropriately named **divers** or "sea ducks," have similar feeding habits. This group prefers open, deep-water habitats, feeding on **submergent plants** (plants that are rooted on the bottom and growing under the water) such as wild celery and widgeon grass. They also eat aquatic insects. The distinguishing characteristic of divers is that they appear to run along the water to gain flight and they dive completely beneath the surface to feed.

Approximately 10 species of diving ducks migrate through the state, with ring-necked ducks and lesser scaup being the most abundant. Lesser numbers of redheads, canvasbacks, and greater scaup are occasionally observed in isolated flocks or with the more abundant lesser scaup. The remaining diving ducks often found migrating through Alabama include the common goldeneye, buffleheads, hooded merganser, common merganser, and ruddy duck.

A shortage of suitable overwintering habitat is the primary reason diving ducks are not common in most of Alabama, except in coastal areas where large bodies of water exist. Historically, except for the major rivers, there were few large bodies of open water in the state and therefore, diving ducks used the major rivers for migration corridors. A few pools and oxbows (old riverbeds) in these river systems that contained an abundance of aquatic foods were chosen as resting and feeding areas during migration. The recent creation of man-made reservoirs in Alabama for flood control, hydroelectric power, and recreation has not changed the traditional migration patterns of diving ducks because most of these reservoirs are not suitable for large concentrations of birds.

The other two groups of ducks are the **dabbling**, or puddle ducks, and the **perching**, or tree ducks. Both of these groups are often called the pond and river species since they have similar habits. Dabbling and perching ducks have brightly colored wing patches. They are usually seen feeding by a "tipping up" motion, and spring straight into flight from the water. The two most common dabbling ducks seen in Alabama are mallards and black ducks. Most of the dabbling ducks are found in small flocks on isolated wetlands; however, each year large numbers of some species are observed on wetlands and reservoirs throughout the state. Northern shovelers and blue-winged teal are early migrating dabbling ducks frequently seen in Alabama during September and October. Other dabbling ducks found throughout Alabama include the gadwall, green-winged teal, northern pintail, and American widgeon or baldpate. The most common perching duck in Alabama is the wood duck.

Other dabbling ducks use habitats similar to those preferred by mallards and black ducks. Often mixed flocks of dabblers that include mallards and black ducks are observed resting together. These species can live in

GLENN D. CHAMBERS

RICHARD T. BRYANT/ARISTOCK

RICHARD T. BRYANT/ARISTOCK

GLENN D. CHAMBERS

Common ducks found in Alabama are (top, left to right) male and female canvasback, male and female mallard, (bottom left to right) male and female green-wing teal, and a wood duck with brood. Canvasbacks are diving ducks, mallard and teal are dabbling ducks, and wood ducks are perching ducks.

the same habitats because they have slightly different feeding patterns. Mallards, black ducks, and pintails frequently feed on agricultural grains and certain native **emergent vegetation** (plants that grow in and above water levels) such as smartweed, bulrush, and sedges; and **floating vegetation** (plants that float on the water) such as duckweed and pondweed. Other species such as gadwall, widgeon, shovelers, and blue-winged teal also prefer native plant seeds, vegetative parts, and invertebrates associated with wetlands. Although there is dietary overlap among the dabblers, the diversity of food items eaten by each species reduces competition for food.

TWO COMMON WATERFOWL SPECIES IN ALABAMA

The Wood Duck

The wood duck (*Aix sponsa*) or "woodie" is a symbol of forested wetlands to hunters, bird watchers, photographers, and artists. By all accounts this bird is one of the most beautiful birds in the world. It can be found in 42 states and five Canadian provinces.

The wood duck is commonly found in sloughs, ponds, and streams in Alabama. This was not the case at the turn of the century when wood duck numbers were low. The recovery of wood duck populations is a classic conservation success story.

In the early part of the 1900s, wood duck numbers were so low that many people feared the wood duck would become extinct. Unregulated hunting during the spring was a primary factor in declining populations. In addition, the birds were losing habitat and suitable nesting cavities throughout their range. The passage of the Migratory Bird Treaty Act of 1918 provided the necessary protection to save the birds from possible extinction.

Regulated hunting and the concerted research and management efforts of waterfowl biologists and managers helped to re-establish woodies across all of their historic range, including Alabama. The increase in beaver populations and subsequent beaver pond impoundments in Alabama since the late 1930s has also helped improve wood duck habitat and contributed to the increase in wood duck populations. Habitat preservation, management, research on nesting box designs, and regulated hunting have all been part of the wood duck's comeback. As a result of these efforts, wood ducks today are the second most common waterfowl species harvested in Alabama, behind mallards.

Waterfowl Species Commonly Seen in Alabama

Species	Period of Peak Abundance
Dabbling Ducks	
Mallard	Dec.–Jan.
Wood Duck	Sept.–Nov.
Black Duck	Dec.–Jan.
Gadwall	Oct.–Dec.
Green-winged Teal	Oct.–Dec.
Blue-winged Teal	Sept.
Northern Pintail	Oct.–Dec.
American Widgeon	Oct.–Dec.
Northern Shoveler	Oct.–Dec.
Diving Ducks	
Lesser Scaup	Nov.–Jan.
Greater Scaup	Nov.–Jan.
Ring-necked Duck	Nov.–Jan.
Canvasback	Nov.–Jan.
Redhead	Nov.–Jan.
Goldeneye	Nov.–Feb.
Bufflehead	Nov.–Feb.
Hooded Merganser	Nov.–Feb.
Ruddy Duck	Dec.–March

BIOLOGY AND LIFE HISTORY OF THE WOOD DUCK

Wood ducks are the most attractive and brightly colored ducks in North America. Male woodies have a large purple and green crested head. The eyes and base of the bill are bright red. The birds also have a pair of white parallel lines running from the bill and back of the eye to the rear of the crest. Male wood ducks have a burgundy-colored chest with white fleckings and a white breast and belly. The purplish-black back sharply contrasts with the breast and belly. The female is also more colorful than other female ducks.

Female woodies are gray-brown with a distinct white, tear-drop-shaped eye ring. Young females may not show this coloration around the eye. They have a sooty-gray crested head. Both sexes have short wings, about 8 to 9 inches long, with iridescent purple wing patches. Adult wood ducks weigh about 1½ pounds and measure about 20 inches long.

The wood duck is the only waterfowl species that nests in Alabama other than mottled ducks, which nest in small numbers in the Mobile/Tensaw River Delta, and Canada

GLENN D. CHAMBERS

Left: Diving or "sea" ducks like redheads prefer open, deep-water habitats and feed on submergent plants such as wild celery and widgeon grass. Redheads are generally found in coastal areas on large bodies of water. Wood ducks prefer to nest in hollow cavities of trees but will readily nest in artificial nesting boxes if cavity trees are unavailable. The hen stays in the nest with the newly born ducklings for about 24 hours and then calls her brood from the nest to the water in search of food.

geese, which have been translocated here by the Alabama Wildlife Federation and other waterfowl conservation groups. The wood duck is technically a puddle duck that prefers to nest in hollow tree cavities. If tree cavities are not available, the birds will nest in artificial cavities or nesting boxes. Either cavity should be near a suitable body of water.

Wood ducks begin to form breeding pairs starting in mid-October. By February most of the birds have paired off and begun searching for a nest site. The birds will remain together through most of the egg incubation period unless it is a late nesting or renesting attempt. In this case the drake will abandon the hen for molting, the period of time when old feathers are lost and replaced by new feathers.

SCOTT NIELSEN/DUCKS UNLIMITED

Male wood ducks are the most attractive and brightly colored ducks in North America. The males (right) are more colorful than the females (left) and have a burgundy-colored chest with white fleckings, a white belly, and a purplish-black back.

Both the drake and hen search for and select a nesting cavity from February to April. Most nesting sites are close to water, but some may be up to 1 mile away. Once a proper site has been selected, the hen begins nesting activities. Wood duck hens have a remarkable tendency to return to the same breeding area year after year.

Hen wood ducks lay one egg a day for 10 to 15 days. An average clutch is 10 to 15 eggs; however, predation and "dump nesting," when several different hens lay eggs in the same nest cavity, make it difficult to accurately determine average nest sizes in the wild. While searching for a nest site, hens sometimes lay eggs in several different nests. Biologists have reported finding dump nests with over 40 eggs.

After the eggs are laid, the hen incubates the eggs for 28 to 37 days. The hen stays in the nest with the newly born ducklings for about 24 hours, allowing the young birds to dry thoroughly. Afterwards, the hen calls her brood from the nest and heads for water. The young birds climb or jump out of their nest and fall to the water or ground, sometimes from staggering heights. The hen cares for the young by herself, taking them to brood-rearing habitat composed of a mixture of aquatic grasses and shrubs.

The first few weeks after hatching are a difficult time for the young birds. Travel over land is hazardous if the nest is located too far from water, and many young birds die in route. Only about 50 percent of the ducklings will live to reach flying age. Predators include raccoons, turtles, alligators, snakes, herons, and even largemouth bass. Recent research indicates that erecting nest boxes on bass ponds with large bass may be condemning the ducklings to a very short life.

The young ducks mature quickly. Ducklings remain covered with down for two to three weeks. By six weeks of age, the ducks are fully feathered, and they can fly at 8 to 10 weeks of age. At this time the adult females and their young separate. The females will then molt their old feathers, which are replaced by a new set of feathers. During this molting phase, the birds are flightless for about three weeks while flight feathers are being replaced. By this time the males have already molted. Most woodies are able to fly by late July to mid-August. By mid-October the drakes will once again be wearing their bright and colorful plumage, and the annual breeding cycle begins again.

HABITAT REQUIREMENTS OF THE WOOD DUCK

Typical wood duck habitat is found in bottomland hardwood swamps, wooded sloughs and marshes, or forested riparian areas. Their optimal habitat usually has an abundance of flooded timber. Ideal woodie habitat in Alabama includes bottomland hardwood forests with shrubs and herbaceous plants in water-impounded areas that have long shorelines. Bottomland forests, either flooded or in close proximity to water, are utilized by wood ducks all year and are critical sites for nesting cover.

Water should be available to the birds three to four weeks before nesting. Mature and overmature trees in these forests are critical habitat components because wood ducks nest in natural tree cavities. In fact, lack of suitable tree cavities for nesting has been one of the primary limiting factors of wood ducks in the past. Bald cypress, sycamore, ash, and blackgum provide cavities in low-lying areas. On drier sites, oaks are important

RICHARD T. BRYANT/ARISTOCK

Typical wood duck habitat is found in bottomland hardwood swamps, wooded sloughs and marshes, and forested riparian areas.

cavity-producing trees. Another beneficial component of wood duck breeding habitat is an abundance of loafing or resting sites. Examples of loafing sites include logs, stumps, muskrat mounds, beaver lodges, buttonbush, or islands in open water and shorelines.

Like most game species, wood ducks require cover for concealment from predators. This type of cover is used for raising broods and for escape cover for molting ducks that cannot fly. A mixture of scrubby shrub habitats, where strong, durable shrubs rise about 2 feet above the water and spread into a thick overhanging canopy, provide cover during the late spring, summer, and early fall. Optimal brood-rearing habitat should be at least 5 acres and consist of shrubs, wetland plants growing in the water, trees, and open water.

The depth of the open water is also important. Shallow water enhances wood duck habitat because the birds normally do not feed in water that is more than 18 inches deep. Plants growing in the shallow water, called **emergent plants**, are a critical food source for growing ducks because they harbor vast numbers of insects and other small animal life that are a major source of dietary protein. Important emergent plants in the open and shrubby water habitats include cattails, water lily, smartweeds, water primrose, pickerel weed, rushes, reeds, canary grass, and sedges.

Flooded bottomland forests with an abundance of mast trees are especially important winter feeding habitat for wood ducks. Acorns of water oak, willow oak, Nuttall oak, white oak, and other oaks provide high energy food for the ducks. The seeds from bald cypress, buttonbush, and various emergent plants are also valuable food items.

Other habitats, such as flooded dead timber, open marshes, lakes, and reservoirs, are frequented by wood ducks only if choice habitats are unavailable. Large open areas, like lakes or reservoirs, provide little food or cover for wood ducks.

THE IMPORTANCE OF WOOD DUCK NESTING BOXES

Man has reduced suitable nesting sites for wood ducks in many areas. Removal of trees with nest cavities, drainage and destruction of wetlands, clearing timber stands for agriculture, and timber harvesting without leaving nesting trees have resulted in reduced nesting sites. To reverse this trend, foresters and other land managers should retain at least three to five trees per acre that provide natural cavities around wetland areas. In the past, because of the accelerated loss of natural nesting cavities and the resulting decline in wood duck numbers, artificial nesting boxes were constructed and placed in wood duck habitat. These boxes are readily accepted as nesting sites by wood ducks and have been responsible in large part for the dramatic increase in wood duck numbers that we enjoy today. The first nest boxes were built and installed by the U.S. Biological Survey, now the U.S. Fish and Wildlife Service, on Chautauqua National Wildlife Refuge in central Illinois during 1937. Since then, hundreds of thousands of nest boxes have been erected by federal and state agencies as well as private landowners and conservation organizations throughout the wood duck's range. The Mobile County Wildlife and Conservation Association has erected hundreds of nest boxes in the Mobile/Tensaw River Delta.

After years of research examining wood duck preferences for nesting boxes, biologists have concluded that the best boxes should be built from seasoned woods such as cypress, cedar, or hemlock. A strip of hardware cloth should be attached to the inside of the box up to the entrance hole to provide traction for the day-old ducklings attempting to leave the box. All boxes must have 3 to 4 inches of wood chips added because the female does not bring nesting material to the cavity. The boxes should be checked each year, and this material should be changed annually. The nest boxes should be placed over water and close to suitable brood habitat.

LEWIS O. ROGERS/SOUTH CAROLINA DNR

Wood duck nesting boxes with predator guards have been successful in hatching wood ducks across Alabama.

The first boxes were built of wood and required frequent replacement. In an effort to reduce maintenance and replacement costs and predation, nest boxes were later built of a variety of materials including plastic, fiberglass, and metal. Boxes not constructed from wood should not be used in full sunlight because they result in elevated temperatures inside the box that can kill the duck embryos. What was intended to provide shelter can become an oven.

Predators are another factor that limit wood duck production. It provides little benefit to put up nest boxes if the result is easy access for predators. Unfortunately, as woodies adapted to using nest boxes, their predators also learned to associate nest boxes with a potential meal. Raccoons, rat snakes, fox squirrels, and starlings commonly destroy wood duck nests.

Research has shown that the raccoon is the major predator of wood duck nests; therefore, predator guards should be installed whenever possible. Equally important is placing boxes in trees that predators cannot reach from another tree. The protection offered by nest boxes and predator guards should be equal to, if not better than, that of natural cavities.

The Mallard

The mallard (*Anus platyrhynchos*) is the most common migratory duck species in Alabama. The U.S. Fish and Wildlife Service estimates that during the annual fall migration, mallards account for about 15 percent of the entire duck population in the U.S. Mallards are the most widely distributed waterfowl species in the northern hemisphere and can be found from the arctic to the subtropics in Asia, Europe, and North America. Many varieties of domestic ducks are descendants of the mallard.

BIOLOGY AND LIFE HISTORY OF THE MALLARD

In breeding color, drake mallards are more recognizable than any waterfowl species. The iridescent green head, white neck ring, brown chest, gray sides, olive-colored bill, and dark rump help in quick identification. As with most ducks, the hens are not very colorful. Hen mallards are a light brown streaked with darker brown. This pattern is more apparent on the back than the belly. A characteristic dark eye stripe and mottled orange bill distinguish hen mallards from females of other duck species. Both sexes have bright orange legs and violet-blue wing patches, which are bordered by white bars on the

PAUL T. BROWN

The mallard is the most common migratory duck in Alabama. Drakes (right) are recognized by their iridescent green head and white neck ring. Hens (left) are a light brown streaked with darker brown.

leading and trailing edges. These wing patches are visible in flight. During the summer when the birds lose their breeding plumage, the sexes look similar. However, males can still be distinguished from females by their olive-colored bills. Mallards nest and produce young in the northern one-third of the United States northwestward to Alaska. Nesting begins in April with hatching of 9 to 10 young about 25 days later.

The fall migration of mallards usually occurs later than that of other dabbling ducks. The birds appear reluctant to migrate south as long as adequate food and open water are available. In fact, some birds have been known to spend the winter in the north and not migrate. Traditionally, the significant mallard migration flights south may not begin to arrive in Alabama until mid-November or later depending on northern weather. Populations in Alabama slowly build through November and December. Because of the prolonged migration window for mallards, there are no sharp migration peaks as seen with other ducks.

Winter concentrations of mallards in the Mississippi Flyway, which includes Alabama, are the largest in North America. The heart of this wintering area extends from Cape Girardeau, Missouri, south to the Gulf of Mexico through the fertile lower Mississippi River Delta region. Waterfowl counts have recorded up to 37 percent of the entire North American mallard population wintering in this region. These tremendous concentrations have led some to nickname the Mississippi Flyway the "mallard flyway." Mallards also winter in much of the United States as far south as the Gulf of Mexico and along the West Coast as far north as Alaska. Recently mallards have expanded their range eastward to include much of the Atlantic Coast from New England to North Carolina. Contrary to the pattern of fall migration, during the spring, mallards appear to rush back to the breeding grounds. Birds start to leave Alabama wintering areas in February and by early April few birds remain.

Adaptability is the key for a species like the mallard that uses a broad range of habitats. Mallards eat both native and agricultural foods, and each of these food groups are equally important during some period of the annual cycle. The mallard's varied diet ensures that the ducks obtain nutrients essential for survival. Native foods are usually nutrient-complete, while cultivated foods may lack some essential requirements like high amounts of protein. Agricultural foods, therefore, can supplement but not totally replace native foods.

Fall and winter foods consist primarily of high energy seeds from aquatic or emergent wetland plants and cultivated sources. Native foods include seeds from sedges, millet, smartweed, coontail, duck potato, duckweed, and mast from nut-producing trees. Some of the cultivated grains that mallards readily consume include corn, rice, sorghum, wheat, barley, and oats. Mallards also feed on tubers and rhizomes of chufa, flatsedge, and bulrush.

In the spring, male and female diets vary because of differing nutrient requirements. Females seek protein-rich foods to obtain nutrients essential for egg production. Their diets consist primarily of aquatic invertebrates such as midges, crustaceans, and mollusks. Males also use these tiny animals but may eat more seeds and other vegetative matter during this time. Male diets shift to animal matter during the molt because the birds need extra protein to replace feathers.

The summer diet of adults and ducklings is dominated by animal matter, although plant foods are eaten when available. Mosquito, dragonfly, and other insect larvae, along with those mentioned previously are readily eaten. By late summer, the diet shifts to include more high energy foods in preparation for fall migration.

MANAGING MIGRATORY WATERFOWL ON A LARGE SCALE

Management of North America's waterfowl resource requires the cooperation of several countries because waterfowl do not recognize the arbitrary boundaries drawn by man. The migratory travels of waterfowl may carry them across the boundaries of two or more nations each year. Most North American ducks winter in the United States and Mexico but breed, raise young, and molt on Canadian wetlands.

International cooperation is critical for waterfowl management. The Migratory Bird Treaty Act of 1918 implemented an agreement between Great Britain, Canada, and the United States for the protection of migratory birds. This treaty established federal jurisdiction for protecting the international migratory bird resource. The act was amended in 1936 and 1974 to include Mexico and Japan, respectively. In 1978 a treaty similar to the Migratory Bird Treaty Act was signed with the Union of Soviet Socialist Republics, giving international protection to waterfowl with global distributions.

Within North America, differences in supply and demand on the waterfowl resource have resulted in managing waterfowl on the **flyway management concept**. There are four flyway councils, and each council's management decisions are based on historic migration patterns. Each council (Atlantic, Mississippi, Central, and Pacific) is composed of the states and Canadian provinces associated with their respective migratory region. This format allows the council to address the interests and concerns of each state or province. Each council has a technical section that is composed of the waterfowl specialists from each state and province. Alabama waterfowl biologists hold a position on the Mississippi Flyway council. Flyway councils meet in the spring and late summer each year to vote on the management recommendations of the technical section. Annual recommendations for harvest regulations are made at the summer meeting. The U.S. Fish and Wildlife Service gives final approval for harvest recommendations.

Management activities can be categorized under four general topics: habitat preservation and enhancement, harvest regulations, and predator control, and disease control. Of the four, habitat preservation and enhancement remains the top priority. Without sufficient breeding, migratory, and wintering habitats, there would be little need for waterfowl harvest regulations.

WATERFOWL HABITAT PROTECTION

Within the last several decades, waterfowl populations have seriously declined in this country. A number of factors are to blame including habitat destruction, chemical contamination of existing wetland habitat, and a series of droughts in the prairie pothole region that have caused duck populations in North America to drop significantly. The federal government and the Alabama Department of Conservation and Natural Resources have been doing midwinter waterfowl surveys since 1955 to count the number of ducks in Alabama and this country. The number of ducks seen in Alabama during these surveys has declined. These declines in waterfowl populations have raised many questions concerning the status of waterfowl habitat in North America. The most critical problem facing North American waterfowl populations today remains the loss of wetland habitat. Nearly 500,000 acres of wetlands are lost annually to agriculture and urban or industrial development in the U.S. Wetlands and potholes in the northern prairie region, coastal marshes, bogs, and bottomland hardwood swamps are essential for breeding, migrating, and wintering waterfowl. More than 50 percent of North America's wetland resources has been lost in some regions, and other regions like the Central Valley of California have lost as much as 80 percent. Alabama has lost 50 percent of its historical wetland acreage.

In addition, upland grass nesting sites, critical to mallards and pintails, are being converted to other uses at a rate of 2 percent a year. Only 12 percent of Canada's natural grasslands, one of the primary waterfowl nesting areas, remain. In the last 10 years, 33 percent of the remaining grasslands in the north-central states have been converted to cropland.

The problem of suitable habitat is worsened by intensive agriculture that results in other environmental problems. Soil erosion, water quality decline, siltation, and chemical contamination result in poorer quality wetland habitat. Poor soil management and wetland drainage can lower the land's productivity for both agriculture and wildlife.

The long-term destruction of wetland habitats may have severely affected the ability of present and future ducks to survive. Although wetlands are lost by both natural and man-made disturbances, the activities of man are ultimately the source of most of the habitat destruction. In the Mississippi Flyway, which historically has been the flyway with the greatest number of migrating and wintering waterfowl, seven states have lost more than 70 percent of their original wetland habitats. Pollution, siltation, and channelization have also affected wetlands by making them unattractive or even deadly to waterfowl that attempt to use these areas. The negative effects of wetland loss are even more severe on waterfowl when combined with the periodic droughts in the prairie breeding grounds.

Ducks in the past have adapted well to the wet and dry cyclic occurrences characteristic of the prairie breeding grounds. Mobility, a long life-span, large clutch sizes, and the ability to re-nest are excellent adaptations for prairie-nesting duck species. Greater mobility of waterfowl permits breeding ducks to seek wetlands suitable for nesting. In higher than average rainfall years when wetlands are abundant, the ducks respond with increased success in reproduction. However, in years of drought when wetlands are reduced, few birds try to reproduce, and predators can find the birds and their nests more easily because they are more concentrated. Most continue to fly farther north to more permanent wetlands.

What has been done to combat the decline in waterfowl populations? Biologists from the U.S. Fish and Wildlife Service and the Canadian Wildlife Service have outlined a plan of action to preserve North America's magnificent waterfowl populations. On May 14, 1986, the U.S. Secretary of the Interior and the Canadian Environment Minister signed a historic agreement to enhance the continued survival of ducks, geese, and swans in North America. That far-reaching document, the North American Waterfowl Management Plan (NAWMP), sets a course of action for both countries to follow for future years. If all agencies, organizations, and individuals interested in waterfowl support the effort, this plan could become a modern conservation success story similar to the restoration of the wild turkey and the whooping crane. Already waterfowl habitat conservation efforts are beginning to pay off with recent increases in some waterfowl populations. In 1995, 83 million ducks packed the flyways compared to 71 million in 1994 and 59 million in 1993.

HOW WATERFOWL HARVEST REGULATIONS ARE DETERMINED

The purpose of hunting regulations is to control the waterfowl harvest. This is accomplished by limiting the harvest to the number of surplus waterfowl of each species. Surplus means that portion of the population in excess of the breeding component necessary to maintain the population at a healthy level. A surplus is based on quantity and quality of breeding habitat and the annual mortality rate. Mortality varies by species population size and is strongly influenced by habitat conditions.

Determining the surplus can be quite difficult. It requires the cooperative efforts of federal, Canadian provincial, and state waterfowl biologists to collect important information. The size of the breeding population, age and sex ratios, survival rates, and habitat conditions must be known to accurately determine the annual surplus. Age and sex ratios, along with survival rates, are estimated from the data collected during the previous hunting season. Estimates of breeding populations, reproductive success, and habitat conditions are obtained from surveys conducted during the current nesting year. All this information is factored together to make a fall flight forecast.

The **fall flight forecast** is an estimate of the total number of ducks available after reproduction and before hunting season. Hunting season length, bag limits, and species restrictions are set using this information.

PREDATOR AND DISEASE CONTROL

The most critical time for waterfowl survival is during nesting and brood-rearing. Just about every mammalian and avian predator will consume waterfowl eggs and chicks. Common predators include raccoons, skunks, opossums, snakes, red and gray foxes, bobcats, crows, hawks, owls, alligators, and largemouth bass. For nesting wood ducks, predator guards can significantly aid in eliminating nest predation. In addition, nesting cavities should be located over water that contains adequate brood-rearing cover composed of a mixture of shrubs and wetland plants.

The U.S. Fish and Wildlife Service estimates that each spring 900,000 adult ducks, mostly nesting hens, are lost to predation on the northern prairie nesting grounds, with the most common predator being the red fox. A recent study by the Delta Waterfowl Foundation found that predator control on the prairie breeding grounds of north-central North Dakota is an effective method of increasing the nesting success of waterfowl. The study found that nesting success was as high as 82 percent in areas where red fox, skunk, and raccoon numbers were controlled. The study suggests that predator control on northern waterfowl breeding grounds may be an effective method of increasing waterfowl populations. More research is needed, however, to fully understand predator and prey relationships as they apply to waterfowl management in North America.

Current knowledge regarding waterfowl disease losses is limited. The number of waterfowl that die each year from disease is unknown. Many times predators and scavengers remove dead birds before they are discovered by man. Other sick birds seek seclusion and die unnoticed. Only when a major die-off occurs are managers able to respond. Most management efforts are concentrated on dispersing birds and disposing of carcasses when a large disease outbreak occurs.

Contaminant poisoning, like lead poisoning, is another factor frequently associated with disease. Contaminant poisoning is something managers may have more control over. Lead poisoning in waterfowl has long been recognized as a major cause of death in some areas. The impact of lead poisoning on reproduction is still unknown. Elimination of lead shot for all waterfowl hunting, while controversial, will go a long way in reducing availability of this contaminant in future years.

HABITAT IMPROVEMENTS FOR WATERFOWL ON FARM AND FOREST LANDS

Alabama farm and forest owners have several options for improving their lands for waterfowl. These options include management of beaver ponds and farm ponds, creating temporary impoundments in hardwood bottomlands, or flooding agricultural fields. The key with any of these alternatives is providing enough food, water, and cover when most migrating waterfowl are present during the fall and winter months. With all of these alternatives, food sources can be enhanced by a practice called **moist soil management**, which encour-

Constructing and Installing a Three-Log Drain

1. Fasten three logs together with nails and short pieces of wood as shown at right. Logs should be 6 to 9 inches in diameter and 10 to 16 feet long.
2. Place a long piece of tin, one that is 24 inches wide and 6 to 8 feet long, in the dam break.
3. Place the logs on top of this tin with the upstream logs completely covered by water and at least 1 foot lower than the downstream end of the drain. Attach the upstream end of the drain by pegging it down with a forked stick.
4. Place another piece of tin, 24 inches wide and 6 to 8 feet long, over the top of the logs and nail it to the bottom two logs. Make several axe blade holes through this piece of tin. Beavers usually cannot repair the dam break once the three-log drain is installed.

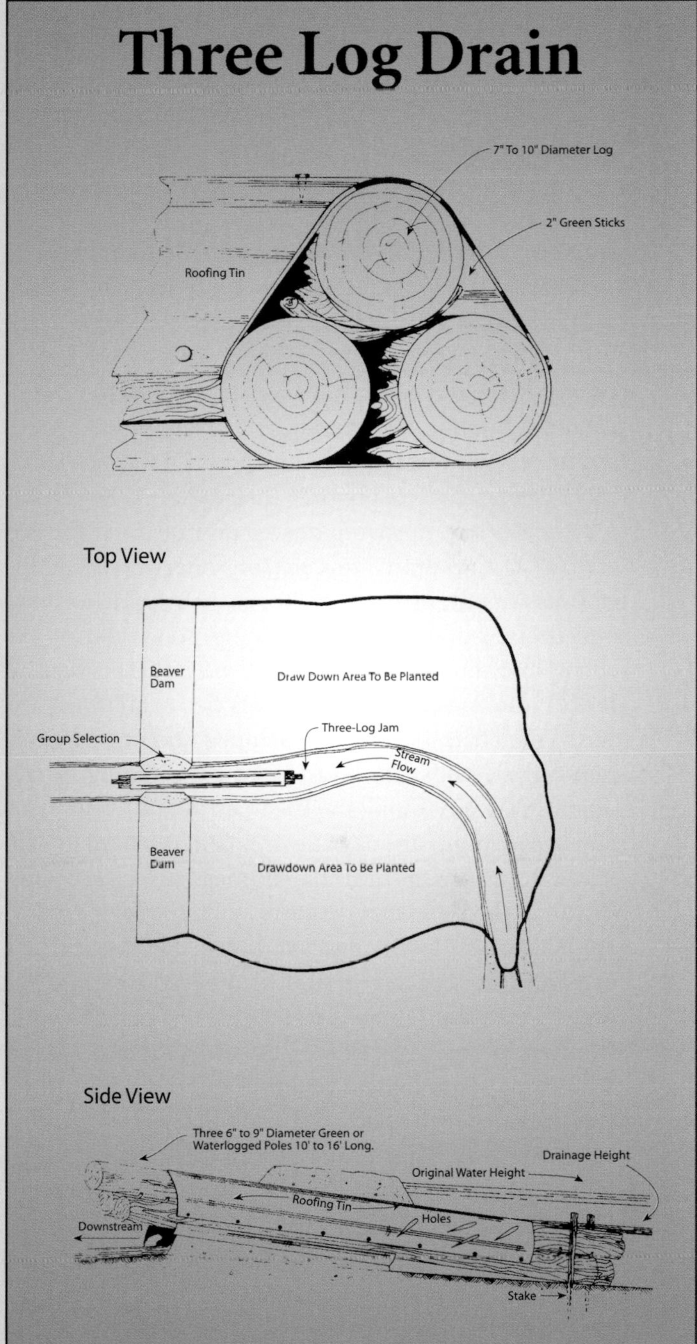

Beaver Pond Leveler

In some instances, beavers can detect the sound of falling water or water movement. They will then plug-up a three-log drain. Another option is installing a beaver pond leveler like those developed by Clemson University researchers. The device consists of a perforated PVC pipe that is encased in a heavy gauge hog wire basket. This part of the device is placed upstream of the dam, in the main run or deepest part of the stream channel. It is connected to a non-perforated section of PVC pipe that is run through the dam to a water control structure downstream. In extensive tests in a variety of situations, the device has worked well, since the beavers cannot hear the sound of falling water as the pond drains; therefore, they don't try to plug up the pipe. Once drained, the pond can then be managed for native plants or established in supplemental food plants. A video that describes the proper construction and installation of this device called "The Clemson Beaver Pond Leveler: One Solution," can be ordered through Clemson University Communications (864-656-5134).

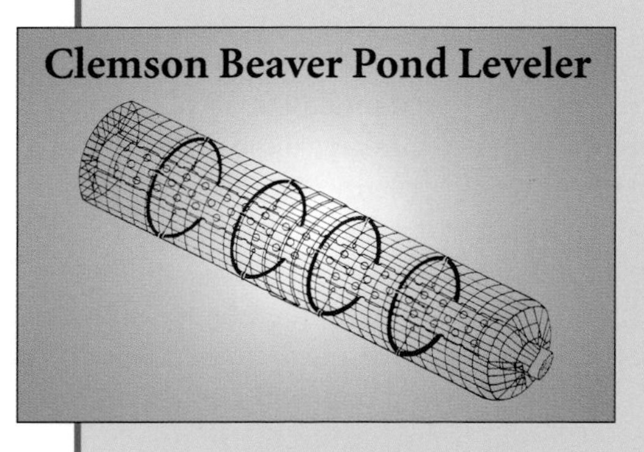
Clemson Beaver Pond Leveler

ages native plants to grow, providing food and cover for waterfowl. Waterfowl foods can also be enhanced by supplemental plantings. Research has shown that waterfowl benefit more from native wetland plants that are higher in protein and other essential nutrients than from most planted agricultural crops. A mixture of both native and planted foods is the best approach for attracting waterfowl and providing all the necessary nutrients that they require.

Beaver Pond Management for Waterfowl

Beaver ponds can provide an attractive habitat for a host of wildlife species including waterfowl. Some of the finest duck hunting in the South occurs over beaver ponds. Managed properly, beaver ponds produce excellent duck hunting opportunities at a relatively low cost. In addition, it is a way that forest owners can turn the liability of having a beaver pond on their land into an asset.

The key to improving beaver pond habitat for waterfowl is the ability to control water levels on the impoundment, or in other words, being able to flood and drain a pond at will. Draining ponds during the spring and summer will allow native vegetation to grow. It will also provide opportunities for establishing exposed mud flats that can be developed into supplemental plantings. Ponds can be drained by breaking beaver dams and maintaining the break until fall. One of several beaver pond drainage devices can be placed in dam breaks to maintain continual drainage of beaver ponds. Without these devices, beavers quickly repair broken dams the night following a break. During the early part of November beavers should be allowed to repair the break in the dam to hold water and re-flood the pond from an incoming stream or by rainfall. Since beavers are mainly active in dam building and repairing during cold weather, this method relies on the beavers' instinct to plug broken dams in time for re-flooding for waterfowl during the fall and winter. Most of the beaver's repair work takes place at night. The only potential weakness in this method is that beavers do not always cooperate and may not repair dams at all if they have moved to a new area.

A more dependable method of draining beaver ponds is to install one of several devices that have been developed specifically for draining beaver ponds for waterfowl habitat enhancement. The two most common devices are a three-log drain and a beaver pond leveler. These devices are effective in draining beaver ponds in preparation for native and supplemental food plant enhancement.

Acorns provide an excellent and preferred source of high energy food for waterfowl. If beaver ponds are fairly new and acorn-producing oaks are still living, beaver ponds should be drained before the onset of the growing season to prevent the trees from dying. In Alabama, water should be completely drained by mid-March. During the dormant season (October through February) ponds can be re-flooded.

If native vegetation is to be enhanced through moist soil management, water should be drained beginning in March and then re-flooded in October. This allows ample time for many of the key waterfowl foods, like smartweeds, to become established before being flooded again in the fall. If areas in beaver ponds are to be planted in supplemental crops, beaver ponds should be drained in mid-June or the first of July to allow mud flats to be exposed and dry out. When mud flats are exposed, Japanese millet seed can be broadcast at a rate of 20 pounds per acre on the moist mud flats. Other millets should not be used in place of Japanese millet since they will not grow well, if at all, in extremely wet soils. The best conditions for broadcasting Japanese millet are during times when the mud flats are about ankle-deep in mud. No further land preparation is needed and fertilization or liming is not necessary in the first two years. While the millet is growing, drainage devices should prevent the pond from re-flooding before the seeds mature. Japanese millet usually matures in about 45 to 60 days, at which time the pond should be re-flooded so that the seed heads are easily accessible to waterfowl. Japanese millet will reseed and grow for several years if ponds are drained in the summer, in June or July, to allow enough time for the seeds to germinate and grow. In some areas, blackbirds strip most of the Japanese millet seeds before ducks arrive on the pond. If this occurs, the Japanese millet should be flooded so that only about 10 inches of the plant and seed head are above water. This will prevent further blackbird feeding damage, since blackbirds do not like to get their tail feathers wet.

Other agricultural crops that are eaten by waterfowl can be grown on drained beaver pond mud flats. Exposed mud flats may be planted in corn, buckwheat, grain sorghum, or soybeans. Waterfowl usually prefer grains more than soybeans and other legumes. Corn and soybeans must be planted in rows. Other crops may be broadcast directly over exposed mud flats. To plant agricultural crops for waterfowl, beaver ponds should be drained in the spring to allow plenty of time for drying. This is especially important if machinery is going to be used during planting. Soil samples should be collected from exposed mud flats that should then be fertilized and limed according to the soil test recommendations. Plantings should be established and timed so that the maturity dates of seeds correspond to the re-flooding period and waterfowl hunting season. Flooded beaver ponds with supplemental plantings should be available to waterfowl approximately two weeks before the opening of waterfowl hunting season.

Grain crops can be planted for ducks and other waterfowl if water levels can be manipulated by flooding in the fall and draining during the spring and summer.

Managing Agricultural Fields for Waterfowl

In some cases farmers that have streams adjacent to agricultural fields, or have another source of water, can flood fields for waterfowl after the crop has been harvested in the fall. Agricultural fields that contain waste grain like corn, grain sorghum, and soybeans can provide a valuable temporary food source for waterfowl. Fields should be flooded shortly after a normal agricultural harvest. To enhance the value of flooded agricultural fields for waterfowl, fallow strips of native plants can be left in the field to increase the availability of food and cover for ducks. These areas should be at least 15 to 20 feet wide, running the entire length of the field. Fallow strips should also be left around field borders. After two to three years, fallow strips should be maintained by disking or burning in late winter to keep these areas productive for waterfowl. The number and size of fallow strips depend upon how much crop and revenue the farmer or landowner is willing to give up to provide quality habitat for waterfowl.

Managing Flooded Woodlands for Waterfowl

Woodlands and bottomland forests can also be managed for waterfowl as **greentree reservoirs**. A greentree reservoir is a forested bottomland that is temporarily flooded during the fall and winter to attract ducks, particularly mallards and wood ducks. Hardwood trees are not killed by winter flooding since the trees are dormant during this time. Most of these reservoirs are created by diverting water from a well or small stream, pond, or reservoir uphill from the site into a hardwood bottom that is bordered by levees. Water flow in and out of the greentree reservoir is regulated by water-control structures. Biologists with the Alabama Department of Conservation and Natural Resources and personnel with the USDA Natural Resources Conservation Service can provide technical assistance as to the type of water-control structures needed, the heights and elevations of levees, and permit requirements (if necessary) from the U.S. Army Corps of Engineers. Usually permits are required to build a levee and to construct a greentree reservoir larger than 10 acres in size. Sites that are less than 10 acres will require a Pre-Discharge Notification from the Corps of Engineers. It is wise to check with the Corps of Engineers, Natural Resources Conservation Service, or Alabama Department of Conservation and Natural Resources prior to any greentree reservoir construction. Failure to do so does not free a land-

GLENN D. CHAMBERS

Hardwood bottoms can be flooded during the fall and winter when trees are dormant to form a greentree reservoir for waterfowl. Water must be drained from the site prior to the growing season to avoid killing valuable mast-producing trees.

JOE MCGLINCY

JOE MCGLINCY

Dikes (left) and water control structures (right) are important components in beaver ponds, greentree reservoirs, or other shallow water areas managed for waterfowl.

owner from any regulations; the restrictions still apply.

Bottomland forests that are potential sites for greentree reservoirs should have clay soils that have good water-holding potential. Ideally, these areas should be at least 10 acres in size on relatively flat ground where a dam or levee can be placed to create water depths of 1 to 18 inches. Greentree reservoirs should have a large number of mast-producing trees including species like cherrybark, water willow, swamp chestnut, and laurel oak, or sweetgum, ashes, bald cypress, elm, maple, pecan, and blackgum. To provide the most benefit, these trees should be at least 40 years old. Ash, elm, and maple are not ideal hard mast–producing trees during the fall, but their winged seeds are an important source of late winter food for ducks when other foods are not available.

After the trees have become dormant in the fall, in mid-October or later, water-control structures are closed to flood the bottomland hardwood stand. Water is retained on the site until late winter. To avoid tree damage or death, water must be removed from the area usually before the buds of trees in the impoundment begin to swell. If water remains during the growing season, flooding will kill or severely stress trees within the reservoir. In addition, attention should be given to the number of consecutive years that the greentree reservoir has been flooded. Research has shown that 20 or more years of annual winter flooding will alter the composition of the forest in the greentree reservoir. Valuable mast-producing species will eventually be replaced by flood-tolerant trees such as bald cypress and water tupelo. One way to avoid or slow this change is to flood greentree areas every other year or in two out of three successive years. The change may also be slowed by flooding at different times during late fall and early winter, drawing the water down slowly in stages. This will mimic the rise and fall of flood water that would have occurred naturally in many of these areas.

During the summer, timber management on greentree reservoirs can improve the area's value for ducks. The timber can be thinned or selectively harvested to remove undesirable trees and promote the growth of mast-producing trees. A timber harvest or thinning should leave about 80 square feet of basal area of the best tree species with diameter at breast height (DBH) of 14 to 30 inches. A variety of tree species should be left so that a mast crop is produced every year. A few snags and live trees with cavities should be left for wood duck nesting sites. As an alternative, wood duck nesting boxes should be installed. Cavity nesting sites should be located on scattered potholes that retain water during the spring and summer after most of the greentree reservoir has been drained. Openings created by selective harvests can be managed for native waterfowl plants or can be planted in corn, sorghum, Japanese millet, or other crops to provide additional food for wintering waterfowl.

Native Plants Versus Supplemental Plantings for Waterfowl

Some landowners have been draining beaver ponds during the spring and seeding the exposed mud flats in grain crops to attract waterfowl for years. While this practice is usually effective, few landowners are aware of the value of native plants to waterfowl. Valuable native waterfowl plants like smartweeds, wild millet, pond

Fig. 1: Habitat and Harvest Management Intensities for Waterfowl

Management Practices	Low Intensity	Medium Intensity	High Intensity
Beaver Ponds	No management	Drain down in early spring and re-flood in early November	Drain down in early spring, broadcast Japanese millet on mudflats, re-flood in early November
Greentree Reservoirs	Flood bottomland hardwoods in November drain in early March	Timber stand improvement to favor mast-producing trees in greentree	Same as low and medium and also create small openings for planted or native waterfowl plants by group selection timber harvest
Agricultural Fields	Flood food crops after harvest in the fall	Leave portions of food crops unharvested and flood in November	Same as medium but also leave fallow strips of native wetland plants within and on the borders of fields
Waterfowl harvest	Follow legal guidelines	Follow legal guidelines and restrict shooting to mornings only	Follow legal guidelines, restrict shooting to one hunt per week only during morning hours

weeds, arrowhead, bulrushes, sedges, and rushes can be encouraged to come up naturally by simply manipulating water levels in wetlands. Waterfowl will concentrate on ponds and other wetlands where natural foods are abundant and frequently prefer these areas more than flooded grain fields.

Although rich in energy, seeds of most cereal grains are nutritionally incomplete. Generally, seeds of naturally occurring plants contain essential nutrients like protein that are deficient or missing entirely in cultivated grains. In addition, seeds of naturally occurring plants often provide as much—or more—energy than do grain crops. Studies have also indicated that waterfowl that feed on seeds of native plants, as opposed to grain crops, were generally in better physical condition for their spring migration and breeding.

Seeds of native plants also persist for extended periods, but those of grain crops spoil and deteriorate rapidly once they are flooded. Unless cultivated intensively, at a considerable management cost, grain crops rarely provide as much food as native plants. In addition, yields of cultivated crops are too dependent on uncontrollable conditions, particularly weather, and often are not as reliable in production as native plants that are adapted to a wide range of moisture conditions. Most grain crops also do not provide adequate protective cover for waterfowl, while native plants may be managed to yield both food and cover.

Another helpful aspect of native plants is that they tend to support high densities and a greater diversity of protein-rich insects and other animal matter, an essential dietary component for most ducks. Cultivated crops attract few aquatic insects, forcing ducks to feed in other areas to obtain their protein requirements.

As a part of moist soil management, native plants can be encouraged to grow on exposed mud flats in wetland areas when the surface water is drained from impoundments during spring and summer months. The types of native food plants that grow following a water drawdown will vary from year to year depending upon the timing and extent of draining as well as the moisture conditions. In general, rapid drainage favors extensive stands of similar vegetation. On the other hand, a drawdown over a longer period or a slow drainage usually provides a greater diversity of vegetation.

After several years without any soil disturbance, perennial plants will soon be replaced by woody shrubs, making impoundments less desirable for waterfowl. Once woody shrubs cover more than one-half of the surface area, disking or burning may be necessary in the spring or summer on drained areas to push back plant succession to herbaceous growth. Generally, disking favors annual plants over perennials and increases the production of seeds eaten by waterfowl.

HARVESTING AND REGULATIONS

Excessive shooting on waterfowl impoundments may cause ducks to abandon an area. As a rule, shooting

Opportunities for Attracting Waterfowl

Keith McCutcheon is a Waterfowl Biologist with the Game and Fish Division, Alabama Department of Conservation and Natural Resources in Hollywood, Alabama. As a waterfowl biologist he sees many opportunities for landowners to improve their property and attract waterfowl.

In Alabama there is a tremendous opportunity to increase overwintering waterfowl habitat by managing beaver ponds, and greentree reservoirs in hardwood bottoms, and by flooding agricultural fields. Technical assistance to improve farm or forest lands for waterfowl can be provided to landowners by our agency, the Natural Resources Conservation Service (NRCS), U.S. Fish and Wildlife Service, or private consultants.

For years the farming community has concentrated on draining farms for cultivation. Many of the drainage structures that were used to remove water from farms are still in place and can now be used to re-flood farmlands for waterfowl. The possibilities for converting some farm acreage into waterfowl habitat are enormous.

Many bottomland hardwoods can also be flooded to provide greentree reservoirs for migrating waterfowl and wood ducks. Wildlife professionals can help determine where to locate a greentree reservoir and decide if the site contains the types of trees (such as oaks) that are valuable for waterfowl. Time of flooding and draining can also be recommended, depending on the objectives of the landowner and site characteristics.

Alabama's abundance of beaver ponds creates a tremendous potential for developing favorable waterfowl habitat. These impoundments can be managed for waterfowl using various methods to manipulate water levels that will favor the growth of preferred herbaceous waterfowl plants and invertebrate foods.

Wildlife professionals can also help landowners obtain necessary permits, if applicable, from the appropriate agency. Permits to divert water to flood areas for waterfowl may be required, depending on the type of stream and volume of water. In some cases the Corps of Engineers requires a 404 permit for diverting water from watersheds and flooding areas. Within the TVA region, an additional 26A permit may also be required. The wildlife professional can help determine what permits may be needed to improve farm or forest land for waterfowl.

should be confined to the early morning hours. This will allow the ducks to return and feed undisturbed before nightfall. Ducks may abandon an area if it is hunted more than once a week, especially if the area is smaller than 20 acres. If an area is larger than 20 acres, part of the impoundment can be set aside as a sanctuary where waterfowl can rest and feed undisturbed.

Federal regulations on managing agricultural crops must be observed where waterfowl hunting occurs. Manipulation of standing crops, other than by flooding, is considered baiting. It is illegal to hunt ducks over crops that have been dragged, mowed, or otherwise deliberately knocked down for the sole purpose of making the food available to ducks. Currently it is legal to deliberately flood a mature standing crop or to harvest the crop in a normal agricultural manner, flood the field, and shoot ducks that are attracted to the area. Since federal regulations are subject to change over time, it is prudent to obtain the latest information on these regulations from a local Alabama Conservation Enforcement Officer, or from a U.S. Fish and Wildlife Service Special Agent.

ORGANIZATIONS DEDICATED TO WATERFOWL

Ducks Unlimited

Ducks Unlimited, Inc. is a non-profit conservation organization dedicated to the protection and management of North American waterfowl. The organization has been active in wetlands acquisition and protection in the breeding grounds of Canada, and to a lesser extent the prairie pothole region of the U.S. Ducks Unlimited also cooperates with government agencies and private organizations by providing matching funding to restore and manage overwintering waterfowl habitat in the U.S. through their M.A.R.S.H. (Matching Aid to Restore State Habitats) Project. To a lesser extent, Ducks Unlimited, Inc. also helps to fund waterfowl research. For more information about Ducks Unlimited, Inc. write the national headquarters at One Waterfowl Way, Memphis, TN 38120-2351, or call (901) 758-3825.

DENNIS HOLT

Excessive hunting of a waterfowl area may cause ducks to abandon it. Shooting should be confined to morning hours only and no more than once or twice a week.

Delta Waterfowl Foundation

The Delta Waterfowl Foundation funds research programs concerning the breeding grounds of the continental waterfowl resource. The Foundation is also active in enhancing the understanding of wetlands ecology and improving marsh management and facilitating the transfer of scientific information to users and managers. For more information on the Delta Waterfowl Foundation, write 102 Wilmot Road, Suite 410, Deerfield, IL 60015, or call (708) 940-7776.

Alabama Waterfowl Association

The Alabama Waterfowl Association, Inc. (AWA) is a private non-profit organization dedicated to enhancing waterfowl habitat in Alabama. The AWA provides technical assistance to landowners interested in improving wetland habitats for waterfowl. To receive AWA's publication *Wetlands and Waterfowl News* or obtain more information, write Alabama Waterfowl Association, Inc., P.O. Box 67, Guntersville, AL 35768, or call (205) 259-2509.

SUMMARY

Waterfowl management is an attractive option for many farm and forest landowners in Alabama. Whether management involves creating and maintaining an agricultural field impoundment, beaver pond, or greentree reservoir, waterfowl can be attracted to many farms and forest lands that have a ready source of water. The key is the ability to manipulate water levels for draining areas for growing native wetland or planted foods in the spring and summer, and flooding these areas in the fall and throughout the winter. This chapter discusses in detail the variety of options that are available for landowners to attract some of the 26 different migratory and resident waterfowl species that frequent Alabama.

Permits Required to Hunt Waterfowl in Alabama

Beginning in 1996, permits are required to hunt waterfowl and other migratory game birds in Alabama. The free permits can be obtained from any sporting license vendors in the state. The permits are part of the federal Migratory Bird Harvest Information Program, which is attempting to obtain waterfowl harvest information from hunters. This information will be used to set future bag limits and hunting seasons for waterfowl. To obtain a permit, hunters must complete a harvest information card with their name, address, and answers to a few short questions about past hunting success. After completing the card, hunters will be given a permit that affixes to their hunting license. Permits must be available for inspection by law enforcement officers when hunting waterfowl or other migratory game birds in Alabama.

Furbearers are an important wildlife resource in Alabama. Among the most common is the raccoon, which can be found throughout the state. The highest raccoon densities are in counties that contain large amounts of water, bottomland hardwood swamps, and fields with grain crops such as corn.

PAUL T. BROWN

CHAPTER 11

Managing for Furbearer Species

CHAPTER OVERVIEW

This chapter features the important furbearer species in Alabama and discusses their biology, life history, and habitat requirements. This information should be helpful in understanding some of the challenges and complexities of managing this diverse resource. The chapter divisions are based upon the habitats that each species commonly uses such as aquatic, semi-aquatic, and terrestrial areas.

CHAPTER HIGHLIGHTS PAGE

AQUATIC FURBEARERS 230
Beaver 230
River Otter 233
Muskrat 235
SEMI-AQUATIC FURBEARERS 237
Mink 237
Raccoon 238
TERRESTRIAL FURBEARERS 240
Opossum 240
Foxes 241
Coyotes 243
Skunks 246
Bobcats 248
FURBEARER CONTROL 252

Furbearers are animals whose skin and hair can be processed into garments. Alabama's furbearer resource includes a diverse group of wildlife that is found throughout the state. Most people enjoy the opportunity to observe a mink, fox, beaver, or any other furbearer in the wild. Trappers, hunters, and photographers spend many hours pursuing these elusive animals. Furbearers also provide a source of income to trappers who sell the hides. Many furbearers are also among nature's best pest control agents because they eat large numbers of mice, rats, and other small mammals. Unfortunately, these same traits make them important predators of game and non-game animals that are valued by man.

Furbearers like beaver, river otter, muskrat, and nutria are considered aquatic because they require surface water of sufficient size to meet their habitat needs. Other furbearers like the raccoon and mink are classified as semi-aquatic because they are usually associated with water, but also spend much of their lives on upland or terrestrial areas. Terrestrial or upland furbearers like the red fox, gray fox, coyote, opossum, striped and spotted skunk, and bobcat require drinking water but are not associated with water. The long-tailed weasel also occurs in Alabama but will not be discussed.

Furbearer management is complicated by several factors:

- The varying amounts of land and habitat needs of different species;
- The diversity of furbearer families;
- The relationship of each species with other furbearers and wildlife;
- Man's relationship to furbearers; and
- The difficulty in obtaining furbearer population estimates and trends.

Currently the Alabama Department of Conservation and Natural Resources (ADCNR) Game and Fish Division conducts annual furbearer surveys to determine indices of abundance. This information, combined with estimated harvest figures reported by pelt dealers, helps to monitor furbearer population trends in Alabama. Within the last few years, estimated harvest figures for most furbearers in the state have declined. Biologists think this decline is caused by the reduction in trapping effort caused by lower pelt prices instead of a decline in furbearer populations. The number of licensed trappers in the state has fallen from a high of 6,094 in the 1979–1980 season to 518 trappers in the 1994–1995 season. Because of the declining numbers of trappers and animals being trapped, actual populations of most furbearers are increasing in Alabama. The fluctuation of trappers, pelt harvests, and prices reflects changes in fashion trends.

AQUATIC FURBEARERS: BEAVER, RIVER OTTER, AND MUSKRAT

Beaver

The beaver (*Castor canadensis*) is one of the most important North American mammals from historical, biological, economic, and aesthetic perspectives. Just like wild turkeys, white-tailed deer, and wood ducks, the restoration of beaver can be considered one of America's great modern wildlife management success stories. Some would argue that we have been too successful. Beavers are considered to be pests in many areas of Alabama due to their ability to dam creeks and streams and flood large areas of farm and forest land.

The pursuit of beavers, primarily for their fur to make hats in Europe, was one factor that led to the exploration and colonization of many portions of North America. The period from 1750 to 1840 has been called the era of beaver trade in North America. The fur trade, based largely on the sale of beaver pelts, greatly influenced the economic and political development of North America. Before the arrival of European settlers in North America, beavers could be found in almost every stream, creek, or river throughout the United States and Canada.

Early explorers and fur trappers over-harvested this resource and sent large numbers of beaver pelts to Europe to be made into hats and coats. As the supply of beavers in an area dwindled, trappers moved to unexplored country to meet Europe's growing demand for finished goods made from beaver pelts.

This pattern of pioneer exploration and local beaver extirpation continued throughout North America. During the 1700s and 1800s, the beaver may have been this continent's most important wildlife species. At this time, they were also abundant in most of Alabama, except in the southern portions of Baldwin and Mobile counties. From about 1870 to 1890, the trapping and uncontrolled harvest of beaver in Alabama was so intense that by 1890 the beaver was scarce in the state. Trapping of beaver was outlawed in Alabama after the close of the 1938 trapping season. At that time only 500 to 600 beavers were left in Alabama. By 1940 the population had increased to about 3,500 and beavers were found in 28 counties in the state.

In 1940, the Alabama Department of Conservation (now the Alabama Department of Conservation and Natural Resources) started trapping beaver in areas where they were abundant and releasing them in suitable habitat where beaver were scarce or absent. The department relocated beaver in 48 counties, but only on lands whose owners requested beavers. At that time, nobody could visualize that the beaver would become a nuisance and cause the economic losses (in damage to timber and crops due to flooding) that it does today. Many biologists believe the beaver is more abundant now than at any other time in modern history. Today the animal's population in Alabama is estimated to be more than 150,000 and it is present in all 67 counties, with the greatest densities in central Alabama.

Beavers are generally considered beneficial, although they may compete with people for the use of the land, water, or timber. Their ponds contain deep holes that are excellent habitat for fish, and also attract a wide variety of other animals including otter, mink, muskrats, raccoons, wading birds, songbirds, waterfowl, amphibians, reptiles, and many other species. The variety of wildlife attracted to these ponds can be used for recreational, scientific, or aesthetic purposes. As mentioned in Chapter 10, a Clemson beaver pond leveler can help manage beaver pond water levels to enhance moist-soil management, which encourages native wetland plants and habitat conditions that are ideal for many species including waterfowl.

Beaver ponds help stabilize water tables, reduce runoff, and serve as catch basins for soil eroded from upstream areas. **Castoreum**, an oily liquid that beavers emit to mark territory, is used in numerous perfumes and cosmetics and in trappers' lures. Finally, beaver meat is excellent table fare if properly prepared. The Mississippi Cooperative Extension Service has a publication entitled *Beaver Burgers* which illustrates how to prepare and cook beavers.

DESCRIPTION

Beavers are the largest rodents in North America and belong to the family *Castoridae*. Adult beavers weigh between 30 and 80 pounds, with an average weight of 33 pounds, and measure 41 to 46 inches in length. They are large-boned and heavily muscled animals with hind legs and feet adapted for swimming. Their back feet are webbed, whereas their forefeet are heavily clawed. They use their nimble forefeet to turn and hold small twigs while they peel the bark with their teeth. The beaver is easily recognized by its large, flat, hairless tail, which is used as a rudder when swimming and for support when standing on land. The tail is also used to slap the water to signal other beavers when danger approaches. Contrary to popular belief, the tail is not used to carry mud used in building dams and lodges.

DR. H. S. BANTON

Beavers, North America's largest rodents, are present in every Alabama county, with the highest densities found in the central portion of the state. Although beavers can cause extensive flood damage to timber, crops, and other areas by diverting water, they also create additional wetland habitat utilized by many wildlife species. Beaver pelts are also valued by the trapping industry.

Beaver fur is comprised of soft, downy underfur and long, coarse guard hairs. The fur is brown, but individual coloration may range from a creamy blond to nearly black. Beavers comb and groom their fur with the aid of the second toenail on their hind feet. The sexes are indistinguishable externally, except for nursing females with swollen mammary glands.

Beavers have large, orange front teeth (incisors) that grow continuously throughout their life. Located directly behind these incisors is a set of lips that seal when the beaver dives under water. The animal also has valves inside its ears and nose that close when it is under water. These features allow the beaver to cut and gnaw on woody or herbaceous plant material under the water.

BIOLOGY AND LIFE HISTORY

Beavers are monogamous, mating in January and February. Usually one to five 1-pound kits are born in March or April. Litters as large as eight have been reported in Alabama, although the average is two. The kits grow rapidly, nursing for about 60 days. By six months of age they weigh between 8 and 10 pounds. The female does all of the kit-rearing. In Alabama and other regions of the South, beavers may produce two litters each year.

Beavers are social animals that live in family units called colonies that range in size from two to eight animals. A colony consists of the adult pair, the current year's offspring or kits, the previous year's offspring, and occasionally a two-and-a-half-year-old offspring. More than one colony may live in a large beaver pond, and each colony defends its own territory. When beavers become sexually mature, at about 18 months, they usu-

RICHARD T. BRYANT/ARISTOCK

Beaver lodges are a common site in many Alabama wetlands. These structures are composed primarily of woody limbs and mud and provide a home for the family unit. Access into the lodge is through an underwater entrance.

ally leave their home colony to form a new colony. Watersheds in Alabama that have established beaver colonies usually contain 10 to 11 beaver per mile.

Beavers have a relatively long life-span of between 10 and 20 years. Alligators and bobcats eat beavers, and historically, wolves, cougars, alligators, and bobcats preyed on the animal. Predation by river otters has been documented, but is rare. Today man is their primary predator, but predation is not considered a limiting factor in most cases. The disease tularemia has been reported to cause widespread population declines of beaver.

HABITAT NEEDS

The beaver is one of few mammals, other than man, capable of modifying its own habitat to suit its needs. When beavers move into an area, they quickly begin building dams to create habitats more to their liking. Beaver dams are usually built on slow, meandering streams that pass through flat, moist, hardwood bottoms. Beavers will use almost any available material to make a dam including logs, dead limbs, bushes, mud, leaves, and other debris. More than one dam may be built along a stream. The number of dams is influenced by stream flow. The greater the flow, the greater the number of dams. Once a dam has been built from surrounding trees and other material, the subsequent and prolonged flooding causes many trees to die. As a result of the open tree canopy, sunlight reaches the ground and water, stimulating the growth of aquatic plants. Often the new plant growth around the edge of a pond (willows, blackgum, and sweetgum) consists of preferred beaver foods.

Beavers are found in a wide variety of wetland habitats ranging from small streams to large lakes or reservoirs that have stable water levels. Any water source that has ample flow for damming activities and a suitable food supply is potential beaver habitat. Appropriate habitat can include streams, rivers, ponds, lakes, swamps, wetlands, and drainage ditches. Beavers living in areas with an abundance of winter foods appear to have larger litters. The availability of winter foods also influences the age when the young move from the den or lodge and therefore affects the age when young beavers breed.

Beavers feed on the cambium layer just under the bark of woody plants as well as on a variety of aquatic and upland herbaceous vegetation. Preferred woody foods in Alabama include sweetgum, cottonwood, dogwoods, sweetbay, blackgum, ironwood, willows, birch, maple, alder, cherry, pine, and poplar. They also feed on the leaves, twigs, and cambium of more than 40 other woody species. An adult may eat about 1½ pounds of food every day. During the summer, they also eat water lilies, pond weeds, and cattails. Sometimes beavers travel substantial distances from the pond or stream to get to corn or soybean fields where they cut down entire plants and drag them back to the water. Beavers do not eat fish.

Beavers are highly territorial and actively defend their colony against outsiders by scent marking. Home ranges usually vary from ½ to 1½ miles of stream length. Home range size, however, varies greatly according to the network of streams and ponds in the vicinity where the beaver lives.

"Busy as a beaver" appropriately describes beaver behavior. They are primarily nocturnal, active for about 12 hours each night while they feed and work on dams and lodges. Most daily movements are centered around the pond and lodge. The adult female in the colony restricts her movements during the spring and summer when she is caring for her young. When young beavers move to another location to establish their own territory, they may travel 5 to 6 miles. Other travels by individual beavers include wanderings by yearlings and adults who have lost their mates.

River Otter

No other North American furbearer inspires such a wide variety of images as does the North American river otter (*Lutra canadensis*). The river otter is one of the most playful members of the weasel family (*Mustelidae*). Many people have seen otters on television, and their playfulness has made them one of America's most popular wildlife species.

Historically, otters could be found in every major river system throughout the United States and Canada, except the arid Southwest and extreme northern Canada and Alaska. Currently, otters are found in

Beaver Control Assistance

Frank Boyd is Alabama's State Director for the USDA Wildlife Services. His agency provides information and technical assistance for a wide variety of nuisance wildlife problems.

Alabama is similar to other southern states in that recreational beaver trapping has decreased, resulting in an increase in beaver populations and damage caused by beavers. Although the economic impact of beaver damage in Alabama is unknown, conservative estimates place damage at more than $20 million annually. Most of the beaver damage in Alabama is found in the central and southern portions of the state. On some individual ownerships, damage has been estimated as high as $250,000. Many times forest owners are unaware of beaver damage on their land until it is too late. This is particularly true with absentee landowners who are not routinely on their land. To reduce losses, forest owners should periodically monitor their land for beaver activity and flooding, and then take appropriate steps to reduce damage.

Our office has a beaver control program that works with other agencies to provide information and technical assistance for beaver and other wildlife problems. The first approach with this program is to provide information and recommendations to landowners so that they can solve beaver problems themselves. This information includes telephone advice, publications, and landowner visits from our agency, the Alabama Cooperative Extension System, and the Alabama Game and Fish Division. We also conduct periodic workshops throughout the state to teach landowners various techniques to trap beavers. These programs are usually advertised through the local county Extension offices. Potential trappers are strongly encouraged to attend these workshops or similar training. A lack of training can increase the odds of personal injury as well as reduce trapping success.

If landowners or managers are unable or unwilling to trap beavers themselves, private contractors are available to provide these services for a fee. Contractors may include local trappers or private pest control companies. Names of these individuals or companies can sometimes be obtained through local Extension or Alabama Game and Fish Division offices.

If these alternatives prove unsuccessful, a third option is to develop a cooperative agreement with our office to conduct and oversee beaver trapping efforts. We can conduct beaver control efforts under a cooperative agreement in which the landowner reimburses our agency for project expenses.

ROGER IDENDEN/ARISTOCK

River otters can be found in most rivers and streams in Alabama. Their diet consists mostly of fish, although they eat a variety of other foods such as crayfish, snakes, frogs, salamanders, and other small aquatic animals.

45 states and all of the Canadian provinces.

In Alabama the river otter appears to be doing well and can be found in every county in the state. Historical records are incomplete, but available information suggests river otters were abundant before the arrival of European settlers. At the peak of land settlement around the late 1800s and early 1900s, river otters were at their lowest population levels. The otter was totally eliminated from a large portion of its historical range as a result of unregulated market hunting and trapping, clearing land for agricultural purposes, and sedimentation.

DESCRIPTION

The North American river otter is a long, cylindrical, aquatic mammal. It has small ears and eyes and a flattened head with a prominent nosepad. The animal's large size surprises most people. Otters weigh between 11 and 30 pounds and measure 36 to 50 inches long. The long, heavy tail, which tapers from the body to its tip, comprises about one-third of the animal's length. They have short legs and feet with fully webbed toes. Males are generally larger than females.

River otters have a thick, durable, and luxurious pelt that was once the standard for all North American furs. Pelt coloration on the back and sides ranges from light to dark chocolate-brown. The belly is usually lighter in color and the neck region appears to have a silvery sheen when observed in bright sunlight.

BIOLOGY AND LIFE HISTORY

Otters become sexually mature at two years of age. Their breeding season in Alabama begins in late winter and early spring. Otters are promiscuous and do not pair bond. Like other members of the weasel family, they exhibit delayed implantation of fertilized eggs. After breeding, the fertilized egg develops slightly and then enters a period of suspended animation. During this period no further development occurs until it implants in the uterine wall 7 to 10 months after fertilization. The egg then resumes development and the active stage of pregnancy lasts about 60 days.

One to five blind young (the average litter size is two) are born between January and May. The young open their eyes 21 to 35 days after birth, and at two months of age they are introduced to the water. Young otters begin eating solid food by two weeks of age and are totally weaned at three months. The mother is their sole caretaker. The family group of mother and young will begin to break up about three months after weaning. Mating occurs immediately after the birth of the young and the cycle begins again.

Otters are primarily fish eaters, although crayfish are also a favored food. Snakes, frogs, salamanders, snails, insects, clams, earthworms, and a variety of mammals and birds have also been found in otter diets. Food preference is usually based on availability and ease of acquisition. As a result otters prey more on slow-moving fish than on faster species. Because of their preference for fish, otters may become economically important pests in commercial catfish ponds and a nuisance in recreational fish ponds.

River otters are solitary animals, although a family group of a female and her young is occasionally seen. Sometimes when food is abundant, otters gather in large concentrations. Otters are most active at night. Their seasonal movements vary between the sexes and among individuals.

The most important factor in determining home range size is the presence of an activity center. **Activity centers** are areas that have adequate water, cover, and food. Otter home range size is determined by the distance between activity centers. Daily travels may range from 1 to 1½ miles of stream. Over the course of a year, they may wander up to 10 miles, and during their lifetime they may travel up to 100 miles. Males generally have larger home ranges due to increased activity during the breeding season.

River otters do not seem to be territorial. Several otters can use the same home range and activity center, but they usually avoid each other by establishing local boundaries using scent marking. When otters defecate, they deposit a musky smelling secretion from a gland located at the base of their tail. They also make a scent mark at the junction of stream tributaries, crossings at beaver dams, and areas where other otters might travel.

HABITAT NEEDS

Otters prefer a wide variety of aquatic and wetland habitats. In general, their habitat consists of any area that has a dependable source of clean water and an ample food supply (primarily fish). Home for an otter may be a beaver lodge, an overhanging bank, a previously excavated burrow, or root mats sticking out from the side of a creek bank. Many otter dens have been constructed by beavers. Otters frequent these sites even when beavers are present in the lodges. They have also been observed using man-made Canada goose nesting platforms as daytime resting sites.

Chemical, industrial, and agricultural pollution has altered the quality of many streams in the state. The major limiting factors for otters are water quality and a decline in wetland habitats. Historically, trapping otters for fur resulted in decreased populations. With regulated trapping stable populations can be maintained.

Muskrat

Muskrats (*Ondatra zibethicus*) are a valuable furbearer in Alabama. Muskrats get their name from the pair of musk glands located at the base of their tails. These glands are used during the breeding season when musk is secreted to mark logs or other areas around their houses, dens, or trails along the bank.

DESCRIPTION

Muskrats are the largest microtine (meadow mice/vole sub-family) rodent in the United States. Muskrats can be found from near the Arctic Circle in Canada to the Gulf of Mexico and are found in every state except Florida. They are stocky animals with a broad head and short legs. Their tail is flattened laterally, is scaly, and has little hair. The pelts consist of soft, thick underfur with long, glossy dark-red to dusky-brown guard hairs. Their unwebbed front feet have four sharp-clawed toes and a small thumb. The large hind feet are webbed or partially webbed with stiff hairs along the toes. The bodies of muskrats measure 10 to 14 inches in length, with the tail measuring an additional 7 to 11½ inches. Adult muskrats weigh between 1¾ and 4¼ pounds. Many people, when first observing muskrats, believe they look like large rats, and muskrats are also easily confused with nutria, a larger fur-bearing exotic that flourishes in the

GLENN D. CHAMBERS

Resembling a giant rat, muskrats can be found in most river systems in Alabama. They are vegetarians and feed on cattails, bulrushes, and other wetland plants.

Mobile/Tensaw River Delta. The appearance, habitat, and habits of nutria are quite similar to those of muskrats.

BIOLOGY AND LIFE HISTORY

Muskrats are prolific breeders. They can produce a litter in about one month. Litter size and the number of litters per year are related to latitude. Muskrats produce more litters the farther south they live, but the litters are smaller. In Alabama, muskrats have three to four young per litter and may have three or more litters per year. They may also breed year-round, but the breeding season usually runs from March through October, with a peak in March through June. Generally these animals are monogamous. The decline in the number of litters throughout the breeding season is due to the decline in sexual activity of the males. Mating usually occurs while the muskrats are under water.

Twenty-eight to 29 days after mating, 3 to 11 blind muskrats are born. The young weigh about ¾ of an ounce and are about 4 inches long. After one week, they are covered with a coarse gray-brown fur. Their eyes open at 14 to 16 days, and at this time they begin to swim, climb, and dive. The kits are weaned in about 24 days and fend for themselves by the end of their first month. The mother is usually ready to give birth again by this time. The first litter may stay in the nest; then the mother will add another nest chamber to accommodate the new litter. Males become sexually mature by six to seven months of age, and females born in early spring may breed in the fall of the same year, although their first litter will usually be small.

Muskrat populations appear to follow a cycle that is influenced by food supply and usually lasts from 6 to 10 years. In general, the cycle follows this pattern:

1) Muskrat numbers are low,

2) A large food supply develops,

3) Overpopulation occurs due to good breeding condition of muskrats,

4) Habitat is damaged by overuse of food sources,

5) Starvation and reduced productivity occur as a large number of muskrats compete for a limited food supply, and

6) Muskrat numbers are low.

Muskrat densities vary depending on the phase of the population cycle, the habitat type and its condition, social pressure by other muskrats, competition, harvest, predation, and geographical area. It appears the amount of shoreline is more important than pond size in determining muskrat habitat and population levels. Breeding densities appear to be one pair per 2½ acres of water.

Muskrats are eaten by a host of predators, including hawks, owls, raccoons, mink, foxes, coyotes, and even largemouth bass and snapping turtles. They also prey upon other muskrats. During periods of overcrowding, some muskrats may kill entire muskrat litters. During a drought year, when overcrowding problems are magnified, muskrats are particularly susceptible to being eaten by other muskrats and predators. Trapping can remove a high percentage of muskrats from the population each year and diseases such as tularemia and hemorrhagic disease can decimate an entire population.

Muskrats are vegetarians and relish cattails, bulrush, smartweed, duck potato, horsetail, water lily, sedges, young willow sprouts, and pickerel weed. They will eat almost any aquatic vegetation, including the bulbs, roots, tubers, stems, and leaves of numerous wetland plants. They will occasionally eat corn, soybeans, grain sorghum, and small grains. At times, particularly during periods of low food supply, muskrats will eat animals, including crayfish, mussels, turtles, frogs, or fish.

Muskrats are not extensive travelers, and the average home range varies from a 66- to 200-foot circle in optimal habitat. During the spring or fall and at times of crisis (flooding, drought, food shortages), muskrats will change their habits and then may move considerable distances. They disperse in the spring, beginning in February or March, and this dispersal lasts about one and a half months. The distances traveled varies, and it appears that all ages of muskrats, not just the young, disperse every spring.

HABITAT NEEDS

Muskrats live in a large variety of habitats which provide a year-round supply of food and water. They can be found in ditches, streams, marshes, lakes, beaver ponds, mine pits, farm ponds, or any other wetland area. For shelter they use bank burrows, houses built of vegetation, and feeding huts. Bank burrows are usually 6 by 8 inches wide and are up to 60 inches long. Muskrats use bank dens mostly during the summer.

Muskrat houses are cone-shaped and measure up to 8 feet in diameter. They are usually constructed of cattail or bulrush stems. Their height varies, and each house will have one or two raised internal chambers. Muskrats usually begin house construction in October, peaking in November. They also eat on feeding huts or platforms that are composed of marsh vegetation. These circular huts are usually smaller than their houses.

The key component of muskrat habitat is slow-moving or non-flowing water that allows the growth of aquatic vegetation. Ideally, the water should be 2 to 3 feet deep. Cattails, bulrush, sedges, and arrowhead (excellent for food and house construction) should be present around the bank.

SEMI-AQUATIC FURBEARERS: MINK AND RACCOON

Mink

Mink (*Mustela vison*) are semi-aquatic furbearers of the *Mustelidae* or weasel family. These weasel-like animals are historically one of North America's most important furbearers. Their pelts are most closely associated with luxury and style. Although no longer as much in fashion, mink is still one of the major wild furs produced in North America. The annual harvest of wild mink in North America is valued at more than $11 million. In Alabama, estimated harvest of mink varies depending on the market price of pelts.

Mink appear in every state except Arizona. In Canada, mink are found everywhere south of the tree line with the exception of a few islands. Alabama mink populations are highest in the coastal marshes and Piedmont areas.

DESCRIPTION

Mink have an elongated body, with the tail comprising one-third to one-half of the total length. Males measure 22 to 28 inches in length and weigh between 2 and 3½ pounds; females are usually about 10 percent smaller than males. Both sexes have a luxurious, dark chestnut-brown to black pelt that consists of a soft, dense underfur concealed by glossy, lustrous guard hairs. Mink usually have white spots and patches on their chin, chest, and belly. The tip of the tail is usually black but is not distinctive.

BIOLOGY AND LIFE HISTORY

Mink are polygamous, and males may roam extreme distances in search of receptive females. Males may fight ferociously among themselves during the breeding season, which extends from late January through March in Alabama. The female remains in heat throughout the entire breeding season, with a peak in receptiveness every 7 to 10 days. This allows the female to be mated several times during the breeding season. Mating activity triggers ovulation, which usually occurs 33 to 72 hours after breeding. It is possible for a female to have her eggs fertilized by two or more males.

Like other members of the weasel family, mink exhibit delayed implantation. Young are born 28 to 30 days after implantation. The total gestation period for mink averages 51 days but varies from 40 to 75 days. In late April and May, one to eight blind young are born. The average litter size is four. Young mink weigh about ½ of an ounce at birth and grow rapidly. By seven weeks of age, they will have obtained 40 percent of their adult weight and 60 percent of their adult length. Their teeth appear at two to three weeks of age, and their eyes open during the third week. The mother raises the young until they leave in late summer or early fall.

GLENN D. CHAMBERS

Mink have historically been an important furbearer in Alabama because of their highly valued pelts. Mink populations have declined in Alabama partly because of a reduction in water quality. Because they are sensitive to contamination in waterways, such as mercury, they are a valuable indicator of the health and quality of aquatic systems.

Mink are carnivorous, and because of their semi-aquatic habits, have adapted to eating both aquatic and terrestrial prey. Feeding studies have indicated that mink are opportunistic predators and will eat mice and rats, fish, crayfish, rabbits, insects, muskrats, amphibians, reptiles, and birds or bird eggs. Availability of any particular food item determines its frequency in the diet, which varies by season and geographic location.

Mink are active mainly at night although they may be seen during the day. They are generally solitary animals, except during the breeding season or when females have kits. Their home range size is variable and depends on habitat quality, especially the abundance of quality food, and the number of denning sites. Males are great travelers. Their home ranges may include 1 to 4 miles of shoreline or an average of 1½ miles of stream. Females move in a small area during the breeding season, and their home ranges may vary from ¼ to 2 miles of shoreline or an average of 1¼ miles of stream.

HABITAT NEEDS

Mink are mainly associated with wetland or aquatic habitats; however, they may move considerable distances from wetlands in search of food. They are shoreline dwellers, and their basic habitat requirements include a permanent source of clean water and shoreline areas free of grazing or development. Shorelines should have adequate vegetation (brushy or grassy areas) to conceal their movements. The availability of den sites appears to limit mink populations. They will construct their own dens but have been known to use beaver lodges or bank burrows, holes, crevices, or log jams. Mink prefer productive waters with high fish, frog, and aquatic invertebrate populations. The water may be turbid or nutrient-rich but must be of acceptable quality. Mercury, polychlorinated biphenyls (PCBs), and pesticides such as DDT, DDE, and dieldrin are known to accumulate in the mink's body tissue and may cause complete reproductive failure. Biologists believe that in areas with high levels of these pollutants, especially mercury, wild mink populations have been adversely affected. Because they are particularly sensitive to mercury in waterways, they are a valuable **indicator species**. An indicator species, as the name implies, alerts man to declines in environmental quality—in this case that mercury contamination in water has reached dangerous levels.

Raccoon

Raccoons (*Procyon lotor*) are among the most adaptable, widespread, and successful wildlife species. They are also important economic and recreational furbearers. These crafty animals provide many hours of enjoyment for raccoon hunters. The annual harvest of wild raccoons in North America for their pelts is estimated to be worth more than $90 million. In Alabama, the annual harvest fluctuates depending on the price of pelts, and number of trappers and hunters.

Every Alabama county has raccoons in both rural and urban settings. Population densities are highest in the Lower Coastal Plain. Raccoons are most abundant in counties with large amounts of water, bottomland hardwood swamps, and grain fields. As urban dwellers can attest, these versatile creatures find their way into city neighborhoods if food is available. Few wild creatures are as cunning when it comes to finding a way into a garbage can.

DESCRIPTION

The raccoon is a very common and easily recognized furbearer. No other mammal in Alabama resembles this black-masked, ringed-tailed animal. Their bodies are stocky, measuring 2 to 3 feet with a broad head and pointed snout. Coloration is generally black above with black-and-white-tipped guard hairs and a pale underfur. Pelts may be variable shades of browns and yellows, with an occasional cinnamon or albino reported. Nearly as rare as the albino is the solid black raccoon, sometimes referred to as a "fisher" raccoon.

Raccoons have five long, slender toes on each foot that make readily identifiable tracks resembling miniature human hand-prints. Males are larger than females and weigh between 15 and 18 pounds. Females weigh between 12 and 16 pounds. The heaviest raccoon ever recorded was a 61-pound male taken in Wisconsin!

BIOLOGY AND LIFE HISTORY

Breeding season for raccoons in Alabama ranges from January to February, with peak activity occurring in February. Although males may mate with several females each spring, there is some evidence of pair bonding be-

PAUL T. BROWN

Raccoons are easily recognized by their black mask and black and white ringed tails. Pelts can range in color from shades of brown and yellow, to cinnamon, to a rare solid black or albino white.

tween males and females. Female raccoons may form a bond with a familiar male about one month before mating. The female will also mate with other males at the peak of estrus. Any type of bond terminates after the breeding season, leaving the female to care for the young alone.

Sixty-three to 65 days after mating, two to eight deaf, blind kittens are born between March and June. The average litter size in Alabama is three. The young weigh less than ½ pound at birth, and their eyes and ear canals open when they are 18 to 24 days old. Infant raccoons feed on the mother's milk until they leave the den at around 10 weeks of age. The mother may move the kittens to ground dens when they can walk and will remain with the young until fall or winter.

Raccoons die from a variety of causes. Canine distemper and rabies are the two diseases that can quickly cause raccoon numbers to decline significantly. Starvation can be a major mortality factor for young raccoons during the winter months. Human-induced mortality is a major cause of death in raccoons. Caution should be used around raccoons that show little fear of man or are unusually aggressive and sluggish. Often this is a sign that the animals are sick and could harbor rabies. No matter how cute the raccoon may seem, **never** approach or pick up a raccoon. If they allow you to get close enough to touch them, then something is wrong.

Raccoons are most active from sunset to sunrise and tend to den up during daylight hours. Peak feeding activity is generally over before midnight. They are least active during the winter months and tend to remain in their dens during extremely cold temperatures.

During the breeding season, adult males may be territorial. At this time home ranges of males usually do not overlap one another. Females are not territorial, and their home ranges often overlap. Several female home ranges may be located within one male's home range. Raccoons may have home ranges of up to 2 miles, although most are about 1 mile. Home ranges vary considerably depending on habitat conditions and population densities. Raccoons will also move considerable distances when food is scarce or populations are low. Densities can vary, depending on food and den availability, from one raccoon per acre to one raccoon per 100 acres.

DENNIS HOLT

Raccoons are most active from sunset to sunrise, primarily foraging for food. They use a variety of dens to raise their young but prefer hollow trees. The lack of suitable dens may be a limiting factor for raccoons in parts of Alabama.

HABITAT NEEDS

This adaptable animal can occupy a wide variety of habitats, although it prefers mature hardwood forest areas with numerous den trees close to water. Since raccoons are water-associated animals, water is a critical habitat requirement. They depend on wetland and aquatic habitats for a large portion of their food and are seldom found far from water.

Raccoons use a variety of den types in which to give birth and raise young. The most common dens are hollow trees. Tree dens may be found in any hollow limb or tree trunk that is large enough to accommodate a female raccoon and her litter. Ground dens are also important for raccoons, especially in areas with few tree dens. Abandoned fox or groundhog burrows are most often used as ground dens. Other types of dens include rock crevices, caves, drains, culverts, abandoned buildings, barns, and brush piles that are located near water.

The lack of suitable den sites may be a limiting factor for raccoons in parts of Alabama. Management practices which include leaving den trees could increase populations in these areas. Management and protection of raccoon habitat, especially wetland habitat, is also necessary for stable populations.

Raccoons also need brush thickets, ground dens, hollow logs, or trees for cover to escape predators and for daytime resting sites. During the day, raccoons may rest on bare tree limbs or in gray squirrel nests.

Feeding areas vary according to the time of year and the types of food available. Since raccoons are omnivorous, their feeding habitats are as varied as the type of food they pursue. They frequent aquatic habitats such as streams, rivers, and ponds when feeding on crayfish, fish, and amphibians. The use of overgrown fields, as well as corn and other grain fields, occurs mainly in the fall and summer when berries, grains, and insects are the mainstay of raccoons' diets. Hardwood forests provide important hard and soft mast, such as persimmons, acorns, and wild grapes. Raccoons usually stay for long periods in these feeding areas. During the spring and summer, raccoons can be major predators on the eggs of ground-nesting birds such as quail and turkeys. If their populations go unchecked, the raccoon can limit the numbers of cavity and ground-nesting birds of many species, including popular game birds.

RACCOON RESTOCKING

Although illegal, raccoons are being imported and released to improve hunting in many areas of the Southeast. Most wildlife biologists do not favor raccoon restocking for a number of reasons. First, this process is not needed if native raccoons are already present in the area, as native raccoons have the ability to increase their population if they are not over-hunted and if there is favorable habitat. Where the land does not have the capacity to support large numbers of raccoons, restocking can only make matters worse.

Second, studies of restocked raccoons have consistently shown that released animals do not increase the raccoon harvest. Over the years, 10 different trial releases have been made in the southeastern states, including Alabama. Of 6,434 marked raccoons that were released, less than 3 percent were recovered by hunters, and radio-telemetry studies have shown that raccoons that are released for restocking have a low survival rate.

A third reason for not restocking raccoons is that the general health of those animals obtained for release is usually not good. Parvovirus and rabies have been found in restocked raccoons. Unfortunately, there is still no way to test a raccoon for rabies without killing the animal and examining the brain.

When all these facts are considered, raccoon hunters should not expect restocking to improve hunting. Protection from illegal hunting and over-harvest, and wise management of raccoon habitat are the best approaches to enhancing raccoon populations.

RICHARD T. BRYANT/ARISTOCK

Opossums are found in every habitat type in Alabama but prefer streams associated with woodlands. Opossums are the only marsupials in North America. Marsupial means that they carry their young in a pouch.

TERRESTRIAL FURBEARERS: OPOSSUM, FOX, COYOTE, SKUNK, AND BOBCAT

Opossum

The Virginia opossum (*Didelphis virginiania*) is the only marsupial and member of the *Didelphidae* family in North America. Being a marsupial, the opossum carries its young in a pouch or on its back and is, in essence, a "moveable nest." This has advantages because the opossum is not restricted to any one area and can move about as available food supplies shift. Opossums can be found across Alabama in every county.

DESCRIPTION

Opossums are about the size of a large house cat, measuring 25 to 33 inches in length and weighing between 6 and 13 pounds. Females are usually smaller than males. The opossum has a whitish to pale gray, cone-shaped head with a white cheek bordered above with a gray or black eye stripe and ring. The first toe on the hind foot is large and opposable (like a human thumb) and lacks a claw. The long tail is almost naked and scaly and is prehensile or grasping. Sex determination is easy because females have an external, fur-lined belly pouch called a marsupium. The teats used for nursing young are enclosed within this pouch.

BIOLOGY AND LIFE HISTORY

The opossum breeding season runs from January through August in Alabama. There are two distinct breeding periods. The first one is in early winter and the second in the spring. Occasionally opossums bear three litters per year.

After a brief 13-day gestation period, the young opossums are born. The tiny young move to the marsupium by grasping the female's belly hair. Once in the marsupium, the young either attach to a teat or die. In this respect the opossums resemble their distant relative the kangaroo, which also bears very small young that are nurtured in a similar manner. The average number of teats for opossums is 13, but only 7 to 10 of these actually produce milk. The number of functional teats determines how many young will survive.

Newborn young are ½ inch long and weigh less than 1 ounce. Young opossums grow rapidly, and by the time they are weaned, at about 96 days of age, they measure 8 inches long and weigh about 6 ounces. Males are sexually mature at age eight and a half months, whereas females are sexually mature at six months of age. When threatened opossums may bare their teeth and act aggressive or feign death by "playing possum."

HABITAT NEEDS

Opossums are found in every habitat type that exists within its range; however, they prefer streams associated with woodlands. They do quite well in a variety of habitats although abundance and distribution appears to be limited by the accessibility of water and the availability of denning sites.

Except during the winter, opossums are active at night. In cold weather they may be observed during intermittent warm spells in the middle of the day. They use a variety of den sites, including holes, stumps, crevices, hollow logs, or any other opening they can find. During the winter they use ground burrows for dens. They often modify abandoned squirrel nests by enlarging them for a suitable home.

There is little need for managing opossums, other than regulating the harvest. Opossums are abundant, have a high reproductive rate, and are very adaptable and live almost anywhere. They do need to be managed however, when unusually high numbers of opossums significantly increase predation on the eggs of ground-nesting birds favored in other wildlife management programs.

Foxes

Like many members of the dog family (*Canidae*), foxes are either beneficial or harmful depending on your perspective. They do prey on some game species, although usually they are not a limiting factor. Exceptions to this, however, are indicated by recent studies that suggest that red fox predation in the prairie-pothole region of North America may be a limiting factor on the success of nesting waterfowl.

In general, foxes are beneficial because they prey on mice, rats, and other agricultural pest animals. They are also highly valued for their luxurious fur. Red fox furs sold in North America during 1978–79 were valued at more than $25 million. This number decreased to $8.5 million in 1983–84 as the number of pelts and the price per pelt dropped. The value of the gray fox harvest in 1982–83 was estimated to be worth more than $13 million dollars. Like other commodities subject to the laws of supply and demand, fox pelt sales are influenced by fashion trends, and at this time the demand has diminished.

There is some question whether or not the red fox is native to North America. Some biologists believe the red fox was native to the northern part of this country but scarce or absent in most of the vast hardwood forests where gray foxes were common. Others believe the North American red fox originated from the European red fox, which was introduced into the southeastern United States around 1750. Nonetheless, the red fox is the most widely distributed carnivore in the world. It occurs throughout the United States, Canada, Europe, Asia, and the former USSR. They have also been introduced in Australia.

The gray fox occurs in much of eastern North America. Its range extends into Mexico, Central America, and Venezuela. They are not found in mountainous areas in the northwestern United States and Canada, parts of the Great Plains, or eastern Central America. Both the red and gray fox are common across Alabama.

DESCRIPTION

The red fox (*Vulpes vulpes*) and the gray fox (*Urocyon cinereoargenteus*) are easily distinguished. The red fox is a small, slender, long-legged canine weighing between 6 and 15 pounds. It measures about 28 inches in length plus a 12-inch tail that it uses for balance and for protecting its face in cold temperatures while sleeping. The white tip on the red fox's tail quickly distinguishes it from the gray fox. The upper body of a red fox is reddish-yellow and is darkest on the shoulders and back. Several color phases of red fox are found, especially in colder regions. Red foxes may be black, silver, "cross" (with a dark cross that appears on the shoulders), or "bastard" (bluish gray).

KAREN LAWRENCE/ARISTOCK

The red fox is one of two fox species found in Alabama, the other being the gray fox. The white tip on the red fox's tail is a key feature that distinguishes it from the gray fox. The upper body of the red fox is reddish-yellow and is darkest on the shoulders and back. Red fox prefer open land and forest edges.

The gray fox, slightly smaller than the red fox, weighs 7 to 11 pounds and is 24 inches long plus a 12-inch tail. There is little color variation in gray foxes. They have a salt-and-pepper coat with buff underfur and rusty yellow sides, legs, feet, and backs of the ears. The reddish sides will occasionally cause the gray fox to be mistaken for a red fox. However, the primary color of the coat is gray, and the bushy tail has a conspicuous black stripe and black tip. A "Samson" fox is a gray fox whose coat does not have guard hairs.

BIOLOGY AND LIFE HISTORY

In Alabama, the breeding season for the red fox occurs between December and February, whereas the breeding season for the gray fox occurs between February and March. The gestation period is between 51 and 54 days for the red fox and 51 to 63 days for the gray fox. One to 10 pups (the average is 5 for red fox, 4 for gray fox) are born April through May. Newly born pups remain at the den site for the first month of their life. Red foxes dig underground dens for pup rearing while gray foxes utilize a wide variety of above ground den sites such as windrows, brush piles, cavities, and thickets. Red fox parents may move the pups from one den to another three or more times before the pups are six weeks old. Litters are sometimes split, with half the litter in one den and half in another. By the 12th week, the pups begin to explore different parts of their parents' home range during daylight hours. By mid-September or early October the young begin to disperse on their own.

During the breeding season, the male and female remain in the territory until the pups are raised. The male defends the territory and brings food to the female until the pups can be left alone for short periods of time. Both parents hunt for food, with the mother returning to nurse the pups during the day. The mother and father will remain with the pups until the pups leave in the fall.

Foxes are opportunistic omnivores and will feed on a wide variety of animals and plant material depending on what is available in a local area. Small mammals, birds, fruits, and insects comprise the bulk of the fox's diet. When rabbits and mice are plentiful, they make up the bulk of the red fox diet. Red foxes often store food in small pockets inside their dens called caches. Caches provide a source of food when prey become scarce.

Besides rabbits and mice, the red fox may eat squirrels, young opossums, raccoons, skunks, housecats, dogs, groundhogs, mink, muskrats, shrews, moles, songbirds, crows, quail, grouse, ducks, turkeys, chickens, geese, woodcock, hawks, owls, bird eggs, turtles and their eggs, and insects ... or in short, anything small or slow enough for them to catch. The red fox also eats plant foods, such as grasses, sedges, nuts, berries, pears, apples, grapes, and other fruits as well as corn, wheat, and many other grains.

Food habits of the gray fox are similar to the red fox, but not as well known. Food availability plays a major role in what is eaten. Animal matter appears to be most important during the winter, whereas insects and fruits are important summer foods. Cottontail rabbits and rodents are the preferred dietary staples of the gray fox.

The area used by a red fox varies between 500 to 2,000 acres and is usually within 1 mile of its den. The den is the focal point of all activities until the end of the denning season. The male red fox is responsible for territorial defense. Scent markings play a role in defining the territory and reinforce the male's familiarity within its own home range.

Gray fox home ranges vary among areas and average between 640 to 1,280 acres. Home ranges are a function of prey availability, population density, and the diversity of habitats present.

Red and gray foxes live about four to five years in the wild. They are subject to a variety of mortality factors including trapping, road-kills, mange, distemper, and rabies. Gray foxes appear to be especially vulnerable to canine distemper which is suspected to be nearly 100 percent fatal to foxes.

HABITAT NEEDS

Red foxes are adaptable to a variety of habitat types. They prefer diverse areas consisting of intermixed cropland, rolling farmland, brush, pastures, mixed hardwood forests, and edges of open areas that provide suitable hunting ground. They select areas where there is a diversity of habitats with plenty of edge. Red foxes may also inhabit suburban areas, particularly parks, golf courses, cemeteries, and large gardens. Red foxes are generally thought of as animals of open land and forest edges.

Gray foxes prefer a diversity of fields and woods rather than a large tract of homogeneous habitat. The basic difference between the habitats of the two species is that gray foxes favor woodlands more than red foxes. Gray foxes prefer 200 to 300 acres of mixed hardwood forests and brush areas interspersed with open fields or croplands. Large, dense stands of timber should have patches of open areas to be useful for gray foxes. The fact that gray foxes prefer woodlands may also be indicative of their uncanny ability to climb trees. Red foxes are not climbers.

An adequate number of denning sites is essential for optimal fox habitat. Red foxes may dig their own dens, but they usually use an abandoned burrow of another animal. The same den may be used for generations. In some cases, red foxes prefer cover strips adjacent to or over den sites. Gray foxes will use woodpiles, rocky outcrops, hollow trees, brush piles, or rock piles for dens.

Coyotes

Coyote (*Canis latrans*) means "barking dog" and is derived from the Aztec word *coyote*. The Native American people called the coyote the "song dog" because of its characteristic howls. In recent years coyotes have been a valuable fur resource in northern states. It is estimated that about 500,000 coyotes are harvested in this country each year. At an average pelt price of $20, this resource would be worth more than $10 million annually. Coyotes are also beneficial in that they catch mice, rats, and other agricultural pests. Although the ability of coyotes to control populations of mice and rats is not proven, they probably help moderate eruptions in prey populations.

Prior to 1900, coyotes were found only west of the Mississippi River, but have now expanded eastward. It is likely that coyote populations expanded in Alabama from local releases as well as eastward migration. Documented releases of coyotes by fox hunters occurred in Barbour County in 1924, Madison County in 1950, and St. Clair County in 1967. In Alabama, the coyote is a resident of every county, and densities are highest in counties having large areas of intermingled pastures, cropland, and woodlands. The highest concentrations occur in the northwestern and southeastern counties of the state.

One disturbing result of coyote population increases is the apparent effect on red fox numbers. Coyotes and red foxes occupy similar habitats. Field studies have shown that red foxes avoid coyote ranges and that coyotes exclude red foxes from the majority of their range. As a result, red fox numbers in Alabama are believed to be declining in areas where coyotes are abundant.

DESCRIPTION

Coyotes are sometimes confused with red or gray foxes, but can be distinguished by their larger body size. They may also be confused with domestic dogs. Coyotes have also been mistaken with the endangered red wolf that once roamed across Alabama. The extirpation of the red wolf from Alabama by early settlers facilitated coyote expansion into the state. Coyotes have a thin muzzle, a bushy tail usually carried at a downward angle, and constantly erect ears. Although coyotes occasional-

Coyotes in Alabama

As a graduate research assistant at Auburn University Chad Philipp conducted research on coyote damage to livestock and agricultural crops. His work helped document the status of coyotes and coyote damage in Alabama. Chad is now a wildlife biologist with USDA/APHIS/Animal Damage Control in Arkansas.

Coyotes have been in Alabama longer than most people think. Early records indicate that coyotes were killed in the state as far back as the 1920s, although it has not been until the early 1970s that we have seen the coyote population take off and expand into all counties across the state.

Two studies at Auburn University have shown that most coyote complaints have centered on livestock and watermelon depredation (damage). In some cases, watermelon producers have gone out of business because they could not keep coyotes from eating watermelons. Most coyote problems are associated with the northwestern and southeastern portions of the state where coyote populations are the highest. Even in these areas, coyote damage is scattered and isolated.

PAUL T. BROWN

Coyotes are larger than foxes. Their coats vary in color from buff-yellow to reddish-yellow to brown. Unlike wolves, they are mostly solitary animals and rarely travel in large packs.

Studies at Auburn and other universities reveal that coyotes will prey upon other wildlife, but the impact is negligible. In one study over 50 percent of coyote scats (droppings) contained deer hair, especially during the peak fawning period. The implications of this study are unclear, because coyotes are known to regularly scavenge on stillborn fawns and deer that are killed by cars and hunters. However, we do know that areas of the state where we have the highest coyote populations are also some of the same areas that have some of the highest deer densities, such as the Blackbelt.

JAMES B. ARMSTRONG

Coyote damage to watermelons is a serious problem for farmers.

Before attempting to control coyote depredation it is important to understand that coyote populations can be reduced but not eradicated. Any damage control program will have to be a continuing process. The type of control method depends on the nature of the problem. Livestock damage can be reduced by husbandry practices that bring calving cows closer to farm houses in smaller pastures. Proper disposal of cattle carcasses, by burying or removal from the property, will also help discourage coyote depredation. Guard dogs and donkeys have also been effective as coyote deterrents. Some watermelon producers have reduced coyote damage by moving fields closer to houses and constructing electric fences around patches.

Trapping and shooting are two techniques that remove coyotes and reduce damage. Leghold trapping with No. 2 traps can be effective if the traps are properly set. Coyotes can also be legally shot during daylight hours in Alabama. There is no closed season or limit on coyotes; however, since they are classified as a furbearer, hunting regulations should be checked. If damage is occurring, hunters can shoot coyotes during other open game or predator seasons.

ly interbreed with domestic dogs, producing "coydog" hybrids, this phenomenon is very rare. Consequently most coyote-looking animals observed in the wild are probably true coyotes.

The size and weight of coyotes are commonly overestimated because their long furry coat masks a smaller body frame. Alabama coyotes measure between 48 and 60 inches long and weigh 20 to 46 pounds; the average female weighs 29 pounds, and the average male weighs 33 pounds.

Wide color variations occur among coyotes. They range in color from almost pure gray to rufous (red) to pure black (melanistic). Coyote coats usually vary from buff-yellow to reddish-yellow to brown. The belly and throat areas are light gray or white. A mane of black-tipped hairs is typical among coyotes, as are black-tipped hairs over the upper tail and rump.

BIOLOGY AND LIFE HISTORY

Coyotes mate in February or early March and give birth to three to seven young 60 to 63 days later. Female coyotes may breed with one or more males but form a pair bond with only one male. The same pair may breed from year to year but not necessarily for life. Litter size varies from 2 to 12 with an average of 4 to 6. Pups are born in late March through May and weigh about ½ pound at birth. Pups are cared for by the mother and possibly other "helpers," usually young of the previous year. The paired male helps by providing food for the young. Pups nurse for five to seven weeks although they begin eating food regurgitated by the mother at three weeks of age. Their eyes open at about 14 days, and their teeth erupt at about the same time. Pups reach adult size by nine months of age.

The young begin leaving the areas of their birth from August through December. Dispersing young may wander miles before suitable habitat is found. During this time, mortality is very high, and as many as 70 percent of the young coyotes will die.

Food supply appears to determine the number of females that breed and particularly affects the number of yearling breeding females. If the food supply is good, more females will breed due to their healthier condition. However, most female coyotes do not breed until their second year.

Coyotes die from many natural and man-induced causes. People are responsible for the majority of deaths of coyotes older than five months of age. Coyotes are also susceptible to a number of canine diseases, including canine distemper, hepatitis, mange, parvovirus, and rabies. Average annual mortality rates of 30 to 40 percent for adults and 70 percent for juveniles are typical.

Coyotes are opportunistic feeders that eat a variety of items depending on food abundance and availability. Cropland, pastures, hayfields, clear-cuts, and overgrown fields are used extensively by coyotes when feeding on mice, rabbits, and insects. Food habit studies of Alabama coyotes have shown that mammals occurred in the diets of 97 percent of those checked; fruits were present in 54 percent and insects in 41 percent.

Rodents, rabbits, and similar-sized mammals and **carrion** (dead animals) make up the bulk of the coyote winter diet. During winter months, rodents and remains of cows can be found in coyote stomachs. Cow remains are usually attributed to scavenging by coyotes. Livestock carrion can be an essential winter food source. Carrion is thought to support higher coyote densities than would exist if livestock producers regularly practiced proper disposal techniques. During the summer, insects (primarily grasshoppers and beetles) comprise a large portion of the diet. It is not uncommon during these periods to see coyotes running and jumping through grassy fields flushing and eating grasshoppers and other insects.

Coyotes also eat vegetable matter and fruits fairly often. It is not uncommon for entire watermelon patches to be decimated by coyotes, particularly if water is scarce. Persimmons are also a favored food during the fall. Other food items they eat include deer, birds, groundhogs, cats, and fish. One study of coyote diets in Alabama and Mississippi found that deer hair was present in over 50 percent of stomachs and scats examined by biologists during peak fawning period. It is unknown, however, whether this was a result of scavenging, predation, or both. Coyotes will drink water from any available source, such as ponds, creeks, rivers, or lakes.

Coyotes are basically solitary animals that do not travel in large packs like wolves. In most cases, coyotes travel and feed as individuals or in pairs, hunting small prey. Often family groups of six to eight coyotes can be seen feeding together, however. Coyote pairs or family groups live in distinct, non-overlapping territories. Territory size is thought to be influenced by coyote densities and prey availability with boundaries that are maintained by scent marking, not fighting. A small percentage of coyotes are nomads and do not respect territorial boundaries. Territory size may vary from 1½ to 10 square miles, but most activity is confined to a much smaller area. In the fall, many young coyotes move from their birth territories in search of a place to settle. These movements often cover 10 to 50 miles.

Coyotes are active day and night but will travel more extensively and for longer periods during darkness. Most activity occurs around sunrise and sunset. This is also the time when most feeding and social interaction occurs. Coyotes usually rest or bed in different locations each day during daylight. In good habitats with plenty of prey, coyote densities during the pup-rearing season can vary from one to five coyotes per square mile. Average population density is usually one adult coyote per square mile.

HABITAT NEEDS

Coyotes are flexible and adaptable in their habitat requirements. They will live almost anywhere adequate food is available. Coyotes use a variety of cover types to escape from domestic dogs or man, to protect their pups, or to use as daytime resting sites. These cover types include brush-thickets, tall grass, or wooded areas. Woodlots in association with creek drainages are often used by coyotes during daylight hours. These woodlots are thought to be centers of activity for coyotes in large areas of open farmland. The number and location of woodlots in this habitat type might determine the population of coyotes.

Coyote dens may be found in a wide variety of places, such as brush-covered slopes, steep banks, rock ledges, thickets, and hollow logs. Denning sites are usually in remote areas away from human activity.

Skunks

We are all familiar with the unmistakable odor that skunks discharge when provoked. This obnoxious odor causes humans to fear and dislike skunks. However, despite our dislike for them, they help man by feeding on insect and rodent pests. The word "skunk" originates from the Algonquin Indians and refers to the spraying of musk.

The striped skunk (*Mephitis mephitis*) can be found throughout the southern half of Canada, the United States except in the desert Southwest, and northern Mexico. Alabama striped skunk populations are probably the highest in the Piedmont area. Open forests and the increase in agricultural lands during the early 1900s benefitted the striped skunk and allowed the animal to expand its range.

The Eastern spotted skunk (*Spilogale putorius*) ranges from northern Mexico throughout the United States, except the Northeast. Spotted skunks in Alabama are less common than striped skunks and are found most frequently in northern Alabama.

DESCRIPTION

The striped skunk and the spotted skunk are often referred to as polecats, civet cats, or hydrophoby cats. The spotted skunk is incorrectly called a civet cat because of its similarity to Old World civets. Skunks are not closely related to either true civets or to cats.

Skunks, along with river otters and mink and other members of the weasel family, have characteristic musk glands that are responsible for their unpleasant odor. The scent is produced by two internal musk glands located at the base of the tail and is usually released for self-defense. Before spraying the oily, sulfur-containing compound, skunks usually stamp their front feet rapidly and growl or hiss. They generally walk a short distance on their front feet and raise their tail as a warning before releasing any scent. The fluid is released in a fine spray that can be accurately dispersed up to 10 feet and less accurately for 20 feet. Skunks can discharge the spray several times within a short period. The fluid is painful if it gets in the eyes and may cause blindness for up to 15 minutes.

Few animals can be confused with these black-and-white spotted or striped animals. Striped skunks are short, stocky mammals about the size of a domestic house cat. They have a triangular-shaped head tapering to a blunt nose, a large bushy tail, and large feet equipped with well-developed claws. Their color pattern is characterized by two prominent white stripes down the back in a coat of jet black fur. The amount of white on the back varies from just a patch on the head to stripes covering the entire back. Adult striped skunks weigh between 4 and 10 pounds, although individuals weighing more than 12 pounds have been recorded. Striped skunks measure between 23 and 28 inches in length.

Eastern spotted skunks are about one-half the size of striped skunks. They measure 10 to 27 inches long and weigh 1 to 4 pounds. Males are generally larger than females. Eastern spotted skunks appear to be much more weasel-like and are readily distinguishable by white spots in front of each ear and on the forehead, and four to six broken white stripes on the back. These animals are much more nervous than striped skunks and are also better climbers.

BIOLOGY AND LIFE HISTORY

The breeding season for skunks in Alabama begins in late January when males begin searching for females near winter dens. Males are polygamous and will mate with several females, sometimes in succession. Females are receptive to males during their one heat period, which lasts 9 to 10 days. Mating triggers ovulation about 42 hours after insemination.

After a gestation period lasting 62 to 66 days, young skunks are born in May and June. Usually 5 to 9 young kits are born in a litter, but there can be as many as 18 or

E. R. DEGGINGER/ANIMALS ANIMALS

PAUL T. BROWN

Alabama has two species of skunks, the spotted (left) and striped (right). The striped skunk is the most common and is found throughout the state, especially in the Piedmont. Spotted skunks are less common and are found primarily in northern Alabama.

as few as 2. Spotted skunks have a gestation period lasting 45 to 60 days and give birth to one to six young (the average is four). The newly born kits are blind, wrinkled, and thinly furred at birth and weigh about 1 ounce. After two to four weeks, their eyes open, and the young skunks are able to discharge fluid from their scent glands. Kits will nurse for six to seven weeks and at this time are able to follow the mother on hunting trips. Young skunks are weaned at about two months of age. Families break up during August and September when the young leave to find their own homes.

Skunks are opportunistic omnivores, consuming both plant and animal material. Favorite skunk foods are grasshoppers, crickets, beetles, yellow jackets, cutworms, and other insect larvae. When insects are not available, skunks will eat mice, rats, shrews, moles, chipmunks, and other small mammals. They will also eat reptiles, amphibians, fish, fruits, and garbage. Occasionally they will feed on poultry and the eggs of ground-nesting birds. Most of a skunk's diet consists of small mammals and insects considered pests by man.

Skunks are nocturnal, becoming active from sunset to slightly after sunrise. Female skunks are not long-distance travelers, whereas male skunks may travel up to 4 to 5 miles a night during the breeding season. Normal skunk home ranges vary from 1 to 1½ miles. During the breeding season, males move slowly, become active during the day, and are reluctant to flee when confronted. This is the time when skunks are often struck by cars. Skunks are not territorial and tolerate other skunks in their range. This non-territoriality allows for high concentrations of breeding skunks in some areas.

Skunks die from a variety of causes. Coyotes, foxes, bobcats, great horned owls, and barred owls all relish skunks. Starvation may also cause some skunks to die during winter. A variety of diseases and parasites are common to skunks. Rabies and leptospirosis are the primary diseases responsible for death. Skunks are a primary source or vector of infection for other species of animals that are susceptible to these two diseases. High skunk populations are a concern for human health and livestock safety.

When a skunk becomes infected with the rabies virus, it may go unnoticed for a period of time. Symptoms may not appear for weeks or months. During this time, the infected animal may transmit the virus to the other animals it contacts. In the final stages of the disease, skunks may seem tame or listless, show signs of excessive salivation, become unusually aggressive or nervous, wander about during the daytime, and show little fear of humans. If a skunk acts strangely—aggressive or nervous, wandering in the daytime, or appearing tame and listless—it should not be approached. Parents should warn children never to approach a skunk or any other wild animal. In areas with large skunk populations, pets and livestock should be vaccinated for rabies. Skunks should not be kept as pets because they cannot be effectively immunized against the disease. Furthermore, they may have contracted rabies at an earlier age and be infected, yet fail to exhibit symptoms of the disease.

HABITAT NEEDS

Because many of the habits of the two species are similar, the discussion on habitat requirements, biology, and natural history of spotted and striped skunks will be combined. Skunks can be found in a variety of habitats throughout Alabama. Favored haunts include rolling hayfields, fencerows, brushlands, woodland edges, weedy fields, rocky outcrops, wooded ravines, stone walls, and drainage ditches. Home to a skunk is an underground den that may be located under buildings and house porches, and in culverts, brush piles, tree stumps, lumber piles, or abandoned fox or groundhog burrows. The dens are lined with leaves, hay, or grasses and are used for loafing, giving birth and raising young, and shelter during periods of inactivity during the winter. During the day, skunks usually sleep in the den, although during the warmer months they may bed in vegetation along fencerows, hayfields, or pastures. In winter skunks may remain inactive in the den for a period of days or weeks. They do not hibernate but become inactive during cold weather, relying on stored body fat to get them through cold months. Several skunks may share the same den during winter to conserve body heat.

Bobcats

Few people ever glimpse a bobcat (*Lynx rufus*) in the wild. Evidence of a bobcat's presence usually consists of remains of prey, tracks along a woods road or field border, or scats (fecal material). Bobcats are beneficial to man in that they eat a wide variety of small rodents and insect pests. On the other hand, bobcats can also become a pest by attacking domestic livestock, poultry, and wildlife favored by man. Pelts of bobcats are valued as fur, although demand and prices fluctuate widely from year to year. In Alabama, fur dealers reported only 26 bobcat pelts sold during the 1994–1995 trapping season.

Historically, the bobcat could be found throughout the lower 48 states, parts of southern Canada, and northern Mexico. It was eliminated from many densely populated and heavily farmed states by the early 1900s. Bobcats are found throughout Alabama but are most abundant in the Coastal Plain and Piedmont.

DESCRIPTION

Often called a bay lynx, barred bobcat, catamount, lynx cat, wildcat, or cat of the mountains, the bobcat (*Lynx rufus*) is one of the most elusive of all furbearing animals. It slightly resembles a housecat but is about twice its size. The bobcat also has longer hind legs and a shorter tail. Male bobcats are larger than females and weigh between 16 and 40 pounds. Males measure about 34 inches long complete with a 5¾-inch tail. Females measure about 31 inches in length and weigh between 8 and 33 pounds. The height of a bobcat at the shoulders varies between 20 and 23 inches.

The bobcat's general coloration is yellowish or reddish brown streaked or spotted with black or dark brown. The guard hairs are black-tipped. Its belly is white with black spots, and there are several black bars along the inside of the forelegs. The tail has several dark bands that become more distinct at the tip. The underside of the tail is whitish. Bobcat fur is dense, short, and very soft. No true color phases occur, but all-black (melanistic) bobcats have been reported. There is tremendous individual color variation within the bobcat's range.

These cats have a short, broad face with ruffs of fur on each side. The ears are prominent and pointed with a tuft of black hair at the tip. The back of the ear is black with a central white spot. The bobcat has sharp, retractable claws. There are four toes on the hind feet and five on the front. The tracks show only four toes on the front feet because the fifth toe is raised.

BIOLOGY AND LIFE HISTORY

Female bobcats generally come into heat once a year from January through March, and sometimes even as late as July. They may breed as early as one year of age, whereas males generally do not become sexually mature until the age of two. About 62 days after breeding, female bobcats give birth to one to four (usually three) kittens in a crude nest made of leaves and moss in the den. The kittens open their eyes in about 10 days. The mother does not leave the young for 2 days, sustaining herself by eating the placenta, feces, and any stillborn kittens. By the end of the fourth week the kittens begin eating solid food. The kittens are weaned by the seventh or eighth week but remain with the mother until autumn.

Bobcats, like other predators, are opportunistic and will attempt to eat almost anything available. Insects, fish, reptiles, amphibians, birds, and mammals have been reported in bobcat diets. Mammalian prey is the most important food group, and the bobcat is best adapted to preying on rabbits. The cottontail rabbit appears to be its principal prey throughout its range.

The bobcat most frequently kills prey weighing between 1½ and 12 pounds, such as rabbits, large rodents, opossum, and similar sized animals. The second most frequently taken prey group includes animals that weigh less than 1 pound, such as squirrels, rats, mice, shrews, and voles. The final group of prey species weighs more than 12 pounds and includes beaver, turkey, and deer.

The size of bobcat home ranges varies according to prey availability and abundance, sex, season, climate, and topography. However, bobcat ranges tend to be large, which accounts for the low densities throughout much of its range. Typical densities average about one animal per 3 to 5 square miles. Home ranges vary from 1 to 80 square miles.

The elusive bobcat is found throughout Alabama and is most abundant in the Piedmont and Coastal Plain. Bobcats are carnivorous and eat a variety of preys such as rabbits, rodents, opossums, turkeys, and other animals. It is not uncommon for an adult bobcat to kill and eat small deer.

Bobcats are territorial, and resident animals confine their movements and activities to specific ranges. Transient bobcats, which are young or sexually immature adults, may exhibit long-range, erratic movements but move into home ranges that are vacant due to the death or removal of a resident bobcat. Male home ranges are usually two to five times larger than those of females and may overlap several females' home ranges. Females tend not to overlap ranges of other females.

The animal's social structure and territorial boundaries are maintained in part by a complex system of scent marking using urine, feces, the anal glands, and scraping with the feet. As part of this system, bobcats use "stretching trees," dry snags without any bark, to help define territories.

Most bobcat mortality is in the juvenile age class. Forty to 50 percent of all juvenile bobcats do not reach their second birthday. The most important factor affecting this mortality is food availability. These young cats are most susceptible to death during the period when they leave their mother and begin establishing their own home range. Bobcats are not commonly preyed upon, although foxes, owls, and adult male bobcats will prey on kittens. Because of the solitary nature of bobcats and their propensity for changing denning and resting areas frequently, bobcat populations rarely succumb to die-offs as a result of disease or heavy parasitic infections.

HABITAT NEEDS

The bobcat is adapted to a wide variety of habitat types, from swamps to deserts and mountain ranges. The only habitat type not used is intensively farmed agricultural land where rocky ledges, swamps, and forested tracts have been eliminated. These reclusive animals appear to prefer forested areas with a dense understory that contains a large number of prey species. Key habitat components include prey abundance, protection from severe weather, availability of rest areas, dense cover, denning sites, and freedom from disturbance.

In some areas, ledges are critical habitat. Ledges appear to be activity centers that provide protective cover, protection from harassment, and an area for courtship activities. These areas may also serve as gathering grounds for solitary cats. Another important feature of bobcat habitat is denning sites. Bobcats use cliffs, rocky ledges, hollow logs and trees, brush, or rock piles as resting and denning sites. The animals use these areas for refuge from harassment, for breeding and raising young, and for shelter. Bobcats need more than one den within their home range because females with young often use more than one denning site.

With habitat protection and annual monitoring of harvests, bobcats will continue to be an elusive but important furbearer species in Alabama. Forest and farm owners can play an active role in maintaining bobcat

Fig. 1: Prevention and Control Techniques for Furbearers

BEAVER	
Exclusion	Small critical areas such as culverts, drains, and other areas may be fenced. Flashing or hardware cloth approximately 3 feet high can be used to protect individual trees in urban settings.
Habitat Modification	Eliminate foods, trees, and woody vegetation where feasible. Continually destroy dams and materials used to build dams. Install a Clemson beaver pond leveler, three-log drain, or other structure in beaver ponds to avoid further pond expansion.
Repellents	None registered; however, there is some evidence that repellents may be useful.
Trapping	The preferred method of control uses one of the following traps: 1) No. 330 Conibear traps, 2) leghold traps No. 3 or larger, 3) basket-type traps primarily used for live trapping, and 4) snares.
Shooting	Rarely effective for complete control efforts and can be dangerous.

RIVER OTTER	
Exclusion	Fence small raceways, tanks, or ponds with 3 x 3–inch mesh wire.
Trapping	The preferred method of control. Use Conibear traps (Nos. 220 and 330), foothold traps (No. 2), and snares. For live release, river otters can be caught in live traps, modified No. 1½ soft-catch traps, and No. 11 long-spring traps.
Shooting	During legal seasons shooting can be effective in damage situations that involve only one or two otters.

MUSKRAT	
Exclusion	Riprap the inside of a pond dam face with rock.
Habitat Modification	Eliminate aquatic vegetation as a food source. Draw down farm ponds during the winter.
Trapping	Most preferred method of control. Use body-gripping traps like the Conibear No. 110. Leghold traps No. 1, 1½, or 2 can also be used.
Shooting	Effective in eliminating some individuals.

MINK	
Exclusion	Usually the best solution to mink predation on domestic animals. Confine animals in fenced areas and seal all openings larger than 1 inch.
Trapping	Most preferred method of control. Use leghold (No. 11 double long–spring or No. 1½ coil-spring) or Conibear-type traps equivalent to No. 120 traps.

RACCOON	
Exclusion	Usually the best method for coping with almost all types of raccoon damage.
Habitat Modification	Remove obvious food sources or shelter. This is usually not practical as a sole method of controlling damage.
Trapping	Live cage traps, body-gripping Conibear-type traps, and foothold traps (No. 1 or No. 1½) are very effective, especially used in conjunction with exclusion and habitat modification.
Hunting	Can be very effective, particularly if trained hounds are used to tree raccoons.

OPOSSUM

Exclusion	Practical where opossums are entering structures.
Habitat Modification	Remove protective cover and plug burrows to reduce the frequency of visits.
Trapping	Most preferred method involves using leghold traps (No. 1½), box traps, cage traps, or body-gripping traps.
Hunting	Can be effective where firearms are permitted.

FOXES

Exclusion	Woven net or electric fences.
Cultural Methods	Protect livestock and poultry during most vulnerable periods such as birth of young.
Frightening	Flashing lights and noise may provide temporary relief. Well-trained guard dogs may be effective in some situations.
Trapping	Most preferred method involves using steel leghold traps (Nos. 1½, 1¾, and 2 double coil–spring trap and the Nos. 2 and 3 double long–spring trap), cage or box traps, and snares.
Hunting	Predator-calling techniques can be effective.
Other Methods	Look for fox dens and remove the young.

COYOTE

Exclusion	Protect livestock in pens at night and use woven wire or electric fences.
Cultural Methods	Select pastures that have a lower incidence of predation. Herding of livestock generally reduces predation due to human presence during the herding period.
Frightening	The use of guard dogs has significantly reduced predation on livestock in some cases. Strobe lights, sirens, propane cannons, and other noise have reduced predation on both sheep and calves.
Trapping	Most preferred method involves using leghold traps; Nos. 3 or 4 are the most effective. Snares are effective where coyotes pass through or under net-wire fences and in trail sets.
Varmint Hunting	Can be effective if sportsmen remove coyotes by shooting during legal hunting season.

SKUNK

Exclusion	Close openings under buildings. In poultry yards install wire mesh fences.
Habitat Modification	Remove garbage, debris, lumber piles, and pet-feeding pans.
Trapping	Most preferred method involves using box or leghold trap. Leghold traps should not be used around residential areas due to the potential to capture domestic pets and the discharge of scent glands.
Shooting	Only practical when animals are not in residential areas.

BOBCAT

Exclusion	Fence poultry and other small livestock located near residential areas.
Cultural Methods	Clear brush and timber in and around farms.
Frightening	Place flashing white lights, loud music, or dogs with livestock.
Trapping	Most preferred method involves using a variety of traps including steel leghold traps (No. 2, preferably No. 3, or No. 4 offset or padded), cage traps (15 x 15 x 40 inches up to 24 x 24 x 48 inches), large body-gripping traps (Victor No. 330 Conibear), or snares. Fur trappers may be willing to trap and remove bobcats year-round in exchange for trapping rights when pelt prices are high.
Hunting	Can use predator calls or experienced trail hounds.

The Decline of Furbearer Trapping

Mike Sievering is a Furbearer Biologist with the Game and Fish Division of the Alabama Department of Conservation and Natural Resources in Northport, Alabama. He routinely gives landowners advice and guidance on how to control nuisance furbearers.

In Alabama, as well as the entire Southeast, the number of trappers has declined. As a result, the number of furbearers that are annually trapped has dropped substantially. Reduced trapping in combination with an apparent increase in furbearer habitat has caused an explosion in many furbearer populations. Corresponding damage caused by nuisance furbearers has increased. Our road-kill surveys have shown a three-fold increase in raccoons in recent years. Consequently we are seeing an unprecedented number of raccoons with rabies and distemper. In the early 1970s the rabies line in Alabama was in the southeastern part of the state; now it lies in north-central Alabama and continues to move northwest at about 10 miles per year. Therefore we are seeing a lot more sick and diseased raccoons in Tuscaloosa, Birmingham, and other central and northern Alabama cities.

Unless market prices increase for furbearer pelts, trapping for profit will continue to decline and we will see increasing problems with nuisance furbearers. The art of trapping as a recreational activity has also declined. Unless youth become involved in trapping at an early age through programs like Hunter Education, trapping for recreation may soon become non-existent. For specific problems with nuisance furbearers, consultants who charge a fee for their services will take the place of recreational trappers. This is already taking place in several counties in Alabama. At least one franchised wildlife nuisance control company is already operating in Birmingham. In the future, we will see more nuisance wildlife problems in Alabama handled by private professionals and business enterprises.

numbers by providing some or all of the habitat components required by bobcats on their lands.

FURBEARER CONTROL

In some cases expanding populations of furbearers can conflict with people. These conflicts include negative interactions that cause damage and economic loss to agricultural commodities, timber, fisheries, and other resources. In some cases, furbearers may also compound human health concerns about diseases such as rabies.

Furbearer control programs can be thought of as having four parts: 1) identifying the problem, 2) understanding the ecology of the problem furbearer, 3) selecting a suitable control option, and 4) evaluating the success of the control method that was used. Identifying the problem involves knowing which furbearer caused the problem as well as the numbers of animals causing the problem, the type of damage, and other biological or social factors related to the problem. Ecology of the problem furbearer refers to understanding the life history and biology of the species so that an effective control method can be chosen and correctly applied. Selecting and implementing the right technique can effectively control most furbearer problems, while choosing the wrong method is a waste of time and money. Understanding which control method to use as well as the correct way to apply it is crucial. Evaluation of the control technique helps to assess any reduction in damage in relation to the costs and impact of the control method. In some situations, a recommended method of control may not be cost effective and other techniques or a combination of techniques will have to be used.

To reduce the problems associated with nuisance furbearers a variety of control techniques have been developed. These control techniques can be broadly categorized as exclusion, habitat modification, frightening, repellents, trapping, shooting, and other methods such as home remedies. Figure 1 highlights damage prevention and control methods for furbearers.

Many control techniques for furbearers can be conducted by landowners and managers; however, if additional help is needed a variety of sources can provide assistance. The Cooperative Extension System is a

PAUL T. BROWN

The gray fox is slightly smaller than the red fox and has a salt-and-pepper coat with buff underfur and rusty yellow sides, legs, feet, and back of the ears. The reddish sides will occasionally cause the gray fox to be mistaken for a red fox. Gray foxes prefer a mixture of woodlands and fields. Unlike the red fox, gray foxes can easily climb trees.

good place to start when there is a problem with nuisance furbearers. The Extension System provides a wide range of information on prevention and control of wildlife damage through local agents in each county and specialists at Auburn (334) 844-9233. In addition, most county Extension offices maintain a list of local trappers who can be contacted to trap nuisance furbearers for a fee. The USDA Wildlife Services also has a wildlife damage management program in Alabama that is based at Auburn University (334) 844-5670. USDA Wildlife Services provides technical assistance for a wide range of wildlife damage problems, especially control of beavers through a cooperative educational and assistance program. Information and technical assistance for furbearer control can also be found through the Alabama Department of Conservation and Natural Resources (ADCNR) at (334) 242-3486. **Often permits are required from ADCNR before any furbearer control method is used.** In cases where furbearers pose a human health threat, local animal control authorities or public health service agencies may also provide assistance. These organizations can usually be found under the government sections of the local public telephone directory.

SUMMARY

Alabama's furbearer resource includes a diverse group of wildlife that can be found in almost any area of the state. Some are valued for their pelts, while others are enjoyed for the uniqueness they contribute to farms and forests across Alabama. Many of the furbearers are among nature's best pest control agents because they eat large numbers of rodents and insects that are considered pests by man. Because some furbearers are so closely associated with aquatic habitats, their general health and abundance are indicators of water quality and environmental health. They can be sentinels for our own survival.

Unfortunately, the survival traits that make many furbearers abundant in Alabama are the same characteristics that occasionally put them at odds with man. As aggressive predators they often prey upon domestic and wild animals that are valued by man. In addition, some pose disease threats to domestic animals, and in rare cases to people. However, the benefits of Alabama's furbearer resource far outweigh the negative aspects.

Today's challenge is to effectively manage the state's furbearers so that their populations are secure and negative interactions with man are held to a minimum. Accomplishing this will be a complex job at best, since some furbearer species may be in decline from habitat degradation and pollution, while others are possibly increasing to high levels as a result of lower trapping and hunting efforts. Future furbearer management will require the collaborative efforts of private landowners, natural resource agencies, and others interested in sustaining healthy furbearer populations for future Alabamians.

Alabama is home to approximately 560 species of non-game wildlife, including the great blue heron.

ROBERT P. FALLS

CHAPTER 12

Managing for Non-game Wildlife

CHAPTER OVERVIEW

Alabama has a diversity of non-game wildlife that provides innumerable benefits, from insect control to aesthetic enhancement. This chapter examines Alabama's non-game wildlife and considerations for their management in woodlands and other areas.

CHAPTER HIGHLIGHTS PAGE

WHAT IS NON-GAME WILDLIFE? 256

NON-GAME WILDLIFE AND FOREST MANAGEMENT 257

ALABAMA MAMMALS 264

BLACK BEARS IN ALABAMA 265

ALABAMA BIRDS 279

ALABAMA REPTILES AND AMPHIBIANS 283

Past wildlife research and management have focused primarily on game animals because of society's historical appreciation of these species. Other species, particularly songbirds, were often overlooked by federal and state agencies and universities and relegated to special interest groups like the Audubon Society. The research that was conducted on birds, reptiles, amphibians, and other non-game species centered on understanding basic biology, with little interest in managing their numbers or habitats.

The environmental movement in the 1960s and 1970s helped awaken the general public to the value of all wildlife species. As a result, the focus of federal and state legislative and administrative efforts was redirected to include non-game species in planning, research, and management. Today, the wildlife profession has responded to the growing interest in and support of non-game animals by initiating active research programs that focus on a broader variety of species. Non-game programs have also been established within many state and federal agencies.

The Alabama Legislature created the Non-game Wildlife Program in 1982, with the recognition that "it is in the public interest to preserve, protect, perpetuate, and enhance non-game wildlife resources of Alabama through preservation of a satisfactory environment and ecological balance." This legislation charged the Alabama Department of Conservation and Natural Resources (ADCNR) with administering the program. After consulting with zoologists and other interested groups throughout the state, ADCNR surmised that there was a lack of information regarding non-game species' occurrence, distribution, life history, biology, and management in Alabama. The ADCNR convened a conference of specialists and lay persons to discuss common interests and concerns for Alabama's non-game resources and to consolidate existing biological information on non-game animals as a foundation for the newly established Non-game Wildlife Program. The "First Alabama Non-game Wildlife Conference" was held in July of 1983 at Auburn University. Four committees emerged from the conference to focus on non-game wildlife: 1) the Fish Committee, 2) the Amphibians and Reptiles Committee, 3) the Bird Committee, and 4) the Mammal Committee. These groups were charged with determining the status of Alabama's non-game wildlife and providing general recommendations to ADCNR. Together the committees produced two publications: *Vertebrate Wildlife of Alabama*, an annotated checklist of many non-game species,

Hummingbirds are one of the most popular species of non-game wildlife in urban and suburban Alabama. The ruby-throated hummingbird is a common site around wildflower gardens and hummingbird feeders.

and *Vertebrate Animals of Alabama in Need of Special Attention*, a technical account of non-game species in the state. Both of these publications are available from the Alabama Agricultural Experiment Station at Auburn University. Alabama's Non-game Program has initiated many efforts aimed at learning more about the state's non-game resources since the 1983 conference.

WHAT IS NON-GAME WILDLIFE?

There is no universal definition of non-game wildlife based upon biological considerations. Instead the term "non-game" is influenced more by political, social, and economic values that are expressed through legislative mandates. As an example, mourning doves are considered game birds in the Southeast, while many northern states regard them as songbirds that are protected from hunting. Non-game status does not necessarily imply that a species will be protected. Many species of nuisance wildlife (animals that cause damage), such as grain-eating blackbirds or root-chewing pine voles, are also non-game wildlife. Because they are pests, they can be legally killed to control the damage they cause. In some cases, special permits are required to kill nuisance non-game animals. Endangered species, on the other hand, are non-game wildlife that need special protection and management to ensure their survival.

The distinction between game and non-game animals is based on whether they are used consumptively (such as game that is hunted, or furbearers that are trapped) or non-consumptively (such as wildlife appreciated by bird watchers, nature photographers, etc.). This has been a convenient way of classifying non-game wildlife, but it has limitations. Many game species are also enjoyed by bird watchers and photographers. In addition, the term "non-consumptive" is a misnomer. Appreciative uses can also impact wildlife through disturbances and habitat degradation. For the purpose of this chapter, we will define non-game as wildlife that are not hunted, not including threatened and endangered species.

NON-GAME WILDLIFE AND FOREST MANAGEMENT

All wildlife require food, cover, water, and space. How these habitat components are arranged determines their value and use by various animals. Forest habitat for non-game species can be enhanced according to a forest level or species approach. A **multiple species approach** is a broad-scale strategy that modifies traditional silvicultural practices to retain critical habitat components (such as riparian habitats, unique areas, and dead wood) that are required by a variety of non-game species. This approach also attempts to maintain biodiversity across the forest landscape. A **single species approach** incorporates considerations on the forest level but also identifies specific habitat characteristics needed by individual species and then provides these features through management. Most non-game management follows a forest level approach because of a lack of information regarding specific habitat characteristics that are required by non-game animals. As continuing research identifies these habitat features, non-game management will begin to shift to species specific management, where habitat improvement practices are refined on a smaller scale to meet the needs of select non-game species.

Multiple Species Approach to Managing Non-game Wildlife

Silvicultural practices can provide a diversity of habitats for a wide range of non-game species. Because of contrasting land ownership patterns, different land-use objectives, and a variety of management practices, the landscape in Alabama is quite diverse, supporting a variety of birds, small mammals, reptiles and amphibians. **To enhance non-game habitats in conjunction with forestry operations, special consideration must be given to retaining and maintaining riparian forests, unique areas, dead wood, and threatened and endangered species**

Alabama's Non-game Program

Bob McCollum is a Non-game Biologist with the Alabama Non-game Program, Alabama Department of Conservation and Natural Resources, Game and Fish Division.

Since 1984 the Alabama Non-game Wildlife Program (ANWP) has been charged with the conservation of the non-game animals of the state: animals that are neither hunted nor trapped. With over 700 non-game vertebrate animal species native to this state, as well as thousands more invertebrates such as butterflies, crayfish, mussels, and snails, this is a formidable task indeed. Projects underway include survey work on animals of concern such as the gray bat, various reptiles and amphibians, a number of shorebirds, fishes, freshwater mussels, and snails. The ANWP also provides technical assistance to other divisions of the Alabama Department of Conservation and Natural Resources as well as to other state and federal agencies. We respond to public requests for information on non-game species and how to attract them, or even to discourage them if they are considered a nuisance. As part of a department-wide educational initiative known as the Basic Outdoor Skills Series (BOSS), the ANWP is developing a backyard wildlife workshop that will provide basic information on attracting wildlife and providing habitat in urban and suburban settings.

Funding for the non-game wildlife program does not include any state-generated funds or any other taxpayers' dollars. The 1982 Alabama legislature created a check-off on the Alabama tax return, enabling citizens to contribute part of their income tax refund to the non-game wildlife program. If taxpayers are not due a refund, direct donations can be made as well.

Non-game Management in Forested Landscapes

Tony Melchoirs is a Wildlife Research Biologist in charge of wildlife research and management on Weyerhaeuser Company lands in the southeast. Most of his work has focused on the effects of silviculture on non-game wildlife species.

During the past 10 years I have been involved in several long-term studies examining the effects of different silvicultural practices on non-game wildlife. Most of my work has concentrated on breeding/wintering birds and small mammal communities in streamside management zones, and on breeding bird communities and nesting success in managed pine stands under different silvicultural systems. Over the years, I have made some observations about non-game wildlife and timber management that may be helpful to non-industrial private forest owners.

First, certain features of a forest provide critical habitat for non-game wildlife, including streamside management zones and other riparian sites, unique ecological areas (such as rock outcrops, cedar glades, seasonal ponds, cypress domes, bogs, beaver ponds, etc.), and dead wood (such as snags and downed logs). These features should be maintained within a managed forest. Secondly, properly conducted silvicultural practices provide suitable habitat for a variety of non-game wildlife. People are often surprised at how many birds, mammals, reptiles, and amphibians occur in managed forest stands. Research is showing that openings created by timber harvests and thinnings, with various stages of vegetative growth, provide ideal nesting habitat for a variety of birds. Obviously, some "area sensitive" species requiring large blocks of mature forest are negatively affected by openings created through silvicultural practices.

Forest owners with a few hundred acres or less should focus their management efforts to increase habitat diversity within forest stands, incorporating critical non-game habitat and unique areas into their management plans. As ownership size increases, forest management for non-game species should concentrate on providing a diversity of habitats by controlling the size, shape, and distribution of managed stands and special areas across the landscape. Protection of rare wildlife, especially threatened and endangered species, should also be an important component of forest non-game management.

habitat. These distinct areas usually comprise only 5 to 10 percent of the forest landscape but support higher concentrations of non-game species than the remaining forest. These sites should be designated as special areas to focus management for non-game wildlife.

PROTECTING RIPARIAN FORESTS

Riparian or streamside forests are areas adjacent to water, usually a stream or river. These are the most productive sites within a forest stand and generally have a greater number and diversity of wildlife than the surrounding upland forests. Streamside forests are "life centers" for aquatic and semi-aquatic communities, critical for many aquatic invertebrates, amphibians, fish, and birds. Riparian hardwood forests provide special habitat for many species of birds, such as the Louisiana waterthrush, pronthonotary warbler, and hooded warbler. At least 49 species of mammals, 160 species of fish, 38 amphibians, and 54 reptiles use riparian forests in the Southeast. They also serve as wintering, resting, feeding, and breeding areas for many migratory birds that depend on these areas as stopover spots during their seasonal movements between North and South America. In pine forests, these are often the only sites that contain hardwoods and consequently help to maintain forest habitat diversity, especially in intensively managed pine stands.

To sustain the benefits of riparian forests, managers often establish streamside management zones adjacent to watersheds. **Streamside management zones (SMZs)** are strips of vegetation (most often hardwoods) adjacent to waterways that are left intact when bordering forest stands are harvested. SMZs maintain riparian habitat for non-game and game species and protect water quality by preventing excessive

BETH MAYNOR YOUNG

Unique areas, like this beaver pond, provide habitat for a host of wetland-associated wildlife species such as ducks, wading birds, reptiles, amphibians, beaver, muskrats, raccoons, and a variety of aquatic life.

runoff and sedimentation.

There is strong evidence that various non-game wildlife prefer different SMZ widths. For example, amphibians and reptiles are reported to be more prevalent in SMZs of 100 to 315 feet in width on each side of a stream. Small mammals (primarily rodents), however, seem to prefer narrow SMZs less than 80 feet in width on each side of a stream. The narrow width may be preferred by small mammals because these areas are often more brushy due to a greater exposure to sunlight. Other studies have found that the minimum SMZ width required to maintain adequate habitat for neotropical migrant songbirds is 150 feet on each side of a stream, however 300 feet or greater is more desirable for increasing songbird diversity. Forest landowners should use these results as a general guideline since more research is needed to determine ideal SMZ widths for each species. SMZ width varies depending on the non-game species requirements and site characteristics (such as size of stream, floodplain width, topography, etc.). Based on a compilation of current research, a minimum SMZ width of 150 feet on each side of a stream would benefit most non-game wildlife.

PROTECTING UNIQUE AREAS

Unique sites are frequently marginal in suitability for forest management but are often ideal habitats for many species of non-game wildlife. Because of the physical characteristics (soil type, topography, poor drainage, etc.) of these areas, they are usually poor sites for growing timber. Consequently, little opportunity cost is foregone from timber production when these areas are reserved for non-game wildlife. In other words, little income is lost when these areas are not managed for timber production and are left undisturbed. Unique areas that should be protected from silvicultural operations include bays, swamps, bogs (such as pitcher plant bogs), cypress domes, rock outcrops, cedar glades, old home sites, upland coves, and threatened and endangered species habitat (see Chapter 13). These areas usually support unique species and a variety of non-game wildlife.

If silvicultural operations are conducted near unique non-game habitats, buffer zones should be established adjacent to these areas and marked off prior to timber sales to prevent disturbance. As an example, land management activities that are conducted close to caves pose a disturbance to hibernating colonies of bats. Caves and other unique sites should have a vegetative buffer zone

Fig. 1: Non-game Forest Management at Two Levels of Intensity

LEVEL 1: MULTIPLE SPECIES APPROACH

- Normal silvicultural practices
- Protection and management of unique features
  - Riparian forest/SMZs
  - Unique areas
    - Wetlands (swamps, bogs, bays)
    - Cypress domes
    - Rock outcrops
    - Caves
    - Cedar glades
    - Upland coves
    - Beaver ponds
    - Threatened and endangered species habitat
  - Dead wood
    - Snags
    - Logs
    - Timber slash

LEVEL 2: SINGLE SPECIES APPROACH

- Includes Level 1 management
- Identifies non-game of interest
- Determines species location(s)
- Investigates habitat features species prefers
- Adjusts management practices to provide essential habitat components required by non-game species
- Incorporates new findings about non-game habitat requirements into forest management plans

at least 150 feet from the edges of these sites.

Existing beaver ponds are valuable for many aquatic invertebrates, amphibians, reptiles, furbearers, and wading birds. Dead trees or snags created by beaver impoundments are heavily utilized by several species of woodpeckers and cavity nesting birds. Whenever possible, beaver ponds should be retained as part of the forest landscape. Landowners should consider their objectives and decide how much beaver activity can be tolerated.

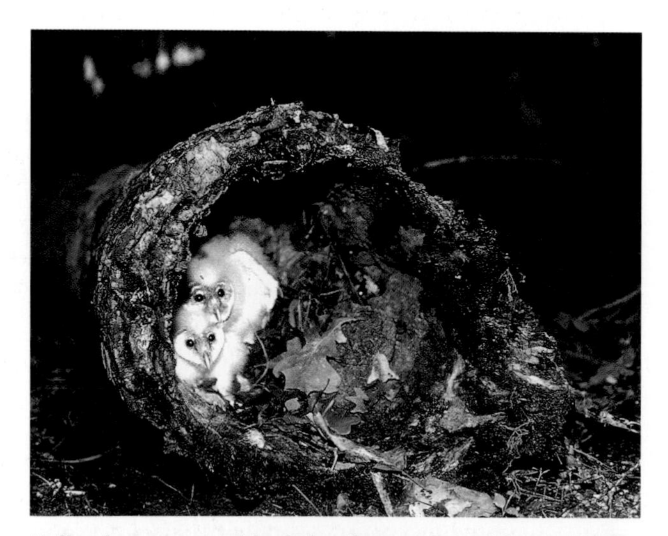

Hollow logs can provide shelter for a variety of forest wildlife such as these immature barn owls.

LEAVING DEAD WOOD

Dead wood is a beneficial habitat component for many non-game species. Standing dead trees or **snags** provide cover for cavity nesting birds and denning mammals. Birds that frequently use snags include the Eastern bluebird, Carolina chickadee, red-bellied woodpecker, red-headed woodpecker, loggerhead shrike, northern flicker, and red-tailed hawk. Snags also provide a ready source of insects for woodpeckers and other insect-feeding wildlife. Woodpeckers utilize snags as drumming sites for territorial marking. Cavity dwelling mammals, such as raccoons and squirrels (gray, fox, and flying) use snags for denning sites. These cavities are also important roosting sites for bats.

A fallen tree, or log, benefits non-game animals. Fallen trees decompose and become ideal hosts for fungi and insects that are eaten by many non-game species. As the tree decays, nutrients are recycled into the soil, and a microhabitat favorable for growing new plants and tree seedlings is created. Insects, salamanders, toads, snakes, mice, shrews, and other smaller animals are attracted to logs and the microhabitat they create. Insect-foraging wildlife, such as skunks and opossums, depend on logs as an abundant source of protein-rich insects. The accumulation of organic matter from rotting wood favors the growth of mushrooms and other fungi that are preferred foods for insects, turtles, birds,

Fig. 2: Effects of Forest Management Practices on Non-game Wildlife

PRACTICE	ADVANTAGES	DISADVANTAGES
Forest Harvest & Regeneration	More intensive methods such as clear-cutting, seedtree, and shelterwood favor early successional species, such as small mammals and reptiles. Group and individual selection is less detrimental to forest interior species.	Snags for cavity-dwelling animals are lost with intensive methods. Interior forest species that require large blocks of mature multi-layered forests suffer. Intensive practices can be detrimental to amphibians and some reptiles, particularly if best management practices are not followed.
Timber Harvest Rotation Length	Long rotations in large forest blocks favor forest interior species. Short rotations favor early successional stage species. Some neotropical migrants and the endangered red-cockaded woodpecker prefer older stands of trees such as those intended for sawtimber.	Both long and short rotations will discourage different species of non-game. For maximum diversity of non-game wildlife, a mixture of long and short rotation stands is preferable. However, this is difficult for the small forest owner.
Thinnings	Encourages the development of shrub & midstory layer ideal for many species of birds. Heavily thinned stands favor species like the gray catbird, rufous-sided towhee, Kentucky warbler and indigo bunting. Other species favored by thinning include the American redstart, black-and-white vireo, hooded warbler, and Eastern wood-pewee. Thinnings conducted in conjunction with prescribed burning are valuable tools in maintaining the endangered red-cockaded woodpecker and threatened gopher tortoise habitats.	Inhibits species that do not require a shrub or midstory layer such as the ovenbird and black-throated green warbler.
Fire	Critical for non-game species that have adapted to fire communities, such as the gopher tortoise and indigo snake.	Detrimental to dead wood (snag, logs, and slash) that provides a vital habitat component for many species of non-game, such as denning sites for foxes and small mammals.
Site Preparation	Less intensive methods are favored to reduce site disturbance. More intensive methods prolong grass and shrub stage favored by small mammal communities.	Intensive methods remove debris and cover for reptiles and amphibians.
Seedling Spacing	Wide seedling spacings (10 x 10 or 12 x 12) delay crown closure and maintain vegetation in an early stage preferred by small mammals, reptiles, and amphibians.	Close seedling spacings (6 x 6 or 6 x 8) accelerate crown closure and loss of early successional grasses and shrubs eliminating food and cover for many small mammals.

and mice. Some ground-nesting birds, such as rufous-sided towhees, prefer to nest adjacent to logs, especially logs that are partially elevated. Depending on their size, hollow logs can shelter a variety of forest mammals such as shrews, ground squirrels, chipmunks, bears, foxes, and coyotes.

Recommendations vary on the amount of dead wood to leave within forest stands for non-game species. Some managers suggest leaving three to five snags per acre for cavity-dwelling wildlife. Additional "green" trees are often left within newly harvested stands to provide future snags. Logs and logging debris (slash) should be scattered in clumps throughout the forest. Ideal places to integrate dead wood into forests are areas that are not routinely burned. These sites include SMZs, wetlands, floodplains, upland coves, and other areas.

Single Species Approach to Managing Non-game Wildlife

A single species approach focuses on particular habitat needs of a non-game species and management practices to enhance its habitat. Few non-game wildlife managers practice this level of management because of the commitment of time and financial resources.

Managers interested in this level of management should determine the species to be favored and the species' habitat requirements. The second step should be to determine distribution and abundance of the species—where they occur and in what numbers. This can be accomplished by visual counts for birds and trapping for small mammals, amphibians, and reptiles. Once a manager has determined where the species occurs, site and habitat characteristics should be examined to learn what distinguishing features make the habitat attractive for this animal. Management practices can then be modified to

Fig. 3: Checklist of Alabama Mammals and Their Distribution

ARMADILLOS	
Nine-banded armadillo (*Dasypus novemcinctus*)	Coastal Plain.
BATS	
Big brown (*Eptesicus fuscus*)	Statewide.
Brazilian free-tailed bat (*Tadarida brasiliensis*)	Distribution unknown.
Eastern pipistrelle (*Pipistrellus subflavus*)	Statewide.
Evening bat (*Nycticeius humeralis*)	Statewide.
Gray myotis (*Myotis grisescens*)	Caves in north Alabama. **Federally endangered species.**
Hoary bat (*Lasiurus cinereous*)	Statewide.
Indiana myotis (*Myotis sodalis*)	Caves in north Alabama. **Federally endangered species.**
Keen's myotis (*Myotis keenii*)	Caves in north Alabama.
Little brown myotis (*Myotis lucifugus*)	North Alabama.
Northern yellow bat (*Lasiurus intermedius*)	One record in Mobile.
Rafinesque's big-eared bat (*Plecotus rafinesquii*)	Distribution unknown.
Red bat (*Lasiurus borealis*)	Statewide.
Seminole bat (*Lasiurus seminolus*)	Statewide.
Silver-haired bat (*Lasionycteris noctivagans*)	Statewide.
Southeastern myotis (*Myotis austroriparius*)	Statewide.
BEARS	
Florida black bear (*Ursus americanus floridanus*)	Main breeding population just north of Mobile.
BEAVER	
Beaver (*Castor canadensis*)	Statewide.
CATS	
Bobcat (*Felis rufus*)	Statewide.
Eastern cougar (*Felix concolor*)	Reliable records are rare. Northern Baldwin County is the only documentation.
Jaguarundi (*Felis yagouaroundi*)	Several sightings on both sides of Mobile Bay.
DEER	
Fallow deer (*Dama dama*)	Found only in Miller's Ferry of Wilcox County.
White-tailed deer (*Odocoileus virginianus*)	Statewide.
DOGS AND ASSOCIATED CANINES	
Coyote (*Canis latrans*)	Statewide.
Gray fox (*Urocyon cinereoargenteus*)	Statewide.
Red fox (*Vulpes vulpes*)	Statewide.
Red wolf (*Canis rufus*)	Have been extirpated from Alabama.
JUMPING MICE	
Meadow jumping mouse (*Zapus hudsonius*)	Piedmont.
MARINE MAMMALS	
Atlantic bottle-nosed dolphin (*Tursiops truncatus*)	
California sea lion (*Zalophus californianus*)	
Spotted dolphin (*Stenella plagiodon*)	
West Indian manatee (*Trichechus manatus*)	
MOLE	
Eastern mole (*Scalopus aquaticus*)	Statewide.

MUSTELIDS	
Long-tailed weasel (*Mustela frenata*)	Statewide.
Mink (*Mustela vison*)	Statewide around wetlands and stream borders.
River otter (*Lutra canadensis*)	Irregular distribution in waterways.
Eastern spotted skunk (*Spilogale putorius*)	Statewide.
Striped skunk (*Mephitis mephitis*)	Statewide.
NATIVE MICE AND RATS	
Alabama beach mouse (*Peromyscus polionotus ammobates*)	Alabama Gulf Coast sand dunes.
Eastern harvest mouse (*Reithrodontomys humulis*)	Statewide.
Eastern woodrat (*Neotoma floridana*)	Statewide.
Golden mouse (*Ochrotomys nuttalli*)	Statewide.
Hispid cotton rat (*Sigmodon hispidus*)	Statewide.
Oldfield mouse (*Peromyscus polionotus polionotus*)	Statewide.
Perdido Key beach mouse (*Peromyscus polionotus trissylepsis*)	Perdido Key area and Gulf Coast.
Prairie vole (*Microtus ochrogaster*)	Found only in Tennessee River valley pasture lands.
White-footed mouse (*Peromyscus leucopus*)	Occurs in northern half of state.
Woodland or pine vole (*Microtus pinetorum*)	Statewide.
NUTRIA	
Nutria (*Myocastor coypus*)	Mobile/Tensaw Delta and upper Mobile Bay.
OLD WORLD MICE AND RATS	
Black rat (*Rattus rattus*)	Occurring mostly around port cities like Mobile.
House mouse (*Mus musculus*)	Statewide.
Norway rat (*Rattus norvegicus*)	Statewide.
OPOSSUM	
Virginia opossum (*Didelphis virginiana*)	Statewide.
POCKET GOPHERS	
Southeastern pocket gopher (*Geomys pinetis*)	Coastal Plain east of Tombigbee and Warrior rivers.
RABBITS	
Eastern cottontail (*Sylvilagus floridanus*)	Statewide.
Marsh rabbit (*Sylvilagus palustris*)	Coastal marshes.
New England cottontail (*Sylvilagus transitionalis*)	Isolated mountain habitats of state.
Swamp rabbit (*Sylvilagus aquaticus*)	Statewide in river bottoms and wetlands.
RACCOONS	
Raccoon (*Procyon lotor*)	Statewide. Most abundant in creek and river bottoms.
SHREWS	
Northern short-tailed shrew (*Blarina brevicauda*)	Restricted to Piedmont.
Southern short-tailed shrew (*Blarina carolinensis*)	Statewide except Piedmont.
Southeastern shrew (*Cryptotis parva*)	Statewide.
SQUIRRELS	
Eastern chipmunk (*Tamias striatus*)	Statewide.
Fox squirrel (*Sciurus niger*)	Statewide.
Gray squirrel (*Sciurus carolinensis*)	Statewide.
Southern flying squirrel (*Glaucomys volans*)	Statewide.
Woodchuck or ground hog (*Marmota monax*)	Mostly occur north of the Fall Line.

RICHARD T. BRYANT/ARISTOCK

DR. JOE CLARK

Alabama is blessed with a variety of non-game wildlife including species like the southern flying squirrel (left), red bat (top right), and black bears. Bears have the smallest young in comparison to the size of the mother of any mammal. Black bear cubs measure about 8 inches in length and weigh 8 to 12 ounces when born.

incorporate these features into the forest landscape.

In some instances non-game species that are of personal interest may also have a high conservation priority, like rare, unique, or endangered species. In these situations, the ADCNR Natural Heritage Section (phone: 334-242-3484), Non-game Program (phone: 334-242-3469), or special interest conservation groups (listed in Appendix O) should be contacted. These agencies and organizations may be willing to dedicate management resources to studying and protecting the non-game wildlife that occur on your land.

ALABAMA MAMMALS

Approximately 63 mammals are known to occur in Alabama. Except for those that are hunted or trapped (game species and furbearers), little is known about the status of these mammals. Determining the status of many of these animals poses challenges, because most are small, secretive, and nocturnal. The lack of information has also been compounded by the fact that there have been few mammalogists conducting field research in Alabama over the course of this century.

Much of Alabama's historical mammal habitat has been dramatically changed due to conversion to agriculture. Over time, the more fertile sites remained in row crops or pasture, but less fertile fields became eroded and were abandoned to revert back to woodlands. These areas lack fertile soils and productive vegetation that support dense populations of small mammals. Some non-game mammals have also been adversely affected by poor land management practices, intensive silviculture, vandalism of bat caves, and widespread application of pesticides. Some silvicultural practices, like clear-cutting, benefit small mammals (such as rats and mice) by providing dense cover and food (such as seeds) associated with grasses, forbs, and shrubs. These species occur in high concentrations for several years following a clear-cut. Raptors (such as hawks and owls), snakes, bobcats, foxes, and coyotes are also common in clear-cuts because of these high prey concentrations.

Non-game wildlife are often at odds with various land-use practices. The expanding human population in the Mobile area, which is the area of highest black bear concentrations in Alabama, and an intolerant attitude toward bears, have severely limited the number of these animals in the state. Black bear need large expanses of undeveloped habitat because of their large home range requirements. Bear habitat has become fragmented and reduced in size by various land-use practices and highways to the point that few areas are capable of supporting populations of these animals. The increased development of the Gulf beaches for human use is also rapidly eliminating the habitat of the Perdido Key and Alabama beach mice. Changing uses of land by man affect large and small creatures alike.

BLACK BEARS IN ALABAMA

The American black bear (*Ursus americanus*) was once found throughout North America from Alaska and northern Canada south to northern Mexico. Currently, 16 sub-species are recognized. Bears found in Alabama are considered to belong to *Ursus americanus floridanus*, generally referred to as the Florida black bear. Accurate data on the historical status and distribution of the Florida black bear in Alabama are generally lacking; however, there have been numerous historic references to the animals being "widespread" and "common." It has been reported that black bears once occupied most forested areas in the Southeast, but reached their peak abundance in the expansive forested bottomlands of the Mississippi, Atchafalaya River, and other major river drainages prior to human settlement in the early 1800s. These areas are rich in legend and lore about the bears that roamed the forest and the men who hunted them. Bears were an important source of food, fur, and oil for early settlers. Early accounts of bear hunts by Indians and early European explorers date to the middle of the 1700s.

The Florida black bear is not currently listed as a threatened subspecies under the Endangered Species Act. The status of the Florida black bear in Alabama is unclear and at the time of publishing is classified as "warranted but precluded." Much of the information on Alabama's remaining black bear population is incomplete or unconfirmed. Within the last few years, five counties have reported black bears either from road-kills or sightings. These counties include **Choctaw County** where recent undocumented reports have come from the Isney area (near the Mississippi border) and near Coffeeville west of the Tombigbee River; **Clarke County** near Chastang Hill, southeast of Jackson; **Washington County** between Basset and Armstrong creeks, and on the Boykin Wildlife Management Area; **Mobile County** in the Celeste and Radcliff Road area (between Citronelle and Creola), Big Creek (south of Big Creek Lake), East Fowl River area, and the Grand Bay area; and **Baldwin County** near Tensaw (north of I-65) and Lillian Swamp on the west side of Perdido River on the Florida border.

Because of the uncertain status of black bears in the state, the Alabama Black Bear Alliance (ABBA) was formed to determine the abundance, ecology, and conservation strategies necessary to protect and maintain black bears in the state. The ABBA is a broad-based cooperative effort lead by conservation organizations like the Alabama Wildlife Federation (AWF) and the Alabama Chapter of The Nature Conservancy (TNC), in cooperation with state and federal agencies, the forest industry, agricultural organizations, the academic community, and landowners. Current efforts include AWF's bear report program and TNC's Alabama Natural Heritage Program, whose goal is to develop and maintain a mapping system for tracking bear locations throughout the state. A study of black bear populations and habitat requirements is also underway. The study is being conducted by the Southern Appalachian Field Laboratory, located at the University of Tennessee. Objectives of the work are to trap and tag bears in and around the Mobile River Delta to determine its status, distribution, basic population parameters (e.g. mortality rates, dispersal, home range size), and habitat requirements.

Black bears are the largest non-game species in Alabama and are only found in isolated areas of the state. Current efforts are underway to determine black bear abundance, distribution, and life history in Alabama.

The decline in black bear abundance can be attributed primarily to human disturbance, illegal kills, and habitat loss. Land drainage and clearing for agriculture reduced the original 25 million acres of bottomland hardwoods in the lower Mississippi River valley to 5 million acres by 1980. Because black bears have a low reproductive rate, the effect of illegally killed adults, especially females, is a serious concern. Habitat loss is the major contributor to declining black bear populations, but poaching may be a primary factor limiting recovery.

Description and Life History

Black bears in Alabama are normally black with a brown muzzle and an occasional white blaze on the chest. Average body weights are 150 to 350 pounds for adult males and 120 to 250 pounds for adult females. Body lengths range from 3 to 6 feet from nose to their short tail. Size typically varies depending on the quality and quantity of available food.

Bears are one of the world's most adaptable carnivores. Their intelligence, omnivorous food habits, dexterity, speed, strength, and elusive behavior contribute to their success through evolutionary time.

Female black bears typically begin having cubs at three to five years of age, but females as young as two years of age may produce young in quality habitats. Conversely, females in poor or marginal habitats may not produce young prior to their seventh year. Limited availability of berries and/or mast during the previous year may decrease litter size, or cause the termination of pregnancy. Mating generally occurs in the summer months and egg implantation is usually delayed for about five months. Cubs are born in winter dens in January and February. Although twins are most common, litter sizes can range from one to five. Cubs are born in a helpless state. Bears have the smallest young in comparison to the size of the mother of any mammal. Although measuring only about 8 inches in length and weighing 8 to 12 ounces when born, they develop and grow rapidly. The sex ratio at birth is usually one male to one female.

Mother and cubs leave the den in April or May when the young weigh from 4 to 8 pounds. The cubs stay with their mother through the first year, which includes sharing the next winter's den. They emerge with her again in the spring, and live with her until the summer when the family unit dissolves. Shortly thereafter, the female goes back into estrus, breeds, and repeats the cycle.

Black bears are not true hibernators. They go through a winter dormancy period termed "carnivorean lethargy," or **torpor.** This is primarily an adaptation to survive periods of food shortages rather than to avoid cold winter weather. In mild winters, with residual food sources available, some bears remain active through the winter. Based on past studies, the onset of denning occurs from late November to early January. Activity, movement, and home range generally decrease rapidly during this period as bears enter "pre-dens" or nests, or enter the den where they will spend the winter.

During the winter "sleep," bears do not eat, drink, urinate, or defecate. Waste products are recycled through their unique metabolic and physiological processes. The nitrogen found in urea is converted to protein. As a result, bears leave dens in the spring with more muscle mass than when they entered. Black bears exhibit varying degrees of lethargy while denning, but most can easily be aroused if disturbed.

Denning activity is influenced by a number of factors such as food availability, age, gender, reproductive condition, photoperiod, and weather conditions. Factors contributing to interruption of the denning period or the changing of den sites during a given winter include human activity, flooding, fluctuating extremes in weather conditions, and the lack of concealment of ground dens. Information collected from monitoring denning behavior indicates bears are more active in winter months in the lower Mississippi River valley than at more northern latitudes. No studies have been done in Alabama.

For some bears, usually males, winter inactivity may be nothing more than bedding for a few days or weeks in one area before moving to new bedding sites. Pregnant females usually seek den sites that are more secure and inaccessible than those typically selected by males. When available, females prefer large, hollow trees, as these provide dry, secure, and well-insulated cover. They also den in brush piles and thickets.

Monitoring bear movements in Louisiana's Tensas Basin has revealed that bears are more active during dusk and dawn, although daytime activity is not unusual. Bears often utilize "daybeds" under forested cover. These sites are usually shallow, unlined depressions scratched in soft ground or leaf litter. Mothers with cubs often bed at the base of a large tree. The female sends the cubs up the tree if she senses danger and either climbs the tree with them, remains at the base of the tree or exits the area alone. Sometimes bears will rest above ground in the crown or lower branches of a tree.

Male black bears move much greater distances than females, often covering two to eight times the area of females. Home range sizes vary from year to year, and from season to season, depending on population density, food availability, sex, age, and reproductive status. Home ranges for males may increase during the mating season in summer. Most bears move extensively in fall when foraging to put on winter fat reserves.

Bear activity revolves around the search for food, water, cover, and mates during the breeding season. Some adult males marked in research studies have ranged up to 35 miles from their capture site. Estimates of home range sizes indicate adult males may utilize over 40,000 acres and adult females up to 18,000 acres. One adult male in the upper Atchafalaya Basin in Louisiana ranged over an 85,000-acre area. Older adult males exert social pressure on younger bears, especially during the spring and summer breeding season, forcing them to disperse to other areas.

Results from studies of radio-collared black bears and observation of bear sign document that uncleared drains, ditches, bayous, and river banks are frequently used by bears to travel between woodlots. Findings indicate that theses travel corridors are important to the movements of adult bears and the dispersal of juveniles through agricultural lands, particularly when they are residing in separate tracts of forested lands or in a severely fragmented forest. Drainage ditches lined with trees and brush, even as narrow as 50 yards in width, have been used by bears to pass through open agricultural areas. However, this may be a minimal width for a viable travel corridor. A good rule of thumb is "the wider the better."

Mobility of bears puts them at considerable risk. Young males are especially prone to long-range movements because they are generally not tolerated by older males. Their movements take them to unfamiliar areas, often those inhabited by humans. This leads to conflict situations where the bear is often the loser. Relocation of nuisance bears is not always successful. Bears have a powerful homing instinct and will often attempt to find their way back to familiar territory. They have been documented to travel up to 400 miles from relocation sites. In their attempt to locate familiar surroundings, they cross roads and highways, increasing the chances of being hit by vehicles. Because of the stress and increased human interaction, the probability of survival of relocated bears is diminished.

Food Habits

Although classified as members of the order *Carnivora* by taxonomists, the diet of black bears is primarily plant material. They are not active predators and only prey on vertebrate animals when the opportunity arises. Bears are better described as opportunistic feeders, as they eat almost anything that is available; therefore, they are more typically classified as "omnivorous."

The growth rate, maximum size, breeding age, litter size, and cub survival rate of black bears are all linked to nutrition. These mammals have a relatively inefficient digestive system thereby forcing them to consume large quantities of food. Consequently, much of their time is spent foraging for food. Feeding signs are usually evident in areas of bear activity. Torn logs, broken saplings, clawed trees, and trampled food plants are indicators of feeding activities. Bears utilize all levels of the forest for feeding. They are excellent tree climbers and can gather foods from treetops and vines. Their keen sense of smell helps them locate food sources.

After emerging from dens in spring, bears may initially be in a "semi-fasting" state as they continue to utilize remaining fat reserves. Food is relatively scarce during this period and weight loss is often more rapid than during denning. Succulent vegetation is eaten first. Then other foods such as residual hard mast (acorns, pecans, etc.), agricultural crops, and insects are consumed. With the arrival of summer, dewberries, blackberries, wild grapes, elderberry, soft mast-producing shrubs, persimmon, pokeweed, devil's walking stick, thistle, and palmetto become staples in the diet. In the fall, hard mast is a particularly important source of fat and carbohydrates, providing the necessary fat reserves for bears to enter the denning period in proper health. Bears exhibit their most rapid weight gain during fall; because of this, hard mast is considered a critical food source at this time.

Agricultural crops can be very important food sources throughout the year, especially in areas of extremely fragmented habitat and high bear density. For example, corn is an important forage crop for the large number of bears inhabiting some agricultural areas in Louisiana. Bears readily take advantage of food opportunities provided by man. Besides crops from both commercial and residential plantings, bears often eat garbage and pet foods.

Habitat Requirements

The remaining black bears that exist in Alabama today do so primarily in relatively large contiguous areas of bottomland hardwood habitat. As mentioned earlier, the ingredients of prime black bear habitat include escape cover, dispersal corridors, abundant and diverse natural foods, water, and denning sites. Because bears are habitat generalists, managed and productive forests can reliably provide the essentials of good black bear habitat.

High quality escape cover is especially critical for bears that live in fragmented habitats and in close proximity to humans. The quality of escape cover can be enhanced when slash and vegetative regrowth, resulting from silvicultural practices such as shelterwood cuts, intermediate thinnings, and small elongated clear-cuts, are combined with "natural" understory thickets. This situation not only offers a refuge for bears but also provides additional opportunities for feeding and denning.

Important food items utilized by bears can be encouraged by forest management. Grasses, thistles, blackberries, pokeweed, and certain fruiting vines are common in well-managed stands. Elderberry, devil's walking stick, French mulberry, red mulberry, and wild grapes all benefit from scattered openings in the forest canopy. Rotting wood from decomposing logging slash harbors protein-rich insects sought by bears during most of the year.

Additional foraging opportunities are made available by the maintenance of small, scattered permanent wildlife openings in or adjacent to the forest. Natural vegetation, cultivated grains and forage crops (such as wheat, oats, rye, corn, and clover), and plants found along the edge of forest openings (blackberries, dewberries, pokeweed, elderberry, and devil's walking stick) are beneficial to bears.

Radio-telemetry studies in other southeastern states have documented the use of heavy cover by bears for day beds and denning sites. These are often in hardwood forests that have been logged within the previous five years. Approximately half of the radio-collared bears (mostly males) used brush piles and ground nests for winter dens. Brush pile dens were made next to discarded logs or in thick briar and vine patches. Use of these areas is encouraged if they are near forest cover or if several large trees are left standing after harvest. This is particularly true for females with cubs who are hesitant to venture into open areas without trees as avenues for escape.

Other forest management practices that benefit the black bear include the identification and protection of potential cavity tree den sites, streamside management zones (SMZs), and corridors linking separate tracts of forest. Cavity trees are especially important in bottomland areas. Most females in seasonally flooded habitats use hollow cypress trees over water for winter dens. A forested buffer zone, such as a SMZ, along major waterways and drainages aids in preventing erosion and siltation and also provides critical elements for quality bear habitat. Thick streamside habitats are utilized by bears for resting, denning and travel areas. Corridors linking separate forested tracts are very important for travel and dispersal of bears.

DR. JOE CLARK

DR. JOE CLARK

Top, a wildlife biologist uses radio-telemetry to locate and track marked black bears; bottom, a wildlife biologist checks a bear denning site in a hollow portion of a tree.

Bottomland Hardwood Management for Timber and Black Bears

Large tracts of mature bottomland hardwood forest composed of a mix of tree species will likely provide for black bear needs and will not require intensive management to maintain good bear habitat. Natural disturbances in the form of tree falls and wind storms typically provide sufficient forest openings needed for forage production and cover.

Species composition and management strategies for bears should be dictated by site characteristics, soils, and the management objectives of the landowner. Stand diversity will be greater using an uneven-aged management system, with single-tree selection, group selection, or small patch harvest cuts. Current research indicates that oaks are most successfully regenerated through group selection or small patch harvest cuts.

One objective of bottomland hardwood manage-

JIM LOW

DR. JOE CLARK

Black bears in Alabama are considered to be a sub-species known as the Florida black bear. Unfortunately much of the information on the state's remaining black bear population is unknown or incomplete. Because of the uncertain status of black bears in the state, the Alabama Black Bear Alliance (ABBA) was formed to determine the abundance, ecology, and conservation strategies needed to protect and maintain bears in Alabama. The Alabama Wildlife Federation (AWF) is an active partner in the ABBA. Current efforts include AWF's bear report program and The Nature Conservancy's Alabama Heritage Program that is developing and maintaining a mapping system for tracking bear locations in the state. Top left, leg snares are one method to capture black bears for marking and recording biological information; left, this black bear has been tranquilized for taking measurements and securing a radio-collar for tracking; and right, black bear tracks are an obvious sign of the presence of bears and should be reported to the Alabama Wildlife Federation at 800-822-9453.

ment is to maintain a diverse forest of maximum growing vigor. Optimal bear habitat forest management should stimulate yield from hard mast-producing trees (oaks, pecan, hickories, etc.) and maintain a diversity of foods. Black bears depend largely on fall and early winter mast crops to provide enough fat reserves to survive winter dormancy. Management of timber stands for oaks and other mast-producing species is important for production of high quality hardwood timber and optimum black bear foraging habitat. Maximizing tree vigor and hard mast production will benefit bears as well as other wildlife species that depend on this carbohydrate-rich food source. Maintaining a diversity of age classes, stand types, and vegetative composition within the forest will provide excellent habitat conditions for black bears.

Rotation length for crop trees should be a minimum of 50 years, with 60 to 100 years for hard mast producers. Stand thinnings (intermediate cuts) should be made when economically or silviculturally feasible, preferably in 5- to 15-year intervals. Intermediate cuts should be designed to improve species composition, remove individual trees of poor quality or vigor, promote regeneration of desirable timber species, encourage food production, and create escape and nesting cover for bears and a variety of other wildlife species. Although regular thinnings are compatible with maintenance of suitable black bear habitat, cuts should carefully consider protection of existing and potential den trees. Midstory timber stand improvement can be accomplished in such a manner as to remove less desirable non-commercial species (such as American hornbeam, box elder, Eastern hop hornbeam) while encouraging those desirable to bear (such as mulberry, swamp dogwood, spicebush, and other fruit-producing shrubs and trees).

Natural regeneration through group selection or small patch clear-cuts will enhance regeneration of shade-intolerant species such as oaks. Early successional food plants, such as dewberry, blackberry, elderberry, and pokeberry, also benefit from these harvest practices.

Key Bottomland Hardwood Species for Black Bears

Crop Trees	Midstory	Understory	Planted
Nuttall Oak	Sassafras	Blackberry	Clovers
Willow Oak	Red Mulberry	Dewberry	Grasses (Bahia)
Water Oak	Dogwood	Pokeweed	Wheat
Cherrybark Oak	Persimmon	Elderberry	Ryegrass
Overcup Oak	Black Gum	Green Briar	Corn
Cow Oak	Mayhaw	Devil's Walking Stick	Milo
White Oak	Rattan	French Mulberry	Sugarcane
Shumard Oak	Muscadine	Palmetto	
Sweet Pecan	Wild Grape		
Hickory	Poison Ivy		
Cypress	Pawpaw		
Swamp Tupelo	Holly		
Live Oak	Gallberry		

A critical component of bear habitat in hardwood bottoms is the availability of denning sites, including cavity trees. Present and potential cavity trees should be identified and maintained regardless of other stand management practices. Any tree species having a large enough cavity for a bear should be maintained; however, special consideration should be given to bald cypress and tupelo gum, especially if cavities are visible, having a minimum diameter at breast height (DBH) of 36 inches, and occurring along rivers, lakes, streams, bayous, sloughs, or other water bodies.

Special or key areas should be identified in hardwood bottoms and management strategies formulated to maintain and enhance them. These areas include cypress stands (important for cavities and den sites), thickets for escape cover (such as canebreaks and palmetto), key food sources, and corridors connecting forested hardwood bottoms with other essential habitats.

SMZs should be located adjacent to all major drainage courses, intermittent and flowing waterways. The SMZs should be as wide as possible, based on site evaluation and landowner objectives.

Construction of logging and other roads in hardwood bottoms that provide access to occupied bear habitat should be limited. Controlling vehicular traffic on these roads with gates and closing them to vehicles after completion of logging will limit disturbances to bears.

Some forest openings in hardwood bottoms should be maintained in early successional natural plant species such as dewberry or pokeweed by annual disking, or planted in foods such as corn in the summer and wheat in winter. Emphasis should be placed on integrating forest and agricultural management by utilizing existing adjacent agricultural fields or rights-of-way as permanent openings.

Forest management practices in natural hardwood stands should incorporate guidelines that retain as many trees as possible that are 36 inches in DBH or greater across a tract or management unit. This will insure that large trees are available for avenues of escape and security for mother bears and their cubs, and for future den trees in the stand.

Black Bear Habitat Management Considerations in Hardwood Plantations

- Timber harvests should be scheduled in such a way as to create optimum habitat diversity. Harvests should not be scheduled on adjacent compartments at the same time.
- The size, shape, arrangement, and proximity of timber harvests should be relative to black bear needs.
- Corridors should be as wide as possible and maintained between plantation fields. Corridors should be managed by selective harvesting, favoring hard mast species and cavity tree den sites. The thicker the cover in these areas, the better.
- Streamside Management Zones (SMZs)—Alabama's Best Management Practices for Forestry

guidelines should be followed, with only selective harvesting of timber allowed within the SMZ. Den trees should be identified and protected in SMZs as well as on other silvicultural sites.

- In intensive, short rotation plantations, coppice (stump sprout) regeneration should be used when feasible to regenerate within one year of harvesting. Intercrop with grains (soybeans, wheat) when possible for at least the first year.
- Females with cubs typically select the base of the largest tree in the vicinity for their day bed site. Leaving a few clumps of large standing trees in each compartment will increase use of these areas by mother bears.

Canebreak Management for Black Bears

A habitat historically associated with the Florida black bear is the canebreak. Extensive stands of switchcane provide bottomland habitat diversity, cover, and a seasonal food supply. Less abundant than in the past, switchcane habitat should be favored in hardwood forest stands when managing for the Florida black bear.

On selected areas where lack of cover is deemed to be a limiting factor, natural regeneration of existing stands of cane should be encouraged to expand through the removal of overstory trees and clearing of competing vegetation. On suitable sites, especially abandoned agricultural fields and public transportation and utility rights-of-way, opportunities exist for artificial regeneration of switchcane. If seeds are available, direct seeding is the favored method of propagation. Another effective method of artificial regeneration, although potentially labor intensive, involves transplanting swithcane.

Switchcane life history traits, including its periodic seeding events and its susceptibility to decline from frequent prescribed burns, intensive cattle grazing, and agricultural clearing, may justify special management procedures for expansion of this habitat type. Areas with existing stands of switchcane deserve protective measures to conserve this valuable component of quality black bear habitat.

Upland Pine Management for Timber and Black Bears

When upland areas are managed intensively for pine production, managers typically use an even-aged management strategy for regeneration of pine stands. Even-aged silvicultural activities can include one of two basic methods: leaving seed trees for natural regeneration, or harvesting trees followed by some form of site preparation for artificial regeneration through seedling planting. Management activities should be dictated by site characteristics, soils, and objectives of the landowner.

Silvicultural practices that impact bear habitat under intensive pine management regimes include regeneration area size and shape, stand age-class distribution, stand thinning, prescribed burning, SMZs, denning and bedding sites, and permanent openings and roadsides. The perimeter shape of timber harvest areas should be as irregular as possible, as determined by site topography, to promote edge habitat and the "edge effect." Although there is no ideal size of regeneration cuts, size should be limited to as small an area as is economically feasible.

To create maximum diversity between stands, there should be at least seven years difference in age classes between adjacent regeneration areas. High between-stand diversity will help ensure a constant supply of soft mast within a relatively small area.

Even-aged pine stands should be thinned as soon as economically feasible (typically by 15 years). This practice will allow sunlight to penetrate to the forest floor and encourage soft mast production and growth of vigorous herbaceous vegetation.

Burning in pine stands should be conducted on a three- to five-year rotation, depending on site conditions. Poor soils and dry sites should be burned less often. Planted pine stands should be burned as soon as practical (7 to 10 years after establishment) and after intermediate thinnings. Hardwood areas should always be protected from fire.

SMZs in uplands should be retained in hardwoods to provide habitat diversity, hard mast production, and travel corridors for bears. Alabama BMPs and site characteristics should be employed during layout of SMZs. If included in the timber base, SMZs should be wide enough to be a silviculturally manageable stand separate from adjacent pine stands. Fire should be excluded from these areas. Herbicides should only be applied in a SMZ on a selective and individual tree basis; otherwise, they should be discouraged.

Mature hollow or cull hardwoods should be left along drainages in upland pine SMZs and travel corridors for denning sites. These trees should be marked and protected during management activities. When feasible, logging slash and treetops should be left for bear bedding areas and for foraging sites for insects, invertebrates, amphibians, reptiles, and small mammals. Deadfall trees should also be left undisturbed for bear insect-foraging sites.

Permanent openings and wide roadsides in upland pine stands should be maintained to promote the production of soft mast. These areas should be burned, bush-hogged, or disked on a three- to five-year cycle. Plantings of corn or clover in summer, and small grains and clover in winter, are beneficial. Vehicle access should be restricted to reduce human disturbance.

Mixed Upland Pine-Hardwood Management for Timber and Black Bears

Upland sites comprised of a mixture of pine and hardwood tend to be managed differently than pure pine stands. Several different types of hardwood trees (oak, cherry, hickory, sweetgum, beech) can be found intermixed with pine. The age classes of the trees may be the same or vary, depending on when previous cuttings took place and what was harvested. These stands can provide good habitat for black bears and still meet landowners' timber objectives.

The biggest challenge in maintaining a mixed pine-hardwood site is managing the lower-canopied or smaller trees in a forest stand. These trees are often the hardwood component of a mixed stand. Since acorns and other hard mast are vital food sources for bears in the fall, forest management activities should favor oaks or other hard mast species as the stand develops.

Harvesting of trees should result in openings large enough that soft mast will be produced as a food source for bears. Normally, as groups of trees or "patches" of the overstory are removed, these provide openings and sunlight that stimulate the growth of fruit-producing plants that are valuable for bears such as dewberry, pokeberry, blackberry, and French mulberry. After a few years these openings become nearly impenetrable thickets. Although availability of natural foods declines over time as openings mature, bears will utilize these sites as denning areas because of the thick cover provided. Also, rotting logs, stumps, and logging slash from timber harvests provide good sources of grubs, insects, and beetles for bears.

If large openings are made at the time of harvest, the landowner should consider leaving some groups of large trees scattered across the tract for sows with cubs to use as a means of escape. Large cavity trees should also be maintained for den sites.

Cypress and Tupelo Management for Black Bears

Management of cypress and tupelo stands is integral to the restoration of the black bear. These stands are generally found in close proximity to bodies of water and have been shown to be a crucial habitat for bears. Bears use this habitat type primarily for escape cover and denning sites. All cypress and tupelo adjacent to water that are 36 inches in DBH (diameter at breast height) or larger, with visible signs of defects, should be protected.

DR. JOE CLARK

Seemingly impenetrable titi thickets, tupelo gum and cypress swamps, as pictured here, and marshes provide refuge for black bears.

Management of both cypress and tupelo stands is similar with a few minor differences. Regeneration generally occurs from stump sprouts of trees up to 14 inches in DBH if sufficient sunlight is available; however, cypress regeneration from sprouts is not reliable. Where there are sources of cypress seedlings they should be hand planted where possible. Tupelo sprouts at higher rates than cypress, so care must be taken to prevent harvested stands from becoming dominated by tupelo. Because a large amount of sunlight is needed to stimulate stump sprouting and sprout survival, harvesting these stands is the preferred method of regeneration.

Regeneration from seed is usually more complex. Both species produce adequate to excellent seed crops yearly, with abundant crops every three to five years. Tupelo is the most consistent and prolific of the two. Cypress needs exposed, wet soil for germination and continued moisture for about two to three years for seedlings to become established. Both species will experience early mortality if they are submerged for any length of time during the first growing season.

Flooding is a constant threat to first-year seedlings as they cannot withstand submergence for any period of time during the early growing season. Once established, growth is vigorous during the early years. Merchantability is usually reached within 20 years, depending upon site quality.

Vigorous growth in dense stands causes natural pruning to occur early on, producing good quality bole development. Thinning should begin by age 20 if markets are available. Thinning to 70 to 80 trees per acre will maintain high rates of growth and bole quality. Subsequent thinnings should be conducted at 10-year intervals to remove poorer quality trees, taking care to retain potential den trees.

Natural stands, where little information is known concerning age and growth, can be maintained at higher stand densities than artificially regenerated stands. Natural stands may be maintained with few or no harvests until the trees reach a marketable size. When the time comes, total harvest is the desired method of regeneration. For periodic income from these natural stands, and to shorten the time to reach a desired diameter, selective harvests may be conducted, removing the poorer quality trees and maintaining a density of 70 to 100 trees per acre.

Retention of small, isolated groups and individual trees of cypress or tupelo within other stands of bottomland hardwoods can be of benefit to bears. When harvesting for regeneration purposes, consideration should be given to these isolated stands or individual trees to promote regeneration and provide escape cover for bears. In areas where water is present much of the year, careful consideration should be taken when harvesting trees. Removal of all trees in permanently flooded areas will almost inevitably lead to conversion of forested wetlands to open water. This is due to regeneration being hindered by the constant presence of water.

When managing stands of tupelo and cypress in occupied bear habitat, loggers should exclude from harvesting trees with visible cavities or defects in the top or trunk, and these trees should also be protected from logging damage. Bears will select these trees for denning sites and escape cover.

Agriculture and Black Bear Management

While forested lands provide optimum bear habitat, agricultural lands can also be managed to enhance overall bear habitat. Use of various habitat management techniques on agricultural lands next to or interspersed with forested tracts can serve to improve and expand occupied habitat.

Agricultural habitat management practices beneficial to bears could be as simple as crop selection or as intensive as the development of wildlife corridors or even the total conversion of agricultural land to hardwood trees. The habitat management options chosen will depend on both the site characteristics and objectives of the landowner.

Habitat quality for bears in agricultural areas can be enhanced by both crop selection and location. Crops such as corn, sugarcane, and winter wheat benefit bears more than soybeans or cotton, not only as forage, but also for cover. Locating preferred crops adjacent to forested lands and travel corridors helps maximize those benefits, as does leaving a portion of the unharvested crop in the field. This benefits bears as well as other wildlife species.

All pesticides and herbicides should be used in accordance with label guidelines and state and federal regulations. Application of chemicals to crops adjacent to forested tracts or travel corridors should be done so that adjacent wildlife habitat is not harmed. A buffer adjacent to forested lands may be left unsprayed, as various plant species in wooded areas adjacent to cropland provide food and cover and could be damaged by drift or inadvertent application. No chemical labeled as harmful to large mammals should be applied to cropland within occupied or potential bear habitat.

For farmers participating in acreage reduction programs, set-aside acreage can be located or used in a manner that provides beneficial wildlife habitat. Set-aside acreage serves to increase availability and diversity of habitat for a wide array of wildlife species, from songbirds to black bear. At least 50 percent of the set-aside lands (not to exceed 5 percent of the crop acreage base) must be planted to an annual or perennial cover crop by November 1 of each year. Locating set-aside acres next to forested lands, ditches, or sloughs can provide additional wildlife habitat or serve as corridors connecting two or more forested tracts. If allowed to go fallow, the land not devoted to a cover crop will develop a cover of native vegetation. This cover, if located next to ditches or in sloughs that join two forested tracts of land, can serve as a wildlife corridor.

Landowners may opt to develop corridors by leaving farmland idle and letting it revert to native vegetative cover. When managed properly, sloughs and lands adjacent to drainage ditches can provide suitable corridors to allow movement of bears among fragmented tracts of otherwise suitable habitat. To allow for adequate cover, these areas should be as wide as possible. If access to drainage ditches is required for periodic maintenance, the corridor could be located on one side of the ditch, leaving the other side open for maintenance and access.

Food Plots and Black Bears

Food plots developed within forested habitat for game species can be beneficial to black bears. If maintained, food plots within a forested tract should be distributed in a manner that minimizes forest fragmentation. For example, they could be grouped fairly close together and close to forest borders rather than distributed evenly throughout the interior.

Black Bears in Alabama

Dr. Mike Pelton is a Professor of Wildlife Science in the Department of Forestry, Wildlife, and Fisheries at the University of Tennessee in Knoxville. For the past 29 years his work has focused on carnivore and black bear ecology in the Southeast, making him internationally known for his work on the black bear. Dr. Pelton describes important black bear management considerations in his own words.

Black bears generally require large continuous forests between 150 to 400 square miles in size with small openings scattered across the landscape. In Alabama the largest known black bear population is in an expansive tract just north of Mobile. Although black bears are considered omnivorous, they are primarily vegetarians, consuming preferred foods such as acorns in the fall, and an assortment of cherries, pokeweed, blueberries, huckleberries, blackberries, and raspberries in the summer. Silvicultural practices that create forest openings are ideal for stimulating berry production, making these areas key summer feeding sites for bears. Large tracts of land that provide adequate acorns in the fall, and plentiful berries during summer, are ideal habitats for the black bear. Black bears are not aggressive predators, although they will consume animal matter such as yellow jackets, bees, and bee nests.

Adequate denning sites are an important habitat component for black bears. Fortunately, they are quite adaptable in terms of denning strategies. Because adult females are particularly vulnerable during the winter, habitat management strategies should include retaining and preserving large trees that provide cavities for denning. Females will use both live upright tree cavities and large logs lying on the ground. Large root masses of fallen trees, and the holes that they make in the ground, are also used for denning. If suitable tree cavities are not available, black bear females will often build a large nest in thick areas that provide protection such as slash or dense vegetation found in newly regenerated forest stands.

MIKE SCOTT

Black bear cubs are born in winter dens during January and February. Twins are most common, although litter size can vary from one to five. Cubs will stay with their mother through their first year.

Access management is also important in maintaining black bear populations. Limiting black bear exposure to people can reduce black bear–vehicle collisions and discourage poaching. Access can be limited by designating areas as refuges for black bears, protecting areas that are naturally inaccessible such as thick swamps, and establishing gates on secondary roads. These inaccessible areas need to be connected by travel corridors to provide cover for bears as they move across the landscape. Corridors facilitate bear movement for feeding, breeding, and dispersal of young who attempt to establish their own territory as a result of being forced out by older bears. Corridors should be wide enough to provide continuous cover and concealment as bears move across the landscape. They should not be interrupted by development or other activity.

Opportunities do exist to reintroduce black bears to other parts of Alabama, particularly on large tracts of private forest land. Industrial forest holdings may offer some of the best opportunities for reintroduction. For black bear reintroduction to be successful in Alabama, public support is essential.

Landscape Management for Black Bears

The goal of landscape management for the black bear is to coordinate habitat management efforts among multiple landowners. Because of the large home ranges of bears, suitable habitat usually cannot be maintained on any one ownership. Landscape management works through a cooperative approach whereby various landowners and user groups work together to promote bear management over a large area. Objectives of landscape management for black bears should include the following:

- Preventing further habitat fragmentation,
- Establishing corridors between existing fragmented habitat,
- Integrating management among tracts to effectively use fragmented resources, and
- Focusing efforts of diverse user groups toward common management objectives.

Successful management at this level is dependent on mutual cooperation and partnerships among landowners. With willing participants, the kinds and extent of management guidelines vary based on considerations such as these:

- Size and shape of individual tracts;
- Landowner objectives and attitudes;
- Number of landowners;
- Forest type and condition;
- Habitat suitability;
- Possibility for corridor development;
- Management strategies already in place;
- Extent and location of non-timbered lands;
- Recreational objectives and support of user groups; and
- Willingness of individual landowners to trade specific objectives for wildlife benefits, if necessary, for the management of the overall area.

Resource managers and private landowners should be aware that, depending on the area or regional conditions, coordinated landscape management may offer the best opportunity for bear management and restoration. The Black Bear Conservation Committee, which operates in Louisiana and Mississippi, has applied the concept of landscape management to bear restoration by forming Bear Management Units (BMUs). Each BMU is administered by a team made up of landowners, agency personnel, and local leaders. The program is entirely voluntary and landowners provide input from planning through implementation of the BMU Management Plan. Landowners interested in establishing a BMU plan can contact the Alabama Black Bear Alliance, 800-822-9453.

Human/Bear Conflicts

Black bears are usually non-aggressive, reclusive animals that are harmless to people unless provoked or threatened. Both humans and bears can coexist peacefully when suitable quality habitat is available and humans modify activities that encourage conflicts. Nevertheless, conflicts between man and bear are inevitable, even if bear numbers are low. The number of conflicts involving the black bear can be expected to increase as restoration efforts proceed.

Reported conflicts between man and black bear in Alabama have been few in recent years, primarily due to low bear densities and the relative isolation of bears from humans. Problems in other areas of the Southeast have included damage to apiaries and crops such as corn, wheat, oats, watermelon, and sugarcane. When compared to other types of agricultural and property losses, those caused by black bears are relatively small; however, some kinds of damage (such as damage to bee hives) can be locally severe.

Black bears have also damaged wooden structures constructed of pressure-treated lumber in parts of its range. Some deer hunters have found the wooden supports of tower stands gnawed so extensively that they were rendered unsafe for use. Bears have also caused extensive damage to wooden signs and out-buildings in some areas. Presumably, chemical salts used to preserve the wood were the main attractant.

Other types of conflicts involving humans and black bears occasionally arise. For example, bears may eat corn and other grains from feeders used by hunters to attract deer. In addition, bears will sometimes wander through or near residential areas, an activity perceived by many people as a threat to human safety. Bears may also become a nuisance, especially when they forage in garbage dumps, cabins, and campsites.

In some states, particularly in the West, black bears can be significant predators of livestock, but this is not common in the East. Livestock predation is not presently a problem in the range of the Florida black bear but may become more frequent if restoration efforts succeed. Though rare, unprovoked attacks on humans in other parts of North America have also been documented. To keep this in perspective though, a person is about 180 times more likely to be killed by a bee and 160,000 more likely to die in a traffic accident than be killed by a bear. Attacks on humans will probably never be commonplace, due to bears' secretive and retreating nature.

Landowners, agricultural producers, and other wildlife resource users have shown a high tolerance for bear damage. For example, many farmers have accepted minor damage to crops as a normal part of farming.

Hunters have bear-proofed stands by placing tin guards around supports or by substituting steel or aluminum in place of wood. Finally, people who see bears near their homes increasingly view the experience as positive. These individuals willingly share their information with wildlife professionals and expect no corrective action to be taken. All of these actions are the result of increased public awareness of the intrinsic value of the black bear.

Continued public education relative to the management of human/bear conflicts is an important key to the future of the black bear in Alabama. Many people in the state are unaware of the bears' presence and are ill-equipped to address problems that arise. The public needs to be provided with factual information about black bears so that human/bear conflicts can be acceptably resolved.

Solutions to Human/Bear Conflicts

Successful restoration of the black bear is partially dependent on immediate and effective responses by wildlife professionals to reported conflicts. Black bears may ultimately be destroyed by individuals who are unaware of available professional solutions to simple problems, who feel that no effective solution for their particular conflict exists, or who think that no one cares. This was the primary mode of addressing human/bear conflicts in the past, a practice that unfortunately still occurs today (such as reports of bear being shot while eating corn in deer feeders). Informing the public about potential conflicts and available solutions is an important strategy in the overall restoration effort.

One goal of black bear management is to promote the natural establishment of a sustainable black bear population in suitable habitat. Consequently, conflict resolution will rely heavily on non-lethal damage control techniques, such as barriers and resource management strategies. Due to low bear densities, destruction of offending animals will only be considered if all other measures have failed, unless human health and safety is jeopardized.

Live trapping and releasing bears into the same general area after aversive conditioning may alter offensive behavior and resolve some conflicts. Releasing trapped animals far from their capture site, however, may cause them to roam over large areas, presumably in search of familiar surroundings. This increases their susceptibility to being killed by vehicles along roads or by humans who perceive a threat to their own well-being. Consequently, bears involved in conflicts with humans should be left in their established territory whenever possible.

Barriers may totally eliminate some problems and offer the greatest immediate relief from conflicts. Barriers, in most cases, are both economically and technically feasible to install and are considered a viable option for controlling many types of bear-related damage.

Management of the resources being damaged or threatened is also applicable to effectively managing bear/human conflicts. In some cases, conflicts may be avoided by keeping susceptible resources away from bear habitat or by removing attractants that lure bears to those resources.

Ideal management plans should emphasize conflict prevention and, when problems arise, the implementation of practical solutions. The following general guidelines and recommendations are presented in support of strategies that can be used to address conflicts involving the black bear.

APIARY PROTECTION

Black bear damage to bees and beehives is the most economically important agricultural problem associated with the Florida black bear in the Southeast. A bear that encounters an unprotected commercial apiary can destroy or badly damage dozens of hives in just one night. Losses to some beekeepers can be a significant financial burden, especially when several apiaries are managed within the home range of a bear that has become a habitual beehive robber. In some cases, individual beekeepers have reportedly sustained as much as $10,000 in damages. Beekeepers, therefore, are wise to initiate damage prevention strategies that preclude or minimize bear damage.

Some bears are especially fond of larval bees and honey and will actively seek out hives in their habitat. Consequently, beehives should be located as far as possible from timber and brush providing bears with cover and travel routes. Honey crops should be harvested as soon as possible after the spring, summer, and fall nectar flows in order to 1) reduce the attractiveness of hives to foraging bears, and 2) prevent the loss of the new honey crop in the event of depredation. When possible, apiaries should be moved to new locations if bear activity is detected nearby. To minimize possible damage to hives and prevent bears from establishing bad habits, apiaries in occupied habitat should be protected using electric fences; bear-resistant platforms; or, with the help of an authorized wildlife professional, aversive conditioning of bears. Fences can also be used to control ongoing damage. Compact apiaries are easier to protect with bear-resistant fencing than those scattered over a larger area. Beekeepers should consolidate hives to form the smallest apiary that can be practically managed.

Plans for various types of bear-resistant fences and

DR. JOE CLARK

DR. JOE CLARK

Black bear damage to bees and beehives is the most economically important agricultural problem associated with black bears in the Southeast. Some bears are especially fond of larval bees and honey and will seek out beehives in their home range.

other types of damage control information can be obtained from the offices of the U.S. Department of Agriculture Animal Damage Control and the Alabama Cooperative Extension System.

CROP AND LIVESTOCK PROTECTION

Some field crops and livestock may occasionally provide food for bears. Like apiaries, crops and livestock should not be located in or close to occupied bear habitat unless the owner is willing to accept occasional losses. Additionally, both should be inspected frequently so that any damage can be discovered quickly, and preventive control measures can be implemented. Gardens, small fields, and pastures should be protected with bear-resistant fences if bear damage is anticipated. Farmers should harvest crops as quickly as possible and consider planting crops that are not attractive to bears.

Gas exploders, noise-making pyrotechnics, strobe lights, electronic sirens and noise generators, guard dogs, and scarecrows can be used to temporarily repel bears from fields or pastures susceptible to damage; however, long-term and repetitious use of these devices may render them ineffective. Harassment with chase dogs may offer some short-term relief from bear problems, but this activity is illegal under most circumstances.

Intensive herding practices can lessen the chance that bears will prey on livestock. Carcasses of dead animals should be hauled to an approved landfill or destroyed by deep burial or incineration to prevent bears from scavenging near susceptible crops and livestock.

PROTECTION OF STRUCTURES

Use of alternative construction materials should be considered where bears regularly damage wooden structures. Steel, aluminum, fiberglass, and other durable materials can be used instead of treated wood to build and/or shield deer stands, signs, and other structures. Foodstuffs that may attract foraging bears should be removed from unoccupied buildings and stored in bear-resistant containers. This includes human and pet food and food items commonly used to attract wildlife, such as molasses and molasses blocks, salt blocks, corn, wheat, and other grains. Bear-resistant doors, window shutters, and fences can also be used to protect unoccupied buildings.

HUNTER EDUCATION AND COOPERATION

Identification, behavior, and management of black bears should be discussed in formal hunter education programs. Likewise, wildlife professionals should promote bear conservation when working with the media, hunters, and other outdoor enthusiasts.

Hunting clubs controlling property in occupied habitat should incorporate bear awareness programs into their annual list of organizational activities so that members can learn facts and dispel myths (such as the mistaken belief that bears and deer cannot live in the same area). Clubs should police their wildlife resources and report any wildlife violation.

Baiting of deer is illegal in Alabama and supplemental feeding during the off-season should be discouraged where the ranges of bears and deer overlap, unless hunters are willing to accept foraging bears at their feeders. This simple action would prevent attracting bears to areas frequented by hunters who may be uninformed about bear behavior and who may kill a bear that is a perceived problem.

Hunting dogs may chase bear instead of legal game in some areas. For this reason use of dogs in occupied habitat should be controlled. Running of dogs outside the hunting season, particularly in late spring, can adversely impact sows and cubs by contributing additional stress. Control of free-ranging dogs in occupied habitat could also reduce the likelihood of potentially dangerous confrontations between dog handlers and bears. In some areas, a program to control feral dogs may be warranted.

ACCESS MANAGEMENT

Unlimited access by humans to occupied habitat increases the chance of contact between man and bear. These contacts may lead to the harassment or destruction of bears under certain conditions. Managing access to bear habitat is one strategy for minimizing this occurrence.

On most privately owned land that provides habitat for black bears, some type of access management program has already been implemented by the landowners or their lessees. For example, most properties are legally posted and patrolled to prevent trespass by unauthorized individuals. In some cases, access points to private property are controlled with a system of locked gates and limited-use roads. Access to abandoned roads is often further restricted with gates and other barriers such as trenches, mounded dirt, and felled trees. Landowners wanting to develop an access management plan for their property can use these same techniques to their advantage.

MINIMIZING ROAD-KILLS AND OTHER HAZARDS

Bears may be killed by vehicles when they wander onto highways, especially those roads that traverse travel corridors historically used by bears. Highway and natural resource management agencies should work to identify these corridors and install culverts crossing under roadbeds. Drift fences can direct bears to culvert entrances and facilitate movement beneath the roadbed.

Collisions between bears and vehicles may also result in human injury or death. Some accidents could possibly be avoided if they were anticipated by informed drivers. Informational billboards and brochures, bear crossing signs, and reduced speed limits at appropriate locations could be used to alert drivers to the potential presence of bears along certain highways.

Electrocution will probably never be a significant cause of bear mortality, but there have been instances where bears have been killed when they touched energized wires while climbing power poles. Electric companies operating in occupied habitat are encouraged to modify poles and conductors so that potential adverse impacts on bears are minimized.

MANAGEMENT OF HUMAN BEHAVIOR

Black bears are normally non-aggressive animals and unprovoked attacks on humans are uncommon throughout the species' range and particularly rare in Alabama. Most attacks on humans have occurred when bears are surprised, cornered, or otherwise threatened. Interaction between humans and out-of-place or "travelling" bears is discouraged because all bears are potentially dangerous and can inflict serious injury. People should avoid bears in all situations where they are encountered.

Bears that are tolerant of human activity may become aggressive, especially if food is expected to be a handout. Feeding bears is not recommended in any situation. Bear cubs should never be approached by humans. A sow with cubs is defiant, defensive, and dangerously aggressive if she perceives that her young are threatened. Even if a person sees a cub who seems to be alone in the woods, the knowledge that the sow is likely to be close by and easily alerted should signal potential danger.

Likewise, thinking ahead can help prevent other types of bear encounters. Hikers should be as noisy as possible in bear habitat, especially when fresh bear sign is encountered. When camping, food and other attractants should be stored far from tents and other sleeping areas. "Friendly" bears should not be tolerated and should be reported to an appropriate wildlife professional as soon as possible.

In a confrontational situation, humans should identify themselves by making noise and moving upwind of the bear. People should remain calm and retreat as soon as possible, especially if cubs are present. Bears that confront humans will often rear on their hind legs to get a better view or smell; this is a non-aggressive behavior. Humans should not climb trees to escape from black bears, which are agile climbers. Natural barriers that block bear movement and allow quick escapes should be used instead.

In seriously threatening situations, bears can be sprayed in the face with capsaicin-based "pepper mace," available in small pressurized canisters that can be easily carried in a pocket, pack, or belt holster. Pepper mace is a strong irritant that reportedly debilitates aggressive bears by burning the mucous membranes of their eyes, nose, and throat. The effects of the spray are temporary and provide an opportunity for a person to move away from a bear to a safer location. Individuals who work or recreate in bear habitat should consider carrying this product if the likelihood of a bear encounter is high.

GARBAGE AND LANDFILL MANAGEMENT

Garbage management is an integral part of managing human/bear conflicts. Bears that obtain meals regularly from landfills may soon become a nuisance because they depend on man for their food. "Garbage dump" bears can begin feeding in dumpsters and residential garbage cans when landfills and dumps no longer supply a dependable source of food. These bears are more likely to be destroyed because this undesirable behavior is almost impossible to change.

Landfills located in occupied habitat should be managed to discourage bears from using them as a food source. When possible, the perimeter of landfills should be enclosed within bear-resistant fences. Additionally, landfill operators should only maintain a small area of exposed garbage and completely cover it with a deep layer of dirt two or more times a day. This reduces odors and makes it difficult for bears to feed. An aversive conditioning program, developed with the assistance of wildlife professionals, should be implemented by landfill operators if problems persist.

Homeowners and campers in or near bear habitat should avoid attracting bears by dumping wastes in closed containers located away from their homes or campsites. Waste collection companies should provide bear-resistant dumpsters in areas where bears are a potential problem. Hunting clubs in bear habitat should remove game carcasses and entrails from skinning sheds as soon as possible. This material should either be hauled to approved landfills or remote dump sites, deeply buried, or completely incinerated. Skinning sheds and other food handling areas should be kept immaculately clean to prevent noxious odors from attracting bears.

Organizations Dedicated to Black Bear Conservation

ALABAMA BLACK BEAR ALLIANCE

The Alabama Black Bear Alliance (ABBA) is a nonprofit conservation consortium formed in 1997 to determine the abundance, ecology, and conservation strategies necessary to protect and maintain black bears in the state. The ABBA is a broad-based cooperative effort composed of conservation organizations like the Alabama Wildlife Federation (AWF) and the Alabama chapter of The Nature Conservancy (TNC), state and federal agencies, the forest industry, agricultural organizations, the academic community, and landowners. Current efforts of ABBA include the AWF's bear report program and TNC's Alabama Natural Heritage Program to develop a mapping system for tracking bear locations throughout the state. A study of black bear populations is presently underway in and around the Mobile River Delta of Alabama. Objectives are to determine bear distribution, obtain estimates of basic population parameters, and evaluate habitat needs. For more information about the ABBA, write the AWF, 46 Commerce Street, Montgomery, Alabama 36104, or call (800) 822-9453.

THE BLACK BEAR CONSERVATION COMMITTEE

The Black Bear Conservation Committee (BBCC) represents a cooperative effort of landowners, representatives from state and federal agencies, private conservation groups, the forest industry, agricultural organizations, and the academic community to work together for management and restoration of black bear in the Southeast. BBCC efforts focus on five main areas: 1) habitat management, 2) information and education, 3) conflict management, 4) research, and 5) funding. Since formation of the BBCC in 1990, a workable strategy has been implemented to achieve realistic goals in each of the five focus areas. To inform landowners, managers, and the general public about black bear conservation and current efforts, the BBCC has written a Black Bear Management Handbook and publishes a periodic newsletter. For more information about the BBCC, write the BBCC Coordinator, P.O. Box 4125, Baton Rouge, Louisiana 70821, or call (505) 338-1040.

ALABAMA BIRDS

Historical accounts indicate Alabama at one time had approximately 412 bird species. Of these, three are now extinct, two no longer exist in Alabama but are found elsewhere, and 22 have been sighted but have not been officially documented. The passenger pigeon, ivory-billed woodpecker, and Carolina parakeet are now extinct. The whooping crane and common raven have been extirpated from Alabama, but they do occur in other parts of their natural ranges. Currently there are 363 bird species on the official state bird list.

Each year Alabama's spring is ushered in with songs and brilliant flights from a diversity of birds. Some of these birds are found year-round in Alabama and are called **residents**. Other birds, called **short-distance migrants**, visit the state only briefly to feed and rest before continuing their migration, or spend the entire winter in Alabama (such as sparrows and some waterfowl). These species breed in northern latitudes and winter in the southern U.S.

A third group of birds, called **long-distance or neotropical migrants**, breed in the spring and summer in North America but spend their winters south of the

Alabama is home to a wide variety of birds, approximately 363 species. Each bird has a unique life history, physical characteristics, and habitat requirements. A few of the birds found in Alabama are pictured above. Top left, American Goldfinch; top middle, red shouldered hawk; top right, red-bellied woodpecker; middle left, pileated woodpecker; middle right, killdeer; bottom left, yellow crowned night heron; and bottom right, indigo bunting.

Checklist of Breeding Birds and Preferred Habitats

EARLY SUCCESSIONAL STAGE[1] FOREST

Bachman's sparrow (*Aimophila aestivalis*)—R
Bachman's warbler (*Vermivora bachmanii*)—NTM. Extremely rare, possibly extinct. Federally endangered species.
Blue grosbeak (*Guiraca caerulea*)—NTM
Brown-headed cowbird (*Molothrus ater*) SDM (nest parasite of other birds)
Common yellowthroat (*Geothlypis trichas*)—NTM
Dickcissel (*Spiza americana*)—NTM
Gray catbird (*Dumetella carolinensis*)—NTM
Indigo bunting (*Passerina cyanea*)—NTM
Painted bunting (*Passerina ciris*)—NTM
Prairie warbler (*Dendroica discolor*)—NTM
Red-winged blackbird (*Agelaius phoeniceus*)—SDM
Rufous-sided towhee (*Pipilo erythrophthalmus*)—SDM
White-eyed vireo (*Vireo olivaceus*)—NTM
Worm-eating warbler (*Helmitheros vermivorus*)—NTM
Yellow-breasted chat (*Icteria virens*)—NTM

MIDSUCCESSIONAL STAGE[2] FOREST

Acadian flycatcher (*Empidonax virescens*)—NTM
Bachman's sparrow (*Aimophila aestivalis*)—R
Blue-gray gnatcatcher (*Polioptila caerulea*)—NTM
Carolina wren (*Thryothorus ludovicianus*)—R
Eastern wood-pewee (*Contopus virens*)—NTM
Hooded warbler (*Wilsonia citrina*)—NTM
Kentucky warbler (*Oporornis formosus*)—NTM
Northern cardinal (*Cardinalis cardinalis*)—R
Pine warbler (*Dendroica pinus*)—NTM
Ruby-throated hummingbird (*Archilochus colubris*)—NTM
Swainson's warbler (*Limnothlypis swainsonii*)—NTM
Wood thrush (*Hylocichla mustelina*)—NTM
Worm-eating warbler (*Helmitheros vermivorus*)—NTM
Yellow-billed cuckoo (*Coccyzus americanus*)—NTM

LATE SUCCESSIONAL STAGE[3] FOREST

American redstart (*Setophaga ruticilla*)—NTM
American swallow-tailed kite (*Elanoides forficatus*)—NTM
American crow (*Corvus brachyrhynchos*)—R. Nest predator of other birds.
Bachman's sparrow (*Aimophila aestivalis*)—R
Barred owl (*Strix varia*)—R. Nest predator of other birds.
Bald eagle (*Haliaeetus leucocephalus*)—R. Federally threatened species.
Blue jay (*Cyanocitta cristata*)—R. Nest predator of other birds.
Broad-winged hawk (*Buteo platypterus*)—NTM
Carolina chickadee (*Parus carolinensis*)—R
Cerulean warbler (*Dendroica cerulea*)—NTM
Common grackle (*Quiscalus quiscula*)—R. Nest predator of other birds.
Cooper's hawk (*Accipiter cooperii*)—NTM
Downy woodpecker (*Picoides pubescens*)—R
Great-crested flycatcher (*Myiarchus crinitus*)—NTM
Hairy woodpecker (*Picoides villosus*)—R
Louisiana waterthrush (*Seiurus motacilla*)—NTM
Mississippi kite (*Ictinia mississipiensis*)—NTM
Northern flicker (*Colaptes auratus*)—SDM
Northern oriole (*Icterus galbula*)—NTM
Northern parula (*Parula americana*)—NTM

[1]Refers to grass, vine, shrub, sapling stage of succession.
[2]Refers to pole (between sapling and small sawtimber) stage of succession.
[3]Refers to sawtimber stage of succession.
R = resident, SDM = short-distance migrants, NTM = Neotropical migrants (long-distance migrants)

United States in Mexico, the Caribbean islands, and other Central and South American tropical countries. Neotropical (New World Tropics) birds are the largest group of birds that migrate into Alabama, comprising 50 percent to 70 percent of the bird species found in the state's forests. They include warblers, vireos, orioles, tanagers, thrushes, flycatchers, hummingbirds, swallows, swifts, shorebirds, and some birds of prey. Unfortunately, over the last decade, there has been a dramatic decline in the number of some neotropical migrants seen each year.

Destruction of native habitat on breeding grounds, wintering areas, and along migration routes is a primary cause of declines in many migrant songbird populations. Habitat loss for these species results from changing land-use practices. Birds thought to require large expanses of unbroken mature forest, called **forest interior species**, have been particularly impacted because most mature forests have been cleared for agriculture, commercial, or residential use or are now in younger forest age classes. Consequently, the remaining forested landscape has been broken into smaller blocks or patches—a process known as **forest fragmentation**. These smaller forest blocks may look like good habitat for forest interior birds, but that can be deceiving. Small patches of forest create edge habitat that is preferred by predators that destroy eggs and young of forest interior bird species. The brown-headed cowbird is a nest parasite that prefers edge habitat, lays eggs in the nests of other birds, and negatively affects nesting.

Some songbirds that require early successional and grassland habitat have also been negatively impacted as a result of changing land-use practices. Conversion of agricultural and grassland habitats to forest may cause a decline of species such as the rufous-sided towhee, prairie warbler, grasshopper sparrow, and Eastern meadowlark. Forest fragmentation and resulting predation/parasitism may be more of a problem in midwestern states that are heavily farmed, and less of a problem in Alabama where large expanses of forest still exist. In general, birds experiencing the largest decline in Alabama are those that require early succession/grassland habitat, not forest interior species.

Partners in Flight: A Cooperative Effort to Save Neotropical Migrant Birds

In 1990 the National Fish and Wildlife Foundation launched Partners in Flight, a comprehensive bird conservation program to stop or reverse declines of migratory bird species. The primary goal of the program is to maintain populations of forest and grassland neotropical birds throughout the Americas through habitat protection, management, professional training, and public education. Partners in Flight brings together more than a dozen federal agencies, 50 state fish and wildlife agencies, 20 conservation organizations, universities, and the forest products industry to work together to enhance migratory bird populations and the vital habitats that sustain them. For more information about Partners in Flight, or to receive a newsletter, write Partners in Flight, National Fish and Wildlife Foundation, 1120 Connecticut Avenue, NW, Suite 900 Bender Bldg., Washington, DC 20036. The newsletter details conservation plans, progress reports, resource listings, and a comprehensive roster of contact people in both government agencies and conservation organizations.

International Migratory Bird Day and the Christmas Bird Count

Each year millions of Americans volunteer their time at the nearest forest, field, marsh, or backyard to observe and count the spring and winter migration of more than 340 species of birds. International Migratory Bird Day is designated to celebrate spring migration and to increase public awareness of the decline of many migrating birds. Bird counts are also conducted each winter on Christmas Day by bird watchers across the U.S. This annual event is commonly referred to as the Christmas Bird Count and serves as an index of bird abundance. International Migratory Bird Day and the Christmas Bird Count are sponsored by many conservation organizations and state wildlife agencies. For more information about these two days, contact the Alabama Department of Conservation and Natural Resources, Non-game Wildlife Program, 64 North Union Street, Montgomery, AL 36130, (334) 242-3861. Readers with access to the Internet's World Wide Web can learn more about International Migratory Bird Day at the National Fish and Wildlife Foundation's web site at http://www.nfwf.org.

Investing in Non-game Wildlife: The Teaming With Wildlife Initiative

For over 50 years, hunters and anglers have paid excise taxes on hunting and fishing equipment to the Sportfish and Wildlife Restoration Fund that was established by the Dingell-Johnson Act (fisheries) and the Pittman-Robertson Act (wildlife). These funds have helped to re-establish wildlife species that were critically low in numbers such as the wild turkey, wood duck, and striped bass. Funds have also been used to purchase and conserve thousands of acres of wildlife habitat not only of benefit to game species but also non-game species. These efforts have provided countless hours of enjoyment for hunters, anglers, and other outdoor enthusiasts.

The Teaming with Wildlife Initiative is an attempt to provide a stable source of funding to state wildlife programs for protection and management of non-game species. It would be similar to current game programs in that it will establish a user fee in the form of a 5 percent surcharge on the manufacturer's price of many outdoor products such as recreation equipment (backpacks, tents, canoes, etc.), photographic equipment (cameras, film, etc.), recreational vehicles, backyard wildlife feeding supplies, and wildlife field or viewing guides. The goal of the program is to raise $350 million dollars annually for non-game wildlife.

The funding initiative provides a way of investing in the future by giving all citizens an opportunity to contribute to conserving the wildlife they deeply appreciate. Three benefits will result from the program. First, it will conserve a wide array of fish and wildlife and their habitats, with an emphasis on preventing species from becoming endangered. Second, it will enhance the recreational opportunities people experience in the outdoors. And third, it will help to foster a responsible stewardship ethic through conservation education efforts. For many years non-game wildlife have lacked equivalent economic research support compared to game species, and what funding has been available for non-game research needs has been drawn from taxes and license fees paid by hunters. This additional outlet for support would go a long way towards a more equitable and balanced arrangement to meet both needs. For more information about the Teaming with Wildlife Initiative, or how you might help, contact the Alabama Department of Conservation and Natural Resources, Non-game Wildlife Program, 64 North Union Street, Montgomery, AL 36130, (334) 242-3861.

ALABAMA REPTILES AND AMPHIBIANS

Alabama has 135 species of reptiles and amphibians, collectively known as **herpetofauna** or **herps** (herp is from the Greek word *herpeton* meaning "creeping thing"). A breakdown of herpetofauna by species includes 3 sirens, 30 salamanders, 28 frogs, 22 turtles (excluding sea turtles and introduced species), 11 lizards, 40 snakes, and the American alligator. The distribution and range of Alabama's herpetofauna correlates closely with the physiographic and natural regions of the state, which suggests that physical features of the landscape play a major role in the occurrence of reptiles and amphibians. A prominent physiographic feature that limits the distribution of about 24 species of reptiles and amphibians is the **Fall Line**, which divides the upland regions from the Coastal Plain. Scientists are unsure of all of the reasons, but changes in the geographic features of the land have clear impacts on the distribution of some animals.

Several aspects of amphibian life history (such as being cold-blooded, having small home ranges, having limited dispersal, and adaptability to aquatic and terrestrial sites) suggest that they may be useful as **indicator species** for monitoring the effects of local environmental disturbances. Land management practices that affect even the small local microclimate and microhabitat features of an area have been shown to impact amphibian populations before they were felt by other wildlife. Generally, forest management practices such as clear-cutting have a negative impact on amphibians and a slightly less negative impact on reptiles. Research is not definitive on the reasons why. Amphibians are most affected by timber cutting because of increased temperature and moisture loss in these areas. Also, amphibians may not be able to disperse as far as reptiles such as snakes and turtles. The threatened Red Hills salamander, found only in Alabama, is one amphibian that has been negatively impacted by forest cutting, which has altered favorable microhabitat and destroyed burrows on steep slopes and coves. Even though they are small in size and easily overlooked, these creatures are very sensitive to alterations.

SUMMARY

Alabama has approximately 560 species of mammals, birds, reptiles, and amphibians that are considered non-game wildlife. Until the early 1980s, little was known about the occurrence, distribution, biology, and land management practices that would enhance the abundance of these species.

Although land managers are trying new and innovative practices for non-game wildlife every day, current woodland efforts consist of two strategies: 1) a multiple species approach that protects special habitat features known to support non-game wildlife (such as riparian forests, upland coves, unique areas), and 2) a single species approach that investigates specific habitat features required by certain non-game wildlife and then incorporates this information into land management programs. With a little planning, non-game wildlife habitat can be a productive component of forest and farm management.

A 27-day-old red-cockaded woodpecker begins its maiden flight from a nesting cavity in a mature longleaf pine. The red-cockaded woodpecker, which is a federally listed endangered species, can be found in some Alabama pine forests that contain mature pine trees 60 years of age or older.

DERRICK HAMRICK

CHAPTER 13

Threatened and Endangered Species

CHAPTER OVERVIEW

Alabama had the fourth highest number of federally threatened and endangered species of any state in 1998, with 68 listed species of animals and 18 listed plant species. Most endangered and threatened species occur on private lands and have some degree of protection under the Endangered Species Act. Its implications for private landowners, as well as innovative approaches to protecting endangered and threatened species are discussed in this chapter.

CHAPTER HIGHLIGHTS PAGE

THE ENDANGERED SPECIES ACT 286

ENDANGERED AND THREATENED SPECIES OF ALABAMA 286

ANIMALS AND PLANTS 287

- Mammals 287
- Birds 289
- Reptiles and Amphibians 292
- Fishes 296
- Mollusks 300

NEW APPROACHES TO PROTECTING ENDANGERED SPECIES 311

The outstanding scientific discovery of the twentieth century is not television, or radio, but rather the complexity of the land organism. Only those who know the most about it can appreciate how little is known about it. "The last word in ignorance is the man who says of an animal or plant: What good is it? If the land mechanism as a whole is good, then every part is good, whether we understand it or not. If the biota, in the course of aeons, has built something we like but do not understand, then who but a fool would discard seemingly useless parts? To keep every cog and wheel is the first precaution of intelligent tinkering."—Aldo Leopold, Sand County Almanac, 1949

The words of Aldo Leopold ring true today as most citizens of the state agree that protecting endangered species is important. However, the issue of how to do this becomes complex and highly emotional when landowners perceive that their rights may be threatened by regulation and economic losses are incurred from land-use restrictions.

The controversy over protecting the northern spotted owl in the Pacific Northwest has highlighted problems with protecting endangered species when economic interests are involved. Closer to home, the red-cockaded woodpecker offers similar challenges in protecting forest owners' rights to use and manage their lands while ensuring the protection and survival of threatened and endangered species. As regulatory agencies, landowners, and conservation groups struggle with this issue, many species continue to decline in number. The success of protecting endangered species will depend upon how well all interested parties work together to develop flexible long-range approaches that protect both endangered animals and landowner rights to use and manage their lands.

THE ENDANGERED SPECIES ACT

The term "endangered species" became a household term in 1973 when the U.S. Congress passed and President Richard Nixon signed the Endangered Species Act (ESA). The act was developed in an effort to reduce the increasing rate of extinction of animals and plants in the United States. The act directs the U.S. Fish and Wildlife Service to identify those species of plants and animals that should be classified based upon scientific evidence as endangered, threatened, or species of concern. **Endangered species** are those that are in danger of becoming extinct unless protected. **Threatened species** are those that are likely to become endangered in the foreseeable future unless conservation measures are taken. While not identified or covered by the ESA, **species of concern** are species that need to be monitored because of imminent threats to habitat, limited range, or because of other physical or biological factors that may cause them to become threatened or endangered within the foreseeable future. **Candidate species** are those species not currently listed as endangered or threatened, but, which may soon be listed as new information is gathered.

The Endangered Species Act of 1973 as amended has 17 provisions. For landowners, the most important of these provisions is **Section 9**, which defines prohibited acts. This section clearly prohibits the possession, import, export, and interstate or foreign sale of listed species. It also is illegal to "take" a listed species from the wild. **Take** is defined as to harass, harm, pursue, hunt, shoot, kill, trap, capture, or collect a listed species. Taking can also include significant habitat modification or degradation that actually kills or injures an endangered species by impairing its survival, which includes disrupting habitats for breeding, feeding, or shelter. If an animal on private property is listed as endangered or threatened, the landowner must avoid a "taking" of that species. Penalties for violations can range from a warning and seizure of illegally held wildlife specimens or products, to civil fines of $25,000, or criminal fines of $50,000 and/or a year in jail for each violation. Plants do not have the same protection as animals under Section 9, except that it is unlawful to "remove and possess any [endangered plant] species from areas under federal jurisdiction; maliciously damage or destroy any species on any area in knowing violation of any state law or regulation, including state criminal trespass law."

Section 7 of the ESA requires federal agencies that manage federally supported programs to ensure that land-use practices do not jeopardize the existence of a listed species or adversely affect critical habitat. **Critical habitat** is an area required for a species' normal needs and survival. Private landowners are not responsible for managing critical habitat unless that habitat is part of a federal action such as a Section 404 wetland permit. Section 7 does **not** require the private landowner to actively manage for a listed species as is required on federal lands.

ENDANGERED AND THREATENED SPECIES OF ALABAMA

Alabama ranks fourth in the nation in the number of federally listed endangered and threatened species.

Endangered Species Act at a Glance

- Enacted by Congress and signed by President Nixon in 1973.
- Preceded by the Endangered Species Preservation Act in 1966, and the Endangered Species Conservation Act in 1969. Since 1973 the ESA has been amended seven times in 1976, 1977, 1978, 1979, 1980, 1982, 1988. Currently the ESA is up for reauthorization by Congress.
- Most powerful law ever enacted to protect animals and plants.
- Its purpose is to conserve the ecosystems upon which endangered and threatened species depend and to provide a program for conserving these species.
- U.S. Fish and Wildlife Service and the National Marine Fisheries Service, in cooperation with state wildlife agencies, are responsible for administering the Act.

At the time of this writing, there are 5 mammals, 4 birds, 6 reptiles, 1 amphibian, 12 fish, 40 invertebrates, and 18 plants that are listed on the federal threatened and endangered species list in Alabama. The high number of endangered species in Alabama, as compared to other states, is probably related to the natural diversity of the state and the large number of mussel species impacted by river impoundments and siltation. With over 300 different soil types and a variety of topography, Alabama has a multitude of unique habitats that are home to numerous rare and uncommon wildlife.

A growing number of states like Alabama have developed programs to help protect a broader spectrum of rare plants and animals. In 1982, the Alabama legislature recognized the importance of native wildlife species that were threatened by habitat loss and passed legislation to create the State's Non-game Wildlife Program. Through this program, species are identified that are in need of special protection or management. These species may or may not be on the federal list.

The following are endangered and threatened species most likely to be impacted by land management practices in Alabama. A complete listing of federally protected endangered and threatened species in Alabama can be obtained from the 1) U.S. Fish and Wildlife Service, Division of Ecological Services, Daphne Field Office, P.O. Drawer 1190, Daphne, AL 36526 (334) 441-5181; 2) Alabama Department of Conservation and Natural Resources, Non-Game Program, 64 North Union Street, Montgomery, AL 36130, (334) 242-3469; or 3) the National Endangered Species Act Reform Coalition (800-457-1013) which maintains lists for all states. State information is also available on the U.S. Fish and Wildlife Service web site: www.fws.gov.

ANIMALS AND PLANTS

Mammals

GRAY BAT (*MYOTIS GRISESCENS*)

Federal Status: Endangered

This small bat uses caves that are normally located within 1 mile of a river or reservoir. Gray bats use warm caves in the summer where they establish maternal and bachelor colonies. In the winter they relocate and hibernate in a few small, cold caves. The gray bat has a wingspread of about 11 to 12 inches and is uniformly dark gray. Gray bats are insect eaters and usually hunt and feed over water.

Gray bats can be adversely affected by forestry operations if their roost sites are disturbed or if wooded corridors that provide them with cover on nightly flights between roosting and feeding sites are removed. Caves that are inhabited by gray bats should be protected by leaving a buffer of undisturbed vegetation around their entrances. Wooded travel corridors between roosting and foraging sites should also be protected. The use of herbicides and pesticides in areas adjacent to foraging and roost sites should be carefully controlled and monitored for unanticipated adverse effects.

Gray bats are known to occur in the following counties: Colbert, DeKalb, Jackson, Laurderdale, Limestone, Madison, Marshall, Morgan, and Shelby. They have also been sighted in Conecuh County and are suspected to occur in other southern Alabama counties.

INDIANA BAT (*MYOTIS SODALIS*)

Federal Status: Endangered

This small bat has fur that ranges from nearly black to

The gray bat uses caves that are normally located within a mile of a river or reservoir. Gray bats use warm caves in the summer where they establish maternal and bachelor colonies. During the winter they relocate and hibernate in a few small, cold caves.

The Indiana bat has fur that ranges from nearly black to light gray or cinnamon on the belly. They hibernate in caves and mines in Alabama during the winter.

Fig. 1: Gray bats are found primarily in northern Alabama, but have also been observed in Shelby and Conecuh counties.

Fig. 2: The Indiana bat is found only in four northern Alabama counties that provide suitable caves for roosting and nesting.

lighter gray or cinnamon on the belly. Individual hairs have dark bases with lighter tips. Its wingspread is about 10 inches. Indiana bats hibernate in caves and mines during the winter. They disperse from their hibernation caves in the spring and form separate male, female, and juvenile colonies. Females form maternal colonies that roost under the loose bark of trees, usually near water. Little is known about where males spend the summer. To date no summer roost sites have been found in Alabama.

The greatest threat land management activities pose to Indiana bats is through the disturbance of hibernating colonies in caves. Although there are large populations of bats, most of them hibernate in only a few sites in the U.S. A buffer area of undisturbed vegetation should be maintained around the entrances of hibernation caves. Harvest of roost trees should also be avoided. If roosting bats are found during timber harvest or site clearing activities, the tree and area around it should be protected until a determination of the bat species is made. Areas that contain Indiana bat roost sites should be managed in a manner that will ensure that an adequate number of roost trees are present, and that a sufficient wooded area is available within the bat's home range to support the colony.

Indiana bats are presently known to occur in Blount, Jackson, Lauderdale, and Morgan counties in Alabama.

Birds

RED-COCKADED WOODPECKER (*PICOIDES BOREALIS*)

Federal Status: Endangered

The red-cockaded woodpecker is a small (7 to 8 inches in length) black and white woodpecker, with **no visible red markings.** It can be distinguished from other black and white woodpeckers by its large white cheek patch and black and white or ladder back. Other small Alabama woodpeckers have either an unstriped white back, a black eye-stripe, or red on the head. The red-cockaded woodpecker, referred to by some as the "RCW", is also the only Alabama woodpecker that makes its home in living pine trees, drilling a round hole approximately 3 inches in diameter through the sapwood and into the heart of the tree. They also peck sap wells, half-dollar-sized wounds that bleed resin onto the tree trunk. The sap encrusted tree stem is often easier to identify than the bird. It can resemble a large wax candle and is easily seen in the open woods the bird usually inhabits. Other woodpeckers and some animals use abandoned red-cockaded dens but often enlarge the entrance. The sap on active trees is clear or amber in color. That on inactive trees turns a dull gray or has the appearance of "icing." Red-cockaded woodpeckers live in small groups in a 1- to 10- acre area called a colony or cluster, using cavity trees for roosting and nesting. A cluster may have from one or two cavities to more than 12. An ideal cluster site is a mature open pine stand with 50 to 80 square feet of basal area per acre. Few, if any, hardwoods should be over 15 feet high. Red-cockaded woodpeckers

The red-cockaded woodpecker is the only Alabama woodpecker that makes its home in living pine trees, drilling a round hole about 3 inches in diameter through the sapwood and into the soft interior of trees infected with red-heart disease.

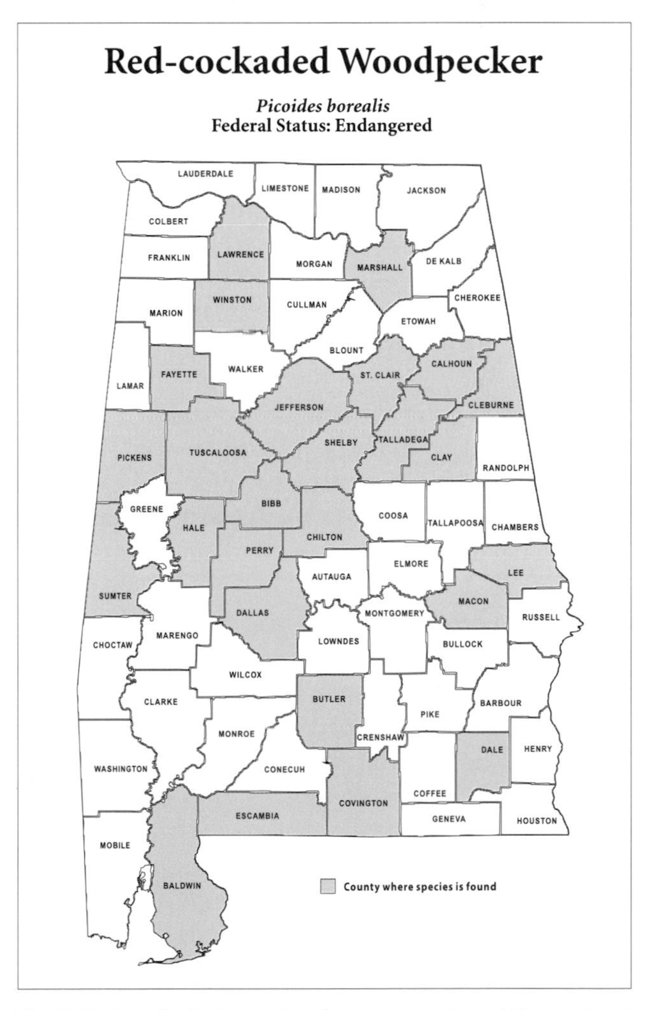

Fig. 3: Red-cockaded woodpeckers are scattered throughout Alabama where suitable habitat exists. The highest densities of red-cockaded woodpeckers are found in the pine belts of the central and southern portions of the state.

Red-cockaded Woodpecker Procedures Manual for Private Woodlands

The U.S. Fish and Wildlife Service is currently developing a "Red-cockaded Woodpecker Procedures Manual for Private Woodlands." For more information about the manual contact the U.S. Fish and Wildlife Service, P.O. Drawer 1190, Daphne, AL 36526, (334) 441-5181 extension 29. An excellent publication, "The Red-cockaded Woodpecker: Information for Alabama Forest Landowners, 1996" is also available from the Alabama Natural Heritage Program, P.O. Box 301455, Montgomery, AL 36130-1455, (334) 242-3033.

forage by prying off the loose bark and feeding on the mites, insects, and larvae underneath, rather than by drilling into dead trees like other woodpeckers.

The red-cockaded woodpecker has three basic requirements: 1) at least some old living pines for roosting, 2) from several dozen to a few hundred acres of pine or mixed pine woods for foraging, and 3) an open understory such as that maintained by fire. The removal of old pines from much of the landscape and the absence of fire from many natural stands have been the major factors in the decline of the red-cockaded woodpecker because of the loss of essential habitat. Red-cockaded woodpeckers prefer mature pine with heartrot disease, which only occurs after pines reach 60 to 80 years of age. The bird has been listed as endangered since the early 1970s and fewer than 4,700 family groups are known to remain across the Southeast in an area comprising about 1 percent of the birds' original range.

When encountered in forestry activities, a determination should be made by an experienced biologist whether or not the site is actively being used. Den trees and the surrounding areas should be left intact until professional advice is obtained. Biologists believe that foraging stands with pines 10 inches in diameter and larger are necessary for successful management of the woodpecker. Logging or other activities near the den trees during the breeding or brood-rearing season may cause red-cockaded woodpeckers to abandon the site or to be unsuccessful in raising their young. The U.S. Fish and Wildlife Service and other agencies listed in this chapter and the appendix can give management advice when red-cockaded woodpeckers are encountered.

In Alabama, fewer than 200 groups are known to exist, with most on national forest lands. Only 25 to 30 groups are known to exist on private lands, scattered primarily across the southern part of the state. However, red-cockaded woodpeckers can occur anywhere in the state where there is old pine timber in open stands. Many biologists believe that the future of red-cockaded woodpeckers in isolated colonies on private lands is bleak and that most will be gone in 15 to 20 years due to the loss of genetic diversity. Counties where they are known to occur include Baldwin, Bibb, Butler, Calhoun, Chilton, Clay, Cleburne, Covington, Dale, Dallas, Escambia, Fayette, Hale, Jefferson, Lawrence, Lee, Macon, Marshall, Perry, Pickens, Shelby, St. Clair, Sumter, Talladega, Tuscaloosa, and Winston.

WOOD STORK (*MYCTERIA AMERICANA*)

Federal Status: Endangered

Wood storks are large wading birds approximately 3½ feet in height with a wingspan of over 5 feet. They are distinguished by a dark unfeathered head and neck, a white body, and a black tail and wing tips. Like most other wading birds, wood storks feed on small fish in shallow freshwater wetlands. They use tall cypresses near the water for colonial nest sites. They occasionally visit Alabama's swamps to forage, but are rarely found nesting in the state.

Land management operations in Alabama complying with Alabama's Best Management Practices for Forestry should not affect wood storks. If nesting should resume here, appropriate care should be given to protect nest sites and the tall cypress trees the storks favor.

Wood storks have been found in Autauga, Baldwin, Barbour, Choctaw, Clarke, Colbert, Conecuh, Covington, Dallas, Elmore, Escambia, Greene, Hale, Henry, Houston, Lauderdale, Lawrence, Limestone, Lowndes, Macon, Marengo, Mobile, Monroe, Montgomery, Morgan, Russell, Sumter, Washington, and Wilcox counties.

BALD EAGLE (*HALIAEETUS LEUCOCEPHALUS*)

Federal Status: Threatened

The adult bald eagle is recognized as the symbol of our country. Adults have dark bodies and wings with the familiar white head, neck, and tail feathers. Young eagles are less distinctive, adding the white feathers gradually after one year of age. Bald eagles are large birds,

RICHARD T. BRYANT/ARISTOCK

The endangered wood stork is a large wading bird about 3½ feet in height with a wingspan of over 5 feet. They are distinguished by a dark unfeathered head and neck, white body, and a black tail and wing tips.

PAUL T. BROWN

The adult bald eagle is easily recognized by its white head. Because of successful conservation efforts, bald eagles have been moved from endangered to threatened status.

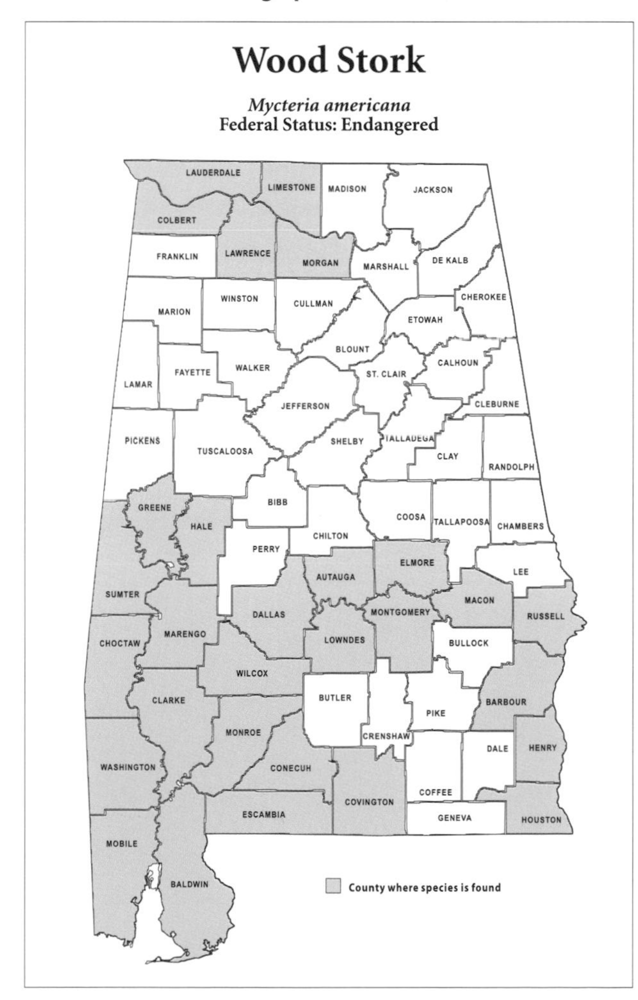

Fig. 4: Wood storks primarily occur in the southern portion of the state.

Fig. 5: The recovery of the bald eagle has resulted in increased numbers and expanded distribution of the species across the state.

with body lengths of 28 to 32 inches and wingspreads of 6 to 7 feet. Eagles catch and eat fish and other prey and will also eat dead animals along lakes, river shores, and roadsides. Bald eagles nest in large trees, often near water. Nests are usually located near the tops of the tallest trees and used year after year. Generally, eagles nest in Alabama from October to May, but may stay on the nest until August. Eagles are increasing in numbers across the nation. Their documented recovery has moved this species from endangered to threatened status.

When land management operations are scheduled in areas where eagles are known or suspected to nest, care should be taken to search the site and to protect the area surrounding the nest tree if one is found. Guidelines have been formulated by the U.S. Fish and Wildlife Service for the Southeast that recommend restrictions of activities around known eagle nests. Although these guidelines are only advisory and voluntary, following them should prevent disturbance that can be detrimental to eagles. Recommended restrictions include no logging or other tree cutting, road building, or use of chemicals toxic to wildlife within a zone ranging from 750 to 1,500 feet around an eagle nest site. Human entry, particularly when eagles are present and nesting, should be restricted. Roost trees and potential replacement nest trees are also important to eagles and should be protected in much the same way as nest trees. If an eagle nest is spotted or suspected, all potentially disturbing activities should be halted immediately and the Alabama Department of Conservation or the U.S. Fish and Wildlife Service notified.

Bald eagles occur in several Alabama counties, usually associated with river systems, lakes, bays, and other bodies of water. Counties with reported eagle presence include Baldwin, Barbour, Cherokee, Choctaw, Covington, Dallas, Elmore, Henry, Jackson, Lauderdale, Limestone, Madison, Marshall, Morgan, Sumter, Tallapoosa, and Winston. Other unconfirmed sightings have been reported in other counties.

The Alabama red-bellied turtle grows to about 13 inches in length and is usually olive, greenish, brown, or black with cream, yellow, orange, or red markings. The head and legs are dark with yellow striping.

Fig. 6: The red-bellied turtle has only been found in the lower Mobile-Tensaw River Delta in Alabama.

Reptiles and Amphibians

ALABAMA RED-BELLIED TURTLE (*PSEUDEMYS ALABAMENSIS*)

Federal Status: Endangered

Also known as the "red-belly," this large freshwater turtle grows to about 13 inches in length. The top of the shell is usually olive, greenish, brown, or black with cream, yellow, orange, or red markings. The bottom is usually cream, yellow, red, or orange with dark markings. The head and legs are dark with yellow striping. The front toenails are long and sharp.

Declines in water quality that threaten the beds of submerged vegetation they feed on are the only land management activities that might threaten these turtles. Adherence to Alabama's Best Management Practices for

Forestry and herbicide labels should protect them from those impacts. Predation on nests and the turtles themselves by collectors, alligators, raccoons, crows, and fire ants is a problem for this species.

The red-bellied turtle is endemic to the lower Mobile-Tensaw River Delta in Alabama and has also been found in Mobile and Baldwin counties, at Dauphin Island, in Weeks Bay, lower Fish River, and in Little River in Monroe County.

FLATTENED MUSK TURTLE (*STERNOTHERUS DEPRESSUS*)

Federal Status: Threatened

This is a small freshwater turtle less than 5 inches in length with a flattened top shell. The top of the shell is brown and the bottom either pink or yellow. The head is greenish with a network of dark markings. The flattened musk turtle feeds on invertebrates such as snails and mussels in small, unimpounded, medium-sized, clear, shallow streams. It can also be found upstream and downstream of impoundments.

These animals are sensitive to changes in streambed habitat and water quality, especially siltation. Adherence to Alabama's Best Management Practices for Forestry should prevent them from being impacted by land management activities.

Currently the flattened musk turtle is found only in the Black Warrior watershed above the Bankhead Dam. The Alabama counties included in its range are Blount, Cullman, Etowah, Fayette, Jefferson, Lawrence, Marshall, Tuscaloosa, Walker, and Winston.

The flattened musk turtle is less than 5 inches in length and has a flattened top shell. The top of the shell is brown and the bottom is pink or yellow.

Fig. 7: Currently, the flattened musk turtle is found only in the Black Warrior River above the Bankhead Dam.

GOPHER TORTOISE (*GOPHERUS POLYPHEMUS*)

Federal Status: Threatened (in portions of its range)

A dry land turtle, the gopher tortoise has a high, domed shell with lengths of up to 15 inches. They have stubby hind feet and flattened front feet with large toenails used for digging. Gopher tortoises favor dry, sandy ridges with open stands of longleaf pine, turkey oak, and other scrub oaks. They also frequent open areas around road shoulders, food plots, and rights-of-way that have well-drained sandy soil. Gopher tortoises dig long, sloping burrows up to 30 feet long and extending as deep as 9 feet below the surface. The burrows almost always have a characteristic mouse-hole shape, with a flat bottom and a rounded arched top and sides, much like the gopher tortoise itself. These dens are used as shelter by gopher tortoises as well as by a variety of other sandhill residents, including the endangered Eastern indigo snake and the diamondback rattlesnake. Gopher tortoises feed on grasses and other plant material near the ground. Feeding trails are often visible leading from the den's sandy apron to foraging areas. Eggs are laid in or near the den entrance in May, June, and July and hatch in about 80 to 100 days. Young tortoises are about the size of silver dollars and are very vulnerable to predation by crows, raccoons, opossums, foxes, skunks, coyotes, and other animals.

JAMES C. GODWIN

The gopher tortoise is a dry land turtle with a high domed shell up to 15 inches long, elephant-like hind feet, and front feet equipped with claws for digging their burrows. They prefer dry, sandy ridges with open stands of longleaf pine and/or scrub oak.

Fig. 8: The gopher tortoise is found in the southern portion of Alabama where sandy soils predominate. It is federally listed as threatened in Choctaw, Mobile, and Washington counties.

Care should be taken with heavy equipment around gopher tortoise dens to avoid collapsing the den and particularly to avoid crushing eggs and young gophers, which dig shallow dens. Fire and/or herbicides may be necessary to maintain gopher tortoise habitat quality when scrub oaks shade out groundcover the gopher tortoise feeds on. Gopher tortoises forced to move to road shoulders and opening edges are vulnerable to predation by animals and humans. Frequent fires and thinnings allow sunlight to reach the forest floor and create good habitat for nesting and feeding.

Gopher tortoises are protected by federal law in the Alabama counties west of the Mobile and Tombigbee rivers and in Mississippi and Louisiana. They are also protected across their range in Alabama by state regulation as non-game species. Counties in Alabama where they occur and are federally protected include Choctaw, Washington, and Mobile. Other counties where they occur are Baldwin, Barbour, Bullock, Butler, Clarke, Crenshaw, Coffee, Conecuh, Covington, Dale, Escambia, Geneva, Henry, Houston, Monroe, Montgomery, Pike, and Wilcox.

RED HILLS SALAMANDER (*PHAEOGNATHUS HUBRICHTI*)

Federal Status: Threatened

The Red Hills salamander has few easily recognizable distinguishing characteristics. It is best found by recognizing potential habitat and searching for burrows rather than the salamander itself. It is a relatively large salamander that grows to 10 inches in length with a dark brown tail and body. It spends almost all its time in its burrow on shady, steep bluff sites, coming to the opening on warm, humid nights to feed on insects and other invertebrate (insect) prey. The shady, moist conditions on the bluffs where the salamander lives are critical to its survival. Loss of shade and cover leads to drying by sunlight and wind that negatively impacts both the salamander and its prey.

Land activities that lead to erosion, such as logging, soil compaction, or loss of canopy cover can be harmful to the Red Hills salamander. Removal of too many trees in the canopy on and immediately above the slopes on which these animals occur can also expose the site to excessive drying. Removal of trees just above the slope can lead to downed trees on the slope, again creating gaps in the canopy which lead to drying of the site. It is recommended that areas known or suspected to contain Red Hills salamanders be investigated by biologists before land management operations begin. If Red Hills salamanders are found on a site, limited logging or other activities are still possible, but consultation with the U.S. Fish and Wildlife Service may be necessary to guide management. Habitat conservation plans, such as those developed by International Paper, have been formulated for private lands in Alabama that permit some

MARK BAILEY

The Red Hills salamander is a secretive amphibian approximately 10 inches in length with a dark brown tail and body. Shady and moist conditions on the bluffs where the salamander lives are critical to its survival.

Gulf sturgeon are anadromous, which means they spend most of their lives in the bays and the Gulf of Mexico, and then journey up coastal rivers to fresh water to spawn. If you catch a sturgeon call the USFWS at (904) 769-0552.

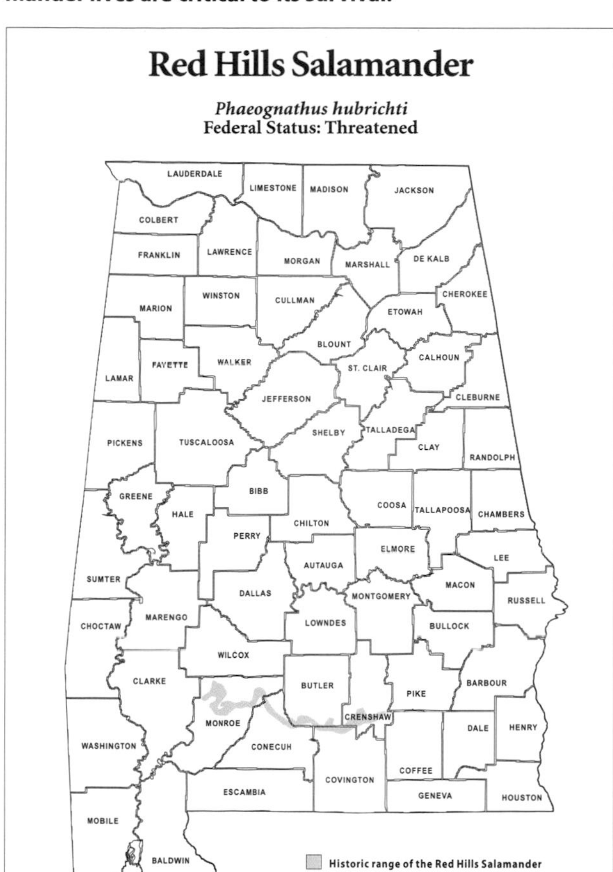

Fig. 9: The Red Hills salamander is found only in Alabama in a narrow belt associated with two siltstone geologic formations called the Tallahatta and the Hatchetigbee.

Fig. 10: The sturgeon is a prehistoric fish that dates back more than 200 million years in fossil records. The gulf sturgeon is only found in the lower Mobile, Tensaw, and Alabama river system and should not be confused with the smaller Alabama sturgeon.

operations while affording protection of the salamander.

The Red Hills salamander is found only in Alabama along a narrow belt associated with two siltstone formations called the Tallahatta and the Hatchetigbee. The salamander's range is bounded on the east by the Conecuh River and on the west by the Alabama River. There are estimated to be less than 55,000 acres of Red Hills salamander habitat remaining within their range. Portions of Butler, Crenshaw, Conecuh, Covington, and Monroe counties contain the entire 55,000 acres.

The blue shiner is endemic to the Mobile basin above the fall line in the upper Coosa River system.

The slackwater darter is distinguished by its black spotted fins and darker upper body.

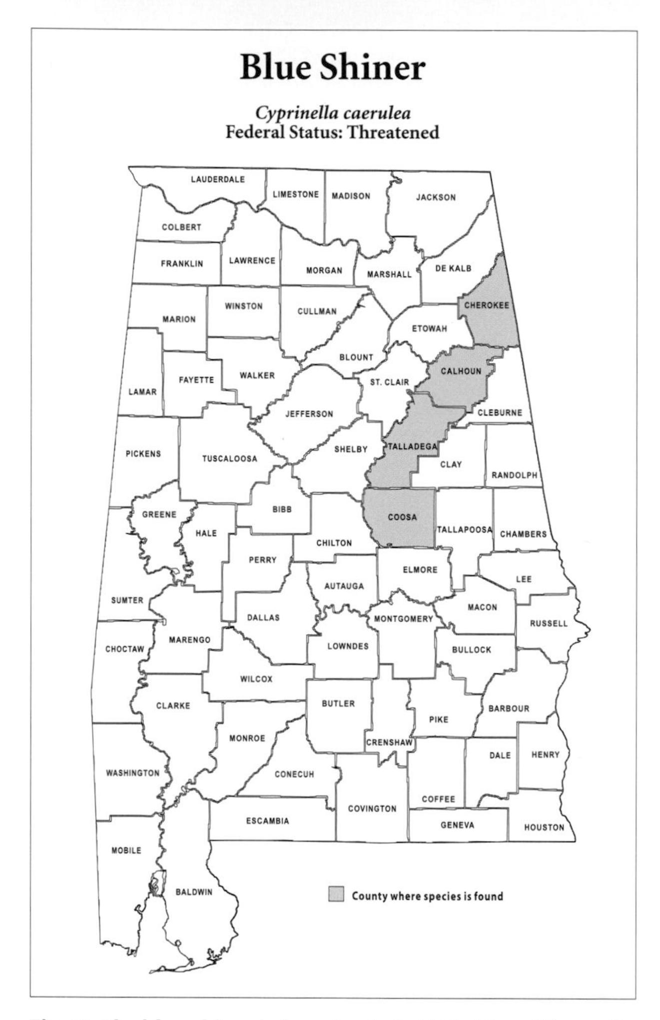

Fig. 11: The blue shiner is found only in Little River, Weogufka Creek, and Choccolocco Creek.

Fig. 12: The slackwater darter is found in streams in the northwestern portion of the state.

Fishes

As of this writing, there are 12 fish species listed as threatened or endangered in Alabama. Most of them occur in northern Alabama rivers and streams. The single biggest problem affecting protected fish that could result from land management practices is sedimentation, which smothers fish eggs and habitat. Particular attention should be given to bridge and road construction that might lead to erosion. Leave wide buffer strips adjacent to streams when cutting timber and avoid disturbing soil on steep slopes. Avoid spraying pesticides near waterways. The best way to avoid impacts on fish is to maintain high water quality and closely follow Alabama's Best Management Practices for Forestry and Agriculture.

Figure 19 is a list of protected fish and the rivers and streams where they occur. Because many of these species are small and not easily caught for identification, they may also occur in additional streams.

Fig. 13: The boulder darter Is only found in Limestone County.

Fig. 14: The watercress darter is only found in several springs and small streams in Jefferson County.

Fig. 15: As the name implies, the Cahaba shiner is only found in portions of the Cahaba River.

Fig. 16: The goldline darter is found in three counties in central Alabama.

RICHARD T. BRYANT/ARISTOCK

The snail dater is brownish in color with black specks located down the middle of the body. It also has a characteristic black marking directly below the eyes. It was the snail darter that halted construction of the Tellico Dam in Tennessee.

Fig. 17: The snail darter is found only in the Paint Rock River in Madison and Marshall counties.

RICHARD T. BRYANT/ARISTOCK

The spotfin chub is a threatened species found in the Tennessee River in Colbert and Lauderdale counties.

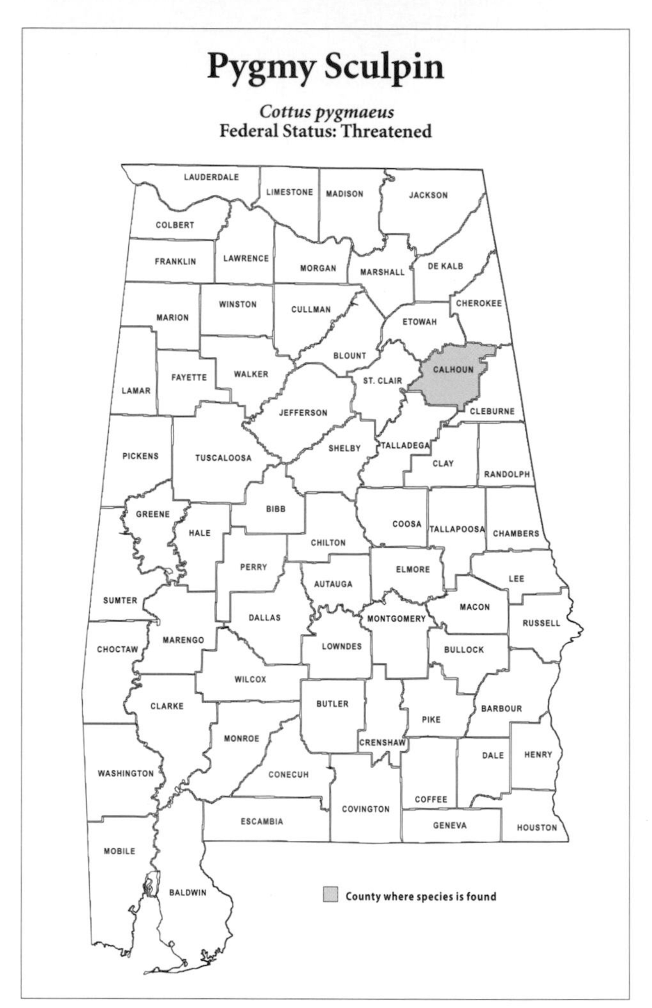

Fig. 18: The pygmy sculpin is found in Coldwater Creek in Calhoun County, Alabama.

FIG. 19: THREATENED AND ENDANGERED FISH SPECIES IN ALABAMA

Species*	Rivers/Creeks	Counties
Gulf Sturgeon (*Acipenser oxyrinchus desotoi*)	Mobile River, Tensaw River, Alabama River below Claiborne Dam, Tombigbee River below Coffeeville Dam	Baldwin, Mobile, Clarke, Monroe, Choctaw, Washington
Pygmy Sculpin (*Cottus pygmaeus*)	Coldwater Creek, Coldwater Spring	Calhoun
Blue Shiner (*Cyprinella caerulea*)	Little River, Weogufka Creek, Choccolocco Creek	Calhoun, Cherokee, Coosa, Talladega
Spotfin Chub (*Cyprinella monacha*)	Tennessee River	Colbert, Lauderdale
Slackwater Darter* (*Etheostoma boschungi*)	Cypress Creek and all its tributaries, except Threet and Little Cypress creeks; Swan Creek; Greenbriar Branch; Flint River	Lauderdale, Limestone, Madison
Watercress Darter* (*Etheostoma nuchale*)	Glenn, Thomas, Roebuck, Tapawingo springs and their spring runs	Jefferson
Boulder Darter* (*Etheostoma wapiti*)	Elk River	Limestone
Cahaba Shiner* (*Notropis cahabae*)	Cahaba River	Bibb, Perry, Shelby
Palezone Shiner* (*Notropis procne*)	Paint Rock River	Jackson
Goldline Darter (*Percina aurolineata*)	Cahaba and Little Cahaba rivers and Schultz Creek	Bibb, Shelby, Jefferson
Snail Darter* (*Percina tanasi*)	Paint Rock River	Madison, Marshall
Alabama Cavefish* (*Speoplatyrhinus poulsoni*)		Lauderdale

* Species names followed by an asterisk (*) are listed as endangered, all others are listed as threatened.

Fig. 20: The palezone shiner is found only in Jackson County.

Endangered Species Assistance

The U.S. Fish and Wildlife Service is the federal agency with regulatory responsibility for the Endangered Species Act. Updated information on endangered and threatened plants and animals occurring in Alabama may be obtained from the following agencies:

U.S. Fish and Wildlife Service
Division of Ecological Services
Daphne Field Office
P.O. Drawer 1190
Daphne, AL 36526 (334) 441-5181

Alabama Department of Conservation and Natural Resources
64 North Union Street
Montgomery, AL 36130
Coordinator, Non-Game Wildlife Program (334) 242-3469
State Lands Division, Natural Heritage (334) 242-3484

Mollusks

As of this writing there are 39 species of freshwater mussels and two aquatic snails that are listed as either threatened of endangered in Alabama. Like the fishes, they can best be protected from negative effects of land management practices by protecting water quality and streambed habitat. Following the guidelines in Alabama's Best Management Practices for Forestry closely will help ensure that impacts are minimized. Below is a list of mussels and counties where they are known to occur.

FIG. 21: THREATENED AND ENDANGERED MUSSEL SPECIES IN ALABAMA

Species	Rivers/Creeks	Counties Known to Occur
Alabama Lamp Pearly Mussel* (*Lampsilis virescens*)	Paint Rock River, Hurricane Creek	Jackson
Fine-rayed Pigtoe Mussel* (*Fusconia cuneolus*)	Paint Rock River	Jackson, Marshall, Madison
Inflated Heelsplitter Mussel* (*Potamilus inflatus*)	Black Warrior and Tombigbee rivers	Choctaw, Clarke, Hale, Greene, Marengo, Sumter, Tuscaloosa, Washington
Orange-footed Pearly Mussel* (*Plethobasus cooperianus*)	Tennessee River	Colbert, Lauderdale, Madison, Morgan, Marshall
Pale Lilliput Pearly Mussel* (*Toxolasma cylindrellus*)	Paint Rock River, Hurricane Creek	Jackson
Pink Mucket Pearly Mussel* (*Lampsilis abrupta*)	Tennessee River, Paint Rock River	Colbert, Jackson, Madison, Morgan, Marshall, Lauderdale, Lawrence, Limestone
Rough Pigtoe Mussel* (*Pleurobema plenum*)	Tennessee River	Colbert, Lauderdale, Madison, Morgan, Marshall, Limestone
Shiny Pigtoe Mussel* (*Fusconaia edgariana*)	Paint Rock River	Jackson, Marshall, Madison
Stirrup Shell Mussel* (*Quadrula stapes*)	Tombigbee and Sipsey rivers	Greene, Pickens
Fine-lined Pocketbook Mussel (*Lampsilis altilis*)	Coosa and Tallapoosa rivers	Calhoun, Cleburne, Coosa, Elmore, Etowah, Hale, Macon, Shelby, St. Clair, Talladega, Lawrence, Winston, DeKalb, Dallas, Clay, Jefferson, Tuscaloosa
Shinyrayed pocketbook (*Lampsilis subangulata*)	Chattahoochee River system	Houston
Alabama Moccasinshell Mussel (*Medionidus acutissimus*)	Alabama, Tombigbee, Cahaba, Coosa, Black Warrior rivers	Greene, Lawrence, Pickens, Tuscaloosa, Winston
Southern Clubshell Mussel* (*Pleurobema decisum*)	Tombigbee, Black Warrior, Alabama, Tallapoosa, and Coosa rivers	Dallas, Greene, Lamar, Macon, Pickens, Shelby, Tuscaloosa
Upland Combshell Mussel* (*Epioblasma metastriata*)	Black Warrior, Cahaba, and Coosa rivers	Bibb, Shelby, Jefferson
Coosa Moccasinshell Mussel* (*Medionidus parvulus*)	Coosa, Cahaba, and Black Warrior rivers	Cherokee, DeKalb, Lawrence, Winston
Triangular Kidneyshell Mussel* (*Ptychobranchus greeni*)	Black Warrior, Cahaba, and Coosa rivers	Bibb, Blount, Jefferson, Lawrence, Shelby, Walker, Winston
Southern Acornshell Mussel* (*Epioblasma othcaloogensis*)	Upper Coosa and Cahaba rivers	Cherokee, St. Clair, Shelby
Ovate Clubshell Mussel* (*Pleurobema perovatum*)	Tombigbee, Black Warrior, Alabama, Tallapoosa, and Coosa rivers	Greene, Jefferson, Lamar, Macon, Pickens, Tuscaloosa, Walker, Winston

FIG. 21: CONTINUED

Species	Rivers/Creeks	Counties Known to Occur
White Wartyback Pearlymussel* (*Plethobasus cicatricosus*)	Tennessee River	Colbert, Lauderdale
Orange-nacre Mucket Mussel (*Lampsilis perovalis*)	Tombigbee, Black Warrior, Alabama, and Cahaba rivers	Bibb, Fayette, Greene, Lawrence, Pickens, Tuscaloosa, Winston
Flat Pigtoe Mussel* (*Pleurobema marshallii*)	Tombigbee River	Greene, Sumter
Heavy Pigtoe Mussel* (*Pleurobema taitianum*)	Tombigbee and Sipsey rivers	Greene, Sumter, Lamar, Pickens
Southern Combshell Mussel* (*Epioblasma penita*)	Tombigbee and Buttahachie rivers	Greene, Lamar, Sumter
Southern Pigtoe Mussel* (*Pleurobema georgianum*)	Coosa River	Calhoun, Cleburne
Tulotoma Snail* (*Tulotoma magnifica*)	Several tributaries of the Coosa River	Calhoun, Coosa, Elmore, Shelby, St. Clair, Talladega
Anthony's Riversnail* (*Leptoxis crassa anthonyi*)	Limestone Creek and Tennessee River	Jackson, Limestone, Lauderdale
Fanshell Mussel* (*Cyprogenia stegaria*)	Tennessee River	
Dromedary Pearly Mussel* (*Dromus dromas*)	Tennessee River	
Yellow-blossum Pearly Mussel* (*Epioblasma florentina florentina*)	Tennessee River	
Purple Cat's Paw Pearly Mussel* (*Epioblasma obliquata obliquata*)	Tennessee River	
Tubercled-blossom Pearly Mussel* (*Epioblasma torulosa torulosa*)	Tennessee River	
Turgid-blossoum Pearly Mussel* (*Epioblasma turgidula*)	Tennessee River	
Cracking Pearly Mussel* (*Hemistena lata*)	Tennessee River	
Ring Pink Mussel* (*Obovaria retusa*)	Tennessee River	
Little-wing Pearly Mussel* (*Pegias fabula*)	Tennessee River	
Clubshell* (*Pleurobema clava*)	Tennessee River	
Black Clubshell Mussel* (*Pleurobema curtum*)	Extirpated	
Dark Pigtoe Mussel* (*Pleurobema furvum*)	Sipsey Fork and North River drainages of the Black Warrior River	
Cumberland Monkeyface Pearly Mussel* (*Quadrula intermedia*)	Tennessee River	

* Species names followed by an asterisk (*) are listed as endangered, all others are listed as threatened.

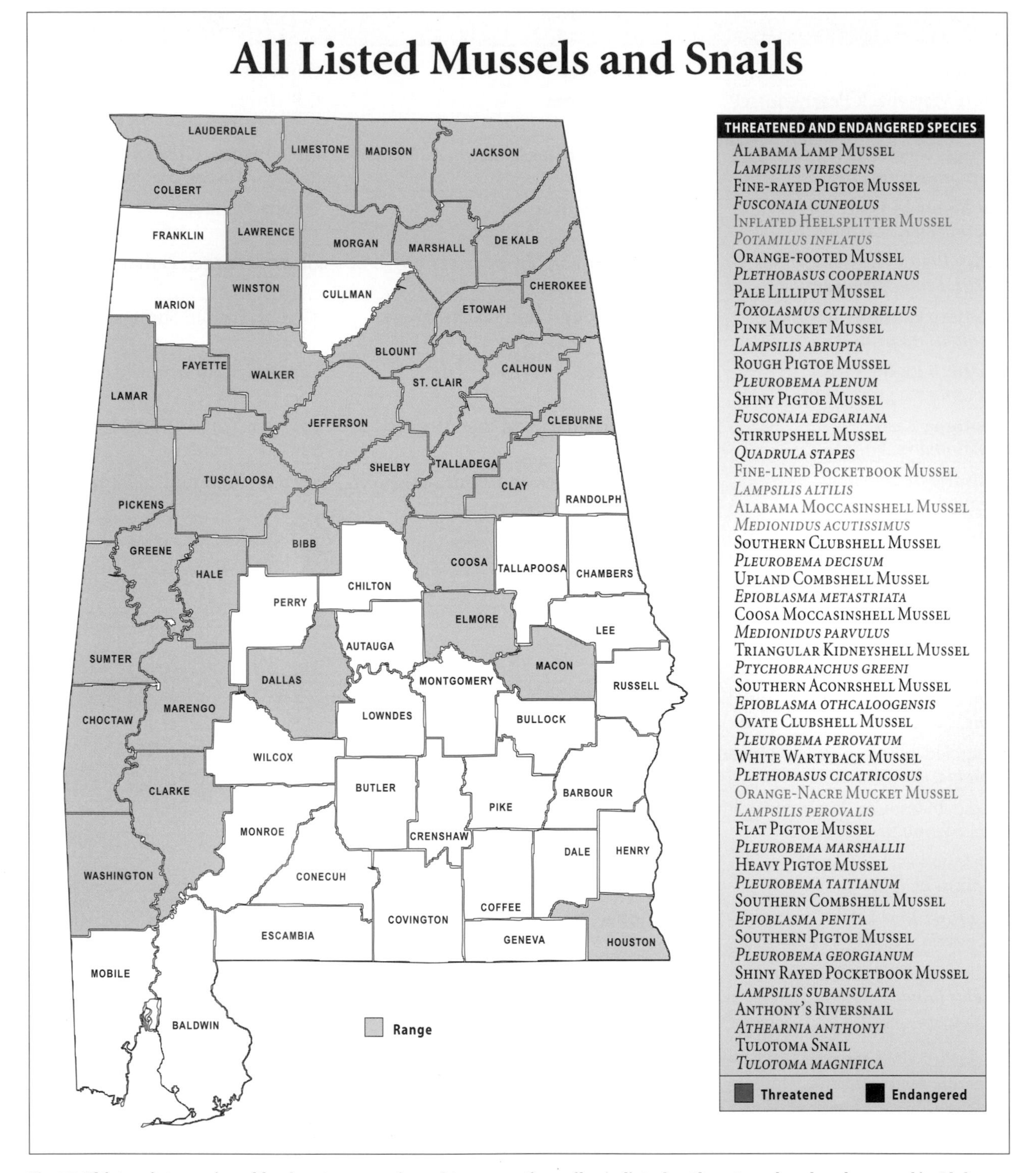

Fig. 22: Thirty-nine species of freshwater mussels and two aquatic snails are listed as threatened and endangered in Alabama. Declining water quality and siltation are the greatest threats to these species.

Plants

ALABAMA CANEBRAKE PITCHER PLANT (*SARRACENIA RUBRA ALABAMENSIS*)

Federal Status: Endangered

Like all pitcher plants, this one is carnivorous, trapping and digesting insects in its tubular leaf. The tube of the Alabama canebrake pitcher plant is 8 to 16 inches tall in the spring and may be curved in shaded conditions. The flower, which appears in April through June, is maroon and droops from a 2-foot stalk. The summer leaves are also tubular and may be up to 27 inches long. They are light green and covered with white "hair." The plant grows in wet areas and seeps along with grasses, sedges, sweetbay, poison sumac, bayberry, and sparkleberry.

Pitcher plants are sun loving. Fire, which releases them from shade and woody brush, is beneficial. They are very dependent on the moist soil conditions where they grow, so any activities that affect the water table or drainage of the site are potentially harmful.

Alabama canebrake pitcher plants are known to occur only in Autauga, Chilton, and Elmore counties.

The endangered Alabama canebrake pitcher plant, like other pitcher plants, traps and digests insects in its tubular leaf. The maroon flower appears from April through June and droops from a 2-foot stalk.

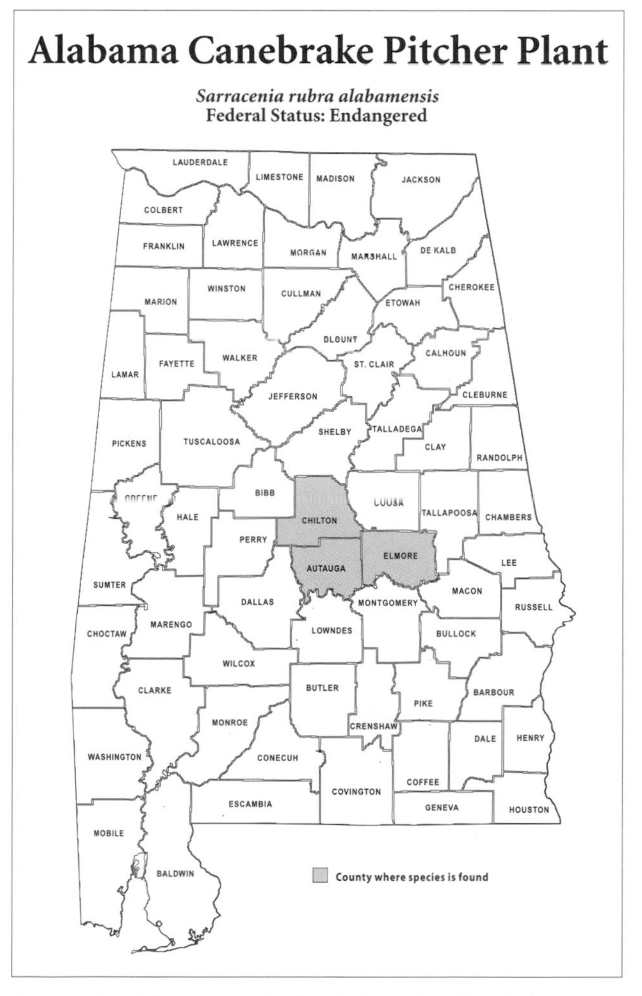

Fig. 23: The Alabama canebrake pitcher plant is only known to occur in Autauga, Chilton, and Elmore counties.

ALABAMA LEATHER FLOWER (*CLEMATIS SOCIALIS*)

Federal Status: Endangered

This is an erect herb about 7 to 12 inches tall with a blue-violet, dangling, bell-shaped flower that appears in April and May. This plant grows in full sunlight near woods edges. It grows in wet silty-clay flats near creeks and streams and is often surrounded by grasses and sedges.

Activities that dry out sites or disturb soils would be detrimental to this plant. The plant is a poor competitor and suffers when the canopy is opened too much, but it should be able to withstand thinning if heavy equipment is not used. Mechanical site preparation would likely destroy it. The plant is only found in Cherokee and St. Clair counties.

HARPERELLA (*PTILIMNIUM NODOSUM*)

Federal Status: Endangered

This is a small herbaceous plant that has narrow, quill-like leaves, which are up to 16 inches long where they grow from the base of the plant, but grow shorter as they get higher on the stem. The flowers are small and inconspicuous and appear in June and July. The plant dies back each year. It grows on rocky banks and shoals near creek beds. Other plants growing in association with harperella include azalea, mountain laurel, and holly.

Land management activities are not likely to take

MALCOLM PIERSON

The Alabama leather flower is about 7 to 12 inches tall with a dangling, blue-violet, bell-shaped flower that appears in April and May.

VERNON BATES

Harperella is a small plant with narrow, quill-like leaves and white flowers that appear in June and July.

Fig. 24: The Alabama leather flower only occurs in Cherokee and St. Clair counties.

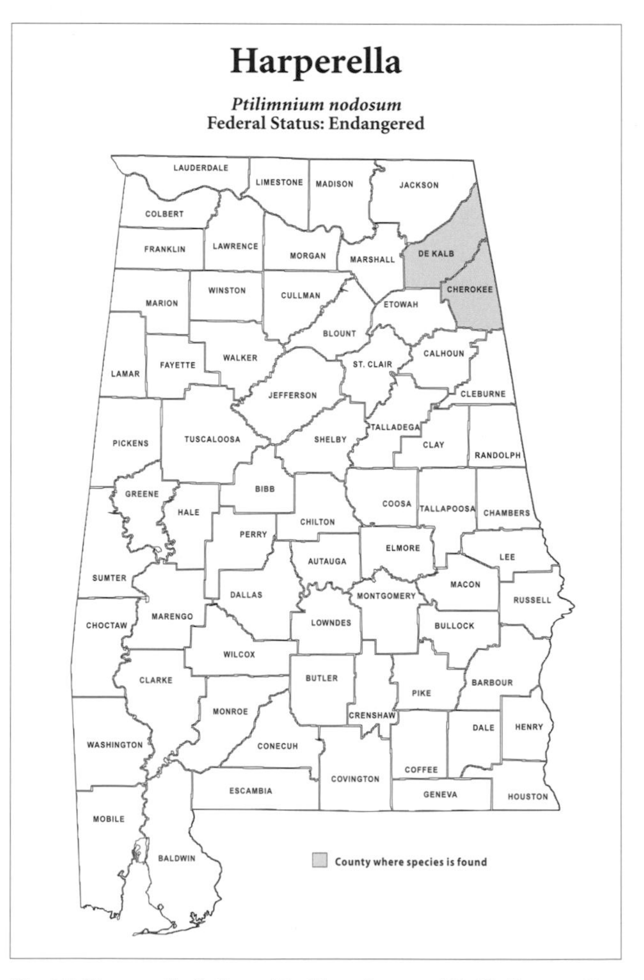

Fig. 25: Harperella is found in Cherokee and DeKalb counties.

place in the areas where this plant grows, if Alabama's Best Management Practices for Forestry are followed. If operations are planned, extra wide streamside zones (SMZs) should be planned to prevent siltation of the site. Timber thinnings and harvest should be done under the supervision of someone familiar with the plant. Herbicides should be used in accordance with labeled instructions and not in SMZs. Currently the plant is found only in Cherokee and DeKalb counties.

KRAL'S WATER PLANTAIN (*SAGITTARIA SECUNDIFOLIA*)

Federal Status: Endangered

This plant is an aquatic growing on or below the water. It grows on rocky creek beds and nearby slopes. It is often found in association with azaleas, mountain laurel, and holly.

Because of where it occurs, this plant should not be affected by land management practices if Alabama's Best Management Practices for Forestry are followed, particularly for SMZs and stream crossings. The plant is found in the Little River drainage in DeKalb and Cherokee counties, and in the West Sipsey Fork in Winston County.

GREEN PITCHER PLANT (*SARRACENIA OREOPHILIA*)

Federal Status: Endangered

This is a rare, carnivorous plant with a tubular, hollow spring leaf and distinct hood common to pitcher plants. It is widest at the top of the tube and tapers to the base. The tube is green or yellow-green with maroon veins. Insects are attracted into the tube where they are trapped and digested. The flowers appear in April and bloom into June. They are yellow and droop from the top of a 2-foot stalk arising from the base of the plant. In the late summer, the tubes dry up and are replaced by flat, sickle-shaped leaves that are pale or reddish at the base. Pitcher plants grow in boggy areas, stream banks, or seeps in a community with grasses, sedges, sphagnum moss, and cinnamon fern.

Fire is essential to the continued survival and vigor of these plants. Fireline construction—where necessary—should be carried out in a way that avoids alteration of the drainage pattern of the site or causes a change in the water table. Timber harvesting, mechanical site preparation, or other land management activities should also be conducted in a manner that avoids those changes.

The pitcher plant is known to occur only on wet sites in Cherokee, DeKalb, Etowah, Jackson, and Marshall counties.

ALABAMA NATURAL HERITAGE PROGRAM

Kral's water plantain is an aquatic plant that is found in the mountainous regions of the state growing on rocky creek beds and nearby slopes.

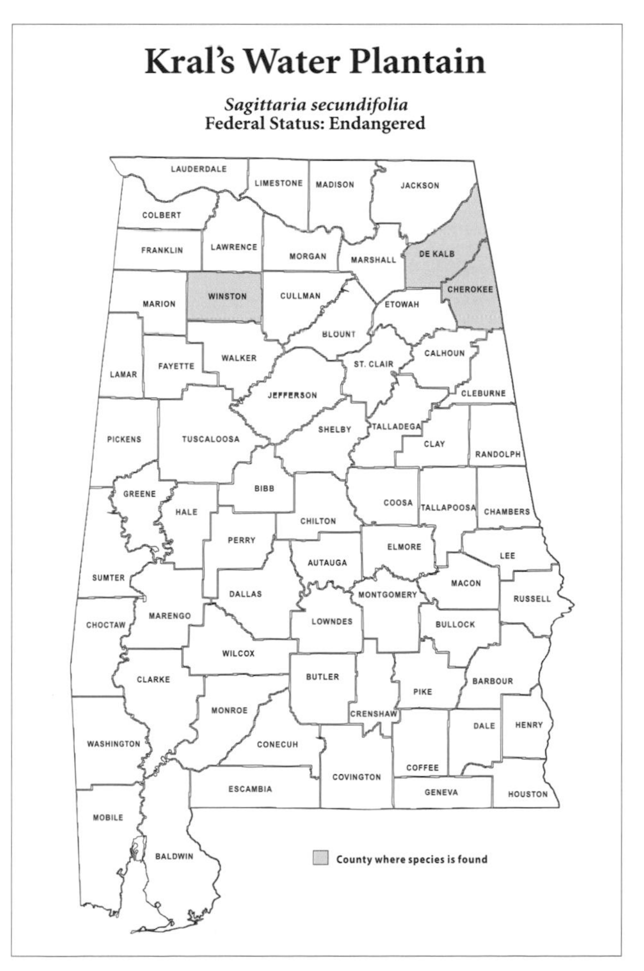

Fig. 26: Kral's water plantain is only found in the Little River drainage in DeKalb and Cherokee counties, and the West Sipsey Fork in Winston County.

RANDY TROUP

The green pitcher plant has yellow flowers that flower from May to early June. These flowers droop from a 2-foot stalk. Pitcher plants grow in boggy areas, stream banks, or seeps.

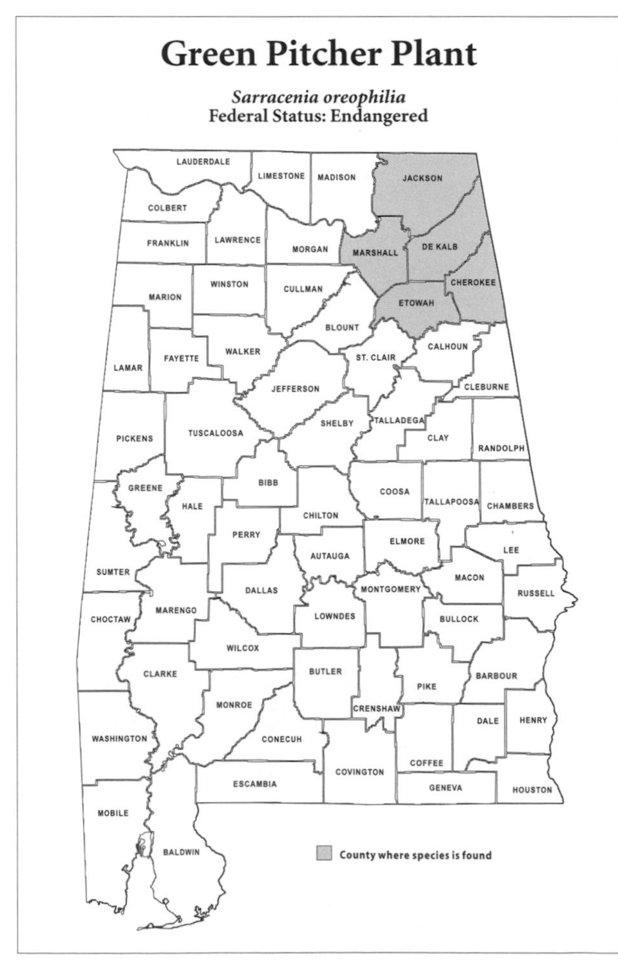

Fig. 27: The green pitcher plant is only known to occur on wet sites in five northwestern counties.

MOREFIELD'S LEATHER FLOWER (*CLEMATIS MOREFIELDII*)

Federal Status: Endangered

This is a vine growing to 16 feet in length and covered with dense white "hairs." The flower is urn shaped and pinkish and blooms in late May through early July. It is often found near seeps under mixed hardwoods on rocky south- and southwest-facing mountain slopes. Trees often found near it include smoketree and chinkapin oak.

Use of fire is unlikely on these sites, but mechanical site preparation and clear-cutting would damage plants. Light thinnings might be tolerated. Care should be taken during thinning operations to avoid direct impacts to plants. This plant is known to occur only in Madison County.

TENNESSEE YELLOW-EYED GRASS (*XYRIS TENNESSEENSIS*)

Federal Status: Endangered

This plant grows in association with ferns, willows, buttonbush, and bulrushes in seeps, springs, and on the banks of small streams. It is grass-like with 5- to 17-inch slender, twisted leaves. The small pale yellow flowers appear in August and September.

The plant occurs in natural openings and prefers sunlight. It can probably tolerate thinnings. However, site preparation, clear-cutting, and other land management practices might be very damaging to this plant. Some herbicides may also adversely affect it. Tennessee yellow-eyed grass is known to occur in Bibb, Calhoun, and Franklin counties.

RELICT TRILLIUM (*TRILLIUM RELIQUUM*)

Federal Status: Endangered

This trillium is a fleshy, low-growing (12 inches or less) plant with waxy, dark green, blotchy leaves. The flower appears to sit right on top of the stalk. It is greenish-brown, purple, or occasionally pure yellow in color and is among the earliest of spring bloomers. The plant dies back to its underground portion in the summer and emerges in the spring. Trillums grow in moist, shady hardwood forests.

Trillium is adversely affected by fire and grows best in the shade. Timber operations within areas inhabited by trillums that remove shade, dry out sites, or use fire will likely adversely impact the plants. Relict trillium is known to occur only in Bullock, Henry, and Lee counties.

ALABAMA STREAK SORUS FERN (*THELYPTERIS PILOSA* VAR. *ALABAMENSIS*)

Federal Status: Threatened

SUSAN WEBER

Morefield's leather flower is a vine that grows up to 16 feet in length and is covered in white "hairs." The flower is urn-shaped, pinkish, and blooms from late May through early July.

JIM ALLISON

Tennessee yellow-eyed grass is a wetland plant that has slender, twisted leaves that are 5 to 17 inches in length, and small, pale yellow flowers that appear in August and September.

Fig. 28: Morefield's leather flower is only known to occur in Madison County.

Fig. 29: Tennessee yellow-eyed grass is known to occur in Alabama only in three counties: Bibb, Calhoun, and Franklin.

CAROLINE R. DEAN

Relict trillium grows low to the ground and has dark green leaves with white blotches. The flower is greenish-brown, purple, or occasionally pure yellow and blooms in early spring.

USFWS

Alabama streak sorus fern is a small, evergreen fern that only grows on sandstone overhangs and cliff faces. It needs cool, shady conditions to survive and is often found in coves.

Fig. 30: Relict trillium is known to occur in only three counties in east-central and southeastern Alabama: Bullock, Henry, and Lee.

Fig. 31: Alabama streak sorus fern is found in Alabama only in Winston County.

This is a small evergreen fern with 4- to 8-inch deeply scalloped leaves. It grows only on sandstone overhangs and cliff faces, often in coves. It needs cool, shady conditions to persist.

The plant is unlikely to be affected by land management practices if Alabama's Best Management Practices for Forestry are implemented. Alabama streak sorus fern is only found in Winston County.

AMERICAN HART'S TONGUE FERN (*PHYLLITIS SCOLOPENDRUM* VAR. *AMERICANA*)

Federal Status: Threatened

This is a leafy fern that grows in humid, deeply shaded conditions near limestone sinks and caves. Since it is unlikely that land management activities will take place in the habitats where this plant occurs, it is not likely that conflicts will arise. The plant can be found only in Morgan and Jackson counties.

American hart's tongue fern is a leafy fern that grows in wet, shaded areas near limestone sinks and caves.

Fig. 32: American hart's tongue fern is found in Alabama only in Morgan County.

MOHR'S BARBARA'S BUTTONS (*MARSHALLIA MOHRII*)

Federal Status: Threatened

This plant may reach a height of 2 feet or more. Its small white-pink flowers are produced in several heads and in a branched arrangement. The flowers bloom in May and June. The plant grows in moist to wet openings in woodlands and along shale-bedded streams. It often grows in association with grasses and sedges and prefers full sunlight or partial shade.

This plant can probably tolerate light thinnings. Mechanical site preparation and clear-cutting could be very damaging to this plant, as would some herbicides. The plant can be found in Bibb, Cherokee, Calhoun, Etowah, and Walker counties.

PRICE'S POTATO-BEAN (*APIOS PRICEANA*)

Federal Status: Threatened

This is a climbing yellow-green vine that grows from a stout, potato-like tuber. The vines may be up to 15 feet long with pale pink or greenish-yellow pea- or bean-type flowers which bloom during July and August. The fruit is a pod about 4 to 6 inches long. The plant grows in forest openings in mixed hardwood stands where ravine slopes grade into creek or stream bottoms. Trees often found growing with this plant include chinkapin oak, basswood, and slippery elm.

Because it occurs in second growth timber, it appears to withstand some timber harvesting. Site preparation using shears and rakes should be avoided where this plant occurs and some herbicides may damage or kill it. This plant is known to occur in Autauga, Madison, and Marshall counties in Alabama.

JIM ALLISON

Mohr's Barbara's buttons grows to a height of 2 feet or more, has small, white-pink flowers in a branched arrangement, and blooms from May through June.

CAROLINE P. DEAN

Price's potato-bean is a climbing yellow-green vine that grows from a potato-like tuber. The vines may be 15 feet long with pale pink or greenish-yellow flowers that bloom during July and August.

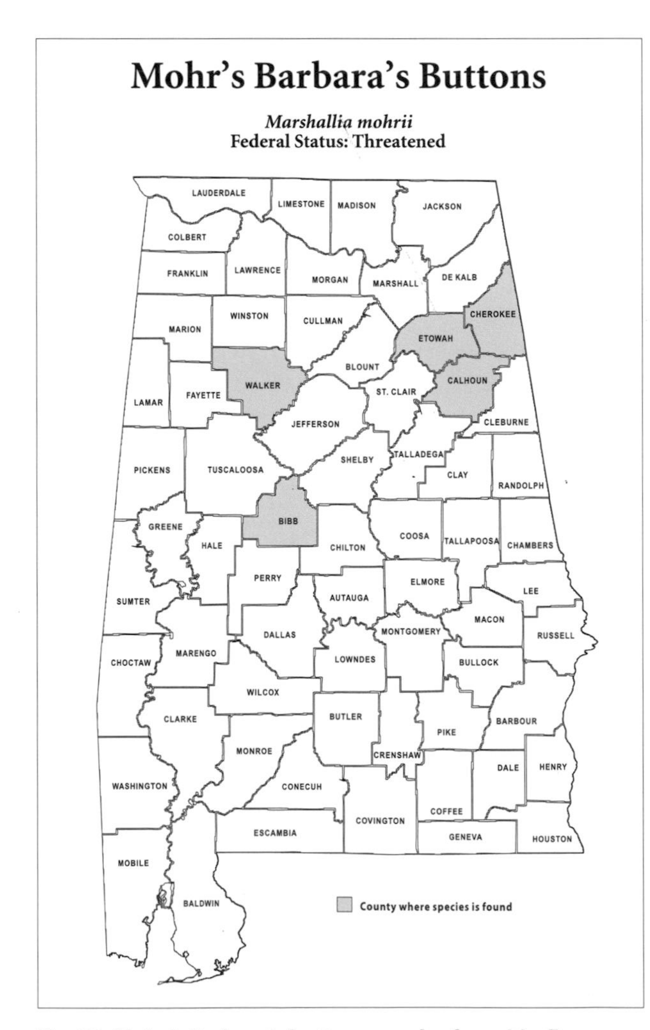

Fig. 33: Mohr's Barbara's buttons can be found in five counties in Alabama: Bibb, Cherokee, Cullman, Etowah, and Walker.

Fig. 34: Price's potato-bean is found in three counties in Alabama: Autauga, Madison, and Marshall.

NEW APPROACHES TO PROTECTING ENDANGERED SPECIES

Realizing the potential economic impact that the Endangered Species Act (ESA) has on private landowners, the U.S. Fish and Wildlife Service (USFWS) has become more flexible in working with private landowners who have endangered or threatened species on their land. Past approaches of implementing the ESA have relied on regulations, enforcement, and fines as a means of protecting species in peril. This approach has met with limited success in long-range protection of species and has polarized private landowners, who own most of the land in the Southeast, and federal and state agencies, who are charged with implementing the ESA.

Recognizing this problem, the USFWS began exploring options in which landowners retain their rights to use and manage their lands, but in a manner that would not jeopardize existing endangered species on their land. Some of the alternatives considered were conservation easements and agreements, economic incentives for landowners, and relocation of listed species from private land to suitable habitat on public lands. These alternatives led to the increased use of Habitat Conservation Plans (HCP), available since 1982, which are flexible agreements between the USFWS and private landowners to protect listed species but at the same time allow the land to be used and managed by the landowner. HCPs have been developed in Alabama to protect the endangered Alabama beach mouse and the threatened Red Hills salamander.

Currently in Alabama there is a cooperative effort between the USFWS and the major conservation organizations and agencies, including the Alabama Wildlife Federation, to develop a statewide HCP for red-cockaded woodpeckers on behalf of landowners. Many states are trying to develop statewide HCPs with multiple options.

One option is called **Safe Harbor**. This strategy is the most attractive for landowners who do not currently have red-cockaded woodpeckers but would like to manage for large sawtimber, poles, or simply an older and more aesthetically pleasing pine forest. As an example, suppose a landowner has a Safe Harbor agreement and does not have red-cockaded woodpeckers on his or her land or has only a few. Then, because of management practices, the red-cockaded woodpeckers either move onto this land or expand their existing populations into a new area. Safe Harbor allows the landowner to convert previously unoccupied habitat to other uses if he or she changes plans later on. The only provisions are that the person must 1) maintain any cavity trees and foraging habitat of red-cockaded woodpeckers that were present on the land prior to the Safe Harbor agreement, 2) avoid converting the land during the nesting season, and 3) allow an opportunity to trap and relocate birds under the direction of the Department of Conservation and Natural Resources, if the habitat is to be adversely altered.

Another option is **translocation of isolated groups**. Under this option, states would obtain a federal permit allowing the relocation of isolated red-cockaded woodpeckers from private lands to other suitable areas. The plan would involve trapping and banding of red-cockaded woodpecker nestlings for one to three years. Before the birds are one year old, they would be moved to Forest Service lands or other approved areas with a substantial red-cockaded woodpecker population and available habitat. After a landowner has allowed red-cockaded woodpeckers to be relocated, he or she would be exempt from any additional legal responsibilities related to red-cockaded woodpeckers and would be free to cut all remaining trees, including nest cavity trees. Under this option, only lands with isolated red-cockaded woodpecker groups will be eligible.

SUMMARY

The Endangered Species Act is one of the most powerful laws to protect and conserve wildlife in the U.S. Although primarily regulatory in nature, innovative habitat conservation plans have emerged as a viable alternative to protecting threatened and endangered species and at the same time guaranteeing the rights of private landowners. The challenge is to develop workable partnerships between private landowners, regulating agencies, and a supportive public to conserve our endangered and threatened species.

Some private property rights groups and business interests would like to see the ESA's restrictions reduced or eliminated. Some strong opponents have even called for the repeal of the act. The National Wildlife Federation, representing a wide range of conservation and environmental groups, has urged Congress to strengthen the ESA in five key areas. These areas include the following:

1. Streamlining the process for listing endangered species.
2. Improving implementation of ESA's requirement for a designated critical habitat.
3. Achieving greater ecosystem protection through better planning or recovery of endangered species.
4. Ensuring adequate funding for authorized activities, especially for development of Habitat Conservation Plans

Red-cockaded Woodpeckers on Private Forest Lands

Ralph Costa is the U.S. Fish and Wildlife Service's Recovery Coordinator for the endangered red-cockaded woodpecker and is located at Clemson University in South Carolina. Because of Ralph's efforts, landmark partnerships have been developed between the federal government and private landowners to protected the endangered red-cockaded woodpecker. These partnerships are unique in that they are flexible enough to protect the interests of private landowners as well as the species. Ralph is coauthor of the book Red-cockaded Woodpecker: Recovery, Ecology and Management *and is also author of the U.S. Fish and Wildlife Service's private lands manual, which provides guidelines for managing property where red-cockaded woodpeckers exist.*

I often hear forest owners express their concerns about having red-cockaded woodpeckers on their land and what they should do if they think they have red-cockaded woodpeckers. The first step is to find out if red-cockaded woodpeckers are indeed on the property. The likelihood of having red-cockaded woodpeckers in pines less than 40 years of age is extremely rare. Generally landowners that have pine stands that are open and park-like and over 60 years of age should be checked before any timber is harvested. More often red-cockaded woodpecker colonies occur in pines that are between 70 and 100 years old. Pine stands that are near established red-cockaded woodpecker populations or adjacent to older stands should also be checked. In most cases pine stands can be surveyed by the landowners themselves with the aid of publications that illustrate what red-cockaded woodpecker cavity trees look like. This information can be obtained from our field office in Daphne (334-441-5181), the Alabama Forestry Commission (334-240-9348), or the Non-Game Program with the Alabama Department of Conservation and Natural Resources (334-242-3469). If landowners are uncomfortable checking for the presence of red-cockaded woodpeckers themselves, professionals with these agencies or private consultants can provide assistance.

Then, if red-cockaded woodpeckers are found on your property you have a legal responsibility to avoid *take*, which is defined as "to harass, harm, shoot, or kill." *Harass* means "to directly threaten the birds themselves," whereas *harm* is "destruction or degradation of the bird's habitat." Harvesting timber where red-cockaded woodpeckers breed, feed, or find shelter is considered *harm*. Fortunately, our private lands manual gives forest owners guidelines to avoid *take* with red-cockaded woodpeckers. The manual is available through my office (864-656-2432) or the field office in Daphne, Alabama.

In a progressive move to eliminate landowner fear about having red-cockaded woodpeckers, the U.S. Fish and Wildlife Service has established a program in several states called **Safe Harbor**. Under this program we issue a permit to the landowner that eliminates any legal responsibility for protecting red-cockaded woodpeckers beyond those already on your property. If you currently have no red-cockaded woodpeckers, then you are not legally responsible for protecting any that might appear in the future. The rationale is to reduce the premature cutting of older timber stands to prevent red-cockaded woodpeckers from becoming established on your land. Under safe harbor you do not have to be afraid to grow longleaf pines to 300 years of age.

Alabama is developing a statewide red-cockaded woodpecker conservation plan that allows any landowner the opportunity to sign up for a safe harbor permit. The program will also provide permits to take demographically isolated red-cockaded woodpeckers from private lands and relocate them to public lands such as the Conecuh or Talladega National Forests. Fortunately for private forest owners, Alabama does not have as many groups of red-cockaded woodpeckers on private lands as compared to other states.

Red-cockaded woodpecker conservation partnerships also exist between the USFWS and six timber companies. Gulf States Paper Company is on the verge of finishing a safe harbor conservation plan for red-cockaded woodpeckers that will include a red-cockaded woodpecker management area dedicated to increasing red-cockaded woodpecker populations to 15 to 20 groups. Clearly we are in a new era of cooperation between federal and state governments and private landowners that will benefit both the landowners and the threatened and endangered species.

DENNIS HOLT

RICHARD T. BRYANT/ARISTOCK

The red-cockaded woodpecker, which has no visible red on its body, prefers open, park-like stands and requires mature pines over 60 years of age for constructing nesting cavities. Cooperative programs, like Safe Harbor, allow for conservation of the red-cockaded woodpecker and protection of the rights of private landowners to manage timber on their lands.

Why Do Species Become Rare, Endangered, and Extinct?

The evolution of new species and their extinction are both natural processes that have occurred throughout time. Scientists estimate that 500 million species have lived at one time on our planet. If this estimate is correct, today's species represent only 2 percent of that total. Scientists also believe that before modern man the percentage of species becoming extinct remained fairly constant, except for dramatic periodic changes when the rate was increased by changing environmental factors. Modern man has also accelerated the rate of extinction as compared to the rate before human intervention. According to scientist E.O. Wilson, today's rate of extinction rivals any of history's extinction episodes.

Species become rare, endangered, or extinct because they cannot adapt to changes in their environment. Environmental changes may be caused by natural factors or by man. The following is a list of factors that increase the likelihood of a species becoming endangered.

- Habitat alteration or destruction.
- Drastic changes in climatic or environmental conditions (like a comet striking the earth).
- Unregulated or commercial hunting (regulated sport hunting has never led to the endangerment of any wildlife species).
- Species having a narrow range of habitat requirements making them vulnerable to changes in their environment, such as the cave-dwelling and endangered Indiana bat.
- Species having limited reproductive rates such as small litter sizes, a long gestation or incubation period, and young that require extensive parental care (e.g. elephants).
- Species that have been exploited or rare species whose numbers have been reduced by collectors.
- Species living in international waters where protection is difficult, including both fish and aquatic mammals.
- Highly specialized species (due to physiological, behavioral, physical, or reproductive specializations).
- Introduction of exotic species, such as kudzu and the fire ant, that compete with or are pests to native wildlife and plants.

by state and local governments.

5. Strengthening the enforcement provisions.

Given the political climate and the highly publicized controversies regarding endangered species, private lands, and economics, it is difficult to determine how the ESA will look in the future. Hopefully the ESA will be reauthorized in such a manner that private landowner concerns and endangered species are both protected.

A bluebird perched on a nest box prepares to feed young hatchlings. The erection of bluebird nest boxes by landowners, homeowners, and bird enthusiasts has been instrumental in the comeback of this strikingly beautiful songbird.

MAC CONE

CHAPTER 14

Attracting Backyard Wildlife

CHAPTER OVERVIEW

Attracting and observing wildlife around homes and residential areas has become a national hobby. Part of the reason for this is that living in close proximity to wildlife gives us a sense of pleasure and connectiveness to nature. As a result more homeowners, as well as businesses, are landscaping their property to attract wildlife by providing habitat for a variety of species. This chapter serves as a guide for improving residential and urban areas for wildlife.

CHAPTER HIGHLIGHTS PAGE

STEPS TO ATTRACTING BACKYARD WILDLIFE 316
ATTRACTING SELECT SPECIES 325
BIRDS 326
REPTILES AND AMPHIBIANS 334
BUTTERFLIES 334
NUISANCE BACKYARD WILDLIFE 336

Attracting wildlife to your backyard is one way to observe many species without ever having to leave home. Watching backyard wildlife has become a national hobby with more than a third of American families annually spending a total of over $500 million on materials and supplies (birdseed, houses, feeders). The exuberant songs of cardinals and mockingbirds, the dazzling display of newly arriving migrants at birdbaths and feeders, and the sight of colorful butterflies fluttering above a wildflower patch are all personal pleasures that aren't easily measured. There is an inherent need for people to live close to the natural world. When we are able to watch natural interactions through our back window, we become more observant and in tune with life's delicate subtleties. A monarch butterfly observed in September marks the beginning of fall migration for many species. Similarly, the appearance of purple martins in February lets us know that spring is not far behind. These gentle heralds of the passing seasons bring meaning to our lives.

Developing wildlife habitat in our own yards is one way to bring us closer to the land and the natural world. Attracting wildlife to our homes fosters a sense of environmental stewardship for those who are ordinarily removed from the land.

STEPS TO ATTRACTING BACKYARD WILDLIFE

With planning and the proper arrangement of food, shelter, and water, many wildlife species can be attracted to urban and suburban settings, providing countless hours of viewing pleasure. There are five steps that need to be taken to attract wildlife to your home: 1) Determine what wildlife to attract, 2) Make an inventory of existing habitat features, 3) Create a backyard habitat plan, 4) Develop a schedule of activities, and 5) Monitor wildlife use of backyard habitat, making adjustments and improvements where needed.

Step 1—Determining What Wildlife to Attract

Alabama is blessed with more than 875 species of vertebrate wildlife (mammals, birds, reptiles, fish, and amphibians) that are either permanent residents or migrants. The species you can expect to attract to your backyard depends on several factors. The most critical factor is how closely you can get your yard to duplicate the habitat requirements of the species you desire. Each species has unique requirements for food, especially in its natural form; water to drink and bathe in; cover or shelter to escape from predators, rest, and build nests; and space or territory in which to live and raise young. Some species, like amphibians, may spend their entire life in your backyard. Other species will visit periodically on a daily (resident birds) or seasonal (migrating birds) basis. Most require a diversity of habitat types.

Another approach in determining which species to attract is to make observations of the wildlife that routinely visit your neighborhood, or nearest natural area. These are the species that you have the best chance of attracting to your yard. The best strategy may be to attract the widest variety of wildlife by providing a diversity of habitat.

RICHARD T. BRYANT/ARISTOCK

With a little planning, backyards can be enhanced to attract visitors such as this red-spotted purple butterfly.

Advantages of Attracting Backyard Wildlife

1. Provides additional wildlife habitat.
2. Adds beauty and pleasure to home surroundings.
3. Can increase real estate values from 3 to 10 percent.
4. Wildlife viewing entertains and educates.
5. Increases awareness of wildlife needs and life history.
6. Certain species control insects and rodents.

A colorful addition to backyard habitats, this male cardinal eats more than 100 kinds of seeds and fruits. Its powerful beak is designed to break apart even the toughest seeds.

Habitat Requirements of Backyard Wildlife

Food: All animals get their energy for survival from plants and other animals. The ideal backyard wildlife plan uses natural vegetation to supply year-round food—from the earliest summer berries to fruits that persist through winter and spring (such as dogwood, Eastern red cedar, and holly).

Space: All animals require space to mate and raise their young. In backyards of limited size, you might only have room to support a limited number of breeding birds and other animals. An animal's requirement for space may be substantially less if food, water, and cover are provided.

Water: Fresh water is essential for all wildlife and is often the factor that limits the presence of some wildlife in backyards. Spring and fall migrants are especially attracted to water during long flights. Frogs and salamanders require standing water to complete parts of their life cycles.

Cover: Backyard wildlife need protective cover during breeding, nesting, hiding, sleeping, feeding, and movement. Try to select cover plants that also double as food sources.

Water is an important component of backyard wildlife habitat. Frogs require standing water to complete their life cycle—from eggs to tadpoles to adult frogs.

Step 2—Making an Inventory of Existing Backyard Habitat

The second step to improving backyard wildlife habitat is to understand what you already have. Often, backyards already provide some of the habitat components readily used by a wide range of wildlife. Start by walking around your property and carefully noting habitat features such as the types of trees and plants that provide cover and food. As you survey your yard, make a rough sketch of all the habitat features you observe. Be sure to include your property's dimensions, the area covered by your house and other structures (such as a garage, storage shed, pool, deck, patio, fence, sidewalk, driveway, etc.). Also note the location of underground water pipes, utilities, septic tanks, irrigation lines, and sprinkler heads. This sketch will serve as the basis for a more complete wildlife landscape map that denotes existing features as well as those to add.

Look at your yard as wildlife would view it, both from a horizontal and vertical perspective. Are there areas that provide adequate food, cover, water, and space? Are there native seed-bearing plants that can provide fruit on a continuous basis? Does your backyard provide adequate cover and safe travel corridors for small mammals and birds? In addition, wildlife use one or more of a combination of five backyard zones: 1) underground (some amphibians and small mammals); 2) ground level, short grasses, and low-lying cover; 3) taller grasses, wildflowers, and forbs; 4) shrubs and vertical vines; and 5) trees. Check for the presence and location of these zones. These levels are used to varying degrees by different wildlife species. Some species may restrict most of their life activities to one level, while others will use several zones to meet their needs. Note the location of various habitat zones on your map.

As you walk around, examine the soil in your yard. Soils determine what will grow well in your yard. Is the soil fill dirt, sand, sandy loam, topsoil, or another soil type? Also note soil moisture and areas that are poorly drained or unusually dry. Soil type is a major consideration when selecting wildlife plants to establish in your yard. To determine the fertility of your soil, take a soil test. This can serve as a basis for selecting plants and for guiding fertilization rates.

As you survey your yard, make a list of the trees, shrubs, and herbaceous plants already in your yard. You may want to note their age, size, health, whether they are exotic or native to your area, their value for food and cover, and any special maintenance requirements. Also notice how the vegetation interacts with the physical characteristics in your yard to form unique microhabitats. For example, large trees may shade a portion of the yard during most of the day, making these areas cooler and moister than the rest of the yard. Other areas receiving full sunlight during the day will probably be drier, with different vegetation than wetter areas. The patterns of sun and shade during the day and throughout the year have an influence on vegetation. Each backyard microhabitat presents different opportunities and constraints in an overall backyard wildlife plan.

As you inventory your yard, do not forget to consider your family's needs and personal use of the property. Consider space requirements for work, play, entertainment, access and traffic patterns, trash collection, security and privacy, and areas for pets. If outdoor cats and dogs are a priority, your expectations for attracting wildlife should be low. Free-ranging cats are a major limiting factor on suburban and rural bird abundance. Think realistically about how much and what type of space you will need for each activity.

After a survey of your property has been completed and important features have been sketched on a rough map, you are ready for the next step in planning your backyard wildlife habitat.

Getting Neighbors Involved

Before planning your backyard habitat, talk to neighbors and discuss your ideas. Ask for their opinions and input. Uninformed neighbors might become concerned as they see your yard being transformed to accommodate wildlife. Discuss your interest in attracting wildlife and the potential that some wildlife may become nuisances or pests. Discussing your intentions early on with your neighbors might prevent future misunderstandings. On the other hand, your neighbors may share your interest and want to jointly plan a combined backyard wildlife effort.

Step 3—Creating a Backyard Habitat Plan

DRAWING WHAT YOU HAVE

Now that you have examined your property as a wildlife manager would and have a rough sketch of what you have found, you are ready to prepare a base map

TES RANDLE JOLLY

Alabama backyard wildlife habitats can be beautiful spaces for both people and wildlife with proper planning and maintenance.

that will guide your wildlife landscaping efforts. From your inventory, you should be able to determine if you need to make major landscape changes or simply modify a reasonably acceptable backyard habitat.

The information on the rough sketch can now be transferred to a base map drawn to scale on graph paper. Be sure that all physical and habitat features are recorded using uniform symbols. For example, one type of circle may be used to represent evergreen trees, another for deciduous trees, a third for shrubs, irregular outlines for beds of plants, squares and rectangles for buildings, straight lines for walkways, and so on. The exact symbols are not as important as using consistent symbols for similar features. Templates for creating these symbols can be obtained from drafting supply stores. The National Wildlife Federation (8925 Leesburg Pike, Vienna, VA 22184-0001. Phone 800-822-9912) also sells an inexpensive planning kit complete with a template of appropriate symbols (trees, shrubs, and other features) for tracing.

MAC CONE

Artificial feeders are an excellent choice for increasing wildlife viewing opportunities and providing supplemental food for a variety of birds such as this house finch and these American goldfinches. A variety of commercial feeders are available.

ENHANCING EXISTING LANDSCAPE

What you already have in your backyard determines how much modification you will have to make to attract wildlife. Since most yards have limited space, begin by removing those plants or trees that have little or no value for wildlife. They take up space that could be utilized by more beneficial plantings. If they don't appear in the planting guide in Appendixes D–J, or wildlife have not been observed using them, take steps for their removal.

Fig. 1: Creating A Backyard Habitat Sketch. A backyard habitat is a landscape plan designed with wildlife in mind. The sketch serves as a blueprint for backyard wildlife habitat components, similar to designs created by landscape architects for commercial and residential grounds. On this sketch, note the diversity of wildlife-attracting plants and placement of water. A variety of planting suggestions are found in Figure 1. For more choices on what to plant turn to the Appendixes D–J. Try to use native trees and plants whenever possible, as they are usually better adapted to the land in a certain area and create fewer long-range problems than some exotic species, like tallow trees.

Backyard wildlife like this turtle need protective cover. Try to select cover plants that also double as food sources.

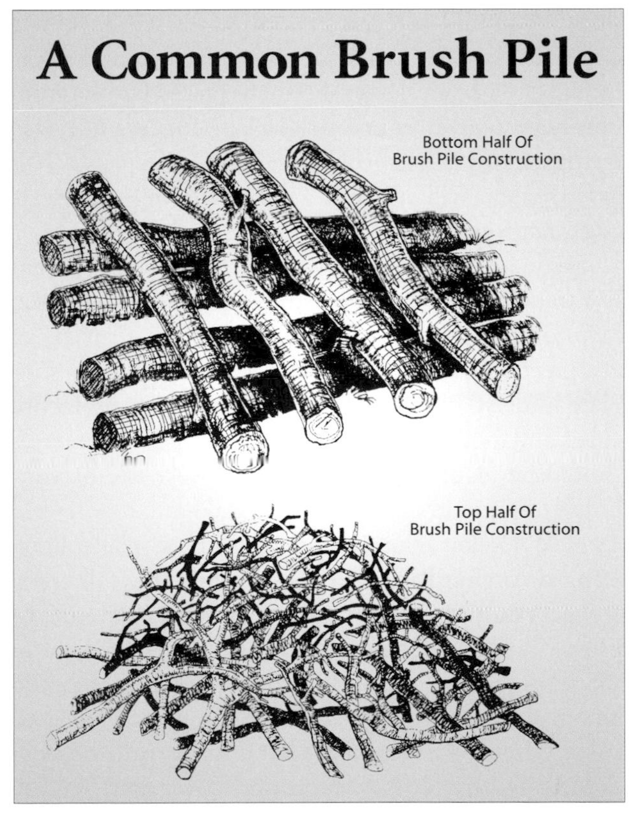

Fig. 2: (Left) Properly constructed bird houses should have ¼ inch ventilation holes under the roof overhang and six ¼ inch drainage holes in the floor. Fig. 3: (Above) Brush piles can provide additional cover for small mammals in backyards. Sticks and branches can be added as the pile rots.

Fig. 4: Trees and Plants Used to Attract Backyard Wildlife

SUGGESTED TREES AND PLANTS

1. Pine
2. Red Mulberry
3. Flowering Dogwood
4. Wax Myrtle
5. Southern Magnolia
6. Blueberry
7. Viburnum
8. Red Maple
9. American Holly
10. Fringe Tree
11. Red Buckeye
12. Black Gum
13. Hawthorne
14. Red Cedar
15. Persimmon
16. Black Cherry
17. American Beauty-Berry
18. Pokeweed
19. Wildflowers
20. Seed-Producing Grasses
21. Wild Azaleas
22. Honeysuckle
23. Salvia
24. Trumpet Creeper
25. Lantana
26. Cypress
27. Oaks
28. Yellow Poplar
29. Beech

Before adding plants to your backyard, remember that they should be located to optimize wildlife viewing opportunities. The layout of plants should give a layering effect. The tallest trees should be placed farthest from observation areas (windows, decks, benches), followed by midcanopy trees, shrubs, and herbaceous plants. This arrangement will enhance wildlife viewing and create edge that is highly attractive to most wildlife.

Begin by framing backyards with a backdrop of native trees that serve as food sources and cover, and also help to screen your yard from neighbors. A variety of evergreen and deciduous trees should be planted to simulate a "forest" canopy, providing nesting sites, protective cover, and food for mammals and birds. Deciduous trees should be planted on the south or west side of your house to provide summer shade.

Next develop an understory by planting smaller flowering or orchard trees in clusters near the tall trees. Stagger plantings so that they are not placed in lines or rows. Clusters of plantings should also be irregular in shape to create more edge for wildlife. Small trees should be surrounded with shrubs, brambles, and other ground cover. Often if these areas are left undisturbed, they will eventually provide native ground cover beneficial to ground-nesting/denning birds and mammals. Shrub borders may also be established that have a mixture of several varieties that vary in shape, height, and density. This creates ideal nesting sites for a wide assortment of birds. Try to choose native trees and shrubs that fruit at different times of the year for a continuous food supply, such as red mulberry in the spring, blackberry in the summer, dogwood in the fall, and holly in the winter.

Some areas of open lawn can be converted to "natural meadows" by simply letting vegetation grow in select sites. These areas are heavily used by small mammals, insects, and butterflies. Natural meadows should be irregular in shape and placed in front of taller vegetation for easy viewing. With a little intentional neglect, most unmanaged lawns (no herbicides and a variety of native grasses) will grow into a diverse grassy and herbaceous meadow in less than a year's time. Expect to see a variety of grasses and "weeds" such as black-eyed susan, thistles, milkweeds, daisies, Queen Anne's lace, and other natural plants described in Appendixes D–J). All that is required for maintenance of natural meadows is two summer mowings to help control tree and shrub invasion. Well-manicured lawns can be transformed into more natural meadows by disking selected sites immediately after the last frost before spring. Natural meadows should be maintained by annual disking in late winter and late summer.

Planted meadows can also be established in select grasses, herbaceous forbs, and wildflowers. There are many commercial seed mixes that are available at local garden centers. A variety of wildflowers should be selected that are well adapted to your area of the state and that will produce blooms in sequence from early spring through fall.

Fig. 5: Select Wildflowers for Meadows

	PLANT SPECIES	COLOR
SPRING	Wild Columbine (*Aquilegia canadensis*)	red and yellow
	Blue False Indigo (*Baptisia australis*)	blue
	Spring Beauty (*Claytonia virginica*)	pink
	Fawn Lily (*Erythronium americanum*)	yellow
	Wild Geranium (*Geranium maculatum*)	magenta
	Crested Iris (*Iris cristata*)	blue, purple
	Foxglove Penstemon (*Penstemon digitalis*)	white
	Beardtongue (*Penstemon smallii*)	purple
	Wild Blue Phlox (*Phlox drummondii*)	blue
	Blue-eyed Grass (*Sisyrinchium angustifolium*)	blue
	Stokes's Aster (*Stokesia laevis*)	purple, blue
SUMMER	Swamp Milkweed (*Asclepias incarnata*)	pink
	Butterfly Weed (*Asclepias tuberosa*)	orange
	Blue False Indigo (*Baptisia australis*)	blue
	Bluebell (*Campanula americana*)	blue
	Lance-leaved Coreopsis (*Coreopsis lanceolata*)	yellow
	Shooting Star (*Dodecatheon meadia*)	pink
	Purple Coneflower (*Echinacea purpurea*)	purple
	Fireweed (*Epilobium angustifolium*)	pink-red
	Blanket Flowers (*Gaillardia* spp.)	red and yellow
	Sneezeweed (*Helenium autumnale*)	yellow-orange
	Common Sunflower (*Helianthus angustifolius*)	yellow
	Standing Cypress (*Ipomopsis rubra*)	red
	Canada Lily (*Lilium canadense*)	orange
	Wild Lupine (*Lupinus perennis*)	blue, purple
	Bee Balm (*Monarda didyma*)	red
	Bergamot (*Monarda fistulosa*)	pink,lavender
	Evening Primrose (*Oenothera biennis*)	yellow
	Showy Evening Primrose (*Oenothera speciosa*)	pink
	Annual Phlox (*Phlox drummondii*)	red
	Black-eyed Susan (*Rudbeckia hirta*)	yellow
	Sage (*Salvia coccinea*)	red
	Sweet Goldenrod (*Solidago odora*)	yellow
	Common Spiderwort (*Tradescantia virginiana*)	purple, white
	Ironweed (*Vernonia altissima*)	purple
	Cardinal Flower (*Lobelia cardinalis*)	red
	Blazing Stars (*Liatris* spp.)	white, pink, lavender
	Giant Coneflower (*Rudbeckia* spp.)	yellow
FALL	Swamp Milkweed (*Asclepias incarnata*)	pink
	Smooth Aster (*Aster laevis*)	pale blue, purple
	Calico Aster (*Aster lateriflorus*)	purple, white
	Stiffed-leaved Aster (*Aster linariifolius*)	violet
	Tickseed Sunflower (*Bidens aristosa*)	yellow
	Purple Coneflower (*Echinacea purpurea*)	purple
	Common Sunflower (*Helianthus annuus*)	yellow
	Evening Primrose (*Oenothera biennis*)	yellow
	Gray Goldenrod (*Solidago nemoralis*)	yellow
	Sweet Goldenrod (*Solidago odora*)	yellow
	Seaside Goldenrod (*Solidago sempervirens*)	yellow
	Ironweed (*Vernonia altissima*)	purple

Helpful Tips for Enhancing Backyard Habitat

- **Lawn:** Convert some open lawn to natural or planted meadows. Wildflowers, butterflies, bees, birds, and small mammals will readily use these areas.

- **Hedges:** Select a variety of plant heights, but maintain a minimum of 3½- to 8-foot-high hedges. The best hedges for bird cover and nesting are evergreens or deciduous shrubs with dense thorny branches such as blackberry. Thorny hedges discourage unwanted predators and disturbance to nesting birds.

- **Hedgerows/Fencerows:** Natural hedgerows and vegetated fencerows provide ideal food and cover as well as travel paths for wildlife. They should consist of shrubs, stunted trees, vines, and small herbaceous plants. The plow-perch method is a natural way to let birds develop a hedgerow or thicket for you. In summer or early fall, till an area adjacent to existing fences. If there is not a fence construct an artificial fence by stringing a double row of wire between posts placed 15 feet apart. Seed-eating birds will perch on the post and wires, and the remains of their meals will fall on the ground beneath the fence. In this way birds will "plant" their own preferred foods, creating an ideal natural hedgerow. Some preferred species include blackberry, blueberry, crabapple, dogwood, elderberry, holly, honeysuckle, mulberry, serviceberry, sumac, and black cherry.

- **Pruning:** Birds prefer unclipped hedges. Pruning should be conducted every few years on azaleas and other early flowering shrubs that bloom from buds formed during the previous summer. Avoid pruning during the nesting and brood-rearing period. Prune after flowering and before bud-set to prevent removing recently formed flower buds.

- **Small Trees:** Be sure orchard and flowering trees receive the amount of sunlight they need. For example, dogwoods prefer some light shade during the day.

- **Large Forest Trees:** Maintain standing dead and dying trees for cavity-nesting and denning wildlife. They also serve as a source of food for a variety of woodpeckers. Dead trees that pose a safety hazard should be removed.

- **Corridors:** Connect thickets, meadows, and other habitat sites by hedgerows, fencerows, or overgrown grass and herbaceous "alleys." Corridors provide protective cover for wildlife, especially small mammals, as they move about in your yard.

- **Water:** Water is a strong attractant for wildlife. Although many mammals and birds satisfy their water needs from the vegetation they eat, they will also readily visit drinking and bathing sites. Reptiles and amphibians must live in or around water. Water sources can be provided from streams, ponds, birdbaths, or shallow water-holding containers sunk in the ground. Depths should vary from ½ inch to at least 2 feet in the center. The sides of the water source should have grasses, shrubs, trees, or rocks for protective cover. Birdbaths are the best way to attract migrating songbirds in the spring, especially in areas where water is a limited resource.

- **Walking Paths:** Walking paths can make visiting your yard more enjoyable, especially when vegetation is wet with rain or dew. Add mulched or stone walkways to your backyard wildlife landscape.

- **Pets:** The number one predators of backyard wildlife are pets—especially cats. Birds and small mammals can be killed or frightened away by the family cat or dogs. Do not let your backyard become a death trap for wildlife. **Without exception, exclude cats from backyard wildlife habitats.**

Step 4—Schedule of Activities

After drawing a base map and choosing the modifications you would like to make, you need to decide where and when to begin. If there are major landscaping changes, it is easy to feel overwhelmed. The best approach is to establish a schedule of activities by season for several years. A schedule will help to break down activities as well as focus efforts during the best times of the year. The schedule should include both establishment and periodic maintenance activities. The following schedule might be a good starting place:

Spring

- Prepare seedbeds and plant select wildlife plants.
- Maintain bird feeders with seed.
- Establish artificial ponds in early spring.
- Maintain hummingbird feeders.
- Plant annuals and perennials from seed.

Summer

- Mow natural meadows.
- Maintain bird feeders with seed.
- Plant annual fall plants.
- Till soil in late summer to establish native grasses.
- Maintain hummingbird feeders.

Fall

- Plant winter greenery such as wheat scattered in wildflower meadows or garden areas.
- Erect bird feeders and fill with seeds.
- Maintain bird feeders with seed.

Winter

- Build and erect nesting boxes.
- Clean and repair existing nesting boxes.
- Plant bareroot and container plants.
- Maintain bird feeders with seed.
- Clean bird baths for spring.
- Trim hedges sparingly on spring flowering hedges such as azaleas.
- Till natural meadows.

Step 5—Monitor and Adjust

As your backyard habitat begins to take shape, record the number and types of wildlife that visit your yard. The best way to do this is in a field notebook that can serve as your diary to document what you have done and the results. Try to note what trees, plants, and other habitat features are being frequented by individual wildlife species and those that are not used. Also take note of those plants that seem to grow well in your yard as well as those that did not fare so well. This information will be useful as you look back to see what changes or modifications are needed to make your yard more attractive to wildlife.

A Brush Pile for Cover

Brush piles can provide additional cover for small mammals in backyards. To build a brush pile, lay four logs (6 feet long and 4 to 8 inches in diameter) parallel to one another about 8 to 12 inches apart on the ground. Then place four more logs the same size across and perpendicular to the first four poles. These will keep pathways open under the pile for small mammals. Next add large limbs and then smaller branches on top until the pile is 4 to 6 feet in height and diameter. Sticks and branches can then be continually added to the top as the pile rots. Backyard brush piles will also provide protective cover for some birds.

ATTRACTING SELECT SPECIES

Small Mammals

More than half of Alabama's terrestrial mammals might visit a well-developed backyard habitat. If you live in a rural or suburban area close to woods or open fields you may be able to attract foxes, bobcats, coyotes, and deer. Residential areas farther removed from cities provide a good opportunity for viewing these animals around homesites. However, if you live in urban or suburban neighborhoods that do not adjoin large tracts of undeveloped land, you probably will not be able to draw these larger mammals to your property.

Some of the mammals that you might be able to see in your backyard include raccoons, opossums, skunks, cottontail rabbits, flying squirrels, gray and fox squirrels, chipmunks, cotton mice, old-field mice, Eastern woodrat, moles, voles, shrews, and bats. Most of these mammals are nocturnal and secretive, and they are dependent on cover to protect them from predators. Suggestions for attracting small mammals include the following:

- Give special attention to cavity trees since they provide denning sites for small mammals. If cavity trees are not available, construct nest boxes as substitutes for natural cavities.
- Plant native trees with edible fruits and nuts such as mulberry, wild cherry, beech, pine and sawtooth oak.

MICHAEL WORTHY/ARISTOCK

Backyards have the potential to attract more than half of Alabama's mammals. Smaller wildlife, such as this secretive Eastern harvest mouse, depend on cover to protect themselves from predators.

- Create maximum habitat diversity and edges in your backyard habitat.
- If cover is scarce, build a brush pile.

BATS

There are 15 different species of bats known to occur in Alabama. There is good potential to attract the little brown bat, big brown bat, red bat, evening bat, and Eastern pipistrelle to backyard habitats. Bats are extremely valuable in residential areas because of their ferocious appetites. A single little brown bat can consume about 500 insects per hour! Little browns and other bats forage for several hours a night, so it is easy to see how they are one of the most efficient natural insect predators.

Most bats roost in caves, in attics, under tree bark, beneath bridges, and in similar dark, covered places. Roosts are often close to open ponds, fields, or other large expanses where bats can forage in the open for flying insects. One method to attract bats is to provide roosting sites by erecting bat boxes. These boxes can be purchased from backyard wildlife specialty stores or from Bat Conservation International, P.O. Box 162603, Austin, Texas 78716. A portion of the money received by Bat Conservation International from bat box purchases goes directly to conserving bats and their habitats.

Bat boxes can also be built of Western red cedar, redwood, or cypress. Rough lumber is best because it makes it easier for the bats to secure a good foothold when roosting. If rough lumber is not available, roughen the inside of the house with a chisel or file. Avoid using treated or painted lumber because bats are susceptible to chemicals.

Bat boxes should be placed on the side of a tree or house approximately 10 to 15 feet above the ground where they are protected from prevailing north and west winds. They should be placed where they can receive maximum heat from the sun, especially during the morning hours. Researchers at Penn State University found that bat houses that were most often used for roosting and raising young received at least seven hours of direct sunlight. Ideal temperatures inside bat houses for raising young should be 80°F to 110°F. Bat houses should be made as airtight as possible to conserve heat by gluing all joints and cracks with silicone caulk. Tar paper or dark shingles placed on the top of bat houses can also help increase heat absorption by the box.

Experiment in your yard to see which direction the bat house should face. South-facing houses will be warmer than north-facing houses, which can also catch the brunt of damaging northerly winds. By testing several choices and recording the high and low temperatures inside the house, you can determine the best position. Some bat enthusiasts will first mount four houses in a group, each facing a different direction, to provide a wide range of box temperatures for bats to select. Because bats forage for flying insects, bat houses should be located near open water sources or open areas such as meadows, large yards, or fields. It may take several years for bats to start roosting in boxes, so try not to be discouraged. Most bat houses are not used the first year they are erected, and some may never be used. If bat houses are not used after the second year, consider relocating them to other sites. Experimenting with different locations and a little patience will eventually pay off.

BIRDS

Of the 363 birds known to occur in Alabama, a good portion of the more common species will visit backyards that offer some of the basic habitat features discussed earlier. Resident birds may appear year-round, while others are only seasonal in their occurrence. On Figure 6 is a list of some of the more common birds, their habitat preferences, and what can be done to make backyard habitats more appealing for each.

KAREN LAWRENCE/ARISTOCK

KAREN LAWRENCE/ARISTOCK

RICHARD T. BRYANT/ARISTOCK

RICHARD T. BRYANT/ARISTOCK

MAC CONE

DR. H. S. BANTON

LEWIS O. ROGERS/SOUTH CAROLINA DNR

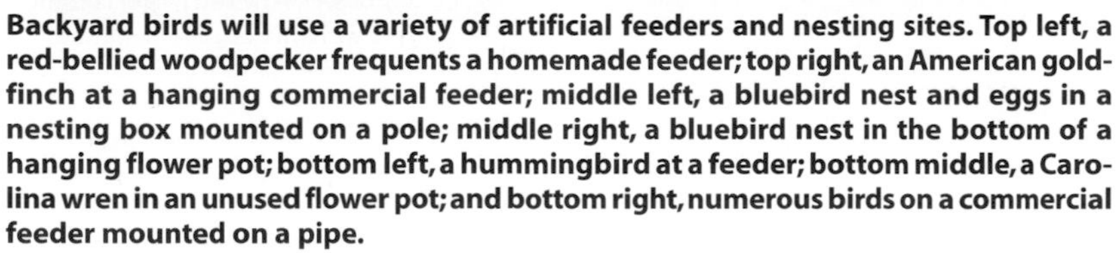

Backyard birds will use a variety of artificial feeders and nesting sites. Top left, a red-bellied woodpecker frequents a homemade feeder; top right, an American goldfinch at a hanging commercial feeder; middle left, a bluebird nest and eggs in a nesting box mounted on a pole; middle right, a bluebird nest in the bottom of a hanging flower pot; bottom left, a hummingbird at a feeder; bottom middle, a Carolina wren in an unused flower pot; and bottom right, numerous birds on a commercial feeder mounted on a pipe.

Fig. 6: Habitat Preferences for Select Birds of Alabama

Birds	Preferred Natural Food	Preferred Nesting Site	Feeders	Nest Boxes	Landscape Preferences
Cardinal Blue Grosbeak Indigo & Painted Bunting	Mostly seeds of wild and cultivated grasses, some insects. Cardinals eat more than 100 kinds of fruit.	Thickets, vines, dense stands of young saplings, other brushy plants.	Yes	No	Cardinals prefer mixed gardens with hedges and lawns backed by a variety of trees. They also have a strong preference for sunflower seeds. Buntings and grosbeaks like brushy pastures and woodland edges, seek an exposed perch to sing on, feed on the ground, favor white proso millet at feeders. Buntings are shy and require heavy cover.
Purple Martin	Vast quantities of insects.	Natural cavities, holes and crevices in sides of bluffs or cliffs. Readily use gourds or martin houses.	No	Yes	Prefer open meadows and lawns near water. Have learned to nest in gourds and special birdhouses placed in suitable habitat. Do not use pesticides in backyard.
Eastern Bluebird	Primarily insects, some fruits and berries.	Natural cavities in trees, old woodpecker holes in	Rarely	Yes	Prefer open areas with scattered trees. No insecticides!
Blue Jay	Acorns, other nuts, and berries, insects, small reptiles and mammals.	Variety of trees 10–13 feet off the ground.	Yes	No	Prefer yards with large numbers of trees, especially oaks, beech, and pines. Water is a major attractant. Peanuts are especially attractive in feeders.
Carolina Wren	Mostly insects.	Cavities or crotches of trees or shrubs.	Yes	Yes	Prefer wooded gardens with dense shrub undergrowth. Will nest in almost any cavity around homes and in gourds hanging from eaves. Prefers peanut butter and suet cakes.
Mocking Bird Catbird Brown Thrasher	Insects, grubs, fruits, and seeds.	Dense, thorny shrubs or vines conceal basket-like nests. Brambles ideal.	Yes	No	Edge situations provided by gardens excellent for mockingbirds. Native berries are important food sources. Catbirds like access to water. Thrashers forage on the ground where leaf litter is plentiful.
Carolina Chickadee Tufted Titmouse	Insects and many plant foods.	Natural cavities and abandoned woodpecker holes.	Yes	Yes	Yards with mature deciduous and evergreen trees supported by dense shrub and small tree understory are best. Chickadees prefer to dig their cavities in partially rotted trucks and stumps, especially pine and birch. Hanging suet feeders and sunflower seeds are especially attractive.
Screech Owl Barred Owl American Kestrel	Mice and insects.	Cavities.	No	Yes	Prefer gardens with many old trees to open, unmowed areas for hunting. Prefer cavities in hardwoods and old woodpecker holes in pines. Readily use appropriate nest boxes. Will use water if provided.
Woodpeckers Red-headed Red-bellied Downy Pileated Flicker Yellow-bellied Sapsucker	Major consumers of forest pest insects, grubs and eggs, ants, beetles; and also berries, nuts, and seeds.	Cavities in dead and dying trees.	Yes	Yes (except pileated)	Pileated and red-bellied prefer old-growth forests with mixed hardwoods. Downy and flicker common in yards with mix of deciduous and evergreen trees and shrubs, some open ground. Optimum habitat for red-headed has open lawns, a few large pines and oaks, and some dead snags nearby. Maintain snags in yards for all woodpeckers. Leave stumps and fallen logs as foraging habitat. Will eat suet; red-headed likes bread on platform feeder.

Fig. 6: Continued.

Birds	Preferred Natural Food	Preferred Nesting Site	Feeders	Nest Boxes	Landscape Preferences
Robin Wood Thrush	Forage on ground for insects; also eats fleshy fruits and berries.	Trees.	Rarely	No	Wooded yards with densely planted understory.
Orchard Oriole Northern Oriole Summer Tanager	Insects, fleshy fruits, especially berries.	Oriole prefers shade and trees near water. Tanager prefers deciduous trees such as oaks.	Yes	No	Prefer high feeding stations with fruit; northern orioles like suet. Attracted to gardens with mixed trees, especially orchard trees, mulberry, tupelos, wild cherry and blackberry. Orioles attracted to fruit at feeders, especially oranges.
Cedar Waxwing	Abundant fleshy fruit on shrubs and trees.	Not in Alabama.	Rarely	No	Manage yards to include many fruiting natives such as dogwoods, buds, and flowers of hardwoods, holly, and red cedar berries in late winter.
White-breasted Nuthatch Brown-headed Nuthatch	Insects and nuts.	Cavities in dead trees or old woodpecker holes.	Yes	No	Leave snags. Many hardwoods and pines are preferred cavity trees. Suet and sunflower seeds are feeder favorites.
Goldfinch Purple Finch Pine Siskin	Buds, soft fruits, seeds, insects in summer.	Not in Alabama.	Yes	No	Sweetgum and sycamore fruits are favored winter foods. Water is one of the best attractants. Most prefer high feeders; goldfinches will feed on the ground. All prefer sunflower and thistle seeds.
Yellow-billed Cuckoo	Caterpillars, grasshoppers, other insects.	8–12 feet high in shrubs or on horizontal tree branches.	No	No	Best natural controller of tent caterpillars. Generally prefer trees with dense canopies, such as oaks.
Ruby-crowned Kinglet	Tiny insects gleaned from foliage high in trees.	Gnatcatchers nest on horizonal limbs 25 feet or higher and use many kinds of trees.	Yes	No	Prefer mature, diverse yards, good mix of evergreen and deciduous trees. Occasionally visit small suet feeders.
Eastern and Gray Kingbird Great Crested & Blue-gray Flycatcher Eastern Phoebe	Mostly catch insects, bees, grasshopper, ants, some fruits.	Often near water. Kingbird likes medium shrubs or trees.	No	Yes (except kingbird)	Like deciduous woods habitat. Attracted by gardens with water. Kingbirds need perch with good view. Great-crested flycatcher will nest in gourds.
Red-winged Blackbird Common and Boat-tailed Grackle	Mostly seeds and grains, some insects.	Wetlands or nearby fields, often in cattails.	Yes	No	Forage in all types of open habitat during non-breeding season. Highly attracted to water sources.
Warblers	Insects, some seeds.	Large trees and shrubs near water.	Suet Feeders	No	Many resident and migrant warblers will be attracted to a diverse garden with many canopy layers, including mature trees. Oaks provide a good source of caterpillars. A water source will bring in seldom-seen species.
Vireos	Insects and spiders, some fleshy berries prior to migration.	All suspend hanging nests in trees from 3–4 feet off the ground to treetops.	No	No	Same as warblers.
Sparrows	Feed on ground, mostly weed and grass seeds, and some insects.	Near ground.	Yes	No	Require mixed garden vegetation with weed, grass, and shrub cover. Will visit feeders regularly. Use water sources often.

Suet Cake Recipe

1 cup of ground suet
1 cup of smooth peanut butter
2–3 cups of yellow corn meal
½ cup of enriched white or whole wheat flour

Melt the suet in a saucepan. Add peanut butter and stir until melted and blended together. In a separate bowl, mix together the dry ingredients. When the suet/peanut butter mix has cooled and begins to thicken, add the dry ingredients and blend thoroughly. Stuff the mixture into a pinecone or form into cakes in muffin tins for use in suet feeders.

RAY DOVE/ARISTOCK

Feeders should be located so that birds like this house finch can be easily observed. Care should be taken to select seeds that have high oil content for energy, such as the sunflower seeds in this feeder.

Artificial Feeders

Artificial feeders can be helpful in supplementing native food for backyard wildlife. Feeders offer a simple way to attract many species of birds and some mammals for backyard viewing. A variety of commercial feeders are available.

Feeders should be located where they are easily visible from windows, porches, or other observation points. Thick shrub or tree cover should be within 10 to 20 feet of the feeder to provide escape cover for birds pursued by predators. Feeders should not be placed close to or within dense shrubbery because these areas can be death traps for feeding birds stalked by cats. Feeders should be positioned far enough away from windows to prevent bird collisions and discourage mock fighting by birds that are threatened and preoccupied with their own reflection in the glass. This is a particular problem in early spring, when birds are beginning to nest. Even the shy bluebird becomes amazingly defensive and will attack any reflection in close proximity.

Each bird species prefers certain foods, but seeds are a preferred food because of their high protein and fat content. Studies have shown that the top grain choices for birds are seeds with a high oil content such as striped and hulled sunflower seeds, finely cracked corn, white proso millet, and thistle seed. Separate feeders should be used for different kinds of grain to reduce competition among several species and prevent grain loss. Avoid using most commercial seed mixes because they contain less-preferred seeds that are eventually wasted as birds pick out the higher quality seeds such as sunflower. Some nuisance birds, such as starlings, brown-headed cowbirds, and house sparrows, can be discouraged by limiting the seeds they prefer. Milo and hulled oats attract starlings, and wheat is preferred by brown-headed cowbirds and house sparrows.

Several kinds of feeders should be placed at various heights and locations in your backyard to accommodate the different eating styles of various birds. For example, millet and cracked corn can be offered in ground feeders for doves, towhees, sparrows, and quail. Sunflower seeds,

Features of a Good Bird Feeder

- Holds enough food for two or three days.
- Easy to fill with seeds.
- Protects food from inclement weather because wet grain spoils quickly. Moldy food is unhealthy for birds.
- Can be easily taken apart and cleaned.
- Discourages unwanted wildlife such as squirrels or predators.
- Keeps spillage and waste grain to a minimum.
- Can be maintained year-round.

mixed grains, and fruit can be placed on a platform or hopper feeder 3 to 4 feet off the ground for perching birds like cardinals, finches, and grosbeaks. A suet feeder can be added and suspended or attached to a tree limb. Suet feeders should be placed in the shade to prevent the food from becoming rancid. At least 12 different bird species will use suet on a year-round basis.

Components of a Good Nest or Denning Box

- Should have ¼-inch ventilation holes under roof overhang.
- Should have six ¼-inch drainage holes in the floor.
- Avoid metals and plastics as nest-box material.
- Avoid using stains and wood finishes that contain pentachlorophenol, green preservative, or creosote.
- Roughen interior wood below entrance hole with a wood chisel to help birds climb out of box.
- Specification and location of box should match requirements of species. Entrance size and cavity depth are important.
- One side of roof should be hinged for easy cleaning and maintenance.
- Roof should extend over all sections for maximum protection. Should extend at least 2 inches over the entrance hole to shelter the interior from rain and to keep predators from reaching in.
- Do not use a perch on the outside next to entrance hole. This will only attract undesirable species such as house sparrows and European starlings.
- For added protection from the rain, the floor should be elevated ¼ inch up from the bottom of the sides to prevent seepage.
- Can provide nesting material such as hair, bits of yarn, or lint from a clothes dryer in a mesh-net bag hung from a tree.
- Provide predator guards or shields to prevent egg predation and the death of mother and young.

TES RANDLE JOLLY

Nesting boxes should be maintained or replaced on a regular basis, unlike the one pictured here. The post supporting this nesting box should have a predator guard to keep out unwelcome visitors such as the egg-consuming black rat snake and other predators.

Nesting Boxes

Nesting and denning boxes provide shelter for some birds and mammals to give birth and raise their young. There are many designs that are commercially available for a variety of birds and mammals. When buying a nest box, be sure that it is designed for the particular bird or mammal in which you have an interest. Commercial birdhouses are often designed for beauty rather than for the birds. What is attractive to people can sometimes be useless for birds. Avoid houses made of plastic or metal. They may suffice in the cooler temperatures of spring, but in summer plastic and metal absorb too much heat and turn a birdhouse into an oven, potentially causing nestling mortality. Birds will abandon these houses, but usually only after the young have died. An exception to the metal box rule is anodized aluminum purple martin houses that have a relatively large entrance and a central ventilating shaft that opens to each compartment, providing sufficient cooling.

Nesting boxes can be easily constructed and should be built with ¾-inch-thick durable wood such as cypress, Western red cedar, or exterior grade plywood. Try to use rough-cut lumber whenever possible because it provides a better foothold for young birds climbing out of the box. Rough lumber also blends into the natural setting.

To avoid nest predation, nesting boxes should be placed out of reach of neighborhood cats, snakes, raccoons, and other invading predators. Ideally,

ISIDOR JEKLIN

Named for the brilliant dark red band around its neck, the ruby-throated hummingbird is readily attracted to artificial feeders. The best designs are those that feature red flowers and can be easily taken apart for cleaning.

bird-nesting boxes should be placed on metal or wooden poles that have a protective predator guard placed 6 inches to 1 foot below the nesting box. Predator guards are metal skirts that prevent predators from reaching the nesting box. Another excellent way of preventing snakes from reaching nest boxes is to encircle the post or tree with a loose skirt of ¾-inch mesh nylon netting.

Hummingbirds

The ruby-throated hummingbird is the only species of hummingbird that nests in Alabama. After nesting in summer, they migrate south of the U.S. to spend their winters. A diverse backyard of evergreen and deciduous trees and shrubs can provide ideal nesting sites.

The main attraction for hummingbirds is nectar-producing flowers, especially tubular flowers like trumpet creeper, columbine, and coral honeysuckle. Hummingbirds are strongly attracted to red and orange flowers. A variety of plants that bloom throughout late spring, summer, and early fall will attract hummingbirds daily. An assortment of flowering native trees, shrubs, vines, and perennials is ideal. Flowering annuals also attract hummingbirds but require more care and replanting each year. Plants that produce single-flowered blossoms are desirable because they provide more nectar than double-flowered plants.

An excellent way to establish a "hummingbird garden" is to layer vegetation, starting with the tallest trees or plants in the back and gradually reducing plant height to the front of the garden. A trellis covered with vines like trumpet creeper, coral honeysuckle, or yellow jasmine can also serve as a back drop for a hummingbird garden. Lower-growing shrubs such as coral bean or firebush, followed by low flowering perennials and annuals should be placed in front of the trellis or taller vegetation. Be sure to consider the best viewing opportunities from windows and porches as you select a location for the garden. Because most flowering plants require almost constant exposure to the sun, the location of hummingbird gardens should be in full sunlight.

Hummingbirds can also be attracted to backyards by using artificial feeders. There are a variety of commercial feeder designs on the market. Some commercial feeders attach directly to windows with suction cups, and these birds will feed at the window's edge with little hesitation. The best designs are those that feature red flowers and can be easily taken apart for cleaning. You may purchase artificial nectar for filling feeders or mix it yourself. The recipe for homemade nectar is one part white granulated sugar with four to six parts water. Do not make the mixture any sweeter than this or substi-

Plants for Hummingbirds

PLANT	PERIOD OF BLOOM
Red Buckeye (*Aesculus sylvatica*)	April–May
Japanese Flowering Quince (*Chaenomeles japanica*)	Mid-May
Wild Columbine (*Aquilegia canadensis*)	May–June
Old-Fashioned Weigela (*Weigela florida*)	Late May
Common Beardtongue (*Penstemon barbatus*)	June–July
Lilies (*Lilium* spp.)	June–Sept.
Scarlet Sage (*Salvia splenders*)	Early Summer–First Frost
Coral Bells (*Heuchera sanguinea*)	All Summer
Fuchia (*Fuchsia* x *hybrida*)	All Summer
Allwood Pink (*Dianthus* x *allwoodii*)	All Summer
Pineapple Sage (*Salvia elegans*)	All Summer
Autumn Sage (*Salvia greggii*)	Summer–Fall
Mealycup Sage (*Salvia farinacea*)	Summer
Jeponeza and Coral Honeysuckle (*Lonicera* spp.)	Summer
Trumpet Creeper (*Campsis radicans*)	Mid-July
Flowering Tobacco (*Nicotiana alata*)	All Summer
Petunia (*Petunia* x *hybrida*)	All Summer
Nasturtium (*Tropaeolum majus*)	All Summer
Garden Phlox (*Phlox paniculata*)	July–Aug.
Bee Balm (*Monarda idyma*)	July–Aug.
Daylilies (*Hemerocallis* spp.)	July–Aug.
Obedient Plant (*Physostegia virginiana*)	Aug.–Oct.
Common Snapdragon (*Antirrhinum majus*)	Mid-Summer–First Frost
Common Rose Mallow (*Hibiscus moscheutos*)	Aug.–Sept.

tute honey. Boil the mixture for about one to two minutes until all the sugar is completely dissolved. After cooling, fill feeders and store unused portions in the refrigerator. Do not add red food coloring because the imitation red flowers on the feeder will attract hummingbirds.

Hummingbird feeders should be placed within an easy viewing distance in early spring to attract hummingbirds as they arrive from their northward migration and should remain up until the late fall. **Feeders should be taken apart every week and cleaned thoroughly with hot water (no soap) and vinegar to remove bacteria and fungus molds.**

Sapsuckers and orioles will also frequent the feeders. Many hummingbird watchers leave feeders out during the winter months to attract other hummingbirds such as the black-chinned and rufous hummingbirds. Other visitors such as ants, bees, and wasps may not be desirable. Some feeders come equipped with "bee guards" to discourage small flying insects. Vegetable oil applied around the feeder openings and wires holding feeders should discourage unwanted insects. Do not

BILL DYER/CORNELL LABORATORY OF ORNITHOLOGY

Pictured here is a female ruby-throated hummingbird with her young. The ruby-throated hummingbird is the only hummingbird species that nests in Alabama. They are known to raise two broods in one season and prefer to nest in isolated areas.

RICHARD T. BRYANT/ARISTOCK

RICHARD T. BRYANT/ARISTOCK

Amphibians and reptiles, such as the tree frog and box turtle, are easily attracted to backyards if cover and water are provided.

use pesticide sprays to control insects on feeders, as birds will occasionally consume the insects and are sensitive to these chemicals.

REPTILES AND AMPHIBIANS

Reptiles and amphibians can be a fascinating addition to backyard habitats. They are also beneficial because they consume thousands of insects during the spring and summer. The American toad, for example, can devour over 200 insects in one night. Most of Alabama's reptiles and amphibians are secretive and require various amounts of low-growing herbaceous plants and shrubs for protective cover. Many of the plants already mentioned in this chapter would meet this need.

Establishing an in-ground freshwater "pond" is probably the best addition for amphibians and reptiles. Artificial ponds should be lined with plastic, irregular in shape, and range from ½ inch to 2 feet in depth. At least two sides of the pond should be bordered by dense native grasses and wetland plants. These plants should appear naturally over time if the pond edges are not mowed. The pond and adjacent vegetation will create a small wetland plant community that is heavily used by many herpetofauna. A few rotting logs can be added to the edges of ponds and other moist sites to provide hiding places for salamanders and other herps. Other moist sites, such as drainage ditches, seeps, and low-lying areas, should also be reserved and maintained for reptiles and amphibians.

Drier sites can provide habitat for many reptiles such as the green anole, Eastern fence lizard, Eastern box turtle, and a variety of snakes. Some of these areas include rock piles, stone walls, brush piles, and sandy areas with protective cover. A portion of these areas should be fully exposed to the sun to provide important warming areas for the cold-blooded reptiles.

BUTTERFLIES

Butterflies can add color, variety, and beauty to any backyard. The most effective method of attracting a wide array of butterflies is to design and plant a wildflower garden. With an understanding of butterfly life cycles, gardens can be designed to provide flowering plants that meet most of the needs of butterflies. A butterfly garden that provides these needs will attract the largest number and variety of butterflies.

Butterflies use various plants for different reasons during their life cycle: egg, caterpillar, chrysalis (the encased "cocoon" containing developing adult butterfly), and adult. After mating, the female butterfly will search for a specific "host plant" on which to lay her eggs. For example, monarchs lay eggs on milkweed, black swallowtails on parsley plants, and Eastern tiger swallowtails on yellow poplar or wild cherry trees. Other butterflies lay eggs on a variety of plants. When the eggs hatch, the caterpillars selectively feed on the surrounding plants. If a preferred plant is not available, caterpillars will starve rather than eat other plants. For a few weeks, caterpillars eat voraciously on their favorite plants and then create chrysalises that they attach to plant stems.

Butterfly gardens should be located in unshaded areas because most butterfly-attracting plants require full sunlight. The garden should be designed so that large splashes of similarly colored flowers are grouped together

Selected Plants for a Butterfly Garden

NECTAR SOURCES

ANNUALS
Cosmos (*Cosmos sulphureus*)
Egyptian Star-cluster (*Pentas lanceolata*)
French Marigold (*Tagetes patula*)
Heliotrope (*Heliotrope arborescens*)
Impatiens (*Impatiens wallerana*)
Mexican Sunflower (*Tithonia rotundifolia*)
Moss Verbena (*Verbena tenuisecta*)
Zinnia (*Zinnia elegans*)

PERENNIALS
Eupatorium (*Eupatorium* spp.)
New England Aster (*Aster novae-angliae*)
Sages (*Salvia* spp.)
Pineapple Sage (*Salvia rutilans*)
Purple Coneflower (*Echinacea purpurea*)
Thrift (*Phlox subulata*)
Vervain (*Verbena bonariensis*)
Verbena (*Verbena canadensis*)

SHRUBS
Azalea (*Rhododendron* spp.)
Butterfly Bush (*Buddleia davidii*)
Glossy Abelia (*Abelia* x *Grandiflora*)
Lantana (*Lantana camara*)
Trailing lantana (*Lantana montevidensis*)

TES RANDLE JOLLY
RICHARD T. BRYANT/ARISTOCK
MAC CONE
TES RANDLE JOLLY

Butterfly houses (left) can be placed in wildflower gardens to provide shelter for butterflies. Top, a gulf fritillary feeds in a wildflower meadow; middle, bananas put out to attract butterflies; bottom, a red-spotted purple butterfly is a common site in Alabama flower gardens.

HOST PLANTS	BUTTERFLIES ATTRACTED
Bloodflower (*Asclepias curassavica*)	Monarch
Butterflyweed (*Asclepias tuberosa*)	Monarch
Parsley (*Petroselinum cripum*)	Black Swallowtail
Dill (*Anethum graveolens*)	Black Swallowtail
Copper Fennel (*Foeniculum vulgare*)	Black Swallowtail
Maypop (*Passiflora incarnata*)	Gulf Fritillary
Blue Passionflower (*Passiflora caerulea*)	Gulf Fritillary
Spicebush (*Lindera benzoin*)	Spicebush Swallowtail
Yellow Poplar (*Liriodendron tulipifera*)	Eastern Tiger Swallowtail
Black Cherry (*Prunus serotina*)	Eastern Tiger Swallowtail

Backyard Wildlife Habitat Program

The National Wildlife Federation's Backyard Wildlife Habitat Program offers homeowners a chance to certify their yards as official NWF habitat, if their property meets specific criteria for food, cover, water, and other features. NWF offers a variety of helpful publications, including a "Gardening with Wildlife Kit." Write to the National Wildlife Federation 1400 16th Street NW, Department HG, Washington, DC 20036-2266 for more information about their backyard wildlife program.

TES RANDLE JOLLY

Butterfly gardens and natural meadows provide an artist's palate of colors to enrich backyards and attract an array of butterflies such as this black swallowtail butterfly.

in the garden. This makes it easier for butterflies to locate preferred plants. Plants should include those that furnish nectar-producing flowers for feeding, function as host plants for laying eggs, and provide food for caterpillars. Many native butterflies prefer purple, yellow, orange, and red blossoms. Clusters of short, tubular flowers or flat-topped blossoms provide ideal shapes for butterflies to land and feed. Plants with single flowers produce more nectar and are more accessible for butterflies than double flowers. Butterfly gardens should be planted with a variety of plants that will provide blooms and nectar when the butterflies are most active from spring until fall. For example, azaleas provide blooms in the spring, butterfly bush in the late spring and the summer, and chrysanthemums in the fall.

Other features can also be added to butterfly gardens to enhance their power of attraction for butterflies. Damp areas or shallow puddles in the garden provide drinking sites and areas where butterflies feed on salts from moist soils. Occasionally large numbers of male butterflies will congregate around a moist area to drink, forming what is often called a "puddle club." Butterflies also like to spread their wings and bask in the sun. Plants can provide some of these sunning areas, and so can large, flat stones or small mounds of bare soil. If you have created a small pond lined in stones or wood, these can serve the dual purpose of securing the pond liner and adding a low sunning area.

To get the most out of your butterfly garden, purchase a field guide to help identify the visitors to your garden. Several excellent sources are listed in Appendix A. To get other ideas about butterfly gardening visit the famed Callaway Gardens butterfly displays in Pine Mountain, Georgia. There you will see one of the greatest varieties of butterfly-attracting plants, demonstration areas of butterfly gardening, as well as the Cecil B. Day Butterfly Center. This is the largest glass-enclosed butterfly conservatory in North America, and it contains more than 1,000 butterflies from all over the world. For more information write Callaway Gardens, Education Department, Pine Mountain, Georgia 31822-2000, or call (706) 663-5153.

NUISANCE BACKYARD WILDLIFE

Backyard wildlife enthusiasts who develop their yards to attract animals can expect to entice some wildlife that will become pests. Depending on your perspective, these animals may be only a minor nuisance, or they may cause absolute havoc. Backyard nuisance wildlife can be divided into three categories: 1) predators that eat wildlife you are trying to attract, 2) aggressive animals that scare off and out-compete other wildlife for food or nest sites, and 3) wildlife that pose a particular problem, such as rabbits eating your garden or raccoons getting into your trash cans.

Most natural predators pose a threat to other wildlife if sufficient cover is not provided. Predation most often occurs around bird feeders and feeding stations. To reduce the potential for this, dense escape cover should be available 8 to 10 feet away from feeding sites to provide for a quick getaway.

Domestic predators, however, are a different story. Cats, even seemingly benign house cats, are efficient predators of backyard wildlife. Feeding stations, even with adequate cover close by, become death traps for wildlife if cats are lurking nearby. If you own a cat, do not allow it to roam freely in your backyard habitat—keep it inside. Neighbors should be informed, diplomatically, of the importance of restraining their cats and the unwelcome impacts that they have on the wildlife you are trying to attract. At the very least, ask neighbors to attach a bell to the cat's collar so that feeding wildlife can be forewarned of approaching danger. Still, cats with bells also take their share of wildlife, and there is no guarantee bells will dissuade an effective cat. Stray cats that have no apparent owner can be captured in a live-trap (such as a Hav-a-Hart or Tomahawk) and taken to the nearest Humane Society or animal pound.

Some wildlife are pests because they dominate feeding sites and prevent other animals from feeding. Gray squirrels, southern flying squirrels, and chipmunks are examples of these types of pests. Gray squirrels cause the most problems associated with bird feeders. Although they are entertaining to watch as they make their way to and around feeders, they often damage feeders and scatter seed everywhere. Several specialized commercial feeders are available just for feeding squirrels, while others have been developed for the purpose of excluding them. If gray squirrels, southern flying squirrels, and chipmunks become a persistent problem, try using modified commercial feeders that are "squirrel proof." These feeders usually have protective metal around feeder openings that prevent squirrels from reaching seeds. Other designs have weight-activated doors that close when a heavy squirrel jumps on a feeder. Feeders should also be positioned away from tree trunks, limbs, roofs, or other points where squirrels can readily access them. Feeders suspended from thin wires or cables that are surrounded by short lengths of garden hose make it difficult for squirrels to climb down to them. Pole-mounted feeders can be protected with squirrel baffles, an inverted cone, or a sheet metal guard attached to the pole 1 to 2 feet below the feeder. Some backyard wildlifers have even tried greasing feeder poles with vegetable shortening (it is biodegradable and harmless), making the poles almost impossible for squirrels to climb.

Raccoons can also be a problem for homeowners. Dog food, pet feeding pans, and trash attract raccoons. Pet food and trash should be stored in cans that have lockable lids. Vegetable gardens and ornamental plants are routinely eaten by rabbits and deer. The most effective method of exclusion for rabbits and deer is by woven wire or electric fences. Several fencing designs are available that can be used in backyard settings. The local county Extension office will have information on preferred designs.

MICHAEL SIEDE/ARISTOCK

Once your backyard is landscaped for wildlife, be prepared to handle nuisance wildlife such as this gray squirrel raiding a bird feeder.

If nuisance wildlife problems do not subside, you can call the Extension Wildlife Damage Management Specialist (334-844-9233) at Auburn University for advice on how to handle a problem animal. Another alternative is to accept the nuisance wildlife problems and read the entertaining book *Outwitting Squirrels—101 Strategies to Reduce Dramatically the Egregious Misappropriation of Seed from Your Bird Feeder* by Bill Adler, Jr., Chicago Review Press, 1988. You may not be able to solve your problem, but at least you will have a good laugh.

SUMMARY

Homeowners have an opportunity to attract some of Alabama's approximately 560 species of mammals, birds, reptiles, and amphibians to their backyards. The key to doing this is planning that includes 1) determining what species to attract and their habitat requirements, 2) making an inventory of existing backyard habitat features, 3) creating and implementing a habitat improvement plan, and 4) monitoring the results of improvement efforts and making modifications where necessary. With planning, backyard wildlife can be attracted to homes that offer food, cover, and water. Providing backyard habitat can attract many species of wildlife that bring hours of viewing enjoyment that enrich our lives, and at the same time help to provide additional habitat for a variety of wildlife species.

Alabama has over 50,000 private ponds scattered across the state. Many of these ponds provide countless hours of recreational fishing as well as habitat for a variety of wildlife species. With a little planning and effort, farm and forest ponds in Alabama can be managed to provide some of the best fishing in the South.

RICHARD T. BRYANT/ARISTOCK

CHAPTER 15

Fish Pond Management

CHAPTER OVERVIEW

More than 1 million acres of rivers, reservoirs, and private impoundments in Alabama support a recreational fishing industry that has a dramatic impact on the state's economy. Each year Alabama anglers spend an estimated $835 million on fishing and fishing-related activities, much of this money going to rural areas. A large portion of recreational fishing is done in private ponds scattered throughout the state. Alabama has about 50,000 private ponds that are used for irrigation, watering livestock, and recreation. They cover more than 134,000 acres. Even though most of these ponds are not used for recreational activities alone, they could provide excellent fishing opportunities with proper management.

CHAPTER HIGHLIGHTS PAGE

UNDERSTANDING HOW PONDS WORK 338

BASIC PRINCIPLES OF FISH POND MANAGEMENT 340

POTENTIAL PROBLEMS 355

ENHANCEMENT TECHNIQUES 356

Small farm ponds are primarily man-made. Farm pond owners must be careful stewards of this resource to provide productive recreational fishing. Think of a pond as a garden or an orchard. It must be selectively laid out, fertilized, seeded (or stocked, in this case), weeded, selectively harvested, and protected from destructive acts of nature (e.g. oxygen depletions) to be bountiful.

UNDERSTANDING HOW PONDS WORK

Good pond management includes several components:

- Enhancing food availability for fish,
- Controlling harvest to maintain a balance of predator-prey populations,
- Weed control,
- Preventing situations that may cause fish kills, and
- Maintaining water quality.

None of these are simple tasks. Ponds are complex ecological systems that require personal commitment and insight for productive, plentiful results.

No two ponds are exactly alike. Ponds close to one another, but on the same **watershed** (the surrounding terrain from which the pond receives rainfall or water drainage), will each be slightly different. These distinctions are not totally understood, but experience has shown that soil characteristics, topography, shading from adjacent trees and vegetation, and localized variations in the watershed are unique for each pond. Other factors critical to managing a pond include the following:

- Plankton availability,
- Water quality, and
- Existing or introduced fish populations.

Plankton

Plankton, the microscopic and near-microscopic organisms that are suspended in the water of a pond, are vital because they are essential to the creation of oxygen in a pond. In considering the overall web of life teeming in the pond's waters, these tiny creatures are the first building blocks. They are needed because fish require oxygen to survive, and oxygen is not freely available in pond water. It must dissolve into the water before the fish can metabolize it. **Dissolved oxygen** comes from the air or through the process of photosynthesis. Aquatic plants, primarily phytoplankton and other algae, release oxygen directly into the water as a by-product of photosynthesis, the primary source of oxygen in water.

Two important factors affecting the health and productivity of fish ponds are oxygen content and forage

RICHARD T. BRYANT/ARISTOCK

Aquatic vegetation around pond edges can provide shelter for fish. Too much aquatic vegetation, however, will eventually "choke-out" a pond. Depth of pond edges is an important factor that determines the presence of aquatic vegetation around a pond's edge. Deep pond edges discourage aquatic plant growth.

base. Both of these factors are influenced by the microscopic-sized organisms in pond water called plankton and occur in two forms. **Phytoplankton** are plants and influence oxygen content in pond water. **Zooplankton** are animals and, along with phytoplankton, they form the base of the food chain on which the quality and quantity of fish populations depend.

Dissolved oxygen in pond water is necessary for fish health and survival. Phytoplankton are the primary source of dissolved oxygen in ponds, as they release oxygen directly into pond water during photosynthesis. A large population of phytoplankton is called a **bloom** and is evidenced by water with a green tint that results from chlorophyll found in plant tissue. When phytoplankton blooms die off rapidly, oxygen is absorbed from the water and fish may become stressed or die. Phytoplankton die-offs can be detected by a change in water color to dark or black and are common in deep hillside ponds or ponds affected by manure or fertilizer runoff.

Both phytoplankton and zooplankton play an important role in the food chain or forage base in a pond. Phytoplankton are eaten by zooplankton and insects which in turn are eaten by minnows and small fish. These same minnows and small fish provide the forage base for larger fish which are the measure most people use to rate the success or failure of a fish pond.

Fig. 1–2: Understanding How Ponds Work

ILLUSTRATION, COURTESY OF RAY SCOTT, IS FROM *GREAT SMALL WATERS*, A 3-VOLUME VIDEO SERIES WHICH CHRONICLES THE CREATION OF A 5-ACRE LAKE HE RECENTLY COMPLETED AND THE REHABILITATION OF A FARM POND. IT IS AVAILABLE BY CALLING (800) 903-2481.

Fig. 1–2: Understanding a pond's ecology and the importance of fertilization is the basis of a successful pond management program.

Water Quality

Color is an indicator of water quality and should be carefully observed. The color or clarity of pond water can be related to plankton populations or to suspended sediments and organic matter. **Productive water** (water that will support healthy fish populations) has a green tint that is produced by chlorophyll pigments contained in the billions of phytoplankton suspended in the water

Sediments (silt, sand, or dirt) washed into ponds after heavy rains can also alter pond color. Normal color should return within a few days as the particles settle. Ponds that receive too much sediment can become unproductive. This situation can cause fish to die because plants become shaded, reducing the amount of sunlight available for photosynthesis and oxygen production. Fish gills can also become clogged with the sediment particles, making it difficult for fish to breathe.

Pond dynamics are also affected by other aspects of water quality. Factors such as pH (whether the water is acidic or basic) and dissolved oxygen affect fish health and pond productivity. Some aspects of water quality fluctuate daily, weekly, or monthly. Dissolved oxygen and pH cycle each day. Alkalinity can change over a period of time, ranging from several weeks to months, depending on the pH of the watershed or soils on the bottom of the pond.

As discussed earlier, photosynthesis is critical to the production of oxygen in a pond. Because photosynthesis is driven by the energy in sunlight, oxygen production does not occur at night. Therefore, dissolved oxygen levels rise throughout the day. After sunset, oxygen slowly declines as plants and animals consume oxygen in their normal respiratory processes. In a well-managed pond, nighttime dissolved oxygen levels should not fall below 3 or 4 parts per million (ppm). Oxygen levels below 3 ppm stress fish, and many species may even suffocate and die when oxygen levels fall below 2 ppm.

Pond pH varies over the course of a day as a result of respiration and photosynthesis. The carbon dioxide released from respiration reacts with water, producing carbonic acid. During nighttime, more carbonic acid is formed because plants are now respiring (more carbon dioxide is produced). The pond becomes more acidic and pH is lowered. Acidic pH levels vary from 1 to 6.9. Basic pH levels vary from 7.1 to 14. The lower the number, the more acidic a compound is as a liquid. During daylight, phytoplankton use carbon dioxide in photosynthesis. This reduces acidity and increases pH. Pond pH normally fluctuates between 6.5 and 9. If the pH drops below 5 (perhaps because of acid runoff in mining areas) or rises above 10 (low alkalinity combined with enhanced carbon dioxide removal by dense phytoplankton or algal blooms), fish may become stressed and die.

Alkalinity is related to pH. The amount of base, a material similar to baking soda, in water defines what is known as alkalinity. These bases, usually bicarbonates, react with acids and minimize pH changes. Alkalinity can increase the availability of carbon dioxide and other nutrients to phytoplankton. A total alkalinity of 20 ppm or more is necessary for effective pond productivity.

BASIC PRINCIPLES OF FISH POND MANAGEMENT

To achieve viable fish populations, pond owners and managers need to keep in mind the cornerstones of healthy fish maintenance. Here are some essential points to consider:

- Proper location, pond construction, and watershed management;
- Fish selection and stocking;
- Removal of undesirable or overpopulated fish;
- Liming and fertilization;
- Harvesting and record-keeping;
- Pond balance; and
- Aquatic weed control.

Pond Construction Considerations

The first step in developing a new pond is identifying a suitable location. This step will directly affect the life, productivity, cost, ease of construction, and overall usefulness of the pond. Proper site selection will save many headaches and expense in the long run.

Soils and the lay of the land are among the first items to consider when evaluating the suitability of a site for a pond. Clay soils retain water while sandy soils allow water to pass through. Sites to avoid include sandy, gravelly, or peaty soils, shale ledges, and areas with rock outcroppings along the area where the dam will be located. These sites have a greater chance of leakage and drastic fluctuations in water level once a pond is constructed. Sites to favor include clay soils, small depressions, valleys, and wet boggy areas, as they typically are well suited to holding water. Soils should have a minimum of 30 percent clay to be suitable for the lining of the pond, including the basin and the levee. A good rule of thumb to use to determine if there is sufficient clay in the soil is to pick up a handful of moist soil and compact it into a ball. If the soil remains intact and does not crumble after considerable handling, there is enough clay in the soil to provide a watertight seal. If there is any doubt as to the quantity and availability of clay on the site, professionals from the Natural Resources Conservation Service can determine if soils have a sufficient amount of clay for constructing a pond by taking several **core samples**, which are representative samples

Fish shelters can be created during pond construction. Slash from clearing operations can be used to make shelters that provide refuge for young fish to escape predation.

of soil that can be analyzed to determine clay content.

Another important consideration in pond location is the size and potential water yield of the watershed present or needed and the size of the pond desired. The drainage area must be able to supply enough water to maintain a reasonably stable water level at all times. Watersheds surrounded by open lands (pastures or fields) will yield a greater water volume than forested watersheds of the same size. The ratio of drained watershed area to pond size should be from 5 to 10 acres of pasture or cropland and from 10 to 20 acres of forest land per surface acre of water impounded. For a watershed composed of a mixture of forest and open land, a general rule of thumb is 7 to 15 acres of watershed per surface area of pond. If the surrounding soil is sandy (well drained), double the acreage required to account for lower runoff.

When desired pond size and water yield from a watershed are out of balance, adjustments will have to be made. When a pond of a given size is desired and the supplying watershed is too large, ponds will have to incorporate features such as additional spillways or diversion ditches to handle excess water. This is especially important to minimize or prevent rapid pond flushing (changing or movement of water through a pond) which can lead to loss of pond fertilizer. Naturally, the necessity for the special water control structures will also increase the construction costs of the pond.

A drained pond reveals circular depressions that were once bluegill spawning beds. Ponds are often drained and restocked when they become out of balance.

Fig. 3: Building a Fish Pond

ILLUSTRATION, COURTESY OF RAY SCOTT, IS FROM *GREAT SMALL WATERS*, A 3-VOLUME VIDEO SERIES WHICH CHRONICLES THE CREATION OF A 5-ACRE LAKE HE RECENTLY COMPLETED AND THE REHABILITATION OF A FARM POND. IT IS AVAILABLE BY CALLING (800) 903-2481.

Fig. 3: Artificial fish shelters are easily created during fish pond construction to provide refuge for young fish to escape predation. These areas are also excellent fishing sites. Shelters can consist of rock piles, stumps, old cars, tire piles, slash debris from logging operations, or other large structures. Some standing trees should be left and protected during pond construction to provide natural fish shelters. A variety of tree shelter types can be created by injecting some trees with a herbicide to promote rapid decay.

Locating a pond where a stream continuously flows presents the same problem as a watershed too large for a desired pond size. Special water control features will be needed to handle the additional water flow and minimize rapid pond flushing. Other problems associated with ponds fed by continuous streams include the introduction of trash fish from upstream and sediment and other potential contaminants from other sources within the watershed. Slow natural springs are the best source of continuous water for ponds.

Livestock and Erosion Considerations

Livestock can cause severe erosion damage on pond banks and levees. The eroded sediments slowly fill the pond and create shallow areas, enhancing weed growth. In addition, animal wastes may wash into the pond during periods of heavy rainfall. This can cause water pollution or nutrient overload problems. To prevent this from happening

- Locate livestock watering areas below the pond;
- Prevent livestock grazing or roaming on watershed land; and
- Fence the pond area to keep cattle away if necessary.

Ponds should also be separated from row crops or fields by a grass barrier. Pesticides, herbicides, and contaminated soils can wash into a pond and kill fish. By placing sod or grass strips 50 to 100 feet wide around the pond, you can reduce soil erosion and chemical runoff from neighboring pastures and fields.

A common problem with farm ponds is leakage due

to improper construction. Soils for pond construction must contain a 30 percent clay mix as mentioned previously, and the dam itself should be constructed with a compacted clay core. Trees or other woody vegetation should not be permitted to grow on the actual pond banks or dam, as the roots will eventually penetrate the soil, causing leakage. Ponds should have a drain so that pond managers can easily regulate water levels. Get help before building or renovating ponds. Contact the local Farm Services Agency office, Natural Resources Conservation Service office, or the Fisheries Section of the Alabama Department of Conservation and Natural Resources, Game and Fish Division (see Appendix O).

SOUTHEASTERN POND MANAGEMENT

Fertilizing and liming ponds increases both food availability throughout the food chain and the total amount of fish a pond can support. Properly limed and fertilized ponds can be stocked with twice as many fish as unfertilized ponds and can also increase fish size.

Pond Liming and Fertilization

Just as you would fertilize fields to increase crop yields, you should fertilize a pond to provide phytoplankton with adequate nutrients for fish growth. Proper fertilization increases food availability throughout the food chain and indirectly increases the total amount of fish a pond can support. Properly fertilized ponds can be stocked with twice as many fish as unfertilized ponds and can increase the size of individual fish. Before fertilizing a pond, it is important to check pH and alkalinity rates. Fertilization will not stimulate a good phytoplankton bloom if alkalinity is below 20 ppm. Adding lime to a pond can increase pH and alkalinity, which helps to make fertilizers more available to phytoplankton. Even without fertilization, liming alone may improve available nutrients that can support a better phytoplankton bloom. The best time to lime a pond is during the winter before ponds are fertilized in early spring.

DAN DUMONT

An automatic feeder can improve fish productivity on well-fertilized ponds. One feeder per 5 acres is sufficient to increase the chance of producing trophy fish.

The process of determining whether or not a pond needs to be limed requires the help of a soil scientist, not guesswork. To have your pond tested, take samples from many places in the mud at the pond's bottom. Combine these and allow them to dry out. Then mix them together and place them in a soil-test box available from your county's Extension office. Mark the sample "fish pond" so that the proper tests can be run. The sample can then be mailed to the Auburn University Soil Testing Lab, where technicians will analyze the sample and return results recommending a liming rate, if needed. Liming rates for most ponds in Alabama, except for ponds in the Black Belt, usually run around 2

Comparison of Fish Pond Fertilizers

Auburn University tested the effectiveness of granular, liquid, powdered water-soluble (WSF), and controlled-release fertilizers (CRF). The following chart summarizes the results of fertilizer comparisons.

Fertilizer	Applications No./Year	Applications Lbs./Acre/Year	Fish Production[1] Lbs./Acre	Cost $/Acre	Comments
None	0	0	100	0	No cost involved but poor fish production.
Granular	10	12	335	$56	Increased fish production three-fold. Disadvantage in application on platforms, slow to dissolve and higher costs.
Liquid	10	40	321	$30	Increased fish production three-fold and instantly available in pond.
WSF	10	2	305	$28	Increased fish production three-fold and rapidly dissolves in water.
CRF	1	69	317	$32.50	Increased fish production three-fold, less applications, and constant rate of nutrients available.

[1]Based on sunfish production. (Source: Alabama Agricultural Experiment Station)

tons per surface acre of pond when required.

If alkalinity and pH are found to be low, the addition of powdered, agricultural limestone should raise pH and alkalinity, increasing the utilization of fertilizers. Do not use quick or slaked lime because these compounds can cause rapid pH changes that may kill fish.

Lime must be spread evenly over the entire pond so it can react with the bottom mud. This can be done from a boat that spreads liquid lime from a water pump or by broadcasting solid lime across the pond evenly. Because limestone dissolves slowly and is washed out of the pond with overflow water, repeat treatments every two to five years. Another alternative is to spread one-fourth the original application of lime into the pond each year. The good news is that adding more than the recommended lime (agricultural lime only) will not harm fish, because limestone simply will not dissolve in the water once the pH reaches the high level of 8.3. Auburn Extension Circular ANR-232, "Liming Fish Ponds," will give you additional tips if you require help, as will your local county Extension agent.

Proper fertilization can double or triple fish production. Infertile ponds will seldom produce more than 200 pounds per acre, but well-managed, fertile ponds will support 300 to 600 pounds of fish per acre. If the pond is not fished often or if the pond receives some natural fertilization, use half the recommended fertilizer rates, or do not fertilize at all. Once you start a fertilization program, it should be continued, or fish growth may become stunted due to reduced food supply and aquatic weed growth will increase. Also, just as trees with poor vigor are more susceptible to diseases and insect infestations, fish without the nutrients available through fertilization are more prone to succumb to disease.

Fertilizers are labeled with N:P:K ratios or the percent composition of nitrogen (N), phosphorus (P), and potassium (K). Phosphorous is usually the nutrient most needed to foster fish production since it is usually tied up in sediment at the bottom of the pond. The equivalent of 8 pounds of phosphate (the form of phosphorus used in ponds) in granular form, or 4 pounds in liquid form, per acre of pond is a commonly recommended treatment rate. Liquid fertilizer is often preferred because it is easier to apply and affects plankton blooms quicker. Powdered water-soluble fertilizers (WSF) are also popular because they dissolve instantly in water, as compared to granular fertilizers. A capsulated granular form of fertilizer, also known as a controlled-release fertilizer (CRF), is also available for fertilizing fish ponds and allows nutrients to be released slowly over time. The advantages of time-released granular fertilizers are that repeated pond fertilizations are reduced and pond nutrient levels are maintained at a fairly constant level.

Recommended Pond Fertilization Rates

Fertilizer Formulation	Pounds/Acre/Application
Granular	
20-20-5	40
16-20-5	40
18-46-0	18
0-46-0	18
Liquid	
13-38-0	10
10-34-0	10
Water Soluble (Powdered)	
10-52-4	2–8
Controlled-Release	
13-13-13	69

Nitrogen is usually the nutrient needed most in soils, but it is rarely needed in ponds, especially older ponds. New ponds may sometimes need nitrogen but once established, most ponds do not require nitrogen fertilization. Potassium (K) is usually not a necessary element in pond fertilization. When applying fertilizers in pond management, it is important to use fertilizer mixes that best provide the fertilizer components (N:P:K) in the amounts recommended by a soil test.

Landowners should understand that pond fertilization is expensive, time-consuming, and labor intensive. Fertilization programs should not be initiated unless the landowner intends to continue fertilizing.

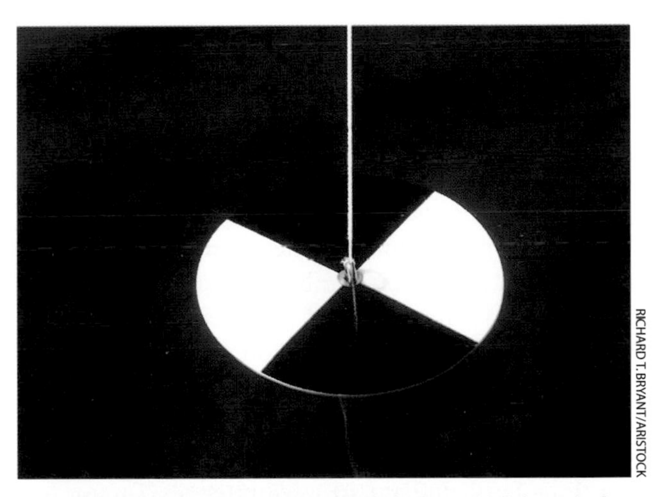

RICHARD T. BRYANT/ARISTOCK

A Secchi disk helps determine pond clarity, which is an index to the amount of phytoplankton available for fish. If the Secchi disk can still be seen when lowered to depths greater than 24 inches, fertilization is needed.

WHEN TO FERTILIZE

One simple method used to determine when to fertilize measures the clarity of pond water. The depth of light penetration in water is a good indicator of the phytoplankton density or bloom. Light penetration can be measured using a **Secchi disk.** A Secchi disk can be made from an 8-inch diameter disk of plywood, metal, or plastic. Mark the disk into four equal sections and paint each set of opposing quarters white and black, respectively, so that the colors can be visible under water. Attach the disk to the bottom of a broomstick, yardstick, or pole with the painted surface facing up toward you. Paint lines on the pole at the distances of 12, 18, and 24 inches from the disk.

Lower the Secchi disk into the water until it just barely disappears from sight and record that depth. Use figure 4 as a fertilization guide based on Secchi disk measurements. Low Secchi disk readings in muddy water (suspended sediments) are not reliable estimates of phytoplankton blooms.

If the Secchi disk disappears at a reading of 12 inches or less, the bloom is too dense and the water contains too many nutrients and oxygen depletion could occur.

Fig. 4: Pond Fertilization Recommendations Based on Secchi Disk Measurements

SECCHI DISK MEASUREMENT	RECOMMENDED MANAGEMENT
24 inches or greater	Fertilize.
18–24 inches	Good Bloom. No Action.
12–18 inches	Dense Bloom. Watch.
6–12 inches	Bloom Too Dense. Find Cause. Prepare to Aerate.
6 inches or less	Likely Oxygen Depletion. Aerate at Night.

Try to determine the source of the excess nutrients such as livestock manures or field fertilizers washing into the pond. If you cannot locate the source, you may have overfertilized the pond. Overfeeding fish can also cause excess nitrogen and phosphorus concentrations in pond water. If this is happening, reduce or stop feeding and be prepared to aerate at night. Low Secchi readings that result from muddiness rather than from excessive algae should not be used as a basis for evaluating pond fertility.

For ponds that do not flood frequently, fertilization should begin in March. If the pond floods during the spring, wait until spring rains are over before fertilizing. The first three applications of fertilizer should be made two weeks apart and then another application should be made based upon water coloration using a Secchi disk. Ponds should generally be fertilized at three to four week intervals from spring through September.

Do not fertilize ponds earlier than late February or early March in Alabama, or before water temperatures have reached a minimum of 60°F depending on your location in the state. Fertilization should stimulate a phytoplankton bloom within two weeks, but the first application does not always stimulate a bloom. If a bloom does not appear, fertilize the pond again and continue fertilizing at two-week intervals. Do not fertilize more than three times. After a bloom has developed, fertilize the pond as necessary to maintain it following the Secchi disk guidelines. Continue managing your phytoplankton until late September or until water temperatures have dropped to 60°F.

Some pond mangers continue to fertilize over the winter months, as they feel that maintaining the bloom provides some extra food and reduces filamentous algae problems. Research suggests that winter fertilization does not increase the growth of fish, however.

HOW TO FERTILIZE

Phytoplankton have no roots and absorb nitrogen, phosphorus, and other required elements directly from the water. Granular fertilizers should not be broadcast directly into the pond because the granules sink to the bottom. The nutrients then become tied up in bottom sediments and are unavailable for phytoplankton uptake.

Granular fertilizers should be placed on platforms (such as a sheet of plywood) that are held by a stand at approximately 12 inches underwater. Place one platform for every 5 to 6 acres of pond surface water. Locate the platform in an area of the pond that receives good wind and wave action to circulate the water. Lay the required number of fertilizer sacks on the platforms and tear off the top layer of each sack. The wave action will distribute the fertilizer in the pond. Granules placed on the platform will slowly dissolve and promote a bloom.

If you are using a liquid fertilizer, you must dilute it with water. If liquid fertilizer is undiluted, it will sink to the bottom and be trapped by sediments. Once diluted, liquid fertilizer can be sprayed or splashed into the pond or applied directly into the prop-wash of a small outboard or electric motor. Apply the fertilizer evenly over as much of the pond surface as possible.

Fertilization is ineffective and should not be attempted in "flushing" ponds that have problems with flooding, large overflows, or flow-throughs because they rapidly lose fertilizer. Some ponds will flush many times in winter and early spring but respond well to fertilization during late spring, summer, and fall.

Muddy ponds (visibility of 12 inches or less) do not usually respond to fertilization. Because of the shading effect, it is difficult to establish phytoplankton blooms in murky water. Therefore, the pond is unproductive and receives little photosynthetically produced oxygen. Contact your county Extension office for information about clearing muddy water.

Do not fertilize ponds infested with an aquatic weed problem because the fertilizer will only stimulate weed growth. The nutrients are absorbed by unwanted vegetation, not by phytoplankton. You must control the weeds first. Starting a fertilization program before weeds appear is one of the best methods to prevent weeds from becoming established. A good phytoplankton bloom can shade out weeds that compete for essential nutrients.

Fish Selection and Stocking

The choice of fish to be stocked depends on the pond owner's goals. The largemouth bass and bluegill combination is the most common strategy for stocking ponds for recreational fishing across the South. (The term "bream," an imprecise generic term commonly used to describe small game fish, won't be used here. Instead, "small sunfish" will be used when referring to the wide variety of small sunfish including bluegill; redear sunfish or "shellcracker"; green, redbreast, longear, dollar, spotted, and orangespotted sunfish; crappies; and others. Large- and smallmouth bass are also in the sunfish family, but aren't included when "small sunfish" is used in the text. When a specific species such as "bluegill" is used, it is referring to that specific species.) The beauty of the largemouth bass–bluegill system is its simplicity. In a well-fertilized pond, zooplankton and insect larvae will be plentiful enough to supply food for young bass and all sizes of bluegill. Bluegill grow rapidly and reproduce repeatedly throughout the spring and summer. The bluegill provide bass with an abundant food supply (prey). With proper harvest techniques, the bass will grow rapidly and prevent bluegill from overcrowding the pond. Enough large small sunfish will survive to reproduce and sustain good bluegill populations.

Channel catfish can be added to a bass-bluegill pond. However, catfish will compete with bass and bluegill for natural foods and lower the number of bass and bluegill caught. Figure 5 gives recommended stocking rates for bass, bluegill, and catfish in new or renovated ponds. Blue catfish may also be stocked instead of channel catfish. Blue catfish are more effective predators than channel catfish and will compete with bass for bluegill.

Fish can be obtained for new or renovated ponds from the Fisheries Section of the Alabama Department of Conservation and Natural Resources. Local district fisheries biologists can be contacted for stocking applications. Private hatcheries offer consulting services and also sell fish and may offer varieties or hybrids selected for rapid growth at times when they are not available from the Fisheries Section of The ADCNR, Game and Fish Division. Contact your county Extension office for a list of live fish suppliers.

Bluegill should be stocked in early autumn (September) to make sure they have grown and matured enough

Ponds That Should Not Be Fertilized

Muddy Ponds.

Mud prevents sunlight from passing through the water. Tiny floating plants must have sunlight to grow. When a pond stays muddy most of the time, do not fertilize the pond until the mud problem is corrected.

Ponds Infested With Unwanted Fish.

If undesirable fish dominate the pond, poison the pond, restock, and then begin fertilization.

Ponds Infested With Weeds.

During warm months, pond weeds use up the fertilizer that the microscopic plants need. In these cases, the pond stays clear even after repeated fertilizer applications.

Ponds Not Fished Heavily.

Fertilizing a large pond is a waste of time and money if the pond is only occasionally fished and the fish are not removed from the pond.

Unbalanced Fish Populations.

If the small sunfish population is overcrowded, it means that there are not enough bass to keep the small sunfish numbers down. It would be unwise to fertilize if this condition existed. Small sunfish should be removed by poisoning or seining and additional fingerling bass added to the pond before fertilizing.

Ponds With High Water Exchange.

Ponds that have a problem with flooding, large overflows, or flow-throughs rapidly lose fertilizers. The ration of drained watershed area to pond size determines the degree of water exchange. Fertilizer would also be flushed into streams, potentially causing eutrophication of the streams.

Alabama's Fish Stocking Program

Owners of new or renovated ponds may apply to stock their ponds with Florida largemouth bass, small sunfish, and redear sunfish. The Fisheries Section of the Alabama Department of Conservation and Natural Resources, Game and Fish Division will stock ponds that are free of fish and larger than ¼ surface acre or ½ surface acre if the pond is not going to be fertilized. Application deadline is usually February 1. To request more information and an application form, contact the District Fisheries Office that serves your county.

BARRY W. SMITH

Bluegill and largemouth bass stocking combinations are the most popular in the South. Bluegill should be stocked in early autumn to make sure they grow and are mature enough to spawn in the spring.

BARRY W. SMITH

Largemouth bass should be stocked near the end of May or June. They will grow rapidly by feeding on young bluegill fry. After the first season, bass should average between ¼ to ½ pound. Where food is plentiful, bass may approach 2 pounds their first year.

District I: P.O. Box 366, Decatur, AL 35601; (205) 353-2634
Colbert, Cullman, DeKalb, Franklin, Jackson, Lauderdale, Lawrence, Limestone, Madison, Marshall, Morgan, and Winston.

District II: P.O. Box 158, Eastaboga, AL 36260; (205) 831-6860
Blount, Calhoun, Chambers, Cherokee, Clay, Cleburne, Coosa, Etowah, Randolph, Shelby, St. Clair, Talladega, and Tallapoosa.

District III: P.O. Box 305, Northport, AL 35476; (205) 339-5716
Bibb, Fayette, Greene, Hale, Jefferson, Lamar, Marengo, Marion, Pickens, Sumter, Tuscaloosa, and Walker.

District IV: 64 N. Union St., Montgomery, AL 36104; (334) 242-3628
Autauga, Bullock, Chilton, Dallas, Elmore, Lee, Landaus, Macon, Montgomery, Perry, and Russell.

District V: P.O. Box 245, Spanish Fort, AL 36527; (334) 626-5153
Baldwin, Choctaw, Clarke, Conecuh, Escambia, Mobile, Monroe, Washington, and Wilcox.

District VI: P.O. Box 292, Enterprise, AL 36330; (334) 347-9467
Barbour, Butler, Coffee, Covington, Crenshaw, Dale, Geneva, Henry, Houston, and Pike.

to spawn in the spring. Bass should be stocked near the end of May or June so they can grow rapidly by feeding on young bluegill fry. This ensures that ample forage is available for bass because bluegill spawn three or four times between spring and fall. After the first season, bass should average ¼ to ½ pound and can approach 2 pounds if food is plentiful. Catfish may be stocked in the fall or spring. When stocking catfish with bass, make sure the catfish are as big as the bass being stocked. Catfish cannot usually reproduce successfully in ponds with bass and bluegill, so be prepared to restock catfish populations as they are gradually fished out.

ALTERNATIVE STOCKING STRATEGIES

It is difficult to manage bass-bluegill populations in ponds less than ½ acre in size. These ponds should be stocked with catfish or other species. Catfish are good fighters when hooked, excellent table fare, and can be stocked at 250 to 500 catfish per acre. At this level you should offer feed to the fish. If stocked alone, catfish may reproduce, and the pond can become overpopulated. Try to prevent spawning activity. Catfish are cavity spawners, and reproduction can be prevented by

- Removing all stumps, rock piles, and other debris from the pond;
- Not allowing muskrats or beavers to colonize the pond (catfish will spawn in the burrows); and
- Not placing containers (tires or milk cans) in the pond that might be used for breeding sites.

Bass stocked at about 20 to 30 fish per acre can also help control catfish spawns. Other fish to consider using in small ponds include blue catfish, redear sunfish, hybrid bluegill, and fathead minnows. Redear sunfish, which are also known as "shellcrackers" because they eat snails, can be stocked with bass and bluegill. Redear sunfish grow larger than bluegill and are excellent sport fish. Shellcrackers are not as prolific as bluegill and do not provide sufficient spawns for bass forage. If redear sunfish are desired, stock 20 to 25 percent redear sunfish in place of bluegill (for example, stock 300 bluegill and 100 redear sunfish per acre). Some pond owners like to stock fathead minnows (1,000 per acre) as a forage fish in channel catfish ponds. These minnows are quickly eliminated if the pond is stocked with bass. Stocking rates may be increased in ponds that are well fertilized.

Many pond owners like to stock hybrid bluegill because they grow rapidly and provide excellent angling, especially if they are fed commercial fish feed, which they can be conditioned to accept. Hybrid bluegill are not sterile like most hybrids. Most of the fish are males,

Fig. 5: Stocking Rates for New and Renovated Ponds Larger than ½ Acre.

SPECIES	FERTILIZE	NUMBER Stocked/Acre
Bass	Yes	100–125
	No	50–75
Bluegill	Yes	1,000
	No	300–500
Catfish	Yes	50
	No	25

but if females are present, they will reproduce. Reproduction will lead to overpopulated ponds. Therefore, predatory fish such as bass should be stocked at 20 to 30 fish per acre to feed on young hybrid bluegill. This combination works best for ponds ½ acre or less.

Species that should not be stocked into farm ponds include black and white crappie, gizzard shad, flathead catfish and bullhead catfish, common carp, and shiners. These species may rapidly overcrowd ponds and/or may reduce populations of desirable fish species. **Pond owners definitely need to consult a qualified fisheries biologist if they stock these species.**

For example, crappie are popular sport fish but are not desirable for small ponds that are less than 50 acres. It takes three years for a crappie to reach a weight of ½ pound. A young, ½-pound female crappie can produce 50,000 eggs in a single spawn. Just a few successful spawns during one season will overcrowd a pond with young crappie. When these young fish mature, they consume all available food and then stop growing and become stunted. Young crappie also compete directly with young bass and bluegills for food. Large crappie will then feed on small bass and bluegill. It is virtually impossible to manage bass and crappie populations together in farm ponds. The end result is poor fishing for all species.

Flathead catfish should not be considered, as they are voracious fish eaters, and can grow large enough to consume even bigger bass. Common carp, another unwise choice, can overpopulate rapidly, eat eggs of other fish, compete for food, and muddy the pond with their bottom-feeding activity.

DOUGLAS P. DARR

The best remedy for removing unwanted fish species or correcting an imbalanced pond is to use Ronenone and restock the pond.

Removal of Unwanted Fish

People do not usually catch many fish from ponds that are poorly managed or ignored. Fish populations often become imbalanced or contaminated with unwanted species. Typically, unmanaged ponds become crowded with small or stunted bass or sunfish populations, stunted green sunfish, bullhead catfish, shiners, or other less desirable species. The best remedy in these situations is to eliminate all the fish and start over. Destroying unwanted fish is easy and inexpensive, and it requires less chemicals if the pond is partially drained and the fish are concentrated. However, fish can survive in small puddles or pools away from the main body of water. Treat all puddles regardless of size, even those that lie within the watershed but not necessarily adjacent to the fish pond.

Ronenone is a registered aquatic chemical that can

Trophy Bass Management Suggestions from American Sport Fish

Don Keller and Barry Smith of American Sport Fish are private fisheries consultants in Montgomery. They provide fingerlings of a variety of fish for stocking ponds and lakes for recreational fishing. As part of their services, they provide technical advice for pond owners who are interested in managing for trophy-sized bass. Both Don and Barry have M.S. degrees in Fisheries Management from Auburn University and 25 years experience managing ponds and lakes.

DON KELLER

Some pond owners define trophy largemouth bass to be in the 6 to 7 pound range, while others put trophy bass in the 10 pounds and over category. Whatever the definition, existing or newly constructed ponds can be managed to produce large bass.

In most cases a sample seining of existing ponds can show if a pond is in balance and that an adequate number of forage fish are available along with bass. If the history of the pond is unknown, electrofishing is the best method of determining relative numbers and condition of adult and sub-adult fish. Ponds that are unproductive often have a low forage base and are characterized by many small-sized bass. Existing ponds that are severely out of balance should be drained, treated with Ronenone, and restocked. With a newly constructed pond, the proper balance of small sunfish and bass can be added to ensure a good start.

Two of the most important considerations with any trophy bass management program are liming and fertilization. Properly limed ponds (a hardness of 20 ppm) enable nutrients in fertilizers to become readily available to plants that provide food for prey fish that are eaten by bass. A good fertilization program can triple the capacity of a pond to produce trophy bass. Ponds should be on a good fertilization program before bass are stocked. To produce trophy-size bass, ponds and lakes should be stocked with Florida largemouth bass, or a cross between the Florida and native bass. These fish grow faster and survive longer than native bass. In the last 20 years, most of the trophy bass in the Southeast have been Florida largemouth bass or a Florida-native cross.

For bass to grow to a large size, landowners need to have a good forage base of prey fish. To provide this forage base for new ponds we recommend that 1,000 small sunfish and 1,000 to 3,000 fathead minnows per acre be stocked in the fall or winter. Stocking small sunfish and minnows during this time allows these fish to spawn in

be used to kill fish. Contact your district fisheries biologist or Extension aquaculture specialist for more detailed information about purchasing and applying Ronenone if the instructions that follow do not answer all your questions. In Alabama this chemical can be purchased from most farm supply or feed-and-seed stores by anyone with a pesticide applicator license, but remember, **Ronenone has been classified as a RESTRICTED USE pesticide.** This means that people who purchase and apply it must attend training for its proper application and storage, to avoid costly mishandling.

Ronenone dissipates from the water within 3 to 20 days depending on water temperature and weather conditions. It comes in both liquid and powder forms at a 5 percent active ingredient concentration. It should be applied at a rate of 2 to 3 ppm; 1 ppm is equal to 2.7 pounds per acre-foot of granular Ronenone. There

JERRY L. MOSS

Ronenone is a registered aquatic chemical that is used to kill fish in ponds that are out of balance. It can be applied as a liquid or powder. It dissipates from the water within 3 to 20 days, depending on water temperatures and weather conditions.

the spring, providing an additional food for bass. Threadfin shad can also be added to the forage base the first summer after the bass have been stocked. Bass may not consume many shad the first summer, but they will take a significant amount the second year. Shad are heavy spring spawners contributing significantly to the total forage base of a pond. A stocking rate of 50 to 100 Florida largemouth bass per acre in mid-May or early June is recommended. The lower the stocking rate, the less competition among bass for food and the faster the growth rate. Stocking rates of bass and forage fish can be manipulated or tailored to account for natural soil fertility and the owner's desire for a particular type of fishing. For example, ponds in the Blackbelt Region are generally more fertile than ponds in the Coastal Plain and can support a higher stocking rate of both forage fish and bass.

For pond owners who have the interest and resources, an artificial feeding program can increase the productivity level of a well-fertilized pond. With this approach, an automatic feeder is installed for each 5 acres. Automatic feeders have been shown to increase the ability of a well-fertilized pond to produce trophy bass.

As bass become established and reproduce, preventing a build-up and overcrowding of smaller bass becomes important. Too many smaller sized bass increase competition and lessen the chances of producing trophy bass. Harvest of bass is necessary to prevent overcrowding. In many situations, bass 13 inches and smaller should be removed when fishing. Bass 14 inches or larger, especially those in the 3- to 5-pound range, should be returned so that they may grow to a trophy size. The number and size of bass to be removed can vary from lake to lake. The owner or manager should consult an experienced fisheries biologist to make this determination. If fishing pressure is not adequate to remove enough small bass, consultants can electrofish a pond to remove the necessary numbers to maintain an adequate forage base for the larger bass that remain in the pond.

Another method to increase the chances of producing trophy bass is to remove by angling a greater number of male bass than female. Past records have shown that females make up the greatest percentage of trophy bass. During spring when a majority of the fishing occurs, fishermen can easily recognize females by the red and swollen area near the vent. Females caught may then be released.

Although there is no minimum pond size for trophy bass management, the number of large bass that can be produced is lower in smaller ponds. The larger the pond or lake, the greater the chance of producing catchable numbers of large bass. For small ponds that are 1 acre or less in size, managing for trophy small sunfish instead of large bass is recommended.

should be 5.4 to 8.1 pounds per acre-foot of the actual product. In order to figure the rate to use, the volume of water in the pond, which is measured in acre-feet, must be approximated. One gallon of the liquid Ronenone with the usual 5 percent active ingredient mix will sufficiently treat an estimated 3 acre-feet of water at 1 ppm, or 1 acre-foot at 3 ppm.

To calculate the acre-feet in any pond, multiply the surface area in acres times the average depth in feet. A 2-acre pond with an average depth of 6 feet would have 12 acre-feet total and would require 4 gallons of the liquid 5 percent formulation to treat at a concentration of 1 ppm; it would require 8 gallons for 2 ppm. Generally, it is safe to stock fish two weeks after applying Ronenone during spring, summer, and autumn. Lower water temperatures during winter may allow Ronenone to remain toxic for as long as a month when the temperatures are very cold. To check for residual Ronenone, place a few small sunfish in a minnow bucket and float them in the pond. If the fish are alive after 24 hours, it is safe to restock fish.

Harvesting and Record-Keeping

Ponds should not be fished for at least one year following stocking. After the first season, bass are often easy to catch. The most common problem in small ponds is removing too many bass. People can catch more than half of a pond's total bass population in one day of intensive fishing. To maintain good fishing, you must carefully control how many pounds of fish are removed.

When bass are overharvested, the pond becomes

Fishing Record-Keeping

Date ______________________

Angler's Initials ______________________

Time Spent Fishing ______________________

SPECIES OF FISH:		Number Harvested	Harvest Weight
	BASS		
	BLUEGILL		
	CATFISH		
	SHELLCRACKER		
	OTHER (*species*)		
	OTHER (*species*)		

Notes ______________________

overpopulated with stunted bluegill. If this happens, it is difficult to restore the balance of predator (bass) and prey (bluegill) in the pond. It may be necessary to poison the fish and start again. As a general rule, fertile ponds can sustain an annual harvest of 25 to 35 pounds of bass per acre. If the pond is infertile, you should not remove more than 10 pounds of bass per acre per year. Do not begin bass fishing in a new pond before bass spawn in the spring, when the water is above 60°F. By practicing catch and release with the bass, you can enjoy successful angling more often. Bluegill should also be harvested. A general rule is to remove 10 to 15 bluegill for each bass taken, or 4 pounds of bluegill for each pound of bass.

The opposite sort of problem from overharvesting of bass is the "bass-crowded" condition where too many bass crowd out smaller sunfish trying to reach maturity. What few small sunfish do compete successfully for food and reach maturity may be of nice size, but numbers are too few for an ample small sunfish catch. Even the bass suffer eventually, because as predators they soon do not have sufficient prey to eat. The bass then become stunted or have a weakened physical condition. This situation can usually be corrected by heavy harvests of the bass. Once the numbers are in a better balance, quality small sunfish fishing can be maintained by regulating the harvest of bass. If fewer than 10 pounds of bass are removed per acre per year, the size of the average small sunfish increases. The harvest numbers of small sunfish, in contrast to bass, has little effect on the balance of populations in the pond. Still, to maintain a good average size of small sunfish, harvest 4 to 6 pounds of sunfish for each pound of bass.

Catfish may be removed when they reach a size that satisfies the pond owner, but usually they attain a size worth catching and eating at 6 to 12 months following stocking. Catfish must be stocked periodically to replace individuals that have been removed. Catfish fry do not usually survive when bass and small sunfish are present in a pond because bass and small sunfish eat them. Large catfish fingerlings (8 inches or longer) should be stocked into ponds with established bass–small sunfish populations as bass will consume the smaller catfish.

Evaluation of Pond Balance

Ponds should be checked every one to two years to ensure that fish populations are in balance or if fishing quality is unsatisfactory. Contact your local district fisheries biologist with the Alabama Department of Conservation and Natural Resources for assistance, but you must request help well in advance to schedule a visit. Private fisheries consultants can also provide assistance in determining if fish populations are in balance. Pond balance can be evaluated from catch records and seine data. When using catch records, do not rely on your memory. Be sure to keep records about the number, species, and size of each fish caught. Record-keeping sheets can be left at a drop-off box attached to a fencepost near the pond so that anglers may leave the information voluntarily without disturbing landowners or managers every time they fish.

Bill Ireland, a conservation leader and past president of the Alabama Wildlife Federation, proudly displays a 15-pound bass he caught in a private lake in Shelby County.

Balance can also be checked with a 10- or 15-foot minnow seine. The best time to seine the pond is early June. Try to seine several shallow areas of the pond that are clear of brush and weeds. If you allow the seine to cup or arch slightly as it is pulled, fish cannot swim easily around it. Record the size, number, and species of fish caught in the seine. Seining samples provide a "snapshot" of reproductive success and indicate the presence of unwanted species.

Sampling with a larger seine (30 feet or larger with ½- to 1-inch mesh) provides even more meaningful results when evaluating pond balance. Seine one or two areas in the pond and record the number of small sunfish in groups of 1) less than 3 inches, 2) 3 to 5 inches, and 3) greater than 5 inches. Bass condition can be

Evaluation of Pond Balance Using Seine or Angler Catch Information

Type Of Fish Caught	Recommendation
Seine data: Small to intermediate bass and small sunfish *Catch data:* Bass and small sunfish of various sizes	No immediate management necessary
Seine data: Small and intermediate small sunfish; few or no bass *Catch data:* Few large bass; few harvestable small sunfish	Remove small sunfish & stock 25, 6–8 inch bass/acre
Seine data: Few intermediate small sunfish; many recently hatched small sunfish *Catch data:* Numerous small, thin bass only; few but large small sunfish	Remove 50–75 bass/ acre (35 lbs.); stock 200, 4–5 inch small sunfish/acre
Seine data: No recent small sunfish hatch, few intermediate small sunfish, unwanted species present *Catch data:* Unwanted species or no fish, few small sunfish of harvestable size	Start over (use Ronenone)

indicated by plumpness, and any undesirable species should be noted.

If you catch both young bass and recently hatched small sunfish fry in the seine, the pond is most likely balanced. The pond is out of balance when few young bass or small sunfish fry, but many intermediate-size small sunfish (3 to 5 inches long), are caught in the seine. If large numbers of undesirable fish species are caught, it is time to poison the pond, with Ronenone, and start over.

Weed Control

Aquatic weeds are a common problem in farm ponds. Some rooted vegetation furnishes habitat for small animals and increases the food available to fish. Vegetation also provides small fish with cover to hide from predators. However, if left unchecked, weeds can take over the entire pond and remove the nutrients required for phytoplankton production. Predators also decline because their smaller prey have too many hiding places.

Aquatic weeds can be controlled using manual, chemical, or biological means. Manual control of plants (hand removal) like cattails can be practical when vegetation first appears. This is similar to pulling weeds from gardens. Woody vegetation along dams can also be successfully controlled by hand.

Another option for aquatic weed control is to use herbicides (chemical control). However, many herbicides are not approved for aquatic use. The weeds in question must be accurately identified. Another problem associated with the use of herbicides is oxygen depletion resulting from the death of planktonic algae. Oxygen depletions often occur after herbicides have been applied during hot weather in ponds with heavy weed over-growth. When considering herbicide control, check with your county Extension office, a fisheries biologist, or an aquaculture specialist for plant identification and treatment recommendations. Whenever applying chemicals, be sure to protect yourself and others by carefully following the label instructions.

One of the simplest and most economical long-term

Aquatic weeds, if left unchecked, can literally choke the life out of a pond by removing the nutrients required for phytoplankton production. Aquatic weeds can be controlled using manual, chemical, and biological methods.

methods of controlling rooted aquatic vegetation in new or recently treated ponds is to stock grass carp. The grass carp or "white amur" is an Asian carp brought into the U.S. to control aquatic weeds. These fish are primarily plant eaters once they reach a length of 10 inches. They do not stir up bottom mud like common carp or disturb the nests of other fish. During warm weather, grass carp can eat 30 to 40 percent of their body weight in weeds daily.

Grass carp prefer flowing water and will swim over a pond spillway if given the opportunity, such as after heavy rains. An escape barrier can be placed across the spillway to prevent this from happening. In Alabama, only sterile Asian grass carp are allowed. A list of certified triploid grass carp suppliers and information about building an escape barrier can be obtained from the Alabama Department of Conservation and Natural Resources, Fisheries Section or the Auburn University Cooperative Extension System.

The number of triploid grass carp that should be stocked depends on which weeds are present and the magnitude of the problem (see Figure 6 for grass carp stocking rates). If large springs flow into your pond, you might have to stock additional grass carp for effective weed control. If the pond contains large bass, you must stock 8-inch or longer grass carp fingerlings, as the bass will eat most smaller fish. After the carp grow to 20 to 30 pounds (five to seven years after stocking), they are less effective in removing weeds and should be removed by seining, bow-fishing, or spearing. They are delicious to eat and are renowned in their native region of the world but lesser-known here.

If weeds have been allowed to get out of hand, it may be necessary to treat the pond with a herbicide first and later stock 10 to 15 grass carp per acre to maintain weed control. For more information on aquatic weed control refer to publication ANR 48 "Chemical Weed Control for Lakes and Ponds" or publication ANR-452 "Grass Carp for Weed Control in Alabama Ponds" which can be obtained through the local county Cooperative Extension System office or the Auburn University Cooperative Extension System publication distribution office (334-844-1592).

POTENTIAL PROBLEMS

Most pond problems are related to improper management. A problem not directly related to management is pond "turnover." Pond turnover is caused by **pond stratification** or layering. Stratification occurs when surface water warms faster than deep water. The warm layer is lighter and does not mix with the cool, deep water. Cool water near the bottom becomes stagnant, does not circulate, and results in low dissolved oxygen. Toxic compounds may also be produced by bacteria and

Trophy Bass Management on Sumter Farm

Dr. Danny Everett is a Wildlife Biologist and Manager of Sumter Farm in Emelle, Alabama. Dr. Everett received his graduate degrees in wildlife from Mississippi State and Auburn University. Through his formal training and years of on-the-ground experience, Dr. Everett has combined practical and innovative approaches that have made Sumter Farm a model for wildlife management in Alabama.

On Sumter Farm trophy bass management encompasses stocking and maintaining a good balance of bass and small sunfish, proper pond fertilization, sufficient fishing pressure to reduce crowding, and record-keeping. To manage for trophy largemouth bass, I like to start off with a "clean lake" that has no fish. The stocking rate that I use is 100 bass and 1,000 small sunfish per acre. Fertilization is a big part of keeping the pond productive, so I usually use 10-34-0 liquid fertilizer applied six to eight times from March or April until September or October depending on the temperature. As long as the water temperature is still warm enough to produce an algae bloom, I will continue to fertilize.

The second year after stocking, the pond or lake is ready for fishing. If the pond is not fished, then it will quickly become crowded with undersized bass. The biggest problem on our lakes is getting enough fishing pressure to remove an adequate number of small bass. Our policy on Sumter Farm is to keep all the bass caught that are less than 16 inches and to release those that are larger than 16 inches, except those to be mounted. Our goal is to remove 30 pounds of bass per acre each year. Removal of smaller bass allows enough food to remain for larger bass. On larger ponds and lakes, I like to stock with thread-fin shad, which helps to maintain a balance between small sunfish and bass for a longer period of time. Shad compete with small sunfish, slowing down small sunfish recruitment. Since bass feed on small sunfish, reduced small sunfish numbers also slows down production of bass and prevents overcrowding. Usually within six years a well-managed pond can routinely produce bass over 10 pounds.

On Sumter Farm we keep records on each pond to track bass productivity. The most important information is the number and weight of fish. Average weights and numbers give us an indication of the productivity of our ponds. Our goal is for the average fish kept that is less than 16 inches to weigh about 1¼ pounds. As an added bonus, our fishing lakes on Sumter Farm also provide us with a reservoir of water to flood our greentree reservoirs for waterfowl in November.

decaying organic matter. A **turnover** occurs when the upper layer cools quickly and mixes with the stagnant layer. The resultant mixture may not contain enough oxygen to support fish. Turnovers usually take place after a cold, heavy rain or the sudden passage of a cold front. Immediate aeration may save the fish. Fish kills can also be caused by oxygen depletions resulting from phytoplankton bloom die-offs or decomposing vegetation killed by herbicide applications including herbicide runoff from upstream areas.

Pond owners should also be informed about the quality of upstream water entering a pond. In some cases industrial and agricultural pollutants such as organochlorides may cause human health and environmental concerns. The risk is usually greatest to subsistence fishermen from these pollutants. In most cases, fish consumption advisories are issued and posted adjacent to waterways that pose a threat to human health. Pond owners concerned about incoming water quality should contact the Alabama Department of Environmental Management (334-271-7700).

ENHANCEMENT TECHNIQUES

In addition to managing a pond correctly, several other techniques can improve farm pond fishing. Some of these techniques include the following:

- Stocking fathead minnows for forage,
- Adding fish shelters/habitat,
- Supplemental feeding,
- Checking and adjusting water levels,
- Aeration and mixing of pond temperature layers, and
- Providing preferred spawning substrate.

Fig. 6: Stocking Rates for Triploid Grass Carp

WEED EVALUATION	GRASS CARP STOCKED/ACRE
New pond or minor weed problem	5
Moderate weed problem (10% to 20% coverage)	10 to 15
Severe weed problem or spring-fed pond	15–20 or more

Fathead Minnows

For newly established or renovated ponds, bass survival and growth can be improved by adding fathead minnows in February or March before the bass are restocked in June. These minnows will spawn in time to produce ample forage for young-of-the-year bass. Unfortunately the benefits of adding fatheads are short-lived, as the bass will eliminate them in a few months and return to foraging on other fish.

Fish Shelters

Artificial fish shelters can be created by anchoring last year's Christmas trees, Eastern red cedars, or assorted slash from forestry operations to the bottom of ponds. These shelters provide refuge for young fish to escape predation. Other good fish structures include stakes driven into the bottom of a pond (stake bed), rock piles, and tire reefs. These structures should be placed no deeper than 2 to 6 feet below the water. Usually only one reef is placed for every 1 to 3 acres.

A good way to increase fish reproduction is to place nesting structures in the pond. Fish spawns can be encouraged by furnishing breeding areas throughout the pond. If your pond has a silty bottom, spawning beds are necessary for successful fish reproduction. Spawning beds allow you to observe the reproductive success of your fish. They can be made by building a frame or box and filling it with around 4 to 6 inches of sand and gravel. Place beds at several locations around the shoreline in 2 to 5 feet of water.

Supplemental Feeding

Providing supplemental, commercial fish feed is a way to increase the growth of small sunfish and catfish.

BARRY W. SMITH

The grass carp or "white amur" is an Asian carp brought into the U.S. to control aquatic weeds. During warm weather, grass carp can eat 30 to 40 percent of their body weight in weeds every day.

Bass do not readily consume artificial feeds but do benefit from the increase in small sunfish that serve as prey. If a pond owner decides to provide supplemental feed, offer the feed in the same area and at the same time each day. Do not overfeed fish. A good rule is to supply what the fish will eat in 10 to 15 minutes. Do not feed more than 15 pounds of feed per acre each day. Fish can be fed from March through May and October through November when most small sunfish growth occurs in Alabama. Late winter feeding is not necessary but will improve small sunfish growth and reproduction. If feeding is continued during the winter months, use a feed that sinks to the bottom of the pond, and do not offer more than 3 pounds of feed per acre daily.

You can feed fish by hand or with a demand or automatic feeder. Floating feeding rings can be made from PVC tubing anchored in place. One feeding station per 3 acres of pond is sufficient. High protein levels in the food are really not essential, as research has demonstrated

JERRY L. MOSS

SOUTHEASTERN POND MANAGEMENT

RICHARD T. BRYANT/ARISTOCK

DAN DUMONT

Artificial fish shelters can be created by anchoring discarded Christmas trees, Eastern red cedars, logging slash, tires, and other large debris to the bottom of ponds. Top right: Fisheries biologists with the Alabama Department of Conservation and Natural Resources or private fisheries consultants can provide assistance in determining if fish populations are in balance. Bottom left: Left unchecked, a leaky pond dike can drain ponds and encourage the filling-in of ponds by encroaching vegetation. This pond has been drained for repair and maintenance. Bottom right: Proper fertilization will stimulate a good phytoplankton bloom if alkalinity is above 20 parts per million (ppm).

that even low levels stimulate excellent growth. Paying more money to purchase feeds with higher levels of protein is a waste of money better used elsewhere. In fact, feeding fish in itself is expensive and can only be justified if there is an obvious need to increase production beyond the levels that ordinary liming and fertilization bring about.

Adjusting Water Levels

Another good way to control aquatic weeds, while improving bass growth and reducing sunfish populations, is to install a drain in the pond. The Alabama Department of Conservation and Natural Resources, FSA (Farm Services Agency) or the Auburn Cooperative Extension System has more information on drains. Ponds with drains have distinct management advantages. In relatively deep ponds, the water can be drawn down 2 to 3 feet in late fall and maintained at that level throughout the winter. Fall drawdown helps control aquatic weeds as a result of freezing and drying on areas of exposed pond bottom. Lowered water levels concentrate fish, which increases forage availability to bass. Bass growth is improved and sunfish populations are reduced. Ponds should be allowed to refill during February and March.

Aeration and Mixing of Pond Temperature Layers

If a pond is deep (8 feet or more) or if it has a history of fish kills, it will benefit from aeration, the mixing of pond temperature layers (called **destratification**), or both. Many varieties of electric aerators can be purchased. Aim for ½ to 1 horsepower of aeration per surface acre of the pond. Turnover or bloom die-offs often call for additional aeration beyond those levels.

Destratification uses blowers, underwater fans, and

RICHARD T. BRYANT/ARISTOCK

Properly constructed and managed farm ponds, like this beautiful lake near Fairhope in Baldwin County, can enhance property values and provide wholesome outdoor recreation for many years to come.

propeller aspirator types of aerators. Using this equipment counteracts the effects of thermal layering (stratification) and usually prevents a fish kill if the situation is caught and remedied in time. It will also increase the area utilized by fish during hot summer months. However, destratification techniques alter algae blooms and may increase low oxygen problems during periods of overcast weather.

Spawning Substrate

Fish can also be encouraged to spawn in specific areas by providing them with the type of spawning substrate they prefer. Sand and gravel beds should be placed in several locations around the shoreline in 2 to 5 feet of water. The sand and gravel should be 4 to 6 inches deep and can be contained in a frame or box if the bottom is particularly silty. These beds allow the pond manager to concentrate seining efforts in areas where spawning can be expected to occur.

SUMMARY

Farm ponds are not creations of nature with their own set of checks and balances. The pond owner must provide the management to carefully monitor fish populations for continued successful harvesting. Once you understand the basics of what to watch for and how to correct problems or imbalances, you can keep fish populations in check. With the information and references contained in this chapter, you can choose a proper location, lime or fertilize the water, stock or remove fish, and protect the area from die-offs when heat or other conditions prove inhospitable. The rewards can be quite bountiful and the pleasure of outdoor fishing opportunities should be well worth the work.

Food plots for white-tailed deer are one of the most popular habitat improvements. Thousands of acres of supplemental plantings provide high quality foods for deer and other wildlife, and also concentrate wildlife for viewing, photography, and recreational hunting.

PAUL T. BROWN

CHAPTER 16

Supplemental Plantings as Wildlife Food Sources

CHAPTER OVERVIEW

Alabamians spend over $35 million annually establishing food plantings for wildlife. This chapter should be useful because it discusses the types and characteristics of various wildlife plants, how these plants meet the nutritional needs of wildlife, and techniques and considerations for conducting a successful and cost-efficient wildlife food plot program. Appendixes D–J provide detailed requirements for establishing and maintaining a wide assortment of wildlife food plants. Additional wildlife planting recommendations for certain species can also be found in other chapters throughout the book.

CHAPTER HIGHLIGHTS PAGE

CATEGORIES OF WILDLIFE FOOD PLANTS 364
NUTRITIONAL NEEDS OF WILDLIFE 366
WILDLIFE PLANTING CONSIDERATIONS 368
- **Choosing the Right Plant** 368
- **Site Selection** 369
- **Size, Shape, and Distribution of Plantings** 371
- **Land Preparation** 371
- **Planting Dates** 372
- **Seeding Rates** 373
- **Fertilization and Liming** 373
- **Inoculation of Legumes** 374
- **Companion Plants** 375
- **Planting Bareroot Tree and Shrub Seedlings** 376
- **Maintenance and Management** 376
- **Costs and Availability** 377
- **Weed and Insect Control** 377
- **Depredation by Domestic Animals and Other Wildlife** 378
- **Record-Keeping** 379

Envision a piece of land from the vantage point of one wildlife species such as a barred owl, pileated woodpecker, mourning dove, white-tailed deer, or quail to name a few. Each animal has particular dietary needs that create a preference for certain types of food. This chapter will focus on enhancing food availability for certain wildlife species by establishing supplemental plantings. In some cases, dramatic differences can result; however, supplemental plantings should not be viewed as a substitute for protecting and enhancing native wildlife foods. Native foods should be inventoried, evaluated, and managed **before** making substantial investments in wildlife food plantings. **In most cases, managing existing native wildlife plants constitutes a more practical and cost-effective method of enhancing wildlife habitat.**

As an example, consider the vision of one Marengo County farmer who walked the low-lying fields on his back 20 acres and began to mumble, "Build it, and they will come." Build what? A levee for an impoundment to create prime waterfowl habitat. Naturally, this took a lot of planning and inquiring about permits. Many months later, an impoundment was created using a levee to flood the field in the fall (for ducks) but leave the area dry in the spring to plant corn or sorghum. Where once there was non-productive pasture land, now wild ducks and other wetland-dependent species congregate. Establishing plantings for wildlife can have profoundly positive effects. This is why food plots are the most popular form of wildlife management practiced on farm and forest lands in the Southeast.

There are three ways that Alabama landowners, managers, and sportsmen can improve the quality and availability of wildlife foods on the lands they own or hunt. One method includes protecting existing high-value native wildlife food plants that already exist. Secondly, managers can enhance and stimulate the growth of native vegetation by **mechanical** (timber thinning, strip disking, mowing, prescribed burning) and **chemical** (herbicides, fertilizers, and lime) means. A third way is to propagate desired wildlife food plants by direct seeding or seedling planting, which is often called **supplemental planting.** Land management plans that list wildlife as an objective should have all three wildlife food enhancement methods included. This provides maximum diversity of quality foods year round.

Planting and cultivation of food plots is the most popular wildlife management technique conducted by sportsmen, managers, and landowners on private lands in the Southeast. This is probably because the results of their efforts are usually visible and tangible in a relatively short period of time. The plants grow rapidly and wildlife readily use the food plots. Since the quantity, quality, and distribution of native wildlife foods are often subject to seasonal and yearly fluctuations, cultivated plantings can supplement native food supplies in many cases. On poor sites with low fertility where native plants are lacking in nutritional value, food plots can have a positive effect on wildlife. In addition, food plots can be used to attract and concentrate animals to facilitate viewing, photography, and harvesting of game animals to accomplish population management objectives.

CATEGORIES OF WILDLIFE FOOD PLANTS

Wildlife plants in the South can generally be classified as grasses or legumes; annuals or perennials; warm season or cool season plants; woody vines, shrubs, or trees. When making distinctions in vegetation types, **grasses** are herbaceous (non-woody) plants that have parallel leaf veins and grow fibrous root systems. Grasses should not be confused with wetland plants, such as sedges or rushes, which are not true grasses. **Legumes** are herbaceous or woody plants that form a pod containing seeds, have netted leaf veins, and usually have a well-developed taproot system. Legumes have the unique ability to produce their own nitrogen, often called "nitrogen-fixing," with the aid of a bacteria of the *Rhizobium* genus that attaches to root nodules (joints). Because of the ability to produce nitrogen, legumes usually do not require nitrogen fertilizers during food plot establishment. To ensure that legumes get the nitrogen-fixing bacteria they need, seeds of legumes should be mixed in with bacteria (**inoculated**) before planting or they can be purchased pre-coated with a bacteria inoculant.

Some wildlife food plants are **annuals** since they germinate, grow, reproduce, and die in one year. Annuals grow only from seeds and most must be established each year by planting. **Perennials**, on the other hand, live for more than one year and can grow from seeds or reproduce from vegetative parts of the plant. Perennials do require some maintenance, such as annual mowing, disking, fertilizing, and liming. Some wildlife plants are also considered **biennials** in that they usually survive at least two growing seasons.

Landowners and wildlife managers interested in establishing wildlife plantings usually do so during two periods of the year: in the spring or early summer, or in the fall. Plants that are normally seeded or begin growth in spring or early summer and grow during these times

PAUL T. BROWN

Flooded agricultural fields, such as this cornfield containing mud flats seeded with Japanese millet and fallow areas managed for native wetland plants, can attract a variety of waterfowl species.

are considered **warm season plants**. Plants that are primarily seeded or begin growing in the fall are classified as **cool season plants.**

Wildlife biologists and ecologists are beginning to make an additional distinction between those wildlife plants considered to be **native** (naturally occurring) and those that are **exotic** (non-native) plants. From an ecological standpoint, more emphasis is being placed on recommending native plants that are better adapted to local site conditions and in many cases are less likely to be **noxious** (invasive and overly competitive with native plants) than exotic plantings. Plants, whether natives or exotics, that could spread as noxious weeds, should be avoided in wildlife plantings.

As an example of a noxious exotic plant, consider the profound effects of kudzu in the South. Kudzu, an Asian import, was utilized initially for erosion control by the USDA Soil Conservation Service (currently the Natural Resource Conservation Service) personnel for plantings along steep banks adjacent to newly constructed roads. No one dreamed what a nightmare it would become once established. Like the dreaded contents of Pandora's box, no one chooses to let kudzu "out" any more.

Cogongrass (*Imperata cylindrica*) has also become invasive in portions of south Alabama. Native to Southeast Asia, cogongrass was first introduced in Gainesville, Florida and Grand Bay, Alabama in the early 1940s as a soil erosion control plant. In Florida it has spread throughout the state, and in Alabama it is also spreading rapidly. Much of the dispersal of cogongrass has been attributed to mowing of interstate medians that enhances growth and reproduction. Cogongrass creates problems with natural regeneration, particularly in longleaf pine stands, and also outcompetes many native grasses and other herbaceous plants that are important for wildlife. Attempts to control cogongrass have largely proven ineffective because of its ability to recover quickly following tillage, burning, or herbicide treatment. Arsenal® is one herbicide that has proven most effective in controlling cogongrass. An integrated management strategy using all available control methods may be the only approach to effectively managing cogongrass.

DAN DUMONT

Arsenal® is one herbicide that has proven most effective in controlling cogongrass.

RICHARD T. BRYANT/ARISTOCK

Some exotic plants have become so invasive that they compete with native vegetation for nutrients and space. Unfortunately, thousands of acres of native vegetation in Alabama have been lost to kudzu.

NUTRITIONAL NEEDS OF WILDLIFE

Wildlife, like all other creatures, have particular nutritional needs. Understanding these needs and the time of year when these nutritional requirements are the greatest is an important aspect in planning wildlife plantings. With proper planning, they will provide the maximum value to the desired species. Wild animals with an adequate supply of quality food will grow larger as a general rule. They also produce more young, look more vigorous and healthy, have a better chance of reaching their genetic potential, and are more resistant to diseases or mortality than animals affected by malnutrition. The lack of quality food can be a major limiting factor for wildlife, affecting growth, reproduction, and survival.

"Quality" wildlife foods are ones that contain adequate amounts of nutrients. Some of the nutrients are **protein** for body growth, development and maintenance; **carbohydrates** as a source of quick energy; **lipids** or **fats** as a main ingredient in physiological processes and a source of stored energy; **vitamins** for maintaining growth and vigor; and an assortment of **minerals** that are needed for bone and tooth formation and mainte-

Exotic Trees and Plants Considered Invasive in Alabama

Exotic Trees	Silktree or Mimosa (*Albizia julibrissin*) Chinaberry (*Melia azedarach*) Tallowtree or Popcorn Tree (*Sapium sebiferum*)
Exotic Shrubs	Bicolor (*Lespedeza bicolor*) Japanese Privet (*Ligustrum japonicum*) Chinese Privet (*Ligustrum sinense*) Multiflora Rose (*Rosa multiflora*)
Exotic Vines	Japanese Honeysuckle (*Lonicera japonica*) Japanese Climbing Fern (*Lygodium japonicum*) Kudzu (*Pueraria lobata*) Chinese Wisteria (*Wisteria sinensis*)
Exotic Grasses and Forbs	Bermuda Grass (*Cynodon dactylon*) Bahia Grass (*Paspalum notatum*) Cogongrass (*Imperata cylindrica*) Fescue (*Festuca* spp.) Japanese Grass (*Microstegium vimineum*) Johnson Grass (*Sorghum halepense*) Tropical Soda Apple (*Solanum viarum*)

For information on control of exotic plants contact your local county Extension system office.

Fig. 1: Categories of Select Wildlife Plants

GRASSES

PERENNIALS		ANNUALS	
Warm Season Perennial Grasses	**Cool Season Perennial Grasses**	**Warm Season Annual Grasses**	**Cool Season Annual Grasses**
Bahia grass	Orchard grass	*Browntop millet*	*Barley*
Big bluestem	Timothy	Corn	*Oats*
Dallis grass		*Foxtail millet*	*Rye*
Indian grass		*Pearl millet*	*Ryegrass*
Switch grass		*Sorghum-sudan hybrids*	*Wheat*
		Sudan grass	

LEGUMES

PERENNIALS		ANNUALS	
Warm Season Perennial Legumes	**Cool Season Perennial Legumes**	**Warm Season Annual Legumes**	**Cool Season Annual Legumes**
Perennial peanut	Alfalfa	Alyceclover	Arrowleaf clover
Sericea lespedeza	*Birdsfoot trefoil*	*Cowpea*	*Ball clover*
White clover	*Red clover*	*Korean lespedeza*	*Berseem clover*
	Soybean	Button clover	Crimson clover
		Striate lespedeza	Subterranean clover
		Velvetbean	Sweetclover
			Common vetch
			Hairy vetch
			Black medic

WOODY PLANTS

VINES	SHRUBS	TREES
Alabama supplejack	American beautyberry	Black cherry
Crossvine	American elder	Flowering dogwood
Greenbriar	Bayberry	Persimmon
Japanese honeysuckle	Dwarf huckleberry	*Sawtooth Oak*
Wild grape	*Shrub lespedezas*	Wild Plum
	Autumn olive	White Oak

Plants considered to be exotics are in italics.

nance of body functions. **Of these nutrients, protein is considered the most essential to wildlife and is often found lacking in native plants.** In fact, scientists at Auburn University consider protein to be the most important factor limiting growth and reproduction of white-tailed deer in Alabama. Crude protein content (a measure of protein) in native wildlife plants usually ranges from a low of 2.4 percent to a high of 38.7 percent; however, most native forages rarely reach a level above 10 percent. Plant protein is concentrated in the growing tips of stems, and in seeds such as beans, grains, and nuts. Legumes, because of their ability to produce nitrogen with root bacteria, are generally excellent sources of protein. Some wildlife species, like white-tailed deer, require 16 to 17 percent protein levels in their diets for body maintenance and levels of 17 percent and above for optimum antler development. Phosphorus, and to a lesser degree calcium, are important nutrients in antler development of deer. Unfortunately, on most soils in the southeast, phosphorus is found lacking and consequently is not readily available to deer in native forages.

Studies have shown that nutrient levels in plants in the Southeast are at their highest level and are more readily digestible to wildlife during the spring and fall growing seasons. Plant nutrient levels drastically decline in late summer and winter months and become less available to wildlife. During these times, digestibility of plants is reduced due to the low moisture content and increased fiber content. Therefore, it is not surprising that the decline in plant quality corresponds, in general, to the nutritional stress periods during late summer and winter months for some wildlife species in the Southeast. Efforts to supplement wildlife diets with plantings should be directed toward these periods of the year when native vegetation is at its lowest nutritional level.

WILDLIFE PLANTING CONSIDERATIONS

Careful planning can mean the difference between success and failure of wildlife food plots. Many factors should be considered when planning food plots, including the following:

- Choosing the right plant,
- Site selection,
- Size, shape, and distribution of food plots,
- Land preparation,
- Planting dates,
- Seeding rates,
- Fertilization and liming,
- Inoculation of legumes,
- Companion plant(s),
- Planting seedlings,
- Maintenance and management,
- Cost and availability of plant materials,
- Weed and insect control,
- Depredation, and
- Record-keeping.

Specific recommendations can be found in Appendixes D–J for each of these factors for many wildlife food plants.

Choosing the Right Plant

Matching the right plant(s) to wildlife that you are interested in enhancing is not an easy task. No single plant will grow in every soil and climate and provide all the nutritional needs of wildlife. For example, there are over 62 different types of plants used in food plots for deer in the Southeast. In reality, wildlife will readily use a variety of plant species for food and cover; however, preferences do exist for certain plants. These preferences depend on the quality and availability of native foods and agricultural crops in the area, and seasonal physiological demands on wildlife, such as gestation, lactation, and antler development. In addition, some species of wildlife are better able to utilize certain plants.

TED DEVOS

There are over 60 different types of plants used in deer food plots across the Southeast. Plant selection for food plots should depend upon demonstrated preferences by wildlife, site characteristics (soil type, climate, and drainage), and the season of the year when supplemental feed is needed.

Some food plots fail because the plants are not adapted to the site. Plant adaptation is limited by soil type, climate, and drainage. Care should be taken to select wildlife plants that are adapted to the climate of your area. For example, some plants are more cold-tolerant than others and will persist longer during the winter months or will have a better chance of survival in regions where the average winter temperature is well below other areas. On the other hand, some plant species are sensitive to prolonged high temperatures and will die out in extreme heat. Obviously, the farther north you go the more you should consider cold-tolerant plants. For example, rye or ryegrass is more cold-tolerant than oats and therefore would have a better chance of survival in northern Alabama than would oats.

To help in your selection of a wildlife food plant, use the quick reference chart found in Appendixes D–J as a guide. In many cases, there are also knowledgeable land managers or natural resource professionals who have experience and successes with plantings for wildlife in your area. Seek them out and ask their advice.

As a final thought, experiment with different plants, mixtures, and seeding rates to see what works best for you on your land. One of the best ways to see how much your field is being used by wildlife, especially deer, is to place a wire fence exclosure in the food plot. An **exclosure** is simply a small fenced-in area, usually 4 to 5 square feet in size, that protects plants inside the exclosure from being eaten by wildlife. Exclosures can be easily constructed with any type of woven-mesh fence material that is held in place by fence posts. The height and vigor of plants inside the exclosure can easily be compared to plants outside the exclosure for an indication of food plot use. After experimenting with several plant varieties and planting styles and recording the results of growth and use by wildlife, you should have a pretty good indication of the most desirable plants for your area.

Site Selection

To ensure successful wildlife plantings, the plant should be well adapted to the soil type of the site chosen for a food plot. For example, some plants are better adapted to moist or wet sites while others prefer well-drained areas. If you are unsure of the soil types on your land,

RICHARD T. BRYANT/ARISTOCK

Powerline or gasline rights-of-way are excellent locations for establishing wildlife food plots. Utility companies should be contacted in advance for policies regarding management activities under rights-of-ways.

LEWIS O. ROGERS/SOUTH CAROLINA DNR

Linear openings in forest stands are excellent locations for wildlife food plots. Multiple plantings that create both food and cover can increase the benefits provided by food plots.

LEWIS O. ROGERS/SOUTH CAROLINA DNR

Exclosures are small, fenced-in areas that protect plants inside the exclosure from being eaten by wildlife. The height and vigor of plants inside the exclosure can easily be compared to plants outside the exclosure for an indication of food plot use by wildlife.

LEWIS O. ROGERS/SOUTH CAROLINA DNR

Mowing is one maintenance practice that can help prolong the availability and quality of some perennial food plots, such as this clover field.

RICHARD T. BRYANT/ARISTOCK

Removal of stumps and roots is a necessary first step when converting forest land to food plots.

you should obtain a copy of your county's soil survey, which can be obtained free of charge from your local county Natural Resources Conservation Service office. These surveys, if available in your county, are an invaluable tool that contain a county soils map on which you can easily identify your property and the prevalent soil types that occur on your land. The soil surveys also have a handy listing of native and propagated plant species that are adapted to each soil type. The cultural requirement section for commonly planted wildlife foods found in Appendixes D–J should also be helpful in determining appropriate wildlife planting sites. In addition, observing wildlife food plantings on similar soil types within the same area can also be helpful in selecting plants that might grow well on your property and be readily utilized by wildlife.

Landowners should avoid wildlife plantings on sites that are considered wetlands to avoid possible violation of the USDA Swampbuster Program or the Clean Water Act. Before establishing food plots in wetland areas you should contact your local Natural Resources Conservation Service office for guidance.

Excellent locations for wildlife plantings include unused corners and edges of agricultural fields, forest openings, forest regeneration areas, idle crop fields and meadows, along fence rows, ditch banks, utility rights-of-way and access roads, logging roads, edges of wide woods roads, logging decks, forest salvage openings, firelanes, abandoned fields, or any other area where enough sunlight reaches the ground. Some wildlife managers have established food plots in strips in pine stands where the basal area has been reduced to 60 square feet or less.

Food plots can be most effective when located along transition zones where several habitat types meet. Food plots adjacent to streamside management zones, travel corridors, and escape cover are also in good locations. For many wildlife species, like bobwhite quail, locating plantings adjacent to escape cover is extremely important in reducing predation. As an added protective measure to reduce unwanted predation of small game by raptors (such as hawks and owls), tree snags and other likely perches adjacent to food plots should be removed if at all possible. Consideration should also be given to alternating locations of small game food plots to prevent possible accumulation of predators from repeated food plot use by small game. To discourage illegal harvest of larger game animals, such as deer and turkey, wildlife plantings should not be located in view of, or be accessible by, a public road or right-of-way. Food plots in plain view of access roads can quickly become drive-by or roadside shooting plots for wildlife poachers. To prevent viewing from access roads, some managers plant rows of trees or shrubs to reduce the visibility of food plots. Examples of good screening tree choices include pine, Eastern red cedar, leyland cypress, and Russian olive.

Location of food plots should be included in forest management plans and determined in advance of harvest layouts and timber sales. Permanent openings on forest land should be easily accessible for food plot cultivation, especially if large mobile contract lime and fertilizer spreaders are to be utilized.

To prevent soil erosion from cultivation activities, food plots should be established on level to near-level ground. If food plots are located on sloping terrain, they should be maintained as natural openings by periodic low intensity disturbances such as prescribed burning, light disking, or mowing. If unlevel openings are planted as food plots, they should be seeded in biennial or perennial plants that require little maintenance and in-

NANCY WEBB

Disking is an important seedbed preparation practice that ensures a well-pulverized seedbed that improves soil to seed contact and seed germination.

RICHARD T. BRYANT/ARISTOCK

When planting wheat, clovers, and other light-seeded plants, consistent seeding depth is important and can be controlled with a grain drill. Seeds will not germinate if planted too deeply.

frequent soil disturbance. This will help reduce soil erosion and provide another variation in the type of food plots available to wildlife.

Size, Shape, and Distribution of Plantings

Size, shape, and distribution of wildlife food plots are important considerations when trying to maximize their effect. Size of plantings determines the degree to which food plots will be used. Large food plots may not be fully utilized by some species, while smaller plots may be overutilized and exhausted in a short time. As a general rule, smaller or less mobile species of wildlife require a proportionally smaller food plot and will benefit from a greater number of plantings per unit area. For example, food plots for quail may be as small as a ¼ acre in size. Plantings for quail should be made adjacent to escape cover so that quail food plots don't become food plots for predators. Plantings for deer and turkeys should range from 1 to 5 acres. Most plantings for dove and waterfowl should be at least 5 acres in size and preferably over 10 acres.

Wildlife, like deer and turkeys, are reluctant to use the center sections of large plots, especially where regular disturbances or hunting are commonplace. Over time deer will eventually begin feeding in the center of large plots, but only at night. To ensure that wildlife have easy access and use of food plots, plantings should be long and irregular in shape. Large, even-sided plantings should be avoided. Vegetative hedgerows can be established through the interior of large openings to provide additional cover.

Wildlife plantings should be evenly distributed across an entire tract of land to ensure access by wildlife. As a general rule, depending on the availability and quality of native wildlife food plants, wildlife plantings should comprise approximately 1 to 5 percent of the total forest land. One plot per 25 acres of forest land is a good ratio of food plots to forest land for deer and turkey.

Land Preparation

Preparing the land before planting is vital in securing a successful stand for most wildlife foods. Proper seedbed preparation enhances the formation of a firm, moist, level seedbed, which is necessary for a successful wildlife planting. A well-pulverized, level seedbed helps maintain a consistent seeding depth that is important for seed germination. This is especially important for small-seeded legumes like many of the clovers, where seed can be wasted if planted too deep. To determine if a seedbed has been properly prepared, walk over your field. If your shoe prints are not much deeper than the sole of your shoe, your preparation is sufficient. If your foot sinks too deep, you should use a cultipacker (a tractor attachment used to compact freshly plowed land) to pack the dirt tighter in the seedbed.

When converting logging decks and logging roads to wildlife food plots, **soil compaction**, also called a **hardpan**, may be a problem. Hardpans are created by weight and pressure from heavy equipment, which compresses the soil so that soil **pore space** (the very small, natural air spaces between soil particles) is reduced. Even the steady traffic of livestock can cause a shallow hardpan. Water and oxygen do not travel well through hardpans or compacted soil, and during periods of low rainfall wildlife plants are doomed to failure. Even with adequate rainfall, water will lie in puddles on the surface of the compacted soil and evaporate before it can seep down to the roots and be utilized by plants.

RICHARD T. BRYANT/ARISTOCK

Soils that have not been in cultivation for some time may require the use of a heavy disk to break apart the soil surface for seedbed preparation. When soil compaction is a problem, subsoiling should be used to improve soil structure.

LEWIS O. ROGERS/SOUTH CAROLINA DNR

Large-seeded wildlife plants like corn or soybeans can be established in rows with an individual hopper-style planter.

To detect a hardpan, use a **push-probe,** a 2- to 2½-foot metal rod with one end sharpened to penetrate through the soil and the other end with a handle to push the probe through the soil. Push-probes can be made or purchased from a forestry equipment supplier. Check for hardpans when the soil is fairly dry (dry enough to operate farm equipment on the site), but not extremely wet or dry. Insert the probe two to three times at various locations across the field. Areas that appear to have little or no surface vegetation could indicate a hardpan. As the probe is inserted, the force required to move it through the soil should remain about the same until a hardpan is reached. At the point of a hardpan, it takes a significantly greater amount of effort to push the probe. When the probe reaches a hardpan, note the spot on the probe to determine the depth of the hardpan. A hardpan may range from 2 to 10 inches thick.

Hardpans detected within the top 6 or more inches of the soil should be broken apart by subsoiling. **Subsoiling** (breaking up the soil to depths of 6 to 12 inches) will fragment compacted soil. To subsoil, a landowner needs a tractor, dozer, or skidder that has at least a 50 horsepower engine. A chisel-plow should be used to break up the hardpan to a depth of 6 or more inches below the soil surface. Sites that are to be established as wildlife food plots and have an obvious hardpan should be subsoiled 30 to 60 days before planting to allow the soil to settle and establish a firm seedbed.

Logging decks and logging roads converted to food plots should be subsoiled to reduce compaction. Subsoiling is also recommended every two or three years on all food plot sites. Subsoiling breaks apart hard subsurface soil, which can reduce water availability and root growth of plants.

Seedbed preparation, like subsoiling, should ideally begin several months, and at the very latest several weeks, before planting. Early seedbed preparation allows time to incorporate lime (if needed) to stabilize soil pH well before planting. Advance seedbed establishment also helps to reduce unwanted weed competition. In fact, if time permits, weed competition can be greatly reduced if the seedbed can be *cultivated* (mixed and then leveled by mechanical means) at one week intervals three to five times before planting. This will improve the quality of the plot significantly.

Planting Dates

Timing of plantings is critical for successful wildlife food plantings. For example, many wildlife food plants have differing optimum planting times that are often only a few weeks in length. Scheduling planting during the range of best planting dates often means the difference between a successful food plot and one that is doomed to failure. Since it seems almost inevitable that bad weather, mechanical failure, and a multitude of other factors delay plantings, wildlife food plots should be planted as early as possible within the recommended period. Late planting of warm season wildlife food plots that are normally planted in the spring may result in increased weed competition. It may also make the planting more susceptible to summer drought because the root system is not fully developed. In comparison, planting a cool season wildlife plant too late in the fall may result in cold-induced stress and ultimately cause a complete planting failure.

RICHARD T. BRYANT/ARISTOCK

A cultipacker can be pulled behind a no-till drill planter to improve seed to soil contact and improve germination. This conservation tillage system may eliminate the need for seedbed preparation.

LEWIS O. ROGERS/SOUTH CAROLINA DNR

Commercial lime spreaders supplied by a co-op or farm supply store can be used to apply lime to wildlife food plots. This is the easiest and most efficient method for liming food plots of a significant number or size.

Staggering two or three plantings of the same plant within the "safe" planting dates can increase the availability of food for wildlife. For example, a wildlife opening that is divided into three smaller fields, with each subfield planted two weeks apart, can extend the length of available food for one month. Caution should be used in staggering planting dates for more than three different time periods since there is the risk of low germination and poor growth. There is no need to stray too far from the optimum planting dates, which are listed by plant species in Appendixes D–J.

Seeding Rates

If seeding rates are too low, weed competition and germination problems can cause a wildlife food plot to fail. On the other hand, extremely heavy seeding rates can be a waste of money. As a rule of thumb, seeding rates for establishing wildlife food plots should be higher than the rates recommended for commercial production of the same plant. Higher seeding rates help to ensure that a significant portion of the seeds germinate. Since seed costs are usually not a major expense in the overall costs of food plot establishment, many wildlife managers routinely exceed the recommended seeding rate to ensure a successful wildlife planting. There are exceptions to this, such as the cost of some species of clover, for example. Wildlife plants can be seeded with a variety of methods including those used to establish agricultural crops. These methods include 1) cyclone or spin-seeders operated from either a tractor power takeoff, batteries, or by hand; 2) band seeders; 3) cultipacker seeders; 4) drills; and 5) broadcasting by hand. Large-seeded plants like grains and vetches should be seeded with a drill. Small-seeded plants can be drilled but care should be taken to ensure that they are not planted too deeply.

Broadcasting seeds can also be effective for small-seeded plants like clovers if a cultipacker is used afterwards to ensure that seeds are in good contact with the soil. Seeding rates vary somewhat depending on the method of planting. For example, with drill-seeding the planting rate is somewhat lower than when broadcasting seeds since planting depth and spacing is more precise when seeds are drilled. In addition, seed lost by broadcasting is greater since wildlife readily feed on exposed seed.

Fertilization and Liming

Very few food plot sites in the Southeast naturally contain the proper amount of nutrients for successful food plot establishment and growth. Most often food plots are initially planted on "new ground" (logging decks, logging roads, forest openings). These areas have not been in cultivation for some time, and consequently lack adequate concentrations of soil nutrients necessary to establish or maintain a wildlife planting. Except on some agricultural lands, most areas chosen for food plots contain low levels of plant nutrients and acidic soils that are not conducive to growing wildlife food plots. The nutritional quality of plants and their value to wildlife are directly related to the fertility (the abundance of nutrients) of the soil in which the plant grows. Numerous studies have demonstrated the positive relationship of soil fertility to the size of animals, reproductive success, and abundance. Deer and turkey prefer forage that has been properly limed and fertilized

Nutrients Important for Plant Growth

MAJOR NUTRIENTS	FUNCTION
Nitrogen (N)	Component of chlorophyll, critical for photosynthesis, often leached from soil.
Phosphorus (P)	Helps plants make food, important for seed and fruit development, root growth and survival, and growth of seedlings.
Potassium (K)	Helps in plant maintenance and in reducing cold temperature stress.
SECONDARY NUTRIENTS	
Calcium (Ca)	Maintain pH of soil.
Magnesium (Mg)	Facilitates plant growth.
Sulphur (S)	
MICRONUTRIENTS	
Manganese (Mn), iron (Fe), boron (B), copper (Cu), molybdenum (Mo), chloride (Cl) and zinc (Zn)	

since it is more nutritious and palatable than unfertilized vegetation.

Nutrient deficiencies in the soil can be corrected by applying the proper rates of fertilizers and lime. The only way to really know how much fertilizer and lime to apply is to have the soil tested. Soil tests should be made several months before planting to allow enough time to receive test results and then apply lime before planting. Tests will determine the nutrient needs of a site for a particular planting as well as help contain costs associated with overfertilization and liming. In some cases overfertilization and liming can reduce plant growth. Fertilization and liming rates should be based, if possible, on the soil test results. Information and help on how to collect soil samples, analyzing soil samples, and interpreting soil test results can be obtained from the local county Cooperative Extension System office. Based upon the results of your soil test, a local farm supply store or co-op will be able to prepare a fertilizer tailored to a particular field's needs.

Fertilizers are sold in many ways but generally in ratios that express the percentage of nitrogen (N), phosphorus (P) or phosphate (P_2O_5), and potassium (K) or potash (K_2O). For example, with triple 20 (20-20-20) fertilizer, the ratio means that for every 100 pounds of fertilizer there are 20 pounds of nitrogen, 20 pounds of phosphorus, and 20 pounds of potassium. Fertilizers with all three nutrients are called "complete" fertilizers. However, if soil tests reveal that only one nutrient is needed, it can be purchased and applied separately.

Soil tests will also determine the pH, which is an indication of the degree of acidity or alkalinity of a soil. This is important in food plot establishment. A pH value of 7 is considered neutral while a pH below this value is acidic. A pH above 7 is alkaline. Most wildlife food plants do best between a pH of 5.8 and 6.5. Soil nutrients are most readily available to plants in this pH range. In addition to determining the pH of a food plot site, soil tests also provide a recommended rate of lime for raising the pH level if the soil is too acidic. Lime should be applied during land preparation and mixed with the soil several months before planting to allow time for the soil pH to change. Since recommended application rates of lime usually run around 1 to 2 tons per acre for acidic soils, lime application is usually done using a commercial spreader or liming truck provided by the dealer or farm supply store. It's wise to make sure that food plots are accessible by lime spreaders and trucks. If not, you may find yourself spreading lime from the back of a pickup truck, which is less effective and much more tiresome.

Inoculation of Legumes

Seeds of legumes should be treated with the proper mixture of live *Rhizobium* bacteria before planting. *Rhizobium* is the scientific name of a bacteria that facil-

DAVID K. NELSON

Wheat is a popular planting that is utilized by a wide variety of wildlife. When left undisturbed it will mature and make seed, providing additional benefits later in the year like brood habitat for young turkeys and quail.

itates nitrogen-fixation (production of nitrogen) and allows nitrogen to become available for plant growth. The technique of mixing bacteria with legume seeds is called inoculation. Some commercial varieties of legume mixtures that are marketed especially for wildlife plantings have bacteria already added, like red clover, birdsfoot trefoil, and white clover. However, for those legume seeds that are not inoculated, you can add *Rhizobium* bacteria with bags of inoculant purchased from your local farm supply store. Directions for mixing the seed with the inoculant are found on the bag. When purchasing a legume inoculant, be sure that you specify the legume that you wish to inoculate, since certain bacteria are specific for certain legumes. Inoculated seed should be protected from high heat or direct sunlight since this tends to kill the bacteria and consequently reduces the effectiveness of inoculation. In addition, you should never directly mix fertilizers with inoculated seed since the salts in fertilizers kill bacteria.

Companion Plants

Plant mixtures are usually more desirable as wildlife plantings. Combination plantings not only provide a diverse food source for wildlife but also lengthen the availability of food. For example, the availability of a wheat-ryegrass mixture will be longer than that of a field of wheat alone. Overseeding winter annuals like wheat and rye on dormant sods of warm season plants like bahia grass, bluestems, or dallis grass is also an excellent way to extend the length of time that a planting is available to wildlife.

For deer and turkey, combination plantings can be used to produce abundant quality forages during all seasons, especially during the late autumn–winter stress period. Some of the most popular mixtures in Alabama include wheat and clover combinations and in northern Alabama, rye and white clover mixtures. Small grain and clover mixtures are the best combinations for deer and turkey, providing forage from fall through early spring.

A mixture of clover and joint vetch can provide high-quality forage from spring to early fall. Mixtures of wheat, oats, and elbon rye are also popular. Ryegrass and arrowleaf clover make good fall and spring combinations and can be maintained by mowing and fertilizing each September. If clover mixtures are desired, blends of ladino, red, and annual clovers might be mixed to derive advantages of each clover type. A combination of plantings can be chosen that supplies quality forage during critical periods year-round.

When planting mixtures, use one-half to two-thirds

DR. MARK DUBOIS

Specialized planting tools that easily and quickly create holes for planting tree and shrub seedlings can be purchased commercially. This planting bar is specifically designed for planting containerized seedlings or "plugs." The extended platform on the left is pushed down with the foot, sending the tip of the planting bar to a pre-determined depth.

DR. MARK DUBOIS

Seedlings should be carefully placed in planting holes and the soil around them compacted by foot to eliminate air pockets near seedling roots. The location of seedlings should be marked with flagging or a wire flag marker for maintenance and protection against mowing or other land management activities.

as much seed of each plant species as you would use if you planted just one species. Companion plants are listed in Appendixes D–J for most of the common wildlife plants.

Planting Bareroot Tree and Shrub Seedlings

Bareroot seedlings are used most often in establishing wildlife shrub or tree plantings. **Bareroot seedlings** are seedlings grown in a nursery bed, lifted after one or two growing seasons, and shipped and planted without any soil around the roots. This is in contrast to seedlings that are potted in soil or balled with burlap. The roots of bareroot seedlings are usually packed with some type of medium like shredded newspapers, sphagnum moss, or sawdust to keep them moist during shipping and until planting. The roots may also be dipped in a gel medium that clings to the roots and keeps them moist. Bareroot seedlings are less expensive than potted or balled plants because less labor is required for growing them in the nursery. They also weigh less and are therefore cheaper to ship.

Care of seedlings before planting is important in order to ensure survival. Poor care of seedlings before planting is one of the main causes of low survival. Seedling roots should be kept moist at all times in a storage area with 85 to 95 percent humidity. Seedlings should be stored in a refrigerated unit at a temperature between 33 to 38°F. If no refrigeration is available, seedlings can be stored at 38 to 50°F for two to three weeks, or at 50 to 75 degrees Fahrenheit for three to five days. Temperatures above 85°F will quickly kill stored seedlings.

Bareroot seedlings can be planted with a planting bar (often called a dibble), shovel, or mattock. Care of seedlings during planting is critical for survival. Seedling roots must be kept moist before planting. Ten minutes of air drying on a warm, sunny, windy day can reduce seedling survival by as much as 50 percent. To prevent damage to the roots, only carry as many seedlings as you can plant in 10 minutes. Keep the remaining seedlings in the shade, with the roots moist and covered.

Seedlings should be planted just slightly deeper than the depth they were planted in the nursery. You can tell the depth in the nursery by a slight color change on the stem. Make sure the hole is big enough to allow the roots to be spread out. Roots should be pointing straight down in the hole and not "J" rooted (roots bent back pointing towards the top of the hole). For more information about planting bareroot tree and shrub seedlings, contact the Alabama Forestry Commission (334-240-9345) or the National Wild Turkey Federation's Project HELP (Habitat Enhancement Land Program) (800-843-6983).

Maintenance and Management

Maintenance and management requirements of plantings are important considerations when choosing a wildlife food. Some plant species are easily maintained while others require a greater degree of care. Perennial and reseeding plant varieties usually involve the least amount of effort. In addition, plants that readily respond to disturbances (disking, mowing, burning) such as partridge pea, clovers, and other reseeding legumes, usually require the least amount of time and effort to establish and maintain.

AUBURN UNIVERSITY

Tree shelters can improve growth and survival of seedlings by reducing plant moisture loss and creating a "greenhouse" effect. They can also protect seedlings from feeding damage by rodents, rabbits, and deer and serve as clear markers for persons operating equipment nearby.

Tree Shelters for Seedlings

Tree shelters are translucent (allows light to filter through) tubes that are placed over young seedlings or small saplings and have been shown to increase their growth and survival. The shelters work by providing a mini-greenhouse effect around individual seedlings, which increases moisture availability for seedlings and also protects them from animal damage. Over time, some tree shelters biodegrade and leave trees free to grow. In many cases seedlings protected by tree shelters have grown more than twice as fast as those not protected with shelters. One drawback to tree shelters is cost, which may range from $1 to $5 per shelter, depending on the size and type of shelter. However, in some cases, cost may outweigh the loss of seedlings or the reduction in growth of seedlings without shelters. For sources of tree shelters consult Appendix K.

Costs and Availability

Establishing wildlife plantings can become expensive in a hurry. As a first step you should decide how much you or your hunting club are willing (or can afford) to spend on food plots. Then you should develop a food plot budget. The cost of establishing food plots depends on several factors but generally runs between $60 to $200 per acre. Maintenance costs also vary but will range from $15 to $45 per acre each year.

The intensity of maintenance and management will affect the long-term cost of food plots. For example, some plant species require little or no maintenance while others require more time, effort, and expense. Generally speaking, perennial and reseeding plant varieties, like many clovers, usually require the least amount of maintenance since they don't have to be re-established each year like annuals. In addition, some wildlife plants are more tolerant to fluctuations in temperature and rainfall. To determine the degree of care required for the plant(s) chosen for food plots, consult the management description found later in the cultural requirements section for wildlife plants found in Appendixes D–J.

To cut costs, compare prices of several plant varieties, fertilizers, and lime. Usually the more common plant varieties are the least expensive. Unless financial constraints are not a concern, annual plant varieties costing over $3 per pound should be avoided unless they are exceptional, with well-documented results. Compare prices among several vendors to obtain the best value. Dealers who sell in bulk tend to have more competitive prices. If you choose varieties that are not readily available, plan several months in advance before planting to allow time to locate, order and receive seeds or seedlings. If you are having trouble locating plant materials from your local farmers' supply store or co-op, refer to the section in Appendix K on seeds and seedling sources.

Weed and Insect Control

Weed and insect control is not the major concern in wildlife food plots that they are for production agriculture. However, in severe infestations, weed and insect problems can cause a wildlife planting to fail. Herbicides may be necessary in some cases to control undesirable weeds, which have little or no value to wildlife. Herbicides are usually applied either before planting, during planting, or after plant emergence. Herbicide labels should be read and carefully followed to ensure applicator safety and prevent unwanted damage to wildlife plants.

Insect damage is most harmful to wildlife plantings during the early stages of growth. Well-established plantings usually have more resistance to insects due to better developed root systems. Effective insect control requires identifying the insect pest and applying a proper insecticide according to label instructions. For help in identifying invading insect pests and advice for pesticides to control infestations, contact your local county Extension agent.

Depredation by Domestic Animals and Other Wildlife

Wildlife food plots will also attract domestic animals and **non-target wildlife species,** wildlife that you are not trying to attract or benefit. Cattle, goats, and swine can easily destroy a wildlife food plot in a matter of a few days. Domestic animals should be fenced out of wildlife plantings. In some situations, non-target wildlife will

Costs of Forage Plantings for White-tailed Deer

Research at Auburn University compared the cost of establishing 39 cool-season and 9 warm-season forages that are commonly planted for white-tailed deer in Alabama. Results of the research showed that small grains (wheat, oats, rye, barley) as a group are the best, most cost-effective forages to attract deer during hunting season from fall through winter. The cost of establishing small grains ranged from $88 to $102 per acre. The study also illustrated that crimson clover and ryegrass were the most cost-effective forages from winter through early spring. Costs of establishing clovers ranged from $44 to $65 per acre. Combining clovers with small grains has the added effect of attracting deer during hunting season and offering high-quality forage for deer during the late winter stress period. The study also indicated that from spring through summer, ladino and red clovers are cost-effective, especially if planted on moderate to fertile sites.

Of the nine warm-season forages tested by Auburn, soybeans, velvetbean, and peas were the most cost-effective forages. These plants produced an abundance of high-quality forage for deer throughout the summer at a relatively low cost.

Fig. 2: Breakdown of Costs for Planting a 1-Acre Sunflower Food Plot

The following example is a breakdown of costs associated with establishing a 1-acre abandoned field in sunflowers. A soil test indicates a low pH and liming needs of 1½ tons per acre, and fertilizer needs of 500 pounds of 4-12-12 per acre. The field has not been cultivated in some time and weed control is necessary.

DESCRIPTION	UNIT	UNIT PRICE	QUANTITY	COSTS
Soil test	each	$5.00	4	$20.00
Sunflower seed	lbs.	0.64	10	6.40
Fertilizer (spread)	100 lbs.	9.00	5	45.00
Lime (spread)	ton	24.00	1.5	36.00
Tractor/Machinery	acre	20.00	2	40.00
Herbicide	pints	20.00	2	40.00
Labor	hour	4.15	2	8.30
				$195.70

Costs vary depending on local prices for seed, fertilizer, lime, herbicides, labor, and equipment requirements.

also become a nuisance and severely damage food plots. For example, in areas where deer densities are high, dove and quail food plots can be completely destroyed by deer browsing. This is especially true for plants like sunflower, peas, and some species of shrub lespedeza. Chufa plots planted for turkeys can also be damaged heavily by raccoons and wild hogs. Options to alleviate unwanted damage to food plots include reducing numbers of offending animals, establishing food plots with plants that are less preferred by non-target wildlife, protecting food plots with electric fences, or a combination of all of the above.

Tree and shrub seedlings that have just come out of a nursery are likely to experience animal damage since they have been heavily fertilized in nurseries. This makes them extremely palatable to browsing animals. Browsing by white-tailed deer and other animals can be a serious threat to the survival of newly planted tree and shrub seedlings. Rabbits and other small rodents can also browse and girdle seedlings. Deer can inflict additional damage to seedlings when antlers are rubbed on the main stem during the rut.

Tree shelters can be used to protect valuable seedlings from browsing animals including beavers and dramatically enhance seedling growth and survival. **Tree shelters** are long, tubular sleeves made of polyethylene or polypropylene that are placed around seedlings for the first three to five years after planting. For more information about tree shelters and their role in protecting tree and shrub seedlings contact the Alabama Forestry Commission (334-240-9348) or the National Wild Turkey Federation (NWTF). Tree shelters can be purchased from the NWTF's Project HELP (Habitat Enhancement Land Program) by calling (800) 843-6983.

Record-Keeping

Many land managers don't bother keeping records on their food plots. Records can be invaluable over time in avoiding past mistakes and in determining the most effective planting strategy on your land. Records should be kept for each food plot and should generally include the following:

- Location and identity (number or name) of food plots,
- Variety of plant(s),

Checklist of Potential Problems in Establishing a Food Plot

Sometimes under what seem to be ideal conditions a wildlife food plot will fail. When a wildlife planting does fail, it's important to find out why to minimize the chance of the same thing happening in the future. The following checklist may help identify why a wildlife planting failed.

PROBLEM	POSSIBLE CAUSES
Seeds did not germinate?	soil type not suitable, dry seedbed, non-viable seed, hard or dormant seed, temperature extremes, herbicide residue, waterlogged soil, too much or too little rain
Seed germinated but did not come up through the soil?	planted too deep, soil crusted at surface, poor seedling vigor, insects or disease, temperature extremes
Seedling came up but did not survive?	soil too acidic or low fertility, insects and disease, drought or flooding, weed competition, no legume inoculation, winter kill, browsed too early, plants pulled up, hard subsoil layer

- Seedbed preparation technique,
- Planting dates (day/month/year),
- Seeding rate,
- Information from soil test results,
- Type and rate of fertilization and liming,
- Planting method,
- Maintenance and management activities,
- Rainfall and temperature during establishment and growing season,
- Use by wildlife,
- Cost of establishment and maintenance, and
- Success of food plots.

DR. RON HAALAND

Combination plantings, such as corn, sunflower, and sorghum pictured here, provide a diversity of food sources and cover for wildlife and also lengthen the availability of food.

Wyncreek Plantation's Food Plot Program

Mickey Easley has long been respected in Alabama as a wildlife biologist and game manager who has built a reputation of excellence by combining practical, on-the-ground experience with his graduate training in wildlife ecology from Mississippi State University. He currently is the wildlife biologist and manager of Wyncreek Plantation in Hardaway, Alabama.

Many times what is *not* happening on a site determines if I'll use an area for a wildlife food plot. For example, many deer plots on Wyncreek Plantation are openings that have been created by fallen and diseased pines killed by southern pine beetles. Once the timber is salvaged from beetle kills these areas make excellent sites for deer and turkey food plots. My "dream" plots for deer are basically composed of four plantings: 1) ryegrass, 2) crimson clover, 3) triticale, or 4) combination plantings of ryegrass and clovers. Sometimes I substitute other clovers for crimson clover, but I have found that they do not withstand the heavy deer grazing pressure that crimson clover can take. I also use wheat for some of my deer plots and choose to plant feed (which comes from lower quality seed) rather than seed wheat. Turkeys will also readily utilize these plots.

For turkeys I like to plant chufa in sandy sites, which makes it a little easier for turkeys to dig up the root tubers to feed on. If the ground is thoroughly broken up, I sometimes plant chufa on other sites. We also have corn plots scattered in small openings for turkeys. These areas provide seed and excellent brood habitat for turkeys. Since I do not use herbicides in my corn patches, grassy areas develop between rows and stalks, which attract insects that are heavily fed upon by turkey hens and poults.

For quail I use a variety of practices to enhance food availability. One of the practices that I routinely incorporate into our management plan is to create "bird strips" with a tractor and disk, usually in pine stands with a basal area between 30 and 60 square feet. These strips zigzag through the woods and create a network of interconnecting food sources for quail. I do this as often as I can during various times of the year, except during the turkey and quail nesting period. I try to cut new strips each year. I often take the path of least resistance through the woods, making sure not to disturb native fruit-producing plants that are valuable food sources for wildlife. My bird strips are either left for native plants to grow or are planted. To encourage

RON BRENNEMAN

Crimson clover is one of the most cost-effective winter and early spring forages for deer in Alabama.

DAVID K. NELSON

Clover and small grain combinations attract deer during the fall and winter hunting season and benefit turkeys, rabbits, and a variety of other wildlife.

native legume production I often fertilize strips with 4-12-12. For planted strips I drill in just about every combination of seeds that have been documented to occur in quail crops. Some of my favorites are foxtail millet, Egyptian wheat, browntop millet, dove proso, and white millet.

A goal should be to provide a diversity of seed sources that remain available for as long as possible. This can be accomplished by selecting plants that have differing seed maturation dates as well as staggering planting dates. For game birds I have found that Egyptian wheat is one of the best individual plants because the seeds remain available for a longer period of time, they are not readily eaten by deer (important in high deer density areas), and the wheat's appearance makes an aesthetically pleasing impression that an area is being planted and managed for wildlife. Another favorite of mine is browntop millet since it is relatively inexpensive and readily available, will grow on most sites, and is eaten by a variety of wildlife from gamebirds to songbirds. Other favorites include partridge pea and annual lespedezas. Although partridge pea is not a high-quality food plant for quail, the seeds are readily available during winter when other seed sources are depleted. Annual lespedezas, such as Kobe and Korean are also excellent choices for quail. Care should be taken in using Kobe lespedeza in woodlands that are routinely burned in March because Kobe sprouts early in the spring and may be killed by fire. All of my planted strips are fertilized with 10-10-10 or 13-13-13 because most of our soils have been depleted of nutrients in the past from years of intensive cotton farming and cattle grazing.

I also plant shrub Lespedeza plots about every five years for quail. I prefer *Lespedeza thunbergii* over *Lespedeza bicolor* because *thunbergii* is fairly resistant to deer browsing and does not become an invasive pest like *bicolor*. Shrub lespedeza strips provide good food sources and overhead protection from avian quail predators. Often I establish bird strips and other food plots in relation to where my *Lespedeza thunbergii* plots are located.

Most of my food plots are not treated with herbicides and contain a mixture of native grasses and weeds. These areas provide excellent insect bugging sites for turkeys and quail. I'm not so interested in producing optimum yield as I am in providing ideal feeding sites.

In my duck impoundments and greentree reservoirs, I like to plant tropical corn, Japanese millet, and Egyptian wheat. I also use chufa in some parts of my waterfowl impoundments that are not frequently flooded. Newly constructed dikes and levees are seeded with *Sericea lespedeza* to stabilize soil and prevent erosion. Although *Sericea lespedeza* is not readily utilized as a food source for quail and other wildlife, it provides excellent cover.

Fig. 3: Wildlife Food Plot Activity Record Sheets

Wildlife food plot activity record sheets provide a detailed account of food plot establishment, maintenance, and success. Record sheets should be 8½ x 11 inches and punched to fit into a three-ringed looseleaf management plan notebook. Below is an example wildlife food plot activity record sheet that can be modified to suit a landowner's/manager's specific needs.

WILDLIFE FOOD PLOT ACTIVITY RECORD

Page ________

1. Plant(s) Grown: ______ **Number of Acres:** ________ **Year:** ____ **Field Number/Name:** ______________

Variety(ies) ______________________ Seed dealer ______________________

Seed: Germination % ______________________ Vigor test ______________________

Seeder type and setting ______________________ Seeding rate ______________________

Planting date(s) ______________________ Depth ______________________

2. FIELD HISTORY

Previous plant(s) ______________________ Planting date ______________________

Variety(ies) ______________________ Seeding rate ______________________

Last year's use by wildlife ______________________

3. SOIL AND SITE PREPARATION INFORMATION

Soil type ______________________ Tillage method ______________________

Describe tillage method(s) ______________________ Date ______________________

Other comments on soil and site preparation: ______________________

____

4. SOIL TEST RESULTS AND RECOMMENDATIONS

	Soil Lab Analysis				Recommendations	
Field Identification	pH	P	K	Other	Lime tons/A	Pounds / Acre N P_2O_5 K_2O Other

Date soil samples were taken ________

5. FERTILIZER AND LIMING PRACTICES *(pounds applied per acre)*

Method	Timing		N	P_2O_5	K_2O	Other Nutrients
Broadcast	Pre-plant					
Topdress	Post-plant	1				
		2				

Type of lime ______ Year Applied ________ Rate __________

Herbicide used ____ Date Applied ________ Rate __________

6. COSTS

Soil test(s)$ ______ Seed(s) $ __________ Fertilizer $ __________

Lime $ __________ Inoculant $ ________ Herbicide $ __________

Labor $ __________ Machinery $ ________ Other $ __________

7. MANAGEMENT ACTIVITIES

____
____
____
____
____

8. SUCCESS OF PLANTING

Field Appearance					*Use by Wildlife*		
Observation Date	Poor	Fair	Good	Excellent	Low	Medium	High

RICHARD T. BRYANT/ARISTOCK

Food plots placed adjacent to woods roads can congregate wildlife for hunting or viewing but should only be established when road access is strictly controlled.

JOE MCGLINCY

Forest openings and other unused areas, such as old logging decks, can be utilized as wildlife food plots in some years or left fallow for brood habitat as shown here.

Ten Steps to Establishing a Successful Wildlife Food Plot

1. Pick out best sites for food plots. Determine size and shape. Get permission if necessary.
2. Select plant varieties according to adaptability and soil type.
3. Check on availability of seed and order if necessary.
4. Obtain soil test several months in advance of planting.
5. Prepare seedbeds beginning several months before planting.
6. If recommended by a soil test, till in 1 to 2 tons of lime per acre in the soil several months in advance.
7. Till soil up to five times beginning several months before planting to reduce weed competition and create a firm, moist seedbed.
8. Apply fertilizer to seedbed based on soil test just before planting.
9. If needed, apply inoculant to legumes prior to planting.
10. Plant seeds or seedlings during recommended planting dates.

SUMMARY

Supplemental plantings can provide an additional source of high-value food for many species of wildlife, especially when native vegetation is scarce or poor in nutritional quality. To establish successful plantings that thrive and are used by wildlife, land managers should plan in advance and follow the guidelines and cultural requirements discussed in this chapter. In addition, there are other helpful sources of information and assistance listed in this chapter that can give forest and farm owners or their lessees guidance on establishing food plots. The listing of sources of wildlife planting material in Appendix K should be useful in locating seeds and seedlings of wildlife plants that may not be available locally. Although wildlife plantings are an excellent method to enhance food quality and availability, food plots should not be viewed as a substitute for forest and farm practices that enhance the availability and quality of native vegetation.

Improving forest and farm lands for wildlife has the potential to provide landowners with supplemental revenue from fee-hunting and other forms of fee-based wildlife recreation such as wildlife viewing and photography. Managing impoundments for waterfowl is one example of how landowners can meet the needs of waterfowl hunters.

DENNIS HOLT

CHAPTER 17

Economic Considerations of Forest and Farm Wildlife Management

CHAPTER OVERVIEW

This chapter discusses economic considerations of managing wildlife on farms and forests in Alabama and in the Southeast. In general good habitat conditions for wildlife can be created while managing for commodities. In many cases, certain forestry and farming practices also improve wildlife habitat with little or no additional cost to the landowner. In providing wildlife habitat, landowners make some concessions in the form of potential lost income from maximum timber or agricultural production. However, an integrated approach to forest, farm, and wildlife management minimizes costs while providing for sustainable production of wildlife, timber, and other land-based resources.

CHAPTER HIGHLIGHTS PAGE

ECONOMICS OF INTEGRATING WILDLIFE INTO FOREST AND FARM MANAGEMENT 387
FOREST VALUES 391
WILDLIFE VALUES 392
FACTORS AFFECTING WILDLIFE MANAGEMENT COSTS 393
ALTERNATIVES TO OFFSET WILDLIFE MANAGEMENT COSTS 396
LANDOWNER ASSISTANCE PROGRAMS 399
INCOME ALTERNATIVES FROM WILDLIFE-RELATED ACTIVITIES 402
STEPS TO DEVELOPING FEE ACCESS RECREATION 409

People place different priorities and values on natural resource use and management. The beauty of a forest, the feeling of being connected to the earth, the satisfaction of knowing you are providing habitat for wildlife, caring for the land in an ecologically sound manner, and practicing good stewardship for future generations are all ways individuals feel about forests and wildlife. For them, maximizing financial returns aren't as important as good stewardship and peace of mind.

Others place more emphasis on the financial productivity of the land. They seek to provide material needs for their family and make a living from the land while protecting their investment. Most landowners are probably somewhere in between these two extremes. Regardless of where you stand, if you are interested in wildlife, you should know that there are costs associated with providing quality habitats for wildlife on farm and forest lands.

Habitat management costs depend on a variety of factors such as landowner objectives, the wildlife species being managed, and the intensity of management. Costs associated with maintaining or improving wildlife habitat can be significantly reduced if it is conducted in conjunction with other land management practices that also enhance wildlife habitat. For example, many forestry and farming practices, like prescribed burning, timber thinnings, and no-till farming, also improve habitat for many wildlife species. Since these activities are normally conducted as a part of forest or farm management, the costs of these practices are partially or totally absorbed as an expense of the forestry or farming operation. As a general rule, the more concessions or management practices that are strictly conducted for wildlife (e.g. maintaining open areas, a lower timber stocking rate, food plot plantings, unharvested row crops, etc.) beyond what would normally be required in a typical timber or farming operation, the higher the opportunity costs. **Opportunity costs** are lost revenue that would have otherwise been received if timber or farm commodities were managed for maximum production. In other words, if timber or farm commodities are managed jointly with wildlife, the same financial return cannot be expected as if every acre was intensively managed for timber or agricultural production.

PAUL T. BROWN

Farm and bobwhite quail management are compatible if considerations are given to providing food and cover for quail. The bonus of having quail as a by-product of agriculture production is the enjoyment of hunting by the landowner and family and friends, or the option of receiving additional income from leasing quail hunting rights.

PAUL T. BROWN

Wild turkey management can be integrated into forestry and farming operations with advance planning and consideration of turkey habitat needs. Leasing lands for turkey hunting is an attractive option for forest and farm owners in Alabama.

Costs of improving wildlife habitat can be reduced significantly if it is conducted in conjunction with other land management practices, and costs may be completely offset if additional revenues can be generated from a recreational fee-access operation like leasing lands for hunting. For example, in some parts of Alabama, the cost of integrating deer and turkey habitat improvement practices with forest (pine) management can range from approximately 82 cents to $5 per acre, depending upon timber stumpage prices. This includes costs for wildlife management practices and opportunity costs (lost revenue) of foregone pine production. Since most annual hunting lease prices in the Southeast range from $2 to $7 per acre ($20 per acre is not unheard of), landowners who can integrate wildlife and other land management practices and at the same time receive supplemental income from hunting leases can more than offset the cost of wildlife management, and in many cases make a profit.

ECONOMICS OF INTEGRATING WILDLIFE INTO FOREST AND FARM MANAGEMENT

To illustrate the potential costs and returns of incorporating wildlife considerations into forest or farming operations it may be helpful to examine two examples where landowners have included wildlife as an objective in their land management plans. Forest and farm owners who are interested in various management alternatives for wildlife should go through an informal financial analysis or budget to get a better idea of management costs and estimated returns on their investments. The following two examples provide general insight into the costs of managing wildlife on woodlands and farms. This first example is based on work done by Dr. Bill McKee, which examines the economics of integrating wildlife and timber management. The second example illustrates a projected budget for implementing select habitat improvements on a farm that is leased for hunting.

Example 1: Forest Wildlife Management

Chester Baker, a forest owner in Coosa County, Alabama, is interested in managing his 160-acre tract of property for loblolly pine and at the same time providing habitat for deer and turkey for a lease hunting operation. In order to provide suitable habitat for deer and turkey, Chester was concerned about the potential loss in revenue from timber due to concessions made to accommodate wildlife. To get a better grasp of his management costs and potential returns, Chester decided that he would treat any potential reduction of revenue from timber as opportunity costs: timber income that is given up to accommodate habitat for deer and turkey.

Loss of timber revenue from maximum timber production then becomes an estimated cost of providing deer and turkey habitat on his land. Obviously, before beginning his economic assessment, Chester had to find current information and make a few assumptions about discount rates, expected costs, timber yields, stumpage prices, and hunting lease rates. To Chester's surprise, his potential net revenues from integrated timber and wildlife management could exceed profits that he would make from timber alone.

The results of Chester's calculations are especially important for small, non-industrial landowners since they illustrate the advantage of receiving annual hunting lease income. This income helps to offset the burden of timber investment for forest owners who normally have to wait years between returns from timber sales.

Example 2: Farm Wildlife Management

Curtis Gibson owns a 500-acre farm in Hale County, Alabama and is interested in knowing the potential cost of habitat improvements for a variety of game species on his farm. Curtis's primary objective is continuing his row crop operation (in soybeans, cotton, and corn). He is also interested in diversifying his farming operation to include a lease hunting program for all game species. Four hundred acres of the 500-acre farm are devoted to row crop production; the remaining 100 acres include a mixed pine–hardwood stand. To improve the farm wildlife habitat, Curtis would like to convert 8 acres of forest to planted wildlife openings and leave 6 acres of unharvested crops along field borders to provide supplemental food and cover. To get a better understanding of the economics of habitat improvements and a lease hunting operation on his farm, Curtis developed a bud-

Chester's Opportunity Costs of Providing Improved Wildlife Habitat

Stumpage Price Range	Type of Management	Present Net Worth[1] ($/acre)	Annual Equivalent[2]	Timber Revenue Foregone $/acre/year
High	Timber Only	175.29	12.09	4.95
	Timber & Wildlife	103.62	7.14	
Medium	Timber Only	113.46	7.83	4.34
	Timber & Wildlife	50.66	3.49	
Low	Timber Only	23.29	1.61	3.52
	Timber & Wildlife	-23.65	-1.91	

Chester's Economic Return from Integrated Timber and Wildlife Management While Receiving Annual Hunting Lease Payments at $5 Per Acre

Stumpage Price Range	Present Net Worth ($/acre)	Annual Equivalent	Income Gain Over Timber Only Management ($/acre/year)
High	181.11	12.49	.40
Medium	128.14	8.84	1.01
Low	49.83	3.44	1.83

[1]Present net worth is discounted value of all revenues minus discounted value of all costs. Positive values are good investments.
[2]Annual equivalent is the annual cash flow that is equivalent to another specific cash flow. Useful for comparing different management strategies.

get. The budget assumes that the entire farm is leased for all game species at $3.50 per acre.

Curtis's budget shows the cost for establishing 8 acres of openings to be $1,130.68 and for 6 acres of unharvested crops to be $180. The total cost for wildlife habitat concessions is $1,310.68. To arrive at a hunting lease price that recovers all costs (break-even), Curtis divided the total cost by the number of acres ($1,310.68 / 500 acres) and came up with a break-even cost of $2.62 per acre. Since Curtis was expecting to receive at least $3.50 per acre for leasing hunting rights, he can expect a net profit of $439.32. Obviously, the higher the lease price,

Curtis's Budget for a Hunting Lease Operation on a 500-acre Farm

	Unit	Price ($)	Quantity	Total Value ($)	Value Per Acre ($)
Receipts From Hunting Lease	acre	3.50	500	1,750.00	3.50
Variable Costs					
Forest Wildlife Openings					
Wheat	lbs.	0.33	20.00	52.80	6.60
Ladino clover	lbs.	3.40	8.00	217.60	27.20
Crimson clover	lbs.	1.00	2.00	16.00	2.00
Lime (spread)	tons	24.00	1.00	288.00	36.00
Fertilizer (spread)	tons	9.00	2.00	144.00	18.00
Machinery	acre	1.11	1.00	8.88	1.11
Tractors	acre	5.30	1.00	42.40	5.30
Labor	hour	4.15	1.35	44.80	5.60
Miscellaneous		1.00	10.00	80.00	10.00
Total Variable Costs				-894.48	-111.81
Fixed Costs					
Machinery	acre	4.57	8.00	36.56	4.57
Tractors	acre	4.83	8.00	38.64	4.83
Total Fixed Costs				-76.20	-9.40
Land Opportunity Costs					
Forest Wildlife Openings	acre	20.00	8.00	-160.00	-20.00
Cost of Openings				-1,130.68	-141.34
Cost of Unharvested Crops	acre	30.00	6.00	-180.00	-30.00
Total Costs				-1,310.68	
Break Even				1,310.68	2.62
Net Profit				439.32	0.88

Economics of Wildlife and Timber Management

Dr. Bill McKee is a wildlife biologist and forester with the Molpus Company in Philadelphia, Mississippi. He received his B.S. in forestry, M.S. in wildlife ecology, and Ph.D. in agricultural economics from Mississippi State University. During his career he has been actively involved in forest operations and management, and wildlife management from both a research and extension perspective. Dr. McKee is most noted for his work on the economics of integrating wildlife and timber management.

Landowners who are considering income from both wildlife and timber management should first define their objectives. Most Alabama farm and forest owners inherently have more than one management objective such as farming, timber, wildlife, or some other alternate operation. Whatever the interest, the key is to first identify your objectives and then develop a management plan that outlines how the objectives will be accomplished. For example, if wildlife is a high priority, consider leaving more streamside management zones, hardwood pockets, food plots, wider road backslopes, natural openings, more prescribed burning, and other habitat improvement practices. The management plan should also include a timber harvesting schedule that identifies which stands will be cut and at what time. The harvest schedule is an integral part of the plan because it helps the landowner understand how much and when timber sale revenue and associated costs will occur. Finally, profitability of the management prescriptions can be acknowledged by summing all costs and revenues by year of occurrence in a cash flow table.

Choosing wildlife habitat management as a primary objective over timber production will reduce some timber revenue because less acreage is devoted to commercial timber production. If the landowner is not satisfied with the projected cash flow, then more emphasis may need to be placed on growing timber. Lost timber revenue caused from intensified wildlife habitat management practices, in many cases, can be recouped from revenue generated from fees received for hunting leases. In some cases hunting lease income exceeds what was lost in timber production because forest owners who emphasize managing for wildlife as their first priority are able to lease their lands at a higher rate. I strongly believe that hunting lease rates in Alabama will continue to increase. Consequently, managing for both wildlife and timber production is and will continue to be an economically viable option for forest owners.

the higher the net profit. For example, a hunting lease rate of $5 per acre would yield a net profit of $1,189.32.

Many of the financial calculations required in a forestry (or forestry and wildlife) operation can now be accomplished quickly and painlessly with the help of computer programs. One such program that works on a stand-by-stand basis is YIELDplus, which was developed by the Tennessee Valley Authority. YIELDplus contains a set of timber growth and yield simulators and a financial analysis package. The financial analysis portion of the program incorporates timber volumes, user-defined stumpage values, and management costs to determine profitability of a potential management plan. One of the major benefits of using a computer program to perform these calculations is the ability to change an estimated timber volume, hunting lease price, cost, or revenue and quickly recompute the analysis. This allows you to fine-tune your potential management options by answering the question "What if...?" This can be important if some of your estimates are subject to change or if profitability is marginal. YIELDplus is available through many county Extension offices or from the Extension Forestry Department at Auburn University. For those who want to use YIELDplus on their own personal computers, copies of the program can be obtained from the Forest Resources Systems Institute (FORS Institute), P.O. Box 1785, Clemson, SC 29633 (864) 656-7723.

RICHARD T. BRYANT/ARISTOCK

Forest and bobwhite quail management are compatible if timber stands are opened to allow sunlight to reach the forest floor, encouraging the growth of herbaceous and shrub vegetation for food and cover. Forest owners have an opportunity to meet the demand for quality quail hunting and, at the same time, receive additional income from hunting leases.

FOREST VALUES

Alabama's forests provide a wide array of benefits to its citizens. Some of these benefits include havens for recreational opportunities, habitats for wildlife, aesthetic beauty, and clean air and water. Although these benefits cannot always be measured in economic terms, their greatest value is in enhancing the quality of life in Alabama. In economic terms though, forestry is big business in Alabama! The 22 million acres of commercial timberland, the majority owned by private forest landowners, supply the raw materials to operate Alabama's largest manufacturing industry and make timber Alabama's most valuable crop. Timber outranks all other commodities as the dominant crop in 34 counties and exceeds the total of all crops combined in 15 other counties. Forestry directly employees over 65,000 people in Alabama, impacts an additional 100,000 jobs, and provides an annual payroll of over $1.7 billion. There is no doubt that Alabama's forests are a major contributor to the state's economic stability and quality of life.

Economic Facts of Alabama's Timber Industry

- Forest industry is the state's largest manufacturing industry, producing an estimated $13.2 billion worth of products annually. The value-added component of this figure is estimated at $6.3 billion annually.
- 70,800 Alabamians are directly and indirectly employed by forestry, with an estimated annual payroll of $2.01 billion.
- 100,000 additional jobs depend on the forest industry.
- $790 million is paid annually to forest landowners for timber.
- Total volume of timber is estimated at 23 billion cubic feet.

Based upon Alabama Forestry Facts 1997, *Alabama Forestry Association*

WILDLIFE VALUES

With few exceptions, wild animals are not sold in a market like timber or farm crops, so market prices cannot be used as a measure of their value. But, wildlife populations and their habitats are often placed in direct competition with agriculture or other land-use alternatives such as urban development—the benefits of which are usually presented in dollars and cents. Society is influenced by both dollar-denominated measures of benefits and costs and non-monetary measures. Wildlife resources contribute economic value and enhance the quality of life. Adequate information is often not available to quantify the non-monetary benefits of wildlife; nevertheless, these benefits are profound.

One of the most commonly recognized values of wildlife is **recreation.** Over 1.8 million Alabamians (56 percent) annually pursue some form of wildlife recreation each year. One of every four Alabamians hunt or fish. There are two categories of wildlife associated recreation, consumptive and non-consumptive. **Consumptive** use of wildlife generally includes those activities that have a direct impact (taking) on wildlife numbers such as hunting, fishing, and trapping. Over 914,000 Alabama sportsmen hunt or fish each year. **Non-consumptive** uses of wildlife are those activities that usually do not directly impact (non-taking) wildlife such as wildlife observation, wildlife photography, wildlife feeding (bird feeders), etc. Approximately 1.2 million people in Alabama participate each year in some form of non-consumptive recreation. Many natural resource professionals and ecologists now recognize that traditional non-consumptive uses do in fact directly impact wildlife in some way. An example would be soil compaction and habitat degradation caused by overuse of natural areas by non-consumptive users. Consequently, the term **appreciative use** is replacing non-consumptive use as a more accurate description.

Most efforts attempting to derive an economic value from wildlife have focused on user expenditures from recreation, primarily from hunting and fishing. To a lesser extent, a few studies have attempted to quantify the economic value of non-consumptive uses. Consumptive and non-consumptive wildlife recreation values are not a complete reflection of the value of wildlife. They should not be used by themselves to make comparisons with competing land-use alternatives (such as urban development) unless it is clear that these economic values represent only a portion of the overall value of wildlife resources. Recreation is only one of at least five categories of wildlife values. Unfortunately, these other values are even more intangible and difficult to quantify in economic terms.

Although the general public is most familiar with the recreational value of wildlife, there are other important categories of use or value. Wild animals are integral parts of biological communities and ecosystems. This contribution to productive ecosystems is referred to as the **biological value** of wildlife. Examples of the biological importance of wildlife include seed dispersal and

Values and Uses of Wildlife

NON-CONSUMPTIVE VALUES

- Hiking and Camping
- Wildlife Photography
- Wildlife Viewing
- Seed Dispersal
- Pollination of Flowers & Crops

CONSUMPTIVE VALUES

- Hunting
- Fishing
- Trapping
- Meat as Food
- Fur Coats

CONTRIBUTIONS

- Biological
- Scientific
- Intrinsic
- Educational
- Economic
- Recreational
- Aesthetic
- Family Traditions

Alabama Sportsmen and the Game They Hunt[1]

GAME HUNTED	NUMBER OF SPORTSMEN	NUMBER OF DAYS AFIELD
Big Game (deer & turkey)	262,000 (73%)	3.7 million
Small Game (rabbits, squirrels, & quail)	160,000 (44%)	1.6 million
Migratory Birds (doves, ducks)	104,000 (29%)	358,000 days
Other Animals (coyote, raccoon, fox, bobcat)	17,000 (5%)	361,000 days

[1]*Based upon 1995 U.S. Fish and Wildlife Service survey. Percentages total more than 100% since sportsmen and women hunt more than one species.*

planting, pollination of agricultural and wild plants, natural population control, regulation of pest species, and nutrient transport and recycling in natural systems.

Secondly, **scientific value** of wildlife is closely related to biological contributions. Each animal and plant retains a genetic and chemical "factory" that has taken centuries to develop. These genetic and chemical secrets have allowed man to develop medicines and pharmaceutical products to cure many common diseases. Ecological studies of wildlife populations have been extremely important in helping us learn more about our planet and ourselves. The health and viability of wildlife is often used as an indicator of environmental quality and ultimately the quality of life for humans. For example, mink *(Mustela vision)* are extremely sensitive to mercury contamination. Decline in mink numbers is thought to be directly related to an increase in watershed pollution from heavy metals like mercury that degrades environmental quality and ultimately threatens human health.

Third, wildlife have an **intrinsic value** or existence value. Several studies have indicated that landowners place a high intrinsic value on the wildlife on their lands, a value often higher than that of generating revenue from their property. Landowners often reserve wildlife viewing and hunting rights exclusively for themselves and their families, even though leasing their lands for hunting and other purposes would be profitable. Past studies have also shown that most people value just knowing that certain wildlife species exist, regardless of whether the individuals will ever get to see them or directly benefit from them. Fourth, **educational** values of wildlife are often overlooked. Environmental education involving wildlife and their habitats vividly illustrates the interconnectiveness of life. Most educators realize the value of using wildlife as a learning tool, one that captivates the attention and imagination of any student. Finally, wild animals have an **aesthetic value.** A portion of this value can be assessed through expenditures for recreational activities such as wildlife observation and wildlife photography, as well as wildlife-related activi ties such as camping and hiking. Recent studies have also shown that the aesthetic value of wildlife is very important to the majority of individuals involved in consumptive activities such as hunting and fishing.

FACTORS AFFECTING WILDLIFE MANAGEMENT COSTS

There is an associated cost or investment in managing wildlife on private farm and forest lands. The amount of investment varies depending on the following factors:

- Landowner objectives.
- Property ownership status, whether the property is paid for or newly purchased or leased.
- Availability of labor and equipment necessary to conduct management practices.
- Condition of the land before wildlife habitat improvements are made.
- Multiple management objectives. Is the land be-

Economic Facts About Wildlife[1]

UNITED STATES

- $83 billion spent annually by 49.7 million hunters and fishermen
- $18.1 billion spent annually by 76.1 million non-consumptive wildlife recreationists

ALABAMA

Hunting industry in Alabama:

- $1.4 billion generated in annual output
- $1.1 billion of annual employee earnings
- 45,952 employed directly or indirectly from hunting related activities
- $1.2 billion generated annually by 1.8 million Alabamians (56% of the population) who participate in wildlife-associated activities
- $31.3 million generated annually from hunting leases on private lands
- The majority of the hunting industry's economic contribution is to rural Alabama. In two rural counties in the Southeast, private land hunters contributed over $6.6 million to the local economy.

[1]*Based upon surveys and studies conducted by the U.S. Fish and Wildlife Service and Auburn University in 1995.*

ing managed solely for wildlife or in conjunction with current land management practices like forestry or row crop production?

Many landowners find that the cost of managing wildlife can be quite high, especially if there is little or no other income being generated from the land. Fortunately, there are some alternatives available to forest landowners that can help defray the cost of wildlife management, and in many cases cover more than just costs associated with improving lands for wildlife.

Landowner Objectives

Landowner objectives have a direct bearing on how much money is invested in wildlife management. For example, the wildlife species that a landowner wants to enhance will require certain types of habitats that have an associated cost to establish or maintain. Some wildlife species require little or no habitat management (low intensity management) while others require constant land management either on a biennial or annual basis (high intensity management). Obviously, management costs will be much greater for those wildlife species that require more intensive management. Some habitats may be provided through traditional forest management practices that, with slight modifications, are already being conducted on the land. Depending on the degree of modification of forest management plans for wildlife, management costs for wildlife can be significantly offset or absorbed as a forest management expense. White-tailed deer and Eastern wild turkey are examples of two wildlife species that generally do well in conjunction with forest management. In most cases, the level or intensity of habitat management for deer and turkey can be low, and the two species will generally do quite well. However, some game species like bobwhite quail require habitats in early successional stages (grasses, legumes, and shrubs). To maintain native quail habitats in grasses, legumes, and shrubs demands a high level of intensive management with frequent disturbances such as prescribed burning, disking, and mowing. Consequently, costs for maintaining quail habitat, on a per acre basis, are considered to be the greatest of any game species in the Southeast. Landowner objectives can therefore determine the intensity of habitat management that will be required and ultimately the costs of wildlife management.

Land Ownership

Most forest landowners in the Southeast who are managing for wildlife and providing recreational fee or lease hunting already own their land. In these cases, wildlife and fee hunting are not primary land management objectives or income-generating activities. The majority of these lands are managed for timber, agriculture, or a combination of the two. Wildlife management and fee hunting then are generally considered supplemental or complementary activities. The cost of land ownership in these cases is not considered a wildlife management cost. However, in some instances individuals have purchased land primarily for managing wildlife and developing a fee hunting operation. In these cases, land costs do become a wildlife management expense. With few exceptions, it is cost-prohibitive to invest in land for the primary purpose of developing a recreational hunting operation. Individuals should be discouraged from purchasing land solely for the purpose of managing and operating a recreational hunting business unless financial considerations are unimportant.

Availability of Labor and Equipment

If landowners and family members are able to perform many of the habitat improvements and other management activities themselves, the costs of hiring additional employees or contracting labor for various land management services can be reduced. Costs associated with landowner and family time spent on wildlife management or other activities must be included in an investment analysis. Likewise, any equipment or facilities present on the property that can be used in the operation will help minimize costs. For instance, a farm tractor can also be used for establishing food plots, firelanes, and disking. A barn or old cabin can be refurbished as a lodge for hunters, fishermen, and other recreationists. Once again, the cost of owning and using equipment must be accounted for in the investment analysis based on the proportion of use in each enterprise. Buying new equipment solely for wildlife habitat management or land enhancement for recreation can be cost prohibitive. Leasing equipment and hiring labor from neighboring farms or clubs is generally more cost effective.

Fig. 1: Typical Equipment Used for Wildlife Habitat Management

All of the land management tools available to farmers, foresters, and construction engineers have been used to manipulate wildlife habitat. The following are some of the most commonly used wildlife habitat enhancement tools.

EQUIPMENT	USE
Tractor & select attachments	Establishing food plots -preparing seedbed -seeding -fertilizing -herbicide
Dibble planting bar	Planting shrub & tree seedlings
Hand-held cyclone seeder	Broadcast seeding and fertilization of food plots
Drip torch and fire beaters	Prescribed burning
Backpack sprayer	Spot application of herbicides
Hypo-hatchet	Herbicide injection into low value wildlife trees
Chainsaw	Removal of low quality wildlife trees and clearing roads
All terrain vehicle (4-wheeler)	Multipurpose
4-wheel-drive truck with winch and trailer	Multipurpose

Fig. 2: Trends of Average Cost of Forest Management Practices that Affect Wildlife from 1976 to 1996[1]

Forestry Practice	1976	1979	1982	1984	1986	1988	1990	1992	1994	1996
Cost in Dollars Per Acre										
Prescribed Burning	3.65	2.95	4.12	7.16	4.84	6.52	8.10	8.14	10.57	14.65
Removing Undesirable Trees (Chemically)	23.41	40.23	40.65	64.82	65.61	57.26	63.70	62.73	67.41	67.65
Timber Cruising	1.18	1.77	2.18	2.26	3.27	2.47	2.02	2.49	2.09	3.06
Marking Trees for Harvesting	8.05	7.14	14.02	14.63	10.57	8.58	8.47	12.72	14.19	12.21
Mechanical Site Preparation	73.36	93.09	114.04	90.23	94.21	92.66	87.45	98.42	100.74	108.05
Planting by Hand*	48.06	41.94	43.56	43.65	47.16	52.56	53.73	51.93	52.83	54.63
Planting by Machine*	34.56	36.81	48.60	49.53	39.51	44.28	40.68	46.71	53.28	58.59
Precommercial Mechanical Thinning	25.97	33.22	49.27	43.18	52.44	55.58	55.43	75.71	79.05	89.22
Fertilization	—	—	38.80	40.35	36.03	35.84	39.29	43.17	41.01	56.52

*Cost figured on planting 900 pine seedlings per acre. (Dubois et al. 1997, *Forest Landowner*)
[1]Costs are based on a weighted average

Condition of the Land

The condition of forest land determines the quality of habitats for wildlife and its potential for supporting adequate wildlife populations. If past land management has degraded the productivity of the land and its ability to produce food and cover for wildlife, then habitat improvement practices will have to be implemented. The poorer the wildlife habitat the greater the need for intensive management practices, which in turn increases wildlife management costs. Landowners interested in a wildlife enterprise should also consider taking additional steps to improve the aesthetic quality of their lands for their potential clientele.

ALTERNATIVES TO OFFSET WILDLIFE MANAGEMENT COSTS

The cost of wildlife habitat improvements can include the expense of hiring natural resource professionals to provide technical assistance in developing recommendations as well as the actual investment of habitat improvement practices. To help offset these costs, there are several alternatives available to forest and farm owners. Alternatives include state and federal cost-sharing programs that partially pay for management practices that directly or indirectly improve wildlife habitat, private and government-sponsored landowner assistance programs that provide technical assistance,

CLEMSON UNIVERSITY

Timber and wildlife management are compatible uses of the land. Thinning pine stands can improve wildlife habitat and provide landowners with additional income from the sale of timber.

and in-kind labor assistance from land user groups. In some cases, there are tax incentives for landowners who improve their lands for wildlife or who establish conservation easements. In addition, landowners who invest in wildlife habitat improvements on their property usually enjoy the benefit of increased land values.

Cost-Sharing Programs

State and federal cost-sharing programs are available to help private landowners defray the costs of land management practices that directly and indirectly benefit wildlife. Although the primary purpose of most of these programs is not wildlife habitat enhancement, wildlife still derive many benefits. State and federal agencies that administer cost-sharing programs are usually located in the city that houses county government agencies, often called the county seat. Figure 3 lists cost-sharing programs that are available for forest and farm owners.

Tax Incentives

To reduce the costs associated with improving farm and forest land for wildlife, some states provide landowners with tax incentives for managing and maintaining wildlife habitat, especially if the general public is given access to private land. To promote and recognize the value of good land stewardship, some states, like Minnesota, have eliminated property taxes on private lands that have protected wetlands and native wildlife habitats. Most state and county governments in the Southeast have an agricultural property tax rate. This is one of the lowest property tax rates and applies to forest land that is being managed and used for both timber and wildlife. Forest and farm owners who derive income from selling access rights to wildlife are also allowed an agricultural property tax rate if their land is not solely managed to produce income from wildlife-based recreation.

Landowners who manage their forest land for timber should take advantage of the reforestation tax credit and seven-year amortization whenever trees are replanted and deduct valid management expenses at the appropriate time. As mentioned earlier, certain forest management activities also improve wildlife habitat. To claim management expenses as deductions, it is important to be able to show a relationship between expenditures and a potential

ART DYAS

Dr. Sam Eichold of Mobile and Dan Dumont, Executive Director of the Alabama Wildlife Federation, sign a conservation easement covering Sturdy Oak Farm in Escambia County, which will be administered by the Alabama Wildlife Federation. Conservation easements allow forest and farm owners the opportunity to conserve wildlife habitat on their land while reducing property, income, and estate taxes.

for profit. To understand what management expenses are deductible as conservation activities under the current IRS guidelines, forest owners would be well advised to seek the advice of an accountant who has expertise in forest taxation. Additional information is available in *Forest Owners' Guide to the Federal Income Tax*, Ag Handbook 708, available from the Superintendent of Documents in Washington, DC, 20402-9328. In addition, forest owners should have an accountant or tax attorney help develop an estate plan to help reduce or eliminate forest estate taxes in the event of death. This planning may prevent the type of crisis management that often occurs when forest landowners pass away. Remember, on a long rotation it takes 70 to 100 years to grow a mature forest in Alabama and land ownership changes every 25 years on average due to the death of the landowner.

CONSERVATION EASEMENTS

Conservation easements allow landowners to conserve wildlife habitat and other valuable land resources on their property and in the process can significantly reduce property, income, and estate taxes. A conservation easement is a deed restriction where the landowner agrees to protect and maintain certain land resources. Conservation easements are flexible enough to be tailored to meet landowner objectives while at the same time providing some degree of protection of land resources. Easements are recorded with land deeds and bind future owners to whatever terms the current owner agrees to convey. Most easements are perpetual, regardless of changes in ownership. Before considering a conservation easement, it is important to know exactly what land-use limitations you are imposing and their impacts. It is wise to consult with an accountant, attorney, and natural resource professional for advice and written terms of an easement. Conservation easements can be administered through public landholding agencies or by established conservation groups like The Nature Conservancy, the Alabama Wildlife Federation at (800) 822-9453, or the Delta Environmental Land Trust Association (D.E.L.T.A.). Conservation easements are recognized in more than 12 states that have adopted versions of the Uniform Conservation Easement Act. Alabama has recently adopted a version of the Uniform Conservation Easement Act. For more information about conservation easements contact the Alabama Wildlife Federation or The Land Trust Alliance, 1319 F Street NW, Suite 501, Washington, DC 2004 (202-638-4725); D.E.L.T.A., 900 West Pine, Suite 7, Vicksburg, MS 39180 or (601-638-8823) and ask for a copy of their publication on conservation easements.

Another way to protect your land for future generations, and gain major tax benefits, is by donating the land to the Alabama Wildlife Federation, the Alabama Forever Wild Land Trust, or to another non-profit land conservation organization. The Alabama Forever Wild Land Trust is a state land protection program to acquire and hold land for recreation and conservation uses. Donations of land to the Forever Wild program can provide double credit on state tax returns for charitable contributions, based on the appraised value of your land, and credit for charitable contributions on your federal tax return. Land donated through estate planning can ease the estate tax burden on heirs. Conservation easements also are accepted as part of the Forever Wild Program. For more information about land donations or the Forever Wild Program, contact the Lands Division of ADCNR, (334) 242-3484, the Nature Conservancy, 2821-C 2nd Ave, S., Birmingham, AL 35233, (205) 251-1155, or the Alabama Wildlife Federation (800) 822-9453.

Increased Land Values

In most cases, properly managed land with abundant wildlife populations has a higher economic and aesthetic value than unmanaged land with low numbers of wildlife. Exceptions may include situations where the presence of wildlife have a negative impact, such as crop damage by deer. Landowners who are not particularly interested in the value of having wildlife on their property are usually concerned about the resale value of their land. In most cases, land that has been managed to improve wildlife habitat has a higher resale value.

In-Kind Assistance

In many cases, land resource users may provide in-kind labor to conduct wildlife habitat improvement practices or other land maintenance activities. For example, sportsmen's groups who lease land for hunting are often willing to provide labor to plant food plots for wildlife, make road and fence repairs, or make other improvements that are beneficial to wildlife and the landowner. To facilitate in-kind labor assistance from recreational users, landowners should consider giving users of the land a copy of the forest or land management plan. This will help identify management activities that could possibly be conducted by user groups. Responsible sportsmen's groups may also help reduce trespassing problems and vandalism by posting, monitoring, and patrolling the landowner's property. A survey of forest products companies in the South found in-kind services are sometimes as valuable as the financial compensation received.

LANDOWNER ASSISTANCE PROGRAMS

Several state and federal agencies offer free technical assistance to forest and farm owners who are interested in improving their lands for wildlife. The type of assistance varies, depending upon the agency, but in general natural resource personnel are available to offer advice and help in developing land management plans that consider wildlife as an objective. Some of the agencies that offer assistance include the following:

- Alabama Department of Conservation and Natural Resources Game and Fish Division (technical wildlife assistance)
- Alabama Forestry Commission (technical forestry assistance)
- Alabama Cooperative Extension System (educational forestry and wildlife assistance)
- USDA Natural Resources Conservation Service (technical soils and wildlife assistance)
- U.S. Fish and Wildlife Service (technical wildlife assistance through the Partners for Fish and Wildlife Program).

Natural resource consultants can also provide technical assistance for a fee. Care should be taken in selecting a consultant since the range of education, experience, and expertise varies greatly between consultants. Consultants who provide wildlife assistance should be certified by The Wildlife Society, the professional organization of wildlife biologists. Consultants who have combined forestry and wildlife training, and are registered foresters and certified wildlife biologists, can provide a wide range of professional services and expertise. For a listing of consultants contact the Alabama Forestry Commission, the Alabama State Board of Registration for Foresters, or the Alabama Chapter

Companies That Offer Landowner Assistance Programs*

Alabama River Woodlands, Inc.
P. O. Box 99
Perdue Hill, AL 36470
(334) 743-8212

Boise Cascade Group
307 Industrial Road
Jackson, AL 36545
(334) 246-2482

Bowater, Inc.
Southern Division Woodlands
Calhoun, TN 37309
(615) 336 2211

Cedar Creek Land & Timber, Inc.
P. O. Box 1769
Brewton, AL 36427
(334) 867-6165

Champion International Corp.
P. O. Box 250
Courtland, AL 35618
(334) 637-2741

Container Corp. of America
P. O. Box 1469
Brewton, AL 36427
(334) 867-3621

Fort James Corp.
P. O. Box 130
Pennington, AL 36916
(205) 654-3000

Georgia-Pacific Corp.
Forest Resources Division
P. O. Box 9576
Columbus, GA 31908
(706) 327-6522

Gulf Lumber Co., Inc.
P. O. Box 1663
Mobile, AL 36601
(334) 457-6872

Gulf States Paper Corp.
P. O. Box 48999
Tuscaloosa, AL 35404
(205) 553-6200

International Paper Co.
3201 International Drive
Selma, AL 36703
(334) 872-3481

MacMillan Bloedel, Inc.
P. O. Box 336
Pine Hill, AL 36769
(334) 963-4391

Mead Coated Board, Inc.
P. O. Box 178
Cottonton, AL 36851
(334) 855-4751

Packaging Corp. of America
Woodlands Division Southern Region
Box 33
Counce, TN 38326
(901) 689-5292

Stone Container Corp.
P. O. Box 21607
Columbia, SC 29221
(803) 359-7232

Tolleson Lumber Co.
Drawer E
Perry, GA 31069
(912) 987-2105

Union Camp Corp.
200 Jensen Road
Prattville, AL 36067
(334) 361-5000

U.S. Alliance
17589 Plant Road
Coosa Pines, AL 35044
(205) 378-5541

**Current at the time of this writing.*

Fig. 3: Cost-Sharing and Other Assistance Programs[1]

COST-SHARING PROGRAM	PURPOSE	ADMINISTERING AGENCY
Alabama Agricultural and Conservation Development Commission Program (AACDCP)	Cost-sharing for reducing soil erosion, reforestation, and water quality improvement practices.	SWCD
Conservation of Private Grazing Land (CPGL)	Technical assistance for improving range and pasture grasslands that can benefit wildlife.	NRCS
Conservation Reserve Program (CRP)	Cost-sharing to retire marginal farmland into permanent vegetation (trees or certain perennial plants) to reduce soil erosion.	FSA and NRCS
Environmental Quality Incentives Program (EQIP)	Cost-sharing to farmers for conservation practices that reduce soil erosion.	NRCS
Farm Loan Conservation Program (FLCP)	Credit given to borrowers from government loan programs based upon environmental and conservation efforts.	FSA
Farmland Protection Program (FPP)	Provides technical and financial assistance for private grazing land, farmland protection, and flood risk reduction. Some practices also benefit wildlife.	FSA
Flood Risk Reduction Program (FRRP)	Encourages farmers of frequently flooded cropland to use the land less intensively. This program is not yet operational.	FSA
Forestry Incentive Program (FIP)	Cost-sharing for certain forest management practices such as site preparation for natural regeneration, tree planting, and timber improvement practices like prescribed burning.	NRCS
Highly Erodible Land Conservation (HELC)	A disincentive program. Contains two provisions, Sodbuster and Conservation Compliance, that discourage the use and conversion of highly erodible forest and grassland into cropland.	FSA
National Natural Resources Conservation Foundation (NNRCF)	A charitable, non-profit corporation to initiate and support conservation activities of USDA.	NRCS

Fig. 3 Continued.

COST-SHARING PROGRAM	PURPOSE	ADMINISTERING AGENCY
Partners for Fish and Wildlife	Technical and financial assistance to private landowners for wildlife habitat restoration.	USFWS and ADCNR
Resource Conservation and Development Program (RC&D)	Authorizes local citizens to organize RC&D Councils to provide leadership and coordination to address a wide variety of rural development and natural resource conservation needs. RC&D Councils compete for federal funds for projects.	NRCS
Rural Abandoned Mine Program (RAMP)	Cost-sharing to reclaim abandoned mines in trees and wildlife cover.	NRCS
Stewardship Incentives Program (SIP)	Cost-sharing to encourage long-term stewardship of non-industrial private forest land for a variety of resources such as forest, wildlife, soils, water quality, recreation, and aesthetics.	AFC and FSA
Water Bank Program (WBP)	A regional wetland protection program that compensates landowners for protecting certain types of existing wetlands and associated uplands.	NRCS
Watershed Protection & Flood Prevention Program (WPFPP)	Locally driven federal program to plan and carry out water management projects on a watershed scale. Provides planning, technical assistance, and cost-sharing assistance to local project sponsors.	NRCS
Wetland Conservation or "Swampbuster" Program (WCP)	A disincentive program. Makes wetland conservation an eligibility requirement before receiving certain agriculture subsidies through USDA.	NRCS and FSA
Wetlands Reserve Program (WRP)	Cost-sharing for restoration and protection of wetlands that in the past were converted to agriculture.	NRCS and ADCNR
Wildlife Habitat Incentives Program (WHIP)	Cost-sharing assistance for wildlife habitat improvement practices as determined by a state committee.	NRCS

ADCNR—Alabama Department of Conservation and Natural Resources. AFC—Alabama Forestry Commission. FSA—USDA Farm Services Agency. NRCS—USDA Natural Resources Conservation Service. SWCD—Soil and Water Conservation District. USFWS—U.S. Fish and Wildlife Service.
[1]*Cost-share and assistance programs available at the time of writing.*

of the Wildlife Society (for addresses see Appendix D).

Many forest products companies also have landowner assistance programs that provide a variety of services. Some of these services include tailored forest management plans, and assistance in forest practices such as site preparation, obtaining tree seedlings, planting, thinning, forest insect control, timber marking, and prescribed burning. Many of these landowner assistance programs offer services at no charge to the landowner. Other companies charge a minimal fee that, in most cases, is much lower than management costs found elsewhere. Some companies require a right of first refusal when timber is marketed, while others do not. Landowners can choose programs that best fit their needs. The obvious benefit to forest owners that participate in an industry landowner assistance program is the available expertise and reduced costs of managing and marketing timber. Since 70 percent of Alabama's forests are owned by non-industrial private forest landowners, the forest industry benefits by having a future supply of timber for their mills.

INCOME ALTERNATIVES FROM WILDLIFE-RELATED ACTIVITIES

THE WILDLIFE MARKET

Current trends in the various segments of the wildlife market may affect income potential for landowners now and in the future. A unique aspect of establishing a market for wildlife recreation is the issue of resource ownership. Landowners do not own or sell wildlife, but they can market access and a recreational experience. For instance, Joe and Jerry McCallister of McCallister Farms in Dothan, Alabama, do not keep an exact tally on the number of deer, quail, or turkey on their land like they keep track of their livestock. Yet they know deer are abundant on their property, and if they choose to sell hunting lease rights, they would be giving permission to enter their land and harvest game animals. The fee-access recreation market differs from traditional forest or agricultural markets in at least two other ways. The fee-access market is not "standardized." Forest products such as timber, or agricultural products such as corn, soybeans, cattle, and hogs have definitive quality standards that determine the prices that forest landowners and farmers receive in an open market. In the case of fee-access, each land parcel and access arrangement is unique. There are no quality standards to help price the various combinations of wildlife, services provided, length of visit or hunt, and physical attributes of the property. In a fee-access operation, the landowner is dealing directly with the consumer during marketing and management. Unlike many forest landowners and farmers, the lease operator must deal directly with the final consumer and often on a continuing basis. Therefore, successful fee-access operations not only demand knowledge of the resource, but also skills in marketing, business management, and customer relations.

A wildlife operation based upon fee-access is one way that forest landowners can partially or fully recoup timber opportunity costs sacrificed by intensifying wildlife habitat management. Forest landowners considering this option should realize that there are also expenses associated with developing a wildlife recreation enterprise. Costs will vary depending on the type and intensity of enterprise. The type of services and amenities provided will depend upon the type of clientele that the landowner is trying to attract. Generally, expenses are directly related to the services and amenities provided. For example, many wildlife enthusiasts would rather not pay for anything other than access to the property. With this arrangement, there are few investment costs for the landowner, such as advertising, and minimal management time. Other clientele may be willing to pay not only for access rights, but also for services and amenities that include lodging, meals, guides, dogs, transportation, and even entertainment. These enterprises usually require additional personnel which adds to the cost. Even by charging higher access fees, landowners face difficulties recovering the costs of a high investment operation. Economic analyses have shown that low-input operations tend to be more profitable on the average, although there are some efficiently managed, high-investment operations that have proven to be quite lucrative.

Standards of Various Resources

Timber	Cattle	Soybeans	Deer
pulp	breed	cracked seed	antler type
pole	weight	moisture	weight
sawtimber	performance	insect damage	sex

Fee-Hunting

Currently, the majority of the fee-access market for wildlife-related recreation centers on hunting. Analysis of hunting license sales over a period of years provides trend information about the demand for hunting. The most recent U.S. Fish and Wildlife Service (USFWS) data indicates that the number of hunters across the country is gradually declining, or at best, has leveled off. Participation in hunting has declined from 16.7 million hunters in 1985 to 14.1 million in 1991. Even if the number of hunters in the U.S. has temporarily stabilized, hunters in America will probably decrease over time as the number of young people involved in hunting declines, and as more lands become unavailable for hunting. The declining number of young people involved in hunting has been documented in the USFWS survey as well as other studies. Two reasons why this may be occurring are urban isolation and single parent homes. In the South, the number of hunters has gradually increased over the past 15 years but appears to have stabilized at 5.1 to 5.2 million during the past five years.

So what is the outlook for the landowner? As the number of hunters stabilizes or quite possibly starts to decline, will the demand for places to hunt decline? The long-term demand may shift downward. However, because people face a decreasing supply of hunting land, it will take a substantial decline in the number of hunters before landowners will be affected. The demand for hunting lands, especially lands that provide a quality recreational experience, will probably continue to be strong. In other words, the forecast is favorable for landowners who currently want to get involved in fee-access hunting. This is especially true in Alabama and in the Southeast as a whole where the hunting lease market is strong and well-managed hunting land commands a premium.

Leases

In the South, hunting leases are by far the most common form of fee access found on private lands. A lease is an agreement between the landowner (lessor) and land-user (lessee) which grants the land-user access rights for specified activities on the landowner's property. Lease agreements are usually made with written contracts outlining provisions that are agreed upon by both the lessee and lessor. A sample lease agreement can be found in Appendix M. A well-written lease protects the interests of both landowner and land-user. Lease payments are usually made on a per acre basis at a rate agreed upon by both landowner and land user. Leases can be seasonal, annual, or on a multi-year basis, depending on the desires of the landowners and the relationship with the land-user. Seasonal hunting leases may be for fall and winter for game species legally hunted during this time of year. Annual leases allow use of the land for the entire year. Under the right conditions, annual leases may extend for several years on a multi-year basis. Some landowners choose to lease access rights for hunting certain species of game. Under this system, a landowner may lease out the hunting rights for several game species to different sportsmen's groups. This system works best when access rights granted to different groups do not overlap during the same time of year.

CLEMSON UNIVERSITY

Skeet ranges can provide landowners with additional income and provide hunters with another activity that does not impact game populations.

Hunting clubs that invest a lot of time and effort into habitat management prefer a long-term (multi-year) agreement to limit the possibility of losing the lease to a higher bidder. When compared to other forms of fee-hunting, leasing has several advantages and disadvantages. Advantages include the following:

- Improved control of land access
- Expected income known in advance
- Fewer management costs
- Familiarity with the individuals using the land.

Disadvantages include:

- Guests possibly present at inopportune times
- Often difficult to monitor the game harvest
- Hunters may feel "ownership ties" (especially with long-term leases) and interfere or disagree with land management practices.

These disadvantages can be avoided or minimized by a well-written lease agreement. Price of hunting leases varies depending upon many factors, most commonly the abundance of game, amenities that are provided (lodging, food plots, etc.), and the aesthetic quality of the land. In Alabama, hunting rights commonly lease for $3 to $7 per acre per year and annual fees up to $20 per acre are not that uncommon. One of the hardest decisions for many landowners is what to charge for a hunting lease. Most landowners wind up charging what their neighbors or others in the area are charging.

Fig. 4: Tract Valuation—Determining a Hunting Lease Price

Placing values on certain aspects of wildlife habitat is one method to determine hunting lease rates. The idea is that certain habitat features have a greater value than other components. Point values are assigned to 15 habitat features. A composite score of all habitat components for a particular land tract is used to assign a lease rate class based on wildlife habitat quality and the components listed below.

TRACT CHARACTERISTICS	POINT VALUE RANGE
Forest stand size & diversity	50 acres or smaller (30)—greater than 300 acres (5)
Streamside management zone widths	300 feet or wider (25)—50 feet or less (10)
Percentage of hardwood mast producers	35% or greater (25)—5% or less (0)
Site index, loblolly pine, age 50	90 or greater (30)—50 or less (0)
Percentage of permanent openings	8% or greater (25)—0.5% or less (0)
Presence of permanent water	5 acres or more (25)—none (0)
Road access	High quality (15)—poor (0)
Tract size	3,000 acres or greater (25)—200 acres or less (10)
Game species diversity	All game species (25)—1 game species (10)
Prescribed burning	33% of tract in 5 years (25)—none (0)
	Late spring/early summer (-15)
Timber thinning	Periodic (25)—none (0)
Adjoining subdivision	-25
Livestock	-25
No habitat diversity	-15
Public road access on two sides or transecting tract	-10

LEASE RATE CLASSES

Class	Total Points	Lease Value Class	Lease Fee ($/acre/year)
1	220+	Exceptional	10
2	165 - 219	Excellent	7
3	110 - 164	Above Average	5
4	55 - 109	Fair	3
5	0 - 54	Poor	2

Methods of Determining Hunting Lease Prices

METHOD	DESCRIPTION
What Neighbors or Others are Charging in the Immediate Area	Lease price is based upon the going rate in the area.
Break-Even Plus 10%	Lease price is based upon management and other costs plus 10%.
Habitat Valuation	Lease price is based upon a subjective rating of the quality and quantity of wildlife habitat.
Baseline Plus Value Added	Base price per acre is charged. Additional fees are assessed on each improvement, amenities, or services provided.
Sealed Bid Approach	Similar to timber sales. A description of the hunting lease is made and a request for offers is solicited.

A Guide to Developing a Hunting Lease

The best protection for landowners and hunters is to have a written contract (lease) for the purpose of preventing misunderstandings. The lease should identify the responsibilities, rules, and restrictions of both parties. Leases should be negotiated and amended as situations change to provide a pleasant experience for both the hunter and the landowner. Landowners should consult an attorney in developing lease agreements for their property. An example lease agreement can be found in Appendix M. The following should be included in a lease agreement:

Introduction. Include the names of all parties involved, date, description of the lease tract indicating boundaries, county, and address.

Purpose of the lease. Include what the lease is for (hunting, fishing, photography, etc.) and what rights are not included, if any.

Terms of the lease. Include the duration of the lease term, including starting and ending dates. Identify the type of lease.

Amount of rent or payments. Include total payment, when rent is due, and any information about advance security deposits.

Transferability of lease. Include the conditions concerning the rights and obligations of either party if the lease is transferred to a third party. Will you allow subleasing or assignment of the lease?

Conditions of the lease. Include how the individuals will gain access to the property, types of game that can be harvested, hunting weapons that may be used, use of facilities, how the gates and keys will be handled, use for non-hunting activities, and any other conditions or activities you will or will not allow on your property.

Statement of compliance with all applicable state and federal game laws.

Breach of contract clause. Include those conditions that authorize a landowner or lessee to cancel the lease.

Landowner's (lessor's) responsibilities. Include what will be provided for the comfort and convenience of the hunters.

An "as is" and "release" clause. State that hunters will take the premises as is in its present form. May also include a brief description of present condition of the property.

Right of renewal. The landowner or hunters may want to make long-term improvements in wildlife habitat or lodging facilities. Such a clause may be inserted if a favorable relationship has been established in the past.

Termination Clause. Include what condition the land will be in when the lease ends or termination if property is sold.

Venue Clause. State that the Lease Agreement will be interpreted according to the laws of Alabama.

Notice Clause. Give the addresses of the lessor and lessee.

Closing. Include formal wording for a notorial acknowledgement.

Liability Insurance. Require the hunting club to obtain liability insurance that lists the landowner(s) on the policy as being an additional insured party. (See Chapter 19 on liability and insurance.)

Waiver of Liability. Attach a liability waiver and a signed sheet identifying what hazardous areas have been witnessed by the hunters or pointed out by the landowner.

Signatures. Make sure to have every individual who hunts on the property sign and date the lease.

DAN DUMONT

Corporate shooting preserves can provide shooting for quail, chukar, pheasant, and mallard ducks. First-class lodging, meals, guides, bird dogs, game cleaning, and other amenities are offered at the better preserves.

Permits

Some landowners sell individual permits or group permits for access to their lands. Permits, in the past, have been popular with large industrial timber companies. Today, however, most timber companies operate under some form of lease system. Permits can be sold on a day-by-day basis to individuals or for specific time periods to groups of hunters. Fees may be charged on a per-person per-day basis, or a flat fee may be assessed for a specific time period. Typical daily permits may range from $4 to $10, and seasonal permits may range from $15 to $20 or more. Advantages of a closely monitored permit program include the following:

- Game harvests can be closely monitored at the exit gate.
- Game harvest levels can be easily adjusted by issuing more or fewer permits.
- Access can be provided to a larger and more varied group of hunters.

Disadvantages include the following:

- Administering permits is time consuming.
- Amount of income is uncertain.
- The landowner does not know the individuals using the property.

Commercial Memberships

Commercial memberships offer sportsmen quality recreation at a high membership fee. Commercial memberships involve selling and buying shares in a hunting club (or other recreational group) that is administered and managed by the landowner. In some operations, memberships may be for life and include a deeded share of the land with costs as high as $100,000 per share. In other operations landowners may require an annual membership fee ranging from $1,000 to $5,000. With this type of enterprise, landowners spend a considerable amount of money on wildlife management, services, and amenities.

Advantages include the following:

- Relatively high income
- Some certainty of income amount.

Disadvantages include the following:

- A large capital investment
- High administrative and operating costs
- Small clientele market.

Wing Shooting Preserves

Shooting preserves offer opportunities for shooting pen-raised fowl (quail, chukar, pheasant, and mallard ducks). Special amenities and services are usually associated with shooting preserves. The most intensively managed preserves offer lodging, meals, guides, bird dogs, and game cleaning and packaging. Other less intensively managed preserves will only provide access and released birds. Shoots are usually booked in advance for either a half or a full-day hunt. Fees for hunting trips can range from $100 to $500 per day, depending upon services and amenities. Old style plantation hunting may cost as much as $1,500 per day. Shooting preserves in Alabama are required to be licensed; however, preserve operators enjoy a longer hunting season.

Advantages of a shooting preserve include the following:

- The potential of being relatively profitable

- A long hunting season
- Instant quality hunting
- Predictable harvest opportunities
- Small acreages to manage.

Disadvantages include the following:

- Large capital investment
- High business risk for the operator
- High management costs
- Relatively small clientele market.

For more information about shooting preserves contact the Alabama Department of Conservation and Natural Resources Game and Fish Division (334-242-3465).

Landowner Cooperatives

Landowner cooperatives are groups of individually owned private lands joined together in a larger unit for the common purpose of managing wildlife and marketing access rights to wildlife. An advantage of landowner cooperatives are that land tracts, otherwise too small for large-scale management of some wildlife, can be joined together to form larger units that can be effectively managed for wildlife. The disadvantage of cooperatives is that landowner objectives may conflict, making the management of the cooperative difficult.

Fee Fishing/Fishing Clubs

Fee fishing and fishing clubs are popular among landowners who have farm ponds, lakes, or streams. Some public waters are receiving heavy fishing pressure and it is sometimes difficult for beginning anglers to experience quality recreation. This trend, combined with declining free access to private waters, has created a strong potential market for fee fishing.

Several types of fee fishing systems work for landowners. Some of these include the selling of day permits, receiving payment for fish that are caught, leasing of exclusive fishing rights, or membership in fishing clubs. Landowners who prefer not to deal with the public would be advised to lease fishing rights to a group of people or fishing club. The most popular fish in these operations include largemouth and smallmouth bass, bluegill, crappie, catfish, and trout. Landowners who participate in fishing clubs agree to provide fishing access in return for an assessed fee based upon the amount of use. Anglers who are members of fishing clubs pay an annual membership fee, which allows access to a number of ponds, lakes, and streams in an area. In most cases, the most popular ponds have been those that have been properly stocked and managed.

What Fisherman are Looking for in Fee Fishing Operations

Good fishing: Assured fish catch; Good fishing weather; Tasty, edible fish; Plentiful ponds; Fair prices; Access to game fish; Big fish; Fish diversity

Aesthetics: Pleasant surroundings; Peaceful, quiet settings; Openness; Clean, shaded streambanks; Country settings; Limited exposure to others

Clean/Safe: Clean surroundings; Limited mosquitoes; Fenced area

Convenient: Handicap access; Boat ramps; Fish-cleaning personnel

Friendly People: Friendly, helpful owners

Fun/Family/Kids: Good for kids and family; Fun and enjoyable

Amenities: Food and bait available; Tackle for rent; No license required; Bathrooms; Bait shop; Dock on premises; No boat needed; Seating and picnic tables; Shade trees; Aesthetic landscaping

DAN DUMONT

RON HAALAND

Sporting clay operations can be developed on farm and forest lands to provide additional income for landowners. Sportsmen pay for the opportunity to walk a designated trail that provides shooting stops simulating hunting scenarios with clay birds as targets rather than game animals.

Sporting Clays

Sporting clay shoots are quickly becoming a popular form of recreation for many sportsmen. Shooters pay for the opportunity to walk a designated forest trail or course that has been designed by the landowner. Along the course, stops are randomly laid out. Each stop represents a particular hunting scenario and the targets are clay birds rather than game animals. At each stop, the clay targets are released by positioned throwers in a variety of ways to simulate flushed quail, flying doves or ducks, a running rabbit, or some other hunting scenario. Sporting clay trails vary in length, depending on the number of stops, terrain, and the landowner's imagination. Advantages of sporting clay enterprises are the following:

- They can be easily developed on most farm and forest land.
- They are not dependent on game populations.
- They can be operated year-round.
- They can provide a reasonable rate of return.

Disadvantages of sporting clay operations include the following:

- They have a high initial investment cost.
- They are labor intensive.

Non-consumptive Wildlife Activities

Rural areas in the Southeast symbolize the outdoors to millions of city-dwelling Americans. Many city families and other suburban and urban residents want more frequent contact with nature and look forward to weekends and vacation opportunities. These folks may be willing to pay for access to an environmental experience. Indeed, surveys have shown that three out of every four Americans are involved in some form of non-consumptive use of wildlife such as observation, photography, and backyard wildlife feeding. This does not include activities where wildlife is part of other recreational endeavors such as camping, hiking, picnicking, and nature study. Only recently have a few landowners or public and private economic development groups recognized the economic potential of non-consumptive wildlife activities. The growing interest and recognition of non-consumptive recreation is part of a larger economic development effort referred to as "nature-based" recreation or "ecotourism."

Since a large portion of the land in the Southeast and in Alabama is privately owned, income opportunities for forest and farm owners from non-consumptive recreation may have some potential. Forest owners can provide viewing access to unique wildlife resources. In these types of endeavors, it is important that landowners take special precautions to avoid injuries to visitors, as well as to prevent disturbances of sensitive habitats and wildlife species. As the human population continues to grow and public lands become even more crowded, the demand for these types of wildlife encounters on private lands should increase. Effective marketing (discussed later in this chapter) of these unique resources may be the key to a successful fee-access recreation enterprise based upon non-consumptive wildlife use.

Non-consumptive wildlife recreation offers another alternative for supplemental income for landowners from a potentially large clientele group. The disadvantages, however, are that there is currently no defined market, financial returns are unsure, and there is a lack of technical assistance.

Non-Wildlife Opportunities

Forest landowners also have the opportunity to produce other income derived from the use of natural

DAN DUMONT

DAN DUMONT

Farm and forest owners have the opportunity to derive income from other land uses such as horseback and trail riding. Lodging in rustic settings can also be offered, utilizing existing structures on the property.

resources. Some of the possibilities include the following:

- Beekeeping
- Bed and breakfast inns
- Campgrounds
- Lodges or cabins
- Swimming
- Horseback riding
- Stagecoach or buggy rides
- Specialty crops like shitake mushrooms
- Ginseng collecting
- Moss collecting
- Pine straw collecting
- Firewood collecting
- Christmas trees
- Geological expeditions
- Ecotourism
- Risk-associated activities:
  - Caving
  - Rock climbing
  - Hang gliding
  - Rafting

STEPS TO DEVELOPING FEE ACCESS RECREATION

Just as with any business venture, the development of a fee-access enterprise involves several critical steps. A number of factors should be considered. These include farm and timber objectives, available resources, market conditions, and requirements for enterprise implementation and management. The following process can help landowners answer important questions: Will the enterprise help to accomplish short- and long-term stewardship objectives as well as personal goals? Is the enterprise compatible with other land management practices and operations? Do you have the necessary resources and expertise? Many nontraditional enterprises have failed because one or more of these important factors were overlooked during the decision-making process.

Step 1—Defining Objectives

The first step prior to selecting an income alternative is probably the most important one. Landowners should clearly define *in writing* their own personal **objectives** for the farm or woodland tract. Why is the landowner interested in an income alternative from wildlife or other land resources? The primary objective may not be income. It may be improving wildlife habitat or increasing the property value. Public relations or control of access may also be an important consideration. Landowner objectives will help determine the best operation as well as the type of management that is required. Financial objectives will help clarify profitability questions such as whether a profit is important and, if so, how much profit is necessary. Many landowners are never quite satisfied with their newly created operation because they never clearly defined where they were and where they wanted to go.

Step 2—The Right Idea

The second step involves **developing an idea** for a recreational enterprise. Ideas for income-producing operations come from a variety of sources, but landowner objectives and the available land resources (such as species of wildlife on the property) will dictate the action taken. Some of the most common sources of ideas for new enterprises are neighboring landowners who are already involved in one or more operations. To some degree, general knowledge about the market demand for an activity will influence a landowner when considering possible options. Both the supply (available resources) and demand (market) must be thoroughly evaluated once an idea has been selected.

Step 3—Determining Demand

No business or enterprise can be successful unless there is a demand for the product or service. A **market study** should be conducted for each proposed recreational activity. Questions to be researched and evaluated include the following:

- Is there sufficient demand for the recreational activity?
- What are the various market segments?
- How can the landowner provide the products and services that will attract the desired clientele?

If there is a market, the landowner should evaluate available resources to determine whether desired products or services can be provided. Marketing will be discussed later in more detail.

Step 4—Resource Inventory

At first, it may seem that this step of **evaluating available resources** is getting more attention than it deserves. After all, most landowners are aware of the resources on their property. Right? Well, yes and no. Indeed they may be "aware" of available resources, but if they have never considered the resources in the context of non-traditional enterprises, they may be unaware of their potential. Also, the landowner may not be aware of the resource **supply** that may or may not be adequate for the expected **demand.** Resource inventory is much more involved than simply determining whether there are enough deer to hunt or birds to photograph. Understanding resources as a fee-access recreation operator is a necessity for a solid marketing plan. Most landowners know how much acreage they have in pasture and woodland, a rough idea of the species that reside on or pass through their land, and about how much money they owe the bank. However, there are many other considerations in operating a successful recreational enterprise. In the marketing process, it is especially important to be aware of all biological and physical resources and to know how they affect the clients' recreational experience. Only then can the landowner use these resources as attributes in marketing the operation. Evaluating the feasibility of a recreational enterprise involves inventorying four resource categories: biological, physical, human (personnel), and financial resources.

AS AN EXAMPLE...

Biological—Determining if the appropriate wildlife species are present normally would have been considered long before this step, but it is also critical that animal numbers are adequate for hunting or observation purposes. Do wildlife numbers need to be increased and can the numbers needed be sustained? Biological or habitat manipulations must be analyzed on a cost and return basis.

Physical—Adequate wildlife populations may be all that is required in certain types of recreational wildlife enterprises. However, could the landowner attract more clientele at a higher price if the property had access roads? Does the property have a cabin that could be rented or a shed in which hunters could camp? Are there streams or ponds that would provide attractive campsites or fishing opportunities?

Personnel—A landowner may have time to handle the management activities of a low-input wildlife enterprise, but this will limit the management time spent on other land management activities. Many operations not only require management assistance but technical and financial assistance as well. Are there any family members or current employees who could assist with the new enterprise? Some farm and timber operations have managers or workers who can take on a few extra management duties. These individuals, however, will have to make sacrifices in how their time is currently being spent. It is extremely important that personnel requirements be evaluated prior to initiating a new wildlife enterprise.

Financial—The development of some enterprises requires more investment capital than others. However, landowners considering any type of new enterprise should inventory all liquid (readily converted to cash) assets and determine how much they are willing to invest in a new enterprise. The specific amount of capital required can be determined by analyzing costs and returns and then evaluating the expected cash flow over time. This process is described in the following section.

Step 5—Developing a Budget

After a landowner has selected an enterprise based upon personal objectives, determined that a market exists for the product(s)/service(s), and evaluated the resources available for the enterprise, the next step should be to develop a budget that will be used to evaluate the economic feasibility of the operation. Total costs can be generated by listing all activities associated with the operation, including time spent by the landowner and family members, and the cost of each activity. The total cost must be compared to the expected gross revenue, which of course, is directly related to the price charged per person or per acre. Landowners often set the price for a recreational enterprise by calculating the gross revenue necessary to cover costs and then determine the additional income necessary to meet financial objectives. In this step a landowner may also want to consider non-monetary costs and indirect benefits of a proposed enterprise. A few things that may be viewed as additional costs are increased risk, stress, loss of privacy or solitude, and having to deal with people. Indirect benefits may include improved public relations, wildlife damage prevention, control of access, and providing work for family members or dependents.

Step 6—Determining Cash Flow

Once a budget has been developed, management activities and income can be recorded by month on a calendar to evaluate the monthly cash flow for the enterprise. Costs are subtracted from income to obtain monthly balances. A negative balance during any month means that at least a portion of the income must be set aside earlier in the year or money must be borrowed. The cash flow calendar can help landowners plan for financial obligations such as lease payments or loans in advance. To maintain a positive cash flow, many landowners who lease their land for hunting receive half the fee during early summer and the other half prior to the arrival of hunters in the fall.

Step 7—Compatibility With Other Land Uses

If the new enterprise appears to meet the financial objectives of the landowner, the final step is to determine if the enterprise is compatible with existing land uses, management practices, and state and federal laws and regulations. For example, a quail hunting operation probably would not be compatible with intensive cotton farming, unless major compromises were made. Likewise, a fishing enterprise would not be compatible with a vegetable crop operation that requires annual draining of the lake for irrigation purposes. Deer hunting leases are fairly compatible with livestock operations if livestock are kept out of the hunting areas during the hunting season.

Step 8—Marketing

The concept of marketing is sometimes confused with advertising. Advertising creates an awareness and favorable impression of a business—in other words, promotion. Actually, it is just one of the key ingredients of successful marketing. **Marketing** is matching the resources of an operation with the needs of the clientele. The marketing process involves a thorough understanding of both available resources and clientele.

UNDERSTANDING CLIENTELE

For the landowner developing a fee-access enterprise for the first time, the most important idea to understand is that the product the client is seeking (with few exceptions) is not something completely tangible—rather it is the entire experience. Successful operators are those who have learned to provide their clientele with a pleasant and memorable experience. Too often the landowner's sole focus is on "the kill" or "the catch." Hunter surveys have repeatedly shown that the overall recreational experience (hospitality, scenery, socializing, camping, cooking, traditions, etc.) is more important to hunters than merely the taking of game. This begins to make sense when you realize that the taking of game usually represents just one moment during several days of enjoyable experiences. Furthermore, when you consider that the time spent hunting is just one of five major phases of an outdoor experience, you may begin to understand why successful operators focus on marketing a hunting experience rather than simply selling a product.

These five phases comprise the recreational experience: **anticipation, travel to, on-site experience, travel back, and recollection**. To use recreational sport hunting as an example, the landowner may use all phases to market a hunting experience.

During the **anticipation** phase, the operator may telephone clients to report on weather conditions, wildlife census results, and the hunting forecast. A few minutes of conversation can elevate the level of enthusiasm. It can also be an opportunity to clarify any details and answer questions.

Landowners who are working with their community can have a real advantage during the **travel to** phase of the hunting experience. Community hospitality, which is viewed as a real bonus by hunters, can be initiated by individual businesses and residents, or can be part of an organized effort. Special events can be economically important in their own right, and can provide an added dimension to the hunting experience. The small town of Pine Apple, Alabama in Wilcox County conducts a very successful Hunter Appreciation Day, for example.

The culmination of the year's planning and anticipation is the **on-site experience.** Landowners have the opportunity to provide their clients with an experience that they will want to repeat. This is the best time to obtain information from the clientele about how to

Considerations in Selecting an Income-Producing Wildlife Management Alternative

1. Why are you interested in an income alternative from wildlife:	Income? Control of trespassers? Other?
2. Define your objectives and expectations:	What type of operation? Managing for what species of wildlife? Expectations for income? Resources needed for initiating a program? Write down objectives. Be specific with objectives.
3. Conduct market study to determine feasibility:	Is there a demand for a particular recreational activity, services, and amenities? Who wants these services? Identify market segments. How can you provide services clientele want?
4. Make an inventory of available resources:	Abundance and species of wildlife on property. Amount of acreage. Evaluation of current habitat conditions. Equipment and labor needed. Productivity of the land. Productivity of wildlife. Habitat improvement practices that are needed. Financial resources needed.
5. Compare costs and returns of alternatives:	Compare alternatives to determine least costly. Determine break-even costs. How to set a price.

6. Select an alternative that is most compatible with existing land management activities.
7. Do you like working with people?
8. Would the timing of the wildlife operation fit into your farming or forestry operation?
9. Are there experts available to help?
10. What are the state and federal laws that would effect the proposed operation?
11. Develop a long-range, comprehensive management plan.
12. Start off small. If the alternative is working satisfactorily, expand slowly.
13. Remember the product you are selling is a recreational experience. The raw materials are innovation, land, equipment, and the wildlife resource.
14. Get advice from your attorney and accountant.

improve their experience. It is important to remember that there are activities other than hunting that can make the experience more complete and enjoyable. The landowner may want to sponsor a first-night social, a campfire cookout, a wildlife management lecture, or some other social event.

Providing memorable experiences has allowed many operators to improve their enterprise from a situation of begging for clients to one where clients must be turned away. During the **recollection** phase, the operator can communicate with clients by phone and newsletters about events of the past season. They can also keep them up to date about improvements that are underway, environmental conditions, and game conditions to keep the interest "smoldering." Some successful operations send their clients a video or photo album of their stay that includes their hunt.

ADVERTISING

Most fee-hunting operators agree that the least expensive and most effective advertisement is by word-of-mouth. Successful and satisfied hunters will spread the word about good hunting opportunities. However, with a new wildlife enterprise, the operator has not had the opportunity to establish a reputation. Initially, landowners must advertise in order to locate and make arrangements with clientele. Probably the best form of advertisement for the beginning operator is printed material such as newspapers, brochures, and magazines. These methods are relatively inexpensive, circulation is widespread, and they can be used to target a particular clientele segment.

For small operations, classified ads in local metropolitan **newspapers** are the most cost-effective means of reaching a large number of potential clients. Major newspapers have sections for advertising hunting lands or sporting equipment. Wording should be short and accurate and sound appealing to prospective clients. Minimum information that should be included are tract size, location, accommodations, general habitat type, wildlife species, and phone number.

Posters and **bulletins** displayed in the community can be effective in attracting local hunters to a new operation. Many business owners allow posters or bulletins to be displayed in their shops. Sporting goods stores, barber shops, supermarkets, and taxidermy shops are effective places to advertise. Posters and bulletins should be attractive and include maps or photographs of the area, a description of what is offered, prices, and a phone number.

Many operations that are medium to large in size develop high quality **brochures** to help promote their unique operation, services, and amenities. When advertisements generate inquiries, the more detailed brochures can be forwarded promptly. Any personal touch like a short, handwritten note is usually appreciated, especially if the inquirer is truly interested. The brochure should be informative but to the point. Including professional-quality artwork and current, tasteful photography as an "eye-catcher" is helpful. Brochure information should include the size and location of the property, a brief description of the resources, services provided, cost of services, type of weather to be expected, how to apply, and who to contact. The brochure should **not** overstate what can be expected.

Hunting or outdoor **magazines** are one of the best ways to solicit a prospective guest. Medium and large operations should advertise in as many magazines as their budget will allow. At a minimum, ads should be in at least four national magazines. Ads should be coded so that the operator knows which magazines produced the most inquiries. The pinnacle of magazine advertisement is to have an outdoor writer of a national magazine do a favorable story on a wildlife enterprise. This will reduce that operations' need to advertise in the future since the article will reach thousands of potential customers.

Local and regional **conservation and hunting groups** often lease hunting areas for their members' use. These organizations provide a built-in clientele for landowners who approach the groups and make them aware of their hunting opportunities. Most sporting organizations meet regularly and would welcome a landowner as a guest.

Major corporations are also potential clients for fee-hunting or other outdoor recreation enterprises. Corporations routinely entertain business clients or their own personnel as a cost of doing business. Operations that provide first-class lodging facilities, outstanding recreational opportunities, and other services and amenities are most attractive to corporate clients.

SUMMARY

There are many alternatives that landowners can choose to integrate wildlife management into farming, forest, or other existing land management operations. Some of these alternatives provide a source of income associated with wildlife and can help defray expenses associated with wildlife management. Landowners should be aware of those alternatives that fit best into their existing land management operations. Current land-use practices that can be easily altered to benefit wildlife require the lowest financial investment. In most cases, this is the most preferred and economically feasible method of managing wildlife on farms and woodlands in the South. Landowners who are interested in managing wildlife on their lands should have as their highest priority the welfare of the wildlife resource. When this occurs the interests of the landowner, the public, and the wildlife resource are best served.

Because of the foresight of a handful of courageous national and state conservation leaders, Alabama is blessed with an abundance of natural resources. As society's demands for use of these resources increase, the challenge for conservation leaders is to manage and protect these resources in a sustainable manner for future generations.

CHAPTER 18

Conservation Ethics and Opportunities

CHAPTER OVERVIEW

This chapter considers important reasons why people have been and are becoming more concerned about conserving forest and wildlife resources. Losses in the ecological web of life, as well as restoration successes, are featured. Conservation leaders are noted for their contributions to shaping attitudes in Alabama and the nation. For those interested in contacting organizations or agencies dedicated to natural resource conservation, a listing is found in Appendix O.

CHAPTER HIGHLIGHTS PAGE

WILDLIFE SUCCESS STORIES 416

CONSERVATION LEADERS 419

ALABAMA CONSERVATION LEADERS: PAST AND PRESENT 426

A CONSERVATION ETHIC 428

HUNTER ETHICS 429

THE ALABAMA WILDLIFE FEDERATION'S LEGACY AND VISION 430

America, by most measures, is the wealthiest nation on earth. It is no coincidence that this country also harbors vast natural resources. The land, water, forests, and wildlife that together make up these resources are a national treasure of benefit to every person. Yet how we as individuals and as a country protect and conserve these resources has changed over the years.

If natural resources were looked upon as if they were money in the bank, how would the money be handled? Some people might spend all the money, enjoy it while it lasted, and not worry about tomorrow. Some wildlife species, like the passenger pigeon which thrived in the 1800s, were decimated by exploitative market hunting and habitat destruction and are now gone forever. Other people might entrust their money to investors who may or may not guard it carefully. Sometimes management mistakes occur due to a lack of understanding of all the potential problems an action can create. One such example was the widespread use of the pesticide DDT in the U.S. earlier in this century. Other management practices can result in habitat changes that could adversely affect wildlife. Whether from greed, ignorance, or economics, a landowner may agree to clear-cut up to the banks of streams, even though professionals advise that erosion will result and water quality and fish populations will decline. Some landowners will go ahead because they believe "that's where the highest value trees are" without regard to conservation values.

Whatever the cause, poor management decisions and mistakes are sometimes made. People who manage land and resources take actions that can hamper future productivity of the land and wildlife. During the first quarter of the twentieth century, for example, a tree disease called chestnut blight was unintentionally brought into this country. It attacked the native American chestnut and wiped out this valuable and attractive species across its entire range. It was a terrible loss to people who enjoyed its versatile wood and aesthetic beauty for everything from furniture to fenceposts. Just as important, many wildlife species that utilized the tree for food and shelter also suffered from the loss of the American chestnut.

More recently, the red fire ant appeared in the United States through the port of Mobile in the 1930s, spreading rapidly. Scientists hypothesize that it came in with bananas or tree seedling nursery stock from Brazil, its native range. Fire ants have now spread to many states in the South. We do not completely understand how the addition of this new species is impacting plants and other native animals. Recent research in Texas indicates that fire ants have a negative impact on wildlife species, especially bobwhite quail.

"It is inconceivable to me that an ethical relation to land can exist without love, respect, and admiration for land, and a high regard for its value. By value, I mean something far broader than mere economic value; I mean value in the philosophical sense."

—Aldo Leopold

A Sand County Almanac, 1949

WILDLIFE SUCCESS STORIES

Not all changes in forests and wildlife populations that have occurred in this century have been negative. Populations of some wildlife species, particularly game species like wood ducks, white-tailed deer, and wild turkey, have increased dramatically. Sportsmen and landowners deserve credit for contributing to these positive changes. Management practices on private lands have helped, and money from hunting licenses, permits, and sporting equipment taxes is distributed to research and management programs that benefit game and non-game wildlife species and their habitats. Laws and regulations have been enacted that protect species from over-hunting.

Alabama, as well as the entire Southeast, has an abundance of renewable natural resources including timber, wildlife, and water. These resources support a strong economy and provide a high standard of living for many citizens. As Alabama and the southeastern region of the U.S. have grown and prospered, agriculture, forestry, mining, urban and industrial expansion, and other developments have impacted natural resources and ultimately affected native plants and animals. While we have become effective at managing some of our game species like deer and turkey, other species, particularly some non-game species, have clearly declined.

In the past we reacted mainly to severe situations when wildlife populations experienced dramatic losses that alarmed the public. People recognized in the 1930s that numerous wildlife species were in trouble. In response, sportsmen, landowners, resource managers, naturalists, other citizens, and Congress rallied together and passed the Federal Aid In Wildlife Restoration Act, also called the Pittman-Robertson Act. This monumental piece of legislation provided state wildlife agencies funding derived from federal excise taxes on sporting arms and ammunition to restore numerous wildlife species including white-tailed deer, wild turkeys, and wood ducks.

During the late 1960s, the plight of some non-game species in danger of extinction, such as bald eagles and whooping cranes, caught the public's attention. Con-

Recognizing that a lack of nesting cavities was causing a decline in wood duck numbers, restoration efforts of wildlife biologists, conservation organizations, and private landowners were initiated to create and erect artificial nesting boxes in wood duck habitat. The wood duck is now abundant across Alabama thanks to these efforts.

gress passed the Endangered Species Act in 1973, following public sentiment in support of the Act. Natural resource agencies and conservation organizations developed programs designed to stem species extinctions. Several endangered species have recovered including the bald eagle, the brown pelican, the peregrine falcon, and the American alligator. While most listed species have not recovered, many feel that real progress has been made.

Along with the successes there have been controversies due to situations in which economic sacrifices have been necessary to protect threatened and endangered species. The controversies began with a small freshwater fish in the Tennessee River called the snail darter and the Tellico Dam Project. Controversies continue today with threatened and endangered species such as the Northern spotted owl in the Pacific Northwest, the red-cockaded woodpecker in the Southeast, the scrub jay in Florida, and the golden-cheeked warbler in Texas. Some biologists argue that endangered species are indicators of the general decline of the ecosystems in which they exist. The decline of the red-cockaded woodpecker, for example, is an indicator of the loss of mature pine forests.

These controversial endangered species occupy habitats that provide natural resources that are economically important. Development has in many cases modified habitats to such a degree that some wildlife species do not have the resources for long-term survival. How do we manage or care for the vast majority of species and keep the ecosystem healthy?

The Wood Duck

One example of a declining species that benefitted from restrictive regulations and the actions of sportsmen and biologists is the wood duck (*Aix sponsa*). Congress passed two laws early in the twentieth century that began to reverse its decline. The Migratory Bird Law of 1913 and the broader Migratory Bird Treaty Act of 1918 initiated legal protection for all birds that cross our national borders in fall-spring migration patterns. The observations of scientists in Illinois that a lack of adequate cavity trees was limiting wood duck nesting and reproductive success led biologists to first begin constructing artificial nesting boxes for wood ducks in the 1930s. In the next decade, word spread across the country. People across the nation began building and erecting nesting boxes. In addition, a 10 percent tax from the sale of sporting arms and ammunition (the Pittman-Robinson Act described earlier) was passed in 1937 and distributed to federal aid projects to support research, land acquisition, and habitat improvement to

GARY W. GRIFFEN

DENNIS HOLT

NWTF

The restoration of the wild turkey in Alabama and other parts of the U.S. is one of the most successful wildlife restoration and conservation efforts in history. From an estimated all-time low of 13,487 birds in 1941, wild turkeys in Alabama are currently estimated to number over 500,000. Conservation partnerships involving the ADCNR, private landowners, and conservation organizations such as the Alabama Wildlife Federation and the National Wild Turkey Federation, are responsible for the dramatic comeback of the wild turkey. Left: A mature gobbler perches on a tree limb. Top right: An ADCNR wildlife biologist transports a captured wild turkey for relocation. Bottom right: A National Wild Turkey Federation biologist releases a captured wild turkey for relocation in an area with low turkey numbers.

benefit wildlife. Money from waterfowl stamps sold to hunters was also used for waterfowl habitat enhancement purposes. By the 1950s, scientists estimated that the population of wood ducks was 10 times greater than at the turn of the century. The combination of laws, biologists, sportsmen's tax money, private landowners, and other individuals working together to promote enhancement and protection of wood duck habitat averted additional declines or a total loss of wood ducks in the U.S.

The Wild Turkey

In some parts of the country, the wild turkey (*Meleagris gallopavo*) was eliminated from its native range by the 1930s. Because the turkey was as popular to eat as it was to hunt, it was one of the earlier species for which bag limits were established. A 1780 New York Colony law prohibited the killing of turkeys between April and August. Nevertheless, the wild turkey was still eliminated from all the New England states, Michigan, Wisconsin, and Minnesota by 1920. To restore its numbers, biologists first tried stocking domestic game-farm varieties of turkeys, but they did not survive well in the wild. Eventually, trapping and relocating wild turkeys, along with protective laws, replenished turkey numbers. All of the lower 48 states have now effectively re-established populations, with over 4 million wild turkeys nationwide.

DENNIS HOLT

Alabama Department of Conservation and Natural Resources wildlife biologists at the Fred T. Stimpson Game Sanctuary in Clarke County use a live box trap to capture and relocate white-tailed deer as part of restoration efforts in Alabama.

Mammals

In 1900 the U.S. population of white-tailed deer *(Odocoileus virginianus)* numbered approximately half a million. Now there are estimated to be over 16 million deer. Elk (*Cervus canadensis*) once ranged over the entire country. By 1900 only about 40,000 remained, most in Yellowstone Park. Now elk numbers are estimated at more than 1 million animals in more than a dozen states. Trappers in the early 1800s nearly eradicated the beaver (*Castor canadensis*), but it now occurs in nearly every state, thriving to such a degree that people often wonder how to coexist with these prolific dam builders. The American bison (*Bison bison*) was overharvested to such an extent that only a few hundred remained in 1900; now the buffalo population exceeds 200,000.

"Conservation is a state of harmony between man and land. By land is meant all things on, over, or in the earth. Harmony with land is like harmony with a friend; you cannot cherish his right hand and chop off his left."

— Aldo Leopold

A Sand County Almanac, 1949

CONSERVATION LEADERS

During this century visionary people have called attention to the need to protect the land and all that's on it from human impacts. Finding a balance between the needs of people, wildlife, and the land that supports us all can be quite challenging. However, as the twentieth century draws to a close, it is important to reflect on how society's thinking has changed over the last 100 years and to note the words and actions of leaders who labored to reverse damaging ecological trends. Visionaries such as Theodore Roosevelt, Aldo Leopold, and Rachel Carson alerted Americans to the dangers of land abuse and of the need for conservation. These three conservationists came from totally different backgrounds. One was a statesman, another a scientist and conservationist, and the third an environmentalist. Their influence on public opinion changed the world of their time and beyond. They became involved with other leaders and established organizations that work to foster wildlife and wild lands. Their examples can be followed by people interested in conservation ethics even today.

Theodore Roosevelt, a U.S. president, statesman, conservationist, hunter, and ardent naturalist, was a driving force in the birth of our country's conservation movement. His vision, energy, and sheer force of will helped shape and define conservation in America.

Theodore Roosevelt and the Onset of Conservation

Theodore Roosevelt, born in 1858, was to the American people in his lifetime much of what Thomas Jefferson had been just over a century before: U.S. President, statesman, conservationist, and ardent naturalist. Jefferson had Lewis and Clark explore the West, gather details, collect specimens, and report back to Washington on the state of the natural world. Roosevelt called for a National Conservation Commission report in 1908, heralded in that period as the most exhaustive inventory of natural resources ever made. Roosevelt also wrote more than 35 books, many detailing his adventures in the wilderness, like *Hunting Trips of a Ranchman.*

Roosevelt took the President's office unexpectedly on September 6, 1901, after the assassination of President William McKinley. Republican statesman Mark Hanna complained: "Now look, that damned cowboy is President of the United States." The Rough Rider never slowed down for Washington bureaucracy. In his next eight years in the White House, he set aside 230 million acres of public lands—more than 50 federal wildlife refuges and 18 national monuments. He doubled the existing number of national parks and quadrupled the standing forest reserves. Among the lands he helped preserve, the most commonly known are Yellowstone, Yosemite Valley, Mesa Verde, the Mariposa Grove of ancient sequoias, and Crater Lake of Oregon.

His zest for wildlife included game and non-game species. Hunting and fishing interests led him to team up with friend George Grinnel, editor of *Forest and Stream* magazine, to organize the Boone and Crockett Club in 1888. Although he had hunted buffalo earlier in his life, his interest in herd restoration led him to help initiate the reintroduction of buffalo to Oklahoma in 1907. He also initiated the reintroduction of elk, turkey, and Texas longhorn cattle in Cache, Oklahoma, with hundreds of their descendants now well established.

Plumed hats for ladies were a fashion vogue of the period. This caused serious loss of some bird populations. Audubon Society President Frank Chapman had awakened Roosevelt to bird losses on Pelican Island, a federally owned property off the coast of Florida. Roosevelt, always dedicated to bird-watching, asked, "Is there any law that will prevent me from declaring Pelican Island a Federal Bird Reservation?" When told that there was none, he announced: "Very well, then I so declare it." While watching the skies in Virginia in 1907, he reported seeing some of the very last of the passenger pigeons, though his sighting could not be substantiated by others. It's possible the person who fought hardest during his time to maintain avian habitat may have been allowed one of the final glimpses of this species.

Like many naturalists, Theodore Roosevelt felt the need to get away to the sanctuary of the wilderness to restore his spirits. One important companion on these excursions was John Muir, founder of the Sierra Club. Another was John Burroughs, who did not share Roosevelt's hunting enthusiasm, but stated, "I have never been disturbed by the President's hunting trips—such a hunter as Roosevelt is as far removed from the game-butcher as day is from night." It was in the support of ethical hunting that Roosevelt gained the nickname "Teddy." While hunting bear in the Mississippi Delta forests, a guide tracked down an aged, lame bear and tied him to a tree to assure Roosevelt a kill. The President was disgusted, and his refusal to shoot prompted a famous Washington Post cartoon. The artist got the story mixed up and created a baby bear instead of a senile one. The Ideal Toy Company was happy for the mistake, and "Teddy" bears were born.

Over the course of his life, Roosevelt balanced his power with a sense of responsibility. He left a legacy of cherished national lands for the betterment of people, wildlife, and the environment. He established and supported many diverse organizations to foster better land management and conservation. He promoted change towards a more ethical use of natural resources. He might best be remembered by the words he used to address the 1908 Conference of Governors, which included members of his cabinet, Supreme Court justices, governors of 38 states and territories, and many eminent citizens, including William Jennings Bryant and Andrew Carnegie. To these he said, "In the past we have admitted the right of the individual to injure the future of the Republic for his own present profit. The time has come for a change. As a people we have a right and the duty... to protect ourselves and our children against the wasteful development of our natural resources."

ALDO LEOPOLD FOUNDATION

Aldo Leopold's combination of training, observation, and poetic writing earned him the title "the father of wildlife management." No other individual has had such a profound ability to convey the meaning of a land and conservation ethic. Here Leopold examines one of the many pines he and his family planted to restore their Wisconsin farm.

Aldo Leopold and the Development of Conservation Ethics

Aldo Leopold, educated with a master's degree from the Yale School of Forestry in 1909, combined science, philosophy, and skilled writing in books like *A Sand County Almanac*, which examined the connection between man, wildlife, and the land. His forceful prose led people to consider issues like game management in a new light. It earned him the title "the father of wildlife management." He held many varied positions during his career, which allowed him the opportunity to understand and view from different perspectives the interactions between people and the land.

Early on, he worked for the U.S. Forest Service cruising timber in the Blue Range of the Arizona Territory, and later, he was a supervisor of the Carson National Forest in northern New Mexico. He nearly died from acute nephritis after surviving a storm in the backcountry. This experience cost him a year's sickbed recuperation.

He had some unfulfilling professional years, beginning in 1924 when he was transferred to Madison, Wisconsin's Forest Product's Laboratory. Even though he felt frustrated in his work, he began writing the book *Game Management*. Published in 1933, it was a landmark publication. Before then, the idea that game populations could be manipulated and enhanced was not widely understood. Because of Leopold's work, the science of wildlife management became more accessible to researchers, land managers, and private citizens alike. The attention this book received encouraged the University of Wisconsin to offer Leopold a position as chair of game management, the first academic role of its kind in the U.S. It was a job he retained for the rest of his career.

To Aldo Leopold, academic education was simply not sufficient in and of itself. Learning was *doing*. Dedication to observation (and record-keeping) of interactions on the land was central to his style of teaching. For instance, when he gave exams to his students of wildlife ecology, he would take them into the field and ask them to explain what they saw. A typical test question might be "Look at the trees and vegetation on this acre of land. Then tell me the sort of soil you would expect beneath them and the predators and prey you expect to live here based on what you see."

"That land is a community is the basic concept of ecology, but that land is to be loved and respected is an extension of ethics."

—Aldo Leopold

A Sand County Almanac, 1949

After his teaching appointment, Leopold bought his own piece of land along a bend in the Wisconsin River as a family weekend retreat. He renovated a very spartan dwelling there with his wife and children's help, and all material for it was made by hand or scavenged from the river, except for the lumber he purchased to give his wife the wooden floor she desired. It is still called "Leopold's shack" and is the focal point of the Leopold Foundation. Leopold's family planted thousands of pines to restock the land. After drought killed some of the pines they tried planting birch, dogwood, tamarack, maple, and more pines. The trees they planted that survived totaled over 36,000. He collected seeds from native flowers and broke through the tough earth to re-establish native vegetation. He added wildlife plantings like sorghum, sudan grass, and millet for quail and pheasants. Leopold was a sportsman and hunter all his life. His efforts to enhance and restore the abused and eroded land around the shack are evidence of his willingness to give back to the land as much—and more—than he would ever take.

ALDO LEOPOLD FOUNDATION

Aldo Leopold's "shack" sits on his farm in Salk County, Wisconsin, much the way it did when he spent time there improving the land and writing about a land ethic. The shack served as a focal point for Leopold's family gatherings, writings, and teachings.

One of his former students, Clay Schoenfeld, writes of the opening session of Professor Leopold's Wildlife Ecology course in 1940. On the board, underneath "Criteria of Conservative Land Use," was recorded "(1) maintain soil fertility, (2) preserve the stability of the water system, (3) yield useful products, and (4) preserve the integrity of the flora and fauna to a reasonable degree." The former student noted, "Conservation, we gathered, was not simply preservation; it was wise use." But "use" in the Leopold lexicon was defined to encompass an ecological imperative at once utilitarian and universal—an ethic of community life—and his community included "the soils, water, fauna, and flora, as well as people." He demanded that his students look beyond the text to broaden their understanding, with

BROOKS PHOTOGRAPHERS

The writings of Rachel Carson in *Silent Spring* awakened the nation to the irresponsible use of chemicals and their potentially harmful effects on humans, wildlife, and the environment.

99 required readings in his class. But rather than being a "classroom czar," they saw him as a patient, attentive listener. Geologist and former student Charles Bradley recalled Leopold "never tired of asking the questions that ended up blowing my mind, revealing me to me, and me to him."

"Instead of learning more and more about less and less, we must learn more and more about the whole biotic landscape."

—Aldo Leopold

A Sand County Almanac, 1949

Leopold's lasting testimony to the idea of a land ethic, *A Sand County Almanac*, was first published posthumously in 1949. In the book he writes of many experiences that make his interaction with the land real: the felling of a tree that reveals the history of the years in the wood's annual rings, a spark in a wolf's eyes, quail roosts among wild plum branches, and the satisfaction of planting trees. He intersperses description with explanations that often sound like poetry, but always, in his own way, he teaches the value of conservation ethics. As one writer noted, "Leopold saw good education and good conservation as inextricably linked, and blamed the growing problem of environmental ruin on our failure to instill an essential attitude of respect." As an example, Leopold wrote, "In short, a land ethic changes the role of *Homo sapiens* from conqueror of the land community to plain member and citizen of it. It implies respect for his fellow-members, and also respect for the community as well." His words were largely read by others in his profession when the first hardbound edition was released. When the environmental movement of the 1960s reawakened interest in conservation, his book was reprinted in paperback. Now followers of Leopold's words come from all walks of life, many varied occupations, and all parts of the world. Regardless of background, Leopold's readers appreciate and applaud his compelling explanations for why a commitment to a land ethic is critical for the present and future generations.

THE ALDO LEOPOLD FOUNDATION

The Aldo Leopold Foundation is a non-profit organization founded by Aldo Leopold's children to perpetuate Leopold's influence in the minds and actions of future generations. The Foundation was originally established as the Aldo Leopold Sand County Trust in 1949 to preserve the family land and shack along the Wisconsin River in Saulk County, Wisconsin. For more information contact the Aldo Leopold Foundation, E12919 Levee Road, Baraboo, Wisconsin 53913-9737, (608) 355-0279.

Rachel Carson and Conservation Awakening

Rachel Carson, born in 1907, spent her early professional years as an employee for the U.S. Fish and Wildlife Service. Her undergraduate and graduate degrees were in marine biology. Not surprisingly, her first works in magazines and a book were about marine environments. Her publications in *Atlantic Monthly* (1937) and *The New Yorker* (1951) magazines helped establish her reputation as a proficient writer. Her book, *The Sea Around Us*, was an immediate success and provided her with the financial means to write of her own interests full time.

A good friend wrote to her about aerial spraying of mosquitoes in Massachusetts that had killed both birds and mammals in her 2-acre backyard wildlife sanctuary. Carson began thinking about this, started investigating common chemicals in use in the 1940s, and submitted an article about this to *Reader's Digest* and several other magazines, all of which promptly rejected it. Rather than quit, she researched more deeply.

A person who trusted less in her experience would have been discouraged and quickly turned away. Instead, she wrote a book called *Silent Spring*. Published in 1962, it awakened the nation to the irresponsible misuse of chemicals and their harmful effects on human health and the environment. One biographer wrote, "Although Carson originally had hoped to complete a manuscript by early 1959, she found the topic complex and controversial. Fortunately, she had established valuable contacts through her years of government service, and her publications opened doors that might have remained closed to someone else. Thus she gained the generous assistance of a broad circle of experts and accumulated an abundance of information."

Carson's proactive approach as a researcher enabled her to pursue an idea she believed in regardless of rejections. She trusted in her conviction that the work was necessary to change the way things were being done. The book caught the attention of the U.S. Congress soon after its publication. CBS aired "The Silent Spring of Rachel Carson" on national television in 1963, bringing her concern over the use of chemicals like DDT to the nation's living rooms. President John F. Kennedy's Science Advisory Committee supported Carson's findings soon afterwards, and in the same year, she won medals from the National Audubon Society and the American Geographical Society. She was elected to the American Academy of Arts and Letters. By the next year, 1964, she had died of cancer. In 1965 *A Sense of Wonder*, which she wrote prior to her death, was published in which she explained the importance of fostering a love of the

Fred T. Stimpson was one of the early wildlife conservation pioneers in Alabama. His passion for restoration and management of the wild turkey and other wildlife helped shape the direction of conservation in Alabama.

natural world in children. Her lasting mark has been in exposing the irresponsible use of damaging chemicals like DDT. In 1980 she was awarded the Presidential Medal of Freedom posthumously.

ALABAMA CONSERVATION LEADERS: PAST AND PRESENT

Fred T. Stimpson

Fred T. Stimpson, or "Mist Fred" as he was often called, began his career in the timber business and later rose to become president of a major timber corporation that controlled a significant amount of forest land in Alabama. Although his profession was timber, his passions were his family and friends, and conservation of wildlife, specifically the restoration, management, and hunting of wild turkey. He was named Alabama's Conservationist of the Year for his tireless efforts in wildlife conservation, including the establishment of research programs, the creation of refuges and sanctuaries, and the initiation of game re-establishment projects. For most of his lifetime, Mr. Stimpson was a driving force who shaped the direction of conservation of Alabama's natural resources.

Mr. Stimpson was founder (in 1933) and second president of the Mobile County Sportsmen's Association, now called the Mobile County Wildlife and Conservation Association. He was an active member of the Alabama Department of Conservation's Advisory Board from 1939 to 1954. He also served as president of the Alabama Wildlife Federation in 1939 and 1940, and again in 1955–56. He used his enormous talents, gregarious personality, and financial resources to keep the fledgling conservation movement alive in Alabama. He loved to hunt wild turkeys and devoted himself to their propagation. He won the first World Championship Turkey Calling Contest in 1940. He passed away in 1957.

Mr. Stimpson served as role model and mentor for many natural resource professionals who have gone on to contribute to the wise use and management of wildlife in Alabama and across the country. His leadership set an example that touched many natural resource professionals and wildlife enthusiasts alike, including his children and grandchildren who have also gone on to make significant contributions to wildlife conservation in Alabama.

Fred Stimpson was an avid fisherman and hunter and he saw firsthand the effects of unchecked water pollution on wildlife and fishery resources. He was a sawmill man of the old school, but he abhorred the practice of dumping sawdust in streams and used his considerable influence among Alabama lumbermen of his day to stop this destructive practice.

As president of the Alabama Wildlife Federation, he led conservationists in what became known as the Stock Law War in the 1930s which removed free-ranging cattle from the land so that private landowners could begin to manage their lands for timber and wildlife. This was a major victory for landowners, wildlife, and conservation.

In 1993 Mr. Fred T. Stimpson was honored posthumously, along with his sons William H. Stimpson and Ben C. Stimpson, and grandsons Fred T. Stimpson, III and William Sandys Stimpson, with the Walter L. Mims Lifetime Achievement Award for their lifelong contributions to wildlife conservation in Alabama.

Walter L. Mims

Walter L. Mims, or "The Judge" as he was affectionately called, was an avid hunter and fisherman from Birmingham who was active in many conservation efforts in Alabama and across the nation. It was often said

Walter L. Mims was a lawyer and an avid hunter and fisherman who was active in conservation efforts in Alabama and across the nation. Because of his efforts, the Alabama Wildlife Federation established in his honor the Walter L. Mims Lifetime Achievement Award to recognize individuals who have made a difference in the conservation of Alabama's natural resources. Mr. Mims was the first recipient.

that he was a "conservationist" long before the term became popular. In the mid-forties he served as president of the Birmingham Chapter of the Izaak Walton League and was recipient of the Woodmen of the World Conservation Award. In 1954 the Alabama Wildlife Federation named him Conservationist of the Year. Mr. Mims served for a number of years as a member of the Alabama Water Improvement Commission, and in 1963 received the Governor's Conservation Achievement Award in recognition of his contributions to water improvement and conservation in Alabama. In 1964 he received the American Motors Conservation Award for his efforts related to conservation of the nation's natural resources.

For 24 years, from 1958 to 1982, Mr. Mims served on the board of directors of the National Wildlife Federation (NWF), the largest conservation organization in the United States. He served as first vice president of the NWF from 1965 to 1974, and in 1972 was elected and served for two years as president of the NWF. He also served for many years on the board of the Alabama Wildlife Federation, serving two terms as its president. At the time of his death in 1993, he was chairman of the Alabama Wildlife Endowment.

The Alabama Wildlife Federation established the Walter L. Mims Lifetime Achievement Award that was first presented to Mr. Mims in 1992 to recognize his lifelong contributions to conservation in Alabama and across the nation. This distinguished award, named in honor of Mr. Mims, is given periodically by the Alabama Wildlife Federation to individuals who have made a significant contribution to the conservation of natural resources and the environment in Alabama. Walter Mims was a great attorney, an avid sportsman, and a skillful and effective conservationist. His friendship and counsel were highly cherished by those fortunate enough to have known him.

William R. Ireland, Sr.

William R. Ireland, Sr. is one of the best friends Alabama's environment and natural resources have ever had. Mr. Ireland served with distinction as President of the Alabama Wildlife Federation in 1992–1993 and has also served on the board of directors for many years. Formerly an executive with his family company, Vulcan Materials, he is now retired and spends a great part of his time in matters involving wildlife conservation.

Mr. Ireland is a Life Member of The Gulf Coast Conservation Association, The Nature Conservancy, the Alabama Wildlife Federation, the National Audubon Society, and the National Wild Turkey Federation. He is a Benefactor Member of Ducks Unlimited, and also serves on the board of directors of Ducks Unlimited, the Alabama Wildlife Endowment, the Cahaba River Society, the Alabama Wildlife Rescue Service, and many other non-profit organizations in Alabama. Mr. Ireland's advice and support have been invaluable to all of these organizations, many of which may not have survived their fledgling years without his leadership and generosity.

Mr. Ireland has been an active supporter and benefactor of the wildlife and fisheries programs at Auburn University and the Department of Conservation and Natural Resources. He recently established the Ireland professorship in wildlife sciences at Auburn. One of his most outstanding accomplishments was his leadership role in the development and successful passage of the Forever Wild Constitutional Amendment. Forever Wild established for the first time in Alabama a state-funded program to acquire land for conservation and recreation purposes.

Mr. Ireland has given much of himself and his personal

Pictured here is William R. Ireland, Sr. and his wife Faye receiving the Alabama Wildlife Federation's Walter L. Mims Lifetime Achievement Award for their outstanding efforts to conserve Alabama's wildlife resources.

Fred T. Stimpson, III, AWF president (1991-92) (left), and Lt. Governor Jim Folsom present Charles D. Kelley (right) with the Conservationist of the Year Award from the Alabama Wildlife Federation.

resources to protecting and enhancing the natural resources of Alabama. Because of this lifelong dedication, Mr. Ireland was awarded the Alabama Wildlife Federation's Walter L. Mims Lifetime Achievement Award in 1994.

Charles D. Kelley

Charles Kelley has been director of the Division of Game and Fish of the Alabama Department of Conservation and Natural Resources for over 37 years, making him the longest tenured Game and Fish director in the country. Throughout his life he has fought to assure a healthy future for the fish and wildlife resources of Alabama.

Mr. Kelley's career did not begin when he assumed the directorship of the Game and Fish Division in 1959; it officially began shortly after his graduation from Auburn University in 1949 when he was recruited by Fred Stimpson as the first executive director of the Alabama Wildlife Federation. In 1956 he was selected as one of the nation's top 10 outstanding professional conservationists for his work in preventing the Corps of Engineers from constructing a high dam near Jackson, Alabama which would have destroyed many thousands of acres of critical river bottom habitat in southwestern Alabama. This was the first project of its kind to be halted. He also worked to stop the channelization of streams in Alabama when it was not popular to do so.

In the late 1950s, Kelley led an aggressive campaign to stop the widespread use of chlorinated hydrocarbons because of their related destruction of non-target wildlife. Charles Kelley and his staff of biologists and enforcement officers are directly responsible for rebuilding Alabama's white-tailed deer and Eastern wild turkey populations. These abundant populations now provide liberal hunting opportunities for sportsmen. He has also worked tirelessly to make low-cost hunting available to all Alabamians.

Charles Kelley has been very successful during his long career in helping to promote the wise use and management of Alabama's wildlife resources. He received the C. W. Watson Award presented by the Southeastern Association of Fish and Wildlife Agencies in 1991 and in the same year was also recognized by the Alabama Chapter of the National Wild Turkey Federation for his outstanding contributions to wildlife conservation. In 1992 he was recognized as Conservationist of the Year by the Alabama Wildlife Federation.

A CONSERVATION ETHIC

What do Roosevelt, Leopold, Carson, and Alabama's conservation leaders have in common? They are people from totally different callings in life. They came from different parts of the country and lived in different times. Yet each person took a stand for what he or she believed

DENNIS HOLT

Water quality in Alabama is continually threatened by pollution and siltation from industry, development, and land-use practices. An understanding of the impacts of land and water usage, in combination with responsible stewardship, is the first step in preventing environmental degradation.

in and changed the course of conservation in America. They were leaders. As such, they became involved in organizations that also fostered a land ethic. They believed that like-minded individuals could work together for the common improvement of habitat for wildlife and man. The example they set encourages others to become involved in organizations that foster wildlife conservation and stewardship of our natural resources.

"Examine each question in terms of what is ethically and aesthetically right, as well as what is economically expedient. A thing is right when it tends to preserve the integrity, stability, and beauty of the biotic community. It is wrong when it tends other wise."

— Aldo Leopold
A Sand County Almanac, 1949

HUNTER ETHICS

As with all human endeavors, recreational sport hunting has its share of unethical participants. Unfortunately these individuals are often the most visible to the non-hunting public who are either uncommitted or non-supportive in their views toward hunting. Attitudes and perceptions by non-hunters about hunting and hunters may quickly become reality when they observe just a few unethical hunters in action and these attitudes most often turn out to be negative.

Recognizing that only about 7 percent of the U.S. population pursues the hobby of sport hunting, and that the future of sport hunting depends upon the support of the non-hunting public, several hunting and conservation communities came together in 1993 to develop a new hunter ethics pledge to serve as a personal compass for ethical hunting. Hunters nationwide now have a concise reminder about responsible behavior outdoors.

The new code of conduct was produced by the Izaak Walton League of America (IWLA) in cooperation with the major national and state conservation organizations in the country. The nine-point code is designed as a pledge for individual hunters, but it may be adopted and adapted by hunting and conservation organizations, hunting clubs, federal and state agencies, manufacturers of hunting-related products and others who are interested in securing a bright future for hunting by promoting responsible behavior. The new code of conduct and pledge serves notice to careless, incompetent, and thoughtless hunters that it's time to clean up their acts. Ethical sportsmen/women who are serious about cleaning up the image of hunting and the hunter support such efforts.

RICHARD T. BRYANT/ARISTOCK

Each hunter must decide for himself what constitutes a "fair chase" hunt. High fences for deer are proliferating in Alabama and raise questions that will have to be addressed in the future.

A Hunter's Pledge

Responsible hunting provides unique challenges and rewards; however, the future of the sport depends on each hunter's behavior and ethics. Therefore, as a hunter, I pledge to:

- Respect the environment and wildlife;
- Respect property and landowners;
- Show consideration for non-hunters;
- Hunt safely;
- Know and obey the law;
- Support wildlife and habitat conservation;
- Pass on an ethical hunting tradition;
- Strive to improve my outdoor skills and understanding of wildlife; and
- Hunt only with ethical hunters.

By following these principles of conduct each time I go afield, I will give my best to the sport, the public, the environment, and myself. The responsibility to hunt ethically is mine; the future of hunting depends on me.

RICHARD T. BRYANT/ARISTOCK

Ethical hunters respect the private property rights of others.

Brochures and wallet cards bearing the Hunter's Code of Conduct and Pledge are available from the Izaak Walton League of America Outdoor Ethics Program; 1401 Wilson Blvd, Level B; Arlington, VA 22209; or from the North American Hunting Club, 12301 Whitewater Dr., Minnetonka, MN 55343. The first 100 copies of the brochure or wallet card are free, with postage and handling costs of $5 per 100 for additional copies.

THE ALABAMA WILDLIFE FEDERATION'S LEGACY AND VISION

The Alabama Wildlife Federation (AWF) is an organization that offers people the opportunity to take part in programs to improve wildlife in the region and to be involved with others who share such goals. As Executive Director Dan Dumont explained in *AWF Past and Present*, one of the original purposes of the AWF was to coordinate the conservation activities of all sportsmen's clubs and conservation groups into one strong, enlightened, and statewide federation.

Organized conservation activities were well underway in Alabama sometime prior to 1925. A forerunner of the AWF was called the Alabama Wildlife Conservation League, which held its first annual meeting in October of 1925 and elected as its President Col. E.F. Allison of Bellamy. The League published a monthly magazine called

What is Ethics?

Ethics comes from the Latin word *ethos* which means guiding beliefs or principles. These inner beliefs serve as the basis for how people conduct themselves in everyday living. They serve to form a person's standard of conduct affecting how he or she interacts with others and the environment each day. Beliefs about what's right and wrong, which form the foundation of ethics, are highly personal and take time to clarify after much thought and reflection, and usually become crystallized with maturity. Personal beliefs are based upon universal truths (what is generally accepted as right and wrong) and individual experiences: people are also influenced by what is generally accepted by society. During the early stages of ethical development, what is accepted is strongly influenced by a mentor who models ethical behavior. With this in mind, it is important that young hunters have a role model or mentor who exhibits ethical hunting behavior while accompanying them in the field.

The following are characteristics of an ethical hunter:

An Ethical Hunter...

1. Has a deep reverence and respect for game and other wildlife.
2. Has a doctrine of fair chase, such as not shooting animals confined in a small enclosure.
3. Values the importance of companionship and the experience of hunting rather than simply the number of game animals bagged.
4. Obeys all game laws.
5. Enjoys sharing a hunt with first-time hunters such as youth, members of the opposite sex, or others who may be new to the sport.
6. Prepares in advance to sharpen shooting skills, sight in rifles, and study game identification, habits, and habitats.
7. Respects the land and those who own it (leaving the land as it was found).
8. Makes every effort to retrieve game after shooting.
9. Takes shots at game that are only within his or her ability and the limits of the hunting equipment being used.
10. Eats the game that is harvested and shares with others, especially landowners, who enjoy game but do not hunt.
11. Is respectful to non-hunters. For example, many individuals are repulsed by the sight of dead game, especially deer in the back of a truck or on the hoods of vehicles. Game should be placed out of sight of the general public when transporting or processing.
12. The true test of an ethical hunter is how he or she acts when alone. There should never be two codes of conduct: one followed when hunting with companions and another when hunting alone. When hunting alone you share your hunt with your conscience. Continue to be an ethical hunter during this time and enjoy the personal satisfaction of doing what is right. This serves to make the hunting experience more rewarding and enriching.

DENNIS HOLT

It goes without saying that obeying game laws is part of being an ethical hunter. However, the true test of an ethical hunter is how he or she acts when alone.

The Alabama Sportsman. At the second annual meeting, a committee was appointed to work with the State Department of Game and Fisheries to work on existing laws affecting wildlife. They also wanted to help draft new legislation to present to the Alabama Legislature.

In 1926, Mr. W. E. Brooks of Brewton, then editor of *The Alabama Sportsman*, wrote of the strength of a collective group of dedicated conservationists: "Scattered

Hunting Ethics

Corky Pugh, Assistant Director of the Alabama Game and Fish Division, is an avid outdoorsman and a forest landowner. He hunts and fishes at every opportunity and founded Chalk Hill Wildlife Reserve, Inc., which he actively managed for hunting for over a decade. He is also author of the book Family and Friends. *We asked Corky, "What is a conservation ethic and how is it developed?"*

Ethics govern behavior and decisions and are based on values and principles and a sense of right and wrong. A conservation ethic deals with behavior and decisions relating to the use and management of natural resources. A strong conservation ethic enables a person to see beyond today in terms of leaving a positive legacy for others in the future. Typically learned by example at an early age, this sense of responsibility for one's own actions results in voluntary adherence to standards of fair chase and self-restraint in hunting and fishing, and of wise use and management of our natural resources. It is based on a high regard for the land and respect for all wild creatures.

I enjoy spring turkey hunting as much as anybody possibly could, and the most rewarding gobbler I've ever taken was not the largest or most challenging. He was the somewhat underweight, mating-worn, opening day tom taken on the small north Monroe County piece of family land passed on to me by my great aunt and then my mother. That old turkey was the most rewarding because of the hard work and countless hours devoted to intensive management of the little forest tract that provided the habitat needed for wild turkeys to thrive.

Alabama has a rich history of wildlife conservation. The original conservationists were avid hunters as well as astute land managers who had the foresight and sense of moral duty and obligation to recognize the importance of leading the way to a new land ethic and the establishment of responsible hunting laws. As a result, sweeping land use changes occurred, and a state game and fish agency was born along with enforceable seasons and bag limits. Now, many decades later, forest products constitute the No. 1 industry in the state, and our game populations are the envy of the nation. Had the people who began this movement not been instilled with a strong conservation ethic, the history of Alabama would not be so bright for wildlife or forestry. Private landowners, given the incentive of harvesting game on their own property, have been the heart of the restoration of deer and turkey populations across Alabama over the past 50 years. It's a success story we can all be proud of.

The future of hunting depends on the social significance attached to it—the cultural value. Fair chase considerations will be pivotal as public policy on hunting will be increasingly determined by the growing non-hunting portion of the population. How well we can manage non-hunted resources so valuable to urban and rural constituents will be a major determining factor. The importance of a strong conservation ethic on the part of sportsmen and forest owners alike becomes more significant in this age of separation from the land and of a "drive-up-window mentality" on the part of people who do not understand the connection of the land, other natural resources, and wildlife.

Scenes like this are a blatant disregard for Alabama's game laws, repulse responsible sportsmen, and send a negative message to the non-hunting public.

individual effort can produce little in the way of tangible results. In union there is strength. Sportsmen in each county can do no better work for the cause of more game and fish than by effecting an organization. The more local clubs there are, the more rapidly conditions will improve. If all the local clubs will affiliate with the state organization, means will be at hand to improve conditions much more rapidly than by scattered and diverse efforts."

What was true of the early beginnings of AWF still holds true today. The group's activities are far-reaching and far-ranging, as has been detailed earlier and elsewhere in this book. Naturally, the Federation has room for new members as this growing body of conservationists moves into the next century. The AWF also wants people interested in affiliating with like-minded individuals to be aware of other opportunities to promote conservation both in the state and the nation. In addition, the AWF works with agency personnel whose activities directly impact our wildlife resources. In Appendix O is a partial listing of agencies and conservation organizations that readers may wish to contact. These organizations and agencies offer opportunities for membership or sources for answers to questions we think will be of interest to our readers. Visit AWF's Web site at alawild.org.

ADCNR Commissioner Jim Martin (left) and Vernon Minton, Jerry Walker, and Walter Tatum of the Marine Resources Division present AWF Executive Director Dan Dumont with a proclamation signed by the governor in recognition of the Federation's role in the passage of a saltwater fishing license.

SUMMARY

Earlier in the history of America natural resources were thought to be inexhaustible and limitless. Consequently, the use of these resources continued at an exponential rate with little effort to manage or conserve what was taken for granted. As a result, land was degraded, wildlife habitat was reduced or eliminated, and many wildlife species populations declined dramatically due either to a loss of habitat and/or unregulated market and subsistence hunting. Alarmed by what they saw, early conservation leaders such as Theodore Roosevelt and Aldo Leopold pushed for management and protection of wildlife and their habitats, and called for a land ethic. Later pioneers such as Rachel Carson awakened the public to the misuse of harmful chemicals and their effect on human health, wildlife, and the environment. While national conservation leaders were making their mark, Alabama conservation visionaries were also at work. The abundance of wildlife and other natural resources in Alabama today is largely due to the foresight and efforts of men like Fred Stimpson, Walter Mims, William Ireland, Charles Kelley, and other dedicated individuals and organizations like the Alabama Wildlife Federation. Abundant natural resources and healthy natural systems have been the key to a successful Alabama economy and quality of life. Today, however, the demand for natural resource use and threats to environmental quality are at an all time high. The future of the state's natural resources and environmental health, like in the past, will depend upon dedicated conservation leaders, the efforts of conservation organizations, and an environmentally literate citizenry. For more information on how to participate in the conservation of Alabama's natural resources and environment, contact the Alabama Wildlife Federation or other organizations listed in Appendix O.

Make an effort to include a young person in your next trip afield. It will enrich you both.

Alabama wetlands are some of the most productive areas in terms of wildlife diversity and abundance. Unfortunately over 50 percent of Alabama's original wetlands have been lost to drainage for development and other land uses. Today federal and state regulations govern the use and management of wetlands to protect these valuable resources for future generations.

RICHARD T. BRYANT/ARISTOCK

CHAPTER 19

Legal Considerations and Natural Resource Management

CHAPTER OVERVIEW

This chapter discusses some of the important legal considerations and guidelines that affect the use and management of private forests and farms in the Southeast, with an emphasis on Alabama. As a result of increasing public concern over land use and management effects on the environment, government regulations have increased over the last quarter century. With this increase in government involvement comes a growing concern among farm and forest owners about their rights as property owners. While these concerns will continue, the fact is that landowners must be aware of existing regulations and resulting legal ramifications.

CHAPTER HIGHLIGHTS PAGE

FEDERAL, STATE, AND LOCAL REGULATIONS ... 439

LANDOWNER RIGHTS AND RESPONSIBILITIES ... 452

RECREATIONAL ACCESS AND LIABILITY: WHAT LANDOWNERS SHOULD KNOW ... 454

OPERATION GAMEWATCH ... 456

Y*esterday*... When the first European settlers came to the New World they saw a seemingly inexhaustible supply of natural resources for the taking. The land was healthy and wildlife, forests, and other natural resources were abundant. Reports from that era depict the eastern and gulf coastal shorelines as teeming with shorebirds, falcons, waterfowl, and other wildlife.

The forests of the Southeast seemed never-ending and were alive with white-tailed deer, Eastern elk, woodland bison, wild turkey, black bear, ruffed grouse, passenger pigeon, cougar, red wolf, and a host of other animals. But things changed as civilization spread across the land. Forests needed to be cleared for cropland and domestic animals. Wildlife were dramatically impacted by the unchecked use of the ax, plowshare, trap, and gun. Forest lands were cleared and either converted to agricultural use or left abandoned and not replanted. Because of the poor silvicultural practices of the day and a lack of understanding of sustainable agricultural practices, much of the land's productive capacity was depleted from overuse and abuse. The settlers' zeal to "conquer the land," along with their poor land-use practices, and the relentless efforts of those who sought to extract vast quantities of natural resources in whatever manner, still haunts the landscape today as evidenced by tracts of remaining marginal land which resulted from erosion.

By the time the twentieth century arrived, much of the nation's natural resources, including those of the Southeast, were exploited and in shambles. Recognizing this alarming trend President Theodore Roosevelt in a special message to Congress in 1909 called for control of private land use to prevent further misuse of the land. In his message Roosevelt's concern is evident:

"The time has fully arrived for recognizing in the law the responsibility to the community, the state, and the nation which rests upon the private owners of private lands. The ownership of forestland is a public trust. The man who would so handle his forest as to cause erosion and to injure streamflow must be not only educated but he must be controlled."

Gifford Pinchot, Roosevelt's first chief of the Forest Service, echoes his close friend's feelings by advocating federal control of timber cutting practices on private forest land to prevent overcutting of the nation's private timberland reserve.

"We saw that only federal control of cutting practices on private [forest] land could assure the Nation the supply of forest products it must have to prosper... that [we] will eventually exercise such control is inevitable, because without it the safety of our forests and consequently the prosperity of our people cannot be assured."

Because of widespread land abuses, leaders like Roosevelt and Pinchot were compelled to advocate private land use control by federal and state governments. Out of this advocacy was born a host of federal and state laws and regulations that today have an effect on the use and management of private forest and farm lands. So began government's involvement in the regulation of private land use and management in America.

As Roosevelt and Pinchot were advocating regulation of private lands to prevent further abuses, Aldo Leopold, an astute forester and naturalist, was attempting to reverse years of land abuse by appealing to the very depths of the human soul. Leopold believed strongly in man's need to live in accordance with an inner set of morally correct principles, principles that could also be extended to the land. Leopold states,...

"A land ethic... reflects the existence of an ecological conscience, and this in turn reflects a conviction of individual responsibility for the health of the land. Health is the capacity of the land for self-renewal. Conservation is our effort to understand and preserve this capacity."

Leopold suggests that it is not only man's responsibility to care for the land; it is also morally right to do so. Implied in Leopold's land ethic is the idea that with the privilege of land ownership comes a responsibility—the responsibility to be wise and thoughtful stewards of the land.

Today...

Over the last 25 years there have been many success stories in the U.S. wherein environmental regulation has resulted in cleaner air and water, and the return of several species, such as the bald eagle, that were on the brink of extinction. Despite these improvements, opinion polls over the past decade indicate a growing perception among the American public that natural resources such as clean water and air, open spaces, and wildlife habitat are threatened and dwindling. A growing majority of Americans indicate their support for greater regulation of what they perceive to be environmentally threatening activities. A study conducted by Dr. John Bliss of Auburn University found that 84 percent of private forest owners in Alabama agreed with the statement "Private property rights are important, but only if they do not hurt the environment." A majority also supported government regulations to protect natural beauty, streams, and wetlands, and threatened and endangered species. However, the majority of Alabama forest owners also supported compensation for landowners when

In earlier times Alabama's forests were heavily harvested for timber products, cleared for cropland and domestic animals, or simply abandoned after cutting. Because of poor forestry and agricultural practices of the past, much of the land's productive capacity was depleted from overuse and abuse.

Past mining operations were unregulated, extracting vast resources from the land with little or no concern for environmental impacts or efforts to restore mines that were abandoned.

Unregulated market hunting, without bag or season limits, combined with radical alterations of habitat had a negative impact on the populations of many wildlife species.

Uncontrolled cattle grazing in the early 1900s had a dramatic impact on Alabama's wildlife habitat and timber production. As president of the Alabama Wildlife Federation, Fred T. Stimpson led conservationists in what became known as the "Stock Law War" in the 1930s which eventually lead to the removal of free-ranging cattle from private lands.

regulations caused them to lose money.

A majority of Americans have also expressed a willingness to sacrifice economic growth, if necessary, to prevent environmental degradation. The same opinion polls also found that nearly two-thirds of Americans think that laws and regulations have not gone far enough in protecting environmental values. Given what some see as the choices between environmental protection and economic growth, or between protecting the environment and preserving private property rights, most Americans side with the environment.

For some private forest or farm owners, complying with government mandated rules and regulations and fearing legal penalties for failure to do so can cause a good deal of anguish. In most cases, private landowners are more than willing to comply with regulations that protect their land. Most of the time it is in their best interest to do so. Unfortunately, because of the complexity of regulations, many private farm and forest owners are left confused. Adding to this confusion are the continuing changes to regulations and the apparent inconsistency in application by state and federal agencies. It is no wonder many landowners are frustrated.

This chapter will address important laws that affect private landowners, agencies that administer the laws, and the responsibility of landowners under the law. This discussion raises the issue of private property rights. Also included is a section on recreational access and liability for landowners who are contemplating providing recreational access to their property.

DAN BROTHERS/ADCNR

The quality of Alabama's rivers and streams is continually threatened by development, industry, and other land-use practices. The federal Clean Water Act helps ensure watershed quality in the state by limiting and/or prohibiting harmful practices that might threaten these valuable resources.

Federal, state and local regulations center primarily around protecting endangered species, and water and air quality. Federal regulations also protect migratory waterfowl, such as the mallard (right), by establishing a framework for hunting season length and bag limits based upon annual surveys.

PAUL T. BROWN

FEDERAL, STATE, AND LOCAL REGULATIONS

Prior to the passage of environmental laws and zoning ordinances, common law nuisance actions were one means of attempting to control environmentally destructive activities. These efforts, however, were usually ineffective. Most of the environmental laws and regulations on the books today were enacted to address environmental problems that were prevalent in the 1960s and 1970s. By the 1960s, many of the nation's waters were unfit for swimming, fishing, or drinking. Air pollution and smog had been demonstrated to cause harmful effects on human health. Development had destroyed large areas of plant, wildlife, and aquatic habitat. Toxic pesticides often killed wildlife in addition to target species and threatened human health as well. Federal laws, many of which are implemented through state plans, have since been enacted to protect the environment.

Environmental laws are designed to protect public health and welfare and the environment. However, the controls and costs they impose on natural resource management or other land uses are most often absorbed by private individuals or firms that own the land. Policy debates have raged among various land-use interest groups. These have included private corporations and landowners who grow timber, environmental groups who seek more land-use regulation, public agencies who are charged with implementing federal and state laws, congressional and legislative members who try to strike a balance between economic and environmental concerns, and judicial systems considering the merits of legislation regulating private property owners.

Today, there are over 100 treaties, international agreements, federal statutes, executive orders, and federal regulations that pertain to land use and wildlife regulation. This does not include state regulations, and in some cases local ordinances, that affect land use activities like forestry or agricultural operations (e.g. fire control ordinances, toxic waste management, wetlands). Currently, legislation centers around three broad topics that affect forest, agriculture, and wildlife management practices: **endangered species, water quality,** and **air quality.** The presence of red-cockaded woodpeckers, gopher tortoises, and other threatened and endangered species affects forest and farm management and operations across portions of the southern landscape. Water quality legislation and regulation to control non-point source pollution affects management practices in wetlands and areas adjacent to streams and rivers. Air quality legislation or guidelines affect smoke management from prescribed burning. This chapter will focus on legislation, regulations, and guidelines that could potentially impact how forest and farm owners use and manage their lands. They include the following:

- Endangered Species Act,
- Clean Water Act,
- Clean Air Act,
- Coastal Zone Management Act,
- Federal Environmental Pesticide Control Act,
- Cost-Share Recipient Responsibilities, and
- State and Local Statutes and Ordinances.

JOE MCGLINCY

The federally threatened Red Hills salamander is only found in Alabama along a narrow belt associated with two siltstone formations called the Tallahatta and the Hatchetigbee in south Alabama. Land management practices are still possible in these areas after consultation with the U.S. Fish and Wildlife Service.

Endangered Species Act

The Endangered Species Act (ESA) was passed in 1973 to prevent the extinction of animals and plants that are drastically declining and exist only in extremely low numbers. The ESA is one of the most far-reaching laws ever enacted. The U.S. Congress established the ESA because it recognized that all species *"are of aesthetic, ecological, educational, historical, recreational, and scientific value to the Nation and its people."*

An **endangered species** is one that is in danger of becoming extinct throughout all or a significant portion of its range. The ESA also protects species that are threatened with extinction within the foreseeable future. Legally, there is no practical difference between endangered and threatened species. In Alabama, there are 68 animals listed as endangered or threatened. Eighteen kinds of plants found in Alabama are considered threatened or endangered at press time.

Very few people would disagree that we need to protect species in danger of being lost forever. But a question often raised is "How much is society willing to sacrifice in order to save from possible extinction an animal or plant that few people have ever seen or even fewer could identify?" This question, being played out in the Pacific Northwest, focuses on how to strike a balance between environmental, social, and economic needs in protecting the threatened northern spotted owl. In the Southeast, the endangered red-cockaded woodpecker poses similar challenges (as well as civil and criminal penalties) to natural resource managers, regulating agencies, and forest landowners.

HOW DOES THE ESA AFFECT PRIVATE LANDOWNERS?

Section 9 of the ESA prohibits the possession, import, export, or interstate or foreign sale of a listed species. It also makes it illegal to "take" a listed species from the wild without an exemption or Section 10 ESA incidental take permit. **Take** is defined "to include harass, harm, pursue, hunt, shoot, wound, kill, trap, capture, or collect a listed species or attempt to engage in any such conduct." Taking can also include significant habitat modification or degradation that actually kills or injures an endangered species by impairing its breeding, feeding, or ability to find shelter. If an animal on private property is listed as endangered or threatened, the landowner must avoid a "taking" of that species. Penalties for violations can range from a warning and seizure of illegally held wildlife specimens and products, to civil fines of $25,000, or criminal fines of $50,000 and/or a year in jail. Plants do not have the same

Why Regulation?

Government regulation of private forest and farm practices is primarily a result of nine concerns:

1. General Public Anxiety Over Natural Resources

Forestry and agricultural operations are highly visible activities. Management practices that alter the landscape can easily become the focus of public concern about the environment, especially from urban populations that have little or no connection with the land or understanding of land management. From the public's perspective, a quick call for regulation may be the only way of addressing concerns.

2. Misapplied Forestry and Agricultural Practices

Some advocates of regulation have witnessed careless or unethical applications of forestry or agricultural practices. Whether these practices are isolated or widespread is irrelevant. Their occurrence is a guaranteed call for regulation.

3. Federal Environmental Laws

Federal laws requiring states to initiate programs to address nationally defined environmental problems have established a climate favorable for regulation.

4. Greater Accountability

Making sure that those who own and manage the land are held more accountable for their management practices has caused some to advocate that performance standards be put into regulations. Legally imposed standards (such as clear-cut size, widths of SMZs) are viewed by some as the most effective means of changing land management practices since the law would require a specific application. Evidence could then prove if a landowner violated a legally specified management practice guideline.

5. Growth of Local Ordinances

Regulation of private forest land is also a response to an expanding number of municipal and county ordinances that restrict forestry practices. Of an estimated 360 ordinances nationwide, over 70 percent were established after 1980, and half since 1985. Local ordinances can become a complex jungle of conflicting regulation.

6. Landscape-Level Concerns

Landscape level effects of site-specific land management practices carried out by numerous landowners have also made regulatory programs appealing. A single administrative authority (such as the state forestry commission) that is responsible for establishing and enforcing forest standards among numerous landowners is viewed by some as an effective way of ensuring the sustainability of multi-owned forest lands.

7. Following the Lead of Others

Regulation of private forest practices is motivated by attitudes such as "We should have it too." Communities may feel deficient because their state or municipality is not regulating private land management practices. Likewise, in any one state, opposing interest groups hurry to present their version of a land use law before their opponents suggest one less to their liking. Securing a political advantage —with limited attention to the substance of a proposed law—can become the name of the game.

8. Increasing Population

Increasing human population puts more pressure on available natural resources and land uses. Regulating the use and management of land resources is seen as a means to ease demand.

9. Level Playing Field

Some large industrial forest products companies who have made the decision to follow Best Management Practices (BMPs) and practice sustainable forestry do not want to lose market opportunities to smaller operators who do not follow BMPs or responsible land management practices.

protection as animals under Section 9, except that it is unlawful to "remove and possess any [endangered plant] species from areas under federal jurisdiction; or maliciously damage or destroy any protected species on any area in knowing violation of any state law or regulation, including state criminal trespass law." If the activity may incidentally take or harm a protected species, landowners may apply for a Section 10 ESA incidental take permit. If approved, this provides limited protection from sanctions in the event of an unintentional and unavoidable take of a protected species, and it also provides protection of the species with a habitat conservation plan that is a required element of an incidental take permit.

If you are involved in any federal program (such as the FIP, CRP, SIP, crop support programs), Section 7 of the ESA requires the federal agency that administers the program to ensure that land-use practices do not jeopardize the existence of a listed species or adversely affect **critical habitat**: areas that are crucial for the survival of the species. It does **not** require the private landowner to actively manage for a listed species as is required on federal lands. However, management activities must not adversely affect critical habitat or the species.

In order to help landowners avoid inadvertent violations of the ESA, resource professionals can determine if any listed species occur on private property. If a listed species is located, arrangements can be made to have a biologist with the Alabama Department of Conservation and Natural Resources (ADCNR) Non-game Program or U.S. Fish and Wildlife Service (USFWS) provide landowners with guidance and management recommendations. Keep in mind that endangered and threatened species are rare and the probability of one being on your land is very low at this time.

The USFWS, the federal agency responsible for implementing the ESA, has indicated a willingness to help landowners find creative ways of continuing use and management activities while at the same time not "taking" an endangered or threatened species. Examples of these efforts include an agreement reached with the USFWS and Georgia-Pacific, a timber company with substantial land ownership in the South, concerning the red-cockaded woodpecker. In Alabama, International Paper has developed a plan that protects the Red Hills salamander (found only in Alabama) which allows forest management activities to continue on some of the impacted lands.

Landowners who have endangered or threatened species on their property are impacted, but their land management objectives may continue with the development of a Habitat Conservation Plan (HCP) with the USFWS. Once a HCP is approved and followed, the landowner is usually not required to make future management changes, even if the needs of the species change over time. The idea is to provide forest owners with an atmosphere of stability and certainty so that they can make the long-term investments necessary to manage private forest lands for profit and at the same time protect endangered species. The HCP is part of the Section 10 process to obtain an incidental take permit. The HCP

Red-cockaded Woodpecker Procedures Manual for Private Woodlands

The USFWS, in an effort to give forest and farm owners guidance about managing red-cockaded woodpeckers on their lands, is currently developing a "Red-cockaded Woodpecker Procedures Manual for Private Woodlands." The manual is more flexible than the Biological Assessment Guidelines, which is the current management policy on federal lands. For more information about the manual contact the nearest USFWS office listed in this chapter.

RICHARD T. BRYANT/ARISTOCK

Timber and red-cockaded woodpecker management can coexist by following management guidelines in the "Red-cockaded Woodpecker Procedures Manual for Private Woodlands" or by participating in the "Safe Harbor" program.

JOE MCGLINCY

The EPAs Endangered Species Pesticide Program is designed to reduce pesticide exposure to endangered species. Before spraying pesticides, landowners should protect endangered species and their habitats by following guidelines published in county bulletins available through the EPA, the U.S. Fish and Wildlife Service, or county Cooperative Extension System offices.

provides protection for the species, while the permit provides protection for the landowner should an incidental take occur.

The ESA also calls for the federal government to encourage, with financial assistance and through incentives, activities by states and others to develop and maintain conservation plans to restore populations of listed species to a point where they no longer are in danger of extinction. The Stewardship Incentive Program (SIP) is the first federal government program to provide private landowners with "positive" incentives to protect, manage, and enhance threatened and endangered species habitat. Work is also underway to develop economic incentives for the landowners through cost-sharing assistance, valuable credits (special credits for managing endangered species habitat), or other creative alternatives that protect private property rights and achieve the goals of the Endangered Species Act.

ENDANGERED SPECIES PESTICIDE PROTECTION PROGRAM

For several years the Environmental Protection Agency (EPA) has been working on ways to protect endangered species from the risks of pesticides. Although not yet completed, EPA is in the process of developing an Endangered Species Pesticide Protection Program with the goal of reducing pesticide exposure to endangered species. The new program will rank each endangered species according to its status, recovery potential, vulnerability to pesticides, potential for exposure, and apparent risks from pesticides. The USFWS will consider all of this information and then issue a biological opinion as to whether the species could be harmed by pesticide exposure. If a species is considered in jeopardy from pesticide use, the EPA and USFWS will, in most cases, prepare a bulletin for each county where the species is found or presently is known to exist. Bulletins will include habitat maps and descriptions, and pesticide use restrictions. Bulletins will be available from local county Extension offices and from pesticide dealers and distributors.

So, what if you pick up a bulletin and find that the area you plan to spray is mapped as a possible habitat for an endangered species? First, read the text under the map. It often explains more about the map and the endangered species. For example, it may state specific habitat descriptions that would eliminate the area you are about to spray as endangered species habitat. The focus of the program is education, not enforcement. Some states are developing creative programs, like landowner

DENNIS HOLT

Over half of the water pollution in the U.S. is caused by non-point source pollution from agriculture, mining, urban development, and forestry. Farm and forest owners should follow Best Management Practices to reduce the potential for damaging water quality as a result of management activities.

agreements, to ease the burden on landowners. If you have questions about the program call the local EPA office or EPA's toll-free number: (800) 447-3813. EPA can tell you if bulletins have been issued for the county where you intend to spray.

Clean Water Act, Non-point Source Pollution, and Wetlands Regulation

Congress passed the Water Pollution Control Act in 1948 to provide technical and financial cooperation to states and municipalities that implement programs to reduce stream pollution from municipal and industrial waste, otherwise known as point source pollution. This act established the policy of Congress to recognize, preserve, and protect the primary responsibilities and rights of the states to control water pollution. The emphasis on public health concerns in the Water Pollution Control Act was altered with the passage of amendments to the act in 1972 called the Federal Water Pollution Control Act. These amendments established a national goal of eliminating all pollutant discharges into the waters of the United States and making all waters safe for fishing and swimming. The 1972 Federal Water Pollution Control Act became known as the Clean Water Act.

The Clean Water Act (the "Act") was the first federal legislation to address pollution caused by storm water runoff from the landscape. Over half of the water pollution in the U.S. is caused by non-point source pollution from agriculture, mining, urban and construction activities, and forestry. Contrary to public perception, forestry is only a minor contributor to the total non-point source pollution in the Southeast. The Act also identified the need to protect wetlands from unwanted human disturbance.

Two sections of the Clean Water Act established the legal framework for non-point source pollution control: Section 208 and Section 404. Section 208 requires all states to assess damages to water quality from non-point source pollution and to develop either regulatory or non-regulatory programs to control them. In the South, most states have chosen to develop non-regulatory programs that contain management guidelines like the Best Management Practices (BMPs). The lead agency in Alabama for developing mandated programs under the Clean Water Act is the Alabama Department of Environmental Management (ADEM), as directed by the Alabama Water Pollution Control Act of 1972. The Alabama Forestry Commission, working in conjunction with the ADEM and EPA, developed the state BMP

ALABAMA FORESTRY COMMISSION

Riparain forests should be retained as streamside management zones to protect water quality and provide habitat and travel corridors for wildlife.

guidelines for forestry in 1993. Alabama's BMPs for Forestry manual can be obtained from the Alabama Forestry Commission (334-240-9305).

Section 404 of the Clean Water Act established a regulatory program for the discharge of dredged or fill materials in navigable waters. This section is regulated by the U.S. Army Corps of Engineers with oversight by the EPA. Much debate and litigation has occurred over what constitutes "navigable waters" and "waters of the United States" for purposes of Section 404 jurisdiction. Following litigation in 1977 the Corps of Engineers expanded the regulatory definition of navigable waters to include wetlands. Wetlands only became regulated in 1977 after the Corps was forced to include them by court order. The following regulatory definition of wetlands is used to administer the Section 404 permit program:

"*...those areas that are inundated or saturated by surface or ground water at a frequency and duration sufficient to support, and that under normal circumstances do support, a prevalence of vegetation typically adapted for life in saturated soil conditions. Wetlands generally include swamps, marshes, bogs, and similar areas.*"

Once the definition was established determining what was being regulated, the debate shifted to how to identify or delineate a wetland. Landowners need to know what parts of their property require permits to carry out certain management activities. Unfortunately, there are no boundary markers telling landowners where a wetland begins or ends. In most cases, wetlands can only be identified and marked by natural resource professionals with sufficient training to recognize vegetation, soils, and hydrology to make wetlands determinations. The delineation process is complex and subjective.

One benefit for forest and farm owners under Section 404 is an exemption for normal silvicultural (forest management), farming, or ranching activities. In order to meet the exemption, the activity must be ongoing—"*normal farming, silvicultural, and ranching activities such as plowing, seeding, cultivating, minor drainage, harvesting for the production of food, fiber, and forest products*"—and must not convert a wetland to an upland site. The rules also state that "*new activities which bring an area into farming, silviculture, or ranching use are not part of an established operation.*" In other words, landowners who want to *begin* farming, forestry, or ranching on lands not previously managed for these activities are not exempt from permit requirements. Additionally, while normal harvesting is exempt, this "*...does not include construction of farm, forest, or ranch roads.*" These activities and other silvicultural practices must comply with mandatory federal BMPs for forested

DENNIS HOLT

Wetland conservation and protection can also provide recreational opportunities for sportsmen when impoundments for waterfowl hunting and fishing are maintained.

wetlands contained in federal regulations. Landowners managing wetlands for timber production must follow appropriate federal and state BMPs in order to be exempt from permit requirements. When constructed in wetlands, roads and skid trails that meet the mandatory federal BMPs do not require permits. A written management plan, records of management activities, and evidence of past silvicultural use will help forest owners demonstrate that their land management activities are indeed ongoing activities. Periodic harvests or other types of timber stand improvement practices also help demonstrate that silviculture is an ongoing activity.

Unfortunately for many forest landowners, EPA has a narrow definition of "normal" silviculture. Landowners may run into problems meeting the exemption. EPA and the Corps of Engineers consider only timber management as silviculture. *If the activity is specifically for wildlife management, recreation, or other land use purposes, it will not qualify under this exemption.* As an example, a forest landowner in Delaware was cited by the EPA for establishing a wildlife food plot in a wetland without a permit. EPA ruled that since the practice was for wildlife, it did not meet the silviculture exemption. The U.S. Forest Service and several other groups are currently working with EPA to widen their definition of silviculture to include multiple-use objectives. Until a final ruling is made, landowners who are planning to implement practices in **jurisdictional wetlands,** wetlands that are protected by law, that are not related to timber management need to contact the Corps of Engineers to see if a permit is required. Even if a management practice is for silvicultural purposes, if it is going to impact a jurisdictional wetland it ***must*** be performed in compliance with BMPs for the landowner's operations to remain exempt. A violation may cause a total loss of exemption, penalties, and additional costs.

If a management activity does not meet the silvicultural exemption, the Corps of Engineers has developed general and nationwide permits for a number of activities that have minimal impacts on wetlands. Nationwide permits were designed to regulate similar activities with little or no delay or paperwork. If the activity does not qualify for authorization under a nationwide permit, it may still be authorized by the U.S. Army Corps of Engineers by an individual or regional general permit.

Increased regulation of forestry operations is a possibility in the future if non-regulatory non-point source pollution programs are not effective in protecting water quality. By practicing good stewardship through the use of BMPs during forestry activities, forest owners are more likely to protect water quality in nearby streams and forestall further regulations. Effective communication with the Corps of Engineers will prevent potentially embarrassing situations that could delay management activities or result in fines. Both the EPA and the Corps of Engineers recognize the need to develop better communications with private landowners. Federal and state governments must also contribute by supporting programs that have a direct impact on wetland resources conservation and protection such as the Wetlands Reserve Program (WRP), Conservation Reserve Program (CRP), and Stewardship Incentive Program (SIP).

ARE WETLANDS ON YOUR PROPERTY?

The USFWS has used aerial photography and satellite imagery to map general wetland areas on U.S. Geological Survey topographical maps. You can obtain a copy of National Wetland Inventory maps by calling 800-USA-MAPS. Ground checks should verify that one or more indicators from each of the three wetland parameters (wetland vegetation, hydric soil, and wetland hydrology) are present before an area can be considered a jurisdictional wetland. If you observe definite indicators of any of the three characteristics, you should seek assistance from either the local Corps District Office or someone who is an expert at making wetland determinations. The Corps office will assist you in defining the boundary of any wetlands on your property, and will provide instructions for applying for a Section 404 permit, if necessary. Legal advice and other professional recommendations should be a part of your planning process.

Normal forestry, farming, and ranching operations are exempt from having to obtain a Section 404 permit from the Corps of Engineers if the activities follow the prescribed BMPs. In the Southeast, EPA has retained authority over forestry, farming, and ranching activities as defined under Section 404. This means that the Corps can determine if wetlands are present in forestry areas and NRCS makes the determination regarding farming and ranching in these areas, but only EPA can make the determination if the activities comply with the conditions necessary to be exempt from permitting. In Alabama's coastal zone, ADEM also has regulatory jurisdiction.

If you intend to conduct forest and wildlife management practices in wetlands, you should contact the local Alabama Forestry Commission (AFC) office for advice. EPA has worked with the AFC to ensure that your local forester is knowledgeable concerning EPA guidelines as they apply to forestry. The advice provided by your local forester could help you stay in compliance with the wetlands portion of the Clean Water Act, which carries penalties for non-compliance. Farming or ranching, on the other hand, in wetlands should be coordinated through the local Natural Resources Conservation Service (NRCS) office.

Clean Air Act

In the Clean Air Act of 1970, as amended, the Environmental Protection Agency was directed to identify and publish a list of air pollutants and to establish air quality standards for those pollutants in order to protect public health. A table of National Ambient Air Quality Standards, published by EPA, did identify primary and secondary pollutants and their maximum acceptable concentrations in the atmosphere. States were then directed, by the act, to submit plans detailing how they intended to achieve and maintain the National Ambient Air Quality Standards.

The Alabama Department of Environmental Management has implemented air quality regulations that basically prohibit outdoor burning. An exemption to this prohibition was granted for "fires set for recognized agricultural, silvicultural, range, and wildlife management practices." A prescribed burn would rarely be a significant source of particulate matter in the air for more that a few hours on a given day. However, if the pollution level already exceeds acceptable standards, a prescribed burn could become a problem. In 1981, because of a growing concern about potential problems caused by smoke from prescribed burning, the forestry community developed voluntary smoke management guidelines. The guidelines were written not to restrict the use of prescribed fire, but to reduce the amount of particulate matter being released into smoke-sensitive locations such as across highways and near residential areas.

The Alabama Forestry Commission has developed a reference handbook entitled *Smoke Screening System for Alabama* which provides guidelines for smoke management when control burning, including identification of situations when smoke may create unsafe conditions. Natural resource professionals use these guidelines when providing burning recommendations to landowners. Prescribed burning for forest or wildlife management should only be conducted by an experienced and qualified professional or manager. State law requires that a permit be obtained **before** any woodland, grassland, field, or new ground is burned that is over ¼ acre in size or lies within 25 feet of natural fuels, such as woods and

LEWIS O. ROGERS/SOUTH CAROLINA DNR

Prescribed burning is a beneficial management tool for timber, wildlife, endangered species, aesthetics enhancement, and improving access to forest stands. Because of increasing air quality concerns, the Alabama Forestry Commission has developed a "Smoke Screening System for Alabama" handbook which will help ensure that burns are conducted safely and that forest owners will be able to continue to use prescribed burning as a wildlife and timber management tool.

grass. Failure to obtain a permit could result in a fine of up to $1,000 and/or up to six months in jail. To obtain a burning permit call the Alabama Forestry Commission toll-free number located on the inside front cover of your phone book. Prescribed fire, when used correctly, is a beneficial tool when a forest is being managed for timber, wildlife, endangered species, aesthetics, and other multiple uses. Following the smoke management guidelines will help ensure that burns are conducted safely and that forest owners will be able to continue to use prescribed burning as a wildlife and timber management tool.

Coastal Zone Management Act

The federal Coastal Zone Management Act of 1972, as amended in 1990, provides a means for federal involvement in coastal zone protection. The definition of a coastal zone includes not only those counties that border the Gulf of Mexico but also those areas that extend several counties inland from the ocean. The act requires that every state with a federally approved program develop a plan to control coastal zone non-point source pollution according to guidelines set by EPA. In Alabama, the Alabama Department of Environmental Management (ADEM) administers the Alabama Coastal Management Program, which regulates use (primarily development) of coastal zones. An additional ADEM effort, the Coastal Non-Point Source Management Program, regulates marine non-point source pollution. In coastal counties, ADEM relies on the current Best Management Practices developed by the Alabama Forestry Commission to serve as a guide for forest management. EPA is in an ongoing process of developing a number of best management recommendations for coastal zone areas. These are extensions of current BMPs that states are to use as guidelines. Some of these practices include those already appearing in the BMP manual, such as protection of streamside management zones (SMZs), road construction and maintenance measures, timber harvesting practices, site preparation and reforestation, prescribed fire, and pesticide and fertilizer applications. Final recommendations will form the basis for further regulatory controls or for stricter voluntary BMPs in the extended coastal zone. For more information about land-use regulations in the coastal zone contact the Alabama Department of Environmental Management's Mobile office at (334) 432-6533.

Federal Insecticide, Fungicide, and Rodenticide Act

After World War II, many new pesticides and herbicides were developed to control undesirable animals and plants. Public fears surrounding the findings that DDT accumulated in the food chain and caused animal mortality, along with the concerns that these chemicals may cause cancer, led to the passage of the Federal Insecticide, Fungicide, and Rodenticide Act (FIFRA). The act authorizes the Environmental Protection Agency to classify and register the use of most herbicides, pesticides, fungicides, and rodenticides. EPA decides on the safety of each proposed chemical and lists specific applications that are allowed for each pesticide. Chemical pesticides that are determined to be hazardous can be banned completely. Approved chemicals can only be used legally in accordance with their EPA label guidelines. The act converts the product label of a pesticide or herbicide into a binding legal document. Landowners should carefully read and follow pesticide label instructions. Manufacturers are liable for damages caused only when a pesticide is used in accordance with label instructions. The Federal Environmental Pesticide Control Act has the following major provisions:

- All pesticides must be registered with the EPA.
- General-use pesticides are available to the general public, while restricted use pesticides are available only to certified individuals.
- Private applicators may use pesticides on their own or leased property, and commercial applicators may apply restricted-use pesticides for a fee.
- Both private and commercial applicators must meet minimum standards of competency.
- Misuse of pesticides is unlawful.
- Enforcement of regulations is delegated to designated state agencies, like the Alabama Department of Environmental Management.

Local Ordinances

In recent years there has been a proliferation of local ordinances regulating land use, especially regarding forestry. According to a recent survey conducted by researchers at the Southern Forest Experiment Station, 461 local ordinances now exist across the U.S. Fortunately for farm and forest owners in the South, most of these local ordinances apply only in the Northeast.

> ## Questions About Pesticides?
>
> The National Pesticide Telecommunications Network (NPTN) is a toll-free telephone service that provides accurate and prompt responses to questions about pesticides. Answers are given on the telephone or sent in the next day's mail. For more information call (800) 856-7378 or write the NPTN, Texas Tech University Health Sciences Center, School of Medicine and Community Health, Lubbock, TX 79430.

Local forestry ordinances have varying objectives that reflect concerns of local governments and their constituents. Often a single ordinance will have multiple objectives. Land-use and zoning ordinances can generally be placed into five categories based on the reason for establishing the ordinance. These include the following:

- Public property, health, safety, and welfare protection ordinances;
- Urban and suburban environmental protection ordinances;
- General environmental protection ordinances;
- Special feature, conservation, and habitat protection ordinances; and
- Forest land preservation ordinances.

The type of local forestry regulatory ordinances that have proven to be the most popular in the South are those directed at the protection of public property and motorists' safety. To find out if your county or city has an ordinance that restricts certain farm or forest management activities, contact your local county or city government office, or your local Alabama Forestry Commission office.

Cost-Share Recipient Responsibilities

Landowners who participate in federal programs are obligated to maintain the management practices for which the funds were received according to the terms of the contract. Cost-share recipients are also responsible for using cost-sharing funds in the manner for which

Sources of Regulatory Information and Assistance

State Agencies

Alabama Cooperative Extension System
Local County Offices
Will provide endangered species bulletins for pesticide spraying in counties where endangered species are known to occur.

Alabama Department of Conservation and Natural Resources, Game and Fish Division
Non-Game Program
64 N. Union Street
Montgomery, AL 36130
(334) 242-3486
Source of technical and educational assistance on endangered species in Alabama.

Alabama Department of Environmental Management
1751 Dickinson Drive
Montgomery, AL 36130
(334) 271-7700 (air and water quality) (334) 450-3222 (coastal zone)
State regulatory authority on air and water quality, and coastal zone land use.

Alabama Forestry Commission
513 Madison Avenue
Montgomery, AL 36130
Provides voluntary guidelines on Best Management Practices for forestry that includes wetlands and water quality protection. Also provides guidance on smoke management for prescribed burning.

City and County Governments

Can provide information on local ordinances that may affect land management for wildlife.

Federal Agencies

U.S. Army Corps of Engineers
Mobile District Permitting Section
(South/Central Alabama)
P.O. Box 2288
Mobile, AL 36628-001
(334) 694-3781

U.S. Army Corps of Engineers
Nashville District Permitting Office
(Northern Alabama)
P.O. Box 1070
Nashville, TN 37202-1070
(615) 736-5181
Federal regulatory authority on wetland delineation and permitting.

U.S. Environmental Protection Agency
345 Courtland Street, N.E.
Atlanta, GA 30365
(404) 347-4015 (wetlands and water quality)
(404) 347-3222 (air quality and coastal regulation)
Provides federal regulatory oversight to the Corps of Engineers for wetlands regulations. Also provides federal guidelines to the Alabama Department of Environmental Management on air and water quality as well as coastal regulation.

U.S. Fish and Wildlife Service
Southeast Regional Office
1875 Century Blvd.
Atlanta, GA 30345
(404) 679-4000

U.S. Fish and Wildlife Service
Division of Ecological Services
Daphne Field Office
P.O. Drawer 1190
Daphne, AL 36526
(334) 441-5181
Provides technical and educational assistance on endangered species.

JOE MCGLINCY

Normal forestry, farming and ranching operations, such as this timber harvest, are exempt from having to obtain a Section 404 permit from the Corps of Engineers if the activities comply with mandatory federal BMPs for forested wetlands.

they were intended, and not for something else. Some farm and forest landowners worry that if they accept cost-sharing monies they are more susceptible to federal regulations. In reality, federal and state laws apply regardless of whether you are receiving cost-sharing assistance or not. The responsibility for regulatory compliance is the same for cost-sharing recipients as it is for landowners who receive no cost-sharing assistance.

The federal government makes clear its intention that cost-share assistance programs are not to be regulatory in nature. The Cooperative Forestry Assistance Act of 1978, as amended in 1990, that authorized the Forestry Incentives Program (FIP), Forest Stewardship Program, the Stewardship Incentive Program (SIP), and the Urban and Community Forestry Assistance Program clearly states that the act does not authorize the federal government to regulate the use of private land or deprive landowners or states of any of their rights. Section 14 of the Cooperative Forestry Assistance Act states the following:

"This Act shall not authorize the Federal Government to regulate the use of private land or to deprive owners of land of their rights to property, unless such property rights are voluntarily conveyed or limited by contract or other agreement. This Act does not diminish in any way the rights and responsibilities of the States and political subdivisions of states."

The Swampbuster and Sodbuster programs are two federal regulatory programs that primarily affect agricultural land. Both programs are administered through the USDA Farm Services Agency and the USDA Natural Resources Conservation Service. The Swampbuster program prohibits all federal commodity program payments to landowners who put wetlands into production for commodity crops. Forest owners who plant wildlife food plots (especially when seeding plants that are used in agricultural production) in forested wetlands may also be considered in non-compliance and may lose cost-sharing assistance from federal programs. Before planting wildlife food plots in what might be a wetland site, be sure to first check with the local Natural Resources Conservation Service office.

The Sodbuster program prohibits all federal commodity program payments to landowners who remove permanent cover from highly erodible land and then put the land in commodity crop production. The intent of this program is to prevent soil erosion, to concentrate row crop production on the most productive lands, and to maintain wildlife habitat. Farm owners who convert marginal lands into agricultural production could find themselves in violation of the Sodbuster program and could forfeit current and future federal cost-sharing assistance. The Sodbuster program, however, does allow a 2 acre per farm exception to plant areas for non-commercial purposes.

The Sustainable Forestry Initiative

The American Forest and Paper Association (AF&PA), representing the forestry industry, has developed a Sustainable Forestry Initiative (SFI) to ensure sound stewardship of industrial forest lands without additional regulation. AF&PA has defined sustainable forestry as "managing our forests to meet the needs of the present without compromising the ability of future generations to meet their own needs by practicing a land stewardship ethic which integrates the growing, nurture, and harvesting of trees for useful products with the conservation of soil, air and water quality, and wildlife and fish habitat." As part of the SFI initiative, AF&PA member companies will follow a 12-point action plan that includes the following:

1. Broaden the practice of sustainable forestry.
2. Ensure prompt reforestation.
3. Protect water quality.
4. Enhance wildlife habitat.
5. Minimize the visual impact of harvesting.
6. Protect special sites.
7. Contribute to biodiversity.
8. Continue to improve wood utilization.
9. Continue the prudent use of forest chemicals to ensure forest health.
10. Foster the practice of sustainable forestry on all forest lands.
11. Publicize progress in reports.
12. Provide opportunities for public outreach and education.

For more information about the SFI, write AF&FA, 1111 19th Street, N.W., Washington, DC 20036 or call (800) 878-8878.

LANDOWNER RIGHTS AND RESPONSIBILITIES

Ownership of private property is a valued concept, especially here in the Southeast where over 75 percent of the forests and farms are owned by nonindustrial private landowners. The proliferation of environmental regulations that potentially limit what a landowner can do on his own land has caused concern among many forest and farm owners and private property rights advocates. For example, the American Farm Bureau (which represents farmers in the U.S.) and other private property rights groups claim that environmental laws, such as the Clean Water Act's Section 404 wetlands protection program and the Endangered Species Act, are eroding constitutionally protected private property rights.

Most landowners enjoy exclusive rights to their property, but these rights are not absolute. The granting of easements, leases, or a mortgage on the property are common examples of instances in which landowners may not enjoy exclusive rights to their property. There are four rights or powers that are reserved to society that are exercised by the government. These rights reserved to the public include the following:

1. The right to tax;
2. The right to take private property (with just compensation) for public use through condemnation;
3. The right to regulate or control the use of private property; and
4. The right to *escheat* (the reverting of property to the state when there are no heirs and no will).

The Fifth Amendment provides protection against the "taking" of private property for public benefit without just compensation. A **taking** occurs when the government usurps or uses private property to the extent that the individual owner can no longer use it. But the loss of some "rights" to a property does not mean in all cases that the property is less valuable or provides fewer gratifications to the owner. Zoning laws may limit ownership rights, but they can also provide security against land uses on neighboring tracts that lower property values. Zoning ordinances are most common in urban and suburban areas.

Increasingly, landowners have depended on the

courts as the last option to fight loss of property rights. Relying upon the expressed limitations contained in the Fifth Amendment that "nor shall private property be taken for public use without just compensation," the majority is precluded from trammeling the rights of the minority. The just compensation clause is used as a shield for protecting private property rights.

The Fifth Amendment, as interpreted for two centuries by the U.S. Supreme Court, provides for a reasonable and fair balance between public good and private property rights. The Supreme Court has recently applied the following guidelines when reviewing the impact of zoning or land-use regulations on private property:

- The government restriction must substantially advance legitimate state interests, and
- The restriction must not deny the owner the economically viable use of his or her land.

The court may also consider the extent to which the restriction frustrates the landowner's reasonable investment-backed expectations or what a landowner could reasonably expect to receive from his or her investment in the property.

Since 1987, the Supreme Court has ruled in four landmark cases that private property rights cannot be eliminated through government regulation without just compensation. The tide has been slowly turning in favor of the property owner. In addition, "takings" laws are being proposed in most states and Congress which seek to protect landowner rights from misapplication of current regulatory laws by overzealous government regulators.

No one can be certain how far the movement to broaden the powers of the public over private property will go or how successful the opposing efforts of proposed private property legislation will be. Legislative actions and court decisions will continue to respond to the public sentiment either for or against programs and regulations to direct land use. If a landowner is concerned about eroding property rights and the impact of land-use regulation, he or she should become informed on the latest developments and express concerns to elected officials or landowner associations and affiliates. Some of these include the Alabama Forestry Association; the Alabama Farmer's Federation; and the Alabama Wildlife Federation. By participation in voluntary programs like the Forest Stewardship program and the TREASURE Forest program, by following Alabama's BMPs, and by avoiding practices that work against the basic rights and interests of others in the community, forest and farm owners can demonstrate responsible land stewardship and should be able to counter further erosion of private property rights.

KEN WILLS/ALABAMA ENVIRONMENTAL COUNCIL

Forested sloughs and other forested wetland areas provide valuable wildlife habitat and should receive special consideration and protection. Before conducting timber and wildlife management practices in and around these areas, contact the Alabama Forestry Commission or other qualified professionals for advice.

RECREATIONAL ACCESS AND LIABILITY: WHAT LANDOWNERS SHOULD KNOW

In the past farmers and forest owners have not been overly concerned about tort laws and regulations that affect the use and management of their property. Increasing demands for recreation by the public have prompted many landowners to develop recreational access programs, such as fee hunting operations, which provide limited access to the public and help supplement landowner income.

One of the primary concerns of many landowners who are interested in establishing a fee hunting operation is liability. Landowners almost always ask, when deciding on a recreational alternative for their lands, "Am I liable for damages if someone gets hurt on my property?" In many cases liability concerns have been the deciding factor in whether or not private lands are opened to the public for recreation. It's understandable that some landowners are reluctant to allow access to their property because they fear liability. However, many of these concerns are more perceived than real since lawsuits against landowners for negligence are rare. This does not diminish the fact that the concerns of landowners are real and should be recognized and understood by those who allow use of their land for recreational purposes by others.

Landowners who allow recreational access can significantly reduce their anxiety and risk exposure by understanding their legal responsibilities to those using their lands, meeting those responsibilities, and developing a sound program of risk reduction that protects both the landowner and land user.

Landowner Responsibility

Landowners have a legal and moral responsibility to ensure that conditions on their property are safe for their guests. Landowners interested in developing recreational operations should be aware of their responsibilities to land-users. In most states in the Southeast, including Alabama, the level of landowner responsibility depends on who comes on the land.

For a landowner to be held liable for personal loss or injury, negligence must be proven. A landowner is most often and easily held liable for gross negligence or willful misconduct, such as setting traps aimed at deterring or harming trespassers. Beyond any intentional misconduct, the landowner must be proven to have breached the duty of reasonable care expected under the law. In determining a landowner's liability for injuries that may occur to someone on his property, the legal status of the visitor must first be determined. The duty of care owed to and expected by the land-user, and thus the landowner's liability, depends on whether the land-user is classified as a **trespasser**, **licensee**, or **invitee**. By law, the greatest degree of landowner responsibility is owed to guests categorized as invitees, with the least responsibility being owed to trespassers.

TRESPASSER

A trespasser enters land uninvited and without the consent of the landowner. Land-users in Alabama, such as hunters, must obtain written permission from the landowner before entering the property. Usually, landowners are only liable for trespasser injuries that result from willful misconduct. An example of willful misconduct would be if the landowner set booby-traps with the intention of causing harm and/or death to the trespasser. Landowner responsibility goes one step further to children who are knowingly trespassing. In this case, landowners are required to exercise reasonable care to eliminate any dangers that pose an unreasonable risk of death or serious bodily harm to trespassing children. For example, if a landowner knows there is an open well on the property and also knows trespassing children are playing near the well, the landowner will be liable for any injuries that occur to the children if they fall into the well. A key element excusing landowner liability is the lack of knowledge of the trespasser's presence.

LICENSEE

A licensee enters property with the permission of the landowner and is not required to pay a fee or render a service for the right of access. In other words, licensees enter property to further their own purposes, not the landowner's. Guests who are friends, business acquaintances, or family members are considered licensees. Landowners have a greater degree of responsibility to licensees than trespassers in that they have a duty to warn them of known dangers. For example, a landowner must warn visitors of a biting dog, an open pit, an abandoned well, or cable gates.

INVITEE

An invitee enters land for the benefit of the landowner by paying a fee or providing a service in exchange for the right of access. However, the hunter in a traditional "hunting lease" situation is often considered more like a licensee than an invitee. A hunter is usually considered to be an invitee where a fee is charged on a per-trip basis (as with shooting preserves). The responsibility of the landowner to the land-user increases, since a fee or service is required and the client assumes that

A Checklist to Reduce Landowner Liability

- Identify all known and potential hazards on the property. Guests should be given a plat (sketch of the property and structures) of the property that marks hazards and identifies property boundaries. If possible, landowners or their representatives should tour the property with guests.
- Develop written rules aimed at preventing accidents and protecting the property. Make sure all visitors are aware of the rules. Have them sign a statement that they have read and understand all rules.
- Have land-users sign a hold-harmless agreement (written release) prior to entering the property, stating that those using the land hold the landowner harmless of any consequences to the land-user while on his or her land. An example of a hold-harmless agreement can be found in the sample lease agreement in Appendix M. Written releases may be helpful as proof that an injured land-user assumed the risks of engaging in an activity. It is important to note, however, that hold-harmless agreements do not relieve landowners of liability associated with demonstrated negligence. **Release statements should not be relied upon as a substitute for providing reasonable care to guests and visitors.**
- Avoid single strand wire or cable gates. Make sure all gates are clearly marked.
- Require that hunters provide references to verify safe behavior and adherence to game laws. Local Conservation Enforcement Officers are a good source to check to see if potential lessees have a history of game law violations.
- Post property against trespassing and prosecute violators.
- Plainly mark and show safety zones ("no hunting" areas) around houses, buildings, livestock, etc.
- Do not tolerate unsportsmanlike behavior or alcohol while hunting.
- Require that guests obey all state and federal game laws and regulations and show proof of having attended an approved hunter safety course.

DENNIS HOLT

Good communications is the key to reducing landowner liability and promoting an enjoyable and mutually beneficial relationship between landowners and sportsmen.

- Encourage guests to exercise good judgement, common sense, and sportsmanlike conduct.
- Keep accurate records of all efforts made to reduce or eliminate known and potential risks to land-users. If a suit is filed, landowners will have an accurate record of the efforts that were taken to make conditions on their property safe.
- Continually monitor risk potentials and make efforts to reduce them.
- Consult legal and professional experts.
- Landowners should require hunting clubs to have liability insurance coverage to minimize exposure to loss from liability and other risks. Landowners should be listed on the policy as an additional insured party. The **Southeastern Wildlife Federation**, a wholly owned subsidiary of the Alabama Wildlife Federation, offers affordable liability coverage for hunting clubs, landowners, and hunt-for-hire operations. For more information write the Southeastern Wildlife Federation, P.O. Box 1109, Montgomery, AL 36102, or call (800) 822-9453 or (334) 832-9453.

Enforcement officers of the ADCNR's Game and Fish Division form the first line of defense against those who would violate the state's game laws. The support of landowners and sportsmen is critical to their success.

the property and other conditions are safe. Landowners engaged in a fee access operation must inspect their property for hidden dangers and make every effort to warn their clientele of all known hazards. If known dangers cannot be removed, the landowner must give adequate warning to the guests and explain where these hazards are located.

Reducing Landowner Liability

Reducing landowner liability involves developing and implementing a sound program of risk reduction. In a fee hunting operation, for example, landowners should inspect their property for hazards. Some hazards may include open wells, abandoned mines, unsafe structures, or dangerous domestic or farm animals. The owner must make every effort to eliminate these hazards. Known dangers that cannot be corrected should be identified and explained to sportsmen. In other words, every effort should be taken to make the hunting conditions and the property safe. Meeting these obligations, as defined by law, will reduce the exposure and potential liability of landowners in fee hunting operations. Liability for personal injury cannot be imposed upon the landowner without proof of negligence.

A key component of a risk reduction program is **"foreseeability"**—being able to anticipate potential problems and acting in advance to reduce or eliminate the occurrence of these problems. Foreseeability is a vital factor in reducing risks. Recognizing that landowners provide a valuable service to the public by allowing public access to their lands, most states like Alabama have enacted recreational use statutes that limit landowner liability for injuries to persons using the land for recreational purposes. These statues do not exempt landowners from injuries caused by willful and malicious activities, or the failure of the landowner to warn against known hazardous conditions.

OPERATION GAMEWATCH

Landowners in Alabama experiencing game law violations on their property have a reliable method of reporting violations to conservation authorities. Operation Gamewatch, a program sponsored by the Alabama Wildlife Federation, is designed to stop fish and game law violations in the state. Landowner and citizen involvement is the key to the success of this program.

The way the program works is quite simple. Anyone who observes a violation can call (800) 272-GAME, a statewide, toll-free, 24-hour hotline. The information reported remains completely anonymous and is passed on to conservation officers in the field. Information that

Regulations should serve as a balance between protecting the habitats of certain wildlife species, such as the rare Florida black bear, and ensuring the property rights of private landowners in Alabama.

Game law violators should be reported to Operation Game-Watch which is sponsored by the Alabama Wildlife Federation in cooperation with the Game and Fish Division of ADCNR.

leads to an arrest in one of six categories makes the reporting individual eligible to receive an anonymous reward. The minimum reward is $50 and the maximum reward is $2,500. Rewards are paid for reports of violations leading to an arrest in the following categories:

1. Big game, deer, and turkey;
2. Small game and furbearers;
3. Fish, both fresh and saltwater species;
4. Protected, endangered, and threatened species like black bear, gopher tortoise, and red-cockaded woodpecker;
5. Selling of fish and game; and
6. Intentional, malicious destruction of fish or game habitat.

Reward amounts are based upon the scope of the violation and are awarded after an arrest and conviction. The Operation Gamewatch Board of the Alabama Wildlife Federation administers reward payments after a careful review of the nature and impact of the poaching violation(s) and bases the amount of the reward on those factors.

Operation Gamewatch is financed by sportsmen and women, clubs, businesses, and private citizens in cooperation with the Alabama Wildlife Federation. Anyone interested in being a part of Operation Gamewatch may make a tax-deductible donation to the Alabama Wildlife Federation/OPERATION GAMEWATCH, 46 Commerce Street, Montgomery, AL 36104. For more information about Operation Gamewatch call the AWF office at (800) 822-WILD or (334) 832-WILD.

SUMMARY

One of the most challenging issues confronting natural resource managers is striking a reasonable balance between voluntary and regulatory approaches concerning the use and management of private forest and farmlands. Should government regulate land management practices that improve timber production and enhance wildlife habitat? If so, how much regulation is necessary to achieve reasonable public expectations without infringing on private property rights? These questions and others will no doubt be debated for some time to come. However, landowners should be aware of current regulations and guidelines that affect how they manage their lands for timber, wildlife, and other resources. In addition, farm and forest owners who are concerned about proposed regulations need to become involved in the process by letting their views be known about how proposed regulation will affect them. Finally, complying with voluntary land-use guidelines such as BMPs, and developing a land stewardship ethic will demonstrate landowners' commitment to and concern for the land and the environment.

Appendixes

Appendix A: References and Suggested Reading

Chapter 1 Alabama's Landscape

Albrecht, W. A. "Soil Fertility and Wildlife—Cause and Effect," *Transactions of the North American Wildlife Conference*, 9 (1944), 19–28.

Bailey, J. A. *Principles of Wildlife Management*. New York: John Wiley & Sons, 1984.

Crawford, B. T. "Wildlife Sampling by Soil Types," *Transactions of the North American Wildlife Conference*, 11 (1946), 357–364.

Dodd, D. B. *Historical Atlas of Alabama*. Alabama University: The University of Alabama Press, 1974.

Alabama Forestry Commission. *Forests of Alabama*. Montgomery, AL.

Hill, E. P. "Litter Size in Alabama Cottontails as Influenced by Soil Fertility," *Journal of Wildlife Management*, 36 (1972), 1199–1209.

Hodgkins, E. J. (ed.). *Southeastern Forest Habitat Regions based on Physiography*. Auburn, AL: Auburn Forestry Departmental Series No. 2, Agricultural Experiment Station, Auburn University, 1965.

Hodgkins, E. J., M. S. Golden, and W. F. Miller. *Forest Habitat Regions and Types on a Photomorphic-Physiographic Basis: A Guide to Forest Site Classification in Alabama-Mississippi*. Southern Cooperative Series No. 210, Auburn, AL:, Alabama Agricultural Experiment Station, 1979.

Joiner, H. M. *Alabama's History: The Past and Present*. Athens, AL: Southern Textbook Publishers, Inc., 1980.

Johnson, W. E and L. R. Sellman. *Forest Cover Photo-interpretation Key for the Piedmont Forest Habitat Region in Alabama*. Auburn, AL: Department of Forestry, Auburn University, 1974.

Johnson, W. E. and L. R. Sellman. *Forest Cover Photo-interpretation Key for the Ridge and Valley Forest Habitat Region in Alabama*. Auburn, AL: Department of Forestry, Auburn University, 1977.

Johnson, W. E and L. R. Sellman. *Forest Cover Photo-interpretation Key for the Cumberland Plateau Forest Habitat Region in Alabama*. Auburn, AL: Department of Forestry, Auburn University, 1979.

Klein, P. H., D. B. Dodd, and W. S. Harris. *Twentieth Century Alabama—Its History and Geography*. Montgomery, AL: Clairmount Press, 1993.

Norrello, R. J. II. *The Alabama Story*. Tuscaloosa, Alabama: The Yellowhammer Press, 1993.

Norrell, R. J. *The Making of Modern Alabama: State History and Geography*. Tuscaloosa, AL: The Yellowhammer Press, 1993.

Chapter 2 Developing a Management Plan

Alabama Environmental Council Publication. *Landowners: Don't Miss Out!—The Many Ways to Benefit from Your Forest*. Birmingham, AL: Alabama Environmental Council.

Decker, D. J., T. W. Seamans, W. Kelley, and R. R. Roth. *Wildlife and Timber from Private Lands: A Landowners' Guide to Planning*. Ithaca, NY: Cornell University, 1990.

Jackson, J. J., G. D. Walker, R. L. Shell, and D. Heighes. *Managing Timber and Wildlife in the Southern Piedmont*. Univ. of Georgia: Cooperative Extension Bulletin, 1981.

Mueller, B. S. *Developing Habitat Management Plans—A Consultant's Perspective*, Quail IV. Tall Timbers Research Station, 1997.

Star, J. and J. Estes. *Geographic Information Systems: An Introduction*. Englewood Cliffs, NJ: Prentice Hall, 1990.

Chapter 3 Wildlife Habitat Management

Barnes, T. G., L. R. Kiesel, and J. R. Martin, (eds.). *Private Lands Wildlife Management: A Technical Guidance Manual and Correspondence Course*. Lexington, Kentucky: Cooperative Extension Service, 1992.

Hassinger, J. and J. Payne. *Dead Wood for Wildlife in Pennsylvania Woodlands*. The Pennsylvania State University Cooperative Extension Service Publ. Number 7.

Hunter, M. L., Jr. *Wildlife, Forests, and Forestry: Principles of Managing Forest for Biological Diversity*. Englewood, Cliffs, NJ: Regents/Prentice Hall, 1990.

Payne, N. F., and F. C. Bryant. *Techniques for Wildlife Habitat Management on Uplands*. New York: McGraw-Hill, Inc., 1994.

Robinson, W. L., and E. G. Bolen (eds.). *Wildlife Ecology and Management*. Second edition. New York: MacMillan Publishing Co.

Yoakum, J. et al. "Habitat Improvement Techniques," S. D. Schemnitz, (ed.) *Wildlife Management Techniques Manual*. Fourth edition. Washington, D.C: The Wildlife Society (1980), 329–404.

Chapter 4 Managing Forests

Adams, N. E. Jr. and C. A. DeYoung. "Response of Deer to Rotation Cattle Grazing Systems," *Proceedings of The Third Annual Meeting of the Southeast Deer Study Group* (1980), 1.

Alabama Forestry Commission Publication. *Seedling Care and Reforestation Standards*, 1991.

Alabama Forestry Planning Committee Publication. *Considerations for Forest Management on Alabama Soils*, 1993.

Allen, J. A., and H. E. Kennedy, Jr. *Bottomland Hardwood Reforestation in the Lower Mississippi Valley*, Slidell, LA: U.S. Dept. of Interior, Fish and Wildlife Service, National Wetland Research Center; New Orleans, LA: USDA Forest Service, Southern Forest Experimental Station, 1989.

Bagwell, C. "Selective Harvesting–The Regeneration Alternative," *Forest Farmer 29th Edition Manual*, 52(3) (1993), 55–56.

Baker, J. B. and W. M. Broadfoot. "A Practical Field Method of Site Evaluation for Commercially Important Southern Hardwoods," *USDA Forest Service General Technical Report SO-26*. New Orleans, LA: Southern Forest Experiment Station, 1979.

Bailey, M. A. "Fire and Our Natural Heritage: Rare Plants, Animals, and Natural Communities." *Alabama's TREASURED Forests*, 11(1) (1992), 28–30.

Barnes, T. G., L. R. Kiesel, and J. R. Martin, (eds.). *Private Lands Management: A Technical Guidance Manual and Correspondence Course*. Lexington: Kentucky Cooperative Extension Service, 1992.

Buckner, J. L. and J. L. Landers. *A Forester's Guide to Wildlife Management in Southern Industrial Pine Forests*. Technical Bulletin No. 10, Bainbridge, GA: International Paper Co., 1980.

Buckner, J. L. et al. "Wildlife Food Plants Following Preparation of Longleaf Pine Sites in Southwest Georgia." *Southern Journal of Applied Forestry*, 3(2) (1979), 56–69.

Burk, J. D. et al. "Wild Turkey Use of Streamside Management Zones in Loblolly Pine Plantations." *Proceedings of the National Wild Turkey Symposium*, 6 (1990), 84–89.

Burns, R. M. and B. H. Honkala. *Silvics of North America*. Agricultural Handbook 654. Washington, DC: USDA Forest Service, 1990.

Bushman, E. S., and G. D. Therres. *Habitat Management Guidelines for Forest Interior Breeding Birds of Coastal Maryland*. Wildlife Technical Publication 88-1. Maryland Forestry, Parks and Wildlife Service, 1988.

Byrd, N. A. *A Forester's Guide to Observing Wildlife Use of Forest Habitat in The South*. 1981edition. USDA Forest Service Southern Region, Forestry Report RB-FR5, 1981.

Bryant, F. C. , F. S. Guthery, and W. M. Webb. "Grazing Management in Texas and its Impact on Select Wildlife," *Symposium on Wildlife—Livestock Relationships*. For., Wildl., and Range Exp. Stn., Moscow: Univ. Idaho (1982), 94–112.

Chabreck, R. H. and R. H. Mills, (eds.). *Integrating Timber and Wildlife Management in Southern Forest*. Annu. For. Symp. 29, Baton Rouge, LA: Louisiana State University, 1980.

Cambre, T. "Let's get started! Hardwood Regeneration." *Alabama's TREASURED Forest*, 4(3), 6–7.

Chritton, C. A., "Effects of Thinning A Loblolly Pine Plantation on Non-game Populations in East Texas," M.S.F. thesis, Stephen F. Austin State University, Nacogdoches, TX , 1988.

Conner, R. N. *Injection of 2.4-D to Remove Hardwood Mid-story within Red-cockaded Woodpecker Colony Areas*. USDA Forest Service Research Paper, SO-251, 1989.

Crawford, H. S., C. L. Kucera, and J. H. Ehrenreich. *Ozark Range and Wildlife Plants*. U.S. Department of Agriculture, Forest Service Agriculture Handbook No. 356, 1969.

Cushwa, C. T. et al. *Burning Clear-cut Openings in Loblolly Pine to Improve Wildlife Habitat*. Georgia Forestry, Research Council Paper. No. 62, 1969.

Decker, D. J. et al. *Wildlife and Timber From Private Lands: A Landowner's Guide to Planning*. Cornell Cooperative Extension Publication, Information Bulletin 193, 1983.

Dickson, J. G. "Streamside Zones and Wildlife in Southern U.S. Forests," R.E. Gresswell, B. A. Barton, and J. L. Kershner, (eds.). *Practical Approaches To Riparian Resource Management: An Educational Workshop*. Billings, MT: U.S. Bureau of Land Management (1989), 131–133.

Dickson, J. G., and J. H. Williamson. "Small Mammals in Streamside Management Zones in Pine Plantations." *Management of Amphibians, Reptiles, And Small Mammals in North America: Proceedings of the Symposium*. Edited by R. C. Szarok, K. E. Severson, and D. R. Patton, tech. coords. USDA For. Serv. Gen. Tech. Rep. RM-166 (1988), 375–378.

Dills, G. G. "Effects of Prescribed Burning on Deer Browse." *Journal of Wildlife Management*, 34 (1970), 540–545.

Dudderer, G. "How Harvest Methods Affect Wildlife." *American Forest*, 93 (9/10) (1987), 23–77.

Duryea, M. and P. M. Dougherty, *Forest Regeneration Manual*. Dordrecht, Boston: Kluwer Academic Publishers, 1991.

Dyess, G. J., M. K. Causey, and H. L. Stribling. "Effects of Fertilization on Production and Quality of Japanese Honeysuckle," *Southern Journal of Applied Forestry*, 18(2) (1994), 68–71.

Easley, J. M. "Plan Your Management Around Control Burning." *Alabama's TREASURED Forests*. 5(1) (1986), 24–25.

Enge, K. M., and W. R. Marion, "Effects of Clear-cutting and Site Preparation on Herpetofauna of a North Florida Flatwoods." *Forest Ecology Management*, 14(3) (1986), 177–192.

Felix, A. C., III, T. L. Sharik, and B. S. McGinnes. "Effects of Pine Conversion on Food Plants of Northern Bobwhite Quail, Eastern Wild Turkey, and White-tailed Deer in the Virginia Piedmont." *Southern Journal of Applied Forestry*, 10(1) (1986), 47–52.

Forestry, Wildlife, and Habitat in the East (an annotated bibliography). Technical Bulletin No. 651, National Council of the Paper Industry for Air and Stream Improvement, Inc., 1986–1990.

Franklin, R. M. *Stewardship of Longleaf Pine Forests: A Guide for Landowners*. Longleaf Alliance Report Number 2. Andalusia, AL: The Longleaf Alliance, Solon Dixon Forestry Education Center, 1997.

Guide to Regeneration of Bottomland Hardwoods. U.S. Forest Service General Technical Report SE-76. Asheville, NC: Southeastern Forest Experiment Station.

Gullion, G. W. "Forest—Wildlife Interactions," *Introduction to Forest Science*. Second edition. R. A. Young and R. L. Giese (eds.) Canada: John Wiley and Sons (1990), 349–383.

Guthery, F. S. et al. "Using Short Duration Grazing to Accomplish Wildlife Habitat Objectives." *Can Livestock Be Used as a Tool to Enhance Wildlife Habitat?* K. E. Severson (ed.) tech. coord., U.S. For. Serv. Gen. Tech. Rep. RM-194 (1990), 41–45.

Guynn, D. G., Jr. "Consider Wildlife Habitat in Forestry Decisions." *Forest Farmers 25th Manual Edition*, 44(5) (1985), 26–28.

Ham, D. L and D. C. Guynn, (eds.). "Integrating Wildlife Considerations and Forest Management." *Proceedings of the Third Annual Forestry Forum*, Clemson University, 1983.

Hamlin, L. W., W. E. Balmer, and D. H. Sims. *Managing the Family Forest in the South*. USDA Forest Service Management Bulletin R8-MB 1, 1993.

Harlow, R. F. and D. H. Van Lear. *Silvicultural Effects on Wildlife Habitat in the South: An Annotated Bibliography 1980–1985*. Technical Paper No. 17, Department of Forestry, Clemson University, 1987.

Harlow, R. F., and D. H. Van Lear. "Effects of Prescribed Burning on Mast Production in the Southeast." *Proceedings of the Workshop on Southern Appalachian Mast Management*. Edited by C. E. McGee, University of Tennessee Department of Forestry and USDA Forest Service, Knoxville, TN (1989), 54–65.

Hobson, S. S., J. S. Barclay, and S. H. Broderick. *Enhancing Wildlife Habitats: A Practical Guide for Forest Landowners*. Northeast Regional Agricultural Engineering Service, 1993.

Howard, R. J. and J. A. Allen. "Streamside Habitats In Southern Forested Wetlands: Their Role and Implication for Management." *Proceedings of the Symposium: The Forested Wetlands of the Southern United States*. D. D. Hook and R. Lea (eds.), USDA For. Serv. Gen. Tech. Rep. SE-50 (1989), 97–106.

Hunter, L. M., Jr. *Wildlife, Forests, and Forestry: Principles of Managing Forests For Biological Diversity*. Englewood Cliffs, NJ: Regents/Prentice Hall, 1990.

Huntley, J. C. "Importance and Management of Mast Species Other Than Oaks," *Proceedings of the Workshop Southern Appalachian Mast Management*. C. E. McGee (ed.) Knoxville, TN: University of Tenneesee Department of Forestry, and USDA Forest Service (1989), 74–82.

Hurst, G. A. "Forestry Chemicals and Wildlife Habitat," *Forest Farmer*, Vol., 48(4) (1989), 10–11.

Hurst, G. A. "SMZs for Wildlife," *Alabama's TREASURED Forests*, 11(1) (1992), 8–9.

Hurst, G. A., J. J. Campo, and M. B. Brooks. "Deer Forage in Burned and Burned—thinned Pine Plantation," *Proceedings of the Annual Conference of the Souteastern Association of Fish and Wildlife Agencies*, 34 (1980), 476–481.

Hurst, G. A. and R. C. Warren. *Enhancing White-tailed Deer Habitat of Pine Plantations by Intensive Management*. Technical Bulletin 107. Mississippi Agricultural and Forestry Experiment Station, 1981.

Hyde, K. J., C. A. DeYoung, and A. Garza. "The Effects of Short Duration Grazing on White-tailed Deer," *Proceedings of The Tenth Annual Meeting of the Southeast Deer Study Group*. Gulf Shores, AL (1987), 21.

Hyman, L. "Growing Loblolly Pine in Alabama." *Alabama's TREASURED Forests*, 5(4) (1986), 13–15.

Jackson, J. J. et al. *Managing Timber and Wildlife in the Southern Piedmont*. University of Georgia Cooperative Extension Publication Bulletin 845, 1981.

Jackson, J. "Wildlife And Timber—Can We Help Small Landowners Manage Them Together?" *Tops*. 17(1) (1984), 26–35.

Jackson, J. "Integrating Wildlife Considerations with Hardwood Forest Management," *Southern Hardwood Management.* D. J. Moorehead and K. D. Coder (eds.), Cooperative Extension Services and USDA For. Serv. Management Bulletin R8-MB 67, 1994.

Jenks, J. A., et al. "Effect of Cattle Stocking Rate on Nutritional Characteristics of White-tailed Deer in Southern Pine Forest," *Proceedings of The Twelfth Annual Meeting of the Southeast Deer Study Group*, 13, Oklahoma City, OK, 1989.

Johnson, A. S. "Pine Plantation as Wildlife Habitat: A Perspective," *Managing Southern Forest for Wildlife and Fish: A Proceedings.* Edited by J. G. Dickson and O. E. Maughan. USDA Forest Service General Technical Report SO-65 (1987), 12–18.

Johnson, R. L. "Oak Seeding—It Can Work," *Southern Journal of Applied Forestry*, 5(1) (1981), 28–33.

Johnson, R. L. and R. Krinard. "Oak Regeneration by Direct Seeding," *Alabama's TREASURED Forest*, 4(3) (1985), 12–15.

Kammermyer, K. E., B. Moser, and J. Wentworth. "Population and Habitat Variables Related to Deer Condition in Northeast Georgia," *Proceedings of the Ninth Annual Meeting of the Southeast Deer Study Group 11*, Gatlinburg, TN, 1986.

Kerpez, T. A., and D. F. Stauffer. "Avian Communities of Pine-hardwood Forest in the Southeast: Characteristics, Management, And Modeling," *Proceedings of Pine-hardwood Mixtures.* T. A. Waldrop. (ed.) USDA Forest Service General. Technical Report SE-58 (1989), 156–169.

Landers, J. L. "Prescribed Burning for Managing Wildlife in Southeastern Pine Forests," *Managing Southern Forest for Wildlife And Fish: A Proceedings.* J. G. Dickson and O.E. Maughan (eds.), USDA For. Serv. Gen. Tech. Rep. SO-65 (1987), 19–27.

Lewis, C. E. and T. J. Harshbarger. "Burning and Grazing Effects on Bobwhite Foods in the Southeastern Coastal Plain," *Wildlife Society Bulletin*, 14(4) (1986), 445–459.

Locascio, C. G. et al. "Effects of Mechanical Site Preparation Intensities on White-tailed Deer Forage in the Georgia Piedmont," *Proceedings of the. Fifth Biennial Southern Silvicultural Research Conference*, J. H. Miller, Comp. USDA For. Serv. Gen. Tech. Rep. SO-74 (1989), 599–602.

Maguire, C. C. "Incorporation of Tree Corridors for Wildlife Movement in Timber Areas: Balancing Wood Production with Wildlife Habitat Management," *Journal of the Washington Academy of Sciences*, 77(4) (1987), 193–199.

Management of Southern Pine Forests for Cattle Production, General Technical Report, R8-GR4, U.S. Forest Service.

Masters, R. E. "Nutrient Response to Forest and Wildlife Management Practices in Oklahoma Quachita Mountains," *Proceedings of The Twelfth Annual Meeting of the Southeast Deer Study Group.* Oklahoma City, OK (1989), 19–21.

Mengak, M. T., D. H. Van Lear, and D. C. Guynn. "Impacts Of Loblolly Pine Regeneration on Selected Wildlife Habitat Components." Tech. Rep. SO-74 (1989), 612–618.

Miller, K. V. et al. *Herbicide And Wildlife Habitat: An Annotated Bibliography On The Effects Of Herbicides On Wildlife Habitat and Their Uses in Habitat Management.* USDA Forest Service Southern Region, Technical Publication R8-TP 13, 1990.

Mitchell, W. A., H. A. Jacobson, and M. Gray. "Preliminary Results on Cattle and Deer Relationships in the Southeastern Pine Forests," *Proceedings of The Second Annual Meeting of the Southeast Deer Study Group.* Mississippi State, MS (1979), 12–13.

Nelson, L. R. and R. L. Cantrell. *Herbicide Prescription Manual for Southern Pine Management.* Clemson University Cooperative Extension Service Publication EC 659, 1995.

Nietfeld, M. T. and E. S. Telfer. *The Effects of Forest Management Practices on Non-game Birds: An Annotated Bibliography.* Technical Report Series No. 112. Canadian Wildlife Service, Western and Northern Region, Alberta, 1990.

Patton, D. R. *Wildlife Habitat Relationships in Forested Ecosystems.* Timber Press, 1992.

Payne, N. F. and F. A. Copes, technical editors. *Wildlife and Fisheries Habitat Improvement Handbook.* USDA Forest Service Wildlife and Fisheries Administrative Report, 1988.

Payne, N. F. *Techniques for Wildlife Habitat Management.* J. H. Miller, Comp., *Proceedings of the. Fifth Biennial Southern Silvicultural Research Conference.* USDA For. Serv. Gen *of Wetlands.* Biological Resource Management Series. McGraw-Hill, Inc., 1992.

Payne, N. F. and F. C. Bryant. *Techniques for Wildlife Habitat Management of Uplands.* Biological Resource Management Series, McGraw-Hill, Inc., 1994.

Payne, N. F. and F. C. Bryant. *Techniques for Wildlife Habitat Management of Uplands.* New York: McGraw Hill, Inc., 1994.

"Popular Herbicides Used in Southern Forest." *Forest Farmer* 48(4) (1989), 6–7.

Putnam, J. A., G. M. Furnival, and J. S. McKnight. *Management and Inventory of Southern Hardwoods*, USDA Agricultural Handbook, 181, 1960.

Robinson, W. L. and E. G. Bolen (eds.) "Forest Management and Wildlife," *Wildlife Ecology and Management.* Second edition. New York: MacMillan Publishing Comp. (1989), 317–336.

Rudolph, D. C. and J. G. Dickson. "Streamside Zone Width and Amphibian and Reptile Abundance," *Southwestern Naturalist*, 35(4) (1990), 472–476.

Salwasser, H. and J. C. Tappeiner II. "An Ecosystem Approach to Integrated Timber and Wildlife Habitat Management," *Proceedings of the Forty-sixth North American Wildlife Conference* (1981), 473–487.

Severson, K. E., and P. J. Urness. "Livestock Grazing: A Tool to Improve Wildlife Habitat." M. Vavra, B. W. A. Laycock, and R. D. Pieper, (eds.), *Ecological Implications of Livestock Herbivory in the West.* Denver, CO: Society of Range Management, 1993.

Sims, D. "Hardwood Management Alternatives," *Alabama's TREASURED Forests*, 11(1) (1992), 14–15.

Sisson, D. C. and D. W. Speake. "Spring Burning for Wild Turkey Brood Habitat: An Evaluation," *Proceedings of the Annual Conference of the Southeastern Association of Fish and Wildlife Agencies*, 48 (1994), 134–139.

Smith, D. R. (ed.). *Proceedings of the Symposium on Management of Forest and Range Habitats for Non-game Birds.* USDA Forest Service General Technical Report WO-1, 1975.

Smith, R. L. "Is Clear-cutting Hardwoods Good for Wildlife?" *Proceedings of the Sixteenth Annual Hardwood Symposium of the Hardwood Research Council.* Hardwood Research Council, Memphis, TN (1988), 16–33.

Smith, D. M. *The Practice of Silviculture.* New York: John Wiley & Sons, Inc., 1962.

Solar, P. G. "Non-game Forest Bird Response to Timber Stand Improvement Practices in a Deciduous Forest." Master's thesis, West Virginia Univ., Morgantown, 1988.

Speake, D. W. "Even-aged Forest Management and Wildlife Habitat," *Proceedings of the Annual Conference of the Southeastern Association of Game and Fish Commissioners*, 24 (1970), 1–4.

Spurr, S. F., and B. V. Barnes. *Forest Ecology.* New York: John Wiley & Sons, 1973.

Stewart, S. "Managing Forests for Wildlife." Part One of a Two-Part Series. *Alabama's TREASURED Forests*, 12(3) (1993), 12–14.

Stewart, S. "Managing Forests for Wildlife." Part two of a Two-Part Series. *Alabama's TREASURED Forests*, 13(2)(1994), 22–24.

Stransky, J. J. and J. H. Roese. "Promoting Soft Mast for Wildlife in Intensively Managed Forests," *Wildlife Society Bulletin*, 12 (1984), 234–240.

Thill, R. E. "Managing Southern Pine Plantations for Wildlife," *Proceedings of the XIXth IUFRO World Congress*, Vol. 2 (1990), 58–68.

Wentworth, J. M. "Deer-Habitat Relationships in the Southern Appalachians." Ph.D. thesis, University of Georgia, Athens, 1989.

Wilson, P. "Planting Acorns by the Ton," *Alabama's TREASURED Forest*, 13(2) (1994), 27.

United States Forest Service. *A Guide for Prescribed Fire in Southern Forests.* USDA Forest Service Technical Publication R8-TP 11, 1989.

Yoakum, J., W. P. Dasmann, H. R. Sanderson, C. M. Nixon, and H. S. Crawford. "Habitat Improvement Techniques," *Wildlife Management Techniques Manual.* S. D. Schemnitz (ed.), The Wildlife Society. (1980), 329–404.

Young, J. A. and C. G. Young. *Seeds of Woody Plants in North America.* Portland, OR: Dioscorides Press, 1992.

Chapter 5–10 Wildlife Species

Alabama Wildlife Federation. *Dove hunting in Alabama: Information for the Hunter and Landowner.*

Allen, R., Jr. *History and Results of Deer Restocking in Alabama.* Alabama Department of Conservation and Natural Resources.

Allen, A. W., Y. K. Bernal, and R. J. Moulton. *Pine Plantations and Wildlife in the Southeastern United States: An Assessment of Impacts and Opportunities.* Information and Technology Rep. 3, Washington, DC: U.S. Dept. Interior, Natl. Biol. Surv., 1996.

Allen, R. H., Jr. and R. E. Waters. *Bobwhite Quail Management: Facts and Fiction.* Alabama Department of Conservation and Natural Resources, Division of Game and Fish, Bulletin No. 5, 1972.

Arner, D. H. *Wild Turkeys on Southeastern Farms and Woodlands.* USDA Leaflet 526, 1963.

Arner, D. H. , J. Baker, and T. Wesley. "Management of Beaver Ponds in the Southeastern United States." M.S. thesis. Mississippi State University, Mississippi State, MS, 1966.

Backs, S. E. *An Evaluation of Releasing First Generation (F1) Bobwhite Quail Produced from Wild Stock.* Indiana Department of Natural Resources, Pittman-Robertson Bulletin No. 14, 1982.

Bailey, R. W. and K. T. Rinell. *History and Management of the Wild Turkey in West Virginia.* West Virginia Department of Natural Resources, Division of Game and Fish. Bulletin No. 6, 1968.

Baskett, T. S., M. W. Sayre, R. E. Tomlinson and R. E. Mirarchi. *Ecology and Management of the Mourning Dove.* A Wildlife Management Institute Book. Harrisburg, PA: Stackpole Books, 1993.

Bateman, H. A. *The Wood Duck in Louisiana.* Louisiana Department of Wildlife and Fisheries, 1977.

Beasom, S. L. and M. D. Springer. "Response of Deer Body Quality to Prescribed Burning in Southeast Texas," *Proceedings of the Fourth Annual Meeting of the Southeast Deer Study Group*, Tallahassee, FL, 1981.

Bellrose, F. C. *Ducks, Geese, and Swans of North America.* Washington, DC: Wildlife Management Institute, 1976.

Bellrose, F. C. *Housing for Wood Ducks.* Illinois Nat. Hist. Survey Circular 45, 1955.

Bellrose, F. C. and D. J. Holm. *Ecology and Management of the Wood Duck.* Mechanicsburg, PA: Stackpole Books, 1994.

Beshears, W. W. *Wood Ducks in Alabama.* Special Report Number 4, Federal Aid Project W-35, Job 1-F, Alabama Department of Conservation and Natural Resources, 1974.

Bevill, V. *Game on Your Land Part 2: Wild Turkey.* South Carolina Wildlife and Marine Resources Department, 1978.

Blackburn, W. E., J. P. Kirk, and J. E. Kennamer. "Availability and Utilization of Summer Foods by Eastern Wild Turkey Broods in Lee County, Alabama." *Proceedings of the National Wild Turkey Symposium,* 3 (1975), 86–96.

Bookhout, T. A. (ed.). *Research and Management Techniques for Wildlife and Habitats.* Bethesda, MD: The Wildlife Society, 1994.

Brennan, L. A. "How Can We Reverse the Northern Bobwhite Decline?" *Wildlife Society Bulletin,* 19 (1991), 544–555.

Brenneman, R. (ed.). *Managing Openings for Wild Turkeys and Other Wildlife—A Planting Guide.* National Wild Turkey Federation, Edgefield, SC, 1991.

Brothers, A. and M. E. Ray, Jr. *Producing Quality Whitetails.* Laredo, TX: Fiesta Publishing Co., 1975.

Buechner, H. K. "The Evaluation of Restocking Pen-Reared Bobwhite," *Journal of Wildlife Management,* 14 (1950), 363–377.

Bull, J. and J. Farrand, Jr. *The Audubon Society Field Guide to North American Birds.* New York: Alfred A. Knopf, 1977.

Burger, L. W., Jr., M. R. Ryan, T. V. Dailey, and E. W. Kuvzejeski. "Reproductive Strategies, Success, and Mating Systems of Northern Bobwhite in Missouri," *Journal of Wildlife Management* 59(3) (1995), 417–426.

Burns, R. M. and B. H. Honkala. *Silvics of North America. Volume 2, Hardwoods.* Agricultural Handbook 654, Washington, DC: USDA Forest Service, 1990.

Cole, C. A., T. L. Serfass, M. C. Brittingham, and R. P. Brooks. *Managing Your Restored Wetland.* University Park, PA: Pennsylvania State University Cooperative Extension, 1996.

Colin, W. F. *Alabama Squirrel Investigations, 1949–1953.* Final Report, Project W-25-R, Montgomery, Alabama: Alabama Department of Conservation and Natural Resources, Game and Fish Division, 1957.

Dasmann, W. *If Deer Are to Survive.* Harrisburg, PA: Stackpole Co., 1971.

Davidson, W. R. and V. F. Nettles. *Field Manual of Wildlife Diseases in the Southeastern United States.* Southeastern Cooperative Wildlife Disease Study. Athens, GA: University of Georgia, 1988.

Davidson, W. R. and E. J. Wentworth. "Population Influences: Diseases and Parasites," *The Wild Turkey: Biology and Management,* J. G. Dickson (ed.) Harrisburg, PA: Stackpole Books, (1992), 101–118.

Davis, J. R. *Game Birds in Alabama: The Eastern Wild Turkey.* Montgomery, AL: Alabama Department of Conservation and Natural Resources, Number 3.

Davis, J. R. *Management for Alabama Wild Turkeys.* Montgomery, AL: Alabama Game and Fish Commission, Alabama Dept. Conservation and Natural Resources. Special Report Number 5, 1976.

Davis, J. R. *Management for Mourning Doves in Alabama.* Special Report Number 6, Montgomery, AL: Alabama Department of Conservation and Natural Resources, 1977.

Davis, J. R. *Gray Squirrel Management in Alabama.* Special Report No. 7, Montgomery, AL: Alabama Department of Conservation and Natural Resources, 1978.

Davis, J. R. *The White-tailed Deer in Alabama.* Special Report Number 8, Alabama Department of Conservation and Natural Resources, 1979.

Davis, J. R. "The Gray Squirrel," *Mammals in Alabama.* Special Report No. 5, Montgomery, AL: Alabama Department of Conservation and Natural Resources.

DeVos, T. "Quail Predators and their Management (part 1)," *Quail Unlimited Magazine.* 9(1) (1990).

DeVos, T. "Quail Predators and their Management (part 2)," *Quail Unlimited Magazine.* 9(3) (1990).

Devos, T. and B. S. Mueller. "Relocating Wild Quail Coveys: A Pilot Study," *Proceedings of Tall Timbers Game Bird Seminar,* C. Kyser, J. L. Landers, and B. S. Mueller (eds.), Tallahassee, FL (1989), 57–59.

Devos, T. and B. S. Mueller. "Reproductive Ecology of Northern Bobwhite in North Florida," *Quail III: National Quail Symposium.* K. E. Church and T. V. Dailey (eds.), Kansas Department of Wildlife and Parks, Pratt (1993), 83–90.

Devos, T. and D. W. Speake. "Effects of Released Pen-Raised Northern Bobwhites on Survival Rates of Wild Populations of Northern Bobwhites," *Wildlife Society Bulletin*, 23(2) (1995), 267–273.

Devos, T. "Quail Management," *Alabama Wildlife.* (Summer 1996), 26–27.

Dickey, C. *Bobwhite Quail Hunting.* Birmingham. AL: Oxmoor House, Inc., 1974.

Dickson, J. G. "Oak and Flowering Dogwood Production for Eastern Wild Turkeys," *Proceedings of the. National Wild Turkey Symposium*, 6 (1990), 90–95.

Dickson, J. G. editor. *The Wild Turkey: Biology and Management.* A National Wild Turkey Federation Book. Harrisburg, PA: Stackpole Books, 1992.

Ellsworth, D. L., R. L. Honeycutt, N. J. Silvy, M. H. Smith, J. W. Bickham, and W. D. Klimstra. "Genetic Structure of Reintroduced Wild Turkey and White-tailed Deer Populations," *Journal of Wildlife Management*, 58(4) (1994), 698–711.

Elman, R. *The Hunter's Field Guide to the Game Birds and Animals of North America.* New York: Alfred A. Knopf, 1974.

Everett, D. D., Jr. "Factors Limiting Populations of Wild Turkeys on State Wildlife Management Areas in North Alabama." Ph.D. thesis: Auburn University, AL (1982).

Everett, D. D., D. W. Speake, and W. K. Maddox. "Wild Turkey Ranges in Alabama Mountain Habitat," *Proceedings of the Annual Conference of the Southeastern Association of Fish and Wildlife Agencies*, 33 (1979), 233–238.

Everett, D. D., D. W. Speake, and W. K. Maddox. "Natality and Mortality of a North Alabama Wild Turkey Population," *Proceedings of the National Wild Turkey Symposium*,4 (1980), 117–126.

Everett, D. D., D. W. Speake, and W. K. Maddox. "Use of Rights-of-Way by Nesting Wild Turkeys in North Alabama," *Proceedings Symposium: Environmental Concerns in Rights-of-Way Management*, 2 (1981), 641–646.

Everett, D. D. , D. W. Speake, and W. K. Maddox. "Habitat Use by Wild Turkeys in Northwest Alabama," *Proceedings of the Annual Conference of the Southeastern Association of Fish and Wildlife Agencies*, 39 (1985), 479–488.

Exum, J. H., J. A. McGlincy, D. W. Speake, J. L. Buckner, and F. M. Stanley. "Evidence Against Dependence upon Surface Water by Turkey Hens and Poults in Southern Alabama," *Proceedings of the National Wild Turkey Symposium*, 5 (1985), 83–89.

Exum, J. H. , J. A. McGlincy, D. W. Speake, J. L. Buckner, and F. M. Stanley. *Ecology of the Eastern Wild Turkey in an Intensively Managed Pine Forest in Southern Alabama.* Tallahassee, FL:Tall Timbers Research Station Bull. 23, 1987.

Halls, L. K. (ed). "White-tailed Deer in the Southern Forest Habitat," *Proceedings of a Symposium.* Nacogdoches, Texas, 1969.

Halls, L. K., R. E. McCabe, and L. R. Jahn (eds.). *White-tailed Deer: Ecology and Management.* Wildlife Management Institute. Harrisburg, PA: Stackpole Books, 1984.

Hamrick, W. J. and J. R. Davis. "Summer Food Items of Juvenile Wild Turkeys," *Proceedings of the Annual Conference of the Southeastern Association of Game and Fish Commission* 25 (1971), 85–89.

Harlow, R. F. and F. K. Jones, Jr. *The White-tailed Deer in Florida.* Fla. Game and Fresh Water Fish Comm. Tech. Bull. No. 9, 1965.

Harlow, Richard F. and D. C. Guynn. *Southeastern Deer Study Group Annotated Bibliography: 1977 to 1990.* Department of Forest Resources, Clemson University, Clemson, SC., 1990.

Harlow, Richard F. and D. C. Guynn. *Southeastern Deer Study Group Annotated Bibliography: 1990 to 1995.* A Publication of the Samuel Roberts Noble Foundation, Ardmore, OK, 1995.

Hewitt, O. H. *The Wild Turkey and Its Management.* Washington, DC:The Wildlife Society, 1967.

Hill, E. P. *The Cottontail Rabbit in Alabama.* Auburn, AL: Agricultural Experiment Station., Auburn University Bull. 400, 1972.

Hillestad, H. O. "Movements, Behavior, and Nesting Ecology of the Wild Turkey in Eastern Alabama." G. C. Sanderson and H. C. Schultz, (eds.) *Wild Turkey Management: Current Problems and Programs.* Columbia: The Missouri Chapter of The Wildlife Society and University of Missouri Press (1973), 109–123

Hillestad, H. O. and D. W. Speake. 1970. "Activities of Wild Turkey Hens and Poults as Influenced by Habitat." *Proceedings of the Annual Conference of the Southeastern Association of Game and Fish Commission*, 24 (1970), 244–251.

Hon, T., D. P. Belcher, B. Mullis, and J. R. Monroe. "Nesting, Brood Range and Reproductive Success of an Insular Turkey Population," *Proceedings of the Annual Conference of the Southeastern Association of Fish and Wildlife Agencies*, 32 (1978), 137–149.

Hudson, P. J. and M. R. Rands (eds.) *Ecology and Management of Game Birds*. BSP Professional Books, London, 1988.

Hurst, G. A. and B. D. Stringer, Jr. "Food Habits of Wild Turkey Poults in Mississippi," *Proceedings of the National Wild Turkey Symposium*, 3 (1975), 76–85.

Hurst, G. A. "Habitat Requirements of the Wild Turkey on the Southeast Coastal Plain," P.T. Bromley and R. L. Carlton (eds.), *Proceedings Symposium: Habitat Requirements and Habitat Management for the Wild Turkey in the Southeast*. Elliston: Virginia Wild Turkey Foundation (1981), 2–13.

Hurst, G. A., and W. E. Poe. "Poult Foods Habits Studied," *Turkitat* 7(2) (1989), 4.

Hurst, G. A. "Foods and Feeding." J. G. Dickson (ed.) *The Wild Turkey: Biology and Management*. Harrisburg, PA: Stackpole Books (1992), 66–83.

Jackson, J. J. *Bring Ducks to Your Land*. Georgia Cooperative Extension. Athens, GA, 1980.

Johnson, B. C. *The Wood Duck*. Montgomery, AL: Alabama Department of Conservation and Natural Resources.

Johnson, A. S. "The Quail Problem, Research, and the International Quail Foundation." *Covey Rise*, 2(3) (1983).

Kennamer, J. E., J. R. Gwaltney, and K. R. Sims. "Habitat Preferences of Eastern Wild Turkeys on an Area Intensively Managed for Pine in Alabama," *Proceedings of the National Wild Turkey Symposium*, 4 (1980), 240–245.

Kennamer, J. E. and M. C. Kennamer. "Status and Distribution of the Wild Turkey in 1994," *Proceedings of the National Wild Turkey Symposium*, 7 (1996).

King, S. L., H. L. Stribling, and D. W. Speake. "Cottontail Rabbits Adapt Well to Cover Left from Prescribed Burning," *Highlights of Agricultural Research*, Vol. 38 No. 3. Auburn, AL: Auburn University, Alabama (1991).

Kroll, J. C. Second edition. *A Practical Guide to Producing and Harvesting White-tailed Deer*. Nacogdoches, TX: Institute for White-tailed Deer Management and Research, Center for Applied Studies in Forestry. College of Forestry. Stephen F. Austin State University, 1992.

Landers, J. L. and A. S. Johnson. *Bobwhite Quail Food Habits in the Southeastern United States with a Seed Key to Important Foods*. Tallahassee, FL: Tall Timbers Research Station Miscellaneous Publication No. 4, 1976.

Landers, J. L. and B. S. Mueller. *Bobwhite Quail Management: A Habitat Approach*. Tallahassee, FL: Tall Timbers Research Station and Quail Unlimited, 1989.

Laramie, H. A. Jr. "Device for Control of Problem Beavers," *Journal of Wildlife Management*, 27 (3) (1963), 471–476.

Lewis, J. C. *The World of the Wild Turkey*. Philadelphia and New York: J. B. Lippencott Co., 1973.

Lewis, M. C., T. L. Shaffer, and K. M. Kraft. "How Much Habitat Management is Needed to Meet Mallard Production Objectives?" *Wildlife Society Bulletin*, 23(1) (1995), 48–55.

Lord, R. D., Jr. *The Cottontail Rabbit in Illinois*. Dept. of Conserv. Tech. Bull., 1963.

Madison, J. *Gray and Fox Squirrels*. East Alton, IL: Bulletin of the Conservation Department, Olin-Mathieson Chemical Corporation, 1964.

Madison, J. *The Cottontail Rabbit*. East Alton, IL: Olin Mathieson Chem. Corp., 1959.

Madison, J. *The White-tailed Deer*. East Alton, IL: Winchester-Western Press, 1961.

Mahan, W. "Cottontail Rabbit," *Game on Your Land: Part 1 Small Game and Wood Duck*. South Carolina Wildlife and Marine Resources Department, Columbia, SC (1978), 50–63.

McBryde, G. L. "Economics of Supplemental Feeding and Food Plots for White-tailed Deer." *Wildlife Society Bulletin*, 23(3) (1995), 497–501.

McGlincy, J. A. *Wild Turkeys and Industrial Forest Management*. Wildlife Management Series No. 1, Bainbridge, GA: International Paper Co. (1991).

Metzler, R. and D. W. Speake. "Wild Turkey Poult Mortality Rates and their Relationship to Brood Habitat Structure in Northeast Alabama," *Proceedings of the National Wild Turkey Symposium*, 5 (1985), 103–111.

Miller, K. V. and L. M. Marchinton (eds.). *Quality Whitetails: Why and How of Quality Deer Management*. Mechanicsburg, PA: Stackpole Books, 1995.

Moore, W. H. "Managing Bobwhites in the Cutover Pinelands of South Florida." J. A. Morrison and J. C. Lewis, eds. *Proceedings of the First National Bobwhite Quail Symposium*. Stillwater, OK: Oklahoma State University (1972), 56–65.

Moore, G. C. and A. M. Pearson. *The Mourning Dove in Alabama*. Auburn, AL: Alabama Cooperative Wildlife Research Unit, 1941.

Moore, G. W. and V. Bevill. *Game on Your Land. Part 2—Turkey and Deer.* Columbia, SC: South Carolina Wildlife and Marine Resources Department, 1978.

Mueller, B. S. *Final Report—An Evaluation of Releasing Pen-raised Quail in the Fall.* Quail Unlimited 2(5), 1984.

Mueller, B. S., J. B. Atkinson, and T. DeVos. "Mortality of Radio-tagged and Unmarked Northern Bobwhites," C. J. Amlaner, Jr. (ed). *Proceedings of the Tenth International Symposium on Biotelemetry.* Fayetteville, AR: University of Arkansas (1988), 236–248.

Neely, W. W. and V. E. Davison. *Wild Ducks on Farmland in the South.* Washington, DC: U.S. Department of Agriculture Farmers Bulletin 2218, 1971.

Owen, C. N. "Food Habits of Wild Turkey Poults *(Meleagris gallopavo silvestris)* in Pine Stands and Fields and the Effects of Mowing Hayfield Edges on Arthropod Populations." Unpublished Master's thesis, Mississippi State University, Mississippi State, MS, 1976.

Payne, N. F. *Techniques for Wildlife Habitat Management of Wetlands.* New York, NY: McGraw-Hill, 1992.

Payne, N. F. and F. C. Bryant. *Techniques for Wildlife Habitat Management of Uplands.* New York, NY: McGraw-Hill, 1994.

Pedersen, E. K., W. E. Grant, M. T. Longnecker. "Effects of Red Imported Fire Ants on Newly-hatched Northern Bobwhite," *Journal of Wildlife Management,* 60(1) (1996), 164–169.

Pelham, P. H. and J. G. Dickson. "Physical Characteristics," *The Wild Turkey: Biology and Management,* (1992), 32–45.

Peoples, J. C., D. C. Sisson, and D. W. Speake. "Mortality of Wild Turkey Poults in Coastal Plain Pine Forests," *Proceedings of the Annual Conference of the Southeastern Association of Fish and Wildlife Agencies,* 49 (1995).

Peoples, J. C., D. C. Sisson, and D. W. Speake. "Wild Turkey Brood Habitat Use and Characteristics in Coastal Plain Pine Forests," *Proceedings of the National Wild Turkey Symposium,* 7 (1996), 89–96.

Quality Deer Management: Proven Techniques for Managing Your Deer Herd. Part I of the Quality Deer Management Association Video Series. Nashville, AR: AAI/Our Gang Productions.

Quality Deer Management: Enhancing Habitat on Your Hunting Lands. Part II of the Quality Deer Management Association Video Series. Nashville, AR: AAI/Our Gang Productions.

Reynolds, J. C., P. Angelstam, and S. Redpath. "Predators, their Ecology and Impact on Gamebird Populations," P.J. Hudson and R. W. Rands (eds.) *Ecology and Management of Game birds.* London: BSP Professional Books (1988), 72–97.

Robel, R. J., R. M. Case, A. R. Bisset, and T. M. Clement, Jr. "Energetics of Food Plots in Bobwhite Management," *Journal of Wildlife Management,* 38(4) (1974), 653–664.

Robinson, B. C. "Gray Squirrel" in *The Hunter's Encyclopedia,* R. R. Camp (ed.). Harrisburg, PA: The Stackpole Company, 1958, 118–122.

Roseberry, J. L., and W. D. Klimstra. *Population Ecology of the Bobwhite.* Carbondale, IL: Southern Illinois Univ. Press, 1984.

Rosene, W. *The Bobwhite Quail: Its Life and Management.* New Brunswick, NJ: Rutgers Univ. Press, 1969.

Rue, L. L. III. *The World of the White-tailed Deer.* New York, NY: J. B. Lippincott Co, 1962.

Russell, D. M. *The Dove Shooter's Handbook.* Tabor City, NC: Atlantic Publishing, 1995.

Sadler, K. C. *Rabbit Management.* Missouri Department of Conservation, 1980.

Sanderson, G. C. and H. C. Schultz, *Wild Turkey Management: Current Problems and Progress.* Columbia, MO: University of Missouri Press, 1973.

Sanderson, G. C. (ed.). *Management of Migratory Shore and Upland Game Birds in North America.* The International Association of Fish and Wildlife Agencies, 1977.

Schorger, A. W. *The Wild Turkey: Its History and Domestication.* Norman, OK: University of Oklahoma Press, 1966.

Seiss, R. S., P. S. Phalen, and G. A. Hurst. "Wild Turkey Nesting Habitat and Success Rate," *Proceedings of the National Wild Turkey Symposium,* 6 (1990), 18–24.

Semel, B. and P. W. Sherman. "Alternative Placement Strategies for Wood Duck Nest Boxes," *Wildlife Society Bulletin,* 23(3) (1995), 463–471.

Sheaffer, S. E. and R. A. Malecki. "Waterfowl Management: Recovery Rates, Reporting Rates, Reality Check," *Wildlife Society Bulletin,* 23(3) (1995), 437–440.

Simpson, R. C. *Certain Aspects of the Bobwhite Quail's Life History and Population Dynamics in Southwest Georgia.* Atlanta, GA: Tech. Bull. WL1, Georgia Dept. Nat. Resources, 1976.

Speake, D. W. "Effects of Controlled Burning on Bobwhite Quail Populations and Habitat on an Experimental Area in the Alabama Piedmont," *Proceedings of the Southeastern Association Game and Fish Commission,* 20 (1966), 19–32.

Speake, D. W., T. E. Lynch, W. J. Fleming, G. A. Wright, and W. J. Hamrick. "Habitat Use and Seasonal Movements of Wild Turkeys in the Southeast," *Proceedings of the National Wild Turkey Symposium*, 3 (1975),122–130.

Speake, D. W. "Predation on Wild Turkeys in Alabama," *Proceedings of the National Wild Turkey Symposium*, 4 (1980), 86–101.

Speake, D. W. R. Metzler, and J. McGlincy. "Mortality of Wild Turkey Poults in Northern Alabama," *Journal of Wildlife Management*, 49 (1985), 472–474.

Speake, D. W. and R. Metzler. *Wild Turkey Population Ecology on the Appalachian Plateau Region of Northeastern Alabama.* P-R Project W-44-6, Montgomery, AL: Alabama Department of Conservation, 1985.

Speake, D. W., and B. Sermons. *Reproductive Ecology of the Bobwhite in Central Alabama.* Fed. Aid in Wild. Restoration Final Rep., Proj. W-44-9, Study XII, 1987.

Stanford, J. A. "Bobwhite Quail Population Dynamics: Relationships of Weather, Nesting, Production Patterns, Fall Population Characteristics, and Harvest." J. A. Morrison and J. C. Lewis (eds.), *Proceedings of the First National Bobwhite Quail Symposium.* Stillwater-OK: Oklahoma State University, (1972), 115–139.

Stoddard, H. L. *The Bobwhite Quail: Its Habits, Preservation and Increase.* New York, NY: Charles Scribners Sons, 1931.

Stoddard, H. L. *Maintenance and Increase of the Eastern Wild Turkey on Private Lands of the Coastal Plain of the Deep Southeast.* Tall Timbers Research Station Bulletin No. l, 1963.

Strange, T. "Wood Duck," *Game on Your Land—Part 1 Small Game and Wood Duck.* Columbia, SC: South Carolina Wildlife and Marine Resources Department (1978), 18–23.

Stribling, H. L. *Mourning Dove Management in Alabama.* Circular ANR-513, Auburn, AL: Alabama Cooperative Extension Service, 1988.

Stribling, H. L. *Bobwhite Quail Management.* Circular ANR-511, Auburn, AL: Alabama Cooperative Extension Service, 1988.

Stribling, H. Lee. *Wild Turkey Management in Alabama.* Circular ANR-512, Auburn, AL: Alabama Cooperative Extension Service, 1988.

Stribling, H. L. *Wood Duck Management in Alabama.* Circular ANR-519, Auburn, AL: Alabama Cooperative Extension Service, 1988.

Stribling, H. Lee. *Growing Chufa for Wild Turkey.* Circular ANR-569, Auburn, AL: Alabama Cooperative Extension Service, 1990.

Stribling, H. L. *Attracting Waterfowl to Beaver Ponds.* Circular ANR-611, Auburn, AL: Alabama Cooperative Extension Service, 1991.

Stribling, H. L. *Bicolor for Bobwhites.* Circular ANR-627, Auburn, AL: Alabama Cooperative Extension Service.

Stribling, H. L. *Cottontail Rabbit Management.* Circular ANR-636, Auburn, AL: Alabama Cooperative Extension Service, 1991.

Stribling, H. L. *Planting Egyptian Wheat for Bobwhite Quail.* Circular ANR-635, Auburn, AL: Alabama Cooperative Extension Service, 1991.

Stribling, H. L. *Planting Partridge Pea for Bobwhite Quail.* Circular ANR-610, Alabama Cooperative Extension Service, Auburn University, AL, 1991.

Stribling, H. L. *Gray Squirrel Management.* Auburn, AL: Alabama Cooperative Extension Service circular ANR-768, 1992.

Tacha, T. C. and C. E. Braun (eds.). *Migratory Shore and Upland Game Bird Management in North America,* Washington, DC: The International Association of Fish and Wildlife Agencies, 1994.

Taylor, W. P. (ed.). *The Deer of North America.* Harrisburg, PA: Stackpole Co, 1956.

Teaford, J. W. *Eastern Gray Squirrel.* Waterways Experiment Station, U.S. Army Corps of Engineers, Technical Report EL-86–6, 1986.

Thill, R. R. "Deer-Cattle Diet Overlap in Louisiana: Preliminary Summer Findings," *Proceedings of the Second Annual Meeting of the Southeast Deer Study Group 12.*

Thill, R. R. and A. Martin, Jr. "Deer Diets on Grazed and Ungrazed Pine—Bluestem Range," *Proceedings of the Seventh Annual Meeting of the Southeast Deer Study Group* 11 (1984).

USDA Forest Service Handbook. *Wildlife Habitat Management Handbook,* Southern Region. U.S. Department of Agriculture. FSH 2609, 1971.

USDA Forest Service Handbook. *Silvics of North America.* Volume 2. Hardwoods. U.S. Department of Agriculture. FSH 654, 1990.

Vanderhoof, R. E. and H. A. Jacobson. "Effects of Agronomic Plantings on White-tailed Deer Antler Characteristics," *Proceedings of the Twelfth Annual Meeting of the Southeast Deer Study Group 11* (1989), 20.

Vangilder, L. D. and E. W. Kurzejeski. *Population Ecology of the Eastern Wild Turkey in Northern Missouri.* Wildlife Monographs No. 130, The Wildlife Society, 1995.

Vangilder, L. D. "Survival and Cause Specific Mortality of Wild Turkeys in the Missouri Ozarks," *Proceedings of the National Wild Turkey Symposium,* 7(1996), 21–31.

Vander Haegen, W. M., W. E. Dodge, and M. W. Sayre. "Factors Affecting Productivity in a Northern Wild Turkey Population," *Journal of Wildlife Management.* 52 (1988), 127–133.

Waer, N. A., H. L. Stribling, and M. K. Causey. "Establishing Cost and Productivity of Forage Planted for White-tailed Deer," *Highlights of Agricultural Research,* Auburn, AL: Auburn University, 1996.

Waters, R. E. *Some Things You Should Know about Wildlife in Alabama.* Alabama Forestry Planning Committee, 1994.

Webster, C. G. *Better Nest Boxes for Wood Ducks.* Washington, D. C.: U.S. Dept. of the Int., Fish and Wildlife Service, 1958.

Wengler, K. F. (ed.). *Forestry Handbook.* New York, NYL: John Wiley & Sons, 1984.

Williams, L. E., Jr. *Managing Wild Turkeys in Florida.* Florida: Real Turkeys Publishers, 1991.

Wheeler, R. J. *The Wild Turkey in Alabama.* Montgomery, AL: Alabama Dept. of Conservation Bull. 12, 1948.

White-tailed Deer Management (video). Auburn AL: Alabama Cooperative Extension Service, 1993.

Wildlife Habitat and the Hunter (video). Anoka, MN: Federal Cartridge Co., 1988.

Whitaker, J. O., Jr. *The Audubon Society Field Guide to North American Mammals.* New York, NY: Alfred A. Knopf, 1980.

Wood, G. W. *The Southern Fox Squirrel.* Brookgreen Journal. Vol. XVIII, No. 3, 1988.

Young, G. L., B. L. Karr, B. D. Leopold, and J. D. Hodges. "Effect of Greentree Reservoir Management on Mississippi Bottomland Hardwoods," *Wildlife Society Bulletin,* 23(3) (1995), 525–531.

Young, J. A. and C. G. Young. *Seeds of Woody Plants in North America.* Portland, OR: Dioscorides Press, 1992.

Chapter 11 Managing Furbearers

Alabama Furbearer Status and Harvest for 1994–1995. Montgomery, AL: Alabama Department of Conservation and Natural Resources.

Baker, O. E., III and D. B. Carmichael, Jr. *South Carolina's Furbearers.* South Carolina Wildlife and Marine Resources Department publication, 1989.

Burt, H. W. and R. P. Grossenheider. *A Field Guide to the Mammals—North America North of Mexico.* Third edition. New York, NY: Houghton Mifflin Company, 1980.

Butler, D. R. "The Reintroduction of the Beaver into the South," *Southeastern Geographer* 31(1991), 39–43.

Fifth Southeastern Furbearer Workshop Minutes, 1991.

Carrinton, R. *The Mammals.* Life Nature Library. New York, NY: Time-Life Books, 1963.

Deems, E. F. and D. Pursley (eds.). *North American Furbearers: A Contemporary Reference.* International Association of Fish and Wildlife Agencies, 1983.

Gorham, J. R., K. W. Hagen, and R. K. Farrell. *Mink: Diseases and Parasites.* Agricultural Handbook No. 175. USDA, 1972.

Henderson, R. F., E. K. Boggess, and B. A. Brown. *Understanding the Coyote.* Manhattan, KS: Cooperative Extension Service, Kansas State University, 1977.

Hill, E. P. *Trapping Beaver and Processing Their Fur.* Zoology—Entomology Departmental Series No. 1. Auburn, AL: Auburn University, 1974.

Hill, E. P., P. W. Sumner, and J. B. Wooding. "Human Influences on Range Expansion on Coyotes in the Southeast," *Wildlife Society Bulletin,* 15 (1987), 521–524.

Hygnstrom, S. E., R. M. Timm, and G. E. Larson (eds.). *Prevention and Control of Wildlife Damage.* Lincoln, NE: University of Nebraska Cooperative Extension Service, 1994.

Johnson, A. S. *Biology of the Raccoon (*Procyon lotor*) in Alabama.* Auburn Univ. Bulletin No. 402, Auburn, AL: Agric. Exp. Sta., 1970.

Minser, W. G., III and M. R. Pelton. *Impact of Hunting on Raccoon Populations and Management Implications.* Bulletin 612, Knoxville, TN: The University of Tennessee, 1982.

Novak, M., J. A. Balser, M. E. Obbard, and B. Malloch. *Wild Furbearer Management and Conservation in North America.* Ontario Ministry of Nat. Res., 1987.

Rue, L. L. *The World of the Beaver.* Philadelphia, PA: J. B. Lippincott Co., Living World Books, 1964.

Sanderson, G. C. "Raccoon." M. Novak, J. Baker, M. Obbard, and B. Mallochs, (eds.). *Wild Furbearer Management and Conservation in North America.* Ontario Ministry of Natural Resources, (1987), 486–499.

Southeastern Cooperative Wildlife Disease Study Publication. *Thoughts about Raccoon Restocking.* 3 pp.

Webster, W. D., J. F. Parnell, and W. C. Biggs, Jr. *Mammals of the Carolinas, Virginia, and Maryland.* Chapel Hill, NC: The University of North Carolina Press, 1985.

Whitaker, J. O., Jr. *National Audubon Society Field Guide to North American Mammals.* New York, NY: Alfred A. Knopf, 1996.

Chapter 12 Managing for Non-game

Behler, J. L. and F. W. King. *The Audubon Society Field Guide to North American Reptiles and Amphibians.* New York, NY: Alfred A. Knopf, 1979.

Black Bear Conservation Committee. *Black Bear Management Handbook for Louisiana, Mississippi, Southern Arkansas, and East Texas.* Baton Rouge, LA: BBCC, 1996.

Blaustein, A. R. "Chicken Little or Nero's Fiddle? A Perspective on Declining Amphibian Populations," *Herpetologica,* 50 (1994), 85–97.

Brainerd, J. W. *The Nature Observer's Handbook.* Chester, CT: Globe Pequot Press, 1986.

Burt, W. H. and R. P. Gorssenheider. *A Field Guide to the Mammals.* Boston, MA: Houghton Mifflin, 1976.

Dodd, K. "The Status of Red Hills Salamander, *Phaeognathus hubrichti,* Alabama, USA, 1976–1988," *Biological Conservation,* 55 (1991), 57–75.

Dunson, W. A., R. L. Wyman, and E. S. Corbett. "A Symposium on Amphibian Declines and Habitat Acidification," *Journal of Herpetology.* 26 (1992), 349–352.

Finch, D. M. and P. W. Stangel (eds.). *Status and Management of Neotropical Migratory Birds,* 1993; 1992 September 21–25; Estes Park, CO. Gen. Tech. Rep. RM-229. Fort Collins, CO: U.S. Department of Agriculture, Forest Service, Rocky Mountain Forest and Range Experiment Station.

Hamel, P. B. *The Land Manager's Guide to Birds of the South.* Chapel Hill, NC: The Nature Conservancy, Southeastern Region, 1992.

Harris, L. "The Faunal Significance of Fragmentation of Southeastern Bottomland Forests." D. D. Hook and R. Lea (eds.). *Proceedings of the Symposium on the Forested Wetlands of the Southeastern United States.* USDA Forest Service GTR SE-50 (1988), 126–134.

Harvey, M. J. *Bats of the Eastern United States.* Arkansas Game and Fish Commission, U.S. Fish and Wildlife Service, and Tennessee Technological University. Cookeville, TN, 1992.

Mount, R. H. (ed.) *Vertebrate Wildlife of Alabama.* Auburn. AL: Alabama Agricultural Experiment Station, Auburn University, 1984.

Mount, R. H. (ed.). *Vertebrate Animals of Alabama in Need of Special Attention.* Auburn AL: Alabama Agricultural Experiment Station, Auburn University, 1986.

Mount, R. H. (reprint). *The Reptiles and Amphibians of Alabama,* Auburn. AL: The University of Alabama Press in cooperation with the Alabama Agricultural Experiment Station, 1996.

Payne, N. F. and F. C. Bryant. *Techniques for Wildlife Habitat Management of Uplands,* New York, NY: McGraw-Hill, Inc., 1994.

Pelton, M. R. 1982. "Black Bear," *Wild Mammals of North America: Biology, Management, and Economics.* J. A. Chapman and G.A. Feldhamer (eds.), 504–14. Baltimore: The John Hopkins University Press (1982).

Peterson, R. T. *A Field Guide to Birds.* Boston, MA: Houghton Mifflin Company, 1980.

Petranka, J. W. "Response to Impact of Timber Harvesting on Salamanders," *Conservation Biology,* 8 (1994), 302–304.

Petranka, J. W., M. P. Brannon, M. P. Hopey and C. K. Smith. "Effects of Timber Harvesting on Low Elevation Populations of Southern Appalachian Salamanders," *Forest Ecology and Management,* 67 (1994), 135–147.

Smith, D. R. Technical Coordinator. *Proceedings of the Symposium on Management of Forest and Range Habitats for Non-game Birds.* [May 6–9, 1975, Tucson, Ariz.] USDA Forest Service General Technical Report WO-1. Washington, DC: USDA Forest Service, 1975.

Staten, M. *Breeding Bird Field Manual: A Guide for Bottomland Hardwood Forest Managers,* Memphis, TN: Anderson Tully Company, 1994.

Van Manen, F. T. and M. R. Pelton. "A GIS Model to Predict Black Bear Habitat Use," *Journal of Forestry,* Vol. 95, No. 8 (1997), 7–12.

Whitaker, J. O. *The Audubon Society Field Guide to North American Mammals.* New York, NY: Alfred A. Knopf, 1980.

Migrant Birds: A Troubled Future?—A slide presentation explaining the plight of neotropical migrants and the Partners in Flight program. Produced by the Information and Education Working Group of Partners in Flight, the program can be ordered form the Crow's Nest Birding Shop, Cornell Lab of Ornithology, 159 Sapsucker Woods Road, Ithaca, New York, 14850 (607-254-2400). Price $50.00 plus $3.95 shipping.

Save the Birds: A Guide to Neotropical Migratory Bird Conservation—An easy-to-understand booklet listing dozens of bird conservation opportunities from the backyard to entire ecosystems. Produced by the Partners in Flight Information and Education Working Group, the guide is helpful to bird clubs and other grassroots groups working for habitat conservation. For a copy write the National Fish and Wildlife Foundation, 1120 Connecticut Avenue, NW, Suite 900, Washington, DC. 20036

Directory of Volunteer Opportunities for Birders—A listing of projects, surveys, and censuses in which birders can participate across the United States. Many of the projects involve neotropical migratory birds. Send $2.00 to American Birding Association, P.O. Box 6599, Colorado Springs, CO 80934.

Chapter 13 Threatened and Endangered Species

Bailey, M. A. "Fire and Our Natural Heritage: Rare Plants, Animals, and Natural Communities," *Alabama's TREASURED Forests*, Vol. 9, No. 1. (1992), 28–30.

Bean, M. J., S. G. Fitzgerald, and M. A. O'Connell. *Reconciling Conflicts Under the Endangered Species Act: The Habitat Conservation Planning Experience.* Washington, DC: World Wildlife Fund, 1991.

Cobb, C. "Living with the Endangered Species Act," *Forest Farmer*, Vol. 52, No. 3. (January/February 1993), 22–23.

Flynn, K. and C. V. Isaacson (compiled materials). *Total Forest Resource Management: Integrating Environmental Concerns. Continuing Education Resource Notebook.* Auburn, AL: Auburn University School of Forestry, 1992.

Greenwalt, L. A. "The Power and Potential of the Act," *in Balancing on the Brink of Extinction: The Endangered Species Act and Lessons for the Future.* K. A. Kohm (ed.). Washington, DC: Island Press (1991), 31–36.

Hoge, D. A. "The Gopher Tortoise," *Alabama's TREASURED Forest*, Vol. 5, No. 2 (1986), 29–30.

Hoge, D. A. "Endangered Species," *Alabama's TREASURED Forest.* Vol. 5, No. (1986), 26–28.

Irwin, L.L. and T. B. Wigley. "Conservation of Endangered Species—The Impact on Private Forestry," *Journal of Forestry*, Vol. 90, Num. 8 (1992), 27–30.

Johnson, R. and B. Wehrle. *Threatened and Endangered Species of Alabama: A Guide to Assist with Forestry Activities.* A cooperative publication between the Solon Dixon Forestry Education Center, Auburn University, U.S. Fish and Wildlife Service, Champion International Corporation, and Canal Wood Corporation, 1995.

Johnston, N. C. and H. S. Pond IV. "Silviculture Practices and Environmental Law," *The Alabama Lawyer* (January, 1992).

Kohm, K. A. (ed.). *Balancing on the Brink of Extinction: The Endangered Species Act and Lessons for the Future.* Washington, DC: Island Press, 1991.

Kulhavy, D. L., R. G. Hooper, and R. Costa (eds.). *Red-cockaded Woodpecker: Recovery, Ecology and Management.* Nacogdoches, TX: A Center for Applied Studies in Forestry Publication, Stephen F. Austin State University, 1995.

Mount, R. H. (ed.). *Vertebrate Wildlife of Alabama.* Auburn, AL: Alabama Agricultural Experiment Station, Auburn University, 1984.

Mount, R. H. (ed.). *Vertebrate Animals of Alabama in Need of Special Attention.* Auburn, AL: Alabama Agricultural Experiment Station, Auburn University, 1986.

Robinson, W. L. and E. G. Bolen. "Non-game and Endangered Wildlife, *Wildlife Ecology and Management.* New York , NY: Macmillan Publishing Co. (1984), 355–372.

The Red-cockaded Woodpecker. Information for Alabama Forest Landowners. Alabama Department of Conservation and Natural Resources Game and Fish Division, and the Alabama Natural Heritage Program publication, 1996.

Woehr, J. R. 1993. "A Proposal to Solve Alabama's Red-cockaded Woodpecker Dilemma," *Southeastern Wildlife* (Winter 1993), 20 and 24.

Chapter 14 Backyard Wildlife

Ajilusgli, G. *Butterfly Gardening for the South*, Dallas, TX: Taylor Publishing Co., 1990.

Barnes, T. G. (unpublished manuscript). *Birds, Bunnies, Butterflies, and Blackeyed Susans: Why Landscape to Enhance Wildlife?* Lexington, KY: Kentucky Cooperative Extension, University of Kentucky.

Barnes, T. G. *Bats: Information for Kentucky Homeowners.* Cooperative Extension publication For-48. Lexington, KY: University of Kentucky, 1991.

Behler, J. L. and F. W. King. *The Audubon Society Field Guide to North American Reptiles and Amphibians*, New York, NY: Alfred A. Knopf, 1979.

Butterfly Gardening. Callaway Gardens fact sheet. Pine Mountain, GA: Callaway Gardens.

Brainerd, J. W. *The Nature Observer's Handbook.* Chester, CN: Globe Pequot Press, 1986.

Brewer, J. and D. Winter. *Butterflies and Moths: A Companion to Your Field Guide.* New York, NY: Prentice Hall Press, 1986.

Burt, W. H. and R. P. Gorssenheider. *A Field Guide to the Mammals.* Boston, MA: Houghton Mifflin, 1976.

Cerulean, S., C. Botha, and D. Legare. *Planting a Refuge for Wildlife: How to Create a Backyard Habitat for Florida's Birds and Beasts.* Florida Game and Fish Commission Non-game Wildlife Program, U.S. Department of Agriculture Soil Conservation Service.

Corley, W. L., W. J. McLaurin, J. T. Midcap, and G. L. Wade. *Wildflowers.* Cooperative Extension Bulletin 994, Athens, GA: University of Georgia, 1994.

Dirr, M. A. *Manual of Woody Landscape Plants.* Champaign: Stipes Publishing Co., 1990.

Druse, Ken. *The Natural Garden*, New York, NY: Clarkson N. Potter, 1988.

Greenhall, A. M. *House Bat Management.* Resource Publication Number 143, Washington, DC: U.S. Fish and Wildlife Service, 1982.

Halpin, A. M. *Foolproof Planting.* Emmaus, PA: Rodale Press, 1990.

Harrison, G. H. *The Backyard Bird Watcher.* New York, NY: Simon and Schuster, Inc., 1979.

Harrison, G. H. and K. Harrison. *America's Favorite Backyard Wildlife.* New York, NY: Simon & Schuster, 1985.

Harvey, M. J. *Bats of the Eastern United States.* Arkansas Game and Fish Commission, U.S. Fish and Wildlife Service, and Tennessee Technological University. Cookeville, TN, 1992.

Henderson, C. L. *Landscaping for Wildlife,* Non-game Wildlife Program—Section of Wildlife, St. Paul, MN: Minnesota Department of Natural Resources, 1987.

Henderson, C. L. *Woodworking for Wildlife,* Non-game Wildlife Program—Section of Wildlife, St. Paul, MN: Minnesota Department of Natural Resources, 1992.

Klots, A. *A Field Guide to the Butterflies of North America.* Boston, MA: Houghton Mifflin, 1951.

Kress, S. *The Audubon Society Guide to Attracting Birds.* New York, NY: Charles Scribner's Sons, 1985.

Logsen, G. *Wildlife in Your Garden.* Emmaus, PA: Rodale Press, 1983.

Mahnken, J. *Hosting the Birds.* Pownal, VT: Storey Communications, 1989.

Martin, A. C., H. S. Zim and A. L. Nelson. *American Wildlife and Plants: A Guide to Wildlife Food Habits.* New York, NY: Dover Publications, 1951.

McElroy, T. P. *The New Handbook for Attracting Birds.* New York, NY: W.W. Norton and Company, 1985.

McKinley, M. *How To Attract Birds.* San Francisco, CA: Ortho Books, 1983.

Merilees, W. J. *Attracting Backyard Wildlife.* Stillwater, MN: Voyageur Press, 1989.

Mitchell, J. H. *A Field Guide to Your Own Back Yard.* New York, NY: W. W. Norton and Company, 1985.

Mount, R. H. (ed.). *Vertebrate Wildlife of Alabama.* Auburn, AL: Alabama Agricultural Experiment Station, Auburn University, 1984.

Peterson, R. T. *A Field Guide to Birds.* Boston, MA: Houghton Mifflin Company, 1980.

Perry, F. *Simon and Schuster's Complete Guide to Plants and Flowers.* New York: Simon and Schuster, 1974.

Philips, H. R. *Growing and Propagating Wild Flowers.* Chapel Hill, NC: University of North Carolina Press, 1985.

Pyle, R. M. *The Audubon Society Handbook for Butterfly Watchers.* New York, NY: Charles Scribner's Sons, 1984.

Robbins, C. S., B. Bruun, and H. S. Zim. *Birds of North America.* New York, NY: Golden Press, 1983.

Schneck, M. *Butterflies: How to Identify and Attract Them to Your Garden.* Emmaus, PA: Rodale Press, 1990.

Schneck, M. *Your Backyard Wildlife Garden: How to Attract and Identify Wildlife in Your Yard.* Emmaus, PA: Rodale Press, 1992.

Smyser, C. A. *Nature's Design.* Emmaus, PA: Rodale Press, 1982.

Still, S. M. *Manual of Herbaceous Ornamental Plants.* Champaign: Stipes Publishing Co., 1994.

Stokes, D., L. Stokes, and E. Williams. *The Butterfly Book.* Little, Brown and Company, 1991.

Stokes, D. and, L. Stokes. *The Bluebird Book.* Little, Brown and Company, 1991.

Tekulsky, M. *The Butterfly Garden.* Boston, MA: Harvard Common Press, 1985.

Terres, J. K. *Songbirds in Your Garden.* New York, NY: Harper and Row Publishers, 1984.

True, D. *Hummingbirds of North America: Attracting, Feeding and Photographing.* Albuquerque, NM: University of New Mexico Press, 1993.

Tuft, C. *The Backyard Naturalist.* Washington, DC: National Wildlife Federation, 1993.

Tuttle, M. D. *America's Neighborhood Bats.* Austin, TX: University of Texas Press, 1988.

Watts, D. L. *Backyard for Butterflies.* Mississippi Outdoors, 1995.

Wilson, W. *Landscaping With Wildflowers and Native Plants.* San Francisco, CA: Ortho Books.

Whitaker, J. O. *The Audubon Society Field Guide to North American Mammals.* New York, NY: Alfred A. Knopf, 1980.

Xerces Society/Smithsonian Institution. *Butterfly Gardening: Creating Summer Magic in Your Garden.* San Francisco: Sierra Club Books, 1990.

Chapter 15 Fish Pond Management

Amos, W. H. *The Life of the Pond (Our Living World of Nature).* New York, NY: McGraw-Hill Book Co., 1967.

Cichra, C. E., and L. T. Cooper. *Fee Fishing as an Economic Alternative for Small Farms.* Mississippi State, MS: Southern Rural Development Center, Mississippi State University, SRDC Series No. 116, 1989.

Everest, J. W., J. W. Jensen, M. Masser, and D. R. Bayen. *Chemical Weed Control for Lakes and Ponds.* Circular ANR-48. Alabama Cooperative Extension System, 1996.

Fishes of Alabama and the Mobile Basin. Fisheries Section of the Alabama Game and Fish Division, Alabama Department of Conservation and Natural Resources, 1997.

Jensen, J. W. *Using Grass Carp for Controlling Weeds in Alabama Ponds.* Circular ANR-452, Alabama Cooperative Extension System, 1996.

Horwitz, E. L. *Our Nation's Lakes.* Washington, DC: U.S. Environmental Protection Agency, 1980.

Masser, M. *Management of Recreational Fish Ponds in Alabama.* Auburn University Circular ANR-577, Alabama Cooperative Extension Service, 1993.

Summers, M. W. *Managing Louisiana Fish Ponds.* Monroe, LA: Louisiana Wildlife and Fisheries Commission, 1976.

Rushton, Y. and C. E. Boyd. "New Opportunities for Sport Fish Pond Fertilization," *Alabama Agricultural Experiment Station Highlights of Agricultural Research,* Vol. 42, No. 3 (1995).

Travnichek, R. J. and H. A. Clonts. "Recreational Fishing is Big Business in Alabama," *Highlights of Agricultural Research.* Alabama Agricultural Experiment Station, Vol. 43, No. 2 (1996).

Chapter 16 Supplemental Plantings

Ball, D. M., C. S. Hoveland, and G. D. Lacefield. *Southern Forages.* Potash & Phosphate Institute and the Foundation for Agronomic Research, 1991.

Brenneman, R. *Planting Bareroot Tree and Shrub Seedlings.* NWTF Wildlife Bulletin No. 12.

Brenneman, R. and T. Mills. *Tree Shelters.* NWTF Wildlife Bulletin No. 11.

Gothard, T. L. "Soil Compaction, Hardpans, and Tree Growth," *Alabama's TREASURED Forests.* 12(1), 24–25.

Halls, L. K. *Southern Fruit-producing Woody Plants Used by Wildlife.* Forest Service General Technical Report SO-16, Southern Forest Experiment Station; U.S. Department of Agriculture, 1997.

Miller, J. H. "Exotic Invasive Plants in Southeastern Forests," *Exotic Pests of Eastern Forests Conference,* Nashville, TN, 1997.

Nagy, G. J. and J. B. Haufler. "Wildlife Nutrition," *Wildlife Management Techniques Manual.* S. O. Schemnitz (ed.), The Wildlife Society, Inc., 1980.

Pedersen, J. F. and D. M. Ball. *Evaluation of Annual Clovers in South Alabama.* Circular 307. Auburn, AL: Alabama Agricultural Experiment Station, Auburn University, 1991.

Rainer, D. "Mineral Supplements Not The Answer," *Southeastern Wildlife.* Southeastern Wildlife Federation, (Summer 1994), 15.

Robinson, W. L. and E. G. Bolen. *Wildlife Ecology and Management.* Macmillan Publishing Company, 1984.

Seedling Care and Reforestation Standards. Alabama Forestry Commission Publication, 1991.

Stribling, H. L. *Wildlife Plantings and Practices.* Alabama Cooperative Extension Circular ANR-485.

Stribling, H. L. *Cool-season Food Plots for Deer.* Alabama Cooperative Extension Circular ANR-592.

Stribling, H. L. *Planting Sawtooth Oak for Wildlife.* Alabama Cooperative Extension Circular ANR-851.

Stribling, H. L. *Fertilizing Honeysuckle for Deer.* Alabama Cooperative Extension Circular ANR-887.

Waer, N. A., H. L. Stribling, M. K. Causey, and W. C. Johnson. "Annual Clover/Red Clover Plantings Provide High Quality Deer Feed," *Highlights of Agricultural Research,* Alabama Agricultural Experiment Station, Vol. 38, No. 2 (1991), 4.

Waer, N. A., H. L. Stribling, and M. K. Causey. "Establishment Cost of Forage Plantings for White-tailed Deer Relative to Production," *Proceedings of The Nineteenth Annual Southeast Deer Study Group Meeting,* Orlando, FL, 1995.

Waters, R. E. *Some Things You Should Know About Wildlife in Alabama.* Alabama Forestry Planning Committee Publication, 1994.

Chapter 17 Economic Considerations

Alabama Environmental Council publication. *Landowners: Don't Miss Out! The Many Ways to Benefit from Your Forest.*

"Companies That Offer Landowner Assistance," *Alabama's TREASURED Forest,* K. Gilliland (ed.), (Summer 1993).

"Comparison of Cost-share Programs," *Alabama's TREASURED Forests,* K. Gilliland (ed.), (Summer 1993), 14.

Decker, D. J. and G. R. Goff (eds.). *Valuing Wildlife: Economic and Social Perspectives.* Westview Press, Inc., 1987.

Dubois, M. R., K. McNabb, and T. J. Straka. "Costs and Cost Trends for Forestry Practices in the South," *Forest Landowner* (March/April), 1997, 7–13.

Forests of Alabama. Alabama Forestry Commission Publication, 1990.

Grafton, W. N., A. Ferrise, D. Colyer, D. K. Smith, and J. Miller. *Conference Proceedings: Income Opportunities for the Private Landowner Through Management of Natural Resources and Recreational Access,* Morgantown, WV: West Virginia University Extension Service (R.D. No. 740), 1990.

McKee, C. W. *Economics of Accommodating Wildlife, Managing Southern Forests for Wildlife and Fish—A Proceedings.* General Technical Report SO-65, New Orleans, LA: Southern Forest Experiment Station, USDA Forest Service, 1987.

McKee, C. W. "Timber and Wildlife Habitat Trade-offs," *Alabama's TREASURED Forests*, K. Gilliland (ed.), (Winter 1987), 11–13.

McKee, C. W. "Valuation of Wildlife Habitat by Forest Industry: One Firm's Approach," G. K. Yarrow and D. C. Guynn (eds.). *Proceedings of Fee Hunting on Private Lands in the South.* Clemson Coop. Ext. Serv., Clemson Univ. (1990), 107–116.

McKenzie, D. F. *A Wildlife Manager's Field Guide to the Farm Bill.* Wildlife Management Institute, Washington, DC, 1997.

Patterson, R. H., J. H. Stevens, and S. K. Nodine. *Forest Stewardship for Outdoor Recreation.* Clemson Extension pub. EB 147. Clemson, SC: Clemson Univ., 1996.

Siegel, W. C., W. L. Hoover, H. L. Haney, Jr., Karen Liu, and H. E. Burghart. *Forest Owners' Guide to the Federal Income Tax.* Agricultural Handbook No. 708. U.S.D.A. Forest Service. Washington, DC: Government Printing Office, 1995.

Small, S. J. *Preserving Family Lands.* Boston: Landowner Planning Center, 1993.

U.S. Department of the Interior, Fish and Wildlife Service and U.S. Department of Commerce, Bureau of Census. *1991 National Survey of Fishing, Hunting and Wildlife—Associated Recreation.* Washington, DC: U.S. Government Printing Office, 1993.

Yarrow, G. K., and D. C. Guynn, Jr. (eds.). *Fee Hunting on Private Lands in the South.* Conference Proceedings, Coop. Extension Publ. Clemson, SC: Clemson Univ.

Chapter 18 Conservation Ethics

Brooks, P. *The House of Life: Rachel Carson at Work.* Boston: Houghton Mifflin, 1972.

Carson, R. *The Sea Around Us.* New York: Oxford University Press, 1951.

Carson, R. *Silent Spring.* Boston: Houghton Mifflin Company, 1962.

Carson, R. *The Sense of Wonder.* New York: Harper and Row, 1965.

Dickson, J. G. (ed.). *The Wild Turkey: Biology and Management.* Harrisburg, PA: Stackpole, 1992.

Dumont, D. *Alabama Wildlife Federation Past and Present.* Alabama Wildlife, Spring 1992.

Flader, S. *Thinking Like a Mountain: Aldo Leopold and the Evolution of an Ecological Attitude Towards Deer, Wolves, and Forests.* Columbia: University of Missouri Press, 1974.

Gartner, C. B. *Rachel Carson.* F. New York: Ungar Publishing, 1983.

Gibbons, Boyd. "Aldo Leopold: A Durable Scale of Values," *National Geographic* 160 (Nov. 1981), 682–708.

Gray, G. G. *Wildlife and People: The Human Dimensions of Wildlife Ecology.* Chicago, IL: University of Illinois Press.

Harbaugh, W. H. *Power and Responsibility: The Life and Times of Theodore Roosevelt.* New York: Farrar, Straus, & Giroux, 1961.

Hodgson, B. "Buffalo Back Home on the Range," *National Geographic,* 186 (5) (1994), 68.

Kellert, S. R. and E. O. Wilson (eds.). *The Biophilia Hypothesis.* Washington, D. C.: Island Press, 1993.

Lee, K. N. *Compass and Gyroscope: Integrating Science and Politics for the Environment.* Washington, D. C.: Island Press, 1993.

Leopold, A. *Game Management.* New York: Charles Scribner's Sons, 1993.

Leopold, A. *Round River.* Oxford University Press, Inc., 1953.

Leopold, A. *A Sand County Almanac.* New York: Ballantine Books, 1966.

Maser, C. *Sustainable Forestry: Philosophy, Science and Economics.* Delray, FL: St. Lucie Press, 1994.

McKnight, B. N. *Biological pollution.* Indianapolis, IN: Indiana Academy of Science, 1993, 121–136.

Mealy, S. P. *Ethical Hunting. Fair Chase:* Spring 1994, 19–20.

Meine, C. *Aldo Leopold: His Life and Work.* Madison: University of Wisconsin Press, 1987.

National Wildlife Federation Conservation Hall of Fame. Fiftieth Anniversary, 1936–1986. National Wildlife Federation Publication.

Phillips, D. "Leopold: Conservation and Education," *Southeastern Wildlife* (Winter 1993), 7–8.

Restoring America's Wildlife. U.S. Fish and Wildlife Service Publication. Washington, D.C.: U.S. Government Printing Office, 1987.

Robinson, W. L. and E. G. Bolen. *Wildlife Ecology and Management.* New York: MacMillan, 1984.

Rossiter, M. W. *Women Scientists in America.* Boston: John Hopkins University Press, 1982.

Scherer, D. and T. Altic. *Ethics and the Environment.* Prentice-Hall, Inc., 1983.

Schoenfeld, C. *Aldo Leopold Remembered.* Audubon, 80 (May 1978), 32–33.

Sterling, P. *Sea and Earth: The Life of Rachel Carson*. New York: Thomas Y. Crowell Publishing, 1970.

Strong, D. H. *Dreamers and Defenders*. Lincoln: University of Nebraska Press, 1988.

Tanner, T., (ed.). *Aldo Leopold: The Man and His Legacy*. Ankeny, IA: Soil Conservation Society of America, 1987.

Whorton, J. *Before Silent Spring: Pesticides and Public Health in Pre-DDT America*. Princeton, NJ: Princeton University Press, 1974.

Chapter 19 Legal Considerations

Alabama Forestry Commission Circular. *There's Something You Should Know Before You Burn*. Montgomery, AL: Alabama Forestry Commission.

Alabama Forestry Commission Circular. *The Smoke Problem*. AFC-FP-01. Montgomery, AL: Alabama Forestry Commission, 1995.

Alabama Forestry Commission Publication. *A Smoke Screening System for Prescribed Fires in Alabama*. Montgomery, AL: Alabama Forestry Commission, 1992.

American Forest and Paper Association. *Sustainable Forestry for Tomorrow's World: First Progress Report on the American Forest and Paper Association's Sustainable Forestry Initiative*. Washington, DC: American Forest and Paper Association, 1996.

Argow, K. A. "Stewardship Begins at Home: The Private Property Responsibility Initiative," *Journal of Forestry*, 92 (1994), 514–17.

Barlowe, R. *Who Owns Your Land?* Michigan State University Cooperative Extension Service.

Bliss, J. C. *Alabama's Nonindustrial Private Forest Owners: Snapshots from a Family Album*. AL Circ. ANR-788, Auburn Univ., 1993.

Bliss, J. C. "Public Opinions—Private Forests," *Highlights of Agricultural Research*. Alabama Agricultural Experiment Station, Auburn, AL: Auburn University, Vol. 40, No. 2 (1993), 5.

Bliss, J. C., S. K. Nepal, R. T. Brooks Jr., and M. D. Larsen. "Forestry Community or Granfalloon?" *Journal of Forestry* 92(9) (1994), 6–10.

Cobb, C. "Living with the Endangered Species Act," *Forest Farmer 29th Manual Edition* 52(3) (1993), 22–23.

Cubbage, F. "Federal Environmental Laws and You," *Forest Farmer 29th Manual Edition* 52(3) (1993), 15–18.

Ellefson, P. V. and A. S. Cheng. "State Forest Practice Programs: Regulation of Private Forestry Comes of Age," *Journal of Forestry* 92(5), 34–37.

Floyd, D. W. and M. A. MacLeod. "Regulation and Perceived Compliance: Nonpoint Pollution Program in Four States," *Journal of Forestry* 91(5) (1993), 41–47.

Goldman-Carter, J. "Takings Issue Could Unravel Wildlife, Habitat Protection," *Alabama Wildlife*. 56(2) (1992), 7–8.

Hickman, C. A. "Local Regulation of Private Forestry." *Forest Farmer 29th Manual Edition* 52(3) (1993), 19–21.

Johnston, N. C. and H. S. Pond, IV. "Silviculture Practices and Environmental Law," *The Alabama Lawyer* (January 1992), 38–43.

Johnston, N. C. "Alabama Coastal Takings," *Water Log*, Mississippi-Alabama Sea Grant Legal Program, Vol. 14, No. 3 (1994), 3–8.

Johnston, N. C. "Habitat Conservation Planning for Private Property under the Endangered Species Act," *Proceedings of the Red Clay Conference*, Environmental Law Association, Georgia Law School, 1995.

Marzulla, N. "Private Property Rights at Risk," *Forest Farmer 29th Manual Edition*, 52(3) (1993), 10–11.

Neal, J. E. "Who Owns Your Land?" *Forest Farmer 29th Manual Edition*, 52(3) (1993), 12–13.

Neal, W. "The Endangered Species Act and the Private Land Owner," *Alabama's TREASURED Forest*, 12(3) (1993), 23–25.

Pinchot, G. *Breaking New Ground*. Seattle: Univ. Wash. Press, 1947.

The Evolution of National Wildlife Law. Council on Environmental Quality by the Environmental Law Institute, Washington, DC: U.S. Government Printing Office, 1977.

Times Mirror Magazines Conservation Council. *Natural Resource Conservation: Where Environmentalism is Headed in the 1990's*. 1992.

Appendix B: Book Steering Committee

Mr. Frank Boyd
USDA - APHIS
Animal Damage Control
Room 118 Extension Hall
Auburn University, AL 36849

Mr. Tommy Counts
USDA
Natural Resource
Conservation Service
P.O. Box 311
Auburn, AL 36830

Mr. Ted DeVos
Natural Resource Officer
Regions Bank
P.O. Box 511
Montgomery, AL 36101-0511

Mr. Dan Dumont
Alabama Wildlife Federation
46 Commerce Street
P.O. Box 2102
Montgomery, AL 36104

Mr. Mickey Easley
Wyncreek Plantation
Rt. 1 Box 32
Hardaway, AL 36039

Dr. Danny Everett
Sumter Farms
Rt. 1 Box 130
Emelle, AL 35459

Dr. Kathryn Flynn
School of Forestry
M. White Smith Hall
Auburn University, AL 36849

Mr. Tim Gothard
Alabama Forestry Commission
513 Madison Avenue
Montgomery, AL 36130

Dr. Harry L. Haney, Jr.
Department of Forestry
Virginia Tech University
Blacksburg, VA 24061

Dr. Jimmy Huntley
654 Huntley Road
Louiseville, MS 39339

Mr. Bill Humphries
Alabama Wildlife Federation
3285 Bankhead Avenue
Montgomery, AL 35501

Mr. Rhett Johnson
Solon Dixon Education Center
Rt. 7 Box 131
Andalusia, AL 36420

Mr. Doug Link
Alabama River Woodlands
P.O. Box 99
Perdue Hill, AL 36470

Dr. Jeff McCollum
Alabama Wildlife Federation
614 Margarete Street
Decatur, AL 35603

Mr. Joe McGlincy
Southern Forestry Consultants
305 West Shotwell Street
Bainbridge, GA 31717

Mr. Mason McGowin
P.O. Box 66
Chapman, AL 36015

Dr. Bill McKee
P.O. Box 688
Molpus Company
Philadelphia, MS 39350

Mr. Dave Nelson
Alabama Department of
Conservation and Natural
Resources
Game and Fish Division
P.O. Box 993
Demopolis, AL 36732

Dr. Dan Speake
Department of Zoology and
Wildlife Sciences
331 Funchess Hall
Auburn University, AL 36849

Mr. Stan Stewart
Alabama Department of
Conservation and Natural
Resources
Game and Fish Division
64 North Union Street
Montgomery, AL 36130

Dr. Emmett Thompson
Dean (Retired) - School of
Forestry
108 M. White Smith Hall
Auburn University, AL 36849

Dr. David Thrasher
Alabama Wildlife Federation
3285 Bankhead Avenue
Montgomery, AL 36106

Dr. Tom Wells
640 Timberlane Road
Pike Road, AL 36064

Appendix C: Nest and Denning Box Dimensions for Some Alabama Birds and Mammals

Species	Floor (Inches)	Depth (Inches)	Height of Entrance Above Floor (Inches)	Diameter of Entrance (Inches)	Height Above Ground (Feet)	Special Notes
Birds						
American Robin	6 x 8	8	no sides		5–15	
Carolina Wren	4 x 4	8	1–6	1¼	6–19	Can use shelf, basket or gourd.
Eastern Bluebird	5 x 5	8	6	1½	5–10	
Crested Flycatcher	6 x 6	10	6	2	8–20	
Purple Martin	6 x 6	6	1–2	2–2¼	10–20	Will also use gourds.
Downy Woodpecker	4 x 4	10	8	1¼	6–20	Put 3–4 inches of sawdust in box.
Red-bellied or Red-headed Woodpecker	6 x 6	15	9	2	8–20	Put 3–4 inches of sawdust in box.
Flicker	7 x 7	18	14	2½	8–20	Put 3–4 inches of sawdust in box.
Tufted Titmouse	4 x 4	8	6	1¼	5–15	
Chickadee	4 x 4	8	6	1⅛	5–15	
House Finch	6 x 6	6	4	2	6–12	
Screech Owl	10 x 10	24	20	3 inches high x 4 inches wide	10–30	
Barred and Barn Owl	12 x 12	25–28	12–16	7 x 7	10–30	
American Kestrel	11 x 11	12	9–12	3 x 4	20–30	
Mammals						
Bats[1]	6 x 8	14	no floor, entrance	¾	12–15	No floor entrance through bottom.
White-footed Mouse	4 x 5	8–11	6	1¼	5–10	
Raccoon	16 x 18	30	25	5 x 9	7–20	
Squirrels	10 x 11	24	20 (on side)	3	12–20	

[1]Includes several species of bats. At the center inside the bat box, an additional upright board equal to the width of the box should extend from the roof to 2 inches above the bottom. This provides a roosting surface for bats.

Appendix D: Quick Reference and Wildlife Usage of Commonly Planted Wildlife Foods

	Wildlife Species											
Cultivated Plants	Black Bear	Deer	Dove	Fox	Quail	Rabbit	Raccoon	Ruffed Grouse	Songbirds	Squirrel	Turkey	Waterfowl
Alfalfa		X				X					X	
Alyceclover		X				X					X	
American Jointvetch		X										
Apple and Crabapple	X	X		X	X		X	X		X		
Coastal panicgrass			X		X				X		X	
Austrian Winter Pea		X	X		X						X	
Autumn Olive	X				X			X	X		X	
Bahia-grass			X		X	X					X	
Beggarweed (Florida)		X	X		X	X			X		X	
Birdsfoot Deervetch		X			X						X	
Black Medic		X				X						
Bluestem (Big and Little)			X		X				X		X	
Buckwheat		X	X		X				X		X	
Chinese Chestnut	X	X					X			X		
Chufa	X	X					X			X	X	X
Clovers												
Arrowleaf Clover		X			X	X		X		X		
Ball Clover		X				X		X			X	
Berseem Clover		X				X					X	
Button Clover		X				X		X			X	
Crimson Clover		X			X	X		X			X	
Red Clover		X				X		X			X	
Subterranean Clover		X			X	X					X	
Sweet Clover		X				X		X			X	
White (Ladino) Clover		X			X	X		X			X	
Corn	X	X	X		X	X	X	X	X	X	X	X
Cowpea		X	X		X	X					X	
Dallis-grass			X		X				X		X	

	Wildlife Species											
Cultivated Plants	Black Bear	Deer	Dove	Fox	Quail	Rabbit	Raccoon	Ruffed Grouse	Songbirds	Squirrel	Turkey	Waterfowl
Egyptian wheat	X	X	X		X				X		X	X
Indiangrass			X		X				X		X	
Lespedezas												
Annual		X			X	X			X		X	
Shrub					X				X		X	
Millets												
Browntop Millet		X	X		X	X		X	X		X	X
Foxtail Millet			X		X			X	X		X	X
Japanese Millet			X		X			X	X			X
Pearl Millet			X		X				X		X	
Proso Millet			X		X				X		X	X
Orchardgrass			X			X		X			X	
Partridge Pea					X				X		X	
Peanut		X	X	X	X		X				X	
Peanut, Perennial		X					X				X	
Rice			X		X				X			X
Ryegrass		X				X						
Sawtooth Oak	X	X		X			X	X		X	X	
Sesame (Benne)			X		X				X			
Singletary Pea		X	X		X	X					X	
Small Grains												
Barley		X	X		X	X			X		X	X
Oats		X	X		X	X		X	X		X	X
Wheat	X	X	X	X	X	X		X	X		X	X
Sorghum	X	X	X		X			X	X		X	X
Soybean		X	X		X	X					X	X
Sunflower		X	X		X			X	X	X	X	
Switchgrass			X		X			X	X		X	
Timothy		X			X	X					X	
Velvet Bean		X			X	X					X	

Cultivated/Native and Naturalized Plants	Black Bear	Deer	Dove	Fox	Quail	Rabbit	Raccoon	Ruffed Grouse	Songbirds	Squirrel	Turkey	Waterfowl
	Wildlife Species											
Vetch		X	X		X	X			X		X	
Weeping Lovegrass									X			
Wild Reseeding Soybean		X	X		X			X			X	X
NATIVE AND NATURALIZED PLANTS Trees, Shrubs & Other Woody Plants												
American Beech	X	X			X		X	X		X	X	X
American Beautyberry	X	X		X	X		X		X	X	X	
American Elder		X	X			X	X	X	X	X	X	
American Hornbeam		X				X			X		X	
Alabama Supplejack		X			X	X	X			X	X	X
Bayberry				X	X				X		X	
Black Cherry	X	X	X	X	X	X	X	X	X	X	X	X
Black Walnut								X		X		
Blackberry, Dewberry & Raspberry	X	X		X	X	X	X	X	X	X	X	
Blueberry	X	X		X	X	X	X	X	X		X	
Butternut	X	X								X		
Chinquapin	X	X				X	X	X	X	X	X	
Dogwood, Flowering	X	X		X	X	X	X	X	X	X	X	X
Elm					X				X	X	X	
Fringetree		X			X						X	
Grape, Wild	X	X		X	X		X	X			X	
Greenbrier	X	X			X		X	X	X	X	X	X
Hackberry	X	X			X		X	X	X	X	X	
Hawthorn		X		X	X	X	X	X	X	X	X	X
Hickory	X	X		X		X	X		X	X	X	
Holly	X	X	X		X		X	X	X	X	X	X
Hophornbeam, Eastern		X			X	X		X	X	X	X	
Hornbeam, American		X				X						
Huckleberry, Dwarf	X	X		X	X		X	X	X	X	X	
Japanese Honeysuckle		X			X	X		X	X		X	

Native and Naturalized Plants	Black Bear	Deer	Dove	Fox	Quail	Rabbit	Raccoon	Ruffed Grouse	Songbirds	Squirrel	Turkey	Waterfowl
	Wildlife Species											
Locust		X			X	X		X		X		
Oak	X	X			X		X	X	X	X	X	X
Magnolia					X				X	X	X	
Maple	X	X			X				X	X		
Palmetto, Dwarf	X	X			X		X		X	X	X	
Persimmon	X	X		X	X	X	X		X	X	X	
Pine	X				X			X	X	X	X	
Poison Ivy		X			X	X		X	X	X	X	X
Poplar, Yellow		X						X	X	X	X	
Plum, Wild		X		X	X		X		X	X	X	X
Red Mulberry	X	X			X		X	X	X	X	X	
Redbay		X			X				X		X	
Redcedar, Eastern		X		X	X	X	X				X	
Sassafras	X	X			X		X		X	X	X	
Saw Palmetto	X	X					X				X	
Serviceberry, Shadblow	X	X				X		X	X	X		
Sugarberry		X			X				X		X	
Sumac		X			X	X		X	X	X	X	
Sweetbay		X							X	X	X	
Trumpet-Creeper		X										
Tupelo	X	X		X	X		X		X	X	X	X
Viburnum		X			X	X	X	X	X	X	X	
Upland Herbaceous Plants												
Bahia-grass			X		X				X		X	
Beggarweed		X	X		X	X		X	X		X	
Lespedeza		X			X	X			X		X	
Milkpea		X			X	X		X			X	
Partridge Pea					X						X	
Pokeweed	X	X	X		X	X	X		X		X	
Ragweed			X		X	X		X	X		X	

Native and Naturalized Plants	Wildlife Species											
	Black Bear	Deer	Dove	Fox	Quail	Rabbit	Raccoon	Ruffed Grouse	Songbirds	Squirrel	Turkey	Waterfowl
Wooly Croton			X		X	X			X		X	
Aquatic and Marsh Plants												
Arrow-Arum												X
Asiatic Dayflower												X
Barnyardgrass			X		X			X	X			X
Bulrush									X			X
Cutgrass, Giant								X				X
Foxtail, Giant			X		X				X		X	X
Pondweed												X
Sedge			X		X			X	X		X	X
Smartweed			X		X			X	X			X
Spikerush												X
Tearthumb												X
Watershield												X
Widgeongrass												X
Wildrice			X		X			X	X			X

Appendix E: Cultivated Plants Important to Wildlife in Alabama

ALFALFA *Medicago sativa*

A deep rooted (long tap root), cool-season perennial legume with a high nutritional quality. Native to Iran and central Asia. Grows during cool and warm seasons. Grows 20 to 36 inches high, has fine stems, and short leafy branches from crown with compound leaves having three leaflets. Flowers are usually a shade of purple. Drought tolerant. Used by **deer, turkey** and **rabbits.** Provides favorable nesting habitat, seeds, insects and foliage for turkey. Alfalfa has the longest production season of any southern legume, from March until November. Common varieties are Apollo, Virginia Cimmaton, Alfagraze and Florida 77.

Planting Rate: Plant 15–20 pounds of seed per acre. Inoculate seeds.

Planting Date: September 1 to October 15.

Planting Depth: Plant seeds less than ½ inch deep.

Companion Plants: Grows best alone.

Site Selection: Requires deep, fertile, well-drained soils. Not suited to wet soils or soils with shallow hardpan. Also not adapted to excessively drained sandy soils. Alfalfa is sensitive to soil acidity. A soil pH of 6.5 or above is required for high yields.

Land Preparation: Break land in June or July and fallow until time to plant. Prepare smooth, firm seedbed before planting. A cultipacker-seeder is the best equipment for planting in prepared seedbeds.

Fertilization/Liming: Fertilize according to soil test, or apply 120 pounds of phosphorus, 120 pounds of potassium and 10 pounds of boron per acre when seedbed is prepared. Apply lime as needed according to soil test for a pH greater than 6.7 or apply at 2 tons per acre. Nitrogen is not needed since alfalfa fixes nitrogen when properly inoculated.

Management: Requires a higher level of management than most plantings. Pest problems may occur with crown and stem rot, alfalfa weevil (controlled by spraying insecticide in the spring), leafhoppers, and nematodes of nonresistant varieties in the sandy coastal plain.

ALYCECLOVER *Alysicarpus vaginalis*

A warm-season tropical annual, erect legume 12 to 24 inches tall with pink flowers that is highly preferred by **deer** in late summer and early fall. Originally from the tropical areas of the Orient. This is an excellent forage for deer, **turkeys** and **rabbits**. Comparable to American jointvetch, but has the advantage of being less expensive. High nutritional quality which is maintained well into late summer. Seasonal production best during July through September. Establishment is slow and weed competition may be a problem.

Planting Rate: Plant 15–20 pounds per acre. Broadcasting by hand is not recommended. Use an inoculant for alyceclover to increase nitrogen fixation, yields and crude protein content. Roll or cultipack after seeding.

Planting Date: May–June.

Planting Depth: 1 inch.

Companion Plants: Does well in combination with forage cowpeas and/or American jointvetch. If planted with forage cowpeas and American jointvetch, plant 40 pounds per acre of cowpeas, 10 pounds per acre of alyceclover and 5 pounds per acre of American jointvetch. Plant peas and then drag in alyceclover and jointvetch. Combinations do best in bottomland soils.

Site Selection: Does well on most sites but best suited for moderate to well-drained sandy loam soils. Grows best in coastal areas with high summer rainfall. Tolerant of soil acidity.

Land Preparation: Old cultivated land should be thoroughly broken and prepared in April or early May and redisked before planting to help destroy competing weeds and grass. New land can be prepared just before planting.

Fertilization/Liming: Follow soil test or apply a fertilizer with low or no nitrogen since the plant is a nitrogen fixing legume. Without a soil test apply 300 pounds of 0-20-20 or an equivalent fertilizer per acre, and 1 ton of lime per acre to obtain a soil pH of 6.2.

Management: Apply lime and fertilize every 2–3 years. Alyceclover will not tolerate competition from grasses and weeds during germination and early growth. To overcome grass and weed problems, it may need to be grazed or clipped off until clover gets established. Stands can be expected to persist for several years if the crop is allowed to produce seed each year. Where natural reseeding is desired allow to produce seed. The following spring, disk and harrow to control weed growth and to provide good soil conditions for germination of seeds that were produced the previous fall. Reseeding is not always dependable. Susceptible to nematodes and not competitive with weeds in seedling stage. Seed production is from 300 to 400 pounds per acre. Seeds are harvested by direct combining. Begin harvest when seed pods begin to shatter and harvest immediately.

AMERICAN JOINTVETCH *Aeschynomene americana*

Also known as Aeschynomene, jointvetch is not a true vetch but a warm-season tropical legume. It is a reseeding summer annual used heavily by **deer** in late summer and early fall. It is slow to establish the first year. Browsing by deer usually begins in June, peaks in August to September, and ends in November. For this reason it is an excellent choice if early season deer bowhunting is planned. The plant is rapidly gaining popularity, but has a disadvantage of being more expensive than other seeds. Weed competition may also present problems. In high deer densities this plant cannot flower and seed out effectively. As a result, it must be planted every year. From central Alabama northward, the growing season is too brief to allow reseeding, requiring annual reseeding.

Planting Rate: Plant 15–20 pounds per acre if hulled, or 20–25 pounds per acre if unhulled. Broadcasting by hand is not recommended since this tends to bunch the seed and promotes weed competition. Proper inoculant, like those used for peanuts, cowpeas, or alyceclover, should be applied to the seed immediately before planting.

Planting Date: April 15–July 4.

Planting Depth: ½ inch.

Companion Plants: Can also be mixed with forage cowpeas and alyceclover. Plant peas first at 40 pounds per acre and then drag in jointvetch at 5 pounds per acre and alyceclover at 10 pounds per acre.

Site Selection: Choose a sandy loam to silt loam surface on a slight slope if possible. Sites too wet or too dry will not produce a quality stand. Well-drained alluvial soils are best suited to jointvetch.

Land Preparation: Prepare by plowing, disking, and packing.

Fertilization/Liming: Soil test first. Prior to the final disking to prepare the seedbed, broadcast 400 pounds of 0-10-20 per acre plus the necessary trace elements. Apply low nitrogen fertilizer at a rate of 15–30 pounds per acre after plants are 15 inches high.

Management: Shallow disking every spring should be followed by application of fertilizer when plants are established. In order to ensure a volunteer crop, plants must be allowed to bloom and seeds to mature. If seed maturation does not occur, additional seed must be replanted. Can use pre-treatment with herbicides such as Treflan® for weed and grass control.

APPLE AND CRABAPPLE *Malus* spp.

Trees in the apple family are small with short trunks and broad, open crowns. **Deer**, **ruffed grouse**, **quail**, **black bears**, and **squirrels** readily feed in orchards after apples mature in the fall. Does best in northern Alabama.

Planting Rate: Plant seedlings which have a root collar of at least ⅜ inch in diameter with spacing at 10 x 10 or 10 x 12 feet apart.

Planting Date: Mid-winter to early spring.

Planting Depth: Plant to root collar.

Companion Plants: After establishment can plant most cool and warm-season grasses with apple or crabapple.

Site Selection: Medium-moist to moist uplands, but some species will tolerate dry sites.

Land Preparation: Planting sites must be open and weed competition must be controlled.

Requires sunlight. Mow to control weeds, brush, vines, and briars until the seedlings outgrow their competitors.

Fertilization/Liming: Except for the year of planting, apply 3 pounds of nitrogen per 1,000 square feet beneath and just beyond the branch spread to improve growth and vigor. Annually apply 3 pounds of nitrogen per 1,000 square feet beneath and just beyond the branch spread.

COASTAL PANICGRASS *Malus* spp.

Coastal panicgrass is a tall, robust, warm-season, perennial grass. Growth habit is upright and the plant looks like a bunch grass, although it produces short rhizomes. Plants are bluish green, leafy, and multi-stemmed. They produce large quantities of viable seed used by **quail**, **turkeys** and select seed-eating **songbirds**. Strong seedling vigor is an outstanding attribute.

Planting Rate: For drilled plantings, 10–15 pounds per acre. Broadcast at 20 pounds per acre and cover seeds with cultipacking or other means.

Planting Date: Since coastal panicgrass seed may display some dormancy, the use of two-year-old seed is recommended. In addition, early spring seeding will increase the rate and uniformity of germination.

Planting Depth: In silty or medium-textured soils, plant seed ½–1 inch deep. In coarse-textured soils plant seed 1–2 inches deep.

Companion Plants: Can grow well with mixtures of annual lespedezas.

Site Selection: Grows best on light-textured, sandy to silt-loam soils. To reduce weed competition use a field that has been cultivated for a few years.

Land Preparation: On sites where tillage equipment can be used, prepare a firm seedbed as for pasture planting. No-till seeding into killed sod (gamoxone or Round-Up®) is the preferred establishment method.

Fertilization/Liming: Follow soil test or apply 100 pounds of nitrogen, phosphorus, and potassium per acre and 200 pound of lime per acre.

Management: Responds well to prescribed burning.

AUSTRIAN WINTER PEA (WINTER PEA) *Pisum sativum* spp. *arvense*

A viney, cool-season annual with stems 2 to 4 feet long and native to the Mediterranean region. Wildlife use of Austrian winter peas is limited, but **doves**, **quail**, **turkeys** will eat some seed. **Deer** will graze the tender foliage when other foods are in short supply. High nutritional value. Seeds mature from May to June.

Planting Rate: Broadcast and harrow in at 30–40 pounds per acre or 20–30 pounds per acre if planted with small grain.

Planting Date: September–October. Best production if peas are planted by October 15.

Planting Depth: ½ inch.

Companion Plants: Can be planted with any of the small grains.

Site Selection: Well-drained loam or sandy loam soils. Does well except on poorly drained soils near the coast.

Land Preparation: No preparation if following a cultivated field or harvest of crop. If in a weedy area prepare a clean seedbed.

Fertilization/Liming: Follow soil test or apply lime on acidic soil for a pH of 5.8–6.2, usually 1–2 tons per acre. Apply 60 pounds of phosphorus and potassium per acre.

Management: Allow peas to produce mature seed every two to three years to ensure a reseeding stand. Remove excessive growth of grass in September to facilitate germination of peas. Disking in the fall every two years will encourage production. A hard seed coat helps ensure reseeding. Repeat annual applications of phosphate and potash. Repeat lime application every five years or as needed to maintain a soil pH of at least 6. Produces 300–400 pounds of seeds per acre. Can be harvested by direct combining. Aphids are the most common pest. Pests may also include downy mildew during warm wet winters, viruses, pea weevil, and nematodes.

AUTUMN OLIVE *Elaeagnus umbellata*

A large, non-native deciduous shrub or small tree that grows to a height of 10 to 15 feet. Fleshy fruits are an excellent source of food for **quail** in late fall and winter when food supplies are scarce. **Turkeys** and **songbirds** readily take the red berries in the early fall. Berry production can begin as early as three to five years after planting. Autumn olive has dark green leaves with silvery undersides and produces an abundance of small yellow sweet flowers each spring, and a heavy crop of berries that ripen throughout August and September. Berries range in color from yellow to dark red and are from ⅛ to ¼ inch in diameter. The dense shrub also provides escape cover for wildlife.

Planting Date: January–March.

Planting Rate: Seedlings which have a root collar of at least ⅜ inch in diameter can be planted at 6–10 feet apart for hedgerows or at least 12 feet apart for individual plants.

Planting Depth: Plant by hand in holes. The hole must be large enough to accommodate roots without crowding. This means the hole will have to be 4–6 inches larger in diameter and 4–6 inches deeper than the actual plant root measurements.

Companion Plants: None.

Site Selection: Nitrogen-fixing plant, therefore it does not require a highly fertile soil. Prefers well-drained, moderately fine-textured soils from deep sandy loam to clay. Best planted as a border or cover strip along the edge of food plots. Tolerant of light shade but does not produce many berries if shaded. pH > 7 recommended.

Land Preparation: Requires sunlight. Mow to control weeds, brush, vines, and briars until seedlings outgrow their competitors.

Fertilization/Liming: Apply a handful of 10-10-10 fertilizer around seedling at time of planting. Apply 15–20 pounds of nitrogen around established trees.

Management: Except for the year of planting, apply annually 3 pounds of nitrogen per 1,000 square feet beneath and just beyond the branch spread. Clipping with a chainsaw is recommended to keep in brushy stage.

BAHIA GRASS *Paspalum notatum*

A deep-rooted, summer perennial that spreads by rhizomes and seed. Plants are approximately 8 to 30 inches tall. Originates from Argentina, Brazil, Uruguay, and Paraguay. **Turkeys** readily graze the foliage and small seedheads. In addition, the grass also supports insects that are an important source of protein for turkeys and **quail**. Will provide forage from May 15 until frost. Pensacola and common are the best varieties. Not well suited for north Alabama.

Planting Rate: Broadcast about 20 pounds of seed per acre.

Planting Date: March–May, September–October

Planting Depth: ¼–½ inch.

Companion Plants: Clovers may be over-seeded on bahia grass alone or in combination with ryegrass. Other winter annuals can also be over-seeded over bahia grass. In a well established bahia grass plot, the sod may need to be disturbed with a light disking in the fall to obtain a good stand of over-seeded winter annuals. Winter peas or annual lespedezas can also serve as companion plants.

Site Selection: Adapted to a wide range of soils from very dry to moderately poorly drained. Two plots of ¼–1 acre in size per 100 acres of forest is usually sufficient for turkeys.

Land Preparation: Seedbed should be similar to that for row crops; however, good stands have been obtained by over-seeding thin stands of small grains (like wheat or rye) in February and March.

Fertilization/Liming: Follow soil test recommendations or 300 pounds per acre of 13-13-13 or 15-15-15. Soil pH should be above 5.7 and tests for phosphorus and potassium should be medium or above. Use 20–60 pounds of nitrogen per acre in May and July. Rate and frequency will depend on forage and on desired seed production. Will tolerate low fertility.

Management: Occasional mowing in spring or early summer can increase seed production and provide multiple seed crops over the season. Responds well to 300–400 pounds of 10-10-10 in early spring and a top-dressing with 30–50 pounds of nitrogen per acre in midsummer. Does well on old roads and roadsides.

BEGGARWEED (FLORIDA) *Desmodium tortuosum*

An annual legume, usually growing 4 to 8 feet high in cultivated ground. Recognized by its small jointed seedpods with hooked hairs that cause it to cling to clothing. The seeds are small, 200,000 seeds per pound. Seeds are a choice food of **game birds**. Seeds are harvested mainly in Georgia, Alabama, and northern Florida.

Planting Rate: Broadcast 10–15 pounds per acre of hulled seed on lightly prepared soil. Will volunteer annually following cultivation or soil disturbance if seeds are allowed to mature each year.

Planting Date: April–May.

Planting Depth: Leave uncovered.

Companion Plants: Grows well with other native legumes.

Site Selection: Most sites, but grows best on well-drained to moderately drained sandy-loam soils.

Land Preparation: Disk lightly on previously harvested cornfields. Otherwise, prepare a clean seedbed.

Fertilization/Liming: Fertilize cultivated crop (corn) in a normal manner or use 300–500 pounds of 5-10-15 per acre or similar fertilizer in growing beggarweed alone.

Management: Allow seed crop to mature and remain on the ground until spring. Prepare a normal seedbed in spring and continue crop rotation, or harrow to form a fresh seedbed. Seeds do not ordinarily remain viable in the soil more than one season. Seasonal disturbance by disking in winter or early spring stimulates volunteer growth.

BIRDSFOOT DEERVETCH *Lotus corniculatus*

Also known as birdsfoot trefoil. A deep-rooted, cool-season, short-lived perennial legume that grows 12 to 30 inches. Native to the Mediterranean region. Flowers are bright yellow and seed pods are brown to purple resembling a bird's foot. Production period April through early October. Varieties include Fergu, Empire, and Viking. Utilized by **turkeys**, **quail**, and **deer**.

Planting Rate: Plant seeds with a cultipacker-seeder at 4–6 pounds per acre or broadcast 12 pounds per acre.

Planting Date: August–September.

Planting Depth: ¼ inch.

Companion Plants: Grows well with bahia grass, ryegrass, clover, and vetch.

Site Selection: Does well on most sites and tolerates drought and moderate soil acidity. Adapted to well-drained soils.

Land Preparation: Well prepared clean seedbed. Disk plot in June and leave fallow until planting date. Redisk and plant seed in firm seedbed.

Fertilization/Liming: Follow soil test or lime if soil pH is under 5.5. Responds well to phosphorus and potassium. Generally 75 pounds of (P) and 150 pounds of (K) per acre are required. Inoculation of seed required.

Management: Plants should be allowed to produce seed for reseeding new plants the following year.

Crown and root rot are serious diseases of birdsfoot deervetch and will reduce stands. Nematodes can also be a problem on sandy soils. Mow in early spring and late summer.

BLACK MEDIC *Medicago lupulina*

A cool-season annual that is 6 to 8 inches tall with small yellow flowers. Native to the Mediterranean region. Short period of production from April to early May. Although utilized by various species of wildlife, availability limits use. Used by **turkeys**, **rabbits** and **deer**.

Planting Rate: Broadcast seed at 10–12 pounds per acre.

Planting Date: September–October.

Planting Depth: ¼ inch.

Site Selection: Grows on most sites but does best on calcareous soils.

Land Preparation: Prepare smooth, clean seedbed.

Fertilization/Liming: Follow soil test. Does well with 300–400 pounds of 0-10-20 or 0-20-20 per acre.

Management: Replant for establishment.

BLUESTEM (BIG AND LITTLE) *Andropogon* spp.

The bluestems, known as "beard grasses" or "sagegrasses," are important, native warm-season forage grasses in the longleaf pine-bluestem range. Southern bluestems are mainly bunchgrasses. Heights range from about 1 foot to 6 feet. Provide excellent cover and feeding sites for ground-nesting wildlife such as **quail** and **turkeys**. Rountree variety is best suited; however, Niagara may prove suitable in Piedmont and mountains.

Planting Rate: Can broadcast at 7–10 pounds per acre with an eze-flo spreader or cyclone spreader and then dragged to lightly cover seed. Can also drill at 8 pounds per acre with special drill. If using a cyclone spreader, try mixing seed with fertilizer or cracked corn for better spreading. A granular soil legume inoculant makes an excellent carrier for the fine seeds. It may require two years to establish a good stand and may have a hard time finding seedlings the first year. Be patient. No-till seeding into killed sod (gamoxone or Round-Up®) is the preferred establishment method. Use 2–3 pounds of pure live seed for old world bluestem or 5–10 pounds for little bluestem.

Planting Date: April–May.

Planting Depth: ¼–½ inch.

Companion Plants: Can no-till plant into corn, milo or soybean stubble or into cool seasons crop such as wheat.

Site Selection: Loam to clay soils.

Land Preparation: Create a tilled, firm seedbed by plowing, disking and cultipacking. If prepared hard seedbed is rained on before planting, harrow and cultipack again before planting.

Fertilization/Liming: Soil test fields and bring fertility up to medium levels for lime, phosphorus and potassium, but do not apply nitrogen at or before planting time. Nitrogen will only stimulate weed competition.

Management: Use 1–1.5 quarts of 2,4-D per acre if broadleaf weeds are problem; then wait 10 days to 2 weeks and apply 1.5 pounds of Atrazine per acre. To reduce weed competition mow weeds down at 6 inch height in May. Warm-season grasses are especially well adapted to management with controlled fire in the spring around April 1.

BUCKWHEAT *Fagopyrum esculentum*

An erect annual with white flowers that produces an abundance of forage and seed for about 10 to 12 weeks for **quail**, **doves**, **turkey**, **ducks** and **deer**. Often found along roadsides and in fields. Buckwheat may be heavily browsed or may not be suitable in areas of high deer density.

Planting Rate: Plant 40–50 pounds per acre of seed.

Planting Date: April–August. Matures in 45 days. Staggered plantings lengthens the period of seed production.

Planting Depth: ½–1 inch.

Companion Plants: Sunflower, millets, and grain sorghum.

Site Selection: Grows on practically any type of soil. Prefers well-drained sandy to sandy loam soils.

Land Preparation: Prepare clean seedbed.

Fertilization/Liming: Broadcast 400–500 pounds of 10-10-10 per acre. Preferred pH of 6–6.5.

Management: Early plowing is advantageous. Cultural requirements are similar to small grains.

CHINESE CHESTNUT *Castanea mollissima*

A medium-sized, deciduous tree with a short, crooked trunk and bur-covered nuts which are readily eaten by **rabbits**, **gray** and **fox squirrels**, **chipmunks**, and **deer**. Chinese chestnuts may be one alternate to fill the void left by the demise of the American chestnut. At the beginning of the twentieth century, the American chestnut ranked as one of the most important wildlife plants of the eastern U.S. The bur-covered nuts of the smaller Chinese chestnut trees are similar to the nuts of the American chestnut and are readily eaten by wildlife. Its form consists of a broad, low-forking trunk with spreading branches.

Planting Rate: Plant seedlings which have a root collar of at least ⅜ inch in diameter spaced 10 x 10 or 10 x 12 feet apart at a minimum. Because of its eventual large crown, plant wider if possible.

Planting Date: Midwinter to early spring.

Planting Depth: Plant seedlings to root collar.

Companion Plants: None.

Site Selection: Dry to medium-moist upland sites and mountain slopes.

Land Preparation: Requires sunlight. Mow and use herbicides to control weeds, brush, vines, and briars until the seedlings outgrow their competitors.

Fertilization/Liming: Apply a handful of 10-10-10 fertilizer around seedlings.

Management: Except for the year of planting, annually apply 3 pounds of nitrogen per 1,000 square feet beneath and just beyond the spread of the branches. Vegetation surrounding seedlings should be controlled for several years by chemical or mechanical means to reduce competition.

CHUFA *Cyperus esculentus*

Also known as nutgrass. Chufa is a sedge with grass-like leaves growing 18 to 24 inches high. While the seeds are eaten by many species of **birds**, the tubers are a choice food for **geese**, **waterfowl** (**mallards** and **pintails**), **turkey**, **deer**, **raccoon**, **squirrels** and **feral hogs**.

Planting Rate: Best if planted 20 or more pounds per acre in rows 2–3 feet apart. If broadcasting plant 40 to 60 pounds per acre.

Planting Date: April 1–July 1.

Planting Depth: 1½ inches.

Companion Plants: Grows best alone.

Site Selection: Well-drained to moderately drained sandy or loam soils. Avoid clay soils.

Land Preparation: A well-prepared seedbed is necessary.

Fertilization/Liming: Use sufficient lime and fertilizer to keep the soil pH above 5.7 and to maintain medium soil test levels for phosphorus and potassium. Normal rate of fertilization is 500 pounds of 10-10-10 per acre.

Management: A top dressing of 50–80 pounds of nitrogen per acre will improve growth. Two or three cultivations to reduce weed competition will be helpful. Chufa sites should be rotated every two to three years since it does best on "new" soils. Rotation also reduces the chance of disease problems from concentrating turkeys year after year in one area. After establishment, disk in March and April to encourage reseeding the second and possibly the third year.

ARROWLEAF CLOVER *Trifolium vesiculosum*

An upright, cool-season, reseeding annual legume that grows to a height of 40 to 50 inches under optimum growing conditions. The non-hairy, arrow-shaped leaves generally have a large white "V" mark. Flowers are predominately white but can have pink and purple shaded flower heads. Native to the Mediterranean region. Seeds germinate in the fall but grow slowly during the winter. Beginning in March arrowleaf clover makes rapid growth until full bloom in June. Seeds mature in late June to early August. Reseeds easily. Provides forage through late spring and early summer for **deer**, **turkeys** and **rabbits**. Common varieties are Yuchi, Meechee, and Amclo.

Planting Rate: Broadcast 10–15 pounds per acre of inoculated seed. Seeds should be scarified since they have a very hard seed coat.

Planting Date: September 1 through October in Piedmont, September 1 through November 15 in coastal regions.

Planting Depth: ¼–½ inch.

Companion Plants: Arrowleaf clover and cereal rye make good fall-to-spring combinations on well-drained soils and can be easily reseeded by disking or mowing and fertilizing each September. Grows well with ryegrass, summer perennial grasses and small grains.

Site Selection: Suited to a wide range of soil conditions, but is best adapted to more fertile, well-drained soils. It is not suited to calcareous and light textured droughty soils of low fertility or on poorly drained soils.

Land Preparation: Prepare a smooth, clean seedbed. Firm with a cultipacker before and after planting. May also be planted in an established summer perennial grass sod by light disking or with a sod seeding machine. Planting in sod should be delayed until about first frost date; however, may get a winter kill in the piedmont if planted at this time.

Fertilization/Liming: Soil test first. Apply 100–200 pounds per acre of 0-20-20 after plants are established. A pH level of 6.5–7.0 is needed for reseeding the following year.

Management: An annual program of shredding in late summer, followed by light disking or late summer burning will help in seed production for the following year. Annual applications of 60–80 pounds of fertilizer per acre should be applied in August or September. Fertilizers should not contain any nitrogen after the first year of planting. Annual seed production is from 200–300 pounds per acre. Can be harvested by direct combining. Crown and stem rot may occur during warm, wet winter weather, especially on loam and clay soils. Thinning stands will help by promoting air movement through stands. Nematodes and virus diseases can also be a problem. Stands turn a distinctive purple-red color in response to stress.

BALL CLOVER *Trifolium nigrescens*

A rapid growing, heavy seeding, cool-season annual legume similar in appearance to white clover. Native to the Mediterranean region. Can be used on droughty or low fertility land where crimson and white clover have a hard time growing. Grows up to 3 feet tall with white fragrant flowers. Hard seed and heavy seed production can result in good natural reseeding. Production primarily during late March through April. Utilized by **deer, turkey** and **rabbits**.

Planting Rate: Broadcast 2–3 pounds per acre. Inoculate with white clover inoculant.

Planting Date: Plant from August 15–October 15.

Planting Depth: ¼–½ inch.

Companion Plants: Ball clover grows well on perennial grass sod (bermudagrass or bahia grass) and with small grains, ryegrass and other clovers.

Site Selection: Adapted to a wide range of soils from loam to clay soils. Tolerates poor drainage and grows on wetter and lighter textured soils than crimson clover. Will also grow on less fertile soils than white clover.

Land Preparation: Prepare a smooth, clean seedbed. Allow seedbed to settle before planting. Ball clover can be planted in established stands of perennial grass with a sod seeding machine, by light disking of sod, or with a grain drill.

Fertilization/Liming: Apply 60 pounds of phosphorus and 60 pounds of potassium per acre at planting time. On soils of low fertility apply 16 pounds of nitrogen per acre at planting time. Maintain soil pH at 6 by liming.

Management: Allow ball clover to mature a good crop of seed each year to ensure reseeding. Use same fertilizer rates annually that were used when planting. There are two methods of harvesting seed: 1) direct combining when clover is completely matured, or 2) clover may be mowed and windrowed when half matured. Allow to dry one or two days, then combine with a pick-up attachment. Yields vary from 50–200 pounds per acre. Dry seed thoroughly after harvest. Clover head weevil may reduce seed production.

BERSEEM CLOVER *Trifolium alexandrinum*

A cool-season annual with white flowers that resembles alfalfa and grows to a height of 2 feet or more. Native to the Mediterranean region. Most of the varieties are not winter hardy. High nutritional quality. Production from November to December and from March to June. Used by **turkey**, **deer** and **rabbits**. Common variety is Big "B."

Planting Rate: Broadcast at 20 pounds per acre or drilled at 10–15 pounds per acre.

Planting Date: September–mid-October.

Planting Depth: ¼ inch.

Companion Plants: Ryegrass, small grains and other clovers.

Site Selection: Does well on non-acidic black belt soils and in high rainfall areas near the coast. More tolerant of alkaline and wet soils than most annuals. Best on loam soils with a pH of 6 or more.

Land Preparation: Prepare a clean seedbed.

Fertilization/Liming: Follow soil test or lime until pH above 6. Usually 1–2 tons of lime per acre is needed. Requires high phosphorus and potassium fertility and 2–4 pounds per acre of boron.

Management: Can be encouraged to reseed by shredding and light disking. Crown rot can be a problem if excess forage is present during freezing.

BUTTON CLOVER *Medicago orbicularis*

A cool-season annual about 2 to 3 feet long. Orange-yellow flowers with coiled seed pods that resemble buttons, hence the name button clover. Native to the Mediterranean region. Productive period from March to May. Used by **deer, turkey** and **rabbits**.

Planting Rate: Broadcast seed at 10 pounds per acre.

Planting Date: September–October.

Planting Depth: ¼ inch.

Companion Plants: Bahia grass, small grains, ryegrass and other clovers.

Site Selection: Adapted to loam and clay soils. Does best on limestone soils.

Land Preparation: Prepare a smooth, clean seedbed. Allow to settle before planting.

Fertilization/Liming: Follow soil test and lime if soil is below a pH of 6. If no soil test is performed, apply 100–200 pounds of low nitrogen fertilizer per acre.

Management: An annual program of shredding in late summer, followed by light disking or late summer burning will help in seed production for the following year. Annual applications of 60–80 pounds of fertilizer per acre in August or September. Seed production from 200–

300 pounds per acre. Can be harvested by direct combining. Crown and stem rot may occur during warm, wet winter weather, especially on loam and clay soils. Thinning stands will help by promoting air movement through clover. Nematodes and virus diseases can also be a problem.

CRIMSON CLOVER *Trifolium incarnatum*

A leafy, reseeding, winter annual legume that grows 10 to 15 inches tall and is foraged by **deer**, **rabbits**, **turkeys**, and **quail**. Native to the Mediterranean region. Distinguished by crimson colored flowers with yellow rounded seeds. Generally available for a shorter period of time than other clovers but is more acid-tolerant than most legumes, and has an excellent soil building capacity. Roots may penetrate to a depth roughly equal to the plants height. Improved varieties are Tibbee, Chief, and Dixie. Seasonal production from late November to December and February to early April.

Planting Rate: Broadcast 20–30 pounds of inoculated seed per acre. Use reseeding variety. Seeds which are harvested by combine do not need to be scarified.

Planting Date: September 15–October 15. Be selective about planting dates. Plant when soil moisture is adequate for seed germination.

Planting Depth: Cover seeds ¼ inch. Use a cultipacker or drag.

Companion Plants: Used alone or in a mixture with wheat and/or ryegrass. If seeding with wheat or ryegrass mixtures, use 20 pounds per acre and plant before September 20. Also grows well with other small grains and clovers.

Site Selection: Best on upland clay soils that are well-drained. Crimson is the best variety for coastal areas.

Land Preparation: A good, firm seedbed is preferred.

Fertilization/Liming: Soil test first. Soil pH should be between 5.8 and 6.5. Phosphorus and potassium soil test levels should be maintained. An application of one pound of boron with a phosphorus and potassium fertilizer will increase seed production. Apply 60–80 pounds of phosphorus and potassium per acre after plants are established. Does well with 300–400 pounds of 0-10-20 or 0-20-20 per acre.

Management: Mow excessive growth in the fall to ensure reseeding. Yields of 150–300 pounds of seed per acre are common. Can combine seed directly. Seed may also be harvested by mowing, and windrowing. Clover head weevils often cause heavy seed losses, resulting in poor natural reseeding. Cool, wet winter weather often results in crown and stem rot when stands are thick. Thinning stands may help reduce this problem.

RED CLOVER *Trifolium pratense*

Medium-rooted perennial upright grass which grows as an annual or biennial. Requires high fertility. Has a characteristic red bloom. Used by **deer**, **turkeys**, and **rabbits**. Common varieties include Redland, Florie, Kenstar, Orbit, and Chesapeake.

Planting Rate: Plant 8–12 pounds per acre on a firm seedbed. Cover seed lightly with harrow or cultipacker. Inoculate seed with red clover inoculant.

Planting Date: August 15–October 15.

Planting Depth: ¼ inch.

Companion Plants: Small grains, ryegrass, and other clovers.

Site Selection: Grows best on fertile, moderately well-drained neutral soils with high organic matter. Poorly adapted to light sandy soils.

Land Preparation: Break land in June or July and then prepare a clean seedbed before planting.

Fertilization/Liming: Follow soil test. If not available, apply 60 pounds phosphorus, 60 pounds of potassium per acre at time of land preparation and work into soil.

Management: Apply 40 pounds phosphorus and 40 pounds of potassium per acre annually.

SUBTERRANEAN CLOVER *Trifolium subterraneum*

A winter annual legume that provides supplemental food for **deer, turkey**, **quail** and **rabbits** in late winter and early spring. Native to the Mediterranean region. Used extensively in Australia and New Zealand. Has a growing season similar to crimson but not as long as arrowleaf. Can tolerate close grazing and produces seed near or just beneath the soil surface; therefore, it gets the name "sub clover." Good shade tolerance and reseeding characteristics. Does well planted under open stands or sawtimber sized pines. In these areas it can be established by burning ground litter followed by broadcasting of seeds and fertilizer. Common varieties include Mt. Barker, Woogenellup, Tallarook, Nangeela, and Meterora. Lower yields than crimson and arrowleaf. Production period during November to December and March to April.

Planting Rate: Broadcast 15–20 pounds per acre over prepared seedbed. In permanent sod, lightly disk or drill with a no-till or sod-seeder on 6–10 inch row spacing. Seeds should be inoculated before planting.

Planting Date: August 15–November 15.

Planting Depth: ½ inch.

Companion Plants: Can be planted in combination with warm–season perennial grasses, cool-season annuals or alone. Does well with ryegrass, small grains and other clovers.

Site Selection: Best suited to well-drained soils. A shallow seedbed is preferred.

Land Preparation: Smooth, firm, well-prepared seedbed.

Fertilization/Liming: Take soil test or apply 300 pounds of 0-20-20 and 10 pounds of 2–3 pounds boron per acre annually in the fall. Maintain a soil pH of 6.2–7.0 with liming.

Management: Apply fertilizer (no nitrogen) after plants established. Relatively free of pests.

SWEET CLOVER YELLOW - *Melilotus officinalis* WHITE - *Melilotus alba*

Generally two warm-season biennial and two annual species are planted. Sweet clover and alfalfa are closely related and have similar lime requirements, but sweet clover is shorter-lived, makes coarser growth and thrives on lower fertility soils. All sweet clovers contain a bitter substance with a vanilla-like odor called "coumarin." Spoiled sweet clover can be very harmful to livestock. Biennial sweet clover does not bloom the first year after planting. Growth starts the second year in February or March from buds that remain alive below the surface of the ground. It is a deep-rooted, drought resistant plant that grows on soils too poor to support other plants. Production period May through August. Used by **deer**, **turkeys** and **rabbits**.

Planting Rate: Broadcast 20 pounds per acre. Use sweet clover inoculant.

Planting Date: September 1–October 30.

Planting Depth: ¼ inch.

Companion Plants: Small grains, ryegrass, and other clovers.

Site Selection: Does best on fertile bottomland sites. Adapted to alkaline soils and other soils that have been heavily limed.

Land Preparation: Break and disk land in advance of planting and allow to settle into smooth, firm seedbed.

Fertilization/Liming: Apply 60 pounds of phosphorus and 60 pounds of potassium per acre at planting time or follow soil test.

Management: Repeat fertilizer (no nitrogen) application annually for maintenance. Sweetclover weevil may become a pest.

WHITE (LADINO) CLOVER *Trifolium repens*

Deer, **rabbits**, **turkeys**, and **quail** will readily use white clover for forage. Ordinarily a long-lived perennial in the northern sections of the South, but in most areas will act as a cool-season annual. Native to the Mediterranean region. White clover produces the bulk of its forage in late winter and early spring. With fertile soil and adequate moisture, it will produce year-round. Frost and drought, however, will stop growth. It is highly nutritious, has the longest grazing season, and the highest production of any pasture clover. It is shallow rooted and spreads by creeping branches and by seeds. Flowers are white and in clusters or heads. Common varieties include Regal, California, Tillman ladino, common white, New Zealand, La-S1 and Osceola.

Planting Rate: Broadcast 8–10 pounds of inoculated seed per acre. Inoculate seed with white clover inoculant.

Planting Date: Best results are obtained if seeded after September when soil moisture is favorable but can be planted in February. Usually planted from September 1–November 15.

Planting Depth: ¼ inch.

Companion Plants: Dallisgrass, bahia grass, ryegrass, small grains and other clovers.

Site Selection: Adapted to a wide range of soils, but performs best in heavier soils such as moist bottom clay or loam soils. Not suited for droughty upland soils of medium to low fertility or to extremely wet soils.

Land Preparation: Prepare a clean, smooth, firm seedbed and allow to settle before planting. If planting on established sod, disk lightly or plant with a sod seeding machine.

Fertilization/Liming: Soil pH should be between 6 and 6.5. Soil test if possible or apply 80 pounds of phosphorus and 80 pounds potassium per acre at planting time.

Management: After testing the soil, the recommended rate of phosphorus and potassium should be applied annually. Fall disking aids in reseeding of stands. Stands of white clover are eventually taken over by grasses but should persist for 3–5 years. Annual applications of 40–60 pounds of phosphorus and 40–60 pounds of potassium per acre should be made in August or September. Intermediate types of white clover usually reseed naturally while giant or ladino types do not reseed very well in the lower coastal areas. Leaf and root diseases may be a problem. Thinning of stands may help. Viruses are a serious problem and may cause clover to die in two or three years. Only solution is to replant. Grass competition from under-utilized plantings maybe a problem. In grass combinations, grasses should be planted in wide rows and clover broadcast to reduce competition.

CORN *Zea mays*

A choice food for many wildlife species including **bear**, **quail**, **dove**, **ruffed grouse**, **deer**, **turkey**, **squirrel**, **raccoon**, **ducks**, **geese**, and **non-game birds**. Originated from southern Mexico. For wildlife choose a variety that resists worms and produces heavy ears close to the ground. Seeds mature in 150 to 180 days depending on variety. Tropical corn is a variety that is early maturing and drought resistant.

Planting Rate: Plant 12–15 pounds per acre in rows 36 to 40 inches apart.

Planting Date: March 1–April 30.

Planting Depth: 1–1½ inch.

Companion Plants: Can be used with browntop millet, field peas, soybeans, reseeding soybeans, and winter legumes.

Site Selection: Well-drained soils are best. Moderately well-drained sites can be chosen if flooding is planned for waterfowl.

Land Preparation: Plant in a well-prepared seedbed or no-till in legumes, small grains, or old crop residues.

Fertilization/Liming: Follow soil test.

Management: For waterfowl, enough corn will probably fall naturally in a flooded condition to provide prolonged use. Rapid deterioration is reduced if allowed to fall naturally. For geese, a mechanical harvest on dry land adjacent to water areas will provide enough grain. For doves and quail, leave corn standing around the edges or harvest mechanically. Can also burn field after harvest to attract doves. For deer and squirrels, leave corn standing in patches around the edge of fields adjacent to wooded areas. Pests can include corn borer, corn rootworm and corn smut.

COWPEA *Vigna sinensis*

An annual, viney, summer legume with weak stems, large leaves and curved pods. With adequate moisture and temperature it will continue to grow indefinitely. Originating from Ethiopia. When available, **quail** and **doves** will feed almost exclusively on cowpea seed. In addition, **deer** and **rabbits** are able to forage from July until frost. Seasonal production from June through August. Other food plants are more economical and dependable for doves and turkeys. Good varieties are red ripper, clay, combine, whippoorwill, and purplehull. Cowpeas may be heavily browsed or may not be suitable in areas of high deer density.

Planting Rate: Drill 1½–2 bushels per acre, or plant in rows 24–30 inches apart using ½–¾ bushel per acre. Can be broadcast at 20–25 pounds per acre.

Planting Date: May 15–June 20, or after all danger of cold weather has passed. Matures in 80–180 days depending on variety.

Planting Depth: 1 inch.

Companion Plants: A mixture of cowpeas and browntop millet is a good choice for deer, quail, and turkeys. Also does well mixed with American jointvetch and alyceclover. Mix 40 pounds per acre of cowpeas, 10 pounds per acre of alyceclover, and 5 pounds per acre of American jointvetch. Plant cowpeas then drag in alyceclover or American jointvetch. Cowpeas also do well with sorghum.

Site Selection: Practically all soil types, but does best on more fertile, well-drained soils. Plantings of ¼ to ½ acre for quail. Recommend larger plantings when deer are present.

Land Preparation: The seedbed should be well prepared for best results.

Fertilization/Liming: Soil test first. Tolerates low fertility and soil acidity. The soil pH should be between 5.8 and 6.5. A soil test will recommend the amount of phosphorus and potassium to apply. On a soil with a low fertility level, use 10–20 pounds of nitrogen per acre at time of planting. Usually 200–300 pounds of 0-20-20 per acre is sufficient.

Management: Management is similar to soybeans. Keep clean if in rows, no cultivation necessary if broadcast. Seed non-reseeding varieties each year. Disk in early April to reestablish stands of the reseeding varieties. Annual applications of 60–80 pounds per acre of fertilizer should be made in August or September. Yields from 10–40 bushels per acre can be expected. Pests include grasshoppers, curculio weevils, southern cornstalk borer, armyworm, rusts and leafspots.

DALLIS GRASS *Paspalum dilatatum*

A long-lived perennial with a fairly deep, strong root system. Originated in Argentina, Uruguay, and Brazil. It grows in bunches 2 to 3 feet high, but forms a sod under heavy use. The leaves are numerous near the ground but few on the stem. Seeds are eaten by a variety of **game** and **non-game birds**. Seedheads are subject to ergot, a black fungus which is toxic to livestock.

Planting Rate: Plant 10 pounds of seed per acre.

Planting Date: February 15–May 30.

Planting Depth: ½ inch.

Companion Plants: White clover, red clover, annual lespedeza and winter peas.

Site Selection: Best suited to fertile, moist, heavy, well-drained to moderately well-drained soils such as clay and loam soils. Survives periodic flooding.

Land Preparation: Prepare a well pulverized, firm seedbed by breaking, disking and harrowing.

Fertilization/Liming: Follow soil test recommendations or apply at planting time 500 pounds of 13-13-13 or its equivalent per acre and 1–2 tons of lime per acre.

Management: Apply 40 pounds of nitrogen, 40 pounds of phosphorus and 40 pounds of potassium per acre annually. When the grass makes seed, it stops growing. Do not allow seed heads to form until fall. Frequent mowing or grazing will keep down ergot fungus. Sugarcane borer can also be a pest problem. Seeds can be harvested directly by combining or by cutting grass with a mower, allowing to dry, and then using a combine with a pickup attachment.

EGYPTIAN WHEAT *Sorghum* spp.

Egyptian wheat is an annual sorghum that grows to heights of 8 feet. It grows in thick stands, and heads will easily fall to the ground (lodge) at maturity. Makes cover and choice seeds for **quail** and **turkeys**.

Planting Rate: Drill 10 pounds per acre in 36 inch rows at ¼ inch or broadcast 6–10 pounds per acre and cover ½ inch.

Planting Date: April 1–June 30.

Planting Depth: ¼–½ inch.

Companion Plants: Can be strip planted alternatively with other warm-season grasses.

Site Selection: Widely adapted to well-drained, light textured soils.

Land Preparation: Plant in well-disked plots. Best to plant in patches 12–15 feet wide and 30–50 feet long. Excellent for providing cover in large fields.

Fertilization/Liming: A soil test is recommended or use 250 pounds per acre of 13-13-13. Apply lime according to soil test to maintain pH level at 5.5–6.5.

INDIANGRASS *Sorghastrum nutans*

A perennial bunchgrass native to tall grass prairies of the eastern Great Plains, and eastern U.S. Spreads by rhizomes and seeds, growing 3 to 6 feet tall. Heat and drought tolerant. Seeds provide food for game and **non-game birds**. Nutritional quality is generally better than most other warm-season perennial grasses.

Planting Rate: Seeds should be planted at 6–10 pounds per acre.

Planting Date: April–May.

Planting Depth: ¼ inch.

Companion Plants: Grows with other grasses and legumes.

Site Selection: Well-drained, fertile soils.

Land Preparation: Prepare a clean seedbed.

Fertilization/Liming: Most fertilizers with a high nitrogen content. Soil pH should be approximately 6.

Management: Difficult to establish due to warm-season weed competition, especially crabgrass. Best to establish in new ground that has just been taken out of trees.

LABLAB *Dolichous lab-lab*

An annual, viney, drought-tolerant, summer legume native to Africa and used extensively in Australia for cattle grazing because of hardiness and high crude protein content. Tests in Texas reveal that Lablab is similar to Iron and Clay cowpea in production, quality and use by **deer**. However, preliminary results indicate that Iron and Clay cowpeas may produce more forage per acre than lablab. Deer readily feed on the foliage of Lablab through the summer and into November if not severely browsed down. If deer feed of plants in the early growth stage plants often die.

Planting Rate: Seeding rate is 40–50 pounds per acre, depending on planting method. Can broadcast or plant in conventional rows with 36 inch row spacing. Better results if planted in rows. It will compete well with weeds if planted in rows, but weeds will often overtake plots if established by broadcasting.

Planting Date: May.

Planting Depth: 1–1½ inches.

Companion Plants: A mixture of cowpeas.

Site Selection: Practically all soil types, but does best on more fertile, well-drained soils.

Land Preparation: The seedbed should be well prepared for best results.

Fertilization/Liming: Follow soil test for best results. The pH should be between 5.8 and 6.5. A high fertility level is essential for good growth. Fertilize with 30 pounds of phosphate and 60 pounds of potash per acre. Commonly used fertilizer is 0-10-20 at 300 pounds per acre.

Management: Management is similar to soybeans. Pests include grasshoppers, curculio weevils, southern cornstalk borer, armyworm, rusts, and leafspots.

LESPEDEZAS (ANNUAL) *Lespedeza* spp.

Kobe, common, Korean and Sericea lespedeza are good reseeding annuals that are low-growing and leafy. Sericea provides good cover but seeds are of low value to most wildlife. Originally from eastern China, Korea and Japan, all other varieties produce abundant seeds that are highly favored by **quail**, but its use by other species is slight. **Deer** occasionally browse annual lespedezas if preferred browse is unavailable. Plot size for quail should be ¼ to 1 acre. Seed yield is usually 100–400 pounds per acre. Seasonal production during July and through September.

Planting Rate: Broadcast 40 pounds of common, Kobe, or Korean seed per acre. Inoculation of seeds is usually not necessary; however, inoculation will ensure that the proper *Rhizobium* bacteria is present.

Planting Date: February 15–March 31. Plant unhulled seed early, hulled seed later.

Planting Depth: Cover seed ¼ inch or less.

Companion Plants: Can be over-seeded on established small grains; however, it is not practical to be used with small grains that are heavily fertilized with nitrogen.

Site Selection: Widely adapted. Especially useful on low-fertility, acidic soils with low phosphorus. Under high fertility, annual lespedezas are often crowded out by more vigorous and higher yielding grasses and legumes. Grows poorly or not at all in wet or deep sands.

Land Preparation: Prepare a firm, clean seedbed by breaking, disking, and harrowing. Allow to settle before seeding.

Fertilization/Liming: A moderate application of phosphorus and potassium is desirable. Nitrogen is not essential, but 10–15 pounds per acre at time of seeding will improve growth.

Management: Reestablish by disking or controlled burning. Care should be taken in fertilizing grass/lespedeza mixtures since high fertilization reduces potential growth of lespedeza. Pests that can reduce yields and stands include bacteria wilt, tar spot, powder mildew and southern blight.

LESPEDEZAS (SHRUB) *Lespedeza* spp.

Perennial shrubs that grow to a height of 5 to 8 feet. Common species are Amquail, japonica, ambrow, appalow, bicolor, and thunbergii attaway. Some hybridization occurs. Shrub lespedezas produce between 65,000 and 85,000 seeds per acre. Bicolor and thunbergii stems remain alive above the ground but japonica dies back to the ground every winter. **Deer** prefer browsing on bicolor to thunbergii or Amquail. Seeds of all three are the choice food of **quail**. Use bicolor or thunbergii where fall frosts occur after October 20. Use japonica where frosts frequently occur before October 20. Lespedeza japonica will drop its seed considerably earlier than either bicolor or thunbergii and therefore is not as valuable as a late winter food source.

Planting Rate: Plant scarified seeds in rows, 15 to 20 seeds per foot or about 5 pounds per acre. Can also broadcast at about 15 pounds per acre. Inoculation is not usually necessary; however, inoculation will ensure that proper *Rhizobium* bacteria is present. To plant seedlings set plants every 18–24 inches in rows 36–42 inches apart to fit cultivating equipment (usually 36–42 inches). Wider spacing is not suitable since bicolor cannot maintain a satisfactory stand unless it shades the ground thoroughly after reaching maturity in 2–

3 years. Recommend planting seed on new openings relatively free of competition.

Planting Date: Plant seeds in early spring, February 15–March, or about expected date of last spring frost. Seedlings should be planted in winter or early spring.

Planting Depth: Plant seeds ¼–½ inch deep. Seedlings to root collar.

Companion Plants: A good companion plant for shrub lespedeza is kobe lespedeza. Can be an excellent way to establish shrub patches in open woodlands or other "new ground." Broadcast a mixture of 20 pounds per acre of shrub seed with 20 pounds per acre of kobe seed on a well disked, fertilized, limed, and packed strip (20' x 300').

Site Selection: Well-drained soils, but not on deep sands.

Land Preparation: A clean, well-disked seedbed is preferred.

Fertilization/Liming: Soils test first. Shrub lespedeza has a moderate to high fertility requirement. Use 500 pounds per acre of 0-20-20 or similar analysis when planting on field borders or other depleted areas. Use 250 pounds per acre in woodland areas.

Management: Shrub lespedezas should be mowed every two to three years to keep from becoming too woody and to stimulate spreading and seed production. Apply high phosphorus fertilizer annually beginning second year. If weeds are a problem control grasses with Poast® at ¾–1 ¼ pints per acre. Apply at a lower rate in sandy soils. Poast® should be mixed into 30 gallons of water. For broadleaf weeds use 2,4-D (Amine formulation) using same rates as Poast®. Seeds harvested commercially by combine as soon as seeds are mature.

BROWNTOP MILLET *Panicum ramosum*

An annual panic grass from Southeast Asia which produces a yellowish-brown, open-panicle seedhead. Early growth is rapid and produces mature seed in 60 to 70 days. Seeds averages 214,000 per pound. A choice food of **game birds** such as **quail**, **doves**, **ducks**, **turkey**, and **non-game birds**.

Planting Rate: Drill or broadcast 20–30 pounds per acre for hay; 8–10 pounds per acre in 2 ½ to 3 ½ foot rows of commercial seed production. For doves broadcast 25 pounds per acre or plant 8–15 pounds seed per acre in rows.

Planting Date: Plant from July to August for waterfowl fields. For doves, plant 60–80 days before the hunting season, usually May 15–July 15. Plant early enough for at least a portion of the field to mature two weeks prior to shooting. Recommend several plantings two to four weeks apart.

Planting Depth: Cover seeds lightly.

Companion Plants: Other millets, corn, cowpeas and sunflowers. Planting a field in strips with various millets that have different maturing times (such as browntop, proso and sunflowers) will attract doves throughout the season.

Site Selection: Almost any upland or bottomland soils with a water table 4 inches or more below the surface from July through September.

Land Preparation: Prepare a good seedbed by disking. A light second disking to kill sprouting weeds helps just before planting. No cultivation for broadcast or drilled planting, clean cultivation for row plantings, especially for doves.

Fertilization/Liming: Follow soil test or use 10-10-10 or similar fertilizer at about 500 pounds per acre. Extra nitrogen usually reduces seed yields. Liming should maintain pH between 5.5 and 6.5.

Management: For waterfowl management flood in the fall and leave flooded until waterfowl go north in the spring. Depth of flooding should be 2–15 inches. Size of plantings should be ¼ acre for quail, 3 acres or more for doves, 5 acres or more for waterfowl, and 1 acre or more for turkeys. For doves keep areas between rows as weed free as possible by cultivating two to three times during growing season or using a pre-emergent herbicide. Broadcast and drilled plantings may need to be raked after mowing to remove stalks and leaving seeds or exposed soil. Fall armyworm can be a pest.

FOXTAIL MILLET *Seteria italica*

An annual, erect plant 3 to 4 feet tall originating from southern Asia. Seeds are an excellent source of food for **game birds** and **non-game birds**. Fairly drought tolerant. Matures in 60 to 70 days. Less productive than pearl millet or sorghum-sudan hybrids. Widely grown at one time but not commonly grown now.

Planting Rate: Seed can be drilled at 15–20 pounds per acre or broadcast at 20–30 pounds per acre.

Planting Date: May–July.

Planting Depth: Cover seeds lightly ¼ inch.

Companion Plants: Other millets, corn, cowpeas and sunflowers.

Site Selection: Well-drained soils.

Land Preparation: Prepare smooth seedbed.

Fertilization/Liming: Follow soil test or use 13-13-13 or similar fertilizer at about 300–400 pounds per acre. Extra nitrogen will usually reduce seed yields. Liming should maintain pH between 5.5 and 6.5.

Management: Size of plantings should be ¼ acre for quail, 2 acres or more for doves, and 1 acre or more for turkeys. Fall armyworm can be a pest.

JAPANESE MILLET *Echinochloa crusgalli* var. *frumentecea*

An annual reseeding grass of Asiatic origin closely related to barnyard grass but produces heavier seed yields. The number of seeds per pound is about 145,000. Seeds are one of the choice foods of **gadwall**, **mallard**, **wood duck**, **teal**, **turkeys**, and **widgeon** and also is used by several **non-game birds**.

Planting Rate: Plant 15–20 pounds of seed per acre broadcasting or drilled.

Planting Date: Preferably from July–August. Seeds matures in 90 days. Maturation of seeds should coincide with flooding for waterfowl.

Planting Depth: Cover seed ¼ inch or less.

Companion Plants: Other millets.

Site Selection: Grows on wetlands, especially those too wet for browntop millet. This plant, like barnyard grass, will not germinate under water but will germinate on exposed muds tolerating shallow flooding during growth.

Land Preparation: For waterfowl, expose mud flats and broadcast seed or disk dry land and broadcast seed. Follow soil test or use 10-10-10 or similar fertilizer at rates of 500 pounds per acre. Do not fertilize when seed is broadcast on exposed muds.

Management: For waterfowl, flood plants in the fall and leave flooded until waterfowl go north in the spring. Depth of flooding should be 2–15 inches. Chipawa variety takes 120 days to mature and is much taller than other varieties and is useful on areas where drawndown conditions are favorable for planting early in the season.

PEARL MILLET (CATTAIL) *Pennisetum glaucum*

An annual, leafy millet 3 to 8 feet tall originating from north-central Africa. Choice food of **dove**, **quail**, **turkeys** and **non-game birds**. Very productive over a short season from June to August.

Planting Rate: Drill seed at 8–10 pounds per acre or broadcast at 25–30 pounds per acre.

Planting Date: April–June.

Companion Plants: Other millets, corn, cowpeas, and sunflowers.

Site Selection: Fairly well adapted. Does not do well on calcareous soils but tolerates drought and soil acidity.

Land Preparation: Prepare a smooth seedbed.

Fertilization/Liming: Follow soil test or use 10-10-10 or similar fertilizer at about 500 pounds per acre. Extra nitrogen will usually reduce seed yields. Liming should maintain soil pH between 5.5 and 6.5.

Management: Size of plantings should be ¼ acre for quail, 2 acres or more for doves, and 1 acre or more for turkeys. Fall armyworm can be a pest.

PROSO MILLET (DOVE PROSO) *Panicum miliaceum*

An annual panic grass native to central Asia. Growth is rapid and produces seed in 80 to 90 days. The seed is the largest of the millets. There are approximately 80,000 seeds per pound. Seeds are a choice food to both **upland game** and **non-game birds** and **waterfowl**.

Planting Rate: Plant 20–35 pounds per acre drilled or 8–10 pounds per acre in 2½-to-3½ foot rows. Row planting is best for dove fields. Can also broadcast at 25 pounds per acre.

Planting Date: From late spring to midsummer. For dove fields, the crop should be planted 90–110 days before hunting season.

Planting Depth: Cover seed lightly ¼–½ inch.

Companion Plants: Other millets, corn, cowpeas and sunflowers. Can be planted in same field at same time as browntop. Different maturity times will keep seed available for doves for a longer period of time.

Site Selection: Almost any soil except those with excessive moisture during the growing season. Tolerates dry conditions.

Land Preparation: A well prepared seedbed is required for best results. A light disking just before planting helps kill sprouting weeds. Cultivation is not required for drilling or broadcasting but is necessary for row planting.

Fertilization/Liming: Fertilizer and lime needs should be based on soil test. If soil test is not conducted use 10-10-10 or similar fertilizer at 500 pounds per acre. A soil pH of 6.0–6.5 is best.

Management: Size of plantings should be ¼ acre for quail, 3 acres or more for doves, and 1 acre or more for turkeys. Fall armyworm can be a pest.

ORCHARDGRASS *Dactylis glomerata*

A cool-season perennial grass originally from Europe. Orchardgrass is moderately used by **deer** and provides good insect-feeding areas for **turkeys**. Orchardgrass is more shade-tolerant than most grasses. In the lower South stands usually do not last more than four years.

Planting Rate: 15–20 pounds per acre.

Planting Date: Plant only in northern Alabama from August 15–November 1.

Planting Depth: ½ inch.

Companion Plants: Orchardgrass does not take competition well. Light seeding rates of legumes (clovers) are an option as companion plants.

Site Selection: Best adapted to well-drained productive soils. Will not tolerate poor drainage.

Land Preparation: Prepare a smooth seedbed.

Fertilization/Liming: Requires high fertility soils and responds well to nitrogen. Follow soil test recommendations.

Management: Annual applications of 80 pounds of nitrogen per acre in September will improve growth. Fall armyworm, rusts and leafspot disease can be a problem. In sandy soils nematodes can be a serious pest. Consult local Extension agent for pesticide recommendations.

PARTRIDGE PEA *Cassia fasciculata* and *nictitans*

Two species of partridge pea are available in Alabama, showy partridge pea (*C. fasciculata*) and sensitive partridge pea (*C. nictitans*). Both are summer annuals with stems 1 to 3 feet tall and many branched. Flowers are bright yellow. Seed pods are flat and hairy with dark black seeds. Seeds are an excellent source of food for **quail**, **turkey**, **doves,** and **songbirds**. Seeds persist through the fall and through much of the winter. Partridge pea is often found in abandoned fields, fencerows and areas that received periodic disturbances.

Planting Rate: Broadcast 2–15 pounds of seed per strip (⅕ acre = 30 feet x 300 feet) and drag in or cultipack. Can plant 5–10 pounds per acre in 36 inch rows. At time of planting, mix inoculant with 1 ounce of water, add to seed, and stir until all seeds are coated. Plant as soon as seeds are dry. Keep inoculated seed in the shade.

Planting Date: In February or early March. Early planting provides time for the seeds to absorb water. Seed scarification is not recommended because the seed embryos are easily damaged.

Planting Depth: ¼–½ inch.

Companion Plants: Annual lespedezas or alone.

Site Selection: Most soils and sites except for continually wet sites.

Land Preparation: Prepare clean seedbed.

Fertilization/Liming: Apply lime and fertilizer according to needs indicated by soil test. If test is not available, apply one ton of lime per acre and 400 pounds of 0-10-20, or 400 pounds of 0-15-30 per acre.

Management: Apply fertilizer annually and lime every 3 years. Will respond to burning or light disking October–February as a maintenance practice in established stands. Ground scarification should be completed before partridge peas begin to emerge in late February. Seasonal disturbances in strips from October 15–March 1 in old fields and woods openings can encourage volunteering. If herbicides are needed, incorporate Treflan® before planting at rates of ¾ to 1¼ pints per acre. Use a lessor rate on sandy soils. Disk shallow 2 times to mix with soil. The herbicide 2, 4-D amine may be sprayed over the top for broadleaf weed control when partridge pea plants are 5–10 inches tall. Use ¾ pint of 2, 4-D amine in 30 gallons of water per acre. Poast® is recommended for grass control at the rate of 1 pint per acre.

PEANUT *Arachis hypogaea*

An erect annual with yellow flowers whose nuts are eaten by **deer**, **turkeys**, **raccoons** and **feral hogs** during July through October. Foliage is eaten by deer. Peanuts may be heavily browsed or may not be suitable in areas of high deer density.

Planting Rate: Plant seeds 3–6 inches apart in 30–38 inch rows.

Planting Date: April 15–May 15.

Planting Depth: 1½–2½ inch.

Companion Plants: None.

Site Selection: Grows best on well-drained sandy soils and in fields that were previously planted in corn. Do not select fields where soybeans or peanuts were planted the previous year or in fields having severe weed problems. Will not tolerate poor drainage.

Land Preparation: Chop up old crop residue and disk lightly as soon as possible after corn or other crop is harvested. In the spring, break land deeply (8–12 inches) so that all plant residue is covered several inches.

Fertilization/Liming: Test soil for proper lime and fertilizer needs. Apply recommended lime and work into the soil several weeks before planting. A soil pH between 5.8 and 6.5 is best. Does well with high levels of residual fertilizer in the soil or apply appropriate amounts of phosphorus and potassium.

Management: Pests such as nematodes, white mold and pod rot might have to be controlled.

PEANUT, PERENNIAL *Arachis glabrata*

A leafy perennial that is long-lived, 12 to 16 inches tall, and spreads by rhizomes. Perennial peanuts have yellow or orange flowers. Originally from Brazil. Readily eaten by **turkeys** and **deer**. Value is more a forage. Produces few seeds and is propagated vegetatively from rhizomes. Can survive temperatures down to 15°F, but a long, cool season greatly reduces potential production. High nutritional quality, but requires one to two years for establishment of a productive stand. Used by wildlife in the fall.

Planting Rate: Rhizomes are planted at about 60 to 80 bushels per acre.

Planting Date: December to early March.

Planting Depth: ½–1 inch.

Companion Plants: Perennial grasses can be interplanted after the peanuts are well established.

Site Selection: Grow best on well-drained sandy soils. Will not tolerate poor drainage.

Land Preparation: Prepare a smooth and clean seedbed.

Fertilization/Liming: Low fertility requirements. Lime application generally necessary at about 1 ton per acre.

Management: Disking in December and early spring to stimulate regrowth every one to two years.

RICE *Oryza sativa*

Lemont, Mars, Labelle varieties are grown for **waterfowl** in a few regions of the Southeast. Difficult to produce where blackbirds are a problem. Valued also for cover and as a substrate and food for invertebrates which many species of waterfowl readily feed upon.

Planting Rate: Broadcast or drill at 100 pounds per acre.

Planting Date: May 20–June 10. Matures in 90–100 days.

Planting Depth: 1 inch.

Companion Plants: None.

Site Selection: Old ricefields in Lower Coastal Plain. Grows on most sites that are sandy loam to loam. Does not do well on sandy sites. Moist areas or edges of standing water/mud flats good.

Land Preparation: Prepare a clean, smooth seedbed by disking.

Fertilization/Liming: Broadcast 200–300 pounds of 5-10-10. Topdress with 60 pounds of nitrogen per acre.

Management: Flood to bring water right up to heads in the milk stage to help reduce blackbird damage. Weeds compete with rice for nutrients, space and sunlight and in severe cases infestations of weeds can completely choke out rice. Rice for harvest must be clean. Rice for waterfowl or crayfish can tolerate some weeds. Cultural practices that help control weeds include good seedbed preparation, planting viable seed to ensure a good stand, and flooding properly.

RYEGRASS (ANNUAL AND PERENNIAL) *Lolium multiflorum*

A dense growing, sod-forming, cool-season annual grass which is hardy. Originally from Europe. Ryegrass is highly competitive and often crowds out other plants in early spring. Makes rapid growth in April and May. Used sparingly by **deer**, **turkey**, and **rabbits** in winter and early spring. Most common varieties are Gulf and Marshall. Marshall is more cold tolerant and higher yielding. Ryegrass can become a pest if fields are to be used later for commercial small grain crops.

Planting Rate: Broadcast or drill 20–40 pounds per acre of seed. Use 10–15 pounds of seed per acre in mixtures.

Planting Date: August 25 through October 15. September or early October are generally the best months to plant, but November overseeding of warm-season grasses can be done along the coast. Natural reseeding is common. Tolerates close continuous grazing. In high rainfall areas of the coast, high production can be expected through the winter from November to May. Farther north, most of the production is concentrated from late February or March through May.

Planting Depth: Cover seed lightly ¼ inch in firm soil.

Companion Plants: Ryegrass and crimson clover make a good fall to spring combination planting on well-drained soils and can be readily maintained by disking or mowing and fertilizing each September. Other annual clovers, winter peas, and small grains also do well with ryegrass. Ryegrass will have a tendency to dominate companion plants in many cases. A light seeding rate may be necessary.

Site Selection: Best on fertile soils. Prefers a moderate to well-drained soil. Tolerates wet, poorly drained soils.

Land Preparation: Prepare a clean, smooth seedbed.

Fertilization/Liming: Apply according to soil test or broadcast 400–500 pounds of 10-10-10 at planting. Top-dress with 40 pounds of nitrogen in February if necessary. Maintain a soil pH of 6 or without a soil test apply 1–2 tons of lime per acre.

Management: Maintained for several years by mowing or disking in September with reapplication of fertilizer. Ryegrass seeds mature in late May and June. Seeds can be harvested by direct combining. Seed production can be hampered by rust attacks especially when within 100 miles of the coast. Armyworms can be a problem.

SAWTOOTH OAK *Quercus acutissima*

A medium-sized, deciduous, exotic tree from China which produces abundant acorns in about five to eight years and used extensively by many species including **ruffed grouse**, **turkey**, **black bear**, **beaver**, **gray** and **red foxes**, **raccoon**, **squirrel**, and **deer**. Acorns are large with annual production from 40 to 80 pounds. Sawtooth oak acorns are commonly undamaged by insects or diseases.

Planting Rate: Plant seedlings spaced 25 x 25 feet or 30 x 30 feet apart. Mature trees will need 100 feet of spacing for good production. Direct seeding of acorns, that are treated with a rodent repellent, is also a method of establishment. Acorns should be planted soon after harvesting or stored in a cool dry place until planting.

Planting Date: Fall and early winter for acorns. February through March for seedlings. Acorns must be harvested soon after falling, planted immediately, or put in cool storage. Seedlings will stand intermittent flooding without damage in the dormant season but not in the growing season.

Planting Depth: Acorns should be planted 1 inch deep. Seedlings planted to root collar.

Companion Plants: Winter cereal grains may be grown in combination with sawtooth oak after seedlings are well established.

Site Selection: Dry to medium-moist upland sites. Will not do well on soils classified as wetlands. Requires sunlight.

Land Preparation: Mow to control weeds, brush, vines, and briars until seedlings outgrow their competitors. To obtain the best results with sawtooth oak cultivate and fertilize for the first 5 years. Sawtooth oak will not compete well with other woody plants and competition must be controlled to prevent seedling mortality. Due to the need for weed control, planting should be made in orchards rather than strips.

Fertilization/Liming: Soil test first. A complete fertilizer will stimulate growth but should not be applied until the second season after seeding or planting. Liming does not seem to be important.

Management: Except for the year of planting, annually apply 3 pounds of nitrogen per 1,000 square feet beneath and just beyond the branch spread. Acorn production is greater with annual application of fertilizer. Remove all overstory competition with adjacent trees to within 6 feet of the nearest branches. Sawtooth oaks must be protected from fire.

SESAME (BENNE) *Sesamum indicum*

An annual that is 4 to 7 feet tall and sometimes referred to as swamp pea. Choice food of **doves**, **quail**, and a variety of **songbirds**. Seed is available for wildlife from September through January and is utilized most heavily after the first killing frost. Sesame may be heavily browsed or may not be suitable in areas of high **deer** density. There are two types of sesame that include shattering varieties where the seed capsules pop open when dry and non-shattering varieties. Shattering varieties include Margo, Oro, Blanco, Eva, Dulee, and Ambia. Non-shattering varieties include Baco, Delco, Rio, and Palmetto.

Planting Rate: Plant seeds in 36 inch rows at 5–6 pounds per acre. Broadcast at a rate of 10–12 pounds per acre.

Planting Date: May 1–June 1. Matures in 120–150 days.

Planting Depth: Depending on soil texture, plant 1–2 inches deep. Heavy soils that dry quickly after being disturbed require deeper planting.

Companion Plants: Has competed well with other small grains in mixtures.

Site Selection: Prefers well-drained, medium-textured soils. Do not plant on same site two years in a row due to wilt.

Land Preparation: Prepare a clean, smooth seedbed. Generally, seedbed preparations used for cotton do well for sesame.

Fertilization/Liming: Soil test first or use 500–600 pounds of 10-10-10 per acre. Best method is to use 500 pounds of 5-10-15 per acre at planting followed by 40–60 pounds of nitrogen per acre after establishment.

Management: Cultivate to keep areas between rows clean. Multiple diskings (three to four) in the middle of May, prior to planting to control weeds.

SINGLETARY PEA *Lathyrus hirsutus* or spp.

Also known as caley peas or rough winter peas, it is an annual reseeding winter legume which makes rapid growth in late winter and early spring. Native to the Mediterranean region. Small red and blue flowers that form seedpods. **Doves**, **quail**, and **turkeys** will eat seeds. **Deer** will graze the tender foliage when other foods are in short supply.

Planting Rate: Broadcast 30–60 pounds per acre. Scarified seeds are the most desirable. Inoculate with a vetch inoculate just prior to seeding.

Planting Date: September 1 through October 15. Usually ready to graze in late January.

Planting Depth: After broadcasting, the seeds can be harrowed in to a depth of ¼–½ inch. Sod seeding machines can also be used in planting winter peas in a grass sod.

Companion Plants: Grows well when overseeded with bahia grass or dallisgrass.

Site Selection: Widely adapted to most soils except poorly drained soils. Best suited to heavier soils of high to medium fertility. Grows on soils too wet for clovers and also does well on both acidic and calcareous soils.

Land Preparation: No preparation needed if following cleanly harvested crops. On sod or weedy fields, a good seedbed should be prepared.

Fertilization/Liming: Soil pH should be between 5.8 and 6.2. Soil test as needed or 60 pounds of phosphorus and potassium per acre.

Management: Allow peas to form mature seeds every two to three years to ensure a reseeding stand. Remove excessive growth of grass in September to facilitate germination of peas. Disking in the fall every two years will encourage production. Hard seed coat helps ensure reseeding. Repeat annual applications of phosphate and potash. Repeat lime application every five years or as needed to maintain a soil pH of at least 6. Produces 300–400 pounds of seeds per acre. Can be harvested by direct combining. Aphids are the most common pest.

BARLEY *Hordeum* spp. RYE *Secale cereale* OATS *Avena* spp.

Small grains are cool-season annuals. Origin is from Iraq, **Turkey** and Europe. Rye is the most cold tolerant of the small grains, followed by barley, wheat, and oats. Oats can be winter-killed in some years. **Deer** are able to graze foliage from December through March while **doves**, **ruffed grouse**, **quail**, and turkey consume seeds in late spring and summer. Provides grazing for **geese** and some species of **ducks**.

Planting Rate: Broadcast 80–120 pounds of seed (about two bushels) per acre where seedbeds are conventionally prepared. Use 100 pounds of seed per acre when planting over fall crop stubble. In mixtures, 60 to 90 pounds of seed per acre are recommended.

Planting Date: September 1–November 15.

Planting Depth: Cover seeds to about 1 inch in depth with a drag or cultipacker.

Companion Plants: Can be interseeded with many of the annual or perennial clovers such as arrowleaf clover to extend the production season and maintain spring forage quality. A mixture of small grains is also popular. Small grains can also be planted adjacent to ryegrass and clovers.

Site Selection: Adapted to most well-drained soils. Best growth is obtained on clay loams and sandy loams. Deep sandy soils and wet, heavy soils are unsuitable. Of the four grains, rye is best suited for sandy soils.

Land Preparation: Plant on a well-prepared seedbed. Can be overseeded on warm-season grass sods.

Fertilization/Liming: Obtain soil test first. Apply 20–50 pounds of nitrogen per acre at planting and again the first of February. Soil pH should be above 5.7. If the soil test for phosphorus and potassium is less than high, the recommended rates of these elements should be applied before planting.

Management: Needs to be reseeded each year. Some natural reseeding will occur if allowed to mature. Ragweed, another important wildlife food, proliferates in the stubble of cut wheat. Fall armyworm and Hessian fly can be a problem. To attract doves wheat fields can be burned after harvest. Full season wheat fields generally provide excellent shooting the following fall after harvest.

SORGHUM *Sorghum* spp.

An annual, small grain crop that is closely related to corn with a similar high nutritional value to wildlife, especially **doves**, **quail**, and **turkeys** as well as **mallards**, **geese**, **pintails**, and **teal**. Originally from northeast Africa. Known for being drought tolerant. The small size of the kernel, which is readily eaten by **birds**, is sorghum's advantage over corn. A variety known as Egyptian wheat is commonly planted for quail and shows some **deer** browsing resistance. Other wildlife varieties include Kafir, Hagair, milo, and small game food sorghum. The latter variety is a dwarf, ranging in height from 18 to 30 inches. A variety known as NK 300 is a tall sorghum while NK 2660 is a grain sorghum. Grain sorghum is used as a feed grain, forage, and silage for livestock, though it is seldom pastured because of the risk of prussic acid poisoning. Deer seldom browse on sorghum but will eat seedheads when mature.

Planting Rate: Plant three to five plants per linear foot in rows 30 to 36 inches apart. This usually requires 4–7 pounds of seed per acre. For wildlife, broadcasting is acceptable.

Planting Date: April 15–July 15, or anytime when the soil temperature reaches 65°F.

Planting Depth: 1 inch deep.

Companion Plants: Corn, soybeans, reseeding soybeans, cowpeas, and singletary peas.

Site Selection: Most well-drained soils, especially clay loams and sandy loams.

Land Preparation: Requires complete seedbed preparation. Do not till into legumes, small grains, or old crop residues.

Fertilization/Liming: Apply approximately 90 pounds of nitrogen per acre either just before or just after planting. A portion may be broadcast, but at least ⅓ should be banded by rows. Add lime if the soil pH is less than 5.8. Does not tolerant highly acidic soils. Broadcast phosphorus and potassium if the soil test reveals that these minerals are low. Commonly used fertilizer is 10-10-10 at 500–600 pounds per acre.

Management: Use cultivation or herbicides to keep weed populations at a minimum. Pest problems could occur with leafspots, lesser cornstalk borer, and sorghum midge.

SORGHUM X SUDAN HYBRIDS AND SUDANGRASS *Sorghum bicolor*

An annual erect plant 4 to 8 feet tall which originated from northeast Africa. Seeds readily eaten by **quail**, **dove**, **turkey**, and **non-game birds**. Productive over a short season from June through September. Very drought tolerant.

Planting Rate: Drill seed at 20–25 pounds per acre or broadcast at 30–35 pounds per acre.

Planting Date: April 1–August 15.

Planting Depth: 1 inch.

Companion Plants: Corn and soybeans.

Site Selection: Most well-drained clay or sandy loams.

Land Preparation: Requires complete seedbed preparation. Do not till into legumes, small grains, or old crop residues.

Fertilization/Liming: Responsive to high nitrogen fertilizers and needs lime on highly acid soils. Commonly used fertilizer is 10-10-10 at 500–600 pounds per acre.

Management: Fall armyworms can be a problem.

SOYBEAN *Glycine max*

A common summer annual legume produced for seed and wildlife food. Originating from China. Soybeans are an important source of winter food for **doves**, **duck**, **rabbit**, **quail**, and **turkeys**, but the principle use among game animals is by **deer**. Deer readily feed on the vegetative parts of the plant all through the growing season. In high deer density areas, deer will decimate soybean plantings. Soybean fields are especially attractive because the plants also provide cover for feeding wildlife. In small food plots of 1 to 2 acres, plant a later maturing, reseeding variety of soybeans or mix with corn or sorghum. Seasonal production July through August.

Planting Rate: Plant 10–12 seeds per foot (about 30–60 pounds per acre) in rows 30–40 inches apart, or broadcast using 1 bushel of seed per acre. Inoculation of seed is essential unless the land has been planted in soybeans within the past few years.

Planting Date: May 1–July 1. Dates vary depending upon variety and location.

Planting Depth: 1 inch.

Companion Plants: Corn or sorghum.

Site Selection: Widely adapted to well-drained and moderately well-drained, medium-textured soils and heavier, textured soils. Grows best in a pH range of 5.8 to 7.0. A high soil fertility level is essential for good growth.

Land Preparation: Prefers a well-prepared seedbed, but may be no-till planted in wheat stubble or old crop residues.

Fertilization/Liming: Follow soil test for best results. The pH should be between 5.8 and 6.5. A high fertility level is essential for good growth. Fertilize with 30 pounds of phosphate and 60 pounds of potash per acre. Commonly used fertilizer is 0-10-20 at 300 pounds per acre.

Management: For food plots, leave plants standing. If growing seed commercially, leave several rows of soybeans around field edges. If possible, plant a later maturing variety. If planting for waterfowl, seeds deteriorates rapidly after flooding (about 30 days). Armyworms may be a pest problem.

SUNFLOWER *Helianthus annuus*

A summer annual whose seeds are rich in oil and readily consumed by **doves**, **ruffed grouse**, **quail**, **songbirds**, **gray squirrels**, and **turkeys**. The small black variety known as Peredovik is a favorite of doves and is widely available. Four to 5 feet tall, it provides good cover for dove shooters when left standing. Another variety, Armavirec, is also recommended and produces seed two weeks earlier than Peredovik. A field with a minimum of 5 acres should be planted to attract a huntable number of doves. May be heavily browsed by **deer** in areas of high deer numbers.

Planting Rate: Space seeds 8–10 inches apart in rows 36 inches apart. Usually requires 6 to 10 pounds of seed per acre. Seeds mature in about 110–120 days.

Planting Date: March 1–August 1. For doves, plant around May 15. Seed matures in 90–120 days.

Planting Depth: 1 inch.

Companion Plants: Browntop planted between sunflowers, or corn adjacent to sunflowers.

Site Selection: Most well-drained soils, but best results have been obtained on clay, clay loam, and sandy loam. Sunflowers do well on cornfields. Sandy soils usually make better yields than heavy, clay soils.

Land Preparation: Prepare a good seedbed by disking the ground thoroughly as for corn or cotton. Incorporate Treflan® at 1–2.5 pints per acre depending on soil type (check label) if weeds are a problem.

Fertilization/Liming: Add sufficient lime to keep the soil pH around 6. Apply nitrogen at planting at a rate of 20 pounds per acre by the row. General recommendations, depending on soil test, are either 400–500 pounds per acre of 4-12-12 or 5-15-15; or 300 pounds of 13–13–13 per acre, or 500–600 pounds of 10-10-10 per acre. Seed production is greatly improved if 1 pound of boron is applied with the fertilizer or as a foliar spray.

Management: Side dress with 80 pounds of nitrogen per acre when plants are 18–24 inches high. Cultivate as needed to ensure that fields are free of weeds and grasses when sunflowers mature so that doves can find seeds. Normally, starlings and blackbirds that feed on mature seed heads will shatter enough seed to attract doves. If not, shred part of the field as the hunting season approaches. Weed control with Treflan® at a rate of ½ pound (1 pint) per acre of sunflowers. Three pounds (2 quarts) of EPTC® (Eptam 6-E) may be used before planting and disk into the soil. If a pre-emergent treatment is not used, plow-out sunflower middles three times during the growing season to keep down weeds. Weed control is important to reduce competition and provide feeding ground for doves. Doves will also land on the seedhead.

SWITCHGRASS *Panicum virgatum*

Switchgrass is a warm-season grass which is one of the true panicums, having no seasonal phases and flowering only in late summer or early fall. It is among the tallest and stoutest panicums in the South, reaching 6 to 7 feet. Common varieties include Cave-in-rock, Pathfinder, Kanlow, and Alamo. Provides excellent cover and feeding sites for **ground-nesting wildlife**. Native to Great Plains and most of eastern U.S.

Planting Rate: Broadcast at 10 pounds per acre or 7 pounds per acre. (See planting rate for bluestem for more information). Seedlings are slow establishing.

Planting Date: April–May.

Planting Depth: ¼–½ inch.

Companion Plants: Can plant with corn at the same time but use moderate rates of atrazine and no other herbicides. Can also no-till into corn, milo or soybean stubble or into cool-season crop such as wheat. Switchgrass can be overseeded with 10 pounds of kobe lespedeza or hairy vetch per acre. Also does well mixed with partridge pea.

Site Selection: Can tolerate poorly drained soils.

Land Preparation: Create a tilled, firm seedbed by plowing, disking, and cultipacking.

Fertilization/Liming: Follow soil test and bring fertility up to medium levels for lime, phosphorus, and potassium but do not apply nitrogen at or before planting.

Management: Responds well to controlled burning in early spring.

TIMOTHY *Phleum pratense*

A cool-season perennial grass originally from northern Europe. Timothy is primarily planted for pastures and is moderately used by **deer** but provides good insect-feeding areas for **turkeys**.

Planting Rate: 8 pounds per acre.

Planting Date: Plant only in northern Alabama from August 15–September 15.

Planting Depth: ½ inch.

Companion Plants: Can be planted with clovers, alfalfa, or birdsfoot trefoil in August or September.

Site Selection: Best adapted to well-drained sites. Will tolerant sites low in fertility but responds well to fertilization.

Land Preparation: Prepare a smooth seedbed.

Fertilization/Liming: Follow soil test recommendations. Does best on productive soils with a pH of 6.0–6.5. Responds well to nitrogen.

Management: Responds well to annual applications of 80 pounds of nitrogen per acre in September. If grown in combination with a legume nitrogen should not be applied annually.

VELVET BEAN *Stitzolobium deeringianum*

A viney annual that extends 20 to 40 feet. Hairs on 2-to-6 inch pods can irritate human skin. Native to India. Seasonal production May through August. Provides browse for **deer** and **rabbits** and seed for **quail**.

Planting Rate: Plant 10 pounds per acre of seed. Plant seeds 15 inches apart.

Planting Date: Seeds planted after last frost.

Planting Depth: ½ inch.

Companion Plants: Grows well with grasses such as bahia grass or dallisgrass.

Site Selection: Plant ¼–½ acre plots in woods and around edges of woods and fields. Tolerant of soil acidity and low fertility.

Land Preparation: Prepare a clean, smooth seedbed or plant in areas cultivated for crops.

Fertilization/Liming: Soil test first. Tolerates low fertility and soil acidity, but grows best if the soil pH is between 6.5 and 7.0. A soil test will recommend the amount of phosphorus and potassium to apply. On a soil with a low fertility level, use 10–20 pounds of nitrogen per acre at time of planting. Usually 100 pounds of 0-20-20 per acre is sufficient.

Management: Velvet bean caterpillar may become a pest.

VETCH *Vicia* spp.

Vetches are viney, cool-season, annual legumes 2 to 4 feet long with seedpods that burst open when ripe. Flowers are white, purple or pale yellow in clusters. Native to the Mediterranean region. **Quail** and **doves** consume both the seeds and the foliage, while **deer**, **turkey**, and **rabbits** will readily browse on the vine and leaves. Vetch is also an excellent soil builder. Hairy vetch is better adapted to cold weather and has the smallest seed of the vetches. Common vetch has the largest seed. Recommended varieties include hairy, smooth, crown, and grandiflora. Crown vetch is a perennial legume that spreads quickly from rhizomes.

Planting Rate: Inoculation of seed is essential. Hairy vetch should be planted at a rate of 25–30 pounds per acre, but common vetch requires 40 to 60 pounds per acre. Seeds may be broadcast and disked lightly to cover, or a sod seeding machine can be used on grass sod.

Planting Date: From September 1 November 1 when soil moisture is favorable.

Planting Depth: Cover to a depth of ½ inch.

Companion Plants: Small grains seeded at 60 pounds per acre or ryegrass and perennial grasses.

Site Selection: Adapted to a wide range of soils, but does best on moderate to well-drained soils and is more tolerant of acidic soils than most legumes.

Land Preparation: When following a clean cultivated crop, little seedbed preparation is needed. If there are considerable weeds or ground cover, the land should be plowed and a seedbed prepared.

Fertilization/Liming: Soil pH should be between 5.8 and 6.5. Add lime to raise the pH if necessary. Nitrogen is not required, but phosphorus and potassium rates should be based on a soil test. Apply a heavy rate for soils testing low and a moderate rate for soils testing medium. Common rates are 60 pounds of phosphorus and potassium per acre at planting time. Plantings following highly fertilized row crops will usually not require additional fertilizer.

Management: Will reseed if allowed to mature. Responds to annual reapplication of fertilizer. The vetch bruchid often attacks hairy vetch pods and destroys the seed and therefore reduces reseeding.

WEEPING LOVEGRASS *Erogrostis curvula*

A perennial bunch grass which makes rapid, vigorous growth. It has a strong seedling vigor, is resistant to heat and drought, and is a heavy seed producer. One of the earliest grasses to start growing in the spring and the last to turn brown in the fall. Its principal use is for cover and for soil stabilization in gullies and severely eroded areas for trees. It is also very effective in establishing quick cover on eroded areas. Provides cover for **ground-nesting birds**. May become a weedy invader of pastures. Is not palatable to cattle.

Planting Rate: Plant 5 pounds of seed per acre.

Planting Date: March–June.

Planting Depth: ¼ inch.

Companion Plants: Annual lespedezas will grow well with lovegrass on critical areas.

Site Selection: Does well on any type of well-drained soil, but prefers sandy loams. Like any other crop, it responds to soil fertility, although it will grow on soils of low fertility much better than most grasses.

Land Preparation: A firm seedbed with sufficient loose soil to give a light seed covering is adequate.

Fertilization/Liming: Apply 13-13-13 at a rate of 600 pounds per acre. Lime is usually not needed.

Management: When planted for erosion control it should be protected from grazing. Seed crop usually matures in early June. Seeds may be harvested with a combine. Yields 100–200 pounds of seed per acre.

WILD RESEEDING SOYBEAN (BOBWHITE) *Glycine ussuriensis*

A viney annual legume native to Asia. Excessive shattering has limited combine yields to about 200 pounds of seed per acre. Seeds are particularly attractive to **quail**; used to some extent by **doves**. Heavier seed production is usually obtained when planted with corn or sorghum. Volunteers well in cultivated corn and in wildlife food patches after disking or without any soil disturbance. Several varieties include bobwhite and MS-128.

Planting Rate: Plant 6–8 pounds of seed per acre in 36–42 inch rows with a corn planter. Use 20–25 pounds per acre if broadcasting. Planting methods in order of preference are 1) plant 6–8 seeds per foot alone in 3-to-4 foot rows or plant soybeans in every third or fourth row in a corn or milo planting, 2) mix by volume ¼ soybean seed with ¾ corn, or 3) ⅓ soybean seed, ⅔ milo and plant together in rows.

Planting Date: April 15–June 1. Size of plots for quail should be ¼–½ acre.

Planting Depth: 1½ inch.

Companion Plants: Corn or sorghum.

Site Selection: Best suited to well to moderately well-drained soils of average or better fertility. Not adapted to deep sands.

Land Preparation: Prepare a good seedbed by disking and harrowing 2 or 3 weeks in advance of planting time.

Fertilization/Liming: Apply 300–400 pounds of 0-14-14 or similar fertilizer per acre at time of planting.

Management: Plantings for wildlife should be protected from domestic animals. Row plantings should be cultivated once or twice to control weeds. To improve volunteer stands, disk the field in early spring. In some cases there may be enough seed on the ground for volunteering; however, where the seed supply is severely depleted by wildlife a light disking in early spring will improve chances of a good stand. Apply 200–400 pounds of 0-14-14 fertilizer annually or as needed to maintain satisfactory production. Plantings in areas with heavy deer populations seldom succeed.

Appendix F: Native and Naturalized Trees and Plants Important to Wildlife in Alabama

AMERICAN BEECH *Fagus grandifolia*

A tree which averages about 60 to 80 feet tall with fruits that are composed of three-angled nuts and are available to wildlife during September to November. Beechnuts are a choice food of **squirrels** and are also eaten by **chipmunks, black bear, feral hogs, ruffed grouse, turkeys,** and occasionally by **quail**. Buds are also a choice food of **turkeys**. Beech grows on fertile bottomlands and uplands and prefers a cool, shady, moist site. Trees do well under partial cuttings or singletree selection cuttings, but do poorly when forest stands are heavily cut or clear-cut. Seeds can be broadcast immediately in the fall or stratified in sand at 41°F for 60 to 90 days for spring broadcasting. Transplanting is usually done with one to three year old seedlings. Seedlings do better under shade rather than in openings.

AMERICAN BEAUTYBERRY *Callicarpa americana*

A deciduous shrub 2 to 8 feet tall that does well on a variety of forest sites. Fruits are a purple berry available in August to September and may persist to January. Fruits are important to **quail** during winter and eaten by **deer** from August to October. Also eaten by **squirrels**, **raccoons**, **opossums**, **turkeys**, and **foxes**. New leaf growth in spring is especially palatable to deer. Seeds collected in the fall can be sown in the field or nursery the next spring. Plants in the nursery can be transplanted the following winter. Plants are also easily rooted from stem cuttings made in September. If cuttings are made in September they can be propagated in the greenhouse and transplanted in winter and will produce fruit the following summer. Best maintained by moderate disturbance such as logging, prescribed burning, and disking in the winter.

AMERICAN ELDER *Sambucus canadensis*

A shrub with many stems whose purple berry-like fruits are an important source of summer and early fall food for many kinds of **songbirds**, **robins**, **mockingbirds**, **catbirds**, **quail**, **turkey**, **ruffed gouse**, and **doves**. **Squirrels**, **rabbits**, **raccoons**, **opossums**, **chipmunks**, and **deer** feed on the fruit and foliage. Because of its thicket-forming capabilities it provides excellent nesting cover for **birds**. Before planting, seeds should be scarified with sulfuric acid for 10 to 20 minutes or stratified in moist sand for 60 days at 68 to 86°F alternating daily, then 120 days at 41°F. Untreated seeds broadcast in the fall do not complete germination until the second year.

AMERICAN HORNBEAM *Carpinus caroliniana*

A small crooked understory tree up to 35 feet tall whose fruit is a nutlet, available in August to October, and is of moderate value as a food source to **squirrels**, **turkey**, and **quail**. Seeds and bark are eaten by **rabbits** and **beaver**. The foliage is seldom eaten by **deer**. Grows best in fertile soils and swamps. Seeds collected in August can be planted in rich moist soil in the fall.

ALABAMA SUPPLEJACK *Berchemia scandens*

A high climbing, large intertwining vine whose bluish-black fruit is eaten by many kinds of **birds**, including **turkey**, **quail**, **mallards**, and **small mammal**s such as **raccoons** and **squirrels**. Often seen growing up trees. Ripened fruits persist on the plants through most of the winter; therefore, they are an important wildlife food. **Deer** readily browse vegetation. Grows well under trees but forage and fruit yields are several times greater when plants are grown in the open. Plants withstand prescribed burning in winter, which is the practice that is usually recommended for keeping the plant growth close to the ground. If not disturbed, most of the foliage will soon grow beyond the reach of deer.

BAYBERRY *Myrica* spp.

Two species found in Alabama are important to wildlife, southern bayberry or wax myrtle (*M. cerifera*) and northern bayberry (*M. pensylvanica*). A shrub or small tree up to 40 feet tall with a fruit that is light green and covered with a bluish-white wax and available to wildlife in September to October. This plant makes excellent cover for game and non-game wildlife. Fruits are generally not considered preferred food of wildlife; however, the extremely hard fruit is very persistent during late winter and often acts as a reserve source of food. Eaten by many kinds of **songbirds**, **quail**, **turkey**, and **foxes**. Fruits, twigs, and foliage are not eaten much by **deer**. Heavy browsing by deer is considered by biologists as an indicator of overpopulation. Grows in thickets, woods, swamps, and sandy areas. Plants survive prescribed burning by resprouting from the root collar. Can be transplanted as seedlings but best grown from seeds in moist, silty, acidic sand.

BLACK CHERRY *PRUNUS SEROTINA*

An important fruit-producing tree in Alabama. The dark purple or black fruits of black cherry are important food sources to a wide variety of **birds** and **mammals** including **songbirds**, **turkey**, **ruffed grouse**, **squirrels**, **raccoons**, **waterfowl**, and **quail**. **Deer** feed on the fruit and along with **rabbits**, feed heavily on the twigs and foliage. Fruits are available during June to August. Black cherry grows on most sites except swampy or dry sites. It is intolerant of shade and tends to dominate secondary succession initiated by fire, windrow, or logging, especially after clear-cutting. For planting, seeds should be stratified in moist sand or peat for 90 to 120 days at 41°F or soaked for 30 minutes in sulfuric acid or a combination of both treatments. Black cherry should be protected from fire; however, after very hot fires, most top-killed trees resprout from the base.

BLACK WALNUT *Juglans nigra*

A tree up to 125 feet tall whose fruit is a nut that is available to wildlife in September to November. The large nuts are favored foods for Eastern **gray squirrels** and **fox squirrels**. Black walnut does best on deep, well-drained but moist soils. Propagation is by seeds in late winter or early spring which are often distributed by squirrels and **rodents**. Seedlings which have a root collar of at least ⅜ inch in diameter can be planted at a spacing of 10 x 10 or 10 x 12 feet. Wider spacings are recommended if space permits, due to the large mature size. Requires sunlight and is shade intolerant.

BLACKBERRY, DEWBERRY, & RASPBERRY *Rubus* spp.

The many species, hybrids, and varieties of *Rubus* spp. are one of the most important sources of fruit for wildlife in Alabama. The green fruits of all three species turn red before they ripen to black. Fruits are readily sought by **deer**, **turkey**, **quail**, **raccoons**, **chipmunks**, **ruffed grouse**, **squirrels** and many **birds** including **woodcock**. The fruits are eaten in June to October mainly when fleshy, but even the dry berries are eaten some in the fall and winter. Leaves are a significant part of deer diets from May to September. *Rubus* spp. is favored by summer soil disturbance, fire, and overstory removal and grows best in full sunlight. Thickets are excellent cover and nesting sites for **birds**, **rabbits**, **small rodents**. Seeds and rootstock persist in most soils. Sandy, well-drained soils as well as river bottoms are good sites for *Rubus* spp. Produces well in full sunlight. When grown in the shade, fruit production is low and stems are nearly thornless. Stands should be stimulated with fire or disking when fruit production declines or they will become too tall and dense for the fruit to be available to wildlife.

BLUEBERRY *Vaccinium* spp.

Five species of blueberries are important to wildlife in Alabama. These include Elliott's Blueberry (*V. elliottii*),tree sparkleberry (*V. arboreum*), dryland blueberry (*V. vacillans*), ground blueberry (*V. myrsinites*) and deerberry (*V. stamineum*) which are found on a wide range of sites. Fruits vary in color to include black, blue, green, and yellow. Fruits ripen in August to October and often persist into winter. Blueberry fruits are eaten by many species of wildlife that include **turkey**, **quail**, **ruffed grouse**, **songbirds** (**scarlet tanager**, **robin**, **cardinal**, **bluebird**, and **thrasher**), **black bear**, **deer**, **chipmunks**, **rabbits**, **foxes**, and **raccoons**. Foliage is readily eaten by deer with browsing greatest in April and May but continuing on into winter. Maximum growth is best in full sunlight. May be propagated from seed or vegetatively. Seeds should be cleaned from the fruit in a household blender and stored in a refrigerator at 40°F until used. Grows well in the shade of other trees.

BUTTERNUT *Juglans cinerea*

A small to medium-sized tree, found primarily in northern Alabama, up to 70 feet tall. The nuts are eaten by **deer**, **squirrels**, and **rodents** and is available during September to October. Prefers rich, moist, well-drained loam soils and also does well on drier rocky sites that have a limestone base. Butternut should be protected from fire.

CHINQUAPIN (CHINKAPIN) *Castanea* spp.

Two species of chinquapin are found in Alabama that are important to wildlife and include Allegheny chinquapin (*C. pumila*) and downy chinquapin (*C. alnifolia*). Chinquapins are a tree or shrub up to 50 feet high that have nuts that are available to a variety of wildlife in late summer and early fall. Nuts are eaten by a number of **birds** as well as **squirrels**, **rabbits**, and **deer**. Chinquapins usually grow in sandy open woodlands and thickets.

DOGWOOD, FLOWERING *Cornus florida*

A shrub or small tree up to 40 feet tall with fruits in bright red clusters. Found on moderately wet to moist slopes, uplands, and coves. Does poorly on excessively dry or wet sites and in full sunlight. Fruits available from September to October and persist through early winter. Virtually all game species use the fruit. It is a preferred food of **turkeys**, and is readily eaten by **deer**, **quail**, **ruffed grouse**, and to a lesser extent by **wood ducks**. The fruit is also eaten by many kinds of **songbirds** such as **cardinals**, **grosbeaks**, **robins**, **brown thrashers**, and **cedar waxwings**. **Mammals** such as **rabbits**, **foxes**, **black bears**, **raccoons**, **squirrels**, and **chipmunks** eat the fruit. Because browse and fruits of dogwood are widespread, it is one of the most important deer food plants. Foliage and twigs are often heavily browsed except in the coastal regions where use is light to moderate. The browse is most palatable in spring and least palatable in winter. Dogwood is also browsed by **rabbits**. Dogwood seeds germinate in the spring. Freshly gathered seed will germinate following moist soil storage for 100 to 130 days at 0 to 10°F. Seed stored above 15°F will not germinate. Except for the year of planting, annually apply 3 pounds of nitrogen per 1,000 square feet beneath and just beyond the branch spread. Native dogwoods are succumbing to a tree disease known as dogwood anthracnose, particularly at higher mountainous elevations.

ELM (AMERICAN, SLIPPERY, WINGED) *Ulmus* spp.

Three species of elm are important to wildlife in Alabama. These species include American elm (*U. americana*), slippery elm (*U. rubra*), and winged elm (*U. alata*). Elms are trees 80 to100 feet in height and are found in a variety of habitats from moist to dry upland soils. Seeds of elm are eaten by **turkeys** and a variety of **songbirds**.

FRINGETREE *Chionanthus virginicus*

A small shrub or tree up to 35 feet high. Purplish date-like fruit is available from August to October and is eaten by many **birds** and **mammals** including **turkey**, **quail**, and **deer**. Leaves are a preferred browse for deer in the Coastal Plain, but in the Piedmont and mountains it is browsed only lightly. Most abundant in the understory of pine-hardwood forest, especially on moist, acidic, sandy loam soils. It does best in semi-open areas but is only moderately tolerant of shade. Hot fires will kill roots but a light fire induces sprouting and helps keep forage available to deer.

GRAPE, WILD *Vitis* spp.

Four species of wild grapes are important to wildlife in Alabama. These include summer grape (*V. aestivalis*) in central and southern Alabama, sweet winter grape (*V. cinerea*), muscadine grape (*V. rotundifolia*), and frost grape (*V. vulpina*) in northern Alabama. A high-climbing woody vine which grows on a variety of sites but grows best in moist fertile soils and is commonly found along stream banks. The dark blue to purple fruits are available in September to October and are preferred by many **songbirds**, **black bear**, **deer**, **raccoons**, **turkeys**, **quail**, **ruffed grouse**, **wood ducks**, **squirrels**, **red foxes**, **coyotes**, and **feral hogs**. Foliage is eaten by deer and **rabbits** and provides cover for a variety of wildlife species, especially escape and nesting cover for songbirds. Wild grape is intolerant of shade and grows best in sunlight. The plants are commonly propagated by seed, cuttings and grafting.

GREENBRIER *Smilax* spp.

Five species of greenbrier in Alabama are important to wildlife. These include common greenbrier (*S. rotundifolia*), laurel greenbrier (*S. laurifolia*), lanceleaf greenbrier (*S. smallii*), saw greenbrier (*S. bona-nox*), and cat greenbrier (S. glauca). Greenbriers are a group of thorny, woody vines that retain their leaves most of the year and grow on a variety of sites such as dense cutover forest, swamps, abandoned fields, and fence rows. They tolerate shade but do best in open areas. Black fruits are available in September to October and are eaten by **turkeys**, **quail**, **ruffed grouse**, **deer**, **black bear**, **opossum**, **raccoon**, **squirrels**, **rats**, and at least 38 species of **songbirds**. Greenbriers are considered one of the most important group of deer food plants. The fast-growing stems and leaves are very palatable and are eaten year-round and can tolerant heavy deer browsing. About 50 to 60 percent of the annual growth of greenbriers may be eaten without root mortality. Greenbriers are also important to **rabbits**. Prescribed burning helps keep greenbriers available to deer and rabbits and increases crude protein levels in older stems.

HACKBERRY *Celtis occidentalis*

A tree which favors a variety of soil types and often found on limestone soils in the blackbelt. Seldom occurs in more acidic soils of pine stands. Fruits are eaten by many **birds** and **small mammals**. Are a preferred food of **turkeys** and **raccoons. Squirrels** occasionally eat the fruit, buds, and bark. Seeds can be sown in fall or stratified in moist sand at 41°F for 60–90 days and planted in spring. Can also be grown from stem cuttings and sprouts will grow from small stumps.

HAWTHORN *Crataegus* spp.

Four species of hawthorn in Alabama are important to wildlife and include blueberry hawthorn (*C. brachyacantha*), parsley hawthorn (*C. marshalli*), riverflat hawthorn (*C. opaca*), and one-flower hawthorn (*C. unflora*). A shrub or small tree 15 to 25 feet high whose fruit is eaten by **songbirds**, **sparrows**, **cedar waxwings**, and **game birds** such as **turkey**, **quail**, **ruffed grouse**, and **wood ducks**. **Squirrels**, **foxes**, **raccoons**, **rabbits**, and **deer** often eat the fruit. Fruits range in color from bright red, yellow, green or black. The browse is fairly palatable to deer and is eaten primarily in May and June. Grown on a wide variety of sites and soils ranging from heavy clay to sandy loam. Some species grow best on dry stony ridges and mountain slopes while others do better in moist river bottoms and along the edges of swamps. Hawthorn is intolerant of shade and seeds into openings quickly. For wildlife purposes, the most practical way to establish is by seeding. Seeds should be cleaned, stratified in sand over winter and planted in spring.

HICKORY *Carya* spp.

Five species of hickories in Alabama are important to wildlife and include water hickory (*C. aquatica*), bitternut hickory (*C. cordiformis*), pignut hickory (*C. glabra*), shagbark hickory (*C. ovata*), and mockernut hickory (*C. tomentosa*). Hickories are slow-growing deciduous trees that may reach a height of 120 feet. Fruits are a hard, bony nut that is available in September to November and is a food source for many kinds of wildlife. Nuts are eaten by **squirrels** and **feral hogs**. Nuts and flowers are eaten by **turkey** and **songbirds**. Nuts and bark are eaten by **black bears**, **foxes**, **rabbits**, and **raccoons. Deer** will occasionally eat the leaves, twigs and even the nuts. Propagation is best by planting seeds in the spring to reduce the chance of rodent damage. Seeds should be stratified in sand or peat at 30 to 35°F for 90 to 150 days. Hickories should be protected from fire.

HOLLY *Ilex* spp.

Seven species of hollies that are valuable to wildlife are found in Alabama. These include dahoon (*I. cassine*), large gallberry (*I. coriacea*), possumhaw (*I. decidua*), gallberry (*I. glabra*), myrtle dahoon (*I. myrtifolia*), American holly (*I. opaca*), and yaupon (*I. vomitoria*). Hollies are small trees or shrubs with fruits that are red, reddish-orange, black, or rarely yellow. Found on most sites in Alabama. For most of the hollies the fruit persists through winter making it extremely valuable as a food source for wildlife. Fruits are eaten by many kinds of **birds** and **mammals**, and the evergreen foliage is an important source of winter food for **deer**. Can be propagated by collecting seeds and germinating in a mixture of sand, soil and peat. Can also be established from cuttings. Terminal twigs are best for use in cuttings. Rootstocks sprout readily after a prescribed fire.

HOPHORNBEAM, EASTERN *Ostrya virginiana*

A small slow-growing tree up to 60 feet tall with nuts that are available during September to October. The nuts, buds and catkins are eaten by **quail**, **squirrels**, **deer**, **rabbits**, **ruffed grouse**, **purple finch**, **rose-breasted grosbeak**, and **downy woodpeckers**. Seeds are a preferred food of **turkeys**. Grows best in rich, moist woods in a wide variety of soils. Hophornbeam can be established by planting 2-foot-high or taller seedlings in forest openings, woodland borders, or old fields. If planted by seeds, broadcasting should be done in fall or seeds should be stratified over winter in sand or peat at 41°F and then broadcast in the spring.

HORNBEAM, AMERICA *Carpinus caroliniana*

Tree which grows in fertile bottomlands and swamps. Fruits and buds are eaten by a variety of wildlife including **quail**. Seeds and bark are eaten by **rabbits** and **beaver**. Forage is eaten by **deer**, but not preferred.

HUCKLEBERRY, DWARF *Gaylussacia dumosa*

A low shrub not usually over 18 inches tall with creeping stems and upright branches. Fruit is blue to black and is eaten by **quail**, **turkey**, **ruffed grouse**, **songbirds**, **foxes**, and **squirrels**. Highly valued by quail during the summer months. Grows best in shaded peaty or sandy soils and is commonly found in flat damp woods that are frequently burned. Fire can be used as a management tool for dwarf huckleberry. Most abundant on areas that have been burned on two-year intervals.

JAPANESE HONEYSUCKLE *Lonicera japonica*

A semi-evergreen trailing and twining woody vine whose black berries persist through winter and are eaten by many kinds of **songbirds**, **turkey**, **quail**, **ruffed grouse**, and **deer**. Honeysuckle browse is highly preferred by deer. In years of low mast (hard and soft fruit) production, honeysuckle foliage is extremely important to deer during the winter. Dense vines and leaves provide excellent cover for quail, turkey, songbirds, and many **mammals**. Originally introduced from Asia, it is found throughout Alabama, mainly on low moist ground, along streams or fence rows, and borders of woods. Grows best in sunlight. Grows on most all moderate to well-drained soils. Firelanes adjacent to creek bottoms make excellent locations for honeysuckle patches. It can be grown in nurseries, but is easily transplanted from old house sites and other areas. Plants are best propagated vegetatively from root and stem cuttings, root stolons, single plants, or even blocks of soil containing roots in late winter to early spring. Sprigs can be planted on a 6-to-12 inch spacing. Plant next to windrows, brushpiles, or any place where vines can climb. Protection from deer browsing is recommended. Seedlings may need to be protected by cages in areas of heavy deer populations. Plantings need cultivation for the first year. Once established will spread as much as 15 feet in a year. Fertilizing honeysuckle increases growth, protein content, and preference by deer. If possible, fertilize annually to increase production and crude protein levels. In some cases honeysuckle will become a pest, out-competing native vegetation.

LOCUST *Gleditsia and Robinia* spp.

Two species of locust are important to wildlife in Alabama, honeylocust (*G. triacanthos*) and black locust (*R. pseudoacacia*). Locust are trees that average 20 to 30 feet tall with fruits that are a flattened linear pods. Fruits of honeylocust are eaten more readily than black locust. Honeylocust fruits are eaten by **rabbits**, **squirrels**, **quail**, **ruffed grouse**, and **deer**. Foliage of young plants are eaten by rabbits. The seeds are high in crude protein. Honeylocust is found on alluvial flood plains while black locust is commonly found on deep, well-drained, fertile loams. Locust can be grown from seeds or from root cuttings. Seeds should be pretreated (scarified) before planting by soaking in water for a short period at 185 to 195°F or by soaking in concentrated sulfuric acid at 60 to 80°F for one to two hours.

MAGNOLIA *Magnolia grandiflora*

A large tree averaging 50 feet tall whose seeds are available in July to October and are a favorite food of **squirrels** and are also eaten by **rodents**, **songbirds**, **opossums**, **quail**, and **turkeys**. Grows best in moist well-drained soils along streams and swamps and in moist upland sites.

MAPLE (RED, SILVER, BOXELDER) *Acer* spp.

Three species of maples are important to wildlife in Alabama and include red maple (*A. rubrum*), silver maple (*A. saccharinum*), and boxelder (*A. negundo*). Maples are medium-sized deciduous trees averaging 50 to 70 feet tall whose leaves and twigs are eaten by **black bear**, **beaver**, and **deer**. The seeds, flowers and buds are eaten by **squirrels**, many **birds**, and other **small mammals**. Maples are shade-tolerant and do well in moist soils of hardwood forest. Should be protected from fire. Propagation is by seed but will sprout readily if burned or cut.

OAK *Quercus* spp.

In Alabama, oaks are the most important hardwood species to wildlife since a number of wildlife species eat the acorns which are available in fall and winter. At least 96 species of wildlife are known to eat acorns. Some of the wildlife that eat acorns include **wood ducks**, **mallards**, **squirrels**, **turkeys**, **deer**, **quail**, **ruffed grouse**, **woodpeckers**, **black bears**, **raccoons**, **bluejay**, and **feral hogs**. Squirrels also eat the catkins in the spring while deer browse on the leaves and twigs of many oaks. Oaks are divided into two broad groups, the white oak and the red oaks. Acorns of the white oaks mature in one year while those of red oaks require two years. White oak acorns are generally more preferred than red oak acorns due to lower tannin content which makes the acorns more palatable. Still, the red oaks do produce acorns every year, but any year's crop depends on weather factors two years prior.

The most common method of regenerating oaks is by clear-cutting small areas with new trees developing from sprouts. Oaks can also be established by seeding or by planting seedlings. Plant seedlings spaced at 10 x 10 or 10 x 12 feet apart. Except for the year of planting, annually apply 3 pounds of nitrogen per 1,000 square feet beneath and just beyond the branch spread. In managing oaks for wildlife, at least half the oak trees in a stand should be at least 40 years of age. A hardwood stand should also be composed of a good mixture of oak species to maintain annual acorn production. It usually takes about 25 years for Alabama's native oaks to start producing a good number of acorns. Oaks regenerate well by clear-cutting. Clear-cuts should be small (less than 25 acres) in size and well distributed over the entire tract. No more than ⅛ of the tract should be regenerated during any cutting cycle.

PALMETTO, DWARF *Sabal minor*

Dwarf palmetto is important to wildlife in Alabama. The dry hard black fruits are eaten by several kinds of **songbirds**, **quail**, **raccoons**, **black bear**, **turkeys**, **squirrels**, and **deer**. Dwarf palmetto is a shrub that grows in dry hills, flatlands, and wet alluvial floodplains. Dwarf palmetto quickly invades areas where the soil is disturbed.

PERSIMMON *Diospyros virginiana*

A tree or shrub usually 30 to 50 feet high whose yellow to orange fruit is available to wildlife in September to November. Fruits are eaten by many species of wildlife and favored by **opossum**, **raccoons**, **foxes**, and **deer**. Fruits are also eaten by **black bears**, **skunks**, **turkeys**, **quail**, **crows**, **rabbits**, **squirrels**, **feral hogs**, **coyotes** and many **songbirds**. Deer readily browse on persimmon sprouts. Grows best on alluvial soils such as clay and heavy loams and will also grow on poorly drained upland soils. A pioneer species in abandoned fields, it is found in the understory of dense stands. Prolonged flooding during the growing season will kill young trees. Optimum seed-bearing age is 20 to 25 years. Seeds remain dormant over winter and germinate in April or May. Before planting, seeds should be stratified under moist conditions for two to three months at 33 to 41°F. Seedlings develop a strong taproot and are difficult to transplant. Persimmon spouts well from stumps and root collars. Plants may be propagated by root cuttings and grafting.

PINE *Pinus* spp.

Seeds are high in fat content and an important food source available from September to December and are eaten by several kinds of **birds** and **mammals**. These include **turkey**, **quail**, **ruffed grouse**, **songbirds** such as **chickadees**, **crossbills**, **grosbeaks**, **jays**, **nutcracker**, **nuthatches**, **warblers**, and **woodpeckers**. Seeds, bark and foliage are eaten by **bears**, **beavers**, and **rabbits**. **Squirrels** will readily eat the cones, seeds, and catkins. Pines are readily established by planting seedlings or by natural seeding.

POPLAR, YELLOW *Liriodendron tulipifera*

A deciduous tree 80 to100 feet in height and found in moist well-drained soils, especially valleys and slopes, often in pure stands. Leaves of sprouts are highly preferred as **deer** browse. Seeds are used by a variety of **birds** and **squirrels** in early fall and winter.

POISON IVY *Rhus radicans*

A shrub or woody-stemmed vine with dull white fruits which are eaten by many kinds of **birds**, including **quail**, **ruffed grouse**, **wood ducks**, and **turkey**. Comprises a significant portion of **squirrel** diets in bottomland hardwoods in years when other mast (hard and soft fruit) is sparse. **Deer** eat the fruit and browse the leaves and twigs.

PLUM, WILD *Prunus spp.*

Wild plum (*P. americana*), Chickasaw plum (*P. angustifolia*) and flatwoods plum (*P. umbellata*) are important fruit-producing trees in Alabama. The dark purple, red, or yellow fruits are important food sources for a wide variety of **birds** and **mammals** including **songbirds, turkey, squirrels, raccoons, foxes, coyotes, ducks**, and **quail**. Fruits are available in late summer. **Deer** feed on the fruit, and along with **rabbits**, feed heavily on the twigs and foliage. Flatwoods plum grows on sandy soils in open pine and hardwood forest on coastal areas. Seed germination is enhanced by stratifying seeds for 90 days at 41°F in moist sand or peat. Seedlings of Chickasaw plum are available commercially.

RED MULBERRY *Morus rubra*

A tree up to 70 feet tall with a dark purple or red fruit resembling a blackberry that is available to wildlife from April to June. Utilized heavily for food by **songbirds**, **squirrels**, **quail**, **turkey**, **ruffed grouse**, **opossums**, and **raccoons**. The foliage is browsed by **deer** mostly during spring and summer, and **rabbits** feed on the bark in the winter. Occurs on moist sites at elevations below 2,000 feet. A minor component of hardwood and coniferous stands. Often occurs alone along stream sides. Easily reproduced by broadcasting seed. Seeds mixed with sand or sawdust may be broadcast or drilled during the fall. Stratified seed or seed soaked in water for one week may be broadcast in the spring and covered with ¼ inch of soil. Red mulberry may be decreasing over much of its range possibly due to a bacterial disease.

REDBAY *Persea borbonia*

An evergreen tree up to 70 feet tall with a dark blue or deep purple egg-shaped fruit that is available in the fall months. The fruit is eaten by several species of **songbirds** and by **turkey**. In the longleaf pine belt, redbay seed may makeup a significant portion of **quail** diets during fall and winter. Redbay is of intermediate palatability to **deer**. At low deer densities it is often not eaten; however, in high deer density areas it is eaten quite often. Heaviest deer browsing takes place in fall and winter. Fire stimulates seed germination and encourages growth of sprouts. Winter burning on three-to-four year intervals is best for making browse available to deer.

REDCEDAR, EASTERN *Juniperus virginiana*

An evergreen tree 40 to 50 feet high whose pale blue berry-like fruits are eaten by many species of wildlife such as **quail**, **turkeys**, **rabbits**, **foxes**, **raccoons**, **skunks**, **opossums**, and **coyotes**. Provides good nesting and roosting cover for many **birds** and dense thickets make good escape cover for **deer**. Foliage provides low quality emergency food for deer in times of stress. Redcedar grows on a variety of soil types and over a wide soil pH range. Appropriate sites vary from dry uplands to limestone to heavy clay. Often found on abandoned farmlands. It is drought and frost resistant but cannot withstand flooding. Should be protected from fire. Because redcedars require sunlight, planting sites should receive at least Ä full sunlight. Control all vegetation until the seedlings outgrow their competitors. Regenerated naturally from bird or mammal droppings on bare soil. Can propagate artificially by storing seed in fruit for one year, then cleaning and scarifying and broadcasting in the fall. Can also store seed in fruit one year, clean and stratify in peat for 100 days at 41°F and broadcast in spring. Seeds have also been stratified outdoors in the shade from May until broadcasting time in fall. ***Do not plant near apple orchards because red cedar is an alternate host for apple rust.***

SASSAFRAS *Sassafras albidum*

A tree which can reach 90 feet in height whose blue fruit is eaten by **songbirds, turkey, quail, raccoons, squirrels, black bear**, and other **mammals**. Fruit production is limited. **Deer** will browse twigs in winter, and foliage in spring and summer. Sprouts are especially palatable after a fire or disturbance. Grows best in open woods and along fence rows on moist, well-drained sandy loam soils. Sassafras is intolerant of shade. For browse availability to deer, sassafras should be knocked back by prescribed burning or disturbance every two to five years. Sassafras seeds remain dormant until spring with some germination in fall. Stratification for 30 days in moist sand at 41°F induces germination.

SAW-PALMETTO *Serenoa repens*

A shrub usually 3 to 7 feet high whose fruit is black to dark brown and eaten by **deer**, **black bear**, **feral hogs**, and **turkey** in winter and spring. Grows on sandy pinelands, prairies, hammocks, or dunes and on acidic or alkaline soils. It is often the dominant shrub in frequently burned flatwoods. Saw-palmetto produces regrowth after burning; however, fruit production is reduced. For maximum fruit production protect from fire.

SUGARBERRY *Celtis laevigata*

A tree which grows up to 100 feet tall with an orange-red to black fruit which ripens in late summer and persists through the winter. Fruits are eaten by many **songbirds**, upland **game birds**, and **small mammals.** It is a preferred fall and winter food of **turkey. Deer** occasionally eat the leaves and twigs. Distributed widely on bottomlands except deep swamps. Most common on clay soils of broad flats or shallow sloughs within the flood plains of major rivers. Seeds may be broadcast in fall or stratified in moist sand at 41°F for 60 to 90 days and broadcast in spring. Sugarberry can also be propagated by cuttings. Trees should be protected from prescribed burning.

SUMAC (WINGED, SMOOTH, STAGHORN) *Rhus* spp.

There are three species of sumac in Alabama that are important to wildlife. These include winged sumac (*R. copallina*), smooth sumac (*R. glabra*) and staghorn sumac (*R. typhina*). Sumacs are slender-branched shrubs or small trees that rarely reach 25 feet tall. Fruits are red or pale green to brown and available in fall and usually persist through winter. Fruits are eaten by many species of **game birds** and **songbirds.** Entire fruiting heads are often eaten by **deer.** The bright red fruit cluster remains on the plant into winter when more desirable foods are scarce and often serve as an emergency food for **turkey. Rabbits** and **squirrels** will use the bark, twigs and fruit. Deer will occasionally browse on the plant. All species grow in dense thickets in old fields, along roadsides, and near powerlines. Responds well to disturbances such as fire and timber cutting. Usually associated with broomsedge and blackberry. Plants will grow well on dry acidic soil and will persist in the understory until canopy closure. Invasion of new areas is by seed, but established sumac reproduces primarily by rootstocks.

SWEETBAY *Magnolia virginiana*

A large shrub or small tree up to 35 feet high whose seeds are available to wildlife in September and eaten by **squirrels, small mammals, songbirds, turkey,** and **quail.** Leaves and twigs are browsed by **deer.** Common on sites that are poorly drained or often flooded.

TRUMPET-CREEPER *Campsis radicans*

A climbing, woody vine with distinctive reddish-brown flowers. New growth of leaves are browsed heavily by **deer** during spring and summer. Grows best on open sites and alluvial soils that are moist but well drained. Also attracts hummingbirds.

TUPELO *Nyssa* spp.

Three species of tupelo are important to wildlife in Alabama. These species include water tupelo (*N. aquatica*), black tupelo (*N. sylvatica*) and swamp tupelo (*N. sylvatica var. biflora*). A large tree up to 100 feet tall with fruits that are dark purple and available to wildlife during September to October. Fruits are eaten by numerous **birds** such as **quail, wood duck, robin, mockingbird, brown thrasher, thrushes, flicker,** and **starling.** Also eaten by **mammals** such as **deer, black bear, fox, beaver, opossum, raccoon** and **squirrels.** Fruits are considered a staple food of **turkeys.** Young trees and sprouts are preferred by deer. Grows in uplands and alluvial bottoms. Swamp tupelo is found on moist sites in the Lower Coastal Plain and does best in coves and swamps. Tupelos require full sunlight for optimum growth. Tupelos should be protected from fire. However, sprouts following fire are highly palatable and nutritious to deer. Untreated seed can be broadcast in the fall. Seed that is broadcast in the spring should be stratified at 30 to 50°F in moist sand for 60 to 90 days.

VIBURNUM *Viburnum* spp.

Four species of viburnums are important to wildlife in Alabama. These species include rusty blackhaw (*V. rufidulum*), mapleleaf viburnum (*V. acerifolium*), possumhaw viburnum (*V. nudum*), and southern arrowwood (*V. dentatum*). Viburnums are shrubs or trees that range in size from 2 to 40 feet in height and whose red or blue-black fruits are available from late summer through fall and are eaten by **songbirds, turkey, quail, squirrels, beaver, rabbits, raccoons, chipmunks,** and **deer.** The twigs, bark and leaves are eaten by deer and beavers. Good seed crops occur every one or two years. Viburnums are understory plants that are found on a variety of sites. Viburnums reproduce vegetatively or by seeds. Invades new areas by seed followed by root suckering or layering of stems. Seed germination is normally delayed until after the second spring after ripening.

Appendix G: Oaks Important to Wildlife in Alabama

WHITE OAKS

White Oak (*Q. alba*)	Does best on deep, well-drained loamy soils.
Post Oak (*Q. stellata*)	Grows on a variety of soils and is found on rocky ridges, sandy outcroppings and southern exposures.
Overcup Oak (*Q. lyrata*)	Grows best on wet clay and silty clay soils, mostly on poorly drained flood plains and borders of swamps. Often found growing in pure stands.
Chestnut Oak (*Q. prinus*)	Grows on dry, sandy, gravelly soils and well-drained coves and bottom sites. Grows on moist to wet bottomlands, but is best suited to well-drained silty clay and loamy sites in the floodplains of streams and rivers.
Swamp Chestnut Oak (*Q. michauxii*)	Occurs on low ground along river bottoms or in ravines with other hardwoods.

RED OAKS

Black Oak (*Q. velutina*)	Commonly found on a variety of sites, primarily on dry uplands. Forms a large portion of hardwood stands in Appalachian foothills.
Blackjack Oak (*Q. marilandica*)	Grows on dry, sandy soils.
Bluejack Oak (*Q. incana*)	Grows on dry sandy ridges, pine-barrens, and dunes in the Coastal Plain.
Chapman Oak (*Q. chapman*)	Does well on sands and other well-drained soils.
Dwarf Live Oak (*Q. minima*)	Grows on deep sands behind beach areas and in seepage areas of sandy-clay uplands on sites that are burned.
Laurel Oak (*Q. laurifolia*)	Occurs along streams, river bottoms and on moist upland soils.
Live Oak (*Q. virginia*)	Found on a wide variety of soils. Most often on sandy soils. Tolerates salt spray and salinity.
Northern Red Oak (*Q. rubra*)	Grows in soils ranging from clay to loamy sands.
Running Oak (*Q. pumila*)	Found in pine flatwoods and open sandy pine-scrub oaks.
Southern Red Oak (*Q. falcata*)	Occurs on dry, sandy or clay soils and on dry ridgetops. Prefers dry to medium-moist upland soils but occasionally appears in fertile river bottoms where it attains its largest size.
Scarlet Oak (*Q. coccinea*)	Dry to medium-moist upland sites and mountain slopes.
Shumard Oak (*Q. shumardii*)	Found mostly on moist well-drained soils including floodplains along streams and also found on dry ridges and limestone hills.
Turkey Oak (*Q. laevis*)	Grows in poor sandy soils.
Water Oak (*Q. nigra*)	Found along streams, edges of swamps, bottomlands, field edges, and on deep or moist upland soils. Grows best on alluvial bottoms, and well-drained silty clay or loam ridges. The most common deep South upland oak.
Willow Oak (*Q. phellos*)	Grows best on alluvial soils.

OAKS OF ALABAMA

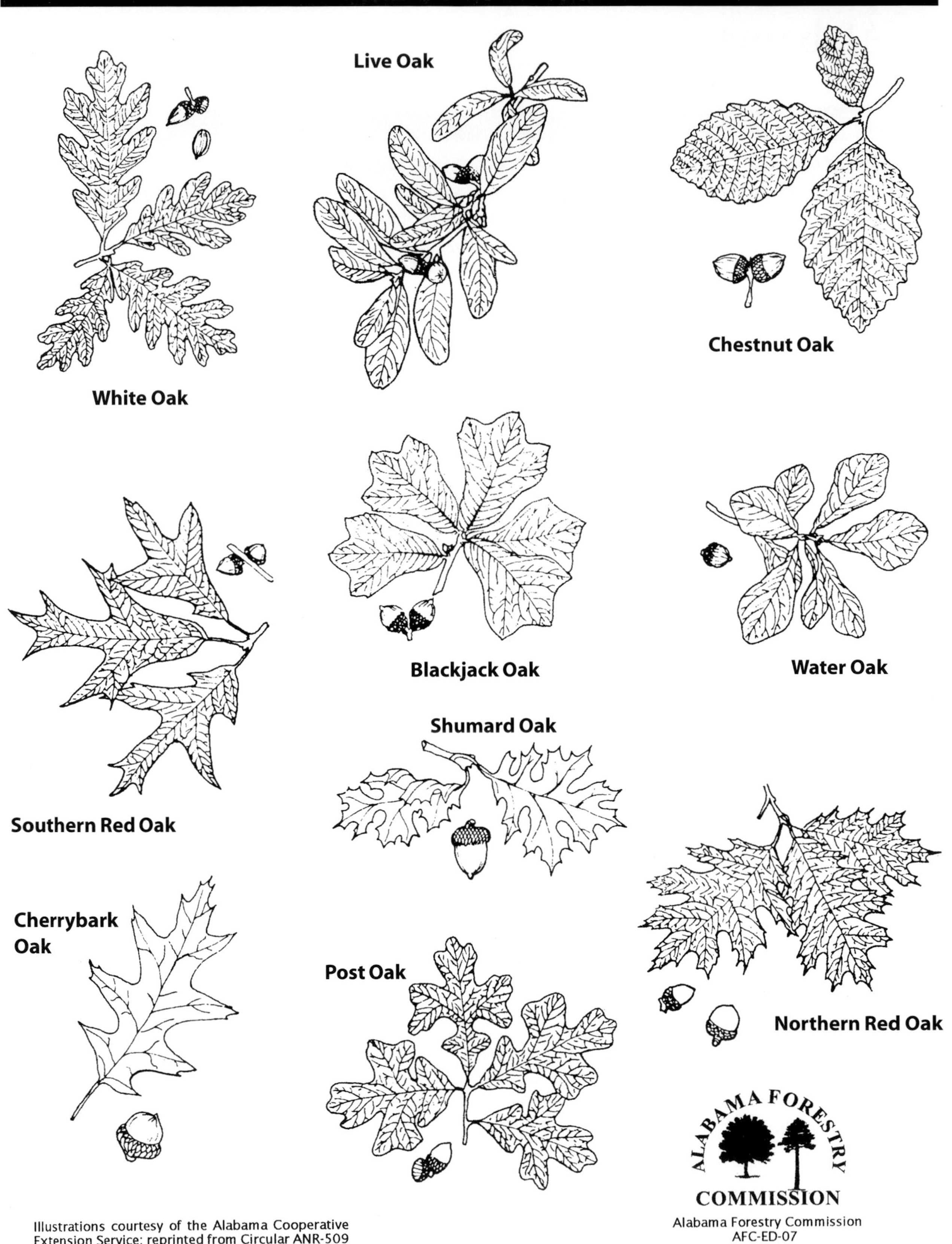

Illustrations courtesy of the Alabama Cooperative Extension Service; reprinted from Circular ANR-509

Alabama Forestry Commission
AFC-ED-07

OAKS OF ALABAMA

Oak trees are an important part of the forests of Alabama. The dependence of man and animal on this family of trees makes them the most widely used hardwood in the United States. The trees are used for many purposes. The wood is used to make furniture, flooring, paneling, lumber, railroad cross ties, firewood and fine paper. Many wildlife species depend on acorns from oaks as a food supply. This family of trees is split into two main groups.

THE WHITE OAK GROUP

The white oak family is identified by having leaves with rounded lobes, light colored bark, and acorns that ripen in one year.

White Oak

This valuable tree, found widely across Alabama, has leaves with many rounded tips. The wood is used for furniture, barrels and paneling. Its acorns are preferred food for wildlife.

Live Oak

This beautiful tree with long, arching branches and thick, evergreen leaves was once heavily cut to supply special timbers for building wooden ships. Today the most common commercial use of this tree is as a street tree where its beauty can be enjoyed by all.

Post Oak

The leaf of this common tree looks like a cross. The wood was once heavily used for fence posts. Sand post oak is a similar but smaller tree that is common on poor, dry sandy soils.

Chestnut Oak

This species grows on upland sites. The leaves have shallow rounded lobes and the bark is rough. The wood is used for cross ties and rough lumber. Swamp chestnut oak is similar, but is found primarily in river and stream bottoms. Its acorns are highly preferred by deer and should be retained for mast production.

THE RED OAK GROUP

The red oak family is marked by leaves that have pointed lobes with bristle tips, dark bark, and acorns that take two years to ripen.

Southern Red Oak

The most common red oak in Alabama has leaves with a long narrow end lobe. The wood is widely used for furniture, paneling and cross ties.

Cherrybark Oak

The largest of the southern red oaks grows in rich bottomlands. It is prized for its high quality wood, used to make furniture and paneling, and its abundant acorns.

Blackjack Oak

This scrub oak is common on poor, dry, and sandy soil and is marked by wide, leathery leaves and stubby branches. The wood is not commercially used, but the tree is an important source of wildlife food. Turkey oak and bluejack oak are also scrub oaks and grow on similar sites.

Northern Red Oak

This tree is common in north and east-central Alabama and has many sharply pointed tips. Its wood is used for furniture and wood flooring. Black oak is a similar tree that grows in the same general range.

Shumard Oak

This large tree with deep pointed lobes is located across Alabama. The wood is used for furniture and crossties. Scarlet oak and Nuttall oak are similar trees that grow in Alabama.

Water Oak

A large tree with rounded crown and variable leaves, some smooth and some with small lobes. The wood is used for rough lumber. The acorns are abundant and are a preferred food for birds and wildlife. The laurel oak and willow oak are similar trees.

Appendix H: Pines Important to Wildlife in Alabama

Loblolly Pine (*P. taeda*)	Produces abundant seeds important to many varieties of wildlife. Susceptible to various insects and diseases including cronartium and Annosus root rot. Grows on a variety of soils. The most common pine in Alabama.
Longleaf Pine (*P. palustris*)	Grows in sandy soils low in organic matter and medium to strongly acidic soils. Produces a large, fatty, winged seed in its large cones. Many species of wildlife depend on longleaf seeds in areas where longleaf is common. Good seed crops occur every 5–7 years. Resistant to most insects, diseases, and windthrow.
Pond Pine (*P. scrotina*)	Found in the southeastern corner of Alabama and prefers wet sites near marshes, swamps, and areas with poor drainage.
Sand Pine (*P. clause*)	Grows mainly in Florida but can be found in coastal Alabama. Grows on light, sandy infertile soil.
Shortleaf Pine (*P. echinata*)	Grows best on fine sandy loam or silt loam soils. Susceptible to nearly all diseases and insects in Alabama including littleleaf disease. Produces large crops of small seeds.
Slash Pine (*P. elliottii*)	Grows on sandy soils that often have poorly drained hardpans. Fast-growing pine that is not very resistant to disease.
Spruce Pine (*P. glabra*)	Grows in lower Alabama on a variety of sites, primarily low areas. Common pine in bottomlands and important to **turkeys** for both seeds and roost sites.
Virginia Pine (*P. virginiana*)	Grows throughout Alabama on a variety of soils but does best on clay, loam, or sandy loam. Small statured and very limby in southern Alabama, grows to larger timber quality in northern Alabama. Poor timber tree.

Appendix I: Upland Herbaceous Plants Important to Wildlife in Alabama

BAHIA GRASS *Paspalum notatum*

See Appendix E.

BEGGARWEED *Desmodium* spp.

See Appendix E.

LESPEDEZA (ANNUAL) *Lespedeza* spp.

See Appendix E.

MILKPEA *Galactia* spp.

A trailing, perennial, herbaceous vine whose seeds are highly palatable to **quail**. Seeds are also valued by **turkeys** and a variety of **songbirds**. **Deer** will feed on the leaves and vines. Found in the ground layer of open woods. Responds well to disturbances such as fire and disking. The bean, or legume family hosts many very important plants for wildlife. High protein browse and quality seeds come from these plants.

PARTRIDGE PEA *CASSIA* SPP.

See Appendix E.

POKEWEED *Phytolacca americana*

A perennial that grows from 3 to 9 feet tall. Dark purple fruits are highly palatable to **dove**, **songbirds**, **raccoons**, and **bear**. Young shoots are eaten by deer. Found along fencerows, cultivated and abandoned land. Also called "Poke Salad."

RAGWEED *Ambrosia* spp.

A highly important summer annual which grows to about 1 to 3 feet tall. Seeds are a main staple of a variety of **birds** including, **dove**, **quail**, **song birds**, and **turkey**. A common weed found in open woods and croplands. Responds well to disturbances such as burning, mowing, and disking. Both "common" and "giant" varieties are common in Alabama. Usually the most common plant in excellent brood-rearing areas for game birds.

WOOLLY CROTON *Croton capitatus*

An annual reseeding forb commonly called goatweed. Seed is a choice food of **quail** and **doves. Deer** do not readily eat this plant. Principally established by natural reseeding. For quail establish a strip 1 to 15 feet wide around a field next to brushy cover. Woolly croton can generally be confined to a strip by mowing once or twice during the season. For doves allow fields to develop. Adapted to almost any soil except extremely wet soils. Soil disturbance is required for good stands to reseed. Disking in early spring is ideal.

Appendix J: Aquatic and Marsh Plants Important to Wildlife in Alabama

ARROW-ARUM *Peltandra virginica*

Native aquatic plant with fibrous rootstocks and light green narrow leaves. The large berries, which contain small seeds, are an important food source for **wood ducks** in **beaver** ponds and several species of **marsh birds**. Found in swamps, marshes, **beaver** ponds, and shores of streams and lakes.

ASIATIC DAYFLOWER *Murdannia keisak*

Annuals with linear leaves whose seeds are one of the most important waterfowl food plants for the larger dabbling ducks (**mallard** and **black duck**), particularly in **beaver** ponds. Found in beaver ponds, marshes and stream banks.

BARNYARD GRASS *Echinochloa crusgalli*

Common annual with thick erect stems up to 3 feet tall. Also called common wildmillet. Seeds are heavily utilized by **waterfowl** and **songbirds** as well as **quail** and **ruffed grouse**. Commonly planted on wildlife refuges. Seeds remain viable in water for long periods of time. Seeds germinate readily after drawdowns forming dense stands. Found in lowland areas, ponds, floodplains, marshes, lake shores, and along roadsides.

BULRUSH *Scirpus* spp.

Cylindrical stemmed wetland plant whose seeds are used by **waterfowl**, **marsh birds,** and **songbirds**. Bulrushes respond well to water drawdown in spring and summer.

CUTGRASS, GIANT *Zizaniopsis miliacea*

A perennial grass which grows over 3 feet tall. Also called southern wild rice and water millet. Giant cutgrass is found growing in mud in several feet of water in fresh or brackish water swamps and marshes. Found along the edges of ponds, lakes, rivers and creeks. Value to **waterfowl** is for nesting and protective cover rather than being an important food source.

FOXTAIL, GIANT *Setaria magna*

A large perennial grass with stiff erect stems which can grow over 6 feet tall. Found in wet soils of swamps, bottomland areas, ditches, and other moist areas. Originally native to Africa, is considered one of the most utilized open field herbaceous grasses by wildlife. Seeds are readily eaten by **waterfowl**, **doves**, **quail**, and **songbirds**. Also provides nesting habitat for **birds** and **small mammals**. Easily propagated by disturbances such as disking and burning.

PONDWEED *Potamogeton* spp.

Submersed aquatic plants with no floating leaves whose fruits, tubers, and roots are highly valued by **waterfowl**. Tubers and seeds have been planted in many areas to improve habitat for waterfowl. Found in alkaline, brackish, or saline water of ponds, slow flowing rivers and marshes. Seeds of other pondweeds are also an important food source for waterfowl.

SEDGE *Carex* and *Cyperus* spp.

Most sedges are perennials with triangular stems. Found mostly in wet soils of marshes, swamps, ditches and along roadsides. Seeds are heavily utilized by **marsh birds**, **soras**, **rails**, **quail**, **ruffed grouse**, **turkey**, **songbirds**, **sparrows**, and **buntings**. **Deer** will readily feed on the stems and leaves of sedges. **Waterfowl** readily utilize the seeds of sedges from marginal areas that have been flooded.

SMARTWEED *Polygonum* spp.

Erect herbaceous perennial (can also be an annual) wetland plants with a jointed-stem and pink to purple flowers. Several species of smartweed are used by wildlife in Alabama and include Pennsylvania smartweed (*P. pensylvanicum*), swamp smartweed (*P. hydropiperoides*), curltop ladysthumb (*P. lapathifolium*) and stout smartweed (*P. densiflorum*). Seeds are used by **waterfowl**, **doves**, **quail**, **ruffed grouse**, **songbirds**, and **marsh birds**. Seeds of swamp smartweed are used the heaviest by wildlife. Smartweeds are found in open swamp forests, sand or clay fields, irrigation ditches, marshes, and along edges of streams, rivers, **beaver** ponds and lakes. Can disk or burn to establish new stands every three years.

SPIKERUSH *Eleocharis* spp.

Aquatic perennials with slender stems that often form a dense mat of vegetation in brackish water. Several species such as dwarf spikerush (*E. parvula*) and squarestem spikerush (*E. quadrangulata*) provide seeds and vegetative shoots for **waterfowl** to feed upon. Found in most of coastal Alabama.

TEARTHUMB *Polygonum sagittatum* and *arifolium*

Annual wetland plants with stems that have downward pointing barbs and leaves that are arrowhead-shaped with pink flowers. Seeds are an important **waterfowl** food in **beaver** swamps. Tearthumb is also found in marshes and swamps and grows equally well in shade or full sunlight.

WATERSHIELD *Brasenia schreberi*

An aquatic perennial plant with long branching stems with floating oval or elliptic leaves that are green on top and purple underneath. Plants are often covered with a slimy coating. The seeds, leaves, and underwater portions of the plant are heavily utilized by **waterfowl**, especially **ring-necked ducks**. Commonly found in acidic ponds and lakes in shallow water.

WIDGEONGRASS *Ruppia maritima*

A submersed aquatic herb with an extensive root system that grows mainly in shallow brackish waters, alkaline lakes, ponds, and streams. The seeds, stems, leaves and roots are eaten by many species of **waterfowl** and **marsh birds**.

WILDRICE *Zizania aquatica*

A perennial wetland grass in the South which grows 7 to 10 feet high with long flat-bladed leaves. Wild rice provides seed and cover for **quail**, **waterfowl**, **rails** and other **marsh birds**. Wild rice, however, is not commonly planted in Alabama. Can plant in fall before freeze or in the spring. Seeds planted in the fall will remain dormant in the soil over winter with growth starting in the spring. Best grown in fresh water streams, sloughs, marshy lakes or ponds having an outlet with soft mud bottom and water from 6 inches to 1 foot in depth. Will also grow well in brackish tidal wetlands. Sheltered bays or coves on larger lakes, streams or rivers make good places to plant. Does not do well in landlocked lakes, in salty waters or strongly alkaline soils. Drain down water to expose mud flats for broadcasting seeds. Broadcast 50 pounds of seed per acre on waters ranging in depth from 6 inches to 1½ feet. Not necessary to cover seeds. Will reseed annually.

Appendix K: Sources of Wildlife Planting Materials

Adams-Briscoe Seed Co.
P.O. Box 18
325 East Second Street
Jackson, GA 30233
404-775-7826
FAX 404-775-7122
shrub lespedeza seed or seedlings,
clovers, bahia grass seed, annual grains

Adams, Tom
Union Springs, AL
334-738-2593
patridge pea

Alabama Crop Improvement Association
South Donahue Street
Auburn University, AL 36849
334-821-7400
shrub lespedeza, seeds for ryegrass,
annual grains

Alabama Forestry Commission
E. A. Hauss Nursery
Route 3, Box 322
Atmore, AL 36502
334-368-4854
FAX 334-368-8624
Native and adapted nursery stock for
wildlife food, habitat enhancement,
and environmental restoration

Augusta Forestry Center
P.O. Box 160
Crimora, VA 24431
703-363-5732
FAX 703-363-5055
shrub lespedeza seedlings,
wildlife mixes, wildflowers

Bailey's
44650 Hwy. 101
P.O. Box 550
Laytonville, CA 95454
707-984-6133
self-staking conical tree shelters

Beersheba Wildflower Garden
P.O. Box 551
Stone Door Road
Beersheba Springs, TN 37305
615-692-3575
FAX 615-692-2120

Ben Meadows
3589 Broad St.
Atlanta, GA 30341
1-800-241-6401
Supertube tree shelters

Botanico
P.O. Box 922
McMinnville, TN 37110
615-934-2868
FAX 615-934-2844
tree/shrub seedlings,
bare root and B&B containers

Boyd Nursery Company
P.O. Box 71
McMinnville, TN 37110
615-668-9898
FAX 615-668-7646
bicolor lespedeza seedlings

Budd Seed, Inc.
P.O. Box 25087
Winston Salem, NC 27114
910-760-9060 or 800-543-7333
FAX 910-765-3168
lespedeza seed, clover, switchgrass,
partridge pea, bahia grass, millets,
crown and hairy vetch, buckwheat

Cartwright Nursery Company
11861 E. Shelby Dr.
Colliersville, TN 38017
901-853-2352
FAX 901-853-2380
tree seedlings

Chestnut Hill Nursery, Inc.
Rt. 3, Box 267
Alachua, FL 32615
800-669-2067 or 904-462-2820
FAX 904-462-2820
fruit tree seedlings

Classic Groundcovers, Inc.
405 Belmont Rd.
Athens, GA 30605
706-543-0145
FAX 706-369-9844

Coastal Gardens
4611 Socastee Boulevard
Myrtle Beach, SC 29575
803-293-2000
FAX 803-293-2000

Cousins Agri-center
P.O. Box 1232
Newberry, SC 29108
803-276-5750
FAX 803-276-5111
lespedezas, clovers, millets,
bahia grass seed, annual grains, ryegrass

Carl R. Gurley, Inc.
P.O. Box 995
Princeton, NC 27569
800-753-3800
FAX 919 936-2200

Cross Seed Company, Inc.
P.O. Box 31881
Charleston, SC 29407
803-766-1687
FAX 803-766-1687
native grasses

C.P. Daniel's Sons Inc.
P.O. Box 119
Waynesboro, GA 30830
706-554-2446
FAX 706-554-4424
millets, clovers, lespedezas
and partridge pea,
shrub lespedeza seedlings

Dabney Nursery
5576 Hacks Cross Rd.
Memphis, TN 38125
901-755-4050
FAX 901-753-8880

Deep South Nursery
Rt. 7 Box 230
Andalusia, AL 36420
334-222-5803 or 334-222-3664
native flowering perennial legumes.
specializes in native legumes in longleaf
pine and hardwood stands, food plot
plantings, landscaping purposes, and
ecological restoration

Dello Nursery
11034 Highway 64
Arlington, TN 38002
901-867-3511
native mountain wildflowers
(trillium, phlox, poppy) and
tree seedlings

Delta View Nursery
Rt. 1 Box 28
Leland, MS 38756
601-686-2352
bare root tree seedlings,
wetland tree species water
tupelo, overcup oak, etc.

Richard Dennis Farms
P.O. Box 1101
St. Stephen, SC 29479
803-567-3071
shrub lespedeza seed, clovers,
bahia grass seed, annual grains

Dixon Gallery and Garden
4339 Park Ave.
Memphis, TN 38117
901-761-5250
wildflower sale once per year,
near Easter, with
unusual specimens

Double Creek Chufa Farm
P.O. Box 6
Highway 29 South
Perote, AL 36061
334-474-3366
chufas and partridge peas

Forestry Suppliers
P.O. Box 8397
Jackson, MS 39284
800-647-5368
Tree Pro and
Tree Pro Junior tree shelters

Gardens of the Blue Ridge
P.O. Box 10
Pineola, NC 28662
704-733-2417
FAX 704-733-8894
woodland plants

Glendale Enterprises, Inc.
Rt. 3 Box 77-P
DeFuniak Springs, FL 32433
904-859-2141
FAX 904-859-2181
chufas, velvet beans,
Florida beggarweed

Heather Farms
P.O. Box 278
Morrison, TN 37357
615-473-7757 or 800-451-3889
FAX 615-645-2375
tree seedlings, sawtooth oak

Hidden Springs Nursery
170 Hidden Springs Lane
Cookeville, TN 38501
615-268-9889
tree seedlings, mulberry trees, chestnut

Hillis Nursery Co., Inc.
Rt. 2 Box 142
Camden, AL 36726
334-668-8941
shrub lespedeza seed, bahia grass seed,
annual grains

Jeffery's Seed Company
Box 88
Goldsboro, NC 27533
919-734-2985
FAX 919-736-7241
bahia grass, sunflower

Kaufman Seed Company
P.O. Box 398
Ashdown, AR 71822
501-898-3328 or 800-898-1082
FAX 501-898-3302
wildlife specialty seeds

Lambert Seed and Supply, Inc.
P.O. Box 128
Camden, AL 36726
334-682-4111 or 800-242-2582
FAX 334-682-5147
chufa, clovers, millets, lespedezas,
partridge pea, bahia grass seed,
annual grains

Lichterman Nature Center
5992 Quince Rd.
Memphis, TN 38119
901-767-7322
FAX 901-682-3050
native grasses, wildflowers,
two plant sales per year (spring and fall)

MacMillian Bloedel Tree Nursery
P. O. Box 336
Pine Hill, AL 36769
334-682-92882 or 800-433-3587
autumn olive, sawtooth oak,
lespedeza thunbergii

Macon Feed and Seed
P.O. Box 3025
Macon, GA 31205
912-746-0291
FAX 912-746-0292
millets, clovers, lespedeza seed,
shrub lespedeza, partridge pea,
wildlife mixes, and grains

Mid-South Seed Company
P.O. Box 5960
North Little Rock, AR 72119
501-945-1474
FAX 501-945-0793

Mississippi Forestry Commission
Waynesboro Nursery
1063 Buckatunna-Mt. Zion Road
Waynesboro, MS 39367
601-735-9512
sawtooth oak, wild pecan

Mississippi Forestry Commission
Winona Nursery
90 Highway 51
Winona, MS 38967
601-283-1456
sawtooth oak

Mixon Seed Company
P.O. Box 1652
Orangeburg, SC 29116
803-531-1777 or 800-822-1377
FAX 801-534-5027
millets, clovers, lespedezas and partridge pea, wildlife seeds, grains, and mixes

Montgomery Seed & Supply Co., Inc.
(mailing) P.O. Drawer 349
Montgomery, AL 36104
(street) 255 Dexter Ave.
Montgomery, AL 36101
334-265-8241
FAX 334-265-8243
WATS 800-633-8700
lespedezas, bahia grass, wildlife seeds, grains, mixes

Mountain Ornamental Nursery
P.O. Box 83
Altamont, TN 37301
615-692-3424
wildlife seeds, wildflowers, tree seedlings

Native Gardens
5737 Fisher Lane
Greenback, TN 37742
615-856-0220
e-mail: clebsch@UTKVX.UTK.edu
wildflowers, tree seedlings

Norplex, Inc.
P.O. Box 814
Auburn, WA 98701
205-735-3431
tree shelter Protex Pro/Gro and other shelters

Norwood Farms
P.O. Box 438
McBee, SC 29101
803-335-6636
FAX 803-335-5197
shrub lespedeza seed, clover, bahia grass, annual grains

Nova Sylva, Inc.
1587, Denalt, Sherbrook, Qc
Canada J1H 2R1
800-567-7318
Tubex tree shelters

Park Seed Company
P.O. Box 46
Greenwood, SC 29648-4206
803-223-7333 or 800-845-3369
FAX 803-941-4206
non-game choices such as wildflowers, buddlia (butterfly bush), bird/butterfly garden seed called "habitat mix"

Pennington Seed, Inc.
AL FAX 334-734-9437
AL 800-768-7333
Due to high volume, they have a different fax/wats/address for every state in the South. Call 800 number first for best number in your area. Bahia grass, birdsfoot trefoil, chufa, buckwheat, deertongue, soil seed sunflower, Egyptian wheat, Florida beggarweed, sesame, wild millets, vetch, clovers, game food sorghum, grains and mixes, wildlife seeds, lespedezas, partridge pea

Pennington Seed, Inc. of Madison
1280 Atlanta Highway
P.O. Box 290
Madison, GA 30650
(706) 342-1234
variety of wildlife plant species

Project HELP
National Wild Turkey Federation
P.O. Box 530
Edgefield, SC 29824
800-843-6983
clovers, lespedezas and partridge pea seed, shrub lespedezas, seedlings, grains, four wildflower mixes, iron clay pea, latcho flatpea, catjang pea—call for brochure

Scott Paper Company
29650 Comstock Rd.
Elberta, AL 36530
334-986-5210
autumn olive, chinese chestnut,
crabapple, dogwood, overcup oak,
persimmon, redbud, sawtooth oak,
lespedeza thunbergii

Sims Brothers
Rt. 2 Box 73
Union Springs, AL 36089
334-738-2619
latest lespedeza varieties from
Auburn University, partridge peas

Spandle Nurseries
RFD 2, Box 125
Claxton, GA 30417
800-553-5771
FAX 912-739-2701
millets, clovers, lespedezas and
partridge pea, shrub lespedeza seedlings,
wildlife seeds, grains, many grasses, chufa

Superior Trees, Inc.
Lee Nursery
P. O. Box 9325
Lee, FL 32059
904-971-5159
American beautyberry, autumn olive,
lespedeza bicolor, chickasaw plum, crabapple,
dogwood, overcup oak, persimmon, redbud,
sawtooth oak, lespedeza thunbergii

Tall Pines Forest Nursery
Rt. 1 Box 110
Cross, SC 29436
803-753-3341
shrub lespedeza seedlings

Tennessee Department of Agriculture
Division of Forestry
East Tennessee Nursery
P. O. Box 306
Delano, TN 37325
901-988-5221 or 615-263-1626
autumn olive, lespedeza bicolor, chinese chestnut,
persimmon, sawtooth oak

Terra Tech
International Reforestation Suppliers
2100 W. Broadway
P.O. Box 5547
Eugene, OR 97405
800-321-1037
tree shelters Protex Pro/Gro,
Tree Pro, and Tree Pro Junior

Trees by Touliatos
2020 E. Brooks Rd.
Memphis, TN 38116
901-346-8065
FAX 901-398-527

Tree Pro
3180 W. 250 N.
West Lafayette, IN 47906
800-875-8071
Tree Pro tree shelters

Tree Sentry
P.O. Box 607
Perrysburg, OH 43552
419-872-6950
self-staking conical tree shelters

Tree Stand Nursery
27 Croft Street
Greenville, SC 29609
864-233-1091 or 912-734-4724
native tree species

Treesentials
Riverside Station
P.O. Box 7097
St. Paul, MN 55107
800-248-8239
Supertube tree shelters

Triangle Nursery
8526 Beersheba Highway
McMinnville, TN 37110
615-668-8022
tree seedlings, wildflowers

Tubex Limited
Tannery House
Tannery Lane
Send, Working, Surrey, GU23 7HB
England
0483 225434
Tubex tree shelters

Warren County Nursery, Inc.
6492 Beersheba Highway
McMinnville, TN 37110
615-668-8941
shrub lespedeza seedlings, sawtooth oak seedlings, small shrubs

Weyerhaeuser
Route 1, Box 108
Eutaw, AL 35462
800-635-0162 or 205-373-8312
autumn olive, lespedeza bicolor, overcup oak, sawtooth oak

Wildwood Nurseries
Rt. 4 Box 616
Walterboro, SC 29488
803-844-2336 or 800-673-7300
FAX 803-844-8990
native and broadleaf evergreens and shrubs, no seeds

Woodlanders, Inc.
1128 Colleton Ave.
Aiken, SC 29801
803-648-7522
native & small shrubs, fruit and nut trees

Wyatt-Quarles Seed Company
P.O. Box 739
Garner, NC 27529
919-772-4243 or 800-662-7591
FAX 919-772-4278
clovers, annual grains
(to wholesale dealers only)

Appendix L: Soil pH Range for Select Southern Tree Species

SOIL pH RANGE FOR SELECT SOUTHERN TREE SPECIES

Common Name	Scientific Name	pH Range
Alder	*Alumus glutinosa*	4.0–7.0
Ash, green	*Fraxinus pennsylvanica*	3.6–7.5
Ash, white	*Fraxinus americana*	4.6–7.5
Baldcypress	*Taxodium distichum*	4.6–7.5
Basswood, American	*Tilia americana*	4.6–8.0
Beech, American	*Fagus grandifolia*	6.0–7.0
Birch, black	*Betula lenta*	5.0–6.0
Birch, river	*Betula nigra*	4.5–6.0
Birch, yellow	*Betula allegahniensis*	5.0–7.0
Blackgum	*Nyssa sylvatica*	4.6–7.0
Buckeye	*Aesculus* spp.	6.0–8.0
Catalpa	*Catalpa* spp.	6.0–8.0
Cherry, black	*Prunus serotina*	4.6–6.2
Cherry, fire	*Prunus pensylvanica*	5.0–6.0
Chinkapin	*Castanea pumila*	5.0–6.0
Cottonwood	*Populus deltoides*	3.6–7.5
Dogwood	*Cornus* spp.	6.0–8.0
Elm	*Ulmus* spp.	5.2–8.0
Eucalyptus	*Eucalyptus* spp.	6.0–8.0
Hackberry	*Celtis occidentalis*	5.0–7.5
Hemlock, eastern	*Tsuga canadensis*	5.0–6.0
Hickory	*Carya* spp.	4.5–5.5
Holly, American	*Ilex opaca*	5.0–6.0
Honeylocust	*Gleditisia triacanthos*	6.0–8.0
Hophornbeam	*Ostrya virginiana*	6.0–7.0
Locust, black	*Robinia pseudoacacia*	4.5–7.5
Magnolia	*Magnolia grandiflora*	5.0–6.0
Maple, red	*Acer rubrum*	4.4–7.5
Mulberry	*Morus* spp.	6.0–8.0
Oak, black	*Quercus velutina*	5.0–5.4
Oak, blackjack	*Quercus marilandica*	5.0–6.0
Oak, bur	*Quercus macrocarpa*	6.0–6.3
Oak, cherrybark	*Quercus falcata* var. *pagodaefolia*	4.5–6.2
Oak, chestnut	*Quercus prinus*	5.0–7.0
Oak, laurel	*Quercus laurifolia*	3.6–5.6
Oak, live	*Quercus virginiana*	6.0–7.5
Oak, northern red	*Quercus rubra*	4.5–6.0

SOIL pH RANGE FOR SELECT SOUTHERN TREE SPECIES, cont.

Common Name	Scientific Name	pH Range
Oak, Nuttall	*Quercus nuttallii*	3.6–6.8
Oak, overcup	*Quercus lyrata*	3.6–5.5
Oak, pin	*Quercus palustris*	6.0–7.0
Oak, post	*Quercus stellata*	5.0–6.0
Oak, scarlet	*Quercus coccinea*	6.0–7.0
Oak, shumard	*Quercus shumardii*	4.4–6.2
Oak, southern red	*Quercus falcata* var	5.0–6.0
Oak, swamp chestnut	*Quercus michauxii*	3.6–6.2
Oak, water	*Quercus nigra*	3.6–6.3
Oak, white	*Quercus alba*	4.5–6.2
Oak, willow	*Quercus phellos*	3.6–6.3
Paulownia	*Paulownia tomentosa*	6.0–8.0
Pecan	*Carya illinoensis*	4.8–7.5
Persimmon	*Diospyros virginiana*	4.4–7.0
Pine, loblolly	*Pinus taeda*	4.5–6.0
Pine, longleaf	*Pinus palustris*	4.5–6.0
Pine, shortleaf	*Pinus echinata*	4.5–6.0
Pine, Virginia	*Pinus virginiana*	4.6–7.9
Pine, white	*Pinus strobus*	4.5–6.0
Redcedar	*Juniperus virginiana*	6.0–7.5
Redbud	*Cercis canadensis*	6.0–8.0
Sassafrass	*Sassafrass albidum*	4.7–7.0
Sourwood	*Oxydendrum arboreum*	4.0–8.0
Sumac	*Rhus* spp.	4.2–7.0
Sweet bay	*Magnolia virginiana*	4.0–5.0
Sweetgum	*Liquidambar styraciflua*	3.6–7.5
Sycamore	*Platanus occidentalis*	4.4–7.5
Tupelo, water	*Nyssa aquatica*	3.6–5.6
Walnut, black	*Juglans nigra*	5.0–7.5
Willow, black	*Salix nigra*	4.6–7.5
Yellow Poplar	*Liriodendron tulipifera*	4.5–7.0

(Source: U. S. Forest Service)

Appendix M: Sample Hunting Lease

This hunting lease agreement is for educational purposes only. It is important to check with your attorney before writing and signing a binding legal agreement. This lease may be more or less inclusive than the parties desire. If the lessor wants to provide other services or rights, such as guides, game cleaning, or allowing the lessee to improve the habitat, they should be included.

HUNTING LEASE AGREEMENT

STATE OF: ______________________________

COUNTY OF: ___________________________

This Lease Agreement (the "Lease") entered into as of the day of _______, by and between _____________ hereinafter referred to as Lessor, and ______________________________ a/an (state whether an individual, a partnership, corporation, or unincorporated association) hereinafter referred to as Lessee.

The Lessor agrees to lease the Hunting Rights, as defined below, on _____ acres more or less, to Lessee for _________________________ ($ ______/ Acre), for a term commencing on ___________________, (the "Commencement Date") and ending on ___________ (the "Expiration Date") on the following described property (the "Land").

SEE ATTACHED DESCRIPTION

The Hunting Rights shall consist of the exclusive right and privilege of propagating, protecting, hunting, shooting, and taking game and waterfowl on the Land together with the right of Lessee to enter upon, across and over the Land for such purposes and none other.

This Hunting Lease Agreement shall be subject to the following terms and conditions:

PAYMENT

1. The Lessee shall pay to the Lessor ______________________, the amount of one (1) year's Rent in full, on or before ____________________ by check payable to Lessor.

COMPLIANCE WITH LAW

2. Lessee agrees for itself, its licensees and invitees to comply with all laws and regulations of the United States and of the State and Local Governments wherein the Land lies relating to the game or which are otherwise applicable to Lessee's use of the Land. Any violation of this paragraph shall give Lessor the right to immediately cancel this Lease.

POSTING

3. Lessee shall have the right to post the Land for hunting to prevent trespassing by any parties other than Lessor, its Agents, Contractors, Employees, Licensees, Invitees, or Assigns provided that Lessee has obtained the Lessor's prior written approval of every sign designed to be so used. Every such sign shall bear only the name of the Lessee. Lessor reserves the right to prosecute any trespass regarding said Land but has no obligation to do so.

LESSOR'S USE OF ITS PREMISE

4. Lessor reserves the right in itself, its Agents, Contractors, Employees, Licensees, Assigns, Invitees, or Designees to enter upon any or all of the Land at any time for any purpose of cruising, marking, cutting or removing trees and timber or conducting any other acts relating thereto and no such use by Lessor shall constitute a

violation of this Lease. This right reserved by Lessor shall be deemed to include any clearing, site preparation, controlled burning and planting or other forestry work or silvicultural practices reasonably necessary to produce trees and timber on the Land. Lessee shall not interfere with Lessor's rights as set forth herein.

GATES/BARRIERS

5. Lessor grants to Lessee the right to install gates or other barriers (properly marked for safety) subject to the written permission of Lessor and the terms and conditions relating thereto as set forth elsewhere in this Lease, on private roads on the Land, and Lessee agrees to provide Lessor with keys to all locks prior to installation and at all times requested by Lessor during the term of this lease.

ROAD OR FENCE DAMAGE

6. Lessee agrees to maintain and surrender at the termination of this Lease all private roads on the Lands in at least as good a condition as they were in on the date first above-referenced. Lessee agrees to repair any fences or other structures damaged by itself, its licensees or invitees.

ASSIGNMENT

7. Lessee may not assign this Lease or sublease the hunting rights the subject of this Lease without prior written permission of Lessor. Any assignment or sublease in violation of this provision will void this Lease and subject Lessee to damages.

FIRE PREVENTION

8. Lessee shall not set, cause or allow any fire to be or remain on the Land. Lessee covenants and agrees to use every precaution to protect the timber, trees, land, and forest products on the Land from fire or other damage, and to that end, Lessee will make every effort to put out any fire that may occur on the Land. In the event that nay fire shall be started or allowed to escape onto or burn upon the Land by Lessee or anyone who derives his/her/its right to be on the Land from Lessee, Lessor shall have the right immediately to cancel this Lease without notice, and any payments heretofore paid shall be retained by Lessor as a deposit against actual damages, refundable to the extent such damages as finally determined by Lessor are less than said deposit. In addition, Lessor shall be entitled to recover from Lessee any damages which Lessor sustains as the result of such fire. Lessee shall immediately notify the appropriate state agency and Lessor of any fire that Lessee becomes aware of on Lessor's lands or within the vicinity thereof.

INDEMNIFICATION AND INSURANCE

9. Lessee shall indemnify, defend and hold harmless Lessor, its directors, officers, employees and agents from any and all loss, damage, personal injury (including death at any time arising therefrom) and other claims arising directly or indirectly from or out of any occurrence in, upon, or at the said Lands or any part thereof relating to the use of said Land by Lessee, Lessee's invitees or any other person operating by, for or under Lessee pursuant to this Lease. Lessee further agrees to secure and maintain a $1,000,000 public liability insurance policy in connection with the use of the Land with Lessor named as an additional insured and with such insurance companies as shall be agreeable to Lessor. This indemnity shall survive the termination, cancellation, or expiration of this Lease.

RULES AND REGULATIONS

10. Lessor's rules and regulations attached hereto as Exhibit "A" are incorporated herein by reference and made an integral part hereof. Lessee agrees that any violation of said rules and regulations is a material breach of this Lease and shall entitle Lessor to cancel this Lease at its option effective upon notice by Lessor to Lessee of such cancellation.

    Lessor reserves the right from time to time, to amend, supplement or terminate any such rules and regulations applicable to this Lease. In the event of any such amendment, supplement, or termination, Lessor shall give Lessee reasonable written notice before any such rules and regulations shall become effective.

MATERIAL TO BE SUBMITTED TO LESSOR

11. If this Lease is executed by or on behalf of a hunting club, Lessee shall provide Lessor, prior to the execution hereof, a membership list including all directors, officers, and/or shareholders, their names and addresses and a copy of Lessee's Charter, Partnership Agreement and By-Laws, if any. During the term of this lease, Lessee shall notify Lessor of any material change in the information previously provided by Lessee to Lessor under this paragraph 11.

LESSEE'S LIABILITY RE: TREES, TIMBER, ETC.

12. Lessee covenants and agrees to assume responsibility and to pay for any trees, timber or other forest products that may be cut, used, damaged or removed from the Land by Lessee or in connection with Lessee's use of the Land or any damages caused thereupon.

NO WARRANTY

13. This Lease is made and accepted without any representations or warranties of any kind on the part of the Lessor as to the title to the Land or its suitability for any purposes; and expressly subject to any and all existing easements, mortgages, reservations, liens, rights-of-way, contracts, leases (whether grazing, farming, oil, gas or minerals) or other encumbrances or on the ground affecting the Land or to any such property rights that may hereafter be granted from time to time by Lessor.

LESSEE'S RESPONSIBILITY

14. Lessee assumes responsibility for the condition of the Land and Lessor shall not be liable or responsible for any damages or injuries caused by any vices or defects therein to the Lessee or to any occupant or to anyone in or on the Land who derives his or their right to be thereon from the Lessee.

USE OF ROADS

15. Lessee shall have the right to use any connecting road(s) of Lessor solely for ingress, egress, or regress to the Land, such use, however, shall be at Lessee's own risks and Lessor shall not be liable for any latent or patent defects in any such road nor will it be liable for any damages or injuries sustained by Lessee arising out of or resulting from the use of any of said Lessor's roads. Lessee acknowledges its obligation of maintenance and repair for connecting roads in accord with its obligation of maintenance and repair under paragraph 6.

SURRENDER AT END OF TERM

16. Lessee agrees to surrender the Land at the end of the term of this Lease according to the terms hereof. There shall be no renewal of this Lease by implication or by holding over.

MERGER CLAUSE

17. This Lease contains the entire understanding and agreement between the parties, all prior agreements between the parties, whether written or oral, being merged herein and to be of no further force and effect. This Lease may not be changed, amended or modified except by a writing properly executed by both parties hereto.

CANCELLATION

18. Anything in this Lease to the contrary notwithstanding, it is expressly understood and agreed that Lessor and Lessee each reserve the right to cancel this Lease, with or without cause, at any time during the Term hereof after first giving the other party thirty (30) days prior written notice thereof. In the event of cancellation by Lessee, all rentals theretofore paid and unearned shall be retained by the Lessor as compensation for Lessor's overhead expenses in making the Land available for lease, and shall not be refunded to Lessee.

APPLICABLE LAW

19. This Lease shall be construed under the laws of the State first noted above.

IN WITNESS WHEREOF, the parties have hereunto caused this Agreement to be properly executed as of the day and year first above written.

WITNESSES:

______________________________ ______________________________

LESSOR

______________________________ ______________________________

LESSEE

(Add notarial acknowledgment for state the land is situated in if other than Alabama).

Appendix N: Sample Timber Sale Contract

STATE OF ___________

COUNTY OF ___________

KNOW ALL MEN BY THESE PRESENTS, that the undersigned ______________, hereinafter referred to as SELLER, whose address is ______________, does hereby GRANT, BARGAIN, SELL, and CONVEY unto ______________, hereinafter referred to as BUYER ________________, designated timber specifically described below lying and being situated in ____________ County, State of _________ to-wit:

SE ___ of the SE ___ of Section ___ and the NE ___ of Section ___, Township ___ North, Range ___ East.

Said sale area plus of minus ___ acres and shown on the attached map (Exhibit A) made a part of this contract.

It is mutually covenanted and agreed between the parties as follows:

1. Timber to be Sold. ***Description of timber to be sold here. Marked or unmarked, sizes, color, species, etc.*** Said marking and basal area to be checked periodically by seller or its agent.

Timber will be harvested in blocks so that the sale area is cut systematically and not in a random fashion.

2. Sales Price. ***If marked by cord/by thousand:*** Buyer, in consideration of the mutual promises and agreements of Seller set forth in this Agreement, agrees to forward its payment to Seller, along with an accurate report, by using weight or scale tickets, showing the volume by species and product cut, delivered, scaled or weighed during the preceding two weeks. Payment will be made by the first Friday following the end of a two week period. Seller retains a lien on timber cut and/or cut and removed for which payment has not been received.

Buyer agrees to pay the following prices:

Pine pulpwood	$___ per cord/ton
Pine CNS	$___ per cord/ton/M
Hardwood pulpwood	$___ per cord/ton
Ply logs	$___ per cord/ton/M
Hardwood logs	$___ per cord/ton/M

If selling by Bid: Buyer, in consideration of the mutual promises and agreements of Seller set forth in this Agreement, agrees to pay ($_____), receipt of which is hereby acknowledge.

3. Length of Contract. The Buyer is to have until ____ in which to cut, manufacture and remove said trees from the property herein above described, and all trees, whether standing, cut, or manufactured, remaining on the above described premises at the date of the termination of this agreement, shall be the property of and belong to the Seller.
4. Access. Buyer, its agents, servants, and employees shall have the right of ingress, egress and regress over, across and along said lands owned by Seller for the purpose of cutting and removing said trees, also free rights of way over and across said lands owned by Seller for such private roads as Buyer may find necessary to construct. Provided, however, that any pastures and game plots located on the herein described premises or on other property belonging to the Seller shall be treated with care and only reasonably necessary roads shall be made across them. Buyer agrees to use existing roads insofar as reasonably possible for the purpose of ingress and egress and agrees to leave all roads now in existence on the premises and used by Buyer in at least as good a condition as they were upon the beginning of logging operations.
5. Care of Property. The Buyer shall cut and remove said trees in a proper, customary and workmanlike manner, doing as little damage as possible to the property of the Seller, fences, and other improvements thereon. The Buyer shall make compensation to the Seller for any negligent, willful or wanton damage to said property,

however, Buyer shall not be responsible or accountable for incidental or unavoidable damages necessarily resulting from the operation of any reasonable timber cutting and logging operation on the above described land.

6. Wet Weather. In the event of prolonged wet weather, Buyer will cease cutting operations to avoid unreasonable damage to Seller's property.
7. Best Management Practices. Buyer agrees to abide by Alabama's Best Management Practices (BMPs) guidelines and mandatory Federal BMPs when roads and skid trails cross streams or other wetlands. Tree tops will be removed immediately and will not be left in any creeks, roads, or food plots on said premises, no harvesting within Streamside Management Zones (SMZs), except for certain marked trees which will be removed perpendicular to the stream. Appropriate stream crossing structures as recommended in the BMP guidelines will be used. SMZs will be marked prior to the sale and shall in no event be less than _____ feet wide from the center of the stream. Buyer will remove all garbage including but not limited to oil cans, trash, and oil filters daily.
8. Indemnification. Buyer hereby covenants and agrees to indemnify, protect and hold harmless the Seller form any and all claims, losses, suits, damages, judgements, expenses including attorney's fees, costs and charges of every kind and nature, whether direct or indirect, on account of or by reason of bodily injuries (including death) to any person or persons, or destruction of property or any person or persons, arising out of or occurring in connection with exercise or failure to exercise by the Buyer of the rights and privileges or obligations herein granted.
9. Attorney's Fees. Should the Seller sue the Buyer to enforce this Agreement or any of its terms, the Buyer shall pay a reasonable attorney fee and all expenses in connection with it.
10. Compliance With Laws. Purchaser shall comply with all laws, rules, regulations and ordinances, whether state, local or federal, including but not limited to all workers compensation laws, in its removal and extraction of timber and in receiving any other benefit or performing any obligation pursuant to this Agreement.
11. Severance Taxes. Purchaser agrees to pay all severance taxes associated with any timber cut in accordance with this Agreement.
12. Notification. Buyer shall notify seller at least 5 days before harvesting begins and 5 days before harvesting ends.
13. CERCLA. Buyer shall indemnify, defend, and hold Seller harmless from any and all liability, claims, costs, fines, fees, actions, or sanctions asserted by or on behalf of any person or governmental authority arising from or in connection with Buyer's use or misuse, handling or mishandling, storage, spillage, discharge, seepage into bodies of water or the groundwater supply, or release into the atmosphere of any hazardous material, pollutants, or contaminant, whether solid, liquid or gas. Buyer shall take all reasonable precautions and safety measures, in accordance with current technology, to prevent the release of hazardous materials, pollutants, and contaminants. In the event Buyer learns of the discharge upon the premises of any hazardous material, pollutant or contaminant, Buyer shall immediately undertake to contain, remove, and abate the discharge. Failure of Buyer to comply with these provisions shall constitute a default.
14. Prevent Forest Fires. Buyer shall be especially bound to exercise reasonable diligence to prevent and extinguish forest fires upon said lands within the time covered by this instrument during which the trees may be removed.

WITNESS our hands and seals in duplicate this the ____ day of _____, 199__.

SELLER:	BUYER:
______________________________	______________________________
__________________________	__________________________
Witness	Witness
By: __________________________	By: __________________________
Its: _________________________	Its: _________________________

Components of a Timber Contract

No two timber-cutting contracts are the same, but all contracts should include the following components:

1. Guarantee of title or right to sell, description of the sale, and location of boundary lines.
2. Specific description of timber being conveyed, method of designating trees to cut, and when, where, and how to determine volume.
3. Terms of payment.
4. Duration and starting date of agreement.
5. Clauses to cover damages to non-designated trees, fences, ditches, streams, roads, bridges, fields, and buildings.
6. Clauses designating road location, construction requirements, maintenance, and condition requirements following harvest.
7. Clauses to cover fire damage where harvesting crews are negligent and to protect seller from liability that may arise in the course of harvesting.
8. Specifications for complete use of the merchantable portion of trees.
9. Clauses to cover protection of soil, establishment of water bars on roads, establishment of filter strips along streams, and protection of drainage systems.
10. Clauses for arbitration in case of disagreement.

Appendix O: Sources of Technical and Educational Assistance

State and Federal Agencies

Alabama Cooperative Extension System
109 Duncan Hall
Auburn University, AL 38649-5612
(334) 844-4444
The ACES, with a statewide network of County Extension offices, conducts informal natural resources educational programs using research-based knowledge and techniques from Auburn and other land grant universities.

Alabama Cooperative Fish and Wildlife Research Unit
331 Funchess Hall
Auburn University, AL 36849
(334) 844-4796
Conducts fish and wildlife research, graduate education, and provides technical assistance.

Alabama Department of Conservation and Natural Resources
64 North Union Street
Montgomery, AL 36130
Main Number (334) 242-3486
Director & Assistant Director, Game and Fish Division (334) 242-3465
Chief, Fisheries Section (334) 242-3471
Chief, Wildlife Section (334) 242-3469
Coordinator, Non-Game Wildlife Program (334) 242-3469
State Lands Division, Natural Heritage Section (334) 242-3484
Provides technical wildlife and fisheries assistance to various groups across the state. Also responsible for management of unique areas and conservation law enforcement.

District Offices of the Alabama Department of Conservation and Natural Resources, Game and Fish Division, Wildlife Section
District I
21348 Harris Station Road
Tanner, AL 35761-9716
(205) 353-2634
District II
4101 Alabama Hwy 21 North
Jacksonville, AL 36265
(205) 435-5422
District III
P.O. Box 305
Northport, AL 35476
(205) 339-5716
District IV
64 North Union Street
Montgomery, AL 36130
(334) 242-3469
District V
P.O. Box 993
Demopolis, AL 36732
(334) 289-8030
District VI
1100 South Three Notch
Andalusia, AL 36420
(334) 222-5415
District VII
P.O. Box 933
Jackson, AL 36545
(334) 246-2165

Alabama Department of Environmental Management
P.O. Box 301463
Montgomery, AL 36130-1463
(334) 271-7700
Responsibilities include monitoring and regulation of water quality, public water supply, underground injection control, solid waste, hazardous waste, air pollution control, water well standards, operator certification, and coastal area functions.

Alabama Forestry Commission
513 Madison Avenue
Montgomery, AL 36130-0601
(334) 241-9400
Provides technical and educational assistance on forest management and fire prevention and suppression.

Alabama State Board of Registered Foresters
513 Madison Avenue
Montgomery, AL 36130
(334) 240-9368
Licenses practicing foresters in Alabama.

Auburn University
Has several colleges and departments that conduct teaching, research, and extension activities in natural resources (forestry, wildlife, and fisheries).

School of Mathematics and Life Sciences

Department of Zoology and Wildlife Sciences
Funchess Hall
Auburn University, AL 36849
(334) 844-4850

Department of Botany and Microbiology
101 Rouse Bldg.
Auburn University, AL 36849
(334) 844-4830

Alabama Cooperative Fish & Wildlife Research Unit
Funchess Hall
Auburn University, AL 36849
(334) 844-4796

College of Agriculture

Fisheries and Allied Aquacultures
Swingle Hall
Auburn University, AL 36849
(334) 844-4786

School of Forestry
108 M. White Smith Hall
Auburn University, AL 36849
(334) 844-1007

Solon Dixon Forestry Education Center
Forestry Education Center
Rt. 7 Box 131
Andalusia, AL 36420
(334) 222-7779

Consortium for Research on Southern Forest Wetlands
U.S. Forest Service
2730 Savannah Highway
Charleston, SC 29414
(803) 724-4271
Provides technical information on forested wetlands.

Geological Survey of Alabama
420 Hackberry Lane
P.O. Box O
Tuscaloosa, AL 35468-9780
(205) 349-2852
Provides topographical maps for land use and management purposes.

Tennessee Valley Authority
400 W. Summit Hill Drive
Knoxville, TN 37902
(615) 632-2101
Federal agency responsible for certain land management and permitting requirements in the Tennessee Valley counties of Alabama.

University of Alabama
Discovering Alabama Series/Alabama Museum of Natural History
Box 870340
Tuscaloosa, AL 35487-0340
(205) 348-2039
Provides educational information about the natural history and natural resources in Alabama.

U.S.D.A. Farm Service Agency
P.O. Box 235013
Montgomery, AL 36123-5013
(334) 279-3504
Provides cost-sharing assistance for farm, forest, and wildlife management. Offices are located in each county.

U.S.D.A. Wildlife Services
Room 118 Extension Hall
Auburn University, AL 36849
(334) 844-5670
Provides wildlife damage management assistance and information.

U.S. Forest Service
2946 Chestnut St.
Montgomery, AL 36107-3010
(334) 832-4470
Responsible for managing national forests in Alabama.

Other U.S. Forest Service Offices:

Conecuh National Forest
Rt. 5, Box 157
Andalusia, AL 36420
(334) 222-2555

Bankhead National Forest
P.O. Box 278
Double Springs, AL 35553
(205) 489-3427

Talladega National Forest
Shoal Creek Ranger District
450 Hwy 46
Heflin, AL 36264
(205) 463-2272

Oakmulgee Ranger District
P.O. Box 67
Centreville, AL 35042
(205) 926-9765

Talladega Ranger District
1001 North Street
Talladega, AL 35160
(205) 362-2909

Tuskeegee National Forest
125 National Forest Road 949
Tuskeegee, AL 36083
(334) 727-2653

U.S.D.A. Natural Resources Conservation Service
P.O. Box 311
Auburn, AL 36830
(334) 887-4560
Provides technical land management assistance for farm and forest owners. Offices are located in each county.

U.S. Department of the Army
Mobile District Corps of Engineers
P.O. Box 2288
Mobile, AL 36628-0001
Provides regulatory guidance and permits for management practices in wetlands and navigable waterways in Alabama.

U.S. Environmental Protection Agency
401 M Street, SW
Washington, D.C. 20460
(202) 260-2090
Federal agency responsible for regulating and enforcing environmental quality in the U.S.

U.S. Fish and Wildlife Service
Division of Ecological Services
P.O. Drawer 1190
Daphine, AL 36526
(334) 441-5181
or
U.S. Fish and Wildlife Service
Division of Ecological Services
6578 Dogwood View Parkway
Jackson, MS 39213 (601) 965-4900
Federal agency responsible for management and regulation of migratory birds and threatened and endangered species. Also responsible for managing National Wildlife Refuges. Provides educational information and limited technical assistance to landowners through special programs.

Conservation and Related Organizations

Alabama Association of Soil and Water Conservation Districts
660 Adams Avenue, Suite 101
Montgomery, AL 26104
(334) 264-4548
Responsibilities include coordination and education of soil and water conservation in Alabama.

Alabama Environmental Council, formerly The Alabama Conservancy
2717 7th Avenue S., Suite 207
Birmingham, AL 35233
(334) 322-3126
Dedicated to the preservation of Alabama's air, water, land, and wildlife.

Alabama Forestry Association
555 Alabama Street
Montgomery, AL 36104
(334) 265-8733
Organization for private forest landowners that promotes sound forest management for products.

Alabama Forest Owners' Association
P.O. Box 104
Helena, AL 35080
(205) 987-8811
A statewide organization affiliated with the National Woodland Owners' Association, providing members with timely information such as legislation, timber markets, environmental issues, and forest taxation.

Alabama Forest Resources Center
P.O. Box 16642
Mobile, AL 36616
(334) 343-9747
Provides educational information for forest landowners and the general public.

Alabama Ornithological Society
5660 Pine Street
McCalla, AL 35115
Organization of bird watchers and others interested in the protection and wise management of birds in Alabama.

Alabama TREASURE Forest Landowners Association
P.O. Box 210476
Montgomery, AL 36121
Association composed of forest owners who are members of the TREASURE Forest program.

Alabama Waterfowl Association, Inc.
P.O. Box 67
Guntersville, AL 35768
(205) 259-2509
To protect, enhance, and create wetlands habitat for all wildlife species and other human values; and to enhance waterfowl population and protect the hunting heritage in Alabama.

Alabama Wildlife Federation
P.O. Box 1109
Montgomery, AL 36102
(334) 832-9453
1-800-822-9453
Alabama's oldest and largest citizen's conservation organization composed of members who share a mutual concern for wildlife and natural resources in the state.

Alabama Wildlife Rehabilitation Center
Oak Mountain State Park
2107 Marlboro Street
Birmingham, AL 35226
(205) 663-7930
Organization dedicated to caring for and rehabilitating injured wildlife.

Alabama Fisheries Society, Alabama Chapter
Department of Fisheries
Auburn University, AL 36849
(334) 844-3474
A professional society to promote the conservation, development, and wise utilization of fisheries, both recreational and commercial.

Alabama Wildflower Society
1808 Epworth Drive
Huntsville, AL 35811
(205) 967-0304
Organization dedicated to the conservation of wildflowers in Alabama.

Aldo Leopold Foundation
E12919 Levee Road
Baraboo, WI 53913-9737
(608) 355-0279
Organization dedicated to promote care of natural resources by fostering an ethical relationship between people and land.

Anglers for Clean Water, Inc.
P.O. Box 17900
Montgomery, AL 36141
(334) 272-9530
A non-profit organization dedicated to educating and informing the American public on conditions of pollution nationwide and to the danger of the failure to halt pollution of the streams, rivers, and lakes of the U.S.

Association of Consulting Foresters of Alabama
P. O. Box 684
Opelika, AL 36803
(334) 745-7530
The Association represents interests of private consulting foresters. They also administer continuing education programs, enforce a code of ethics, and promote the use of private consulting foresters.

Ducks Unlimited, Inc.
One Waterfowl Way
Memphis, TN 38120-2351
(901) 758-3825
The mission is to fulfill the annual life cycle needs of North American waterfowl by protecting, enhancing, restoring, and managing important wetlands and associated uplands.

Izaak Walton League of America, Inc., The
707 Conservation Lane
Gaithersburg, MD 20878
(800) 453-5463
Promotes means and opportunities for educating the public to conserve, maintain, protect, and restore the soil, water, forest, air, and other natural resources of the U.S. and promotes the enjoyment and wholesome utilization of those resources.

National Audubon Society
700 Broadway
New York, NY 10003-9501
(212) 979-3000
The Society's goals are to protect the air, water, land, and habitat that are critical to health and the health of the planet. To accomplish these goals the Society uses solid science, policy research, forceful lobbying, litigation, citizen action, and education.

National Wild Turkey Federation, Inc.
The Wild Turkey Building
P.O. Box 530
Edgefield, SC 29824-0530
(803) 637-3106
A non-profit organization dedicated to the wise conservation and management of the American wild turkey.

National Wildflower Research Center
2600 FM 973 N.
Austin, TX 78725
(512) 929-3600
A non-profit organization devoted to the re-establishment and conservation of native plants by promoting their use in public and private landscape designs. Serves as a national clearinghouse of information on native plants, their sources, landscaping with them, and appropriate resource organizations and agencies.

National Wildlife Federation
1400 16th Street N.W.
Washington, DC 20036-2266
(202) 797-6800
The mission of the National Wildlife Federation is to educate, inspire, and assist individuals and organizations of diverse cultures to conserve wildlife and other natural resources and to protect the earth's environment in order to achieve a peaceful, equitable, and sustainable future.

The Nature Conservancy of Alabama
Pepper Place 2821 C
Second Avenue South
Birmingham, AL 35233
(205) 251-1155
Non-profit organization committed to preserving biological diversity by protecting natural lands through "natural heritage programs" that identify ecologically significant natural areas.

The North American Bluebird Society
Box 6295
Silver Spring, MD 20906

Quail Unlimited, Inc.
P.O. Box 10041
Augusta, GA 30903
(803) 637-5731
A non-profit conservation organization dedicated to improving quail and upland game bird populations through habitat management and research.

The Wildlife Society, Alabama Chapter
118 Extension Hall
Auburn University, AL 36849
(334) 844-5670
The professional organization of wildlife biologists, managers, and scientists in Alabama.

Southeastern Cooperative
Fish Disease Project
Dept. of Fisheries and Allied Aquacultures
Auburn University, AL 36849-5419
(334) 844-4786
Provides fish-kill diagnostic services, training in fish diseases, and research on fish diseases.

Southeastern Society of American Foresters
Association Services Group
1961 Lower Big Springs Road
LaGrange, GA 30240
(706) 845-9085
Southeastern organization representing all segments of the forestry profession. Objectives are to advance the science, technology, education and practice of forestry.

Southeastern Cooperative Wildlife Disease Study
University of Georgia
Athens, GA 30602
(706) 542-1741
Provides wildlife disease and diagnostic services to southeastern states.

For additional listings of conservation organizations, see The National Wildlife Federation's Conservation Directory, which may be purchased by calling toll-free (800-432-6564) or writing The National Wildlife Federation, 1400 16th Street, N.W., Washington, DC, 20036-2266.

Appendix P: Glossary of Terms

absentee landowner: A landowner who does not live on the land (and often not in the county) where his or her property is located.

acid soils: Soils with a pH value below 7.0, which are better suited for pines than hardwood trees.

acre: An area of land equal to 43,560 square feet or 10 square chains. A square acre would be approximately 209 feet by 209 feet, and a circular acre would have a radius of 117.75 feet.

ad valorem tax: An annual tax assessed on the basis of the land value.

aesthetics: Pleasurable benefits, mental or physical, such as natural beauty, which humans may experience as a result of environmental resources.

all-aged stand: (See uneven-aged stand.)

allowable cut: The volume of timber that can be cut from a forest during a given period without exceeding the forest's net growth for that period.

altricial: Helpless and naked when hatched or born.

annual: A plant that grows, reproduces, and dies within one year.

appreciative uses: (See non-consumptive uses.)

approaches: The entry and exit of a road or skid trail through a stream crossing.

aquatic ecosystem: An interacting community of plants and animals (i.e. amphibians, birds, fish, insects, and vegetation) requiring an abundance of water during some part of their life cycle.

artificial regeneration: Establishing a forest by planting or direct seeding instead of allowing the trees to reseed themselves. (See natural regeneration also.)

aspect: The compass direction toward which a slope faces. Aspect is important to know when determining how sunshine and climate will affect the growth of trees and other vegetation on a hillside.

backblade: To move soil by dropping a bulldozer blade into the soil and operating the tractor in reverse.

back slope: The soil profile in the side of a hill that is exposed during cut and fill type road construction.

backyard wildlife: Wildlife that are attracted to and visit suburban and urban homesites.

banks: The sides of a channel which hold or carry water, as in stream banks.

bareroot seedlings: Seedlings grown in a nursery bed which are lifted and shipped for replanting without any soil around the roots.

basal area: Cross sectional area of a tree, in square feet, measured at breast height. Used as a method of measuring the stocking of timber in a given stand.

bed: The bottom of a water channel or stream, also called the stream bed.

bedding: A mechanical site preparation technique where top soil is mounded into rows. Trees planted on top of the rows will be well drained and will benefit from a concentration of nutrients and organic matter during initial stages of growth.

Best Management Practices (BMPs): Management practices that limit negative impacts on the land. BMPs for forestry are voluntary guidelines for silvicultural practices that minimize soil erosion and water quality problems during timber harvest and other management practices.

biennial: A plant that usually survives at least two growing seasons.

Biltmore stick: (See tree scale stick.)

blowdown (windfall): An uprooted tree that has been toppled by the wind.

bloom: An increase in phytoplankton growth. Large increases can deplete oxygen in ponds and lakes when phytoplankton die, causing fish to die. Water that is green in tint has healthy phytoplankton levels, while black water is an indication of phytoplankton dying and decaying causing a rapid drop in pond oxygen levels.

board foot: A unit of wood 1 foot by 1 foot by 1 inch (or 144 cubic inches), often used in describing the volume of lumber that logs would produce when sawed.

bole: The main trunk of a tree.

breast height: (See DBH.)

browse: Leaves, buds, twigs, etc. of shrubs or trees that are eaten by wildlife.

buffer: A designated strip or zone of standing or felled trees or other vegetation used as wildlife shelter, windbreaks, visual screening, or for other purposes.

buffer species: Prey species that serve as alternative food sources for predators.

canopy (crown or overstory): The upper leafy branches of trees and shrubs which intercept light and shade the forest floor.

castoreum: An oily liquid exuded by beavers to mark territory. Also used to make perfumes, cosmetics, and trappers lures.

chain: A unit of measure used by foresters equal to 66 feet; 10 square chains equals 1 acre.

chemical site preparation: Application of chemical

herbicides to reduce plant competition. Used when converting mixed stands to pine plantations. Often the area is control burned afterwards, from which the term "brown and burn" arises.

Clean Water Act: A law passed by Congress to protect water quality.

clear-cutting: A method of forest harvesting which removes all trees regardless of size or species from any stand of timber.

climax community: Vegetation that should appear in the natural succession of plant communities as the final stage in a long progression of changes. Example: from grassy plants and light-seeded trees (early succession) to dominant, closed-canopy trees (climax community); sometimes biologists use the term to also mean the vegetation *and* the associated wildlife in such a community.

codominant: Trees with medium-sized crowns forming the general level of the crown cover, but not as extensive as dominant trees. They receive full light from above but are crowded on the sides and therefore receive comparatively little light from the sides.

commercial: Viewed with regard to profit, such as a commercial thinning operation, wherein a landowner would benefit from the sale of harvested trees.

compaction: The compression of air and moisture-holding spaces between soil particles by the operation of heavy equipment or by excessive human or livestock trampling, or similar forces. In general, soils are more quickly compacted when wet. After compaction, soil productivity decreases and erosion increases.

compartment: Areas that have similar characteristics such as vegetation, soils, topography, productivity, and other features. Compartments are useful in dividing up large and diverse land tracts into manageable units.

compartment burning: Prescribed burning conducted over an entire management unit.

competition: The struggle for survival that occurs when organisms, trees, vegetation, or wildlife make similar demands on environmental resources.

conifer: A tree belonging to the order *Coniferales*, such as pine, spruce, fir, or cedar, that is usually evergreen and cone-bearing. (One exception is bald cypress, which is a deciduous conifer.) Often these are also referred to as "softwood" trees.

conservation: The wise use and management of natural resources.

conservation easement: A legal agreement that places restrictions on property uses. The agreement is usually between a landowner and a non-profit group.

controlled burning: (See prescribed burning.)

cord: (See standard cord.)

core sample: Soil samples taken when constructing a pond that help determine the percent clay content. High clay content is important for retaining water in a pond.

corridor: A pathway which serves as a conduit for wildlife to move from one patch of land to another, which can be as small as a brushy fencerow or as large as a streamside management zone.

cost-sharing: A policy whereby one party or agency agrees to pay a percentage of the specific expenses of a second party.

cove: Area surrounding a spring, stream, or creek head on upland sites that usually supports a diversity of plant species because of high soil fertility and moisture content.

crawler: A tractor with continuous treads instead of wheels.

critical habitat: Defined by the Endangered Species Act as an area required for a species' normal needs and survival.

critical shading of water: Shading that occurs when water receives the greatest protection from overheating and ultraviolet exposure caused by solar radiation.

crop tree: A tree which has been selected for future timber harvesting, usually based on its relative timber market value.

crown: The upper portion of a tree consisting of the trunk, canopy, and expanding branches.

cruise: (See timber cruise.)

cull: A tree or log of no commercial timber value which is sometimes removed during thinning practices to reduce competition with crop trees.

culverts: Materials to convey water, usually metal or plastic pipe but also wooden trough constructions.

cultipacker: A tractor attachment which is pulled over newly prepared seedbeds to firm up and compact the ground to facilitate seed and soil contact which enhances germination.

cut and fill: The practice of removing earthen material from a hill or construction site and placing down a slope to provide a relatively level road bed.

cutting contract: A written, legally binding document used to accomplish the sale of standing timber which specifies provisions covering expectations of either the buyer or the seller.

cutting cycles: The schedule for harvesting or felling operations in a forest stand.

DBH: Diameter at breast height. The diameter of a tree outside of the bark at roughly breast height or 4½ feet from the ground.

deciduous: A tree which loses its leaves during the year, usually in the fall, such as an oak, ash, maple or hickory.

deck: An area cleared to provide a site for loading logs onto transport vehicles. Also called a ramp or loading deck.

decking: Rough or unfinished lumber used to provide a stable surface for roads, stream crossings, or landings.

defects: Tree defects are portions of the tree which are unmerchantable and detract from its commercial value such as rot, crookedness, cavities, excessive limbs, etc.

definable bank: The bounds of a water body at or below its normal flow level. This area usually has no terrestrial plants growing on it.

deposition: The act of depositing, as when soils are deposited by rising flood waters.

depredation: The damage that one species causes to another, such as deer destruction of agricultural crops, or herons (birds) preying upon commercial catfish.

destabilize: Term, here used in reference to the soil, to expose and/or loosen the earth, thus making it more susceptible to erosion.

diameter: The length of a straight line passing through the center of a circle to the periphery. Tree diameter is usually measured at 4½ feet above ground level (see DBH) with instruments such as a tree scale stick, whereas log diameter is measured at the smallest end of the log.

dibble: Also called dibble bar or planting bar. A tool for hand planting bare-rooted seedlings.

direct seeding: Artificially placing seed by hand, farm equipment machinery, or aircraft onto a germination surface.

disking: Breaking up plants (either above or below the soil surface), organic matter and soil with a harrow to improve the ground for replanting and to reduce plant competition.

dissolved oxygen: The primary source of oxygen to fish and other aquatic organisms which is derived from plant and phytoplankton photosynthesis.

diversion device: A structure, also called drainage device, to intercept and reroute water from a road surface.

dominant tree: A tree which has a crown extending above the general level of the crown cover and which receives full sunlight from above.

dredge: Earthen material that is dug and removed from a channel or the bottom of a water body, often to improve drainage.

duff: (See litter layer.)

easement: An interest or right to limited land use granted by the owner to another party.

ecology: The branch of science dealing with the interrelationships of plants and animals with their environment.

ecosystem: An interacting, interdependent community of living organisms together with their physical environment.

ecosystem management: Management which seeks to blend social, economic, and scientific principles to achieve healthy ecosystems over long periods of time, while at the same time allowing production of the many valued resources our society seeks from components of the ecosystem.

ecotone: The zone of transition between two (or more) plant communities that shares the characteristics of both communities.

edge: An area where one type of habitat meets and blends with another.

EHD: Epizootic hemorrhagic disease (EHD) is a viral disease that kills white-tailed deer by causing extensive internal hemorrhaging.

endangered species: Animals and plants that will probably become extinct unless protected.

Endangered Species Act (ESA): Federal act passed by Congress in 1973 aimed to reduce the rate of extinction of animals and plants in the U.S. The ESA directs the U.S. Fish and Wildlife Service to identify and protect plant and animal species determined to be endangered, threatened, or species of concern.

endemic: Plant or animal species that is native to a particular area.

entomology (forest entomology): The science concerning insects and their relationship to forests and forest products.

ephemeral streams: Low places in the landscape where water flows only after significant rainfall. These stream beds often do not have a well-defined channel.

erosion: The dislodging and removal of soil particles by wind or water.

even-aged forest: Trees which are essentially the same age. Tree age does not vary more than 10 to 20 percent.

even-aged forest management: Forest management with periodic harvesting of all trees on part of the forest at one time, or in several cuttings over a short period, that result in stands containing trees which are all the same age or nearly the same age. This type of management is usually applied to conifers, and to a lesser extent, some hardwoods.

evergreen: A tree which retains some or all of its leaves throughout the year, such as most conifers (one noticeable exception is the deciduous bald cypress). Although most hardwoods are deciduous, some exceptions include the evergreen live oak, magnolia, and holly.

exclosure: A small, fenced-in area, usually 4–5 square feet in size, that protects plants inside the exclosure from being eaten by wildlife. Used to give an indication of plant use by wildlife.

featured species approach: A type of wildlife management which focuses on enhancing habitat for one or a few select species rather than many or all species.

fecundity: The quality or power of producing offspring in large numbers. The inherent reproductive potential of a species.

fell: To cut or knock down standing trees or other vegetation.

fell and burn method: A site preparation method used primarily in steep terrain in which trees are cut in the spring and the site is control burned in the summer. This method of site preparation lessens soil disturbance in comparison to other harvesting techniques.

fill (Also see "cut and fill"): To raise the elevation of a surface by depositing dredged or excavated material onto it. Sometimes "fill dirt" or "fill soil" is used to refer to the soil material that is deposited.

filtration strip (filter strip): A strip of land on which vegetation is maintained, or where mulch, or a special fabric is placed so that upland sediment and pollutants can be intercepted and prevented from flowing into the water.

financial maturity: The age at which a tree is no longer increasing in monetary value at a profitable rate.

firebreaks (firelane): Natural or artificially constructed barriers to the spread of fire.

fire danger rating: A numerical classification of the measurement of weather and fuel (or potential of a wooded area to burn) factors on a scale of 1 (lowest) to 5 (highest) used to determine optimal days for prescribed burning in the forest.

floodplain: Areas adjacent to bodies of water that are most prone to flooding when the water overflows its banks.

food plot: An area of land planted as a source of food for wildlife.

forb: Any herbaceous plant other than a grass, or legume.

forest floor: The earth beneath a stand of trees, usually with accumulations of organic debris and low vegetation.

forest fragmentation: The breaking up of large expanses of forest land into smaller blocks or patches.

forest interior species: Birds and other wildlife that require large expanses of unbroken mature forest to thrive.

forest inventory: A survey to assess the characteristics and future timber productivity of a stand.

forest resource managers: This group, also called natural resource managers, includes foresters, wildlife biologists, recreation planners, and other natural resource professionals.

form: The shape of a tree or log, usually in conjunction with timber quality and value.

fragile areas: Environmentally sensitive areas that are easily altered physically, biologically, or chemically and difficult to restore.

furbearer species: Animals whose skin and hair can be processed into pelts that are used by man. Include such species as beaver, river otter, muskrat, mink, raccoon, fox, and bobcat.

game species: Wildlife that are managed and hunted for recreation.

girdling: Cuts in a tree's bark made by man, animals, insects or sometimes by disease damage which completely encircles the tree's bark and cambium, penetrating the sapwood. Girdling usually kills the tree by stopping the flow of nutrients between the roots and the crown.

Geographic Information System (GIS): A method for entering, storing, manipulating, analyzing, and displaying information on a computer about land characteristics, use, and management.

grade: The steepness of rise or fall in a road surface.

ground cover: Low-growing vegetation such as grasses, forbs, vines, or shrubs.

grading: Evaluating and sorting trees, logs, or lumber according to quality.

ground water: Water stored and/or flowing out of sight under the ground surface.

group selection: A method of harvesting that removes clusters of trees of all sizes and ages in small areas. Usually an area less than two acres.

growing stock: All live trees (except rough and rotten trees) in a forest or stand, including sawtimber, pole timber, saplings, and seedlings.

habitat: A place where a plant or animal naturally lives and grows.

hand planting: Planting seed or seedlings by hand.

hardpan: An area of severely compacted soil which does not provide conditions conducive to tree or plant growth. May be up to 5 feet below the surface.

hardwoods: Broadleaf, trees such as oaks, maples, ashes, and elms. These trees do not necessarily produce hard wood.

harvests: Gathering merchantable portions of trees for commercial and domestic use. Also refers to the number of game killed during hunting seasons.

headquarters: also called "coverts" or "prime habitat corners" are points where more than three vegetation or habitat types meet. Normally these areas are ideal for meeting most wildlife needs.

herbicide: A natural or synthetic chemical applied specifically to control vegetation.

herpetofauna (herps): Reptiles and amphibians collectively grouped together.

high-grading: A practice of removing trees of higher commercial timber value, leaving a stand of poor quality trees of lesser value or depleted timber productivity.

high flow: The increased volume and speed of water that exceeds a stream's normal rate of flow.

high water mark: Physical evidence that can be seen of past flooding, such as discoloration of vegetation and debris suspended in branches or off the ground.

home range: An area where an animal spends most of its time from birth until death.

implementation: The carrying out of instructions contained in a management plan, harvesting plan, or reforestation plan.

impoundments: An accumulation of water into pools or ponds formed by blocking the natural drainage.

improvement cut: A type of intermediate cut with the primary objective being to improve the growth and quality of the remaining trees.

indicator species: Plants or animals whose abundance and health are a reflection of environmental quality.

increment borer: A hollow, auger-like tool used to extract a core of wood from a tree to determine its age from tree rings without felling the tree.

in-kind assistance: A type of compensation for the benefits of the use of land that includes volunteer activities such as planting wildlife food plots, patrolling for trespassers or vandalism, and making road or fence repairs.

inoculation: The process of mixing live *Rhizobium* bacteria with legume seeds to facilitate the production of nitrogen, necessary for growth.

intermediate cut: Removing immature trees from the forest at a period between early growth and maturity to improve the quality of the remaining forest stand, a cut which may or may not generate income.

intermediate trees: Trees of lesser height than dominant or codominant trees but with crowns extending into the overall crown cover formed by the dominant and codominant trees. These trees receive little light from above and none from the sides; their crowns are considerably crowded on the sides.

intermittent bodies of water: Water pathways which contain water in well defined channels but only during certain parts of the year.

inventory (tree inventory): (See timber cruise.)

invitee: A person who enters another person's land for his or her own purposes, such as recreational sports, and who is required to pay a fee or perform a service for this privilege.

jurisdictional wetlands: Wetlands that are protected by the government.

KG blade: A blade on a crawler (or tractor with continuous treads instead of wheels) which is used to clear unwanted vegetation in preparation for planting tree seedlings.

label restrictions: Explicit instructions from the manufacturer with approval from federal and state authorities on when, where, and how a particular chemical (such as a herbicide or pesticide) may be applied. Directions usually include worker and environmental safety precautions.

landing (log deck or yard): A site where logs are sorted and loaded onto trucks for hauling to handling or processing facilities.

landowner cooperatives: Individually owned private lands which have been joined together in a larger unit for the common purpose of managing wildlife.

land trust: A private organization with a common goal of protecting farmland or natural areas, such as forests, from development.

lease: An agreement between the landowner (leasor) and land user (leasee) which grants the land user rights for specified activities, such as hunting or farming, on the property.

legume: A family of plants that produce a pod, such as that of a pea or bean, that splits into two halves with seeds inside.

licensee: A person who enters another person's property with the permission of the landowner for purposes such as hunting, but who is not required to pay a fee.

litter layer (duff): The natural accumulation of dead leaves, branches and stems of dead trees and other forest vegetation which builds up on the ground and decays over time.

log: Dead portion of a fallen tree that provides food and shelter for wildlife. In forestry, also considered a 16.3 foot section of newly harvested tree.

log deck: (See landing.)

logger: An individual whose occupation is harvesting timber. Loggers are usually in business for themselves, and often employ other people.

logging: The practice of harvesting timber.

log rule: A method used to determine the board foot contents for logs of various diameters and lengths.

matrix: A large landscape comprised of many different types of habitat patches.

mature tree: A tree that has reached the desired size or age for its intended use, which varies considerably by species and intended use.

MBF: An abbreviation for "thousand board feet," which is a unit of measurement for tree volume or sawed timber.

mechanical planter: A tree planting machine pulled by a

tractor and operated by a person who places trees into the ground.

mechanical site preparation: Use of heavy equipment such as bulldozers with attachments to clear debris or incorporate vegetative material into the soil to improve planting and growing conditions for new forest trees and plants.

mensuration: Traditionally, the branch of forestry which addresses the measurement of present and future volume, growth, and development of individual trees and stands and their eventual timber products; the measurement of forest lands also pertains.

merchantable (or merchantable timber): Trees which have the potential to be managed as a product and sold for a profit.

merchantable height: A term which refers to the height (or length) of a tree trunk for commercial value purposes. It is measured to a point where the diameter is too small to obtain a particular product.

microclimate: A small area (such as a cove) that has different physical characteristics (such as soils, soil moisture, soil fertility, exposure to sunlight) than its surrounding landscape. Because of these differences, microclimates create unique habitats that support plant and animals species that are uncommon in the surrounding landscape.

midstory: The middle layer of a forest stand between the tree canopy and the herbaceous layer. Composed of shrubs and small trees.

minimum residual cover: The fewest number of trees necessary to provide shade, organic material and soil-holding capacity for the protection of the biological integrity of aquatic ecosystems.

mulch: Coarse material which protects soil from rainfall impacts and erosion losses and which improves germination and vegetative growth. Examples of mulches include hay, bark, wood chips, and geotextile fabric.

multiple species management: management approach that provides an assortment of habitat types for a variety of wildlife species.

multiple use: Land management for more than one purpose. Management for some combination of the following: wood production, water, wildlife, recreation, forage, aesthetics, or clean air.

natural barrier: Areas that are without fuel or food, such as a rocky cliff or a barren field, which provide resistance to the spread of fire, insects, or disease.

natural drainage: Stream pathways which collect and expel runoff water.

natural regeneration: Young trees that originate from seed or sprouts of trees present on or near the site area.

natural stand: A stand of trees resulting from natural regeneration.

neotropical migrants (long-distance migrants): Birds that breed in the spring and summer in North America but spend their winters south of the U.S. in Mexico, the Caribbean islands, and other Central and South American tropical countries.

net growth: The net increase in volume of timber for a certain area of land for a certain period of time. Net increase is determined by the gross increase in the volume of trees from the beginning to the end of the time period, plus the volume of trees which become merchantable during the period, less the tree volume from dead, rough, or rotten trees.

non-consumptive uses (appreciative uses): Recreational activities that usually do not directly impact resources on the landscape. Examples: photography and scenic viewing.

non-game wildlife: Wildlife that are not hunted.

non-point-source pollution: Water pollution which cannot be traced to one identifiable facility but comes from a broader area.

normal passage of water and/or aquatic animals: Movement of water or animals which has not been obstructed as the result of man-made activity.

nutrients: Substances that nourish soil and vegetation such as nitrogen, potassium, and phosphorus. These nutrients, added as fertilizer, can sometimes cause problems for aquatic ecosystems.

opportunity costs: Lost income that would ordinarily have been received if another practice had not been given up, such as the income from raising cotton that might be foregone to plant timber in the same location.

organic debris: Refuse such as tree tops, limbs, or severely damaged tree stems which are left following road construction, logging or site preparation.

organic matter: Dead material from plants or animals, an excess of which depletes oxygen availability in streams.

overstocked: Trees growing too closely for optimal timber growth and wildlife value.

overstory: (See canopy.)

overtopped trees (suppressed trees): Trees with crowns which are entirely below the general level of crown cover and receive no direct light either from above or from the sides.

pathology (forest pathology): The science of diseases of forest trees, stands, and products.

partial cut: A selective timber harvest method where particular trees are usually designated to remain in the stand.

patch: Small units of land designated by similar landscape features, flora and fauna. When combined, patches form a landscape matrix.

patch burning: Prescribed burning conducted for wildlife habitat improvement where burns are small in size (less than 10 acres) and scattered over an entire stand. The effect of patch burning is to leave a mosaic of burned and unburned sites across a stand.

perennial: A plant that lives for more than one year and grows from seeds or reproduces from vegetative parts of a plant.

perennial bodies of water: Bodies which contain water within well defined areas virtually year round under normal climate conditions.

permanent road: A road constructed, utilized, and maintained beyond the time period of a single operation such as a timber harvest.

pesticide: A collective term which refers to chemicals used to control pests such as insects or rodents.

photogrammetry: The science of making reliable measurements by the use of aerial photographs.

physical integrity of waters of the state: The retention of a body of water in its natural condition without alteration of a stream course, depth, or clarity and with freedom from obstructions that might occur as the direct result of man-made activity.

pioneer species: Plant species which first colonize disturbed sites. Usually light-seeded weeds, grasses, and trees such as ash, poplar, loblolly pine, and sweetgum.

pioneer successional stage: The successional stage composed of a collection of plants that colonize bare ground.

plankton: Microscopic organisms in ponds and lakes that provide oxygen and food for fish. Can be either plants (phytoplankton) or animals (zooplankton).

plantation: An artificially forested area established by planting or by direct seeding. Plantations are usually made up of a single tree species.

plant succession: The natural progression of forms of vegetation (including trees, shrubs, forbs, and grasses) from ones that usually appear first after any disturbance to ones that appear last and seem to be permanent, or persistent until disturbance sets the cycle in motion again.

pole stage: Intermediate age trees with a DBH of 4–12 inches. Also marketable product 12" or larger, tall, straight and limbless.

point-source pollution: Pollutants that can be identified as coming from an identifiable source.

pollutants: Elements which can enter streams in quantities exceeding the water's natural ability to neutralize them before negative changes occur in the physical, chemical, or biological integrity of the waters. Examples include sediments, organic debris, increased temperature, nutrients, chemicals, and trash.

portable logging mats: Temporary road or stream-crossing surface constructed of rough-cut lumber for temporary use during a logging operation.

precocial: Newly hatched birds that are covered with down and are capable of moving around when first hatched, as opposed to altricial which are born naked and helpless.

prescribed burning (controlled burning): Preplanned fire that is deliberately set in a time and manner to meet specific objectives.

prescriptions: Forest management recommendations.

preservation: 1) With respect to wood, treating wood products with chemicals to prevent damage by insects or decay organisms; 2) With respect to land management, maintaining an environment undisturbed by human activities, influence, or intervention.

pulpwood: Wood cut primarily to be converted into wood pulp for the manufacture of paper, fiberboard, or other wood fiber products. Pulpwood trees are usually a minimum of 4 inches DBH.

puddling: The destruction of root systems and soil structure by the tearing and churning action of heavy equipment operating on saturated soils. Puddled soils are also more susceptible to erosion than are undisturbed soils.

push-probe: An implement used to detect compacted hardpan soils.

reforestation (regeneration): The restocking, regrowth, or revitalization of a forest through natural regeneration or through artificially planted seeds or seedlings, resulting in a stand of newly established trees.

regeneration cut: Either the partial or complete harvest of trees to encourage the sprouting of desirable new tree growth.

release: 1) A written document (often called a hold-harmless agreement) stating that a land-user assumes the risks of engaging in an activity, signed by the land-user and the landowner; 2) Removing some trees from a stand to allow optimal growth of the remaining trees, also called a release cutting.

remote sensing: A means of acquiring information using airborne equipment and techniques to determine the characteristics of a landscape area, commonly using aerial photographs from aircraft and satellites.

resident birds: Birds that are found year-round in an area.

residual trees: Trees that remain in a stand after a harvesting practice.

right-of-way: The legal right of passage over another person's land.

rip-rap: Large stones which are arranged over loose soil to ward off erosion.

riparian zone: The land adjacent to a body of water, usually a stream or river but also other bodies of water. This zone is influenced by periodic flooding.

roller chopping: A technique for site preparation in which a large water-filled drum surrounded by sharpened fins is pulled behind a bulldozer, crushing and cutting remaining trees and brush.

rotation: The period of years required to grow timber crops to a specified point of maturity. At that point the trees will be harvested.

rutting: Impressions left in the ground after soil is compacted by the wheels or tracks of heavy equipment operating in soft earth. Deep rutting can disrupt surface and subsurface hydrology on flat lands and cause soil erosion on steep lands by concentrating surface runoff.

safe harbor: A management strategy for red-cockaded woodpeckers (RCWs) on private lands that allows continued forest management and provisions for protecting existing RCWs.

salvage cut: Harvesting dead or damaged trees which have been damaged or killed by natural causes such as storms, disease, insect infestation, flooding, etc.

sampling: Taking a detailed measurement of a smaller section of a population in order to project information about the whole.

sanitation cut: Harvesting or killing trees that are infected, highly susceptible, or threatened by insects or diseases. Conducted to protect the rest of the forest stand.

sapling: A young tree with a DBH of less than 4 inches and from 3–10 feet in height.

sawlog: A log of suitable size and quality for sawing into lumber; the minimum inside bark diameter at the small end is usually 11 inches for hardwood and 10 inches for softwood but varies by sawmill.

scats: Fecal droppings that are often used to identify wildlife in an area. Close examination of scat contents is used to determine the food habits of some wildlife species.

Secchi disk: An 8 inch diameter disk with alternating white and dark bands that is lowered by a rope or stick into ponds or lakes to determine light penetration, phytoplankton density, and if a pond or lake needs to be fertilized.

Section 404: Section 404 of the Clean Water Act usually requires that a permit be obtained from the U. S. Army Corps of Engineers before a discharge of dredged or fill materials can be made into waters of the United States. There are several exemptions, and details on the act or questions regarding this can be directed to U.S. Army Corps of Engineers, Mobile District, P.O. Box 2288, Mobile, AL 36628-0001, or (205) 690-2581.

Section 9: One of the most important sections of the Endangered Species Act for landowners that defines prohibitive acts such as to take, harass, or harass federally listed species.

sediment: Soil particles that are suspended or deposited in waterways and can cause pollution.

seedling: A young tree, usually with a DBH of less than 2 inches, that has grown from a seed.

seed tree method: Removing all trees from the forest at one time except for a few scattered trees left to provide seed to establish a new forest stand. There are guidelines which must be followed for this method of regeneration to be successful.

selection method: Harvesting individual trees or small groups of trees at periodic intervals (usually 5 to 15 years) based on their physical condition or degree of maturity, a process which produces an uneven-aged stand.

select species management: Also called "featured species approach" where management concentrates on providing habitat for one or two select or "featured" species.

severance tax: A tax on forest products after they are harvested.

shade tolerance: The ability to thrive in the shade of other trees or plants. Trees that are shade tolerant are also called *understory* tree species.

shearing and raking: A site preparation technique using one large tractor equipped with a cutting blade to remove trees just above the ground surface and a second tractor and raking blade unit that pushes felled trees into windrow piles.

shelterbelt: A windbreak or barrier composed of living trees, brush, or other vegetation.

shelterwood harvest: Removing trees on the harvest area in a series of two or more cuttings over time so new seedlings can become established from the seeds of older trees, a process which eventually produces an even-aged stand.

shooting preserve: A facility which offers opportunities for shooting pen-raised birds such as quail, pheasant, and ducks.

short-distance migrants: Birds that breed elsewhere and winter in the southern United States.

side bank: See back slope.

silviculture: The establishment, care and cultivation of stands of forest trees, often used interchangeably with the word *forestry*.

single-tree selection: A harvesting method favoring uneven-aged stands in which only individual trees are removed. May be particularly desirable in site-sensitive areas like wetlands and riparian zones, as site disturbance is usually minimal.

sink: An area which attracts or holds wildlife populations where the habitat is not conducive to sustain the species' numbers. Can cause some species to be extirpated from local areas.

site capabilities: A descriptive term referring to the potential of a specific piece of land to nurture and sustain particular types of growth of vegetation (including trees), wildlife, or combinations.

site index: A measure of the relative productivity of a land area to grow a particular tree species, expressed as an expected height of dominant trees in an even-aged stand at an index age, such as 25 years or 50 years.

site preparation: The use of machines, fire, herbicides, or combinations of treatments to remove slash, improve planting conditions, and provide initial control of weed competition.

sketch map: A descriptive map, that is included in wildlife management plans, of land and other features important for planning and conducting wildlife habitat improvement practices.

skid (skidding): To drag logs with a specialized tractor to a landing.

skid trails: Paths where logs have been dragged in a timber harvesting operation.

slash: Tree stems, tops, branches, leaves, and other debris which for various reasons cannot be sold and are left behind after a commercial timber operation.

slough: An open water inlet from a larger body of water.

snag: A standing dead tree sometimes left to remain in a forest stand for wildlife perches, shelter, or nesting sites.

soil stabilizing materials: Silt fencing, geotextile fabric, straw blankets, and other materials applied to ward off erosion.

softwoods: Trees belonging to the order *Coniferales*, usually evergreen, cone bearing, and with needles or scalelike leaves such as pines, spruces, firs, and cedars.

soil texture: The relative proportions of sand, silt, and clay in soil.

soil type: A subdivision of a soil series based upon differences in soil texture.

source: Permanent populations of an animal species which are viable and which contribute individuals that can move to other less populated areas, called sinks; this term usually refers to wildlife population dynamics.

species of concern: Animal and plant species that must be continually monitored because of imminent threats to habitat, limited range, or because of other physical or biological factors that may cause them to become threatened or endangered within the foreseeable future.

stand: A group of trees with similar characteristics in age and species that are treated as a single unit in a forest management plan.

standard cord: A measure of wood volume, usually for pulpwood or firewood, described as wood stacked 4 feet high, 4 feet wide, and 8 feet long (or 128 cubic feet). Because logs are irregularly shaped, the actual wood content of a standard cord varies from about 75 to 100 cubic feet.

stocking density: The number or basal area of trees per unit of area in a stand, compared with a desirable number for best growth and management. (See also understocked and overstocked.)

strategic grazing: The use of livestock as a tool in habitat management with a system of short-duration grazing alternated with periods of no grazing, to allow for vegetative reproduction.

stream shading: The degree of shade from tree canopies that are adjacent to and over streams and creeks. Stream shading is important because the removal of shade-producing trees and other vegetation from streambanks and shorelines will directly raise the water temperature and indirectly result in lower levels of dissolved oxygen in the water. This change could place fish and organisms under stress.

streamside management zone (SMZ): A strip of vegetation along watersheds which is left intact when adjacent forests are harvested. These areas are left intact because of their value in preventing erosion, maintaining water quality, and enhancing wildlife habitat.

stumpage: The amount paid to a timber owner based on the value of a tree prior to harvest. Often called "on the stump" value.

subsoiling: Breaking up soils to depths of 6–12 inches in order to alleviate compaction.

succession: The natural process of ecological change on a site. The site changes from one form of vegetation (and the related wildlife species it best supports) to another over time.

sunscald: Damage to a tree or plant from too much exposure to direct sunlight.

supplemental planting: Propagating desired wildlife food plants by direct seeding or planting of seedlings.

surface waters: Exposed water above the ground surface.

sustainable use: Providing desired ecological conditions as well as economic benefits from land over extended periods of time.

sustained yield: Management of a forest to yield a relatively constant supply of trees and/or wildlife over time.

take: 1) To harass, harm, pursue, hunt, shoot, wound, kill, trap, capture, or collect an endangered species. Taking can also include significant habitat modification or degradation that will eventually kill an endangered species by impairing its breeding, feeding, or sheltering. As an example, removing the nest of an endangered bird constitutes a "take" because it impairs its breeding capabilities. 2) An act of the government to usurp or utilize private property to the extent that the private individuals can no longer use it, which requires just compensation by law.

temporary access roads: Roads not expected to be maintained much longer than the activity they support. For example roads constructed for logging operations.

territory: A defended portion of animal's home range.

tertiary roads: Little used roads at least 25 feet wide which can be beneficial sites for planted and native vegetation.

threatened species: Animals and plants that are likely to become endangered in the foreseeable future.

thinning: Cutting trees in an immature forest stand in order to reduce the stocking density and to concentrate site productivity on fewer, higher quality trees. These can be pre-commercial (cutting non-merchantable trees) thinnings or commercial (cutting merchantable trees) thinnings.

timber cruise: The process of estimating the volume and value of timber standing on any given piece of property.

timber purchasers (or buyers): Agents who locate commercial timber and negotiate terms of purchase.

topor: A state or period of dormancy or inactivity found in some animals during temperature extremes.

transpiration: The process by which plants release water as water vapor.

tree scale stick (Biltmore stick): A stick used to estimate the height and diameter of standing trees, which commonly has gradations and scales marked on it to estimate the volume of standing trees.

tree shelters: Long, tubular sleeves made of polyethylene or polypropylene placed around newly established trees for protection during the first few years of growth.

topography: The physical and natural features of the landscape, including characteristics such as elevation, slope, and surface area configuration.

tract: A parcel of land considered separately from adjoining land because of differences in ownership, tree types, management objectives, or other characteristics.

trash: Unnaturally occurring, man-made discarded substances, such as bottles or petroleum wastes, which may be carried into waters by storm runoff.

understocked: Insufficient number of plants or seedlings to achieve the desired objectives.

understory vegetation: Small trees, shrubs or other plants which grow beneath the canopy of larger, dominant trees.

uneven-aged stand (all-aged stand): A stand containing trees of three or more viable and well-established age classes with the difference in ages being at least 20 years.

upland runoff: Surface drainage water which flows from higher elevations of a landscape into the natural drainage system of a watershed.

vendors: Contractors who provide tree harvesting, site preparation, tree planting, or other forestry services for a fee.

veneer: A thin sheet of wood sliced or peeled on a veneer machine and often used for plywood or surfacing furniture. Veneer logs are large (usually more than 18 inches in diameter), high-quality logs.

vertical zones (vertical structure): levels of growth stated in relation to the forest floor ranging from underground, ground level, taller grasses, shrub level, midstory, and tree canopy layers.

volume table: Tables that list the volume of individual trees of specific size, often by species.

washouts: Removal of natural or man-made obstructions in a drainage system due to high stream flows.

water bar: A long mound of dirt constructed to prevent soil erosion and water pollution by diverting drainage from a road or skid trail into a filtration strip.

water bodies: Bays, branches, creeks, lakes, ponds, rivers, etc.

water diversions: Structures or devices which change the direction of drainage flow, some as simple as wood pilings strategically placed but others more complicated.

watershed: Terrain surrounding large rivers or small streams.

water quality impairment: The reduction of water quality below established standards.

waters of the state: This includes every watercourse, stream, river, wetland, pond, lake, coastal, ground, or surface water, wholly or partially in the state, natural or artificial, which is not entirely confined and retained on the property of one single landowner.

waters of the United States: These include all waters such as lakes, rivers, streams (including intermittent streams), mudflats, sandflats, wetlands, and sloughs which are susceptible to use in interstate or foreign commerce, recreation, fish and shellfish production and industrial use; impoundments and tributaries of waters just described; and wetlands that are adjacent to waters just described (other than waters that are themselves wetlands).

wetlands of the United States: Those areas that are inundated or saturated by surface or ground water at a frequency and duration sufficient to support, and that under normal circumstances do support, a prevalence of vegetation typically adapted for life in saturated soil conditions. Wetlands generally include swamps, marshes, bogs, and similar areas.

wildfire: Fires burning out of control and not set for a purpose.

wildlife management plan: Written guides for how and when to implement wildlife habitat and population improvement practices.

windfall: (See blowdown.)

windrows: Long stretches of accumulated forest debris such as branches, usually constructed after a logging operation.

wing ditch: A smaller, secondary "turn out" ditch that diverts drainage water from primary roadside ditches so that it is filtered out into the surrounding area.

woodland grazing: The combination of both timber and cattle management on the same lands.

yard: An area where logs, sections of logs, pulpwood bolts, and other timber harvest products are collected and stored prior to being processed or transported to the mill.

YIELDplus: A computer program that analyzes timber growth and yield projections.

yield tax: A tax levied on the sale price of a commodity such as trees.

Appendix Q: Threatened and Endangered Species List

Threatened and endangered species list for southeastern states. Below we provide a list of species residing in the southeastern United States (Alabama, Arkansas, Florida, Georgia, Kentucky, Louisiana, Maryland, Mississippi, Missouri, North Carolina, South Carolina, Tennessee, Virginia, West Virginia) that are federally listed as Threatened or Endangered (as of July 31, 1997) by the U.S. Fish and Wildlife Service (USFWS). Taxon status: E = endangered, T = threatened. Basic habitat requirements are: A = aquatic, T = terrestrial. Taxa were designated aquatic if they rely primarily on aquatic or semiaquatic habitats. Summary data for each state detail the total number of species listed, number of aquatic species listed, number of terrestrial species listed, and percent of total listed species which are aquatic. This list was constructed using information provided on the web page of the USFWS. The USFWS periodically updates the web page listing of threatened and endangered species as new information becomes available. To consult this list use the web address: http://www.fws.gov

State (summary data): Taxon	Common Name	Taxon Status	Habitat
Alabama			
89 listed species: 68 aquatic, 17 terrestrial (76% of listed species are aquatic)			
Plants (19 species)			
Amphianthus pusillus	Little amphianthus	T	A
Apios priceana	Price's potato-bean	T	T
Arabis perstellata	Rock cress	E	T
Asplenium scolopendrium var. *americanum*	American hart's-tongue fern	T	T
Clematis morefieldii	Morefield's leather-flower	E	T
Clematis socialis	Alabama leather-flower	E	T
Dalea foliosa	Leafy prairie-clover	E	T
Helianthus eggertii	Eggert's sunflower	T	T
Lesquerella lyrata	Lyrate bladderpod	T	T
Lindera melissifolia	Pondberry	E	A
Marshallia mohrii	Mohr's Barbara's buttons	T	T
Ptilimnium nodosum	Harperella	E	A
Sagittaria secundifolia	Kral's water-plantain	T	A
Sarracenia orephila	Green pitcher-plant	E	A
Sarracenia rubra alabamensis	Alabama canebrake pitcher-plant	E	A
Spigelia gentianoides	Gentian pinkroot	E	T
Thelypteris pilosa var. *alabamaensis*	Alabama streak-sorus fern	T	T
Trillium reliquum	Relict trillium	E	T
Xyris tennesseensis	Tennessee yellow-eyed grass	E	A
Mollusks (45 species)			
Athearnia anthonyi	Anthony's riversnail	E	A
Cyprogenia stegaria	Fanshell	E	A
Dromus dromas	Dromedary pearlymussel	E	A
Elliptio chipolaensis	Chipola slabshell	T	A
Elliptoideus sloatianus	Purple bankclimber	T	A
Epioblasma brevidens	Cumberlandian combshell	E	A
Epioblasma capsaeformis	Oyster mussel	E	A
Epioblasma florentina florentina	Yellow blossom pearlymussel	E	A
Epioblasma metastriata	Upland combshell	E	A
Epioblasma obliquata obliquata	Purple cat's paw pearlymussel	E	A
Epioblasma othcaloogensis	Southern acornshell	E	A
Epioblasma penita	Southern combshell	E	A
Epioblasma torulosa torulosa	Tubercled-blossom pearlymussel	E	A
Epioblasma turgidula	Turgid-blossom pearlymussel	E	A
Fusconaia cuneolus	Fine-rayed pigtoe	E	A
Fusconaia edgariana	Shiny pigtoe mussel	E	A
Hemistena lata	Cracking pearlymussel	E	A
Lampsilis abrupta	Pink mucket pearlymussel	E	A

Threatened and endangered species list. Continued.

State (summary data): Taxon	Common Name	Taxon Status	Habitat
Lampsilis altilis	Fine-lined pocketbook	T	A
Lampsilis perovalis	Orange-nacre mucket	T	A
Lampsilis subangulata	Shinyrayed pocketbook	T	A
Lampsilis virescens	Alabama lampmusssel	E	A
Medionidus acutissimus	Alabama moccasinshell	T	A
Medionidus parvulus	Coosa moccasinshell	E	A
Medionidus penicillatus	Gulf moccasinshell	E	A
Obovaria retusa	Ring pink mussel	E	A
Pegias fabula	Little-wing pearlymussel	E	A
Pleurobema clava	Clubshell	E	A
Plethobasis cicatricosus	White wartyback pearlymussel	E	A
Plethobasus cooperianus	Orange-foot pimple back pearlymussel	E	A
Pleurobema curtum	Black clubshell	E	A
Pleurobema decisum	Southern clubshell	E	A
Pleurobema furvum	Dark pigtoe	E	A
Pleurobema georgianum	Southern pigtoe	E	A
Pleurobema marshalli	Flat pigtoe	E	A
Pleurobema perovatum	Ovate clubshell	E	A
Pleurobema plenum	Rough pigtoe	E	A
Pleurobema pyriforme	Oval pigtoe	E	A
Pleurobema taitianum	Heavy pigtoe	E	A
Potamilus inflatus	Inflated heelsplitter	T	A
Ptychobranchus greeni	Triangular kidneyshell	E	A
Quadrula intermedia	Cumberland monkeyface pearlymussel	E	A
Quadrula stapes	Stirrupshell	E	A
Toxolasma cylindrellus	Pale lilliput pearlymussel	E	A
Tulotoma magnifica	Tulotoma snail	E	A
Crustaceans (1 species)			
Palaemonias alabamae	Alabama cave shrimp	E	A
Fishes (12 species)			
Acipenser oxyrhynchus desotoi	Gulf sturgeon	T	A
Cottus pygmaeus	Pygmy sculpin	T	A
Cyprinella caerulea	Blue shiner	T	A
Cyprinella monacha	Spotfin chub	T	A
Etheostoma boschungi	Slackwater darter	T	A
Etheostoma nuchale	Watercress darter	E	A
Etheostoma wapiti	Boulder darter	E	A
Notropis albizonatus	Palezone shiner	E	A
Notropis cahabae	Cahaba shiner	E	A
Percina aurolineata	Goldline darter	T	A
Percina tanasi	Snail darter	T	A
Speoplatyrhinus poulsoni	Alabama cavefish	E	A
Amphibians (1 species)			
Phaeognathus hubrichti	Red Hills salamander	T	A
Reptiles (5 speices)			
Caretta caretta	Loggerhead sea turtle	T	A
Drymarchon corais couperi	Eastern indigo snake	T	T
Gopherus polyphemus	Gopher tortoise	T	T
Pseudemys alabamensis	Alabama redbelly turtle	E	A
Sternnotherus depressus	Flattened musk turtle	T	A
Birds (5 species)			
Charadrius melodus	Piping plover	T	A
Falco peregrinus anatum	American peregrine falcon	E	T
Haliaeetus leucocephalus	Bald eagle	T	A
Mycteria americana	Wood stork	E	A
Picoides borealis	Red-cockaded woodpecker	E	T

Threatened and endangered species list. Continued.

State (summary data): Taxon	Common Name	Taxon Status	Habitat
Mammals (5 species)			
Myotis grisescens	Gray bat	E	A
Myotis sodalis	Indiana bat	E	T
Peromyscus polionotus ammobates	Alabama beach mouse	E	A
Peromyscus polionotus trissyllepsis	Perdido Key beach mouse	E	A
Trichechus manatus	West Indian manatee	E	A
Arkansas			
25 listed species: 16 aquatic, 9 terrestrial (64% of listed species are aquatic)			
Plants (5 species)			
Geocarpon minimum	(no common name)	T	T
Lindera melissifolia	Pondberry	E	A
Platanthera leucophaea	Eastern prairie fringed orchid	T	T
Ptilimnium nodosum	Harperella	E	A
Trifolium stoloniferum	Running buffalo clover	E	T
Mollusks (7 species)			
Arkansia wheeleri	Ouachita rock-pocketbook	E	A
Epioblasma florentina curtisi	Curtis' pearlymussel	E	A
Lampsilis abrupta	Pink mucket pearlymussel	E	A
Lampsilis powelli	Arkansas fatmucket	T	A
Lampsilis streckeri	Speckled pocketbook	E	A
Mesodon magazinensis	Magazine Mountain shagreen	T	T
Potamilus capax	Fat pocketbook	E	A
Crustaceans (2 species)			
Cambarus aculabrum	(no common name)	E	A
Cambarus zophonastes	(no common name)	E	A
Insects (1 species)			
Nicrophorus americanus	American burying beetle	E	T
Fishes (3 species)			
Amblyopsis rosae	Ozark cavefish	T	A
Percina pantherina	Leopard darter	T	A
Scaphirhynchus albus	Pallid sturgeon	E	A
Birds (4 species)			
Falco peregrinus anatum	American peregrine falcon	E	T
Haliaeetus leucocephalus	Bald eagle	T	A
Picoides borealis	Red-cockaded woodpecker	E	T
Sterna antillarum	Least tern	E	A
Mammals (3 species)			
Myotis grisescens	Gray bat	E	A
Myotis sodalis	Indiana bat	E	T
Plecotus townsendii ingens	Ozark big-eared bat	E	T
Florida			
92 listed species: 32 aquatic, 60 terrestrial (36% of listed species are aquatic)			
Plants (54 species)			
Amorpha crenulata	Crenulate lead plant	E	T
Asimina tetramera	Four-petal pawpaw	E	T
Bonamia grandiflora	Florida bonamia	T	T
Campanula robinsiae	Brooksville bellflower	E	T
Cereus eriophorus var. *fragrans*	Fragrant prickly-apple	E	T
Chamaesyce deltoidea deltoidea	Deltoid spurge	E	T
Chamaesyce garberi	Garber's spurge	T	T
Chionanthus pygmaeus	Pygmy fringe-tree	E	T
Chrysopsis floridana	Florida golden aster	E	T
Cladonia perforata	Florida perforate cladonia	E	T
Clitoria fragrans	Pigeon wings	T	T

Threatened and endangered species list. Continued.

State (summary data): Taxon	Common Name	Taxon Status	Habitat
Conradina brevifolia	Short-leaved rosemary	E	T
Conradina etonia	Etonia rosemary	E	T
Conradina glabra	Apalachicola rosemary	E	T
Crotalaria avonensis	Avon Park harebells	E	T
Cucurbita okeechobeensis okeechobeensis	Okeechobee gourd	E	T
Deeringothamnus pulchellus	Beautiful pawpaw	E	T
Deeringothamnus rugelii	Rugel's pawpaw	E	T
Dicerandra christmanii	Garrett's mint	E	T
Dicerandra cornutissima	Longspurred mint	E	T
Dicerandra frutescens	Scrub mint	E	T
Dicerandra immaculata	Lakela's mint	E	T
Eriogonum longifolium gaphalifolium	Scrub buckwheat	T	T
Eryngium cuneifolium	Snakeroot	E	A
Euphorbia telephioides	Telephus spurge	T	T
Galactia smallii	Small's milkpea	E	T
Harperocallis flava	Harper's beauty	E	T
Hypericum cumulicola	Highlands scrub hypericum	E	T
Jacquemontia reclinata	Beach jacquemontia	E	A
Justicia cooleyi	Cooley's water-willow	E	A
Liatris ohlingerae	Scrub blazingstar	E	T
Lindera melissifolia	Pondberry	E	A
Lupinus aridorum	Scrub lupine	E	T
Macbridea alba	White birds-in-a-nest	T	T
Nolina brittoniana	Britton's beargrass	E	T
Paronychia chartacea	Papery whitlow-wort	T	T
Pilosocereus robinii	Key tree-cactus	E	T
Pinguicula ionantha	Godfrey's butterwort	T	T
Polygala lewtonii	Lewton's polygala	E	T
Polygala smallii	Tiny polygala	E	T
Polygonella basiramia	Wireweed	E	T
Polygonella myriophylla	Sandlace	E	T
Prunus geniculata	Scrub plum	E	T
Rhododendron chapmanii	Chapman rhododendron	E	T
Ribes echinellum	Miccosukee gooseberry	T	T
Schwalbea americana	American chaffseed	E	T
Scutellaria floridana	Florida skullcap	T	A
Silene polypetala	Fringed campion	E	T
Spigelia gentianoides	Gentian pinkroot	E	T
Thalictrum cooleyi	Cooley's meadowrue	E	A
Torreya taxifolia	Florida torreya	E	T
Warea amplexifolia	Wide-leaf warea	E	T
Warea carteri	Carter's mustard	E	T
Ziziphus celata	Florida ziziphus	E	T
Mollusks (1 species)			
Orthalicus reses	Stock Island tree snail	T	T
Crustaceans (1 species)			
Palaemonetes cummingi	Squirrel Chimney Cave shrimp	T	A
Insects (1 species)			
Heraclides aristodemus ponceanus	Schaus swallowtail butterfly	E	T
Fishes (2 species)			
Acipenser oxyrhynchus desotoi	Gulf sturgeon	T	A
Etheostoma okaloosae	Okaloosa darter	E	A
Reptiles (9 species)			
Caretta caretta	Loggerhead sea turtle	T	A

Threatened and endangered species list. Continued.

State (summary data): Taxon	Common Name	Taxon Status	Habitat
Chelonia mydas	Green sea turtle	E	A
Crocodylus acutus	American crocodile	E	A
Dermochelys coriacea	Leatherback sea turtle	E	A
Drymarchon corais couperi	Eastern indigo snake	T	T
Eretmochelys imbricata	Hawksbill sea turtle	E	A
Eumeces egregius lividus	Bluetail mole skink	T	T
Neoseps reynoldsi	Sand skink	T	T
Nerodia clarkii taeniata	Atlantic salt marsh snake	T	A
Birds (11 species)			
Ammodramus maritimus mirabilis	Cape Sable seaside sparrow	E	A
Ammodramus savannarum floridanus	Florida grasshopper sparrow	E	T
Aphelocoma coerulescens	Florida scrub jay	T	T
Charadrius melodus	Piping plover	T	A
Falco peregrinus anatum	American peregrine falcon	E	T
Haliaeetus leucocephalus	Bald eagle	T	A
Mycteria americana	Wood stork	E	A
Picoides borealis	Red-cockaded woodpecker	E	T
Polyborus plancus audubonii	Audubon's crested caracara	T	T
Rostrhamus sociabilis plumbeus	Everglade snail kite	E	A
Sterna dougallii dougallii	Roseate tern	T	A
Mammals (13 species)			
Felis concolor coryi	Florida panther	E	T
Microtus pennsylvanicus dukecampbelli	Florida salt marsh vole	E	A
Myotis grisescens	Gray bat	E	A
Neotoma floridana smalli	Key Largo woodrat	E	A
Odocoileus virginianus clavium	Key deer	E	T
Oryzomys palustris natator	Rice rat	E	A
Peromyscus gossypinus allapaticola	Key Largo cotton mouse	E	A
Peromyscus polionotus allophrys	Choctawahatchee beach mouse	E	A
Peromyscus polionotus niveiventris	Southeastern beach mouse	T	A
Peromyscus polionotus phasma	Anastasia Island beach mouse	E	A
Peromyscus polionotus trissyllepsis	Perdido Key beach mouse	E	A
Sylvilagus palustris hefneri	Lower Keys rabbit	E	A
Trichechus manatus	West Indian manatee	E	A
Georgia			
48 listed species: 33 aquatic, 15 terrestrial (69% of listed species are aquatic)			
Plant (22 species)			
Amphianthus pusillus	Little amphianthus	T	A
Baptisia arachnifera	Hairy rattleweed	E	T
Echinacea laevigata	Smooth coneflower	E	T
Helonias bullata	Swamp pink	T	A
Isoetes melanospora	Black-spored quillwort	E	A
Isoetes tegetiformans	Mat-forming quillwort	E	A
Isotria medeoloides	Small whorled pogonia	T	T
Lindera melissifolia	Pondberry	E	A
Marshallia mohrii	Mohr's Barbara's buttons	T	T
Oxypolis canbyi	Canby's dropwort	E	A
Ptilimnium nodosum	Harperella	E	A
Rhus michauxii	Michaux's sumac	E	T
Sagittaria secundifolia	Kral's water-plantain	T	A
Sarracenia oreophila	Green pitcher-plant	E	A
Schwalbea americana	American chaffseed	E	T

Threatened and endangered species list. Continued.

State (summary data): Taxon	Common Name	Taxon Status	Habitat
Scutellaria montana	Large-flowered skullcap	E	T
Silene polypetala	Fringed campion	E	T
Spiraea virginiana	Virginia spiraea	T	A
Torreya taxifolia	Florida torreya	E	T
Trillium persistens	Persistent trillium	E	T
Trillium reliquum	Relict trillium	E	T
Xyris tennesseensis	Tennessee yellow-eyed grass	E	A
Mollusks (9 species)			
Epioblasma metastriata	Upland combshell	E	A
Epioblasma othcaloogensis	Southern acornshell	E	A
Lampsilis altilis	Fine-lined pocketbook	T	A
Medionidus acutissimus	Alabama moccasinshell	T	A
Medionidus parvulus	Coosa moccasinshell	E	A
Pleurobema decisum	Southern clubshell	E	A
Pleurobema georgianum	Southern pigtoe	E	A
Pleurobema perovatum	Ovate clubshell	E	A
Ptychobranchus greeni	Triangular kidneyshell	E	A
Fishes (7 species)			
Cyprinella caerulea	Blue shiner	T	A
Etheostoma etowahae	Etowah darter	E	A
Etheostoma scotti	Cherokee darter	T	A
Percina antesella	Amber darter	E	A
Percina aurolineata	Goldline darter	T	A
Percina jenkinsi	Conasauga logperch	E	A
Percina tanasi	Snail darter	T	A
Reptiles (2 species)			
Drymarchon corais couperi	Eastern indigo snake	T	T
Caretta caretta	Loggerhead sea turtle	T	A
Birds (5 species)			
Charadrius melodus	Piping plover	T	A
Falco peregrinus anatum	American peregrine falcon	E	T
Haliaeetus leucocephalus	Bald eagle	T	A
Mycteria americana	Stork wood	E	A
Picoides borealis	Red-cockaded woodpecker	E	T
Mammals (3 species)			
Myotis grisescens	Gray bat	E	A
Myotis sodalis	Indiana bat	E	T
Trichechus manatus	West Indian manatee	E	A
Kentucky			
43 listed species: 31 aquatic, 12 terrestrial (72% of listed species are aquatic)			
Plants (9 species)			
Apios priceana	Price's potato-bean	T	T
Arabis perstellata	Rock cress	E	T
Arenaria cumberlandensis	Cumberland sandwort	E	T
Conradina verticillata	Cumberland rosemary	T	T
Helianthus eggertii	Eggert's sunflower	T	T
Solidago albopilosa	White-haired goldenrod	T	T
Solidago shortii	Short's goldenrod	E	T
Spiraea virginiana	Virginia spiraea	T	A
Trifolium stoloniferum	Running buffalo clover	E	T
Mollusks (21 species)			
Alasmidonta atropurpurea	Cumberland elktoe	E	A
Cyprogenia stegaria	Fanshell	E	A
Dromus dromas	Dromedary pearlymussel	E	A
Epioblasma brevidens	Cumberlandian combshell	E	A

Threatened and endangered species list. Continued.

State (summary data): Taxon	Common Name	Taxon Status	Habitat
Epioblasma capsaeformis	Oyster mussel	E	A
Epioblasma obliquata obliquata	Purple cat's paw pearlymussel	E	A
Epioblasma torulosa rangiana	Northern riffleshell	E	A
Epioblasma torulosa torulosa	Tubercled-blossom pearlymussel	E	A
Epioblasma walkeri	Tan riffleshell	E	A
Hemistena lata	Cracking pearlymussel	E	A
Lampsilis abrupta	Pink mucket pearlymussel	E	A
Obovaria retusa	Ring pink mussel	E	A
Pegias fabula	Little-wing pearlymussel	E	A
Plethobasus cicatricosus	White wartyback pearlymussel	E	A
Plethobasus cooperianus	Orange-foot pimple back pearlymussel	E	A
Pleurobema clava	Clubshell	E	A
Pleurobema plenum	Rough pigtoe	E	A
Potamilus capax	Fat pocketbook	E	A
Quadrula cylindrica strigillata	Rough rabbitsfoot	E	A
Quadrula fragosa	Winged mapleleaf mussel	E	A
Villosa trabalis	Cumberland bean pearlymussel	E	A
Crustaceans (1 species)			
Palaemonias ganteri	Kentucky cave shrimp	E	A
Fishes (4 species)			
Etheostoma chienense	Relict darter	E	A
Notropis albizonatus	Palezone shiner	E	A
Phoxinus cumberlandensis	Blackside dace	T	A
Scaphirhynchus albus	Pallid sturgeon	E	A
Birds (5 speices)			
Charadrius melodus	Piping plover	T	A
Falco peregrinus anatum	American peregrine falcon	E	T
Haliaeetus leucocephalus	Bald eagle	T	A
Picoides borealis	Red-cockaded woodpecker	E	T
Sterna antillarum	Least tern	E	A
Mammals (3 species)			
Myotis grisescens	Gray bat	E	A
Myotis sodalis	Indiana bat	E	T
Plecotus townsendii virginianus	Virginia big-eared bat	E	T
Louisiana			
21 listed species: 14 aquatic, 7 terrestrial (67% of listed species are aquatic)			
Plants (4 species)			
Geocarpon minimum	(no common name)	T	T
Isoetes louisianensis	Louisiana quillwort	E	A
Lindera melissifolia	Pondberry	E	A
Schwalbea americana	American chaffseed	E	T
Mollusks (3 species)			
Lampsilis abrupta	Pink mucket pearlymussel	E	A
Margaritifera hembeli	Louisiana pearlshell	T	A
Potamilus inflatus	Inflated heelsplitter	T	A
Fishes (2 species)			
Acipenser oxyrhynchus desotoi	Gulf sturgeon	T	A
Scaphirhynchus albus	Pallid sturgeon	E	A
Reptiles (3 species)			
Caretta caretta	Loggerhead sea turtle	T	A
Gopherus polyphemus	Gopher tortoise	T	T
Graptemys oculifera	Ringed map turtle	T	A
Birds (7 species)			
Charadrius melodus	Piping plover	T	A
Falco peregrinus anatum	American peregrine falcon	E	T

Threatened and endangered species list. Continued.

State (summary data): Taxon	Common Name	Taxon Status	Habitat
Haliaeetus leucocephalus	Bald eagle	T	A
Pelecanus occidentalis	Brown pelican	E	A
Picoides borealis	Red-cockaded woodpecker	E	T
Sterna antillarum	Least tern	E	A
Vireo atricapillus	Black-capped vireo	E	T
Mammals (2 species)			
Ursus americanus luteolus	Louisiana black bear	T	T
Trichechus manatus	West Indian manatee	E	A
Maryland			
15 listed species: 10 aquatic, 5 terrestrial (67% of listed species are aquatic)			
Plants (6 species)			
Aeschynomene virginica	Sensitive joint-vetch	T	A
Agalinis acuta	Sandplain gerardia	E	T
Helonias bullata	Swamp pink	T	A
Oxypolis canbyi	Canby's dropwort	E	A
Ptilimnium nodosum	Harperella	E	A
Scirpus ancistrochaetus	Northeastern bulrush	E	A
Mollusks (1 species)			
Alasmidonta heterodon	Dwarf wedge mussel	E	A
Insects (2 species)			
Cicindela puritana	Puritan tiger beetle	T	T
Cicindela dorsalis dorsalis	Northeastern beach tiger beetle	T	A
Fishes (1 species)			
Etheostoma sellare	Maryland darter	E	A
Birds (3 species)			
Charadrius melodus	Piping plover	T	A
Falco peregrinus anatum	American peregrine falcon	E	T
Haliaeetus leucocephalus	Bald eagle	T	A
Mammals (2 species)			
Myotis sodalis	Indiana bat	E	T
Sciurus niger cinereus	Delmarva Peninsula fox squirrel	E	T
Mississippi			
32 listed species: 24 aquatic, 8 terrestrial (75% of listed species are aquatic)			
Plants (3 species)			
Apios priceana	Price's potato-bean	T	T
Lindera melissifolia	Pondberry	E	A
Schwalbea americana	American chaffseed	E	T
Mollusks (11 species)			
Epioblasma penita	Southern combshell	E	A
Lampsilis perovalis	Orange-nacre mucket	T	A
Medionidus acutissimus	Alabama moccasinshell	T	A
Pleurobema curtum	Black clubshell	E	A
Pleurobema decisum	Southern clubshell	E	A
Pleurobema marshalli	Flat pigtoe	E	A
Pleurobema perovatum	Ovate clubshell	E	A
Pleurobema taitianum	Heavy pigtoe	E	A
Potamilus capax	Fat pocketbook	E	A
Potamilus inflatus	Inflated heelsplitter	T	A
Quadrula stapes	Stirrupshell	E	A
Fishes (3 species)			
Acipenser oxyrhynchus desotoi	Gulf sturgeon	T	A
Etheostoma rubrum	Bayou darter	T	A
Scaphirhynchus albus	Pallid sturgeon	E	A
Reptiles (5 species)			

Threatened and endangered species list. Continued.

State (summary data): Taxon	Common Name	Taxon Status	Habitat
Caretta caretta	Loggerhead sea turtle	T	A
Drymarchon corais couperi	Eastern indigo snake	T	T
Gopherus polyphemus	Gopher tortoise	T	T
Graptemys flavimaculata	Yellow-bloched map turtle	T	A
Graptemys oculifera	Ringed map turtle	T	A
Birds (7 species)			
Charadrius melodus	Piping plover	T	A
Falco peregrinus anatum	American peregrine falcon	E	T
Grus canadensis pulla	Mississippi sandhill crane	E	A
Haliaeetus leucocephalus	Bald eagle	T	A
Pelecanus occidentalis	Brown pelican	E	A
Picoides borealis	Red-cockaded woodpecker	E	T
Sterna antillarum	Least tern	E	A
Mammals (3 species)			
Myotis sodalis	Indiana bat	E	T
Trichechus manatus	West Indian manatee	E	A
Ursus americanus luteolus	Louisiana black bear	T	T
Missouri			
22 listed species: 13 aquatic, 9 terrestrial (59% of listed species are aquatic)			
Plants (7 species)			
Asclepias meadii	Mead's milkweed	T	T
Boltonia decurrens	Decurrent false aster	T	T
Geocarpon minimum	(no common name)	T	T
Lesquerella filiformis	Missouri bladderpod	E	T
Lindera melissifolia	Pondberry	E	A
Platanthera praeclara	Western prairie fringed orchid	T	T
Trifolium stoloniferum	Running buffalo clover	E	T
Mollusks (4 species)			
Epioblasma florentina curtisi	Curtis' pearlymussel	E	A
Lampsilis abrupta	Pink mucket pearlymussel	E	A
Lampsilis higginsi	Higgins' eye pearlymussel	E	A
Potamilus capax	Fat pocketbook	E	A
Fishes (4 species)			
Amblyopsis rosae	Ozark cavefish	T	A
Etheostoma nianguae	Niangua darter	T	A
Noturus placidus	Neosho madtom	T	A
Scaphirhynchus albus	Pallid sturgeon	E	A
Birds (4 species)			
Charadrius melodus	Piping plover	T	A
Falco peregrinus anatum	American peregrine falcon	E	T
Haliaeetus leucocephalus	Bald eagle	T	A
Sterna antillarum	Least tern	E	A
Mammals (3 species)			
Myotis grisescens	Gray bat	E	A
Myotis sodalis	Indiana bat	E	T
Plecotus townsendii ingens	Ozark big-eared bat	E	T
North Carolina			
49 listed species: 26 aquatic, 23 terrestrial (53% of listed species are aquatic)			
Lichens (1 species)			
Gymnoderma lineare	Rock gnome lichen	E	T
Plants (25 species)			
Aeschynomene virginica	Sensitive joint-vetch	T	A
Amaranthus pumilus	Seabeach amaranth	T	A
Cardamine micranthera	Small-anthered bittercress	E	A

Threatened and endangered species list. Continued.

State (summary data): Taxon	Common Name	Taxon Status	Habitat
Echinacea laevigata	Smooth coneflower	E	T
Geum radiatum	Spreading avens	E	T
Hedyotis purpurea var. *montana*	Roan Mountain bluet	E	T
Helianthus schweinitzii	Schweinitz's sunflower	E	T
Helonias bullata	Swamp pink	T	A
Hexastylis naniflora	Dwarf-flowered heartleaf	T	T
Hudsonia montana	Mountain golden heather	T	T
Isotria medeoloides	Small whorled pogonia	T	T
Liatris helleri	Heller's blazingstar	T	T
Lindera melissifolia	Pondberry	E	A
Lysimachia asperulaefolia	Rough-leaved loosestrife	E	A
Oxypolis canbyi	Canby's dropwort	E	A
Ptilimnium nodosum	Harperella	E	A
Rhus michauxii	Michaux's sumac	E	T
Sagittaria fasciculata	Bunched arrowhead	E	A
Sarracenia oreophila	Green pitcher-plant	E	A
Sarracenia rubra jonesii	Mountain sweet pitcher-plant	E	A
Schwalbea americana	American chaffseed	E	T
Sisyrinchium dichotomum	White irisette	E	T
Solidago spithamaea	Blue Ridge goldenrod	T	T
Spiraea virginiana	Virginia spiraea	T	T
Thalictrum cooleyi	Cooley's meadowrue	E	A
Mollusks (6 species)			
Alasmidonta heterodon	Dwarf wedge mussel	E	A
Alasmidonta raveneliana	Appalachian elktoe	E	A
Elliptio steinstansana	Tar river spinymussel	E	A
Lasmigona decorata	Carolina heelsplitter	E	A
Mesodon clarki nantahala	Noonday snail	T	T
Pegias fabula	Little-wing pearlymussel	E	A
Insects (1 species)			
Neonympha mitchellii francisci	Saint Francis' satyr butterfly	E	T
Arachnids (1 species)			
Microhexura montivaga	Spruce-fir moss spider	E	T
Fishes (3 species)			
Cyprinella monacha	Spotfin chub	T	A
Menidia extensa	Waccamaw silverside	T	A
Notropis mekistocholas	Cape Fear shiner	E	A
Reptiles (1 species)			
Caretta caretta	Loggerhead sea turtle	T	A
Birds (5 species)			
Charadrius melodus	Piping plover	T	A
Falco peregrinus anatum	American peregrine falcon	E	T
Haliaeetus leucocephalus	Bald eagle	T	A
Picoides borealis	Red-cockaded woodpecker	E	T
Sterna dougallii dougallii	Tern roseate	T	A
Mammals (6 species)			
Canis rufus	Red wolf	E	T
Glaucomys sabrinus coloratus	Carolina northern flying squirrel	E	T
Myotis sodalis	Indiana bat	E	T
Plecotus townsendii virginianus	Virginia big-eared bat	E	T
Sorex longirostris fisheri	Dismal Swamp southeastern shrew	T	A
Trichechus manatus	West Indian manatee	E	A

South Carolina
30 listed species: 16 aquatic, 14 terrestrial (53% of listed species are aquatic)
Plants (19 species)

Threatened and endangered species list. Continued.

State (summary data): Taxon	Common Name	Taxon Status	Habitat
Amaranthus pumilus	Seabeach amaranth	T	T
Amphianthus pusillus	Little amphianthus	T	A
Echinacea laevigata	Smooth coneflower	E	T
Helianthus schweinitzii	Schweinitz's sunflower	E	T
Helonias bullata	Swamp pink	T	A
Hexastylis naniflora	Dwarf-flowered heartleaf	T	T
Isoetes melanospora	Black-spored quillwort	E	A
Isotria medeoloides	Small whorled pogonia	T	T
Lindera melissifolia	Pondberry	E	A
Lysimachia asperulaefolia	Rough-leaved loosestrife	E	A
Oxypolis canbyi	Canby's dropwort	E	A
Ptilimnium nodosum	Harperella	E	A
Rhus michauxii	Michaux's sumac	E	T
Ribes echinellum	Miccosukee gooseberry	T	T
Sagittaria fasciculata	Bunched arrowhead	E	A
Sarracenia rubra jonesii	Mountain sweet pitcher-plant	E	A
Schwalbea americana	American chaffseed	E	T
Trillium persistens	Persistent trillium	E	T
Trillium reliquum	Relict trillium	E	T
Mollusks (1 species)			
Lasmigona decorata	Carolina heelsplitter	E	A
Reptiles (2 species)			
Caretta caretta	Loggerhead sea turtle	T	A
Drymarchon corais couperi	Eastern indigo snake	T	T
Birds (6 species)			
Charadrius melodus	Piping plover	T	A
Falco peregrinus anatum	American peregrine falcon	E	T
Haliaeetus leucocephalus	Bald eagle	T	A
Mycteria americana	Wood stork	E	A
Picoides borealis	Red-cockaded woodpecker	E	T
Sterna dougallii dougallii	Roseate tern	T	A
Mammals (2 species)			
Myotis sodalis	Indiana bat	E	T
Trichechus manatus	West Indian manatee	E	A
Tennessee			
87 listed species: 66 aquatic, 21 terrestrial (76% of listed species are aquatic)			
Plants (19 species)			
Apios priceana	Price's potato-bean	T	T
Arabis perstellata	Rock cress	E	T
Arenaria cumberlandensis	Cumberland sandwort	E	T
Astragalus bibullatus	Pyne's ground-plum	E	T
Conradina verticillata	Cumberland rosemary	T	T
Dalea foliosa	Leafy prairie-clover	E	T
Echinacea tennesseensis	Tennessee purple coneflower	E	T
Geum radiatum	Spreading avens	E	T
Gymnoderma lineare	Rock gnome lichen	E	T
Hedyotis purpurea var. *montana*	Roan Mountain bluet	E	T
Helianthus eggertii	Eggert's sunflower	T	T
Isotria medeoloides	Small whorled pogonia	T	T
Lesquerella perforata	Spring Creek bladderpod	E	T
Pityopsis ruthii	Ruth's golden aster	E	T
Sarracenia oreophila	Green pitcher-plant	E	A
Scutellaria montana	Large-flowered skullcap	E	A
Solidago spithamaea	Blue Ridge goldenrod	T	T
Spiraea virginiana	Virginia spiraea	T	A

Threatened and endangered species list. Continued.

State (summary data): Taxon	Common Name	Taxon Status	Habitat
Xyris tennesseensis	Tennessee yellow-eyed grass	E	A
Mollusks (43 species)			
Alasmidonta atropurpurea	Cumberland elktoe	E	A
Alasmidonta raveneliana	Appalachian elktoe	E	A
Anguispira picta	Painted snake coiled forest snail	T	A
Athearnia anthonyi	Anthony's riversnail	E	A
Conradilla caelata	Birdwing pearlymussel	E	A
Cyprogenia stegaria	Fanshell	E	A
Dromus dromas	Dromedary pearlymussel	E	A
Epioblasma brevidens	Cumberlandian combshell	E	A
Epioblasma capsaeformis	Oyster mussel	E	A
Epioblasma florentina florentina	Yellow-blossom pearlymussel	E	A
Epioblasma metastriata	Upland combshell	E	A
Epioblasma obliquata obliquata	Purple cat's paw pearlymussel	E	A
Epioblasma othcaloogensis gubernaculum	Southern acornshell	E	A
Epioblasma torulosa	Green-blossom pearlymussel	E	A
Epioblasma torulosa torulosa	Tubercled-blossom pearlymussel	E	A
Epioblasma turgidula	Turgid-blossom pearlymussel	E	A
Epioblasma walkeri	Tan riffleshell	E	A
Fusconaia cor	Shiney pigtoe	E	A
Fusconaia cuneolus	Fine-rayed pigtoe	E	A
Hemistena lata	Cracking pearlymussel	E	A
Lampsilis abrupta	Pink mucket pearlymussel	E	A
Lampsilis altilis	Fine-lined pocketbook	T	A
Lampsilis virescens	Alabama lampshell	E	A
Medionidus acutissimus	Alabama moccasinshell	T	A
Medionidus parvulus	Coosa moccasinshell	E	A
Obovaria retusa	Ring pink mussel	E	A
Pegias fabula	Little-wing pearlymussel	E	A
Plethobasus cicatricosus	White-wartyback pearlymussel	E	A
Plethobasus cooperianus	Orange-foot pimple back pearlymussel	E	A
Pleurobema decisum	Southern clubshell	E	A
Pleurobema georgianum	Southern pigtoe	E	A
Pleurobema gibberum	Cumberland pigtoe	E	A
Pleurobema perovatum	Ovate clubshell	E	A
Pleurobema plenum	Rough pigtoe	E	A
Ptychobranchus greeni	Triangular kidneyshell	E	A
Pyrgulopsis ogmoraphe	Marstonia royal snail	E	A
Quadrula cylindrica strigillata	Rough rabbitsfoot	E	A
Quadrula fragosa	Winged mapleleaf mussel	E	A
Quadrula intermedia	Cumberland monkeyface pearlymussel	E	A
Quadrula sparsa	Appalachian monkeyface pearlymussel	E	A
Toxolasma cylindrellus	Pale lilliput pearlymussel	E	A
Villosa perpurpurea	Purple bean	E	A
Villosa trabalis	Cumberland bean pearlymussel	E	A
Crustaceans (1 species)			
Orconectes shoupi	Nashville crayfish	E	A
Arachnids (1 species)			
Microhexura montivaga	Spruce-fir moss spider	E	T
Fishes (15 species)			
Cyprinella monacha	Spotfin chub	T	A
Cyprinella caerulea	Blue shiner	T	A
Erimystax cahni	Slender chub	T	A
Etheostoma sp.	Bluemask darter	E	A
Etheostoma boschungi	Slackwater darter	T	A

Threatened and endangered species list. Continued.

State (summary data): Taxon	Common Name	Taxon Status	Habitat
Etheostoma percnurum	Duskytail darter	E	A
Etheostoma wapiti	Boulder darter	E	A
Noturus baileyi	Smoky madtom	E	A
Noturus flavipinnis	Yellowfin madtom	T	A
Noturus stanauli	Pygmy madtom	E	A
Percina antesella	Amber darter	E	A
Percina jenkinsi	Conasauga logperch	E	A
Percina tanasi	Snail darter	T	A
Phoxinus cumberlandensis	Blackside dace	T	A
Scaphirhynchus albus	Pallid sturgeon	E	A
Birds (4 species)			
Falco peregrinus anatum	American peregrine falcon	E	T
Haliaeetus leucocephalus	Bald eagle	T	A
Picoides borealis	Red-cockaded woodpecker	E	T
Sterna antillarum	Least tern	E	A
Mammals (4 species)			
Canis rufus	Red wolf	E	T
Glaucomys sabrinus coloratus	Carolina northern flying squirrel	E	T
Myotis grisescens	Gray bat	E	A
Myotis sodalis	Indiana bat	E	T
Virginia			
49 listed species: 37 aquatic, 12 terrestrial (76% of listed species are aquatic)			
Plants (10 species)			
Aeschynomene virginica	Sensitive joint-vetch	T	T
Arabis serotina	Shale barren rock-cress	E	T
Betula uber	Virginia round-leaf birch	T	A
Echinacea laevigata	Smooth coneflower	E	T
Helonias bullata	Swamp pink	T	A
Iliamna corei	Peter's mountain mallow	E	T
Isotria medeoloides	Small whorled pogonia	T	T
Platanthera leucophaea	Eastern prairie fringed orchid	T	T
Scirpus ancistrochaetus	Northwestern bulrush	E	A
Spiraea virginiana	Virginia spiraea	T	A
Mollusks (20 species)			
Alasmidonta heterodon	Dwarf wedge mussel	E	A
Condradilla caelata	Birdwing pearlymussel	E	A
Cyprogenia stegaria	Fanshell	E	A
Dromus dromas	Dromedary pearlymussel	E	A
Epioblasma brevidens	Cumberlandian combshell	E	A
Epioblasma capsaeformis	Oyster mussel	E	A
Epioblasma torulosa gubernaculum	Green-blossom pearlymussel	E	A
Epioblasma walkeri	Tan riffleshell	E	A
Fusconaia cor	Shiny pigtoe	E	A
Fusconaia cuneolus	Fine-rayed pigtoe	E	A
Hemistena lata	Cracking pearlymussel	E	A
Lampsilis abrupta	Pink mucket pearlmussel	E	A
Pegias fabula	Little-wing pearlymussel	E	A
Pleurobema collina	James River spinymussel	E	A
Pleurobema plenum	Rough pigtoe	E	A
Polygyriscus virginianus	Virginia fringed mountain snail	E	A
Quadrula cylindrica strigillata	Rough rabbitsfoot	E	A
Quadrula intermedia	Cumberland monkeyface pearlymussel	E	A
Quadrula sparsa	Appalachian monkeyface pearlymussel	E	A
Villosa perpurpurea	Purple bean	E	A

Threatened and endangered species list. Continued.

State (summary data): Taxon	Common Name	Taxon Status	Habitat
Crustaceans (2 species)			
Antrolana lira	Madison Cave isopod	T	A
Lirceus usdagalun	Lee County cave isopod	E	A
Insects (1 species)			
Cicindela dorsalis dorsalis	Northeastern beach tiger beetle	T	A
Fishes (5 species)			
Cyprinella monacha	Spotfin chub	T	A
Erimystax cahni	Slender chub	T	A
Etheostoma percnurum	Duskytail darter	E	A
Noturus flavipinnis	Yellowfin madtom	T	A
Percina rex	Roanoke logperch	E	A
Amphibians (1 species)			
Plethodon shenandoah	Shenandoah salamander	E	A
Birds (4 species)			
Charadrius melodus	Piping plover	T	A
Falco peregrinus anatum	American peregrine falcon	E	T
Haliaeetus leucocephalus	Bald eagle	T	A
Picoides borealis	Red-cockaded woodpecker	E	T
Mammals (6 species)			
Glaucomys sabrinus fuscus	Virginia northern flying squirrel	E	T
Myotis grisescens	Gray bat	E	A
Myotis sodalis	Indiana bat	E	T
Plecotus townsendii virginianus	Virginia big-eared bat	E	T
Sciurus niger cinereus	Delmarva Peninsula fox squirrel	E	T
Sorex longirostris fisheri	Dismal Swamp southeastern shrew	E	A
West Virginia			
19 listed species: 12 aquatic, 7 terrestrial (63% of listed species are aquatic)			
Plants (5 species)			
Arabis serotina	Shale barren rock-cress	E	T
Ptilimnium nodosum	Harperella	E	A
Scirpus ancistrochaetus	Northeastern bulrush	E	A
Spiraea virginiana	Virginia spiraea	T	A
Trifolium stoloniferum	Running buffalo clover	E	T
Mollusks (8 species)			
Cyprogenia stegaria	Fanshell	E	A
Epioblasma torulosa rangiana	Northern riffleshell	E	A
Epioblasma torulosa torulosa	Tuberculed-blossom pearlymussel	E	A
Lampsilis abrupta	Pink mucket pearlymussel	E	A
Obovaria retusa	Ring pink mussel	E	A
Pleurobema clava	Clubshell	E	A
Pleurobema collina	James River spinymussel	E	A
Triodopsis platysayoides	Flat-spired three-toothed snail	T	T
Amphibians (1 species)			
Plethodon nettingi	Cheat Mountain salamander	T	A
Birds (2 species)			
Falco peregrinus anatum	American peregrine falcon	E	T
Haliaeetus leucocephalus	Bald eagle	T	A
Mammals (3 species)			
Glaucomys sabrinus fuscus	Virginia northern flying squirrel	E	T
Myotis sodalis	Indiana bat	E	T
Plecotus townsendii virginianus	Virginia big-eared bat	E	T

Index

acadian flycatcher 281
aerial photographs 27
aflatoxins 145, 162
Alabama beach mouse 263
Alabama Black Bear Alliance 47, 279
Alabama canebrake pitcher plant 75, 303
Alabama cavefish 299
Alabama Conservation Enforcement 226
Alabama Cooperative Extension System 233, 277, 399, 450
Alabama Department of Conservation and Resources 18, 106, 208, 219, 223, 226, 230, 252, 253, 256, 257, 283, 287, 292, 299, 345, 349, 355, 357, 360, 399, 407, 442, 450
Alabama Department of Environmental Management 444, 447, 448, 450
Alabama Farmer's Federation 453
Alabama Forestry Association 453
Alabama Forestry Commission 18, 76, 91, 119, 198, 312, 376, 379, 399, 401, 444, 445, 447, 448, 449, 450, 453
Alabama Forever Wild Land Trust 398
Alabama game 393
Alabama Game and Fish Division 233, 432
Alabama lamp pearly mussel 300
Alabama leather flower 303, 304
Alabama mammals
  distribution 262
Alabama moccasinshell mussell 300
Alabama Natural Heritage Program 76, 265, 279
Alabama Nature Conservancy 47
Alabama Non-game Wildlife Program 257
Alabama red-bellied turtle 292
Alabama streak sorus fern 306, 308
Alabama supplejack 6, 89, 113, 367, 511
Alabama Waterfowl Association 227
Alabama Wildlife Federation 11, 47, 68, 96, 135, 161, 208, 265, 279, 311, 398, 426, 427, 453
Alabama's fish stocking program 350
Alabama's landscape 1
Alabama's Non-game Program 257
Albany Area Quail Management Project 183
alder 532
Aldo Leopold Foundation 425
alfalfa 197, 367, 484
Allen, Durward L. 4
alligator 148, 215, 220, 232, 283, 293, 417
Allison, E.F. 138
allwood pink 333
alyceclover 367, 484
American basswood 532
American beautyberry 6, 41, 71, 113, 150, 153, 171, 322, 367, 511
American beech 6, 511, 532
American bison 419
American crow 281
American elder 6, 367, 511
American Farm Bureau 452
American Forest and Paper Association 452
American hart's tongue fern 309
American holly 6, 63, 322, 532
American hornbeam 6, 188, 269, 511, 515
American jointvetch 121, 125
American kestrel 329, 478
American redstart 5, 74, 261, 281
American robin 478
American Sport Fish 352
American swallow-tailed kite 281
American widgeon 212, 214
American woodcock 59
annual phlox 323
annual plants 40, 364
Anthony's riversnail 301
Appalachian Plateau 7
apple 485
appreciative use of wildlife 392
aquatic and marsh plants important to wildlife in 524
aquatic furbearers 230
  beaver 230
  beaver ponds 231
  river otter 233
aquatic furbears
  muskrat 235
arrangement of habitat components 38
arrow-arum 524
arrowhead 225, 236
arrowleaf clover 367, 491
arsenal 365
artificial regeneration 54
ash 15, 53, 57, 71, 89, 94, 117, 188
asiatic dayflower 524
aspect 2
Atlantic bottle-nosed dolphin 262
attracting backyard wildlife 315
Auburn University 68, 72, 76, 78, 82, 119, 122, 124, 126, 135, 142, 145, 152, 169, 183, 198, 244, 253, 256, 337, 345, 346, 352, 357, 358, 367, 377, 378, 390, 394, 427, 428, 436
Austrian winter pea 486
autumn olive 367, 487
autumn sage 333
avian pox 144, 147
azalea 303, 335

Bachman's sparrow 281
Bachman's warbler 281
backyard habitat
  enhancing 324
backyard wildlife
  advantages of attracting 316
  artificial feeders 330
  attracting select species 325
  bats 326
  birds 326
  butterflies 334
  cover 325
  enhancing existing landscape 319
  habitat requirements 317
  hummingbirds 332

involving neighbors 318
nesting boxes 331
nuisance backyard wildlife 336
reptiles and amphibians 334
small mammals 325
steps to attracting 316
suggested trees and plants 322
Backyard Wildlife Habitat Program 336
bahia grass 94, 96, 99, 100, 148, 150, 158, 160, 171, 178, 197, 270, 366, 367, 375, 487, 523
Baker, Chester 387
bald cypress 9, 15, 188, 215, 216, 224, 270, 532
bald eagle 281, 290, 291
baldpate 212
Baldwin County 265
ball clover 367, 491
Bankhead National Forest 106, 107
bareroot seedlings 376
barley 121, 197, 204, 206, 218, 367, 378, 505
barn owl 478
barnyard grass 94, 204, 206, 524
barred owl 148, 187, 247, 281, 329, 364, 478
basal area 62
Basic Outdoor Skills Series 257
basswood 188, 309
bat 478
big brown 262
Brazilian free-tailed bat 262
Eastern pipistrelle 262
evening bat 262
gray myotis 262
hoary bat 262
Indiana myotis 262
Keen's myotis 262
little brown myotis 262
northern yellow bat 262
Rafinesque's big-eared bat 262
red bat 262
Seminole bat 262
silver-haired bat 262
Southeastern myotis 262
bat boxes 326
Bat Conservation International 326
bayberry 303, 367, 511
bear management units 275
beardtongue 323
beaver 229, 230, 231, 232, 233, 235, 236, 238, 248, 250, 253, 258, 259, 260, 262, 351, 379, 419
biology and life history 231
castoreum 231
control 233
description 231
habitat needs 232
prevention and control techniques 250
beaver pond 221
beds 69
bee balm 323, 333
beech 7, 10, 15, 65, 74, 85, 88, 143, 150, 153, 154, 188, 189, 192, 272, 322
beggarweed 94, 150, 170, 171, 174, 175, 488, 523
benne 206
Bent Creek Research and Demonstration Forest 63
bergamot 323
bermuda grass 32, 71, 99, 170, 175, 178, 366
berseem clover 367, 492
Best Management Practices for Forestry 84, 91, 191, 270, 290, 292, 293, 300, 309, 441, 444, 445, 448, 453
bicolor 366
bicolor lespedeza 32
biennials 364
big brown bat 326
biodiversity 45
bird feeder 330
birds-in-a-nest 75
birdsfoot deervetch 488
birdsfoot trefoil 367, 375
bitternut hickory 7, 15
bittersweet 96
bitterweed 96
black bear 43, 46, 91, 93, 255, 262, 264, 265, 266, 267, 268, 269, 270, 271, 272, 273, 274, 275, 276, 277, 278, 279, 436, 457
access management 278
agriculture and black bear management 273
apiary protection 276
Black Bear Conservation Committee 275
bottomland hardwood management 268
canebreak management 271
crop and livestock protection 277
cypress and tupelo management 272
description and life history 266
food habits 267
food plots 273
garbage/landfill management 278
habitat management in hardwood plantations 270
habitat requirements 267
human/bear conflicts 275
hunter education and cooperation 277
in Alabama 274
key bottomland hardwood species 270
landscape management 275
management of human behavior 278
minimizing road-kills/hazards 278
mixed upland pine-hardwood management 272
protection of structures 277
torpor 266
upland pine management 271
Black Bear Conservation Committee 279
Black Belt 2, 3, 9, 10, 13, 51, 108, 115, 116, 134, 345
black birch 532
black cherry 6, 57, 71, 85, 86, 88, 96, 150, 153, 171, 172, 188, 322, 324, 335, 367, 512, 532
black clubshell mussel 301

black duck 212, 213, 214
black gum 85, 154, 270, 322
black locust 57, 88, 532
black medic 367, 489
black nightshade 96
black oak 7, 8, 88, 519, 532
black pine snake 75
black rat 263
black swallowtail 334, 335, 336
black tupelo 113, 150, 153, 171
black walnut 88, 188, 512, 533
black willow 533
black-and-white vireo 261
black-and-white warbler 74
black-eyed susan 322, 323
black-throated green warbler 74
blackberry 6, 33, 41, 69, 70, 71, 74, 82, 90, 94, 113, 114, 118, 119, 150, 151, 154, 171, 172, 174, 175, 178, 188, 194, 195, 196, 197, 268, 269, 270, 272, 322, 324, 512
blackgum 6, 71, 74, 188, 215, 224, 232, 532
blackhead 144, 147
blackjack 89
blackjack oak 153, 519, 532
blanket flowers 323
blazing stars 323
Bliss, John 436
bloodflower 335
blue catfish 349, 351
blue false indigo 323
blue grosbeak 281, 329
blue jay 281, 329
blue passionflower 335
blue shiner 298, 299
blue-eyed grass 323
blue-gray gnatcatcher 281
blue-winged teal 212, 213, 214
bluebell 323
blueberry 6, 72, 112, 113, 150, 171, 188, 322, 324, 512
bluegill 349
bluejack oak 153, 519
bluestem 94, 96, 99, 367, 375, 489
bobcat 145, 147, 148, 168, 187, 193, 220, 229, 230, 232, 240, 247, 248, 249, 251, 252, 262, 264, 325, 393
  biology and life history 248
  description 248
  habitat needs 249
  prevention and control techniques 251
bobwhite quail 4, 9, 32, 33, 35, 44, 45, 51, 59, 66, 76, 89, 91, 102, 164, 165, 166, 167, 168, 169, 170, 171, 178, 180, 182, 183, 195, 199, 370, 386, 391, 394, 416
  Amquail 176
  bevy 181
  biology and life history 166
  brood-rearing cover 170
  buffer foods 168
  cover 170
  coveys 166
  developing cover 177
  escape cover 172
  estimating numbers 181
  factors limiting populations 167
  factors that affect populations on farms 174
  farm management 174
  food 170
  forest management 178
  habitat and harvest management intensities 180
  habitat improvements 173
  habitat requirements 170
  hunting 181
  myths and misconceptions 168
  nesting 166
  nesting cover 170
  pen-raised 169
  pen-released 182
  prescribed burning 179
  releasing pen-raised birds for shooting 182
  roosting cover 171
  seasonally important foods 171
  strip disking 179
  thermal cover 172
  transition zones 177
Book Steering Committee 477
bot fly 188, 193
boulder darter 297, 299
boxelder 15, 269, 515
Boyd, Frank 233
bracken fern 96
breeding birds and preferred habitats 281
bristlegrass 204, 206
broad-winged hawk 148, 281
broomsedge 71
brown thrasher 329
brown-headed cowbird 281, 330
brown-headed nuthatch 5, 329
browntop millet 499
browse 7
brush piles 88, 89, 325
buckeye 71, 88, 96, 532
buckwheat 206, 222, 489
buffer 61
bufflehead 212, 214
bullgrass 204
bullhead catfish 351, 352
bulrush 218, 225, 236, 306, 524
bunting 329
bur oak 532
Burroughs, John 421
butterfly bush 335
butterfly garden
  selected plants 335
butterfly pea 171
butterflyweed 323, 335
butternut 63, 88, 512
butterwort 75
button clover 367, 492
buttonbush 216, 306

cahaba shiner 297, 299
calico aster 323
California sea lion 262
Callaway Gardens 336
Canada goose 214, 235
Canada lily 323
canary grass 216
candidate species 286
canebreak 271
canvasback 212, 214
cardinal 32, 35, 329, 331
cardinal flower 323

Carolina chickadee 87, 260, 281, 329
Carolina cranebill 204, 206
Carolina parakeet 279
Carolina silverbell 188
Carolina wren 281, 329, 478
Carson, Rachel 419, 425
catalpa 88, 532
catbird 329
categories of select wildlife plants 367
cattail 216, 233, 236
cattle 94, 95, 96, 98, 99, 100, 101, 102, 103, 123, 124, 144, 147, 158, 166, 191, 202, 244, 271, 344, 378, 381, 402, 421, 426, 437
Causey, Keith 126
Cecil B. Day Butterfly Center 336
cedar waxwing 329
cerulean warbler 281
chaffseed 75
Chalk Hill Wildlife Reserve 432
Champion International Corporation's forest patterns land classification 92, 95
channel catfish 349
chapman oak 519
Chapman's aster 75
Chautauqua National Wildlife Refuge 216
cherrybark oak 224, 270, 532
chestnut oak 7, 8, 88, 153, 519, 532
chestnut-sided warbler 74
chickadee 478
chinaberry 366
Chinese chestnut 490
Chinese privet 366
Chinese wisteria 366
chinkapin 532
chinquapin oak 188, 309, 513
chipmunk 43, 325, 337
Choctaw County 265
Christmas Bird Count 282
chufa 32, 146, 148, 150, 158, 159, 160, 161, 162, 163, 218, 379, 380, 381, 490
chukar 406
Chunnenuggee Hills 10
cinnamon fern 305
Clarke County 265
Clean Air Act 439, 447
Clean Water Act 85, 118, 370, 438, 439, 444, 445, 447, 452
clear-cutting 54, 59, 156, 190, 264
  daylighting 158
  regeneration 53
  small patch 53
  strip or patch 55
  wildlife habitat 58
Clemson beaver pond leveler 250
Clemson University 62, 312
climate 4
climax community (forest) 41
clover 82, 100, 121, 122, 123, 124, 148, 150, 151, 152, 158, 159, 160, 171, 196, 197, 198, 199, 268, 270, 272, 367, 370, 371, 373, 375, 376, 377, 378, 380, 381
Coastal Non-Point Source Management Program 448
coastal panicgrass 486
Coastal Plain 2, 3, 5, 9, 10, 13, 14, 66, 69, 75, 76, 79, 96, 99, 126, 138, 156, 178, 248, 262, 263, 283
Coastal Zone Management Act 439, 448
coccidiosis 144
coffeeweed 71
cogongrass 365, 366
commercial thinnings 73
common beardtongue 333
common carp 351
common goldeneye 212
common grackle 281
common merganser 212
common raven 279
common rose mallow 333
common snapdragon 333
common spiderwort 323
common sunflower 323
common vetch 367
common yellowthroat 281
commonly planted wildlife foods 479
Conecuh National Forest 96
conservation ethics/opportunities 415
  Alabama Conservation Leaders 426
  Alabama Wildlife Federation's legacy and vision 430
  conservation leaders 419
  hunter ethics 429
  hunter's pledge 430
  hunting ethics 432
  mammals 419
  wild turkey 418
  wildlife success stories 416
  wood duck 417
Conservation Reserve Program 144, 147, 447
constructing/installing a three-log drain 221
consumptive use of wildlife 392
controlled burning 74
  effects 76
cool season plants 365
Cooperative Extension System 252, 357, 360, 374
Cooperative Forestry Assistance Act 451
Cooper's hawk 148, 187, 281
Coosa moccasinshell mussel 300
copper fennel 335
coral bean 332
coral bells 333
coral honeysuckle 332, 333
corn 143, 144, 145, 147, 148, 150, 158, 160, 162, 171, 174, 175, 176, 182, 188, 194, 204, 205, 206, 207, 218, 222, 223, 224, 228, 233, 236, 240, 243, 267, 268, 270, 272, 273, 275, 276, 277, 278, 330, 342, 364, 365, 367, 370, 372, 380, 381, 388, 402, 495
Corps of Engineers 445, 446, 447
cosmos 335
cost-sharing 400
Costa, Ralph 312
cotton 273, 388
cotton mice 325
cotton rat 168, 263
cottontail rabbit 35, 41, 51, 192, 243, 248, 325

cottonwood 15, 88, 532
cougar 93, 436
county soil surveys 28
coves 2
cow oak 270
cowpeas 148, 171, 204, 206, 367, 496
coyote 32, 33, 41, 43, 145, 146, 148, 168, 193, 195, 229, 230, 236, 240, 243, 244, 245, 246, 247, 251, 261, 262, 264, 293, 325, 393
  biology and life history 245
  carrion 245
  description 243
  habitat needs 246
  prevention and control techniques 251
coyote-thistle aster 75
crabapple 118, 324, 485
crabgrass 150, 161, 175, 204
cracking pearly mussel 301
crappie 349, 351
crayfish 236, 237
crested flycatcher 478
crested iris 323
crimson clover 121, 122, 160, 197, 367, 378, 380, 381, 389, 493
crossvine 367
croton 71, 206
cull 80
cultivated plants important to wildlife in Alabama 484
Cumberland monkeyface 301
cypress 154, 270, 322, 326

dallis grass 367, 375, 496
dark pigtoe mussel 301
day-lighting 90
daylily 333
Delta Environmental Land Trust Association 398
Delta Waterfowl Foundation 220, 227
Department of Conservation and Natural Resources 311
destratification 360
determining what wildlife eat 35
developing a wildlife habitat management plan 17
devil's walking stick 267, 268, 270
DeVos, Ted 169
dewberry 33, 70, 155, 267, 269, 270, 272, 512
diamondback rattlesnake 293
dickcissel 281
dill 335
direct seeding 54
dissolved oxygen 340
diversity 11
dogwood 63, 71, 72, 74, 85, 86, 112, 113, 118, 143, 148, 150, 151, 153, 154, 155, 156, 171, 178, 186, 187, 188, 202, 232, 269, 270, 322, 324, 367, 423, 513, 532
Douglas, Howard 138
dove proso millet 204
doveweed 40, 180, 204, 205
downy woodpecker 281, 329, 478
dromedary pearly mussel 301
dropseed 99
duck potato 236
Ducks Unlimited 226
dusky gopher frog 75, 96
Dutchman's breeches 96
dwarf huckleberry 6, 515
dwarf live oak 153, 519
dwarf palmetto 516

Easley, Mickey 182, 380
Eastern bluebird 32, 87, 260, 329, 478
Eastern box turtle 334
Eastern chipmunk 263
Eastern coachwhip 148
Eastern cottontail rabbit 192
  biology and life history 192
  cover 194
  description and historical overview 192
  developing food 197
  developing transition zones 195
  diet 194
  forest management 197
  habitat improvements 194
  habitat requirements 194
  harvest 199
  limiting factors 193
Eastern cougar 262
Eastern elk 436
Eastern fence lizard 334
Eastern gammagrass 99
Eastern harvest mouse 263
Eastern hemlock 532
Eastern hophornbeam 6, 153, 269, 514
Eastern indigo snake 75, 293
Eastern meadowlark 282
Eastern mole 262
Eastern phoebe 329
Eastern pipistrelle 326
Eastern red cedar 6, 7, 71, 359, 370, 517
Eastern spotted skunk 246, 263
Eastern tiger swallowtail 334, 335
Eastern wood-pewee 74, 261, 281
Eastern woodrat 263, 325
economics of forest/farm wildlife management
  advertising 413
  aesthetic value 393
  alternatives to offset costs 396
  availability of labor and equipment 395
  biological value 392
  commercial memberships 406
  condition of the land 396
  conservation easements 398
  cost-sharing programs 397
  developing fee access recreation 409
  educational value 393
  equipment used for habitat management 395
  factors affecting costs 393
  fee fishing/fishing clubs 407
  fee-hunting 402
  forest values 391
  hunting lease 404
  in-kind assistance 399
  income alternatives 402
  income-producing wildlife management alternatives 412
  increased land values 398

integrating wildlife 387
intrinsic value 393
land ownership 394
landowner assistance programs 399
landowner cooperatives 407
landowner objectives 394
leases 403
non-consumptive wildlife activities 408
non-wildlife opportunities 408
permits 406
scientific value 393
sporting clays 408
tax incentives 397
understanding clientele 411
values and uses of wildlife 392
wildlife and timber management 390
wildlife market 402
wildlife values 392
wing shooting preserves 406

ecosystem management 46, 47
ecotone 39
edge 39
edge-adapted species 39
Egyptian star-cluster 335
Egyptian wheat 160, 182, 381, 497
elderberry 96, 267, 268, 269, 270, 324
elm 15, 71, 88, 89, 224, 513, 532
endangered and threatened species 286, 313
Alabama canebrake pitcher plant 303
Alabama leather flower 303
Alabama red-bellied turtle 292
Alabama streak sorus fern 306
American hart's tongue fern 309
bald eagle 290
fishes 296
flattened musk turtle 293
gopher tortoise 293
gray bat 287
green pitcher plant 305
harperella 303
Indiana bat 287
Kral's water plantain 305
Mohr's Barbara's buttons 309
mollusks 300
Morefield's leather flower 306
mussel species 300
plants 303
Price's potato-bean 309
Red Hills salamander 294
red-cockaded woodpecker 289
relict trillium 306
Tennessee yellow-eyed grass 306
wood stork 290
endangered species 286, 440
list 557
new approaches to protecting 311
Endangered Species Act 75, 76, 265, 285, 286, 287, 299, 311, 417, 439, 440, 443, 452
critical habitat 286
Section 7 286
Section 9 286
take 286
endangered species assistance 299
Endangered Species Pesticide Protection Program 443
enhancing wildlife habitat
choosing the right plant 368
companion plants 375
costs and availability 377
costs of forage plantings for white-tailed deer 378
depredation by domestic animals/other wildlife 378
establishing a successful wildlife food plot 383
fertilization and liming 373
inoculation of legumes 374
land preparation 371
maintenance and management 376
nutritional needs of wildlife 366
planting bareroot tree, shrub seedlings 376
planting considerations 368
planting dates 372
record-keeping 379
seeding rates 373
site selection 369
size, shape, and distribution of plantings 371
weed and insect control 377
wildlife food plants 364
Environmental Protection Agency 443, 450
eucalyptus 532
eupatorium 335
European starling 32
evapotranspiration 2
even-aged stands 53
even-aged versus uneven-aged management 54
evening bat 326
evening primrose 323
Everett, Danny 358
exclosure 369
exotic plants 365
exotic trees and plants considered invasive 366

Fall Line 9, 263, 283
Fall Line Hills 9
fallow deer 262
fanshell mussell 301
Farm Services Agency 345, 360
farmland matrix 93
fathead minnow 351, 358
favoring wildlife plants with herbicides 71
fawn lily 323
Federal Aid In Wildlife Restoration Act 416
Federal Environmental Pesticide Control Act 439
Federal Insecticide, Fungicide, Rodenticide Act 449
fee fishing operations 407
feral hog 145, 146, 148
fertilizing honeysuckle for deer 82
fertilizing oaks to increase acorn production 121
fescue 99, 170, 175, 178, 197, 366
fibroma 193
fine-lined pocketbook mussel 300
fine-rayed pigtoe mussel 300
fire adapted landscapes 75
fire ant 146, 148, 293, 313, 416
fire cherry 532
fire-dependent animals 75
firebreaks 90
firebush 332
firelanes 90

firelines 90
fireweed 323
fish pond management 339
adjusting water levels 360
aeration, mixing pond temperature layers 360
alternative stocking strategies 351
basic principles 342
construction considerations 342
enhancement 358
evaluation of pond balance 355
fathead minnows 359
fertilizers 346
fish selection and stocking 349
fish shelters 359
harvesting and record-keeping 354
how to fertilize 348
liming and fertilization 345
livestock and erosion considerations 344
phytoplankton 341
plankton 340
potential problems 357
removal of unwanted fish 352
spawning substrate 361
supplemental feeding 359
water quality 341
weed control 356
when to fertilize 347
zooplankton 341
fishing record-keeping 354
flat pigtoe mussel 301
flathead catfish 351
flattened musk turtle 293
flatwoods plum 6
flatwoods salamander 75
flicker 478
flicker woodpecker 329
flies 111
Florida black bear 262, 265
Florida pine snake 75
Florida pussley 174
flowering dogwood 6, 322
flowering tobacco 333
flying squirrel 87, 325, 337
Folkerts, George 72
foreseeability 456
forest and farm wildlife management 385
forest fertilization 81
forest fragmentation 282
forest harvest and regeneration 53
forest interior species 282
forest management
costs 396
multiple species approach 257, 260
single species approach 257·
forest roads 90
forest stand management
forested corridors 91
forest stand management considerations 91
forest stand value for wildlife 50
forest stewardship 453
forest types 5
forested wetlands 14
Forestry Incentives Program 451
forests
bald cypress 9
bottomland 147, 215, 223
bottomland hardwood 117, 268
early successional stage 281
hardwood 191
late successional stage 281
loblolly 8, 9
loblolly-shortleaf pine 8, 9, 10
mid successional stage 281
mixed hardwood-pine 191
mixed pine-hardwood 117
mixed-pine 5
oak 9
oak-hickory 7
oak-pine 8, 9, 10
oak-tupelo gum-bald cypress 15
pine 58, 155
pine-hardwood 117, 155, 272
pine-oak 7
riparian 80, 260, 445
riparian hardwood 258
single species approach 260
water tupelo 9
fox
biology and life history 242
description 241
habitat needs 243
prevention and control techniques 251
fox squirrel 87, 185, 186, 187, 189, 190, 192, 199, 217, 263, 325
management intensities 190
foxglove 323
foxtail 94, 204, 367, 381
foxtail grass 40, 204
foxtail millet 500
fragmentation 93
Fred T. Stimpson Game Sanctuary 138
French marigold 335
French mulberry 268, 270, 272
freshwater marsh 14
fringetree 113, 322, 513
fuchia 333
fungicides 449
furbearer control 252
furbearer trapping 252

gadwall 212, 213, 214
gallberry 6, 113, 150, 153, 171, 270
garden phlox 333
Geographic Information Systems 24
Geological Survey of Alabama 28
Georgia-Pacific 442
giant coneflower 323
giant cutgrass 524
giant foxtail 524
Gibson, Curtis 388
gizzard shad 351
glossary of terms 546
glossy abelia 335
golden eagle 148
golden mouse 263
golden-cheeked warbler 417
goldeneye 214
goldfinch 329
goldline darter 297, 299
gopher tortoise 3, 75, 76, 261, 293, 294, 439, 457
goshawk 148
government regulation 441
grasshopper sparrow 282
gray bat 287, 288
gray catbird 74, 261, 281
gray fox 148, 187, 220, 230, 241, 243, 262
gray goldenrod 323

gray squirrel 41, 51, 87, 185, 186, 187, 189, 190, 191, 192, 199, 240, 263, 337
- biology and life history 186
- cover requirements 189
- food habits 188
- habitat improvements 189
- habitat requirements 188
- harvest requirements 191
- hunting 191
- limiting factors 187
- management intensities 190
- reproductive cycle 186
- water requirements 189

grazing system 98
great horned owl 147, 148, 187, 247
great-crested flycatcher 281
greater scaup 212, 214
green anole 334
green ash 532
green pitcher plant 75, 305, 306
green sunfish 352
green-winged teal 212, 214
greenbriar 6, 71, 89, 112, 113, 114, 118, 172, 178, 188, 270, 299, 367
Grinnel, George 421
grosbeak 331
ground cherry 96
group selection 64
Gulf Coast 160, 212, 263
gulf fritillary 335
gulf sturgeon 295, 299
gum 117
gypsy moth 33

habitat conservation plans 311, 442
habitat improvements
- openings 83
- riparian forests, streamside management zones 84

habitat preferences for select birds 329
habitat succession 40
hackberry 6, 15, 88, 89, 94, 188, 514, 532
Hadaway, Jeff 199
hairy vetch 367
hairy woodpecker 281
half-cuts 89
harperella 303, 304
hawthorn 6, 113, 514
headquarters 39, 40
heavy pigtoe mussel 301
heliotrope 335
hemorrhagic disease 236
herbicides 66, 70, 71, 72, 80, 81, 87, 89, 101, 118, 157, 205, 271, 273, 287, 294, 305, 306, 309, 322, 344, 356, 364, 377, 378, 380, 381, 395, 449
- application methods 81
- applying 81
- managing red-cockaded woodpeckers 80

herbivores 32
heron 215
herpetofauna 9, 283, 334
hickory 5, 6, 7, 8, 10, 57, 71, 74, 87, 88, 154, 188, 189, 192, 270, 272, 514, 532
high grading 53, 65
hispid cotton rat 263
holly 113, 270, 303, 322, 324, 514
honeylocust 63, 88, 94, 113, 532
honeysuckle 71, 112, 113, 114, 118, 119, 120, 172, 174, 178, 194, 195, 197, 322, 324, 366, 367
hooded merganser 212, 214
hooded warbler 74, 258, 261, 281
hophornbeam 150, 188, 532
- habitat and harvest management intensities 159

horizontal habitat diversity 38
horsetail 96, 236
house finch 478
house mouse 263
house sparrow 32, 330
huckleberry 72, 113, 150, 153, 154, 171, 367
hummingbirds 332, 333
hunting ethics 431
hunting lease 404, 405
hunting lease operation 389
Huntley, Jimmy 96
Hurst, George 147
hybrid bluegill 351
hydrilla 33

impatiens 335
importance of dead wood 43
Indian grass 94, 99, 367
Indian hemp 96
Indiana bat 287 288
Indiangrass 497
indigo bunting 74, 261, 281
individual-tree selection 64
inflated heelsplitter mussel 300
Interior Low Plateau 5
Intermediate forest stand practices 72
International Migratory Bird Day 282
International Paper 442
Interspersion 38
Interspersion index 39
Ireland, William R. 427
ironweed 323
ivory-billed woodpecker 279
Izaak Walton League of America 429

Jacobson, Harry 133
jaguarundi 262
Japanese climbing fern 366
Japanese flowering quince 333
Japanese grass 366
Japanese honeysuckle 6, 82, 366, 515
Japanese millet 206, 222, 500
Japanese privet 366
Jeponeza honeysuckle 333
jimson weed 96
Johnson grass 71, 96, 204, 366
Johnson, Rhett 68
jointvetch 375
jurisdictional wetlands 446

Keller, Don 352
Kelley, Charles D. 428
Kentucky warbler 74, 261, 281
king snake 148
kingbird 329
Kral's water plantain 305
Kral's yellow-eyed grass 75
kudzu 32, 365, 366

lablab 123, 497
ladino clover 389
lance-leaved coreopsis 323
land classification 5
  general management/high timber yield 95
  protected/restricted 95
landowner assistance programs 399
landowner liability 455
landscape level management 92
lantana 322, 335
large gallberry 6
large swamp rabbit 192
largemouth bass 215, 220, 236, 349, 350
larkspur 96
Lauderdale Wildlife Management Area 107
laurel oak 87, 153, 224, 519, 532
legal considerations: natural resource management 435
  Clean Air Act 447
  Clean Water Act 444
  Coastal Zone Management Act 448
  cost-share recipient responsibilities 449
  Endangered Species Act 440
  endangered species pesticide protection program 443
  ESA's affect on private landowners 440
  Federal Insecticide, Fungicide, Rodenticide Act 449
  federal, state, and local regulations 439
  invitee 454
  landowner responsibility 454
  landowner rights and responsibilities 452
  licensee 454
  local ordinances 449
  non-point source pollution 444
  Operation Gamewatch 456
  recreational access and liability 454
  reducing landowner liability 456
  regulatory information and assistance 450
  trespasser 454
  wetlands 447
  wetlands regulation 444
legumes 21, 41, 59, 62, 66, 69, 70, 74, 75, 78, 79, 82, 112, 156, 170, 174, 175, 176, 177, 180, 194, 195, 196, 197, 204, 222, 363, 364, 367, 368, 371, 374, 375, 376, 383, 394
Leopold, Aldo 29, 45, 46, 168, 286, 416, 419, 422, 423, 425, 429, 433, 436
lespedeza 94, 113, 170, 171, 172, 175, 176, 180, 182, 195, 206, 366, 367, 379, 381, 523
lespedezas (annual) 498
lespedezas (shrub) 498
lesser scaup 212, 214
leyland cypress 370
limestone rough pigtoe mussel 300
little brown bat 326
little-wing pearly mussel 301
live oak 89, 153, 270, 519, 532
living brush piles 89
loblolly pine 5, 7, 9, 10, 14, 53, 59, 61, 65, 74, 78, 156, 178, 387, 522, 533
locust 71, 515
loggerhead shrike 87
long-distance or neotropical migrants 279
long-tailed weasel 263
longleaf pine 5, 9, 10, 14, 45, 62, 63, 65, 75, 76, 77, 78, 80, 98, 156, 178, 522, 533
Louisiana waterthrush 258, 281
Lower Coastal Plain Region 14
lymphoproliferative 147

magnolia 10, 63, 515, 532
mallard 212, 213, 214, 217, 218, 219, 223, 406, 439
  biology and life history 217
  diet 218
  migration 218
management by land classification 95
managing for bobwhite quail 165
managing for furbearer species 229
managing for mourning dove 201
managing for non-game wildlife 255
managing for rabbits 192
managing for squirrels 185
managing for waterfowl 211
managing for white-tailed deer 105
managing for wild turkey 137
managing forests for wildlife and timber productio 49
maple 15
marsh hawk 187
marsh rabbit 192
mast trees and shrubs 85
maypop 335
McCallister Farms 402
McCallister, Jerry 402
McCallister, Joe 402
McCollum, Bob 257
McCutcheon, Keith 226
McGlincy, Joe 19
McKee, Bill 390
McMillan, Ed 123
meadow beauty 75
meadow jumping mouse 262
mealycup sage 333
measuring interspersion and edge 39
Melchoirs, Tony 258
Mexican quail 168
Mexican sunflower 335
microclimates 2
midstory 39
Migratory Bird Harvest Information Program 209
Migratory Bird Law 417
Migratory Bird Treaty Act 213, 218, 417

milk vetch 96
milkpea 171, 523
milkweed 96, 322
millet 150, 158, 160, 171, 176, 197, 204, 206, 207, 218, 222, 224, 225, 330, 365, 367, 381, 423
milo 205, 207, 270, 330
mimosa 366
Mims, Walter L. 426
mink 229, 230, 231, 236, 237, 238, 242, 246, 250, 263, 393
  biology and life history 237
  description 237
  habitat needs 238
  indicator species 238
  prevention and control techniques 250
Mississippi Flyway 218, 219
Mississippi kite 281
Mississippi State University 133, 147, 182, 380, 390
Mobile County 265
Mobile County Wildlife and Conservation Associatio 216, 426
mockernut 7
mocking bird 329
Mohr's Barbara's buttons 309, 310
Mohr's goat's-rue 75
Molpus Company 390
monarch 335
monitoring and revision of forest stand management 95
Morefield's leather flower 306, 307
morning glory 71, 204
moss verbena 335
mottled duck 211, 212, 214
mountain laurel 303
mourning dove 58, 200, 201, 202, 203, 205, 206, 209, 256, 364
  biology and life history 202
  common diseases 203
  food needs 203
  habitat and harvest management intensities 209
  habitat improvements 204, 206
  nesting 203
  November-January fields 205
  permits required 209
  preferred native and cultivated grasses 204
  preferred native and planted foods 206
  protection 208
  September and October fields 204
  small grain planting recommendations 208
  what is a legal dove field? 206
Muir, John 46, 421
mulberry 269, 324, 325, 532
multiflora rose 366
muscadine 270
mushroom 43, 113, 188
muskrat 3, 229, 230, 231, 235, 236, 237, 242, 250, 259, 351
  biology and life history 236
  description 235
  habitat needs 236
  prevention and control techniques 250

Nasturtium 333
National Ambient Air Quality Standards 447
National Fish and Wildlife Foundation 282
National Marine Fisheries Service 287
National Pesticide Telecommunications Network 449
National Wetland Inventory 447
National Wild Turkey Federation 138, 152, 163, 376, 379
National Wildlife Federation 319, 336, 427
native forage yield 102
native grazing ranges 99
  loblolly/shortleaf pine–hardwood 99
  longleaf/slash pine–wiregrass 99
  upland hardwood–bluestem 99
native plants 365
native versus exotic species 33
natural regeneration 55
natural regions of Alabama 5
  coastal plain region
    upper coastal plain region 9
    lower coastal plain region 14
  piedmont region 8
  plateau region 5
  ridge and valley region 8
  the fall line 9
  wetlands 14
Natural Resources Conservation Service 342, 345, 370, 447
natural versus artificial regeneration 58
naturalized plants 33
nest and denning box dimensions 478
nest/denning box 331
New England Aster 335
New England cottontail 192
nine-banded armadillo 262
Nixon, Richard 286
non-consumptive use of wildlife 392
non-game forest management 260
non-game wildlife 256
  Alabama mammals 264
  birds 279
  black bear 265
  effects of forest management practices 261
  forest management 257
  indicator species 283
  leaving dead wood 260
  management in forested landscapes 258
  multiple species approach to managing 257
  protecting riparian forests 258
  protecting unique areas 259
  reptiles and amphibians 283
  single species approach to managing 261
Non-game Wildlife Program 256, 287
non-target wildlife species 378
North American Waterfowl Management Plan 219
Northern cardinal 281
Northern flicker 87, 281
Northern oriole 281
Northern parula 281
Northern pintail 212, 214
Northern red oak 7, 8, 519, 532

Northern short-tailed shrew 263
Northern shoveler 212, 214
Northern spotted owl 417
Norway rat 263
noxious plants 365
nutria 230, 235, 236, 263
nutrients for plant growth 374
nuttall oak 15, 56, 216, 270, 533

oak 6, 10, 15, 53, 113, 117, 188, 322, 516
oak regeneration for wildlife 62
oaks important to wildlife 519
oats 505
obedient plant 333
Odum, Eugene 43
old-fashioned weigela 333
old-field mice 263, 325
opossum 3, 35, 43, 145, 146, 147, 148, 168, 220, 229, 230, 240, 241, 242, 249, 251, 260, 263, 293, 325
  biology and life history 241
  description 240
  habitat needs 241
  prevention and control techniques 251
opportunity costs of improved wildlife habitat 386, 388
orange-footed pearly mussel 300
orange-nacre mucket mussel 301
orangespotted sunfish 349
orchard grass 94, 367, 501
orchard oriole 329
osage orange 188
osceola clover 121
ovate clubshell mussel 300
ovenbird 74
overcup oak 15, 270, 519, 533
owl 41, 145, 187, 195, 220, 264

painted bunting 281
pale lilliput pearly mussel 300
palezone shiner 299
palmetto 172, 178, 188, 267, 270
panhandle lily 75
panic grass 75, 94, 99, 150, 171, 172, 174, 180
parsley 335
Partners in Flight 282
partridge pea 171, 172, 174, 175, 176, 177, 180, 376, 381, 502, 523
paspalum 94, 99, 150, 204, 206, 366
passenger pigeon 279, 416, 421, 436
paulownia 533
pawpaw 270
peanut 502
peanut, perennial 367, 503
pearl millet 500
pecan 150, 154, 224, 267, 269, 270, 533
pecan nut 113
Pelton, Mike 274
Peoples, Chuck 152
Perdido Key beach mouse 263
perennial plants 41, 364
persimmon 7, 71, 74, 85, 86, 113, 118, 150, 153, 171, 188, 240, 245, 267, 270, 322, 367, 516, 533
pesticides 84, 238, 264, 273, 287, 296, 329, 344, 378, 439, 443, 449
petunia 333
pheasant 406
Philipp, Chad 244
Phillips, Doug 11
physiographic regions of Alabama 13
pickerel weed 216, 236
Piedmont 5, 8, 9, 13, 79, 154, 175, 237, 246, 247, 248, 249, 262, 263
pignut 7
pileated woodpecker 329, 364
pin oak 533
Pinchot, Gifford 46, 436
pine 53, 516
pine beetle 41
pine pastures 99
pine siskin 329
pine snake 148
pine warbler 5, 281
pineapple sage 333, 335
pineland threeawn 99
pines important to wildlife in Alabama 522
pink mucket pearly mussel 300
plant succession 51
planting acorns for wildlife 56
plants important to wildlife 6
  alfalfa 484
  alyceclover 484
  American jointvetch 485
  apple/crabapple 485
  arrowleaf clover 491
  Austrian winter pea 486
  Autumn olive 487
  bahia grass 487
  ball clover 491
  barley 505
  beggarweed 488
  berseem clover 492
  birdsfoot deervetch 488
  black medic 489
  bluestem 489
  browntop millet 499
  buckwheat 489
  button clover 492
  chinese chestnut 490
  chufa 490
  coastal panicgrass 486
  corn 495
  cowpea 496
  crimson clover 493
  dallis grass 496
  Egyptian wheat 497
  foxtail millet 500
  Indiangrass 497
  Japanese millet 500
  lablab 497
  lespedezas (annual) 498
  lespedezas (shrub) 498
  oats 505
  orchardgrass 501
  partridge pea 502
  peanut 502
  peanut, perennial 503
  pearl millet 500
  proso millet 501
  red clover 493
  rice 503
  rye 505
  ryegrass 503
  sawtooth oak 504

sesame (benne) 504
singletary pea 505
sorghum 506
sorghum x sudan hybrids 507
soybean 507
subterranean clover 494
sudangrass 507
sunflower 507
sweet clover 494
switchgrass 508
timothy 508
velvet bean 509
vetch 509
weeping lovegrass 510
white (ladino) clover 495
wild reseeding soybean (bobwhite) 510
plants of concern in Alabama 75
plants toxic to cattle 96
plateau Region 5, 7
plum 79, 85, 86, 91, 113, 153, 171
plum fruit 113
poison ivy 270, 516
poison oak 150, 171
poison sumac 303
pokeberry 206, 269, 272
pokeweed 69, 204, 267, 268, 270, 274, 322, 523
pond balance 356
pond fertilization 347, 348
pond pine 522
pond stocking rates 351
pond stratification 357
ponds that should not be fertilized 349
pondweed 524
poor-joe 174
poplar 53
pore space 371
possumhaw 7, 113
post oak 10, 89, 153, 519, 533
Powell, Ed 138
prairie vole 263
prairie warbler 281, 282
pre-commercial thinning 72
precipitation 4
predators 33
prescribed burning 9, 74, 100, 101, 156, 191, 197, 271, 364, 447
compartment burn 157
compartment or block 35
oak regeneration 62
patch 35
planning and conducting 78
Price's potato-bean 309, 310
problems in establishing a food plot 379
productive water 341
Project HELP 379
pronthonotary warbler 258
proso millet 501
protection of high-value habitats 91
pruning 83
Pugh, Corky 432
Purdue, Billy 207
purple cat's paw pearly mussel 301
purple coneflower 323, 335
purple finch 329
purple martin 329, 478
purpletop 94
push-probe 372
pygmy sculpin 296, 299

Quail Unlimited 183
Quality Deer Management Association 135
Queen Anne's lace 322

rabbit 3, 34
Eastern cottontail 263
habitat and harvest management intensities 198
marsh rabbit 263
New England cottontail 263
swamp rabbit 263
raccoon 3, 41, 87, 145, 146, 147, 148, 160, 162, 168, 188, 215, 217, 220, 228, 229, 230, 231, 236, 237, 238, 239, 240, 242, 250, 252, 259, 260, 263, 293, 325, 331, 336, 337, 379, 393, 478
biology and life history 238
description 238
habitat needs 239
prevention and control techniques 250
restocking 240
ragweed 40, 66, 171, 174, 177, 180, 204, 205, 206, 523
raspberry 512
rat snake 148, 187, 217
rattan 270
recreation 392
red bat 326
red buckeye 322, 333
red cedar 88, 322, 326, 533
red clover 125, 367, 375, 378, 493
red fox 34, 41, 148, 187, 220, 230, 241, 243, 262
Red Hills 10
Red Hills salamander 59, 283, 294, 295, 442
red maple 15, 45, 57, 62, 71, 322, 515, 532
red mulberry 7, 63, 88, 113, 188, 270, 322, 517
red oak 51, 56, 85, 87, 88, 117, 119, 153, 154, 156, 189
red wolf 262, 436
red-bellied turtle 292, 293
red-bellied woodpecker 87, 260, 329, 478
red-cockaded woodpecker 5, 9, 74, 75, 76, 80, 81, 261, 284, 286, 289, 290, 311, 312, 313, 417, 439, 440, 442, 457
Red-cockaded Woodpecker Procedures Manual 290, 442
red-cockaded woodpeckers on private forest lands 312
redear sunfish 350
red-eyed vireo 5, 74
red-headed woodpecker 87, 260, 329, 478
red-shouldered hawk 148, 187
red-tailed hawk 87, 148, 187, 260
red-winged blackbird 281, 329
redbay 517
redbud 63, 71, 533

redear sunfish 349, 351
redhead 212, 214
Redland II clover 121
redroot 94
redwood 326
reed 216
references and suggested reading 460
regal ladino 121, 122, 123
relict trillium 306, 308
residents—birds 279
rice 503
Ridge and Valley region 1, 8
ring pink mussel 301
ring-neck pheasant 33
ring-necked duck 214
river birch 532
river otter 229, 230, 232, 233, 234, 246, 250, 263
  activity centers 234
  biology and life history 234
  description 234
  habitat needs 235
  prevention and control techniques 250
robin 329
rodenticides 449
Ronenone 352
Roosevelt, Theodore 419, 421, 433, 436
rough-leaved yellow-eyed grass 75
royal catch-fly 75
ruby-crowned kinglet 329
ruby-throated hummingbird 281, 332
ruddy duck 212, 214
ruffed grouse 436
rufous-sided towhee 43, 74, 281, 282
runner oak 153
running oak 519
rushes 216
Russian olive 370
rye 148, 160, 367, 505
ryegrass 197, 270, 367, 503

Safe Harbor 311, 312
sage 323, 335
salt and brackish marsh 14
salvia 322
sample hunting lease 534
sample timber sale contract 538
sand pine 522
sandhill crane 145
sassafras 7, 63, 71, 88, 112, 113, 172, 178, 270, 517, 533
saw palmetto 150, 517
sawtooth oak 32, 325, 367, 504
scarlet oak 8, 519, 533
scarlet sage 333
scarlet tanager 74
Schoenfeld, Clay 423
screech owl 329, 478
scrub jay 417
seaside goldenrod 323
seasonally important native deer foods 113
Secchi disk 347, 348
Section 404 of the Clean Water Act 445, 447
sedge 94, 160, 162, 170, 194, 213, 216, 218, 225, 236, 243, 303, 305, 309, 364, 524
seed-in-place method 55
seed-tree 61
seed-tree regeneration 53
seedling-in-place method 55
semi-aquatic furbearers 237
  mink 237
  raccoon 238
sensitive brier 94
sericea lespedeza 367
serviceberry 74, 86, 324
sesame (benne) 504
shade-tolerant and shade-intolerant trees 63
shagbark 7
shelterwood 61
  oak regeneration 62
shelterwood regeneration 53
shiners 351, 352
shiny pigtoe mussel 300
shinyrayed pocketbook 300
shooting star 323
short-distance migrants 279
shortleaf pine 5, 7, 8, 9, 10, 78, 156, 522, 533
shoveler 213
showy evening primrose 323
shrew 43
shumard oak 15, 270, 519, 533
sicklepod 71
side-oats grama 99
Sierra Club 421
Sievering, Mike 252
silktree 366
silver maple 515
silvicultural practices 28, 41, 48, 49, 50, 51, 53, 54, 59, 83, 86, 87, 90, 95, 103, 116, 120, 154, 178, 257, 258, 260, 264, 267, 271, 274, 436, 445
  effects on wildlife habitat 51
single-tree selection 64
singletary pea 505
site index 50, 51
site preparation 66
  band spraying 72
  brown and burn 70
  chemical 70
  fell and burn 69
  mechanical 66
skunk 43, 145, 147, 148, 168, 193, 220, 229, 230, 240, 242, 246, 247, 251, 260, 263, 293, 325
  description 246
  habitat needs 247
  prevention and control techniques 251
slackwater darter 296, 299
Slade, Tony 138
slash 44
slash pine 5, 9, 14, 78, 156, 522
small sunfish 350
smallmouth bass 349
smartweed 216, 236, 525
Smith, Barry 352
smooth aster 323
smooth sumac 518
snag 43, 87, 88, 260
snail darter 298, 299, 417
snapping turtle 236
sneezeweed 323
soil compaction 371
soil pH range for select southern tree species 532
soils 2
  acidic 3
  alluvial 7

chemistry 3
clay/rocky 8
fertility 3
types 2
sorghum 158, 160, 171, 175, 176, 182, 204, 205, 206, 218, 222, 223, 224, 236, 364, 366, 367, 380, 423, 506
sorghum x sudan hybrids 507
sources of technical and educational assistance 541
sources of wildlife planting materials 526
sourwood 533
southeastern pocket gopher 263
southeastern shrew 263
Southern acornshell mussel 300
Southern Appalachian Field Laboratory 265
Southern clubshell mussel 300
Southern combshell mussel 301
Southern flying squirrel 199, 263
Southern Forest Experiment Station 449
Southern Hardwood Laboratory 57
Southern magnolia 322
Southern pigtoe mussel 301
Southern red oak 5, 519, 533
Southern short-tailed shrew 263
soybean 144, 147, 148, 150, 159, 171, 174, 175, 194, 204, 206, 208, 222, 223, 233, 236, 271, 273, 367, 372, 378, 388, 402, 507
spacing pine seedlings 55
sparkleberry 303
species of concern 286
sphagnum moss 305
spicebush 335
spicebush swallowtail 335
spikerush 525
Sportfish and Wildlife Restoration Fund 282
spotfin chub 299
spotted dolphin 262
spotted skunk 246, 247
spring beauty 323
spruce pine 522
spurge 96
staghorn sumac 518
standing cypress 323
stands 50
even-aged 53
mixed age 53
uneven-aged 53
starling 330
Stewardship Incentive Program 443, 447, 451
Stewart, Stan 37
stiffed-leaved aster 323
Stimpson, Fred T. 138, 426
stirrup shell mussel 300
stocking rates for triploid grass carp 359
stocking rates for woodland grazing and wildlife 102
stokes's aster 323
strawberry 7
strawberry bush 113
streamside management zones 26, 37, 61, 84, 85, 90, 92, 155, 156, 191, 198, 258, 259, 260, 261, 268, 270, 271, 305, 370, 390, 441, 445, 448
striped skunk 246, 263
subsoiling 372
subterranean clover 367, 494
sudan grass 423, 507
suet cake recipe 330
sugarberry 7, 518
sugarcane 270, 273, 275
sumac 63, 94, 172, 174, 178, 303, 324, 533
summer tanager 329
Sumter Farm 358
sunflower 171, 204, 205, 206, 330, 378, 379, 380, 507
supplemental plantings as wildlife food sources 363
sustainable development 46
Sustainable Forestry Initiative 452
sustainable use 46
Swainson's warbler 281
swamp chestnut 224
swamp chestnut oak 153, 519, 533
swamp milkweed 323
swamp tupelo 270
sweet bay 303, 518, 533
sweet clover 94, 113, 494
sweet goldenrod 323
sweetclover 367
sweetgum 5, 88, 188, 204, 206, 224, 232, 272, 533
switchgrass 99, 367, 508
sycamore 15, 188, 533

take 440
taking 452
Tall Timbers Research Station 168, 169, 174, 179, 181, 183
Talladega National Forest 107
tallowtree 366
Teaming With Wildlife Initiative 282
tearthumb 525
temperature 5
Tennessee Valley 7
Tennessee Valley Authority 390
Tennessee yellow-eyed grass 306, 307
terrestrial furbearers 240
bobcats 248
coyotes 243
foxes 241
opossum 240
skunks 246
The American Forest and Paper Association 452
The Nature Conservancy 265, 279, 398
The Wild Turkey in Alabama 138
The Wildlife Society 399
thistle 267, 322
threatened and endangered fish in Alabama 299
threatened and endangered mussel species in Alabama 300
threatened and endangered species 285, 286
three-awn grass 75
thrift 335
tickseed sunflower 323
timber and wildlife stand improvement practices 80
timber industry
economics 391
timber stand improvement (TSI) 80, 81
timothy 508
timothy grass 94, 367

topographical maps 28
topography 2, 5
floodplains 2
upland 2
traditional approaches to wildlife management 44
trailing lantana 335
translocation of isolated groups 311
TREASURE Forest 453
tree shelters 377
trees and plants important to wildlife 511
Alabama supplejack 511
American beautyberry 511
American beech 511
American elder 511
American hornbeam 511, 515
bayberry 511
black cherry 512
black walnut 512
blackberry 512
blueberry 512
boxelder 515
butternut 512
chinquapin 513
dewberry 512
dogwood 513
dwarf huckleberry 515
dwarf palmetto 516
Eastern hophornbeam 514
Eastern redcedar 517
elm 513
fringetree 513
greenbrier 513
hackberry 514
hawthorn 514
hickory 514
holly 514
Japanese honeysuckle 515
locust 515
magnolia 515
oak 516
persimmon 516
pine 516
poison ivy 516
raspberry 512
red maple 515
red mulberry 517
redbay 517
sassafras 517
saw-palmetto 517
silver maple 515
smooth sumac 518
staghorn sumac 518
sugarberry 518
sweetbay 518
trumpet creeper 518
tupelo 518
viburnum 518
wild grape 513
wild plum 517
winged sumac 518
yellow poplar 516
triangular kidneyshell mussel 300
trophy bass management 352, 358
tropical croton 71
tropical soda apple 366
trumpet creeper 71, 322, 332, 333, 518
tubercled-blossom pearly mussel 301
tufted titmouse 329, 478
tularemia 193, 194, 232, 236
tulotoma snail 301
tupelo 518
tupelo gum 188
turgid-blossoum pearly mussel 301
turkey oak 10, 519
turnover 358

U.S. Army Corps of Engineers 223, 450
U.S. Biological Survey 216
U.S. Department of Agriculture Animal Damage Control 277
U.S. Fish and Wildlife Service 32, 76, 208, 216, 217, 219, 220, 226, 286, 287, 290, 292, 299, 311, 312, 399, 403, 425, 442, 450
U.S. Forest Service 93, 96, 100
U.S. Geological Survey 447
University of Tennessee 274
upland combshell mussel 300
upland herbaceous plants important to wildlife in 523
Upper Coastal Plain 9, 10
Urban and Community Forestry Assistance Program 451
USDA Farm Services Agency 28, 451
USDA Natural Resources Conservation Service 18, 28, 223, 399, 451
USDA Soil Conservation Service 365
USDA Swampbuster Program 370
USDA Wildlife Services 233, 253
USDA/APHIS/Animal Damage Control 244

Valley and Ridge 13
Van Lear, David 62
velvetbean 367, 509
verbena 335
vertical habitat diversity 39
vertical layering 39
vervain 335
vetch 148, 158, 197, 509
viburnum 7, 113, 172, 178, 322, 518
Virginia opossum 240, 263
Virginia pine 7, 8, 522, 533
vole 33, 41, 235, 248, 256, 263, 325
volunteer chufa 161

warm season plants 364
Washington County 265
water hemlock 96
water hyacinth 33
water lily 216, 236
water oak 5, 56, 153, 216, 270, 519, 533
Water Pollution Control Act 444
water primrose 216
water tupelo 7, 9, 15, 224, 533
water willow 224
watercress darter 297, 299
waterfowl
attracting 226
beaver pond management 222
biology and life history 212
commonly seen in Alabama 214
dabbling 212
divers 212
emergent vegetation 213
fall flight forecast 220
floating vegetation 213

flyway management concept 218
flyways 212
greentree reservoirs 223
habitat and harvest management intensities 225
habitat improvements: farm/ forests 220
habitat protection 219
harvesting and regulations 225
how harvest regulations are determined 220
managing agricultural fields 223
managing flooded woodlands 223
managing migratory waterfowl on a large scale 218
moist soil management 220
molts 212
native versus supplemental plants 224
perching 212
permits required 227
predator and disease control 220
submergent plants 212
wetlands destruction 219
watermelon 275
watershed 340
watershield 525
wax myrtle 71, 150, 178, 322
weeping lovegrass 510
West Indian manatee 262
wetlands 14, 260, 273
Wetlands Reserve Program 447
Weyerhaeuser Company 258
wheat 121, 122, 148, 150, 151, 158, 171, 197, 204, 206, 207, 218, 270, 275, 367
Wherry's sweet pitcher plant 75
white (ladino) clover 495
white ash 532
white clover 367, 375
white crappi 351
white oak 5, 7, 8, 45, 56, 85, 87, 88, 117, 119, 153, 154, 216, 270, 367, 519, 533
white pine 88, 533
white snakeroot 71, 96
white wartyback mussel 301
white-breasted nuthatch 329
white-eyed vireo 281
white-footed mouse 32, 41, 263, 478
white-tailed deer 3, 10, 32, 33, 34, 41, 45, 66, 82, 85, 104, 105, 106, 107, 110, 111, 116, 120, 126, 133, 135, 230, 262, 362, 364, 367, 378, 379, 394, 416, 419, 428, 436
Alabama records 135
Alabama's Deer Management Assistance Program 134
anthrax 110
antlers 109
area requirements 116
biology and life history 107
cool-season planting mixtures 124
cover 112
crop and ornamental damage 134
deer herd management 126
diet 112
disease and parasites 110
effects of an over-populated herd 127
epizootic hemorrhagic disease 110
establishing food plots 120
estimating population trends 128
fertilization of forage 119
forest management 116
gestation 107
habitat and hunting management intensities 120
habitat improvements 116
habitat requirements 111
harvest 127
health of a herd 128
how to age deer jaw bones 131
keys to healthy deer 108
management considerations 133
managing openings 122
measuring, recording antler characteristics 130
nasal bots 111
parasitic worms 111
prescribed burning 118
quality deer management 129
record-keeping 129
salt or mineral blocks 124
seasonal deer forage plantings 125
skin tumors 110
social carrying capacity 127
supplemental feeding 123
supplemental feeding using soybeans 126
supplemental plantings 120
white-topped pitcher plant 75
whooping crane 279
widgeon 213
widgeongrass 525
wild azalea 322
wild bean 94
wild blue phlox 323
wild boar 32
wild cherry 143, 148, 154, 155, 172, 178, 325, 334
wild columbine 323, 333
wild geranium 323
wild grape 7, 69, 74, 82, 85, 89, 113, 118, 150, 153, 171, 188, 240, 267, 268, 270, 367, 513
wild hog 7, 85, 145, 162, 379
wild indigo 96
wild lettuce 94
wild lupine 323
wild plum 188, 367, 517
wild reseeding soybean (bobwhite) 510
wild strawberry 151, 155
wild turkey 32, 33, 41, 43, 45, 59, 61, 66, 76, 82, 85, 91, 99, 118, 136, 137, 138, 139, 141, 143, 144, 145, 146, 147, 148, 149, 150, 151, 152, 153, 154, 155, 156, 157, 158, 159, 162, 163, 212, 220, 230, 282, 387, 394, 416, 418, 426, 428, 432, 436
annual population levels 143
brood-rearing 142, 149
cover 151
description and life history 138
direct feeding 162
diseases and parasites 144
Eastern wild turkey 138
escape cover 149
factors that limit populations 144

fall and winter period 143, 149
Florida wild turkey 138
foods and feeding 148
habitat improvement practices 152
habitat loss 144
habitat requirements 148
hard mast producers 153
hardwood management 153
hunting 162
key habitat types 149
limiting factors 147
managing openings 157
mating season 139
nesting 141, 149, 151, 152
nesting and brood habitat 152
pine management 155
planting chufas 162
poaching 144
predation 145
predators 148
prescribed burning 156
roosting 149
seasonally important foods 150
supplemental plantings 158
trapping and relocation 138
water 151
weather 144
wildflowers for meadows 323
wildlife
definition of 32
economics 394
exotic 32
wildlife food plot activity 382
wildlife habitat
cover 35
food 32
food preferences 34
home range 38
requirements 32
space 38
travel corridors 37
water 37
wildlife habitat management 31
basic biology and habitat requirements 37
strategic grazing 101
wildlife considerations 101
wildlife habitat management plan
compartment information 26
compartment record sheets 23
sketch map 22
sources of information and assistance 27
wildlife management 44
ecosystem approach 45
featured species management 44
limiting factor management 45
multiple species management 44
preservation 44
sustained yield 44
wildlife management plan
additional considerations 28
assistance 18
designating management compartments 22
developing and implementing 20
equipment and facilities 21
financial considerations 22
format for a simple plan 26
identifying objectives 18
record-keeping and evaluation 29
resource inventory 19
selecting habitat improvement practices 26
the property 19
tools for developing a habitat 27
writing the plan 18
wildlife stand improvement (WSI) 80, 81
wildrice 525
willow 15, 71, 87, 88, 89
willow oak 153, 216, 270, 519, 533
Wilson, E.O. 313
windrow management for wildlife 68
winged sumac 518
winter pea 197
wiregrass 75, 99
witchhazel 188
wood duck 211, 212, 213, 214, 215, 216, 217, 220, 223, 224, 226, 230, 282, 416, 417, 418
biology and life history 214
emergent plants 216
habitat requirements 215
nesting 215
nesting boxes 216
predators 217
wood sorrel 150, 171
wood stork 290, 291
wood thrush 5, 281, 329
woodchuck 263
woodland bison 436
woodland forage plants utilized by cattle and wild 94
woodland grazing and wildlife 96
woodland grazing considerations 100
woodland or pine vole 263
woolly croton 523
worm-eating warbler 281
Wyncreek Plantation 182, 380
food plot program 380

yaupon 7, 113
yellow birch 532
yellow jasmine 332
yellow poplar 5, 7, 57, 61, 62, 63, 71, 88, 113, 188, 322, 334, 335, 516, 533
yellow-bellied sapsucker 329
yellow-billed cuckoo 281, 329
yellow-blossum pearly mussel 301
yellow-breasted chat 281
YIELDplus 390
Youngblood, Lee 161

zebra mussel 33
zinnia 335